Environment

Investment

Science

Sports

Statistics/Demographics

Elementary Algebra

EDITION 4

Elementary Algebra

EDITION 4

Mark Dugopolski

Southeastern Louisiana University

Higher Education

Boston Burr Ridge, IL Dubuque, IA Madison, WI
New York San Francisco St. Louis Bangkok
Bogotá Caracas Kuala Lumpur Lisbon London Madrid
Mexico City Milan Montreal New Delhi Santiago
Seoul Singapore Sydney Taipei Toronto

The McGraw-Hill Companies

ELEMENTARY ALGEBRA, FOURTH EDITION

Published by McGraw-Hill, a business unit of The McGraw-Hill Companies, Inc., 1221 Avenue of the Americas, New York, NY 10020.

This book is printed on acid-free paper.

1 2 3 4 5 6 7 8 9 0 VNH/VNH 0 9 8 7 6 5 4 3

ISBN 0–07–254628–X

Publisher: *William K. Barter*
Senior sponsoring editor: *David Dietz*
Developmental editor: *Erin Brown*
Executive marketing manager: *Marianne C. P. Rutter*
Senior marketing manager: *Mary K. Kittell*
Senior project manager: *Vicki Krug*
Lead production supervisor: *Sandy Ludovissy*
Lead media project manager: *Audrey A. Reiter*
Media technology producer: *Jeff Huettman*
Coordinator of freelance design: *Rick D. Noel*
Cover designer: *Todd Damotte/Rokusek Design*
Cover image: *@Corbis, Volume 622, Architectural Details, No.1*
Lead photo research coordinator: *Carrie K. Burger*
Supplement producer: *Brenda A. Ernzen*
Compositor: *Interactive Composition Corporation*
Typeface: *10.5/12 Times Roman*
Printer: *Von Hoffmann Corporation*

Photo Credits
Page 17, p. 150, p. 254, p. 338, p. 349, p. 460: © Susan Van Etten; p. 90, p. 216, p. 395: © The McGraw-Hill Companies, Inc.; p. 164: © Ann M. Job/AP/Wide World Photos; p. 201 top: © Gary Conner/PhotoEdit; p. 249, p. 475: © Digital Vision; p. 419: © Bobby Lanier/Martin Marietta. All other photos © PhotoDisc, Inc.

Library of Congress Cataloging-in-Publication Data

Dugopolski, Mark.
Elementary algebra / Mark Dugopolski. — 4th ed.
p. cm.
Includes index.
ISBN 0–07–244391–X (hard copy : alk. paper)
1. Algebra. I. Title.

QA152.2 .D84 2004
512.9—dc21

2002152181
CIP

www.mhhe.com

To my wife and daughters,
Cheryl, Sarah, and Alisha

CONTENTS

Chapter 3 Linear Equations in Two Variables and Their Graphs 143

Chapter 4 Systems of Linear Equations and Inequalities 205

Chapter 5 Polynomials and Exponents 249

Chapter 9 Quadratic Equations, Parabolas, and Functions 457

PREFACE

Elementary Algebra, Fourth Edition, is designed to provide students with the algebra background needed for further college-level mathematics courses. The unifying theme of this text is the development of the skills necessary for solving equations and inequalities, followed by the application of those skills to solving applied problems. My primary goal in writing the fourth edition of *Elementary Algebra* has been to retain the features that made the third edition so successful, while incorporating the comments and suggestions of third-edition users. As always, I endeavor to write texts that students can read, understand, and enjoy, while gaining confidence in their ability to use mathematics.

Content Changes

While the essence of previous editions remains, the topics have been rearranged to reflect the current needs of instructors.

- *Systems of Linear Equations and Inequalities*, previously found in Chapter 7 of the third edition, has moved to Chapter 4 in the new edition.
- Chapter 3 has been condensed from seven sections to five. Content from Section 3.5 *Applications of Linear Equations* of the third edition has been incorporated into Section 3.4 *The Point-Slope Form* of the new edition. Section 3.6 *Introduction to Functions* has been moved to Chapter 9, Section 7.
- *Combining Functions* has been newly added to Chapter 9 and appears as the last section.
- The *Geometry Review*, found in Appendix A, now includes a set of Review Exercises.
- Unit conversion problems have been added to Section 2.6.

In addition to these changes, the text and exercise sets have been carefully revised where necessary. New, applied examples have been added to the text and new, applied exercises have been added to the exercise sets. Particular care has been given to achieving an appropriate balance of problems that progressively increase in difficulty from routine exercises in the beginning of the set to more challenging exercises at the end of the set. As in earlier editions, fractions and decimals are used in the exercises and throughout the text discussions to help reinforce the basic arithmetic skills that are necessary for success in algebra.

Features

- Each chapter begins with a Chapter Opener that discusses a real application of algebra. The discussion is accompanied by a photograph and, in most cases by a real-data application graph that helps students visualize algebra and more fully understand the concepts discussed in the chapter. In addition, each chapter contains a Math at Work feature, which profiles a real person and the mathematics that he or she uses on the job. These two features have corresponding real data exercises.

- The fourth edition continues to emphasize real-data applications that involve graphs. Applications appear throughout the text to help demonstrate concepts, motivate students, and to give students practice using new skills. Many of the real-data exercises contain data obtained from the Internet. Internet addresses are provided as a resource for both students and teachers. An Index of Selected Applications listing applications by subject matter is included at the front of the text.
- Every section begins with In This Section, a list of topics that shows the student what will be covered. Because the topics correspond to the headings within each section, students will find it easy to locate and study specific concepts.
- Important ideas, such as definitions, rules, summaries, and strategies, are set apart in boxes for quick reference. Color is used to highlight these boxes as well as other important points in the text.
- The fourth edition contains margin features that appear throughout the text:

  Calculator Close-Ups give students an idea of how and when to use a graphing calculator. Some Calculator Close-Ups simply introduce the features of a graphing calculator, where others enhance understanding of algebraic concepts. For this reason, many of the Calculator Close-Ups will benefit even those students who do not use a graphing calculator. A graphing calculator is not required for studying from this text.

  Study Tips are included in the margins throughout the text. These short tips are meant to continually reinforce good study habits and to remind students that it is never too late to make improvements in the manner in which they study.

  Helpful Hints are short comments that enhance the material in the text, provide another way of approaching a problem, or clear up misconceptions.

- At the end of every section are Warm-up exercises, a set of ten simple statements that are to be answered true or false. These exercises are designed to provide a smooth transition between the ideas and the exercise sets. They help students understand that every statement in mathematics is either true or false. They are also good for discussion or group work.
- Most section-ending exercise sets in the fourth edition begin with about six simple writing exercises. These exercises are designed to get students to review the definitions and rules of the section before doing more traditional exercises. For example, the student might simply be asked what properties of equality were discussed in this section.
- The end-of-section Exercises follow the same order as the textual material and contain exercises that are keyed to examples, as well as numerous exercises that are not keyed to examples. This organization allows the instructor to cover only part of a section if necessary and easily determine which exercises are appropriate to assign. The *keyed exercises* give the student a place to start practicing and building confidence, whereas the *nonkeyed exercises* are designed to wean the student from following examples in a step-by-step manner. *Getting More Involved exercises* are designed to encourage *writing, discussion, exploration,* and *cooperative learning. Graphing Calculator Exercises* require a graphing calculator and are identified with a graphing calculator logo. Exercises for which a scientific calculator would be helpful are identified with a scientific calculator logo. Please refer to page xxi for a visual guide of the icons.

- Every chapter ends with a four-part Wrap-up, which includes the following:

  The Chapter Summary lists important concepts along with brief illustrative examples.

  Enriching Your Mathematical Word Power appears at the end of each chapter and consists of multiple-choice questions in which the important terms are to be matched with their meanings. This feature emphasizes the importance of proper terminology.

  The Review Exercises contain problems that are keyed to the sections of the chapter as well as numerous miscellaneous exercises.

  The Chapter Test is designed to help the student assess his or her readiness for a test. The Chapter Test has no keyed exercises, thus encouraging the student to work independently of the sections and examples.
- **UPDATED!** At the end of each chapter is a Collaborative Activities feature that is designed to encourage interaction and learning in groups. Many of the Collaborative Activities for the fourth edition have been updated. Instructions and suggestions for using these activities and answers to all problems can be found in the Instructor's Solutions Manual.
- The Making Connections exercises at the end of Chapters 2–9 are designed to help students review and synthesize the new material with ideas from previous chapters, and in some cases, review material necessary for success in the upcoming chapter. Every Making Connections exercise set includes at least one applied exercise that requires ideas from one or more of the previous chapters. In essence, the Making Connections are cumulative reviews.

Supplements for the Instructor

ANNOTATED INSTRUCTOR'S EDITION

This ancillary includes answers to all section ending exercises, review exercises, Making Connections exercises, and chapter tests. Each answer is printed next to each problem on the page where the problem appears. The answers are printed in a second color for ease of use by instructors.

INSTRUCTOR'S TESTING AND RESOURCE CD-ROM

This CD-ROM contains a computerized test bank that utilizes Brownstone Diploma® testing software. The computerized test bank enables instructors to create well-formatted quizzes or tests using a large bank of algorithmically generated and static questions. When creating a quiz or test, the user can manually choose individual questions or have the software randomly select questions based on section, question type, difficulty level, and other criteria. Instructors also have the ability to add or edit test bank questions to create their own customized test bank. In addition to printed tests, the test generator can deliver tests over a local area network or the World Wide Web, with automatic grading.

Also available on the CD-ROM are preformatted tests that appear in two forms: Adobe Acrobat (pdf) and Microsoft Word files. These files are provided for convenient access to "ready-to-use" tests. Preformatted chapter tests and final tests can also be downloaded as a Word (.doc) file or can be viewed and printed as a (.pdf) file at www.mhhe.com/dugopolski.

INSTRUCTOR'S SOLUTIONS MANUAL

Prepared by Mark Dugopolski, this supplement contains detailed worked solutions to all of the exercises in the text. The solutions are based on the techniques used in the text. Instructions and suggestions for using the Collaborative Activities feature in the text are also included in the Instructor's Solutions Manual.

PageOut

PageOut is McGraw-Hill's unique point-and-click course website tool, enabling instructors to create a full-featured, professional quality course website without knowing HTML coding. With PageOut instructors can post a syllabus online, assign McGraw-Hill Online Learning Center content, add links to important off-site resources, and maintain student results in the online grade book. Instructors can also send class announcements, copy a course site to share with colleagues, and upload original files. PageOut is free for every McGraw-Hill user. For those instructors who are short on time, there is a team on hand, ready to help build a site!

Learn more about PageOut and other McGraw-Hill digital solutions at www.mhhe.com/solutions.

Supplements for the Student

STUDENT'S SOLUTIONS MANUAL

Prepared by Mark Dugopolski, the *Student's Solutions Manual* contains complete worked-out solutions to all of the odd-numbered exercises in the text. It also contains solutions for all exercises in the Chapter Tests. It can be purchased from McGraw-Hill.

DUGOPOLSKI VIDEO SERIES (Videotapes or CD-ROMs)

The videos are text-specific and cover all chapters of the text. The videos feature an instructor who introduces topics and works through selected problems from the exercise sets. Students are encouraged to work the problems on their own and to check their results with those provided.

DUGOPOLSKI TUTORIAL CD-ROM

This interactive CD-ROM is a self-paced tutorial specifically linked to the text that reinforces topics through unlimited opportunities to review concepts and practice problem solving. The CD-ROM contains algorithmically generated chapter- and section-specific questions. It requires virtually no computer training on the part of students and supports Windows and Macintosh computers.

ONLINE LEARNING CENTER

The Online Learning Center (OLC), located at www.mhhe.com/dugopolski, contains resources for students and instructors. The OLC consists of the Student Learning Site, the Instructor Resource Site, and the Information Center.

Through the Instructor Resource Site, instructors can access links to professional resources, a PowerPoint presentation (transparencies), printable tests, a link to PageOut, and group projects. To access the Instructor Resource Site, instructors must have a passcode that can be obtained by contacting a McGraw-Hill Higher Education representative.

The Student Learning Site is also passcode-protected. Passcodes for students can be found at the front of their texts when newly purchased. Passcodes are available free to students when they purchase a new text. Students have access to algorithmically generated "bookmarkable" practice exercises (including hints), section- and chapter-level testing, audiovisual tutorials, interactive applications, and a link to NetTutor™ and other interesting websites.

The Information Center can be accessed by students and instructors without a passcode. Through the Information Center users can access general information about the text and its supplements.

ALEKS®

ALEKS® (**A**ssessment and **LE**arning in **K**nowledge **S**paces) is an artificial intelligence-based system for individualized math learning, available over the World Wide Web. ALEKS® delivers precise, qualitative diagnostic assessments of students' math knowledge, guides them in the selection of appropriate new study material, and records their progress toward mastery of curricular goals in a robust classroom management system. It interacts with the student much as a skilled human tutor would, moving between explanation and practice as needed, correcting and analyzing errors, defining terms, and changing topics on request. By sophisticated modeling of a student's "knowledge state" for a given subject matter, ALEKS® can focus clearly on what the student is most ready to learn next, building a learning momentum that fuels success.

To learn more about ALEKS® including purchasing information, visit the ALEKS® website at www.highedmath.aleks.com.

NetTutor

NetTutor is a revolutionary system that enables students to interact with a live tutor over the World Wide Web by using NetTutor's Web-based, graphical chat capabilities. Students can also submit questions and receive answers, browse previously answered questions, and view previous live chat sessions.

NetTutor can be accessed on the Online Learning Center through the Student Learning Site.

Acknowledgments

First of all I thank all of the students and professors who used the previous editions of this text, for without their support there would not be a fourth edition. I sincerely appreciate the efforts of the reviewers who made many helpful suggestions for improving the third edition:

Viola Lee Bean, *Boise State University*
Steve Boettcher, *Estrella Mountain Community College*
Annette M. Burden, *Youngstown State University*
Linda Clay, *Albuquerque Technical Vocational Institute*
John F. Close, *Salt Lake Community College*
Donna Densmore, *Bossier Parish Community College*
Linda Franko, *Cuyahoga Community College*
Laura L. Hoye, *Trident Technical College*
Krystyna Karminska, *Thomas Nelson Community College*

Angela Lawrenz, *Blinn College*
Vivian Dennis-Monzingo, *Eastfield College*
Michele Olsen, *College of the Redwoods*
Sandra L. Spain, *Thomas Nelson Community College*
Brenda Weaver, *Stillman College*
Kevin Yokoyama, *College of the Redwoods*
Limin Zhang, *Columbia Basin College*
Deborah Zopf, *Henry Ford Community College*

I also want to express my sincere appreciation to my wife, Cheryl, for her invaluable patience and support.

Hammond, Louisiana M.D.

What is ALEKS®?

ALEKS is...

- A comprehensive course management system. It tells you exactly what your students know and don't know.

- Artificial intelligence. It totally individualizes assessment and learning.

- Customizable. Click on or off each course topic.

- Web based. Use a standard browser for easy Internet access.

- Inexpensive. There are no set up fees or site license fees.

ALEKS (Assessment and LEarning in Knowledge Spaces) is an artificial intelligence-based system for individualized math learning available via the World Wide Web.

http://www.highedmath.aleks.com

ALEKS delivers precise, qualitative diagnostic assessments of students' math knowledge, guides them in the selection of appropriate new study material, and records their progress toward mastery of curricular goals in a robust course management system.

ALEKS interacts with the student much as a skilled human tutor would, moving between explanation and practice as needed, correcting and analyzing errors, defining terms and changing topics on request. By sophisticated modeling of a student's "knowledge state" for a given subject matter, ALEKS can focus clearly on what the student is most ready to learn next, thereby building a learning momentum that fuels success.

The ALEKS system was developed with a multimillion-dollar grant from the National Science Foundation. It has little to do with what is commonly thought of as educational software. The theory behind ALEKS is a specialized field of mathematical cognitive science called "Knowledge Spaces."

Knowledge Space theory, which concerns itself with the mathematical dynamics of knowledge acquisition, has been under development by researchers in cognitive science since the early 1980s. The Chairman and founder of ALEKS Corporation, Jean-Claude Falmagne, is an internationally recognized leader in scientific work in this field.

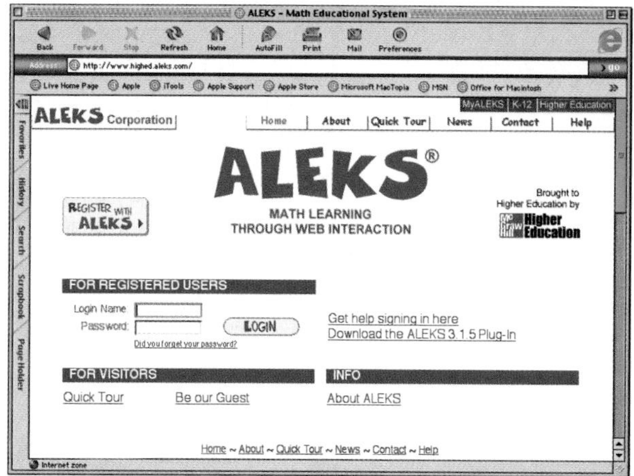

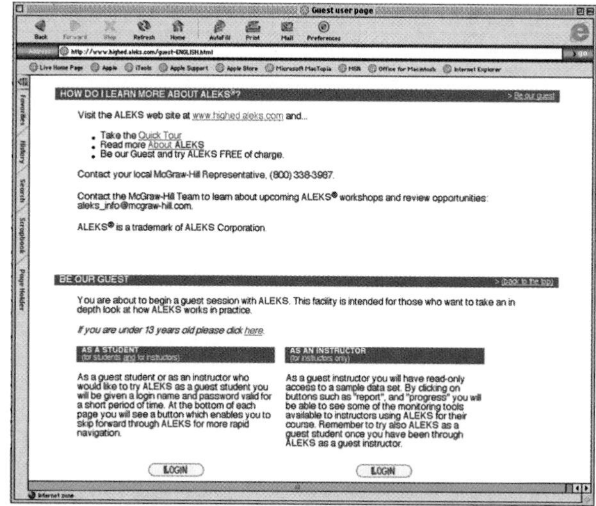

Please visit http://www.highedmath.aleks.com for a trial run.

CHAPTER 3

Linear Equations in Two Variables and Their Graphs

If you pick up any package of food and read the label, you will find a long list that usually ends with some mysterious looking names. Many of these strange elements are food additives. A food additive is a substance or a mixture of substances other than basic foodstuffs that is present in food as a result of production, processing, storage, or packaging. They can be natural or synthetic and are categorized in many ways: preservatives, coloring agents, processing aids, and nutritional supplements, to name a few.

Food additives have been around since prehistoric humans discovered that salt would help to preserve meat. Today, food additives can include simple ingredients such as red color from Concord grape skins, calcium, or an enzyme. Throughout the centuries there have been lively discussions on what is healthy to eat. At the present time the food industry is working to develop foods that have less cholesterol, fats, and other unhealthy ingredients.

Although they frequently have different viewpoints, the food industry and the Food and Drug Administration (FDA) are working to provide consumers with information on a healthier diet. Recent developments such as the synthetically engineered tomato stirred great controversy, even though the FDA declared the tomato safe to eat.

In Exercise 79 of Section 3.4 you will see how a food chemist uses a linear equation in testing the concentration of an enzyme in a fruit juice.

Chapter Opener

Each **chapter opener** features a real-world situation that can be modeled using mathematics. Each chapter contains exercises that relate back to the chapter opener.

3.4 The Point-Slope Form (3-45) **187**

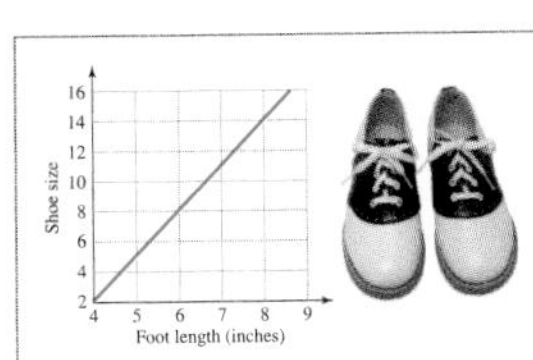

FIGURE FOR EXERCISE 71

72. ***Celsius to Fahrenheit.*** Water freezes at 0°C or 32°F and boils at 100°C or 212°F. There is a linear equation that expresses the number of degrees Fahrenheit (F) in terms of the number of degrees Celsius (C). Find the equation and find the Fahrenheit temperature when the Celsius temperature is 45°.

73. ***Velocity of a projectile.*** A ball is thrown downward from the top of a tall building. Its velocity is 42 feet per second after 1 second and 74 feet per second after 2 seconds. There is a linear equation that expresses the velocity v in terms of the time t. Find the equation and find the velocity after 3.5 seconds.

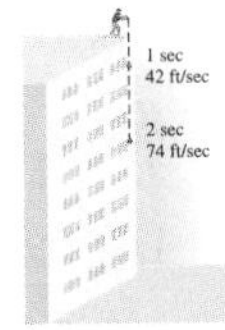

FIGURE FOR EXERCISE 73

74. ***Natural gas.*** The cost of 1000 cubic feet of natural gas is $39 and the cost of 3000 cubic feet is $99. There is a linear equation that expresses the cost C in terms of the number of cubic feet n. Find the equation and find the cost of 2400 cubic feet of natural gas.

75. ***Expansion joint.*** When the temperature is 90°F the width of an expansion joint on a bridge is 0.75 inch. When the temperature is 30°F the width is 1.25 inches. There is a linear equation that expresses the width w in terms of the temperature t.

a) Find the equation.

b) What is the width when the temperature is 80°F?

c) What is the temperature when the width is 1 inch?

76. ***Perimeter of a rectangle.*** A rectangle has a fixed width and a variable length. Let P represent the perimeter and L represent the length. $P = 28$ inches when $L = 6.5$ inches and $P = 36$ inches when $L = 10.5$ inches. There is a linear equation that expresses P in terms of L.

a) Find the equation.

b) What is the perimeter when the $L = 40$ inches?

c) What is the length when $P = 215$ inches?

d) What is the width of the rectangle?

77. ***Stretching a spring.*** A weight of 3 pounds stretches a spring 1.8 inches beyond its natural length and weight of 5 pounds stretches the same spring 3 inches beyond its natural length. Let A represent the amount of stretch and w the weight. There is a linear equation that expresses A in terms of w. Find the equation and find the amount that the spring will stretch with a weight of 6 pounds.

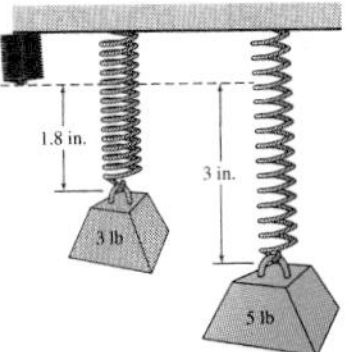

FIGURE FOR EXERCISE 77

78. ***Velocity of a bullet.*** A gun is fired straight upward. The bullet leaves the gun at 100 feet per second (time $t = 0$). After 2 seconds the velocity of the bullet is 36 feet per second. There is a linear equation that gives the velocity v in terms of the time t. Find the equation and find the velocity after 3 seconds.

79. ***Enzyme concentration.*** The amount of light absorbed by a certain liquid depends on the concentration of an enzyme in the liquid. A concentration of 2 milligrams per milliliter (mg/ml) produces an absorption of 0.16 and a concentration of 5 mg/ml produces an absorption of 0.40. There is a linear equation that expresses the absorption a in terms of the concentration c.

a) Find the equation.

b) What is the absorption when the concentration is 3 mg/ml?

Margin Notes

Margin notes include: **helpful hints, study tips,** and **calculator close-ups.** The **helpful hints** point out common errors or reminders. The **study tips** provide practical suggestions for improving study habits. The optional **calculator close-ups** provide tips on using a graphing calculator to aid in your understanding of the material. They also include insightful suggestions for increasing calculator proficiency.

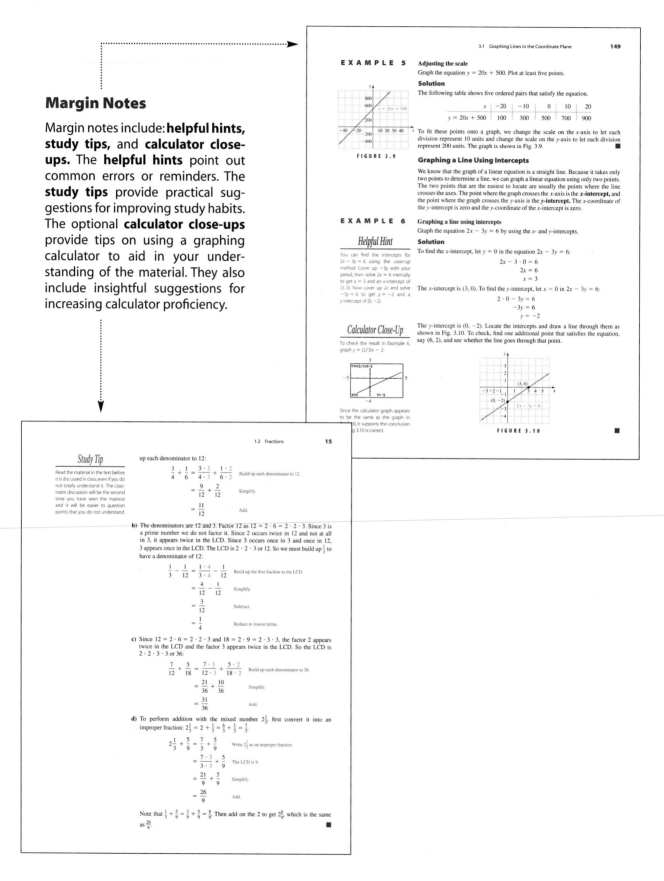

EXAMPLE 5 **Adjusting the scale**

Graph the equation $y = 20x + 500$. Plot at least five points.

Solution

The following table shows five ordered pairs that satisfy the equation.

x	-20	-10	0	10	20
$y = 20x + 500$	100	300	500	700	900

To fit these points onto a graph, we change the scale on the x-axis to let each division represent 10 units and change the scale on the y-axis to let each division represent 200 units. The graph is shown in Fig. 3.9. ■

FIGURE 3.9

Graphing a Line Using Intercepts

We know that the graph of a linear equation is a straight line. Because it takes only two points to determine a line, we can graph a linear equation using only two points. The two points that are the easiest to locate are usually the points where the line crosses the axes. The point where the graph crosses the x-axis is the ***x*-intercept,** and the point where the graph crosses the y-axis is the ***y*-intercept.** The x-coordinate of the y-intercept is zero and the y-coordinate of the x-intercept is zero.

EXAMPLE 6 **Graphing a line using intercepts**

Graph the equation $2x - 3y = 6$ by using the x- and y-intercepts.

Helpful Hint

You can find the intercepts for $2x - 3y = 6$ using the *cover-up method.* Cover up $-3y$ with your pencil, then solve $2x = 6$ mentally to get $x = 3$ and an x-intercept of $(3, 0)$. Now cover up $2x$ and solve $-3y = 6$ to get $y = -2$ and a y-intercept of $(0, -2)$.

Solution

To find the x-intercept, let $y = 0$ in the equation $2x - 3y = 6$:

$$2x - 3 \cdot 0 = 6$$
$$2x = 6$$
$$x = 3$$

The x-intercept is $(3, 0)$. To find the y-intercept, let $x = 0$ in $2x - 3y = 6$:

$$2 \cdot 0 - 3y = 6$$
$$-3y = 6$$
$$y = -2$$

The y-intercept is $(0, -2)$. Locate the intercepts and draw a line through them as shown in Fig. 3.10. To check, find one additional point that satisfies the equation, say $(6, 2)$, and see whether the line goes through that point.

Calculator Close-Up

To check the result in Example 6, graph $y = (2/3)x - 2$:

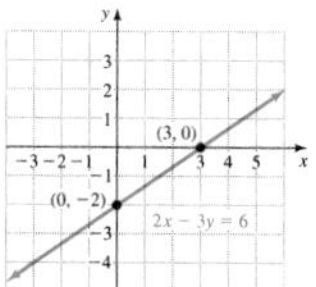

Since the calculator graph appears to be the same as the graph in [Fig. 3.1]0, it supports the conclusion [that Fi]g. 3.10 is correct.

FIGURE 3.10 ■

Study Tip

Read the material in the text before it is discussed in class, even if you do not totally understand it. The classroom discussion will be the second time you have seen the material and it will be easier to question points that you do not understand.

up each denominator to 12:

$$\frac{3}{4} + \frac{1}{6} = \frac{3 \cdot 3}{4 \cdot 3} + \frac{1 \cdot 2}{6 \cdot 2} \quad \text{Build up each denominator to 12.}$$
$$= \frac{9}{12} + \frac{2}{12} \quad \text{Simplify.}$$
$$= \frac{11}{12} \quad \text{Add.}$$

b) The denominators are 12 and 3. Factor 12 as $12 = 2 \cdot 6 = 2 \cdot 2 \cdot 3$. Since 3 is a prime number we do not factor it. Since 2 occurs twice in 12 and not at all in 3, it appears twice in the LCD. Since 3 occurs once in 3 and once in 12, 3 appears once in the LCD. The LCD is $2 \cdot 2 \cdot 3$ or 12. So we must build up $\frac{1}{3}$ to have a denominator of 12:

$$\frac{1}{3} - \frac{1}{12} = \frac{1 \cdot 4}{3 \cdot 4} - \frac{1}{12} \quad \text{Build up the first fraction to the LCD.}$$
$$= \frac{4}{12} - \frac{1}{12} \quad \text{Simplify.}$$
$$= \frac{3}{12} \quad \text{Subtract.}$$
$$= \frac{1}{4} \quad \text{Reduce to lowest terms.}$$

c) Since $12 = 2 \cdot 6 = 2 \cdot 2 \cdot 3$ and $18 = 2 \cdot 9 = 2 \cdot 3 \cdot 3$, the factor 2 appears twice in the LCD and the factor 3 appears twice in the LCD. So the LCD is $2 \cdot 2 \cdot 3 \cdot 3$ or 36:

$$\frac{7}{12} + \frac{5}{18} = \frac{7 \cdot 3}{12 \cdot 3} + \frac{5 \cdot 2}{18 \cdot 2} \quad \text{Build up each denominator to 36.}$$
$$= \frac{21}{36} + \frac{10}{36} \quad \text{Simplify.}$$
$$= \frac{31}{36} \quad \text{Add.}$$

d) To perform addition with the mixed number $2\frac{1}{3}$, first convert it into an improper fraction: $2\frac{1}{3} = 2 + \frac{1}{3} = \frac{6}{3} + \frac{1}{3} = \frac{7}{3}$.

$$2\frac{1}{3} + \frac{5}{9} = \frac{7}{3} + \frac{5}{9} \quad \text{Write } 2\tfrac{1}{3} \text{ as an improper fraction.}$$
$$= \frac{7 \cdot 3}{3 \cdot 3} + \frac{5}{9} \quad \text{The LCD is 9.}$$
$$= \frac{21}{9} + \frac{5}{9} \quad \text{Simplify.}$$
$$= \frac{26}{9} \quad \text{Add.}$$

Note that $\frac{1}{3} + \frac{5}{9} = \frac{3}{9} + \frac{5}{9} = \frac{8}{9}$. Then add on the 2 to get $2\frac{8}{9}$, which is the same as $\frac{26}{9}$. ■

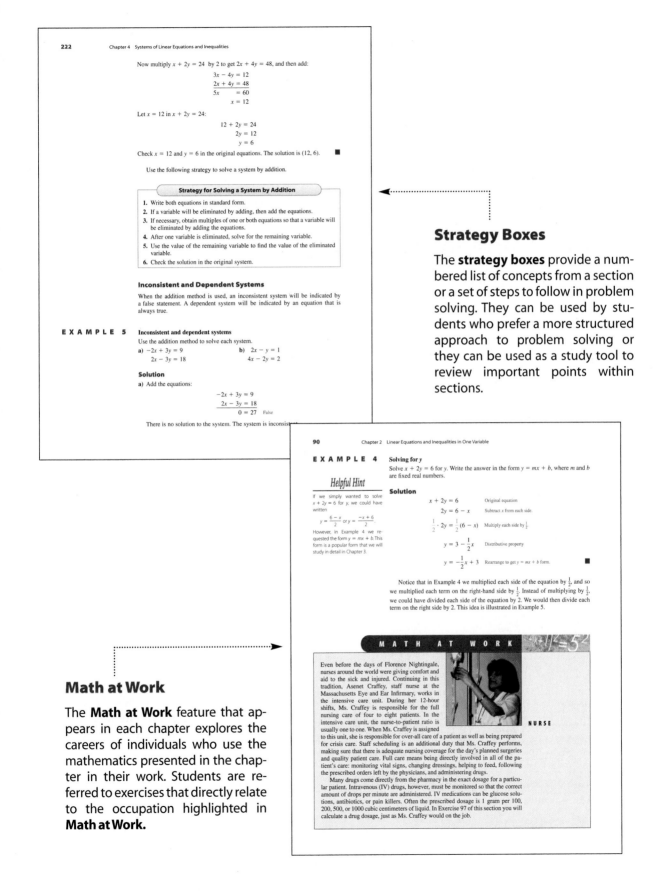

Now multiply $x + 2y = 24$ by 2 to get $2x + 4y = 48$, and then add:

$$\begin{aligned} 3x - 4y &= 12 \\ 2x + 4y &= 48 \\ \hline 5x \quad &= 60 \\ x &= 12 \end{aligned}$$

Let $x = 12$ in $x + 2y = 24$:

$$\begin{aligned} 12 + 2y &= 24 \\ 2y &= 12 \\ y &= 6 \end{aligned}$$

Check $x = 12$ and $y = 6$ in the original equations. The solution is (12, 6). ■

Use the following strategy to solve a system by addition.

Strategy for Solving a System by Addition

1. Write both equations in standard form.
2. If a variable will be eliminated by adding, then add the equations.
3. If necessary, obtain multiples of one or both equations so that a variable will be eliminated by adding the equations.
4. After one variable is eliminated, solve for the remaining variable.
5. Use the value of the remaining variable to find the value of the eliminated variable.
6. Check the solution in the original system.

Inconsistent and Dependent Systems

When the addition method is used, an inconsistent system will be indicated by a false statement. A dependent system will be indicated by an equation that is always true.

EXAMPLE 5 **Inconsistent and dependent systems**

Use the addition method to solve each system.

a) $-2x + 3y = 9$
$2x - 3y = 18$

b) $2x - y = 1$
$4x - 2y = 2$

Solution

a) Add the equations:

$$\begin{aligned} -2x + 3y &= 9 \\ 2x - 3y &= 18 \\ \hline 0 &= 27 \quad \text{False} \end{aligned}$$

There is no solution to the system. The system is inconsist

Strategy Boxes

The **strategy boxes** provide a numbered list of concepts from a section or a set of steps to follow in problem solving. They can be used by students who prefer a more structured approach to problem solving or they can be used as a study tool to review important points within sections.

EXAMPLE 4 **Solving for *y***

Solve $x + 2y = 6$ for y. Write the answer in the form $y = mx + b$, where m and b are fixed real numbers.

Helpful Hint

If we simply wanted to solve $x + 2y = 6$ for y, we could have written

$$y = \frac{6 - x}{2} \text{ or } y = \frac{-x + 6}{2}.$$

However, in Example 4 we requested the form $y = mx + b$. This form is a popular form that we will study in detail in Chapter 3.

Solution

$x + 2y = 6$ Original equation

$2y = 6 - x$ Subtract x from each side.

$\frac{1}{2} \cdot 2y = \frac{1}{2}(6 - x)$ Multiply each side by $\frac{1}{2}$.

$y = 3 - \frac{1}{2}x$ Distributive property

$y = -\frac{1}{2}x + 3$ Rearrange to get $y = mx + b$ form. ■

Notice that in Example 4 we multiplied each side of the equation by $\frac{1}{2}$, and so we multiplied each term on the right-hand side by $\frac{1}{2}$. Instead of multiplying by $\frac{1}{2}$, we could have divided each side of the equation by 2. We would then divide each term on the right side by 2. This idea is illustrated in Example 5.

MATH AT WORK

Even before the days of Florence Nightingale, nurses around the world were giving comfort and aid to the sick and injured. Continuing in this tradition, Asenet Craffey, staff nurse at the Massachusetts Eye and Ear Infirmary, works in the intensive care unit. During her 12-hour shifts, Ms. Craffey is responsible for the full nursing care of four to eight patients. In the intensive care unit, the nurse-to-patient ratio is usually one to one. When Ms. Craffey is assigned to this unit, she is responsible for over-all care of a patient as well as being prepared for crisis care. Staff scheduling is an additional duty that Ms. Craffey performs, making sure that there is adequate nursing coverage for the day's planned surgeries and quality patient care. Full care means being directly involved in all of the patient's care: monitoring vital signs, changing dressings, helping to feed, following the prescribed orders left by the physicians, and administering drugs.

NURSE

Many drugs come directly from the pharmacy in the exact dosage for a particular patient. Intravenous (IV) drugs, however, must be monitored so that the correct amount of drops per minute are administered. IV medications can be glucose solutions, antibiotics, or pain killers. Often the prescribed dosage is 1 gram per 100, 200, 500, or 1000 cubic centimeters of liquid. In Exercise 97 of this section you will calculate a drug dosage, just as Ms. Craffey would on the job.

Math at Work

The **Math at Work** feature that appears in each chapter explores the careers of individuals who use the mathematics presented in the chapter in their work. Students are referred to exercises that directly relate to the occupation highlighted in **Math at Work.**

CAUTION When dividing polynomials by long division, we do not stop until the remainder is 0 or the degree of the remainder is smaller than the degree of the divisor. For example, we stop dividing in Example 6 because the degree of the remainder -6 is 0 and the degree of the divisor $x - 2$ is 1.

WARM-UPS

True or false? Explain your answer.

1. $y^{10} \div y^2 = y^5$ for any nonzero value of y.
2. $\frac{7x+2}{7} = x + 2$ for any value of x.
3. $\frac{7x^2}{7} = x^2$ for any value of x.
4. If $3x^2 + 6$ is divided by 3, the quotient is $x^2 + 6$.
5. If $4y^2 - 6y$ is divided by $2y$, the quotient is $2y - 3$.
6. The quotient times the remainder plus the dividend equals the divisor.
7. $(x + 2)(x + 1) + 3 = x^2 + 3x + 5$ for any value of x.
8. If $x^2 + 3x + 5$ is divided by $x + 2$, then the quotient is $x + 1$.
9. If $x^2 + 3x + 5$ is divided by $x + 2$, the remainder is 3.
10. If the remainder is zero, then (divisor)(quotient) = dividend.

Warm-Ups

Warm-ups appear before each set of exercises at the end of every section. They are true or false statements that can be used to check conceptual understanding of material within each section.

5.5 EXERCISES

Reading and Writing After reading this section, write out the answers to these questions. Use complete sentences.

1. What rule is important for dividing monomials?
2. What is the meaning of a zero exponent?
3. How many terms should you get when dividing a polynomial by a monomial?
4. How should the terms of the polynomials be written when dividing with long division?
5. How do you know when to stop the process in long division of polynomials?
6. How do you handle missing terms in the dividend polynomial when doing long division?

Simplify each expression. See Example 1.

7. 9^0 8. m^0 9. $(-2x^3)^0$
10. $(5a^3b)^0$ 11. $2 \cdot 5^0 - 3^0$ 12. $-4^0 - 8^0$
13. $(2x - y)^0$ 14. $(a^2 + b^2)^0$

Find each quotient. Try to write only the answer. See Example 2.

15. $\frac{x^8}{x^2}$ 16. $\frac{y^9}{y^3}$
17. $\frac{6a^7}{2a^{12}}$ 18. $\frac{30b^2}{3b^6}$
19. $-12x^5 \div (3x^9)$ 20. $-6y^5 \div (-3y^{10})$
21. $-6y^2 \div (6y)$ 22. $-3a^2b \div (3ab)$
23. $\frac{-6x^3y^2}{2x^2y^2}$ 24. $\frac{-4h^2k^4}{-2hk^3}$
25. $\frac{-9x^2y^2}{3x^5y^2}$ 26. $\frac{-12z^4y^2}{-2z^{10}y^2}$

Find the quotients. See Example 3.

27. $\frac{3x - 6}{3}$

b) Use the accompanying graph to determine whether the sky diver (with no air resistance) falls farther in the first 5 seconds or the last 5 seconds of the fall.

c) Is the sky diver's velocity increasing or decreasing as she falls?

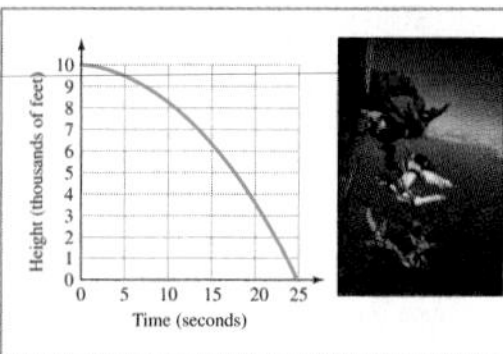

FIGURE FOR EXERCISE 67

68. ***Skydiving.*** If a sky diver jumps from an airplane at a height of 8256 feet, then for the first five seconds, her height above the earth is approximately given by the formula $h = -16t^2 + 8256$. How many seconds does it take her to reach 8000 feet.

69. ***Throwing a sandbag.*** If a balloonist throws a sandbag downward at 24 feet per second from an altitude of 720 feet, then its height (in feet) above the ground after t seconds is given by $S = -16t^2 - 24t + 720$. How long does it take for the sandbag to reach the earth? (On the ground, $S = 0$.)

70. ***Throwing a sandbag.*** If the balloonist of the previous exercise throws his sandbag downward from an altitude of 128 feet with an initial velocity of 32 feet per second, then its altitude after t seconds is given by the formula $S = -16t^2 - 32t + 128$. How long does it take for the sandbag to reach the earth?

71. ***Glass prism.*** One end of a glass prism is in the shape of a triangle with a height that is 1 inch longer than twice the base. If the area of the triangle is 39 square inches, then how long are the base and height?

72. ***Areas of two circles.*** The radius of a circle is 1 meter longer than the radius of another circle. If their areas differ by 5π square meters, then what is the radius of each?

73. ***Changing area.*** Last year Otto's garden was square. This year he plans to make it smaller by shortening one side 5 feet and the other 8 feet. If the area of the smaller garden will be 180 square feet, then what was the size of Otto's garden last year?

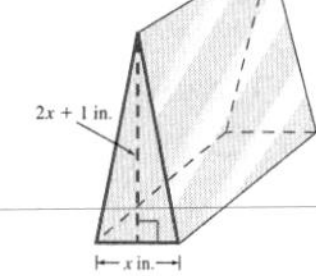

FIGURE FOR EXERCISE 71

74. ***Dimensions of a box.*** Rosita's Christmas present from Carlos is in a box that has a width that is 3 inches shorter than the height. The length of the base is 5 inches longer than the height. If the area of the base is 84 square inches, then what is the height of the package?

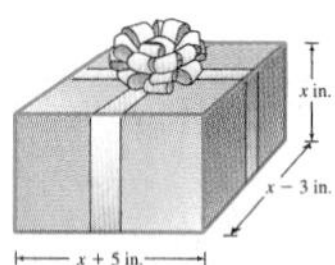

FIGURE FOR EXERCISE 74

75. ***Flying a kite.*** Imelda and Gordon have designed a new kite. While Imelda is flying the kite, Gordon is standing directly below it. The kite is designed so that its altitude is always 20 feet larger than the distance between Imelda and Gordon. What is the altitude of the kite when it is 100 feet from Imelda?

76. ***Avoiding a collision.*** A car is traveling on a road that is perpendicular to a railroad track. When the car is 30 meters from the crossing, the car's new collision detector warns the driver that there is a train 50 meters from the car and heading toward the same crossing. How far is the train from the crossing?

77. ***Carpeting two rooms.*** Virginia is buying carpet for two square rooms. One room is 3 yards wider than the other. If she needs 45 square yards of carpet, then what are the dimensions of each room?

Exercises

The theme of mathematics in everyday situations is carried over to the exercise sets. Applications based on real-world data are included in each set. The **Index of Selected Applications** can help students to quickly identify exercises that associate the mathematics that may be used in their areas of interest.

310 (6-8) Chapter 6 Factoring

Solve each problem by factoring.

87. ***Uniform motion.*** Helen traveled a distance of $20x + 40$ miles at 20 miles per hour on the Yellowhead Highway. Find a binomial that represents the time that she traveled.

88. ***Area of a painting.*** A rectangular painting with a width of x centimeters has an area of $x^2 + 50x$ square centimeters. Find a binomial that represents the length.

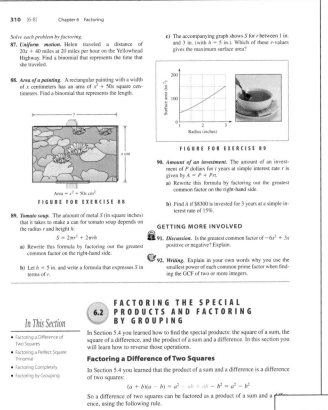

FIGURE FOR EXERCISE 88

89. ***Tomato soup.*** The amount of metal S (in square inches) that it takes to make a can for tomato soup depends on the radius r and height h:

$$S = 2\pi r^2 + 2\pi rh$$

a) Rewrite this formula by factoring out the greatest common factor on the right-hand side.

b) Let $h = 5$ in. and write a formula that expresses S in terms of r.

c) The accompanying graph shows S for r between 1 in. and 3 in. (with $h = 5$ in.). Which of these r-values gives the maximum surface area?

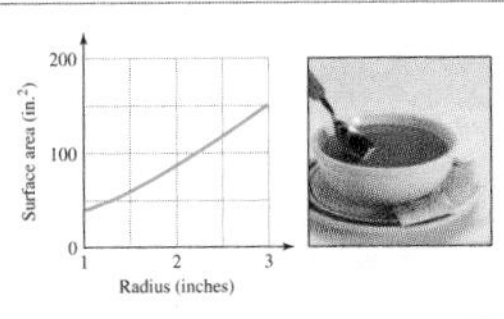

FIGURE FOR EXERCISE 89

90. ***Amount of an investment.*** The amount of an investment of P dollars for t years at simple interest rate r is given by $A = P + Prt$.

a) Rewrite this formula by factoring out the greatest common factor on the right-hand side.

b) Find A if \$8300 is invested for 3 years at a simple interest rate of 15%.

GETTING MORE INVOLVED

91. ***Discussion.*** Is the greatest common factor of $-6x^2 + 3x$ positive or negative? Explain.

92. ***Writing.*** Explain in your own words why you use the smallest power of each common prime factor when finding the GCF of two or more integers.

6.2 FACTORING THE SPECIAL PRODUCTS AND FACTORING BY GROUPING

In This Section

- Factoring a Difference of Two Squares
- Factoring a Perfect Square Trinomial
- Factoring Completely
- Factoring by Grouping

In Section 5.4 you learned how to find the special products: the square of a sum, the square of a difference, and the product of a sum and a difference. In this section you will learn how to reverse those operations.

Factoring a Difference of Two Squares

In Section 5.4 you learned that the product of a sum and a difference is a difference of two squares:

$$(a + b)(a - b) = a^2 - ab + ab - b^2 = a^2 - b^2$$

So a difference of two squares can be factored as a product of a sum and a diff-ence, using the following rule.

Getting More Involved appears within selected exercise sets. This feature may contain

 **Writing,**

 **Cooperative Learning,**

 **Exploration,** and/or

Discussion exercises. Each of these components is designed to give students an opportunity to improve and develop the ways in which they express mathematical ideas.

The exercise sets contain exercises that are keyed to examples, as well as exercises that are not keyed to examples.

406 (7-58) Chapter 7 Rational Expressions

Show a complete solution to each problem. See Example 7.

45. ***Apples and bananas.*** Bertha bought 18 pounds of fruit consisting of apples and bananas. She paid \$9 for the apples and \$2.40 for the bananas. If the price per pound of the apples was 3 times that of the bananas, then how many pounds of each type of fruit did she buy?

46. ***Running backs.*** In the playoff game the ball was carried by either Anderson or Brown on 21 plays. Anderson gained 36 yards, and Brown gained 54 yards. If Brown averaged twice as many yards per carry as Anderson, then on how many plays did Anderson carry the ball?

47. ***Fuel efficiency.*** Last week, Joe's Electric Service used 110 gallons of gasoline in its two trucks. The large truck was driven 800 miles, and the small truck was driven 600 miles. If the small truck gets twice as many miles per gallon as the large truck, then how many gallons of gasoline did the large truck use?

FIGURE FOR EXERCISE 46

48. ***Repair work.*** Sally received a bill for a total of 8 hours labor on the repair of her bulldozer. She paid \$50 to the master mechanic and \$90 to his apprentice. If the master mechanic gets \$10 more per hour than his apprentice, then how many hours did each work on the bulldozer?

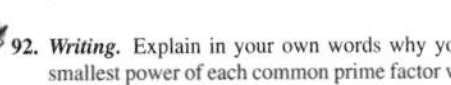

COLLABORATIVE ACTIVITIES

How Do I Get There from Here?

Grouping: Three students per group
Topic: Distance formula

Suppose that you want to decide whether to ride your bicycle, drive your car, or take the bus to school this year. The best thing to do is to analyze each of your options. Read all the information given below and then have each person in your group pick one mode of transportation. Working individually, answer the questions using the given information, the accompanying map, and the distance formula $D = RT$. Present a case to your group for your type of transportation. Compute costs for one entire 32-week school year (two 16-week semesters).

If you travel by car, you will need to pay for a parking permit, which costs \$150/year. Traffic has increased, so it takes 12 minutes to get to school. What is your average speed? Determine the cost (use current gas prices) of your gasoline for a full 32-week academic year assuming your car will get 30 mpg. Are there any other costs associated with using a car?

If you travel by bicycle, then you will need to buy a new bike lock for \$25 and two new tubes at \$5.00 apiece. As well as cost, determine how fast you would have to bike to beat the car.

If you travel by bus, then it will cost \$7.50 per month for a student bus pass. The bus stops at the end of your block. It leaves at 8:30 A.M. and will get to the college at 8:55 A.M. On Mondays and Wednesdays you have a 9:00 A.M. class, four blocks from the bus stop. As well as cost find the average speed of the bus and figure the cost for the full 32-week school year.

When you present your case, include the time needed to get there, speed, total cost, convenience, and any other expenses. Have at least three reasons why this would be the best way to travel. Consider unique features of your area such as traffic, weather, and terrain.

After each of you has presented your case, decide as a group which type of transportation to choose.

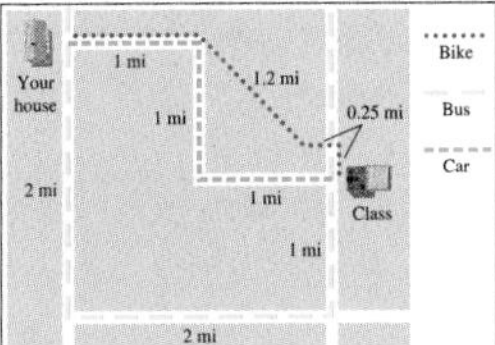

Collaborative Activities

Collaborative Activities appear at the end of each chapter. The activities are designed to encourage interaction and learning in a group setting.

Calculator Exercises

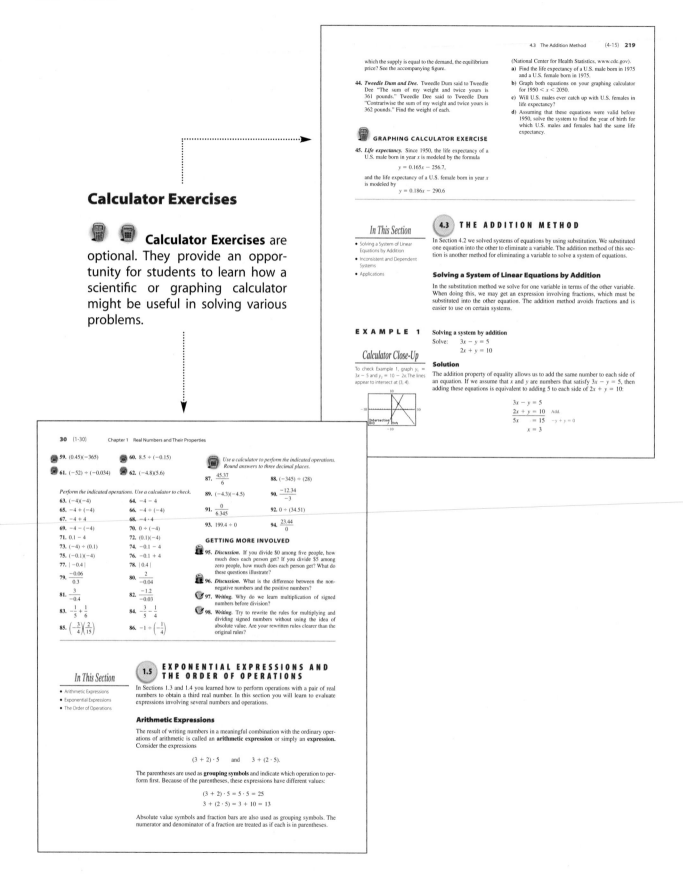

Calculator Exercises are optional. They provide an opportunity for students to learn how a scientific or graphing calculator might be useful in solving various problems.

which the supply is equal to the demand, the equilibrium price? See the accompanying figure.

44. ***Tweedle Dum and Dee.*** Tweedle Dum said to Tweedle Dee "The sum of my weight and twice yours is 361 pounds." Tweedle Dee said to Tweedle Dum "Contrariwise the sum of my weight and twice yours is 362 pounds." Find the weight of each.

GRAPHING CALCULATOR EXERCISE

45. ***Life expectancy.*** Since 1950, the life expectancy of a U.S. male born in year x is modeled by the formula

$$y = 0.165x - 256.7,$$

and the life expectancy of a U.S. female born in year x is modeled by

$$y = 0.186x - 290.6$$

(National Center for Health Statistics, www.cdc.gov).

a) Find the life expectancy of a U.S. male born in 1975 and a U.S. female born in 1975.

b) Graph both equations on your graphing calculator for $1950 < x < 2050$.

c) Will U.S. males ever catch up with U.S. females in life expectancy?

d) Assuming that these equations were valid before 1950, solve the system to find the year of birth for which U.S. males and females had the same life expectancy.

In This Section

- Solving a System of Linear Equations by Addition
- Inconsistent and Dependent Systems
- Applications

4.3 THE ADDITION METHOD

In Section 4.2 we solved systems of equations by using substitution. We substituted one equation into the other to eliminate a variable. The addition method of this section is another method for eliminating a variable to solve a system of equations.

Solving a System of Linear Equations by Addition

In the substitution method we solve for one variable in terms of the other variable. When doing this, we may get an expression involving fractions, which must be substituted into the other equation. The addition method avoids fractions and is easier to use on certain systems.

EXAMPLE 1 **Solving a system by addition**

Solve: $3x - y = 5$
$2x + y = 10$

Calculator Close-Up

To check Example 1, graph $y_1 = 3x - 5$ and $y_2 = 10 - 2x$. The lines appear to intersect at (3, 4).

Solution

The addition property of equality allows us to add the same number to each side of an equation. If we assume that x and y are numbers that satisfy $3x - y = 5$, then adding these equations is equivalent to adding 5 to each side of $2x + y = 10$:

$$\begin{aligned} 3x - y &= 5 \\ 2x + y &= 10 \quad \text{Add.} \\ 5x &= 15 \quad -y + y = 0 \\ x &= 3 \end{aligned}$$

59. $(0.45)(-365)$ **60.** $8.5 \div (-0.15)$

61. $(-52) \div (-0.034)$ **62.** $(-4.8)(5.6)$

Perform the indicated operations. Use a calculator to check.

63. $(-4)(-4)$ **64.** $-4 - 4$

65. $-4 + (-4)$ **66.** $-4 \div (-4)$

67. $-4 + 4$ **68.** $-4 \cdot 4$

69. $-4 - (-4)$ **70.** $0 \div (-4)$

71. $0.1 - 4$ **72.** $(0.1)(-4)$

73. $(-4) \div (0.1)$ **74.** $-0.1 - 4$

75. $(-0.1)(-4)$ **76.** $-0.1 + 4$

77. $|-0.4|$ **78.** $|0.4|$

79. $\frac{-0.06}{0.3}$ **80.** $\frac{2}{-0.04}$

81. $\frac{3}{-0.4}$ **82.** $\frac{-1.2}{-0.03}$

83. $-\frac{1}{5} + \frac{1}{6}$ **84.** $-\frac{3}{5} - \frac{1}{4}$

85. $\left(-\frac{3}{4}\right)\left(\frac{2}{15}\right)$ **86.** $-1 \div \left(-\frac{1}{4}\right)$

Use a calculator to perform the indicated operations. Round answers to three decimal places.

87. $\frac{45.37}{6}$ **88.** $(-345) \div (28)$

89. $(-4.3)(-4.5)$ **90.** $\frac{-12.34}{-3}$

91. $\frac{0}{6.345}$ **92.** $0 \div (34.51)$

93. $199.4 \div 0$ **94.** $\frac{23.44}{0}$

GETTING MORE INVOLVED

95. ***Discussion.*** If you divide \$0 among five people, how much does each person get? If you divide \$5 among zero people, how much does each person get? What do these questions illustrate?

96. ***Discussion.*** What is the difference between the nonnegative numbers and the positive numbers?

97. ***Writing.*** Why do we learn multiplication of signed numbers before division?

98. ***Writing.*** Try to rewrite the rules for multiplying and dividing signed numbers without using the idea of absolute value. Are your rewritten rules clearer than the original rules?

In This Section

- Arithmetic Expressions
- Exponential Expressions
- The Order of Operations

1.5 EXPONENTIAL EXPRESSIONS AND THE ORDER OF OPERATIONS

In Sections 1.3 and 1.4 you learned how to perform operations with a pair of real numbers to obtain a third real number. In this section you will learn to evaluate expressions involving several numbers and operations.

Arithmetic Expressions

The result of writing numbers in a meaningful combination with the ordinary operations of arithmetic is called an **arithmetic expression** or simply an **expression.** Consider the expressions

$$(3 + 2) \cdot 5 \quad \text{and} \quad 3 + (2 \cdot 5).$$

The parentheses are used as **grouping symbols** and indicate which operation to perform first. Because of the parentheses, these expressions have different values:

$$(3 + 2) \cdot 5 = 5 \cdot 5 = 25$$
$$3 + (2 \cdot 5) = 3 + 10 = 13$$

Absolute value symbols and fraction bars are also used as grouping symbols. The numerator and denominator of a fraction are treated as if each is in parentheses.

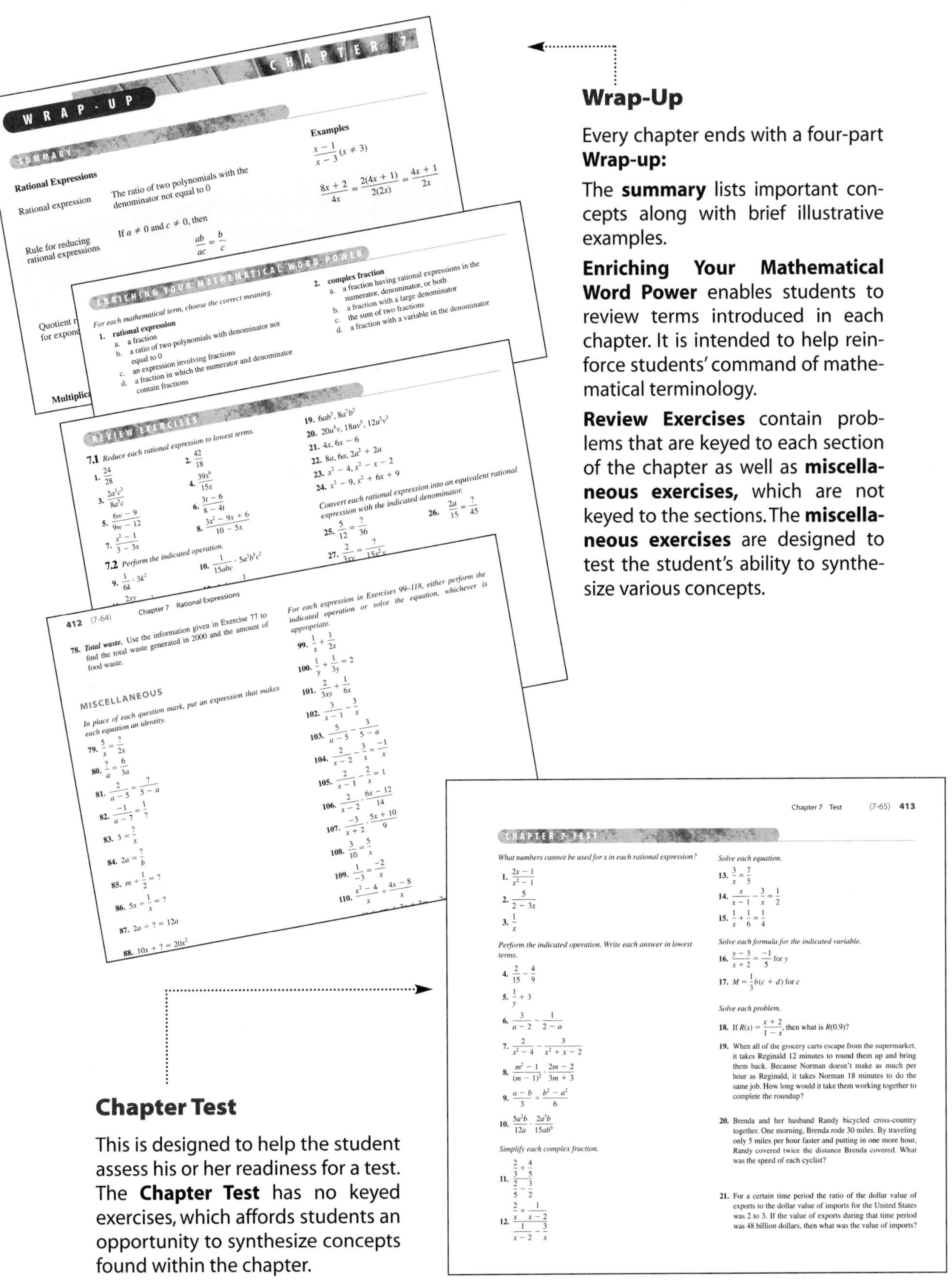

Wrap-Up

Every chapter ends with a four-part **Wrap-up:**

The **summary** lists important concepts along with brief illustrative examples.

Enriching Your Mathematical Word Power enables students to review terms introduced in each chapter. It is intended to help reinforce students' command of mathematical terminology.

Review Exercises contain problems that are keyed to each section of the chapter as well as **miscellaneous exercises,** which are not keyed to the sections. The **miscellaneous exercises** are designed to test the student's ability to synthesize various concepts.

Chapter Test

This is designed to help the student assess his or her readiness for a test. The **Chapter Test** has no keyed exercises, which affords students an opportunity to synthesize concepts found within the chapter.

MAKING CONNECTIONS CHAPTERS 1–7

Solve each equation.

1. $3x - 2 = 5$
2. $\frac{3}{5}x = -2$
3. $2(x - 2) = 4x$
4. $2(x - 2) = 2x$
5. $2(x + 3) = 6x + 6$
6. $2(3x + 4) + x^2 = 0$
7. $4x - 4x^3 = 0$
8. $\frac{3}{x} = \frac{-2}{5}$
9. $\frac{3}{x} = \frac{x}{12}$
10. $\frac{x}{2} = \frac{4}{x - 2}$
11. $\frac{w}{18} - \frac{w - 1}{9} = \frac{4 - w}{6}$
12. $\frac{x}{x + 1} + \frac{1}{2x + 2} = \frac{7}{8}$

Solve each equation for y.

13. $2x + 3y = c$
14. $\frac{y - 3}{x - 5} = \frac{1}{2}$
15. $2y = ay + c$
16. $\frac{A}{y} = \frac{C}{B}$
17. $\frac{A}{y} + \frac{1}{3} = \frac{B}{y}$
18. $\frac{A}{y} - \frac{1}{2} = \frac{1}{3}$
19. $3y - 5ay = 8$
20. $y^2 - By = 0$
21. $A = \frac{1}{2}h(b + y)$
22. $2(b + y) = b$

Calculate the value of $b^2 - 4ac$ for each choice of a, b, and c.

23. $a = 1, b = 2, c = -15$
24. $a = 1, b = 8, c = 12$
25. $a = 2, b = 5, c = -3$
26. $a = 6, b = 7, c = -3$

Perform each indicated operation.

27. $(3x - 5) - (5x - 3)$
28. $(2a - 5)(a - 3)$
29. $x^7 \div x^3$
30. $\frac{x - 3}{5} + \frac{x + 4}{5}$
31. $\frac{1}{2} \cdot \frac{1}{x}$
32. $\frac{1}{2} + \frac{1}{x}$
33. $\frac{1}{2} \div \frac{1}{x}$
34. $\frac{1}{2} - \frac{1}{x}$
35. $\frac{x - 3}{5} - \frac{x + 4}{5}$
36. $\frac{3a}{2} \div 2$
37. $(x - 8)(x + 8)$
38. $3x(x^2 - 7)$
39. $2a^5 \cdot 5a^9$
40. $x^2 \cdot x^8$
41. $(k - 6)^2$
42. $(j + 5)^2$
43. $(g - 3) \div (3 - g)$
44. $(6x^3 - 8x^2) \div (2x)$

Solve.

45. ***Present value.*** An investor is interested in the amount or present value that she would have to invest today to receive periodic payments in the future. The present value of $1 in one year and $1 in two years with interest rate r compounded annually is given by the formula

$$P = \frac{1}{1 + r} + \frac{1}{(1 + r)^2}.$$

   a) Rewrite the formula so that the right-hand side is a single rational expression.
   b) Find P if $r = 7\%$.
   c) The present value of $1 per year for the next 10 years is given by the formula

$$P = \frac{1}{1 + r} + \frac{1}{(1 + r)^2} + \frac{1}{(1 + r)^3} + \cdots + \frac{1}{(1 + r)^{10}}.$$

   Use this formula to find P if $r = 5\%$.

414

Making Connections

These nonkeyed exercises are designed to help students synthesize new material with ideas from previous chapters and, in some cases, review material necessary for success in the upcoming chapter. They may serve as a cumulative review.

CHAPTER 1

Real Numbers and Their Properties

It has been said that baseball is the "great American pastime." All of us who have played the game or who have only been spectators believe we understand the game. But do we realize that a pitcher must aim for an invisible three-dimensional target that is about 20 inches wide by 23 inches high by 17 inches deep and that a pitcher must throw so that the batter has difficulty hitting the ball? A curve ball may deflect 14 inches to skim over the outside corner of the plate, or a knuckle ball can break 11 inches off center when it is 20 feet from the plate and then curve back over the center of the plate.

The batter is trying to hit a rotating ball that can travel up to 120 miles per hour and must make split-second decisions about shifting his weight, changing his stride, and swinging the bat. The size of the bat each batter uses depends on his strengths, and pitchers in turn try to capitalize on a batter's weaknesses.

Millions of baseball fans enjoy watching this game of strategy and numbers. Many watch their favorite teams at the local ball parks, while others cheer for the home team on television. Of course, baseball fans are always interested in which team is leading the division and the number of games that their favorite team is behind the leader. Finding the number of games behind for each team in the division involves both arithmetic and algebra. Algebra provides the formula for finding games behind, and arithmetic is used to do the computations. In Exercise 97 of Section 1.6 we will find the number of games behind for each team in the American League West.

In This Section

- The Integers
- The Rational Numbers
- The Number Line
- The Real Numbers
- Absolute Value

1.1 THE REAL NUMBERS

The numbers that we use in algebra are called the real numbers. We start the discussion of the real numbers with some simpler sets of numbers.

The Integers

The most fundamental collection or **set** of numbers is the set of **counting numbers** or **natural numbers.** Of course, these are the numbers that we use for counting. The set of natural numbers is written in symbols as follows.

The Natural Numbers

$$\{1, 2, 3, \ldots\}$$

Braces, { }, are used to indicate a set of numbers. The three dots after 1, 2, and 3, which are read "and so on," mean that the pattern continues without end. There are infinitely many natural numbers.

The natural numbers, together with the number 0, are called the **whole numbers.** The set of whole numbers is written as follows.

The Whole Numbers

$$\{0, 1, 2, 3, \ldots\}$$

1ST NATIONAL BANK
Account No.
36207
STATEMENT OF ACCOUNT
PREVIOUS BALANCE $ 50
CHECKS PAID $ 70
NEW BALANCE −$ 20
Please call the bank at your earliest convenience.

FIGURE 1.1

Although the whole numbers have many uses, they are not adequate for indicating losses or debts. A debt of \$20 can be expressed by the negative number -20 (negative twenty). See Fig. 1.1. When a thermometer reads 10 degrees below zero on a Fahrenheit scale, we say that the temperature is $-10°$F. See Fig. 1.2. The whole numbers together with the negatives of the counting numbers form the set of **integers.**

100
90
80
70
60
50
40
30
20
10
0
−10
−20

Degrees Fahrenheit

FIGURE 1.2

The Integers

$$\{\ldots, -3, -2, -1, 0, 1, 2, 3, \ldots\}$$

The Rational Numbers

A **rational number** is any number that can be expressed as a ratio (or quotient) of two integers. The set of rational numbers includes both the positive and negative fractions. We cannot list the rational numbers as easily as we listed the numbers in the other sets we have been discussing. So we write the set of rational numbers in symbols using **set-builder notation** as follows.

The Rational Numbers

$$\left\{\frac{a}{b} \,\middle|\, a \text{ and } b \text{ are integers, with } b \neq 0\right\}$$

↑ The set of ↑ such that ↑ conditions

Helpful Hint

Rational numbers are used for ratios. For example, if 2 out of 5 students surveyed attend summer school, then the ratio of students who attend summer school to the total number surveyed is 2/5. Note that the ratio 2/5 does not tell how many were surveyed or how many attend summer school.

We read this notation as "the set of numbers of the form $\frac{a}{b}$ such that a and b are integers, with $b \neq 0$." Note how we use the letters a and b to represent numbers here. A letter used to represent some numbers is called a **variable.**

Examples of rational numbers are

$$\frac{3}{1},\ \frac{5}{4},\ -\frac{7}{10},\ \frac{0}{6},\ \frac{5}{1},\ -\frac{77}{3},\ \text{and}\ \frac{-3}{-6}.$$

Note that we usually use simpler forms for some of these rational numbers. For instance, $\frac{3}{1} = 3$ and $\frac{0}{6} = 0$. The integers are rational numbers because any integer can be written with a denominator of 1.

If you divide the denominator into the numerator, then you can convert a rational number to decimal form. As a decimal, every rational number either repeats indefinitely $\left(\text{for example, } \frac{1}{3} = 0.333\ldots\right)$ or terminates $\left(\text{for example, } \frac{1}{8} = 0.125\right)$.

The Number Line

The number line is a diagram that helps us to visualize numbers and their relationships to each other. A number line is like the scale on the thermometer in Fig. 1.2. To construct a number line, we draw a straight line and label any convenient point with the number 0. Now we choose any convenient length and use it to locate other points. Points to the right of 0 correspond to the positive numbers, and points to the left of 0 correspond to the negative numbers. Zero is neither positive nor negative. The number line is shown in Fig. 1.3.

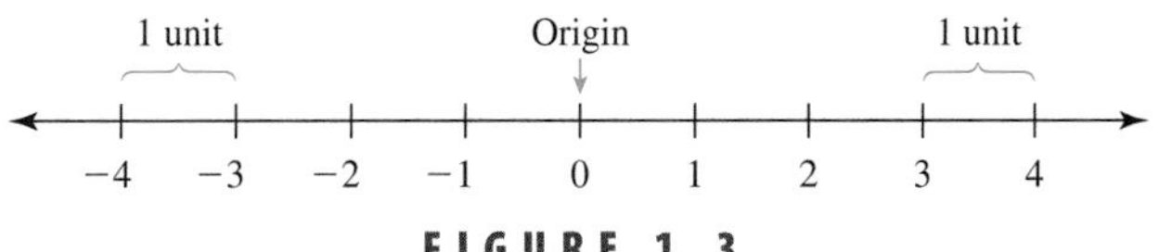

FIGURE 1.3

The numbers corresponding to the points on the line are called the **coordinates** of the points. The distance between two consecutive integers is called a **unit** and is the same for any two consecutive integers. The point with coordinate 0 is called the **origin.** The numbers on the number line increase in size from left to right. *When we compare the size of any two numbers, the larger number lies to the right of the smaller on the number line.* Zero is larger than any negative number and smaller than any positive number.

EXAMPLE 1 **Comparing numbers on a number line**

Determine which number is the larger in each given pair of numbers.

a) $-3, 2$ **b)** $0, -4$ **c)** $-2, -1$

Solution

a) The larger number is 2, because 2 lies to the right of -3 on the number line. In fact, any positive number is larger than any negative number.

b) The larger number is 0, because 0 lies to the right of -4 on the number line.

c) The larger number is -1, because -1 lies to the right of -2 on the number line. ■

The set of integers is illustrated or **graphed** in Fig. 1.4, on the next page by drawing a point for each integer. The three dots to the right and left below the number line and the blue arrows indicate that the numbers go on indefinitely in both directions.

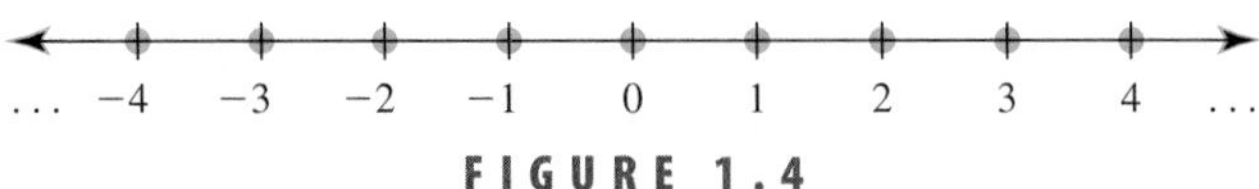

FIGURE 1.4

EXAMPLE 2 **Graphing numbers on a number line**

List the numbers described, and graph the numbers on a number line.

a) The whole numbers less than 4

b) The integers between 3 and 9

c) The integers greater than -3

Solution

a) The whole numbers less than 4 are 0, 1, 2, and 3. These numbers are shown in Fig. 1.5.

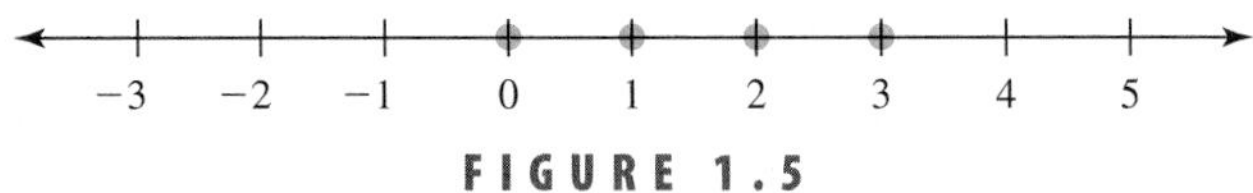

FIGURE 1.5

b) The integers between 3 and 9 are 4, 5, 6, 7, and 8. Note that 3 and 9 are not considered to be *between* 3 and 9. The graph is shown in Fig. 1.6.

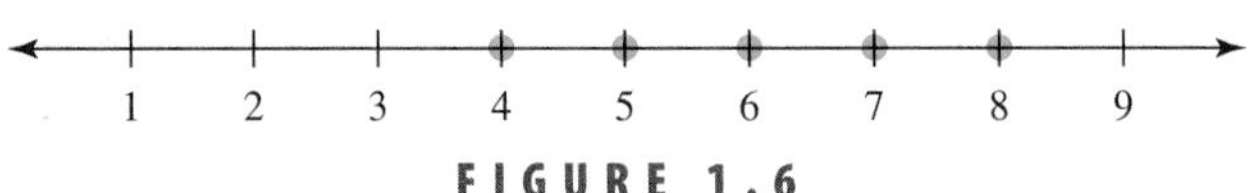

FIGURE 1.6

c) The integers greater than -3 are -2, -1, 0, 1, and so on. To indicate the continuing pattern, we use three dots on the graph shown in Fig. 1.7.

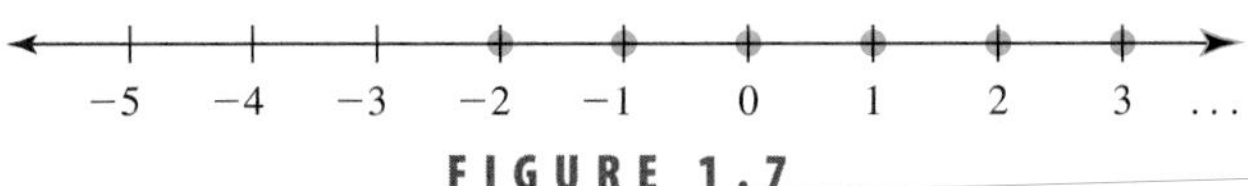

FIGURE 1.7

The Real Numbers

For every rational number there is a point on the number line. For example, the number $\frac{1}{2}$ corresponds to a point halfway between 0 and 1 on the number line, and $-\frac{5}{4}$ corresponds to a point one and one-quarter units to the left of 0, as shown in Fig. 1.8. Since there is a correspondence between numbers and points on the number line, the points are often referred to as numbers.

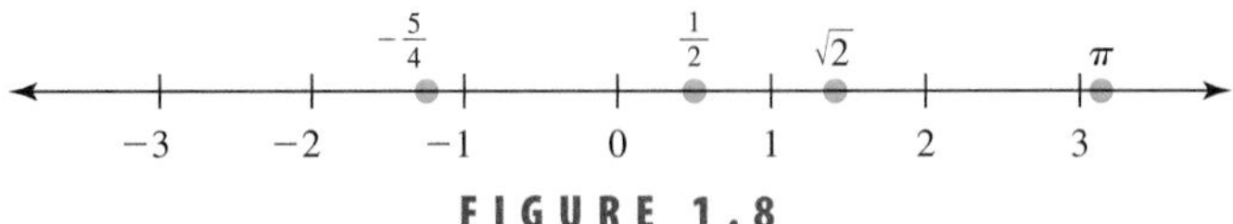

FIGURE 1.8

The set of numbers that corresponds to *all* points on a number line is called the set of **real numbers** or R. A graph of the real numbers is shown on a number line by shading all points as in Fig. 1.9. All rational numbers are real numbers, but there are

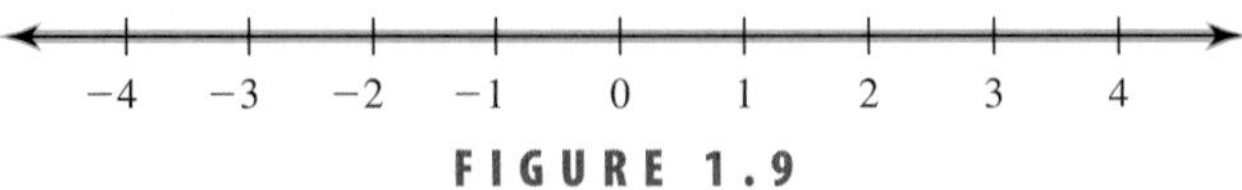

FIGURE 1.9

Study Tip

Start a personal library. This book as well as other books that you study from should be the basis for your library. You can also add books to your library at garage-sale prices when your bookstore sells its old texts. If you need to reference some material in the future, it is much easier to use a familiar book.

Calculator Close-Up

A calculator can give rational approximations for irrational numbers such as $\sqrt{2}$ and π.

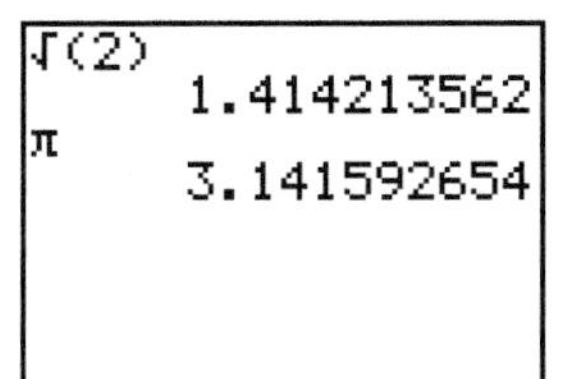

The calculator screens in this text may differ from the screen of the calculator model you use. If so, you may have to consult your manual to get the desired results.

points on the number line that do not correspond to rational numbers. Those real numbers that are not rational are called **irrational.** An irrational number cannot be written as a ratio of integers. It can be shown that numbers such as $\sqrt{2}$ (the square root of 2) and π (Greek letter pi) are irrational. The number $\sqrt{2}$ is a number that can be multiplied by itself to obtain 2 $(\sqrt{2} \cdot \sqrt{2} = 2)$. The number π is the ratio of the circumference and diameter of any circle. Irrational numbers are not as easy to represent as rational numbers. That is why we use symbols such as $\sqrt{2}$, $\sqrt{3}$, and π for irrational numbers. When we perform computations with irrational numbers, we sometimes use rational approximations for them. For example, $\sqrt{2} \approx 1.414$ and $\pi \approx 3.14$. The symbol $\approx$ means "is approximately equal to." Note that not all square roots are irrational. For example, $\sqrt{9} = 3$, because $3 \cdot 3 = 9$. We will deal with irrational numbers in greater depth when we discuss roots in Chapter 8.

Figure 1.10 summarizes the sets of numbers that make up the real numbers, and shows the relationships between them.

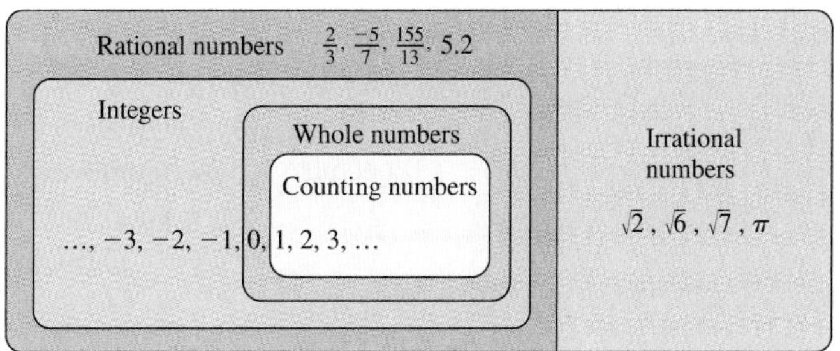

FIGURE 1.10

EXAMPLE 3 **Types of numbers**

Determine whether each statement is true or false.

a) Every rational number is an integer.

b) Every counting number is an integer.

c) Every irrational number is a real number.

Solution

a) False. For example, $\frac{1}{2}$ is a rational number that is not an integer.

b) True, because the integers consist of the counting numbers, the negatives of the counting numbers, and zero.

c) True, because the rational numbers together with the irrational numbers form the real numbers. ■

Helpful Hint

In mathematics, every statement is either true or false. If there is one exception to the statement, then the statement is false. If there is an exception and that is part of the statement, then the statement is true.

Absolute Value

The concept of absolute value will be used to define the basic operations with real numbers in Section 1.3. The **absolute value** of a number is the number's distance from 0 on the number line. For example, the numbers 5 and -5 are both five units away from 0 on the number line. So the absolute value of each of these numbers is 5. See Fig. 1.11. We write $|a|$ for "the absolute value of a." So

$$|5| = 5 \quad \text{and} \quad |-5| = 5.$$

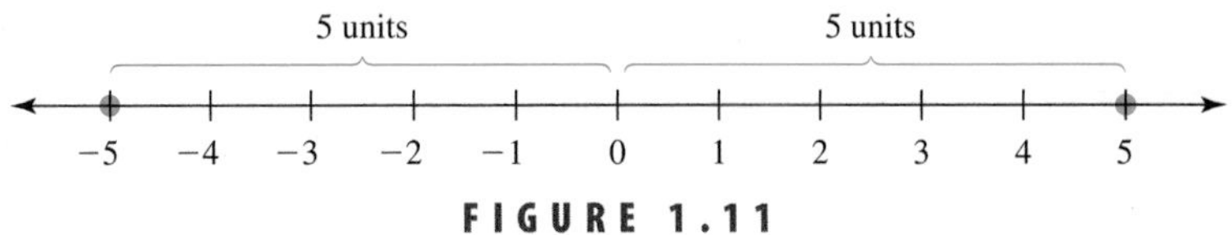

FIGURE 1.11

The notation $|a|$ represents distance, and distance is never negative. So $|a|$ is greater than or equal to zero for any real number a.

EXAMPLE 4 **Finding absolute value**

Evaluate.

a) $|3|$ **b)** $|-3|$ **c)** $|0|$

d) $\left|\frac{2}{3}\right|$ **e)** $|-0.39|$

Study Tip

Exercise sets are designed to gradually increase in difficulty. So start from the beginning and work lots of exercises. If you get stuck, go back and study the corresponding examples. If you are still stuck, move ahead to a new type of exercise.

Solution

a) $|3| = 3$ because 3 is three units away from 0.

b) $|-3| = 3$ because -3 is three units away from 0.

c) $|0| = 0$ because 0 is zero units away from 0.

d) $\left|\frac{2}{3}\right| = \frac{2}{3}$

e) $|-0.39| = 0.39$ ■

Two numbers that are located on opposite sides of zero and have the same absolute value are called **opposites** of each other. The numbers 5 and -5 are opposites of each other. We say that the opposite of 5 is -5 and the opposite of -5 is 5. The symbol "$-$" is used to indicate "opposite" as well as "negative." When the negative sign is used before a number, it should be read as "negative." When it is used in front of parentheses or a variable, it should be read as "opposite." For example, $-(5) = -5$ means "the opposite of 5 is negative 5," and $-(-5) = 5$ means "the opposite of negative 5 is 5." Zero does not have an opposite in the same sense as nonzero numbers. Zero is its own opposite. We read $-(0) = 0$ as the "the opposite of zero is zero."

In general, $-a$ means "the opposite of a." If a is positive, $-a$ is negative. If a is negative, $-a$ is positive. Opposites have the following property.

Opposite of an Opposite

For any real number a,

$$-(-a) = a.$$

Remember that we have defined $|a|$ to be the distance between 0 and a on the number line. Using opposites, we can give a symbolic definition of absolute value.

Absolute Value

$$|a| = \begin{cases} a & \text{if } a \text{ is positive or zero} \\ -a & \text{if } a \text{ is negative} \end{cases}$$

EXAMPLE 5 **Using the symbolic definition of absolute value**

Evaluate.

a) $|8|$

b) $|0|$

c) $|-8|$

Solution

a) If a is positive, then $|a| = a$. Since $8 > 0$, $|8| = 8$.

b) If a is 0, then $|a| = a$. So $|0| = 0$.

c) If a is negative, then $|a| = -a$. So $|-8| = -(-8) = 8$. ■

WARM-UPS

True or false? Explain your answer.

1. The natural numbers and the counting numbers are the same. True
2. The number 8,134,562,877,565 is a counting number. True
3. Zero is a counting number. False
4. Zero is not a rational number. False
5. The opposite of negative 3 is positive 3. True
6. The absolute value of 4 is -4. False
7. $-(-9) = 9$ True
8. $-(-b) = b$ for any number b. True
9. Negative six is greater than negative three. False
10. Negative five is between four and six. False

1.1 EXERCISES

Reading and Writing *After reading this section write out the answers to these questions. Use complete sentences.*

1. What are the integers?
The integers are the numbers in the set $\{\ldots, -3, -2, -1, 0, 1, 2, 3, \ldots\}$.

2. What are the rational numbers?
The rational numbers are numbers of the form $\frac{a}{b}$ where a and b are integers and $b \neq 0$.

3. What is the difference between a rational and an irrational number?
A rational number is a ratio of integers and an irrational number is not.

4. What is a number line?
A number line is a line on which there is a point corresponding to every real number.

5. How do you know that one number is larger than another?
The number a is larger than b if a lies to the right of b on the number line.

6. What is the ratio of the circumference and diameter of any circle?
The ratio of the circumference and diameter of any circle is the number π, which is approximately 3.14.

Determine which number is the larger in each given pair of numbers. See Example 1.

7. $-3, 6$ 6

8. $7, -10$ 7

9. $0, -6$ 0

10. $-8, 0$ 0

11. $-3, -2$ -2

12. $-5, -8$ -5

13. $-12, -15$ -12

14. $-13, -7$ -7

List the numbers described and graph them on a number line. See Example 2.

15. The counting numbers smaller than 6

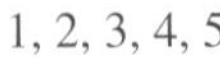
1, 2, 3, 4, 5

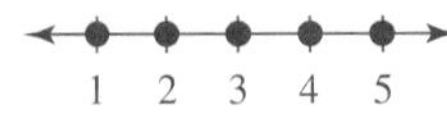

16. The natural numbers larger than 4

5, 6, 7, 8, 9, . . .

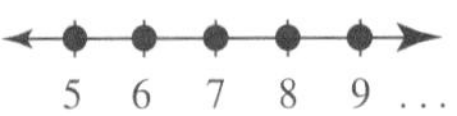

17. The whole numbers smaller than 5

0, 1, 2, 3, 4

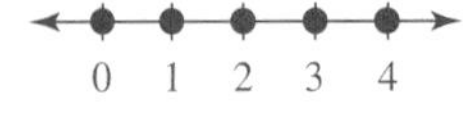

18. The integers between −3 and 3

−2, −1, 0, 1, 2

−2 −1 0 1 2

19. The whole numbers between −5 and 5

0, 1, 2, 3, 4

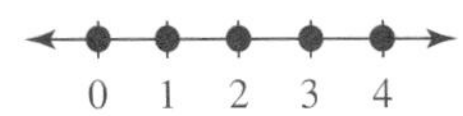

20. The integers smaller than −1

−2, −3, −4, −5, . . .

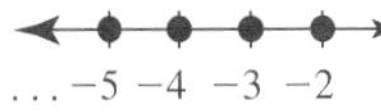

21. The counting numbers larger than −4

1, 2, 3, 4, 5, . . .

22. The natural numbers between −5 and 7

1, 2, 3, 4, 5, 6

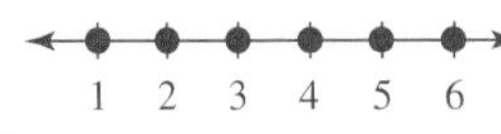

23. The integers larger than $\frac{1}{2}$

1, 2, 3, 4, 5, . . .

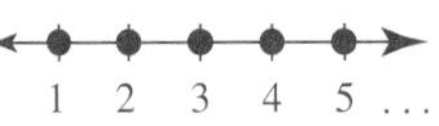

24. The whole numbers smaller than $\frac{7}{4}$

0, 1

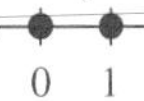

Determine whether each statement is true or false. Explain your answer. See Example 3.

25. Every integer is a rational number. True

26. Every counting number is a whole number. True

27. Zero is a counting number. False

28. Every whole number is a counting number. False

29. The ratio of the circumference and diameter of a circle is an irrational number. True

30. Every rational number can be expressed as a ratio of integers. True

31. Every whole number can be expressed as a ratio of integers. True

32. Some of the rational numbers are integers. True

33. Some of the integers are natural numbers. True

34. There are infinitely many rational numbers. True

35. Zero is an irrational number. False

36. Every irrational number is a real number. True

Determine the values of the following. See Examples 4 and 5.

37. $|-6|$ 6

38. $|4|$ 4

39. $|0|$ 0

40. $|2|$ 2

41. $|7|$ 7

42. $|-7|$ 7

43. $|-9|$ 9

44. $|-2|$ 2

45. $|-45|$ 45

46. $|-30|$ 30

47. $\left|\frac{3}{4}\right|$ $\frac{3}{4}$

48. $\left|-\frac{1}{2}\right|$ $\frac{1}{2}$

49. $|-5.09|$ 5.09

50. $|0.00987|$ 0.00987

Select the smaller number in each given pair of numbers.

51. −16, 9 −16

52. −12, −7 −12

53. $-\frac{5}{2}, -\frac{9}{4}$ $-\frac{5}{2}$

54. $\frac{5}{8}, \frac{6}{7}$ $\frac{5}{8}$

55. $|-3|, 2$ 2

56. $|-6|, 0$ 0

57. $|-4|, 3$ 3

58. $|5|, -4$ −4

Which number in each given pair has the larger absolute value?

59. −5, −9 −9

60. −12, −8 −12

61. 16, −9 16

62. −12, 7 −12

Determine which number in each pair is closer to 0 on the number line.

63. −4, −5 −4

64. −8.1, 7.9 7.9

65. −2.01, −1.99 −1.99

66. 2.01, 1.99 1.99

67. −75, 74 74

68. −75, −74 −74

Consider the following nine integers:

$$-4, -3, -2, -1, 0, 1, 2, 3, 4$$

69. Which of these integers has an absolute value equal to 3? −3 and 3

70. Which of these integers has an absolute value equal to 0? 0

71. Which of these integers has an absolute value greater than 2? −4, −3, 3, 4

72. Which of these integers has an absolute value greater than 1? −4, −3, −2, 2, 3, 4

73. Which of these integers has an absolute value less than 2? −1, 0, 1

74. Which of these integers has an absolute value less than 4? −3, −2, −1, 0, 1, 2, 3

True or false? Explain your answer.

75. If we add the absolute values of −3 and −5, we get 8. True

76. If we multiply the absolute values of −2 and 5, we get 10. True

77. The absolute value of any negative number is greater than 0. True

78. The absolute value of any positive number is less than 0. False

79. The absolute value of -9 is larger than the absolute value of 6. True

80. The absolute value of 12 is larger than the absolute value of -11. True

GETTING MORE INVOLVED

81. ***Writing.*** Find a real-life question for which the answer is a rational number that is not an integer.
What is the probability that a tossed coin turns up heads?

82. ***Exploration.*** **a)** Find a rational number between $\frac{1}{3}$ and $\frac{1}{4}$.
b) Find a rational number between -3.205 and -3.114.
c) Find a rational number between $\frac{2}{3}$ and 0.6667.
d) Explain how to find a rational number between any two given rational numbers.
a) $\frac{7}{24}$ b) -3.115 c) 0.66669
d) Add them and divide the result by 2.

83. ***Discussion.*** Suppose that a is a negative real number. Determine whether each of the following is positive or negative, and explain your answer.
a) $-a$ **b)** $|-a|$ **c)** $-|a|$ **d)** $-(-a)$ **e)** $-|-a|$
If a is negative, then $-a$ and $|-a|$ are positive. The rest are negative.

84. ***Discussion.*** Determine whether each number listed in the table below is a member of each set listed on the side of the table. For example, $\frac{1}{2}$ is a real number and a rational number. So check marks are placed in those two cells of the table.

	$\frac{1}{2}$	-2	π	$\sqrt{3}$	$\sqrt{9}$	6	0	$-\frac{7}{3}$
Real	✓	✓	✓	✓	✓	✓	✓	✓
Irrational			✓	✓				
Rational	✓	✓			✓	✓	✓	✓
Integer		✓			✓	✓	✓	
Whole					✓	✓	✓	
Counting					✓	✓		

1.2 FRACTIONS

In This Section

- Equivalent Fractions
- Multiplying Fractions
- Dividing Fractions
- Adding and Subtracting Fractions
- Fractions, Decimals, and Percents
- Applications

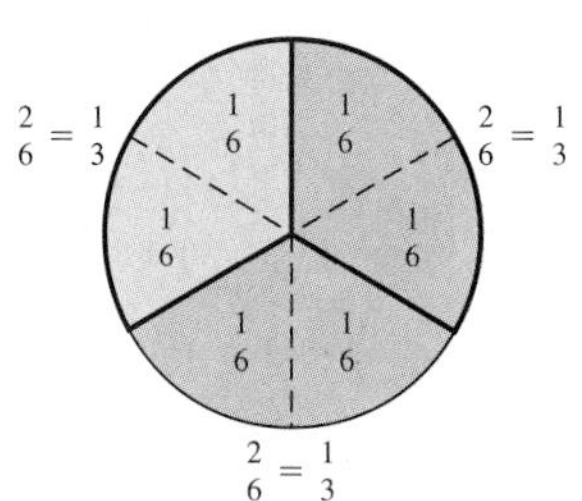

FIGURE 1.12

In this section and Sections 1.3 and 1.4 we will discuss operations performed with real numbers. We begin by reviewing operations with fractions. Note that this section on fractions is not an entire arithmetic course. We are simply reviewing selected fraction topics that will be used in this text.

Equivalent Fractions

If a pizza is cut into 3 equal pieces and you eat 2, you have eaten $\frac{2}{3}$ of the pizza. If the pizza is cut into 6 equal pieces and you eat 4, you have still eaten 2 out of every 3 pieces. So the fraction $\frac{4}{6}$ is considered **equal** or **equivalent** to $\frac{2}{3}$. See Fig. 1.12. Every fraction can be written in infinitely many equivalent forms. Consider the following equivalent forms of $\frac{2}{3}$:

$$\frac{2}{3} = \frac{4}{6} = \frac{6}{9} = \frac{8}{12} = \frac{10}{15} = \ldots$$

The three dots mean "and so on."

Notice that each equivalent form of $\frac{2}{3}$ can be obtained by multiplying the numerator (top number) and denominator (bottom number) of $\frac{2}{3}$ by a nonzero number. For example,

$$\frac{2}{3} = \frac{2 \cdot 5}{3 \cdot 5} = \frac{10}{15}.$$ The raised dot indicates multiplication.

Converting a fraction into an equivalent fraction with a larger denominator is called **building up** the fraction. A fraction can be built up by multiplying the numerator and denominator by the same number, or by using multiplication of fractions (which will be discussed later in this section). For example, to build up $\frac{2}{3}$ to a denominator of 15 we could multiply $\frac{2}{3}$ by the number 1, using $\frac{5}{5}$ as an equivalent form of 1:

$$\frac{2}{3} = \frac{2}{3} \cdot 1 = \frac{2}{3} \cdot \frac{5}{5} = \frac{10}{15}$$

Building Up Fractions

If $b \neq 0$ and $c \neq 0$, then

$$\frac{a}{b} = \frac{a \cdot c}{b \cdot c}.$$

Multiplying the numerator and denominator of a fraction by a nonzero number changes the fraction's appearance but not its value.

EXAMPLE 1 **Building up fractions**

Build up each fraction so that it is equivalent to the fraction with the indicated denominator.

a) $\frac{3}{4} = \frac{?}{28}$ **b)** $\frac{5}{3} = \frac{?}{30}$

Helpful Hint

In algebra it is best to build up fractions by multiplying both the numerator and denominator by the same number as shown in Example 1. So if you use a different method, be sure to learn this method.

Solution

a) Because $4 \cdot 7 = 28$, we multiply both the numerator and denominator by 7:

$$\frac{3}{4} = \frac{3 \cdot 7}{4 \cdot 7} = \frac{21}{28}$$

b) Because $3 \cdot 10 = 30$, we multiply both the numerator and denominator by 10:

$$\frac{5}{3} = \frac{5 \cdot 10}{3 \cdot 10} = \frac{50}{30}$$ ■

Helpful Hint

In algebra it is best to reduce fractions as shown here. First factor the numerator and denominator and then divide out (or cancel) the common factors. Be sure to learn this method.

Converting a fraction to an equivalent fraction with a smaller denominator is called **reducing** the fraction. For example, to reduce $\frac{10}{15}$, we *factor* 10 as $2 \cdot 5$ and 15 as $3 \cdot 5$, and then divide out the *common factor* 5:

$$\frac{10}{15} = \frac{2 \cdot \cancel{5}}{3 \cdot \cancel{5}} = \frac{2}{3}$$

The fraction $\frac{2}{3}$ cannot be reduced further because the numerator 2 and the denominator 3 have no factors (other than 1) in common. So we say that $\frac{2}{3}$ is in **lowest terms.**

Reducing Fractions

If $b \neq 0$ and $c \neq 0$, then

$$\frac{a \cdot c}{b \cdot c} = \frac{a}{b}.$$

Dividing the numerator and denominator of a fraction by a nonzero number changes the fraction's appearance but not its value.

EXAMPLE 2 **Reducing fractions**

Reduce each fraction to lowest terms.

a) $\frac{15}{24}$ b) $\frac{42}{30}$

c) $\frac{13}{26}$ d) $\frac{35}{7}$

Calculator Close-Up

To reduce a fraction to lowest terms using a graphing calculator, display the fraction and use the fraction feature.

```
15/24▸Frac
                 5/8
42/30▸Frac
                 7/5
123456/222222▸Fr
ac
        .5555525556
```

If the fraction is too complicated, the calculator will return a decimal equivalent instead of reducing it.

Solution

For each fraction, factor the numerator and denominator and then divide by the common factor:

a) $\frac{15}{24} = \frac{\cancel{3} \cdot 5}{\cancel{3} \cdot 8} = \frac{5}{8}$

b) $\frac{42}{30} = \frac{7 \cdot \cancel{6}}{5 \cdot \cancel{6}} = \frac{7}{5}$

c) $\frac{13}{26} = \frac{1 \cdot \cancel{13}}{2 \cdot \cancel{13}} = \frac{1}{2}$ The number 1 in the numerator is essential.

d) $\frac{35}{7} = \frac{5 \cdot \cancel{7}}{1 \cdot \cancel{7}} = \frac{5}{1} = 5$ ■

Strategy for Obtaining Equivalent Fractions

Equivalent fractions can be obtained by multiplying or dividing the numerator and denominator by the same nonzero number.

Multiplying Fractions

Suppose a pizza is cut into three equal pieces. If you eat $\frac{1}{2}$ of one piece, you have eaten $\frac{1}{6}$ of the pizza. See Fig. 1.13. You can obtain $\frac{1}{6}$ by multiplying $\frac{1}{2}$ and $\frac{1}{3}$:

$$\frac{1}{2} \cdot \frac{1}{3} = \frac{1 \cdot 1}{2 \cdot 3} = \frac{1}{6}$$

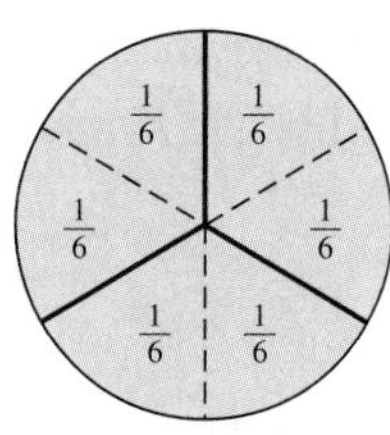

FIGURE 1.13

This example illustrates the definition of multiplication of fractions. To multiply two fractions, we multiply their numerators and multiply their denominators.

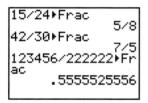

Multiplication of Fractions

If $b \neq 0$ and $d \neq 0$, then

$$\frac{a}{b} \cdot \frac{c}{d} = \frac{a \cdot c}{b \cdot d}.$$

EXAMPLE 3 **Multiplying fractions**

Find the product, $\frac{2}{3} \cdot \frac{5}{8}$.

Solution

Multiply the numerators and the denominators:

$$\frac{2}{3} \cdot \frac{5}{8} = \frac{10}{24}$$

$$= \frac{\not{2} \cdot 5}{\not{2} \cdot 12} \quad \text{Factor the numerator and denominator.}$$

$$= \frac{5}{12} \quad \text{Divide out the common factor 2.}$$

■

It is usually easier to reduce before multiplying, as shown in Example 4.

EXAMPLE 4 **Reducing before multiplying**

Find the indicated products.

a) $\frac{1}{3} \cdot \frac{3}{4}$ **b)** $\frac{4}{5} \cdot \frac{15}{22}$

Calculator Close-Up

A graphing calculator can multiply fractions and get fractional answers using the fraction feature. Note how a mixed number is written on a graphing calculator.

```
1/3*3/4▸Frac
            1/4
4/5*15/22▸Frac
           6/11
(3+1/4)*8/5▸Frac
           26/5
```

Solution

a) $\frac{1}{3} \cdot \frac{3}{4} = \frac{1}{\not{3}} \cdot \frac{\not{3}}{4} = \frac{1}{4}$

b) Factor the numerators and denominators, and then divide out the common factors before multiplying:

$$\frac{4}{5} \cdot \frac{15}{22} = \frac{2 \cdot \not{2}}{\not{5}} \cdot \frac{3 \cdot \not{5}}{\not{2} \cdot 11} = \frac{6}{11}$$

■

Dividing Fractions

Suppose that a pizza is cut into three pieces. If one piece is divided between two people $\left(\frac{1}{3} \div 2\right)$, then each of these two people gets $\frac{1}{6}$ of the pizza. Of course $\frac{1}{3}$ times $\frac{1}{2}$ is also $\frac{1}{6}$. So dividing by 2 is equivalent to multiplying by $\frac{1}{2}$. In symbols:

$$\frac{1}{3} \div 2 = \frac{1}{3} \div \frac{2}{1} = \frac{1}{3} \cdot \frac{1}{2} = \frac{1}{6}$$

The pizza example illustrates the general rule for dividing fractions.

Division of Fractions

If $b \neq 0$, $c \neq 0$, and $d \neq 0$, then

$$\frac{a}{b} \div \frac{c}{d} = \frac{a}{b} \cdot \frac{d}{c}.$$

In general if $m \div n = p$, then n is called the **divisor** and p (the result of the division) is called the **quotient** of m and n. We also refer to $m \div n$ and $\frac{m}{n}$ as the quotient of m and n. So in words, *to find the quotient of two fractions we invert the divisor and multiply.*

EXAMPLE 5 **Dividing fractions**

Find the indicated quotients.

a) $\frac{1}{3} \div \frac{7}{6}$ **b)** $\frac{2}{3} \div 5$ **c)** $\frac{3}{8} \div \frac{3}{2}$

Calculator Close-Up

When the divisor is a fraction on a graphing calculator, it must be in parentheses. A different result is obtained without using parentheses. Note that when the divisor is a whole number, parentheses are not necessary.

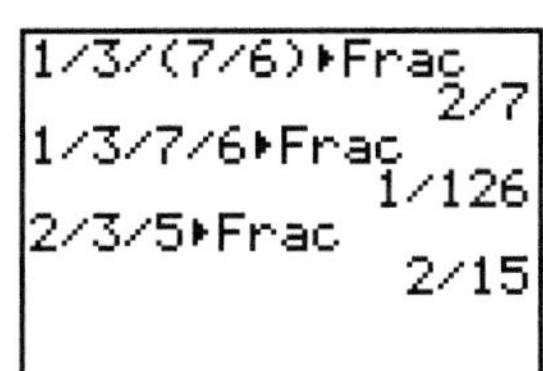

Try these computations on your calculator.

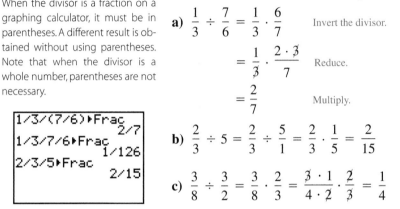

Solution

In each case we invert the divisor (the number on the right) and multiply.

a) $\frac{1}{3} \div \frac{7}{6} = \frac{1}{3} \cdot \frac{6}{7}$ Invert the divisor.

$= \frac{1}{\not{3}} \cdot \frac{2 \cdot \not{3}}{7}$ Reduce.

$= \frac{2}{7}$ Multiply.

b) $\frac{2}{3} \div 5 = \frac{2}{3} \div \frac{5}{1} = \frac{2}{3} \cdot \frac{1}{5} = \frac{2}{15}$

c) $\frac{3}{8} \div \frac{3}{2} = \frac{3}{8} \cdot \frac{2}{3} = \frac{\not{3} \cdot 1}{4 \cdot \not{2}} \cdot \frac{\not{2}}{\not{3}} = \frac{1}{4}$ ■

Adding and Subtracting Fractions

To understand addition and subtraction of fractions, again consider the pizza that is cut into six equal pieces as shown in Fig. 1.14. If you eat $\frac{3}{6}$ and your friend eats $\frac{2}{6}$, together you have eaten $\frac{5}{6}$ of the pizza. Similarly, if you remove $\frac{1}{6}$ from $\frac{6}{6}$ you have $\frac{5}{6}$ left. To add or subtract fractions with identical denominators, we add or subtract their numerators and write the result over the common denominator.

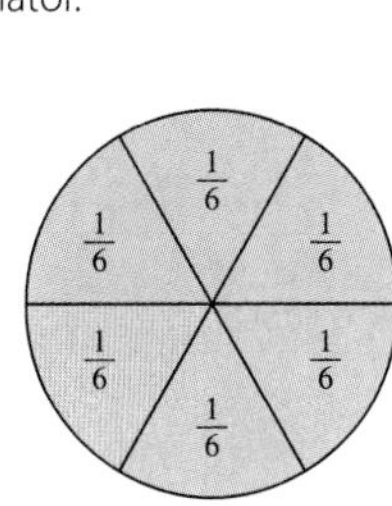

FIGURE 1.14

Addition and Subtraction of Fractions

If $b \neq 0$, then

$$\frac{a}{b} + \frac{c}{b} = \frac{a + c}{b} \quad \text{and} \quad \frac{a}{b} - \frac{c}{b} = \frac{a - c}{b}.$$

EXAMPLE 6 **Adding and subtracting fractions**

Perform the indicated operations.

a) $\frac{1}{7} + \frac{2}{7}$

b) $\frac{7}{10} - \frac{3}{10}$

Helpful Hint

A good way to remember that you need common denominators for addition is to think of a simple example. If you own 1/3 share of a car wash and your spouse owns 1/3, then together you own 2/3 of the business.

Solution

a) $\frac{1}{7} + \frac{2}{7} = \frac{3}{7}$ **b)** $\frac{7}{10} - \frac{3}{10} = \frac{4}{10} = \frac{\not{2} \cdot 2}{\not{2} \cdot 5} = \frac{2}{5}$ ■

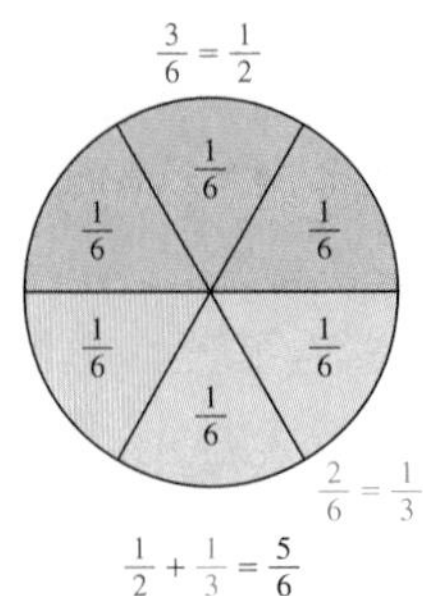

$\frac{3}{6} = \frac{1}{2}$
$\frac{1}{6}$
$\frac{2}{6} = \frac{1}{3}$
$\frac{1}{2} + \frac{1}{3} = \frac{5}{6}$

FIGURE 1.15

To add or subtract fractions with different denominators, we must convert them to equivalent fractions with the same denominator and then add or subtract. For example, to add $\frac{1}{2}$ and $\frac{1}{3}$, we build up each fraction to a denominator of 6. See Fig. 1.15. Since $\frac{1}{2} = \frac{3}{6}$ and $\frac{1}{3} = \frac{2}{6}$, we have

$$\frac{1}{2} + \frac{1}{3} = \frac{3}{6} + \frac{2}{6} = \frac{5}{6}.$$

The smallest number that is a multiple of the denominators of two or more fractions is called the **least common denominator (LCD).** So 6 is the LCD for $\frac{1}{2}$ and $\frac{1}{3}$. Note that we obtained the LCD 6 by examining Fig. 1.15. We must have a more systematic way.

The procedure for finding the LCD is based on factors. For example, to find the LCD for the denominators 6 and 9, factor 6 and 9 as $6 = 2 \cdot 3$ and $9 = 3 \cdot 3$. To obtain a multiple of both 6 and 9 the number must have two 3's as factors and one 2. So the LCD for 6 and 9 is $2 \cdot 3 \cdot 3$ or 18. If any number is omitted from $2 \cdot 3 \cdot 3$, we will not have a multiple of both 6 and 9. So each factor found in either 6 or 9 appears in the LCD the maximum number of times that it appears in either 6 or 9. The general strategy follows.

Helpful Hint

The *least* common denominator is *greater than* or equal to all of the denominators, because they must all divide into the LCD.

Strategy for Finding the LCD

1. Factor each denominator completely.
2. The LCD contains each distinct factor the maximum number of times that it occurs in any of the denominators.

Note that a **prime number** is a number 2 or larger that has no factors other than itself and 1. If a denominator is prime (such as 2, 3, 5, 7, 11) then we do not factor it. A number is **factored completely** when it is written as a product of prime numbers.

EXAMPLE 7 **Adding and subtracting fractions**

Perform the indicated operations.

a) $\frac{3}{4} + \frac{1}{6}$

b) $\frac{1}{3} - \frac{1}{12}$

c) $\frac{7}{12} + \frac{5}{18}$

d) $2\frac{1}{3} + \frac{5}{9}$

Solution

a) First factor the denominators as $4 = 2 \cdot 2$ and $6 = 2 \cdot 3$. Since 2 occurs twice in 4 and once in 6, it appears twice in the LCD. Since 3 appears once in 6 and not at all in 4, it appears once in the LCD. So the LCD is $2 \cdot 2 \cdot 3$ or 12. Now build

Study Tip

Read the material in the text before it is discussed in class, even if you do not totally understand it. The classroom discussion will be the second time you have seen the material and it will be easier to question points that you do not understand.

up each denominator to 12:

$$\frac{3}{4} + \frac{1}{6} = \frac{3 \cdot 3}{4 \cdot 3} + \frac{1 \cdot 2}{6 \cdot 2} \quad \text{Build up each denominator to 12.}$$
$$= \frac{9}{12} + \frac{2}{12} \quad \text{Simplify.}$$
$$= \frac{11}{12} \quad \text{Add.}$$

b) The denominators are 12 and 3. Factor 12 as $12 = 2 \cdot 6 = 2 \cdot 2 \cdot 3$. Since 3 is a prime number we do not factor it. Since 2 occurs twice in 12 and not at all in 3, it appears twice in the LCD. Since 3 occurs once in 3 and once in 12, 3 appears once in the LCD. The LCD is $2 \cdot 2 \cdot 3$ or 12. So we must build up $\frac{1}{3}$ to have a denominator of 12:

$$\frac{1}{3} - \frac{1}{12} = \frac{1 \cdot 4}{3 \cdot 4} - \frac{1}{12} \quad \text{Build up the first fraction to the LCD.}$$
$$= \frac{4}{12} - \frac{1}{12} \quad \text{Simplify.}$$
$$= \frac{3}{12} \quad \text{Subtract.}$$
$$= \frac{1}{4} \quad \text{Reduce to lowest terms.}$$

c) Since $12 = 2 \cdot 6 = 2 \cdot 2 \cdot 3$ and $18 = 2 \cdot 9 = 2 \cdot 3 \cdot 3$, the factor 2 appears twice in the LCD and the factor 3 appears twice in the LCD. So the LCD is $2 \cdot 2 \cdot 3 \cdot 3$ or 36:

$$\frac{7}{12} + \frac{5}{18} = \frac{7 \cdot 3}{12 \cdot 3} + \frac{5 \cdot 2}{18 \cdot 2} \quad \text{Build up each denominator to 36.}$$
$$= \frac{21}{36} + \frac{10}{36} \quad \text{Simplify.}$$
$$= \frac{31}{36} \quad \text{Add.}$$

d) To perform addition with the mixed number $2\frac{1}{3}$, first convert it into an improper fraction: $2\frac{1}{3} = 2 + \frac{1}{3} = \frac{6}{3} + \frac{1}{3} = \frac{7}{3}$.

$$2\frac{1}{3} + \frac{5}{9} = \frac{7}{3} + \frac{5}{9} \quad \text{Write } 2\tfrac{1}{3} \text{ as an improper fraction.}$$
$$= \frac{7 \cdot 3}{3 \cdot 3} + \frac{5}{9} \quad \text{The LCD is 9.}$$
$$= \frac{21}{9} + \frac{5}{9} \quad \text{Simplify.}$$
$$= \frac{26}{9} \quad \text{Add.}$$

Note that $\frac{1}{3} + \frac{5}{9} = \frac{3}{9} + \frac{5}{9} = \frac{8}{9}$. Then add on the 2 to get $2\frac{8}{9}$, which is the same as $\frac{26}{9}$. ■

Helpful Hint

Recall the *place value* for decimal numbers:

```
tenths
|┌hundredths
||┌thousandths
|||┌ten thousandths
0.2635
```

So $0.2635 = \frac{2635}{10,000}$.

Fractions, Decimals, and Percents

In the decimal number system, fractions with a denominator of 10, 100, 1000, and so on are written as decimal numbers. For example,

$$\frac{3}{10} = 0.3, \quad \frac{25}{100} = 0.25, \quad \text{and} \quad \frac{5}{1000} = 0.005.$$

Fractions with a denominator of 100 are often written as percents. Think of the percent symbol (%) as representing the denominator of 100. For example,

$$\frac{25}{100} = 25\%, \quad \frac{5}{100} = 5\%, \quad \text{and} \quad \frac{300}{100} = 300\%.$$

Example 8 illustrates further how to convert from any one of the forms (fraction, decimal, percent) to the others.

EXAMPLE 8 **Changing forms**

Convert each given fraction, decimal, or percent into its other two forms.

a) $\frac{1}{5}$ **b)** 6% **c)** 0.1

Solution

a) $\frac{1}{5} = \frac{1 \cdot 20}{5 \cdot 20} = \frac{20}{100} = 20\%$ and $\frac{1}{5} = \frac{1 \cdot 2}{5 \cdot 2} = \frac{2}{10} = 0.2$

So $\frac{1}{5} = 0.2 = 20\%$. Note that a fraction can also be converted to a decimal by dividing the denominator into the numerator with long division.

b) $6\% = \frac{6}{100} = 0.06$ and $\frac{6}{100} = \frac{\cancel{2} \cdot 3}{\cancel{2} \cdot 50} = \frac{3}{50}$

So $6\% = 0.06 = \frac{3}{50}$.

c) $0.1 = \frac{1}{10} = \frac{1 \cdot 10}{10 \cdot 10} = \frac{10}{100} = 10\%$

So $0.1 = \frac{1}{10} = 10\%$. ■

Calculator Close-Up

A calculator can convert fractions to decimals and decimals to fractions. The calculator shown here converts the terminating decimal 0.333333333333 into 1/3 even though 1/3 is a repeating decimal with infinitely many threes after the decimal point.

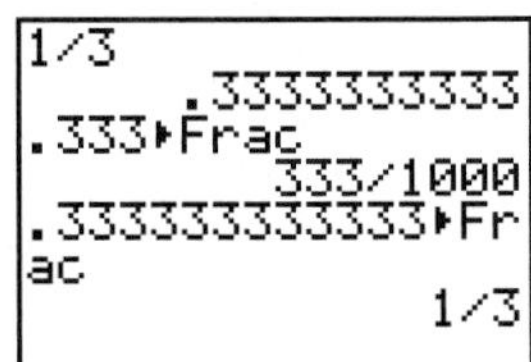

Applications

The dimensions for lumber used in construction are usually given in fractions. For example, a 2 × 4 stud used for framing a wall is actually $1\frac{1}{2}$ in. by $3\frac{1}{2}$ in. by $92\frac{5}{8}$ in. A 2 × 12 floor joist is actually $1\frac{1}{2}$ in. by $11\frac{1}{2}$ in.

EXAMPLE 9 **Framing a two-story house**

In framing a two-story house, a carpenter uses a 2×4 shoe, a wall stud, two 2×4 plates, then 2×12 floor joists, and a $\frac{3}{4}$-in. plywood floor, before starting the second level. Use the dimensions in Fig. 1.16 to find the total height of the framing shown.

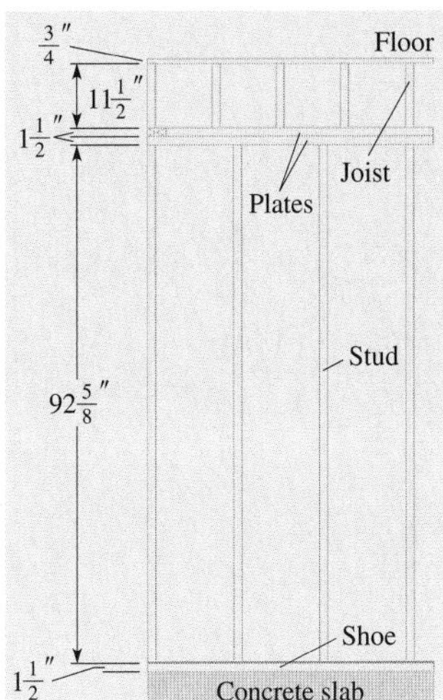

FIGURE 1.16

Solution

We can find the total height using multiplication and addition:

$$
\begin{aligned}
3 \cdot 1\frac{1}{2} + 92\frac{5}{8} + 11\frac{1}{2} + \frac{3}{4} &= 4\frac{1}{2} + 92\frac{5}{8} + 11\frac{1}{2} + \frac{3}{4} \\
&= 4\frac{4}{8} + 92\frac{5}{8} + 11\frac{4}{8} + \frac{6}{8} \\
&= 107\frac{19}{8} \\
&= 109\frac{3}{8}
\end{aligned}
$$

The total height of the framing shown is $109\frac{3}{8}$ in. ■

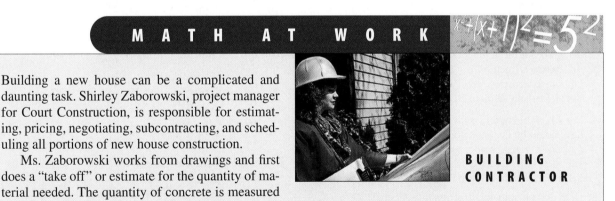

MATH AT WORK

BUILDING CONTRACTOR

Building a new house can be a complicated and daunting task. Shirley Zaborowski, project manager for Court Construction, is responsible for estimating, pricing, negotiating, subcontracting, and scheduling all portions of new house construction.

Ms. Zaborowski works from drawings and first does a "take off" or estimate for the quantity of material needed. The quantity of concrete is measured in cubic yards and the amount of wood is measured in board feet. If masonry is being used, it is measured in bricks or blocks per square foot.

Scheduling is another important part of the project manager's responsibility and it is based on the take off. Certain industry standards help Ms. Zaborowski estimate how many carpenters are needed and how much time it takes to frame the house and how many electricians and plumbers are needed to wire the house, install the heating systems, and put in the bathrooms. Of course, common sense says that the foundation is done before the framing and the roof. However, some rough plumbing and electrical work can be done simultaneously with the framing. Ideally the estimates of time and cost are accurate and the homeowner can move in on schedule.

In Exercise 103 of this section you will use operations with fractions to find the volume of concrete needed to construct a rectangular patio.

WARM-UPS

True or false? Explain your answer.

1. Every fraction is equal to infinitely many equivalent fractions. True
2. The fraction $\frac{8}{12}$ is equivalent to the fraction $\frac{4}{6}$. True
3. The fraction $\frac{8}{12}$ reduced to lowest terms is $\frac{4}{6}$. False
4. $\frac{1}{2} \cdot \frac{2}{3} = \frac{1}{3}$ True
5. $\frac{1}{2} \cdot \frac{3}{5} = \frac{3}{10}$ True
6. $\frac{1}{2} \cdot \frac{6}{5} = \frac{6}{10}$ True
7. $\frac{1}{2} \div 3 = \frac{1}{6}$ True
8. $5 \div \frac{1}{2} = 10$ True
9. $\frac{1}{2} + \frac{1}{4} = \frac{2}{6}$ False
10. $2 - \frac{1}{2} = \frac{3}{2}$ True

1.2 EXERCISES

Reading and Writing *After reading this section write out the answers to these questions. Use complete sentences.*

1. What are equivalent fractions?
   If two fractions are identical when reduced to lowest terms, then they are equivalent fractions.
2. How can you find all fractions that are equivalent to a given fraction?
   Reduce the fraction to lowest terms and then multiply the numerator and denominator by every counting number.
3. What does it mean to reduce a fraction to lowest terms?
   To reduce a fraction to lowest terms means to find an equivalent fraction that has no factor common to the numerator and denominator.
4. For which operations with fractions are you required to have common denominators? Why?
   Common denominators are required for addition and subtraction, because it makes sense to add $\frac{1}{3}$ of a pie and $\frac{1}{3}$ of a pie and get $\frac{2}{3}$ of a pie.
5. How do you convert a fraction to a decimal?
   Convert a fraction to a decimal by dividing the denominator into the numerator.
6. How do you convert a percent to a fraction?
   Convert a percent to a fraction by dividing by 100, as in $4\% = \frac{4}{100}$.

Build up each fraction or whole number so that it is equivalent to the fraction with the indicated denominator. See Example 1.

7. $\frac{3}{4} = \frac{?}{8}$ $\frac{6}{8}$
8. $\frac{5}{7} = \frac{?}{21}$ $\frac{15}{21}$
9. $\frac{8}{3} = \frac{?}{12}$ $\frac{32}{12}$
10. $\frac{7}{2} = \frac{?}{8}$ $\frac{28}{8}$
11. $5 = \frac{?}{2}$ $\frac{10}{2}$
12. $9 = \frac{?}{3}$ $\frac{27}{3}$
13. $\frac{3}{4} = \frac{?}{100}$ $\frac{75}{100}$
14. $\frac{1}{2} = \frac{?}{100}$ $\frac{50}{100}$
15. $\frac{3}{10} = \frac{?}{100}$ $\frac{30}{100}$
16. $\frac{2}{5} = \frac{?}{100}$ $\frac{40}{100}$
17. $\frac{5}{3} = \frac{?}{42}$ $\frac{70}{42}$
18. $\frac{5}{7} = \frac{?}{98}$ $\frac{70}{98}$

Reduce each fraction to lowest terms. See Example 2.

19. $\frac{3}{6}$ $\frac{1}{2}$
20. $\frac{2}{10}$ $\frac{1}{5}$
21. $\frac{12}{18}$ $\frac{2}{3}$
22. $\frac{30}{40}$ $\frac{3}{4}$
23. $\frac{15}{5}$ 3
24. $\frac{39}{13}$ 3
25. $\frac{50}{100}$ $\frac{1}{2}$
26. $\frac{5}{1000}$ $\frac{1}{200}$
27. $\frac{200}{100}$ 2
28. $\frac{125}{100}$ $\frac{5}{4}$
29. $\frac{18}{48}$ $\frac{3}{8}$
30. $\frac{34}{102}$ $\frac{1}{3}$
31. $\frac{26}{42}$ $\frac{13}{21}$
32. $\frac{70}{112}$ $\frac{5}{8}$
33. $\frac{84}{91}$ $\frac{12}{13}$
34. $\frac{121}{132}$ $\frac{11}{12}$

Find each product. See Examples 3 and 4.

35. $\frac{2}{3} \cdot \frac{5}{9}$ $\frac{10}{27}$
36. $\frac{1}{8} \cdot \frac{1}{8}$ $\frac{1}{64}$
37. $\frac{1}{3} \cdot 15$ 5
38. $\frac{1}{4} \cdot 16$ 4
39. $\frac{3}{4} \cdot \frac{14}{15}$ $\frac{7}{10}$
40. $\frac{5}{8} \cdot \frac{12}{35}$ $\frac{3}{14}$
41. $\frac{2}{5} \cdot \frac{35}{26}$ $\frac{7}{13}$
42. $\frac{3}{10} \cdot \frac{20}{21}$ $\frac{2}{7}$
43. $\frac{1}{2} \cdot \frac{6}{5}$ $\frac{3}{5}$
44. $\frac{1}{2} \cdot \frac{3}{5}$ $\frac{3}{10}$
45. $\frac{1}{2} \cdot \frac{1}{3}$ $\frac{1}{6}$
46. $\frac{3}{16} \cdot \frac{1}{7}$ $\frac{3}{112}$

Find each quotient. See Example 5.

47. $\frac{3}{4} \div \frac{1}{4}$ 3
48. $\frac{2}{3} \div \frac{1}{2}$ $\frac{4}{3}$
49. $\frac{1}{3} \div 5$ $\frac{1}{15}$
50. $\frac{3}{5} \div 3$ $\frac{1}{5}$

51. $5 \div \frac{5}{4}$ 4

52. $8 \div \frac{2}{3}$ 12

53. $\frac{6}{10} \div \frac{3}{4}$ $\frac{4}{5}$

54. $\frac{2}{3} \div \frac{10}{21}$ $\frac{7}{5}$

55. $\frac{3}{16} \div \frac{5}{2}$ $\frac{3}{40}$

56. $\frac{1}{8} \div \frac{5}{16}$ $\frac{2}{5}$

Find each sum or difference. See Examples 6 and 7.

57. $\frac{1}{4} + \frac{1}{4}$ $\frac{1}{2}$

58. $\frac{1}{10} + \frac{1}{10}$ $\frac{1}{5}$

59. $\frac{5}{12} - \frac{1}{12}$ $\frac{1}{3}$

60. $\frac{17}{14} - \frac{5}{14}$ $\frac{6}{7}$

61. $\frac{1}{2} - \frac{1}{4}$ $\frac{1}{4}$

62. $\frac{1}{3} + \frac{1}{6}$ $\frac{1}{2}$

63. $\frac{1}{3} + \frac{1}{4}$ $\frac{7}{12}$

64. $\frac{1}{2} + \frac{3}{5}$ $\frac{11}{10}$

65. $\frac{3}{4} - \frac{2}{3}$ $\frac{1}{12}$

66. $\frac{4}{5} - \frac{3}{4}$ $\frac{1}{20}$

67. $\frac{1}{6} + \frac{5}{8}$ $\frac{19}{24}$

68. $\frac{3}{4} + \frac{1}{6}$ $\frac{11}{12}$

69. $\frac{5}{24} - \frac{1}{18}$ $\frac{11}{72}$

70. $\frac{3}{16} - \frac{1}{20}$ $\frac{11}{80}$

71. $3\frac{5}{6} + \frac{5}{16}$ $\frac{199}{48}$

72. $5\frac{3}{8} - \frac{15}{16}$ $\frac{71}{16}$

Convert each given fraction, decimal, or percent into its other two forms. See Example 8.

73. $\frac{3}{5}$ 60%, 0.6

74. $\frac{19}{20}$ 95%, 0.95

75. 9% $\frac{9}{100}$, 0.09

76. 60% 0.6, $\frac{3}{5}$

77. 0.08 8%, $\frac{2}{25}$

78. 0.4 40%, $\frac{2}{5}$

79. $\frac{3}{4}$ 0.75, 75%

80. $\frac{5}{8}$ 0.625, 62.5%

81. 2% $\frac{1}{50}$, 0.02

82. 120% $\frac{6}{5}$, 1.20

83. 0.01 $\frac{1}{100}$, 1%

84. 0.005 $\frac{1}{200}$, 0.5%

Perform the indicated operations.

85. $\frac{3}{8} \div \frac{1}{8}$ 3

86. $\frac{7}{8} \div \frac{3}{14}$ $\frac{49}{12}$

87. $\frac{3}{4} \cdot \frac{28}{21}$ 1

88. $\frac{5}{16} \cdot \frac{3}{10}$ $\frac{3}{32}$

89. $\frac{7}{12} + \frac{5}{32}$ $\frac{71}{96}$

90. $\frac{2}{15} + \frac{8}{21}$ $\frac{18}{35}$

91. $\frac{5}{24} - \frac{1}{15}$ $\frac{17}{120}$

92. $\frac{9}{16} - \frac{1}{12}$ $\frac{23}{48}$

93. $3\frac{1}{8} + \frac{15}{16}$ $\frac{65}{16}$

94. $5\frac{1}{4} - \frac{9}{16}$ $\frac{75}{16}$

95. $7\frac{2}{3} \cdot 2\frac{1}{4}$ $\frac{69}{4}$

96. $6\frac{1}{2} \div \frac{7}{2}$ $\frac{13}{7}$

97. $\frac{1}{2} + \frac{1}{3} + \frac{1}{4}$ $\frac{13}{12}$

98. $\frac{1}{2} + \frac{1}{3} - \frac{1}{6}$ $\frac{2}{3}$

99. $\frac{1}{2} \cdot \frac{1}{2} \cdot \frac{1}{2}$ $\frac{1}{8}$

100. $\frac{2}{3} \cdot \frac{2}{3} \cdot \frac{2}{3}$ $\frac{8}{27}$

Solve each problem. See Example 9.

101. ***Inheritance.*** Marie is entitled to one-sixth of an estate because of one relationship to the deceased and one-thirty-second of the estate because of another relationship to the deceased. What is the total portion of the estate that she will receive?
$\frac{19}{96}$

102. ***Diversification.*** Helen has $\frac{1}{5}$ of her portfolio in U.S. stocks, $\frac{1}{8}$ of her portfolio in European stocks, and $\frac{1}{10}$ of her portfolio in Japanese stocks. The remainder is invested in municipal bonds. What fraction of her portfolio is invested in municipal bonds? What percent is invested in municipal bonds?
$\frac{23}{40}$, 57.5%

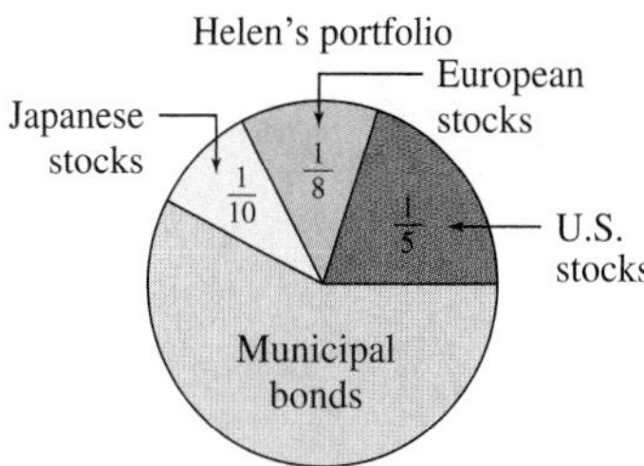

FIGURE FOR EXERCISE 102

103. ***Concrete patio.*** A contractor plans to pour a concrete rectangular patio.

a) Use the table to find the approximate volume of concrete in cubic yards for a 9 ft by 12 ft patio that is 4 inches thick. 1.3 yd^3

b) Find the exact volume of concrete in cubic feet and cubic yards for a patio that is $12\frac{1}{2}$ feet long, $8\frac{3}{4}$ feet wide, and 4 inches thick. $36\frac{11}{24}$ ft^3 or $1\frac{227}{648}$ yd^3

Concrete required for 4 in. thick patio

L (ft)	W (ft)	V (yd^3)
16	14	2.8
14	10	1.7
12	9	1.3
10	8	1.0

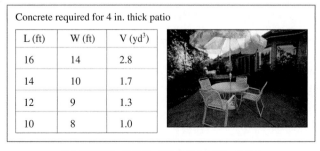

FIGURE FOR EXERCISE 103

104. ***Bundle of studs.*** A lumber yard receives 2×4 studs in a bundle that contains 25 rows (or layers) of studs with 20 studs in each row. A 2×4 stud is actually $1\frac{1}{2}$ in. by $3\frac{1}{2}$ in. by $92\frac{5}{8}$ in. Find the cross-sectional area of a bundle in square inches. Find the volume of a bundle in cubic feet. (The formula $V = LWH$ gives the volume of a rectangular solid.) 2625 in.^{2}, 140.7 ft^3

GETTING MORE INVOLVED

105. ***Writing.*** Find an example of a real-life situation in which it is necessary to add two fractions.

106. ***Cooperative learning.*** Write a step-by-step procedure for adding two fractions with different denominators. Give your procedure to a classmate to try out on some addition problems. Refine your procedure as necessary.

107. ***Fraction puzzle.*** A wheat farmer in Manitoba left his L-shaped farm (shown in the diagram) to his four daughters. Divide the property into four pieces so that each piece is exactly the same size and shape. Each daughter gets 3 km^2 $\div$ 4 or a $\frac{3}{4}$ km^2 piece of the farm. Divide the farm into 12 equal squares. Give each daughter an L-shaped piece consisting of 3 of those 12 squares.

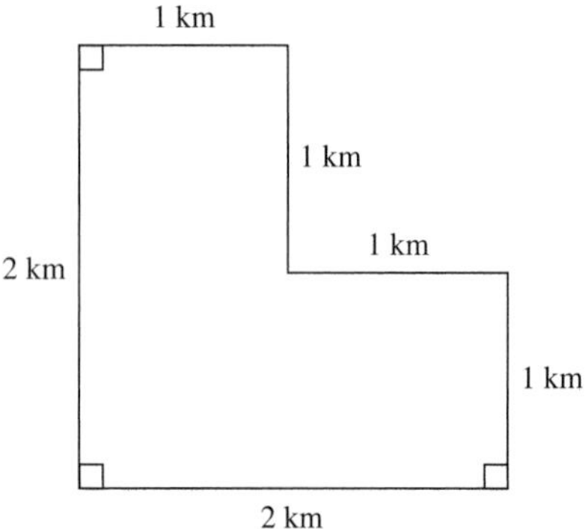

FIGURE FOR EXERCISE 107

1.3 ADDITION AND SUBTRACTION OF REAL NUMBERS

In This Section

- Addition of Two Negative Numbers
- Addition of Numbers with Unlike Signs
- Subtraction of Signed Numbers

In arithmetic we add and subtract only positive numbers and zero. In Section 1.1 we introduced the concept of absolute value of a number. Now we will use absolute value to extend the operations of addition and subtraction to the real numbers. We will work only with rational numbers in this chapter. You will learn to perform operations with irrational numbers in Chapter 8.

Addition of Two Negative Numbers

A good way to understand positive and negative numbers is to *think of the positive numbers as assets and the negative numbers as debts.* For this illustration we can think of assets simply as cash. For example, if you have \$3 and \$5 in cash, then your total cash is \$8. You get the total by adding two positive numbers.

Think of debts as unpaid bills such as the electric bill or the phone bill. If you have debts of \$70 and \$80, then your total debt is \$150. You can get the total debt by adding negative numbers:

$$\underset{\text{\$70 debt}}{\underset{\uparrow}{(-70)}} \quad \underset{\text{plus}}{\underset{\uparrow}{+}} \quad \underset{\text{\$80 debt}}{\underset{\uparrow}{(-80)}} \quad = \quad \underset{\text{\$150 debt}}{\underset{\uparrow}{-150}}$$

We think of this addition as adding the absolute values of -70 and -80 ($70 + 80 = 150$), and then putting a negative sign on that result to get -150. These examples illustrate the following rule.

Sum of Two Numbers with Like Signs

To find the sum of two numbers with the same sign, add their absolute values. The sum has the same sign as the given numbers.

EXAMPLE 1 **Adding numbers with like signs**

Perform the indicated operations.

a) $23 + 56$ **b)** $(-12) + (-9)$ **c)** $(-3.5) + (-6.28)$ **d)** $\left(-\frac{1}{2}\right) + \left(-\frac{1}{4}\right)$

Solution

a) The sum of two positive numbers is a positive number: $23 + 56 = 79$.

b) The absolute values of -12 and -9 are 12 and 9, and $12 + 9 = 21$. So

$$(-12) + (-9) = -21.$$

c) Add the absolute values of -3.5 and -6.28, and put a negative sign on the sum. Remember to line up the decimal points when adding decimal numbers:

$$\begin{array}{r} 3.50 \\ \underline{6.28} \\ 9.78 \end{array}$$

So $(-3.5) + (-6.28) = -9.78$.

d) $\left(-\frac{1}{2}\right) + \left(-\frac{1}{4}\right) = \left(-\frac{2}{4}\right) + \left(-\frac{1}{4}\right) = -\frac{3}{4}$ ■

Study Tip

Exchange phone numbers, cellular phone numbers, pager numbers, and e-mail addresses with several students in your class. If you miss class and you can't reach your instructor, then you will have someone who can tell you the assignments. If you are stuck on a problem, you can contact a classmate for help.

Addition of Numbers with Unlike Signs

If you have a debt of \$5 and have only \$5 in cash, then your debts equal your assets (in absolute value), and your net worth is \$0. **Net worth** is the total of debts and assets. Symbolically,

$$\underset{\text{\$5 debt}}{\underset{\uparrow}{-5}} \quad + \quad \underset{\text{\$5 cash}}{\underset{\uparrow}{5}} \quad = \quad \underset{\text{Net worth}}{\underset{\uparrow}{0.}}$$

For any number a, a and its opposite, $-a$, have a sum of zero. For this reason, a and $-a$ are called **additive inverses** of each other. Note that the words "negative," "opposite," and "additive inverse" are often used interchangeably.

Additive Inverse Property

For any number a,

$$a + (-a) = 0 \quad \text{and} \quad (-a) + a = 0.$$

EXAMPLE 2 **Finding the sum of additive inverses**

Evaluate.

a) $34 + (-34)$ **b)** $-\frac{1}{4} + \frac{1}{4}$ **c)** $2.97 + (-2.97)$

Solution

a) $34 + (-34) = 0$

b) $-\frac{1}{4} + \frac{1}{4} = 0$

c) $2.97 + (-2.97) = 0$ ■

Helpful Hint

We use the illustrations with debts and assets to make the rules for adding signed numbers understandable. However, in the end the carefully written rules tell us exactly how to perform operations with signed numbers, and we must obey the rules.

To understand the sum of a positive and a negative number that are not additive inverses of each other, consider the following situation. If you have a debt of \$6 and \$10 in cash, you may have \$10 in hand, but your net worth is only \$4. Your assets exceed your debts (in absolute value), and you have a positive net worth. In symbols,

$$-6 + 10 = 4.$$

Note that to get 4, we actually subtract 6 from 10.

If you have a debt of \$7 but have only \$5 in cash, then your debts exceed your assets (in absolute value). You have a negative net worth of $-\$2$. In symbols,

$$-7 + 5 = -2.$$

Note that to get the 2 in the answer, we subtract 5 from 7.

As you can see from these examples, the sum of a positive number and a negative number (with different absolute values) may be either positive or negative. These examples help us to understand the rule for adding numbers with unlike signs and different absolute values.

Sum of Two Numbers with Unlike Signs (and Different Absolute Values)

To find the sum of two numbers with unlike signs (and different absolute values), subtract their absolute values.

- The answer is positive if the number with the larger absolute value is positive.
- The answer is negative if the number with the larger absolute value is negative.

EXAMPLE 3 **Adding numbers with unlike signs**

Evaluate.

a) $-5 + 13$ **b)** $6 + (-7)$ **c)** $-6.4 + 2.1$

d) $-5 + 0.09$ **e)** $\left(-\frac{1}{3}\right) + \left(\frac{1}{2}\right)$ **f)** $\frac{3}{8} + \left(-\frac{5}{6}\right)$

Calculator Close-Up

Your calculator can add signed numbers. Most calculators have a key for subtraction and a different key for the negative sign.

```
-5+13
                8
-5+.09
            -4.91
3/8+-5/6▸Frac
           -11/24
```

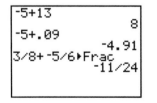

You should do the exercises in this section by hand and then check with a calculator.

Solution

a) The absolute values of -5 and 13 are 5 and 13. Subtract them to get 8. Since the number with the larger absolute value is 13 and it is positive, the result is positive:

$$-5 + 13 = 8$$

b) The absolute values of 6 and -7 are 6 and 7. Subtract them to get 1. Since -7 has the larger absolute value, the result is negative:

$$6 + (-7) = -1$$

c) Line up the decimal points and subtract 2.1 from 6.4.

$$\begin{array}{r} 6.4 \\ -2.1 \\ \hline 4.3 \end{array}$$

Since 6.4 is larger than 2.1, and 6.4 has a negative sign, the sign of the answer is negative. So $-6.4 + 2.1 = -4.3$.

d) Line up the decimal points and subtract 0.09 from 5.00.

$$\begin{array}{r} 5.00 \\ -0.09 \\ \hline 4.91 \end{array}$$

Since 5.00 is larger than 0.09, and 5.00 has the negative sign, the sign of the answer is negative. So $-5 + 0.09 = -4.91$.

e) $\left(-\frac{1}{3}\right) + \left(\frac{1}{2}\right) = \left(-\frac{2}{6}\right) + \left(\frac{3}{6}\right) = \frac{1}{6}$

f) $\frac{3}{8} + \left(-\frac{5}{6}\right) = \frac{9}{24} + \left(-\frac{20}{24}\right) = -\frac{11}{24}$ ■

Study Tip

The keys to success are desire and discipline. You must want success and you must discipline yourself to do what it takes to get success. There are a lot of things that you can't do anything about, but you can learn to be disciplined. Set your goals, make plans, and schedule your time. Before you know it you will have the discipline that is necessary for success.

Subtraction of Signed Numbers

Each subtraction problem with signed numbers is solved by doing an equivalent addition problem. So before attempting subtraction of signed numbers be sure that you understand addition of signed numbers.

Now think of subtraction as removing debts or assets, and think of addition as receiving debts or assets. If you have \$100 in cash and \$30 is taken from you, your resulting net worth is the same as if you have \$100 cash and a phone bill for \$30 arrives in the mail. In symbols,

$$100 \underset{\text{Remove}}{\underset{\uparrow}{-}} \underset{\text{Cash}}{\underset{\uparrow}{30}} = 100 \underset{\text{Receive}}{\underset{\uparrow}{+}} \underset{\text{Debt}}{\underset{\uparrow}{(-30)}}.$$

Removing cash is equivalent to receiving a debt.

Suppose you have \$15 but owe a friend \$5. Your net worth is only \$10. If the debt of \$5 is canceled or forgiven, your net worth will go up to \$15, the same as if you received \$5 in cash. In symbols,

$$10 \underset{\text{Remove}}{\underset{\uparrow}{-}} \underset{\text{Debt}}{\underset{\uparrow}{(-5)}} = 10 \underset{\text{Receive}}{\underset{\uparrow}{+}} \underset{\text{Cash}}{\underset{\uparrow}{5}}.$$

Removing a debt is equivalent to receiving cash.

Notice that each subtraction problem is equivalent to an addition problem in which we add the opposite of what we want to subtract. In other words, *subtracting a number is the same as adding its opposite.*

Subtraction of Real Numbers

For any real numbers a and b,

$$a - b = a + (-b).$$

EXAMPLE 4 **Subtracting signed numbers**

Perform each subtraction.

a) $-5 - 3$

b) $5 - (-3)$

c) $-5 - (-3)$

d) $\frac{1}{2} - \left(-\frac{1}{4}\right)$

e) $-3.6 - (-5)$

f) $0.02 - 8$

Solution

To do *any* subtraction, we can change it to addition of the opposite.

a) $-5 - 3 = -5 + (-3) = -8$

b) $5 - (-3) = 5 + (3) = 8$

c) $-5 - (-3) = -5 + 3 = -2$

d) $\frac{1}{2} - \left(-\frac{1}{4}\right) = \frac{2}{4} + \frac{1}{4} = \frac{3}{4}$

e) $-3.6 - (-5) = -3.6 + 5 = 1.4$

f) $0.02 - 8 = 0.02 + (-8) = -7.98$ ■

WARM-UPS

True or false? Explain your answer.

1. $-9 + 8 = -1$ True
2. $(-2) + (-4) = -6$ True
3. $0 - 7 = -7$ True
4. $5 - (-2) = 3$ False
5. $-5 - (-2) = -7$ False
6. The additive inverse of -3 is 0. False
7. If b is a negative number, then $-b$ is a positive number. True
8. The sum of a positive number and a negative number is a negative number. False
9. The result of a subtracted from b is the same as b plus the opposite of a. True
10. If a and b are negative numbers, then $a - b$ is a negative number. False

1.3 EXERCISES

Reading and Writing *After reading this section write out the answers to these questions. Use complete sentences.*

1. What operations did we study in this section?
   We studied addition and subtraction of signed numbers.
2. How do you find the sum of two numbers with the same sign?
   The sum of two numbers with the same sign is found by adding their absolute values. The sum is negative if the two numbers are negative.
3. When can we say that two numbers are additive inverses of each other?
   Two numbers are additive inverses of each other if their sum is zero.
4. What is the sum of two numbers with opposite signs and the same absolute value?
   The sum of two numbers with opposite signs and the same absolute value is zero.
5. How do we find the sum of two numbers with unlike signs?
   To find the sum of two numbers with unlike signs, subtract their absolute values. The answer is given the sign of the number with the larger absolute value.
6. What is the relationship between subtraction and addition?
   Subtraction is defined in terms of addition as $a - b = a + (-b)$.

Perform the indicated operation. See Example 1.

7. $3 + 10$ 13
8. $81 + 19$ 100
9. $(-3) + (-10)$ -13
10. $(-81) + (-19)$ -100
11. $-0.25 + (-0.9)$ -1.15
12. $-0.8 + (-2.35)$ -3.15
13. $\left(-\frac{1}{3}\right) + \left(-\frac{1}{6}\right)$ $-\frac{1}{2}$
14. $\frac{2}{3} + \frac{1}{12}$ $\frac{3}{4}$

Evaluate. See Examples 2 and 3.

15. $-8 + 8$ 0

16. $20 + (-20)$ 0

17. $-\frac{17}{50} + \frac{17}{50}$ 0

18. $\frac{12}{13} + \left(-\frac{12}{13}\right)$ 0

19. $-7 + 9$ 2

20. $10 + (-30)$ -20

21. $7 + (-13)$ -6

22. $-8 + 20$ 12

23. $8.6 + (-3)$ 5.6

24. $-9.5 + 12$ 2.5

25. $3.9 + (-6.8)$ -2.9

26. $-5.24 + 8.19$ 2.95

27. $\frac{1}{4} + \left(-\frac{1}{2}\right)$ $-\frac{1}{4}$

28. $-\frac{2}{3} + 2$ $\frac{4}{3}$

Fill in the parentheses to make each statement correct. See Example 4.

29. $8 - 2 = 8 + (?)$ $8 + (-2)$

30. $3.5 - 1.2 = 3.5 + (?)$ $3.5 + (-1.2)$

31. $4 - 12 = 4 + (?)$ $4 + (-12)$

32. $\frac{1}{2} - \frac{5}{6} = \frac{1}{2} + (?)$ $\frac{1}{2} + \left(-\frac{5}{6}\right)$

33. $-3 - (-8) = -3 + (?)$ $-3 + 8$

34. $-9 - (-2.3) = -9 + (?)$ $-9 + (2.3)$

35. $8.3 - (-1.5) = 8.3 + (?)$ $8.3 + (1.5)$

36. $10 - (-6) = 10 + (?)$ $10 + (6)$

Perform the indicated operation. See Example 4.

37. $6 - 10$ -4

38. $3 - 19$ -16

39. $-3 - 7$ -10

40. $-3 - 12$ -15

41. $5 - (-6)$ 11

42. $5 - (-9)$ 14

43. $-6 - 5$ -11

44. $-3 - 6$ -9

45. $\frac{1}{4} - \frac{1}{2}$ $-\frac{1}{4}$

46. $\frac{2}{5} - \frac{2}{3}$ $-\frac{4}{15}$

47. $\frac{1}{2} - \left(-\frac{1}{4}\right)$ $\frac{3}{4}$

48. $\frac{2}{3} - \left(-\frac{1}{6}\right)$ $\frac{5}{6}$

49. $10 - 3$ 7

50. $13 - 3$ 10

51. $1 - 0.07$ 0.93

52. $0.03 - 1$ -0.97

53. $7.3 - (-2)$ 9.3

54. $-5.1 - 0.15$ -5.25

55. $-0.03 - 5$ -5.03

56. $0.7 - (-0.3)$ 1

Perform the indicated operations. Do not use a calculator.

57. $-5 + 8$ 3

58. $-6 + 10$ 4

59. $-6 + (-3)$ -9

60. $(-13) + (-12)$ -25

61. $-80 - 40$ -120

62. $44 - (-15)$ 59

63. $61 - (-17)$ 78

64. $-19 - 13$ -32

65. $(-12) + (-15)$ -27

66. $-12 + 12$ 0

67. $13 + (-20)$ -7

68. $15 + (-39)$ -24

69. $-102 - 99$ -201

70. $-94 - (-77)$ -17

71. $-161 - 161$ -322

72. $-19 - 88$ -107

73. $-16 + 0.03$ -15.97

74. $0.59 + (-3.4)$ -2.81

75. $0.08 - 3$ -2.92

76. $1.8 - 9$ -7.2

77. $-3.7 + (-0.03)$ -3.73

78. $0.9 + (-1)$ -0.1

79. $-2.3 - (-6)$ 3.7

80. $-7.08 - (-9)$ 1.92

81. $\frac{3}{4} + \left(-\frac{3}{5}\right)$ $\frac{3}{20}$

82. $-\frac{1}{3} + \frac{3}{5}$ $\frac{4}{15}$

83. $-\frac{1}{12} - \left(-\frac{3}{8}\right)$ $\frac{7}{24}$

84. $-\frac{1}{17} - \left(-\frac{1}{17}\right)$ 0

Use a calculator to perform the indicated operations.

85. $45.87 + (-49.36)$ -3.49

86. $-0.357 + (-3.465)$ -3.822

87. $0.6578 + (-1)$ -0.3422

88. $-2.347 + (-3.5)$ -5.847

89. $-3.45 - 45.39$ -48.84

90. $9.8 - 9.974$ -0.174

91. $-5.79 - 3.06$ -8.85

92. $0 - (-4.537)$ 4.537

Solve each problem.

93. ***Overdrawn.*** Willard opened his checking account with a deposit of \$97.86. He then wrote checks and had other charges as shown in his account register. Find his current balance. $-\$8.85$

Deposit		97.86
Wal-Mart	27.89	
Kmart	42.32	
ATM cash	25.00	
Service charge	3.50	
Check printing	8.00	

FIGURE FOR EXERCISE 93

94. ***Net worth.*** Melanie has a \$125,000 house with a \$78,422 mortgage. She has \$21,236 in a savings account and has \$9,477 in credit card debt. She owes \$6,131 to the credit union and figures that her cars and other household items are worth a total of \$15,000. What is Melanie's net worth? \$67,206

95. ***Falling temperatures.*** At noon the temperature in Montreal was 5°C. By midnight the mercury had fallen 12°. What was the temperature at midnight? -7°C

96. ***Bitter cold.*** The overnight low temperature in Milwaukee was -13°F for Monday night. The

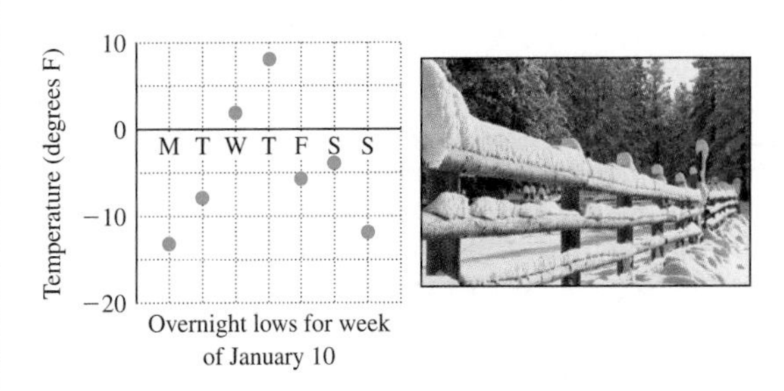

FIGURE FOR EXERCISE 96

temperature went up 20° during the day on Tuesday and then fell 15° to reach Tuesday night's overnight low temperature.

a) What was the overnight low Tuesday night? $-8°F$

b) Judging from the accompanying graph, was the average low for the week above or below 0°F? Below zero

GETTING MORE INVOLVED

97. ***Writing.*** What does absolute value have to do with adding signed numbers? Can you add signed numbers without using absolute value?
When adding signed numbers, we add or subtract only positive numbers which are the absolute values of the original numbers. We then determine the appropriate sign for the answer.

98. ***Discussion.*** Why do we learn addition of signed numbers before subtraction?
Subtraction is defined as addition of the opposite, $a - b = a + (-b)$.

99. ***Discussion.*** Aimee and Joni are traveling south in separate cars on Interstate 5 near Stockton. While they are speaking to each other on cellular telephones, Aimee gives her location as mile marker x and Joni gives her location as mile marker y. Which of the following expressions gives the distance between them? Explain your answer.

a) $y - x$ **b)** $x - y$

c) $|x - y|$ **d)** $|y - x|$

e) $|x| + |y|$

The distance between x and y is given by either $|x - y|$ or $|y - x|$.

1.4 MULTIPLICATION AND DIVISION OF REAL NUMBERS

In This Section

- Multiplication of Real Numbers
- Division of Real Numbers
- Division by Zero

In this section we will complete the study of the four basic operations with real numbers.

Multiplication of Real Numbers

The result of multiplying two numbers is referred to as the **product** of the numbers. The numbers multiplied are called **factors.** In algebra we use a raised dot between the factors to indicate multiplication, or we place symbols next to one another to indicate multiplication. Thus $a \cdot b$ or ab are both referred to as the product of a and b. When multiplying numbers, we may enclose them in parentheses to make the meaning clear. To write 5 times 3, we may write it as $5 \cdot 3$, $5(3)$, $(5)3$, or $(5)(3)$. In multiplying a number and a variable, no sign is used between them. Thus $5x$ is used to represent the product of 5 and x.

Multiplication is just a short way to do repeated additions. Adding together five 3's gives

$$3 + 3 + 3 + 3 + 3 = 15.$$

So we have the multiplication fact $5 \cdot 3 = 15$. Adding together five -3's gives

$$(-3) + (-3) + (-3) + (-3) + (-3) = -15.$$

So we should have $5(-3) = -15$. We can think of $5(-3) = -15$ as saying that taking on five debts of \$3 each is equivalent to a debt of \$15. Losing five debts of \$3 each is equivalent to gaining \$15, so we should have $(-5)(-3) = 15$.

These examples illustrate the rule for multiplying signed numbers.

Helpful Hint

The product of two numbers with like signs is positive, but the product of three numbers with like signs can be positive or negative. For example,

$$2 \cdot 2 \cdot 2 = 8$$

and

$$(-2)(-2)(-2) = -8.$$

Product of Signed Numbers

To find the product of two nonzero real numbers, multiply their absolute values.

- The product is *positive* if the numbers have *like* signs.
- The product is *negative* if the numbers have *unlike* signs.

EXAMPLE 1 **Multiplying signed numbers**

Evaluate each product.

a) $(-2)(-3)$ **b)** $3(-6)$ **c)** $-5 \cdot 10$

d) $\left(-\frac{1}{3}\right)\left(-\frac{1}{2}\right)$ **e)** $(-0.02)(0.08)$ **f)** $(-300)(-0.06)$

Calculator Close-Up

Try finding the products in Example 1 with your calculator.

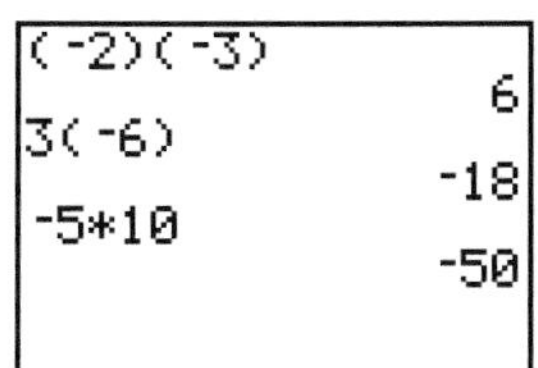

Solution

a) First find the product of the absolute values:

$$|-2| \cdot |-3| = 2 \cdot 3 = 6$$

Because -2 and -3 have the same sign, we get $(-2)(-3) = 6$.

b) First find the product of the absolute values:

$$|3| \cdot |-6| = 3 \cdot 6 = 18$$

Because 3 and -6 have unlike signs, we get $3(-6) = -18$.

c) $-5 \cdot 10 = -50$ Unlike signs, negative result

d) $\left(-\frac{1}{3}\right)\left(-\frac{1}{2}\right) = \frac{1}{6}$ Like signs, positive result

e) When multiplying decimals, we total the number of decimal places in the factors to get the number of decimal places in the product. Thus

$$(-0.02)(0.08) = -0.0016.$$

f) $(-300)(-0.06) = 18$ ■

Division of Real Numbers

We say that $10 \div 5 = 2$ because $2 \cdot 5 = 10$. This example illustrates how division is defined in terms of multiplication.

Division of Real Numbers

If a, b, and c are any real numbers with $b \neq 0$, then

$$a \div b = c \quad \text{provided that} \quad c \cdot b = a.$$

Using the definition of division, we get

$$10 \div (-2) = -5$$

because $(-5)(-2) = 10$;

$$-10 \div 2 = -5$$

because $(-5)(2) = -10$; and

$$-10 \div (-2) = 5$$

because $(5)(-2) = -10$. From these examples we see that the rule for dividing signed numbers is similar to that for multiplying signed numbers.

Division of Signed Numbers

To find the quotient of two nonzero real numbers, divide their absolute values.

- The quotient is *positive* if the two numbers have *like* signs.
- The quotient is *negative* if the two numbers have *unlike* signs.

Zero divided by any nonzero real number is zero.

EXAMPLE 2 **Dividing signed numbers**

Evaluate.

a) $(-8) \div (-4)$ **b)** $(-8) \div 8$ **c)** $8 \div (-4)$

d) $-4 \div \frac{1}{3}$ **e)** $-2.5 \div 0.05$ **f)** $0 \div (-6)$

Helpful Hint

Do not use negative numbers in long division. To find $-378 \div 7$, divide 378 by 7:

$$\begin{array}{r} 54 \\ 7\overline{)378} \\ \underline{35} \\ 28 \\ \underline{28} \\ 0 \end{array}$$

Since a negative divided by a positive is negative

$$-378 \div 7 = -54.$$

Solution

a) $(-8) \div (-4) = 2$ Same sign, positive result

b) $(-8) \div 8 = -1$ Unlike signs, negative result

c) $8 \div (-4) = -2$

d) $-4 \div \frac{1}{3} = -4 \cdot \frac{3}{1}$ Invert and multiply.

$= -4 \cdot 3$

$= -12$

e) $-2.5 \div 0.05 = \frac{-2.5}{0.05}$ Write in fraction form.

$= \frac{-2.5 \cdot 100}{0.05 \cdot 100}$ Multiply by 100 to eliminate the decimals.

$= \frac{-250}{5}$ Simplify.

$= -50$ Divide.

f) $0 \div (-6) = 0$ ■

Division can also be indicated by a fraction bar. For example,

$$24 \div 6 = \frac{24}{6} = 4.$$

If signed numbers occur in a fraction, we use the rules for dividing signed numbers. For example,

$$\frac{-9}{3} = -3, \quad \frac{9}{-3} = -3, \quad \frac{-1}{2} = \frac{1}{-2} = -\frac{1}{2}, \quad \text{and} \quad \frac{-4}{-2} = 2.$$

Note that if one negative sign appears in a fraction, the fraction has the same value whether the negative sign is in the numerator, in the denominator, or in front of the fraction. If the numerator and denominator of a fraction are both negative, then the fraction has a positive value.

Study Tip

If you don't know how to get started on the exercises, go back to the examples. Cover the solution in the text with a piece of paper and see if you can solve the example. After you have mastered the examples, then try the exercises again.

Division by Zero

Why do we exclude division by zero from the definition of division? If we write $10 \div 0 = c$, we need to find a number c such that $c \cdot 0 = 10$. This is impossible. If we write $0 \div 0 = c$, we need to find a number c such that $c \cdot 0 = 0$. In fact, $c \cdot 0 = 0$ is true for any value of c. Having $0 \div 0$ equal to any number would be confusing in doing computations. Thus $a \div b$ is defined only for $b \neq 0$. Quotients such as

$$8 \div 0, \quad 0 \div 0, \quad \frac{8}{0}, \quad \text{and} \quad \frac{0}{0}$$

are said to be **undefined.**

WARM-UPS

True or false? Explain your answer.

1. The product of 7 and y is written as $7y$. True
2. The product of -2 and 5 is 10. False
3. The quotient of x and 3 can be written as $x \div 3$ or $\frac{x}{3}$. True
4. $0 \div 6$ is undefined. False
5. $(-9) \div (-3) = 3$ True
6. $6 \div (-2) = -3$ True
7. $\left(-\frac{1}{2}\right)\left(-\frac{1}{2}\right) = \frac{1}{4}$ True
8. $(-0.2)(0.2) = -0.4$ False
9. $\left(-\frac{1}{2}\right) \div \left(-\frac{1}{2}\right) = 1$ True
10. $\frac{0}{0} = 0$ False

1.4 EXERCISES

Reading and Writing *After reading this section write out the answers to these questions. Use complete sentences.*

1. What operations did we study in this section?
   We learned to multiply and divide signed numbers.
2. What is a product?
   A product is the result of multiplication. The product of a and b is ab. The product of 2 and 4 is 8.
3. How do you find the product of two signed numbers?
   To find the product of signed numbers, multiply their absolute values and then affix a negative sign if the two original numbers have opposite signs.
4. What is the relationship between division and multiplication?
   Division is defined in terms of multiplication as $a \div b = c$ provided $c \cdot b = a$ and $b \neq 0$.
5. How do you find the quotient of nonzero real numbers?
   To find the quotient of nonzero numbers divide their absolute values and then affix a negative sign if the two original numbers have opposite signs.
6. Why is division by zero undefined?
   Division by zero is undefined because it cannot be made consistent with the definition of division: $a \div b = c$ provided $c \cdot b = a$.

Evaluate. See Example 1.

7. $-3 \cdot 9$ -27
8. $6(-4)$ -24
9. $(-12)(-11)$ 132
10. $(-9)(-15)$ 135
11. $-\frac{3}{4} \cdot \frac{4}{9}$ $-\frac{1}{3}$
12. $\left(-\frac{2}{3}\right)\left(-\frac{6}{7}\right)$ $\frac{4}{7}$
13. $0.5(-0.6)$ -0.3
14. $(-0.3)(0.3)$ -0.09
15. $(-12)(-12)$ 144
16. $(-11)(-11)$ 121
17. $-3 \cdot 0$ 0
18. $0(-7)$ 0

Evaluate. See Example 2.

19. $8 \div (-8)$ -1
20. $-6 \div 2$ -3
21. $(-90) \div (-30)$ 3
22. $(-20) \div (-40)$ $\frac{1}{2}$
23. $\frac{44}{-66}$ $-\frac{2}{3}$
24. $\frac{-33}{-36}$ $\frac{11}{12}$
25. $\left(-\frac{2}{3}\right) \div \left(-\frac{4}{5}\right)$ $\frac{5}{6}$
26. $-\frac{1}{3} \div \frac{4}{9}$ $-\frac{3}{4}$
27. $\frac{-125}{0}$ Undefined
28. $-37 \div 0$ Undefined
29. $0 \div \left(-\frac{1}{3}\right)$ 0
30. $0 \div 43.568$ 0
31. $40 \div (-0.5)$ -80
32. $3 \div (-0.1)$ -30
33. $-0.5 \div (-2)$ 0.25
34. $-0.75 \div (-0.5)$ 1.5

Perform the indicated operations.

35. $(25)(-4)$ -100
36. $(5)(-4)$ -20
37. $(-3)(-9)$ 27
38. $(-51) \div (-3)$ 17
39. $-9 \div 3$ -3
40. $86 \div (-2)$ -43
41. $20 \div (-5)$ -4
42. $(-8)(-6)$ 48
43. $(-6)(5)$ -30
44. $(-18) \div 3$ -6
45. $(-57) \div (-3)$ 19
46. $(-30)(4)$ -120
47. $(0.6)(-0.3)$ -0.18
48. $(-0.2)(-0.5)$ 0.1
49. $(-0.03)(-10)$ 0.3
50. $(0.05)(-1.5)$ -0.075
51. $(-0.6) \div (0.1)$ -6
52. $8 \div (-0.5)$ -16
53. $(-0.6) \div (-0.4)$ 1.5
54. $(-63) \div (-0.9)$ 70
55. $-\frac{12}{5}\left(-\frac{55}{6}\right)$ 22
56. $-\frac{9}{10} \cdot \frac{4}{3}$ $-\frac{6}{5}$
57. $-2\frac{3}{4} \div 8\frac{1}{4}$ $-\frac{1}{3}$
58. $-9\frac{1}{2} \div \left(-3\frac{1}{6}\right)$ 3

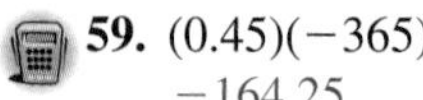
59. $(0.45)(-365)$ -164.25

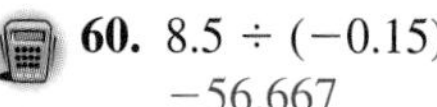
60. $8.5 \div (-0.15)$ -56.667

61. $(-52) \div (-0.034)$ 1529.41

62. $(-4.8)(5.6)$ -26.88

Perform the indicated operations. Use a calculator to check.

63. $(-4)(-4)$ 16

64. $-4 - 4$ -8

65. $-4 + (-4)$ -8

66. $-4 \div (-4)$ 1

67. $-4 + 4$ 0

68. $-4 \cdot 4$ -16

69. $-4 - (-4)$ 0

70. $0 \div (-4)$ 0

71. $0.1 - 4$ -3.9

72. $(0.1)(-4)$ -0.4

73. $(-4) \div (0.1)$ -40

74. $-0.1 - 4$ -4.1

75. $(-0.1)(-4)$ 0.4

76. $-0.1 + 4$ 3.9

77. $|-0.4|$ 0.4

78. $|0.4|$ 0.4

79. $\dfrac{-0.06}{0.3}$ -0.2

80. $\dfrac{2}{-0.04}$ -50

81. $\dfrac{3}{-0.4}$ -7.5

82. $\dfrac{-1.2}{-0.03}$ 40

83. $-\dfrac{1}{5} + \dfrac{1}{6}$ $-\dfrac{1}{30}$

84. $-\dfrac{3}{5} - \dfrac{1}{4}$ $-\dfrac{17}{20}$

85. $\left(-\dfrac{3}{4}\right)\left(\dfrac{2}{15}\right)$ $-\dfrac{1}{10}$

86. $-1 \div \left(-\dfrac{1}{4}\right)$ 4

Use a calculator to perform the indicated operations. Round answers to three decimal places.

87. $\dfrac{45.37}{6}$ 7.562

88. $(-345) \div (28)$ -12.321

89. $(-4.3)(-4.5)$ 19.35

90. $\dfrac{-12.34}{-3}$ 4.113

91. $\dfrac{0}{6.345}$ 0

92. $0 \div (34.51)$ 0

93. $199.4 \div 0$ Undefined

94. $\dfrac{23.44}{0}$ Undefined

GETTING MORE INVOLVED

95. ***Discussion.*** If you divide \$0 among five people, how much does each person get? If you divide \$5 among zero people, how much does each person get? What do these questions illustrate?

96. ***Discussion.*** What is the difference between the non-negative numbers and the positive numbers?

97. ***Writing.*** Why do we learn multiplication of signed numbers before division?

98. ***Writing.*** Try to rewrite the rules for multiplying and dividing signed numbers without using the idea of absolute value. Are your rewritten rules clearer than the original rules?

1.5 EXPONENTIAL EXPRESSIONS AND THE ORDER OF OPERATIONS

In This Section

- Arithmetic Expressions
- Exponential Expressions
- The Order of Operations

In Sections 1.3 and 1.4 you learned how to perform operations with a pair of real numbers to obtain a third real number. In this section you will learn to evaluate expressions involving several numbers and operations.

Arithmetic Expressions

The result of writing numbers in a meaningful combination with the ordinary operations of arithmetic is called an **arithmetic expression** or simply an **expression.** Consider the expressions

$$(3 + 2) \cdot 5 \qquad \text{and} \qquad 3 + (2 \cdot 5).$$

The parentheses are used as **grouping symbols** and indicate which operation to perform first. Because of the parentheses, these expressions have different values:

$$(3 + 2) \cdot 5 = 5 \cdot 5 = 25$$
$$3 + (2 \cdot 5) = 3 + 10 = 13$$

Absolute value symbols and fraction bars are also used as grouping symbols. The numerator and denominator of a fraction are treated as if each is in parentheses.

EXAMPLE 1 **Using grouping symbols**

Evaluate each expression.

a) $(3 - 6)(3 + 6)$

b) $|3 - 4| - |5 - 9|$

c) $\dfrac{4 - (-8)}{5 - 9}$

Calculator Close-Up

One advantage of a graphing calculator is that you can enter an entire expression on its display and then evaluate it. If your calculator does not allow built-up form for fractions, then you must use parentheses around the numerator and denominator as shown here.

```
(3-6)(3+6)
                -27
abs(3-4)-abs(5-9
)
                 -3
(4--8)/(5-9)
                 -3
```

Solution

a) $(3 - 6)(3 + 6) = (-3)(9)$ Evaluate within parentheses first.

$= -27$ Multiply.

b) $|3 - 4| - |5 - 9| = |-1| - |-4|$ Evaluate within absolute value symbols.

$= 1 - 4$ Find the absolute values.

$= -3$ Subtract.

c) $\dfrac{4 - (-8)}{5 - 9} = \dfrac{12}{-4}$ Evaluate the numerator and denominator.

$= -3$ Divide. ■

Exponential Expressions

An arithmetic expression with repeated multiplication can be written by using exponents. For example,

$$2 \cdot 2 \cdot 2 = 2^3 \qquad \text{and} \qquad 5 \cdot 5 = 5^2.$$

The 3 in 2^3 is the number of times that 2 occurs in the product $2 \cdot 2 \cdot 2$, while the 2 in 5^2 is the number of times that 5 occurs in $5 \cdot 5$. We read 2^3 as "2 cubed" or "2 to the third power." We read 5^2 as "5 squared" or "5 to the second power." In general, an expression of the form a^n is called an **exponential expression** and is defined as follows.

Study Tip

If you need help, do not hesitate to get it. Math has a way of building upon the past. What you learn today will be used tomorrow, and what you learn tomorrow will be used the day after. If you don't straighten out problems immediately, then you can get hopelessly lost. If you are having trouble, see your instructor to find out what help is available.

Exponential Expression

For any counting number n,

$$a^n = \underbrace{a \cdot a \cdot a \cdot \ldots \cdot a.}_{n \text{ factors}}$$

We call a the **base** and n the **exponent.**

The expression a^n is read "a to the nth power." If the exponent is 1, it is usually omitted. For example, $9^1 = 9$.

EXAMPLE 2 **Using exponential notation**

Write each product as an exponential expression.

a) $6 \cdot 6 \cdot 6 \cdot 6 \cdot 6$ **b)** $(-3)(-3)(-3)(-3)$ **c)** $\dfrac{3}{2} \cdot \dfrac{3}{2} \cdot \dfrac{3}{2}$

Solution

a) $6 \cdot 6 \cdot 6 \cdot 6 \cdot 6 = 6^5$

b) $(-3)(-3)(-3)(-3) = (-3)^4$

c) $\dfrac{3}{2} \cdot \dfrac{3}{2} \cdot \dfrac{3}{2} = \left(\dfrac{3}{2}\right)^3$ ■

EXAMPLE 3 **Writing an exponential expression as a product**

Write each exponential expression as a product without exponents.

a) y^6 **b)** $(-2)^4$ **c)** $\left(\frac{5}{4}\right)^3$ **d)** $(-0.1)^2$

Solution

a) $y^6 = y \cdot y \cdot y \cdot y \cdot y \cdot y$

b) $(-2)^4 = (-2)(-2)(-2)(-2)$

c) $\left(\frac{5}{4}\right)^3 = \frac{5}{4} \cdot \frac{5}{4} \cdot \frac{5}{4}$

d) $(-0.1)^2 = (-0.1)(-0.1)$ ■

To evaluate an exponential expression, write the base as many times as indicated by the exponent, then multiply the factors from left to right.

EXAMPLE 4 **Evaluating exponential expressions**

Evaluate.

a) 3^3 **b)** $(-2)^3$ **c)** $\left(\frac{2}{3}\right)^4$ **d)** $(0.4)^2$

Solution

a) $3^3 = 3 \cdot 3 \cdot 3 = 9 \cdot 3 = 27$

b)
$$\begin{aligned}(-2)^3 &= (-2)(-2)(-2)\\ &= 4(-2)\\ &= -8\end{aligned}$$

c)
$$\begin{aligned}\left(\frac{2}{3}\right)^4 &= \frac{2}{3} \cdot \frac{2}{3} \cdot \frac{2}{3} \cdot \frac{2}{3}\\ &= \frac{4}{9} \cdot \frac{2}{3} \cdot \frac{2}{3}\\ &= \frac{8}{27} \cdot \frac{2}{3}\\ &= \frac{16}{81}\end{aligned}$$

d) $(0.4)^2 = (0.4)(0.4) = 0.16$ ■

Calculator Close-Up

You can use the power key for any power. Most calculators also have an x^2 key that gives the second power. Note that parentheses must be used when raising a fraction to a power.

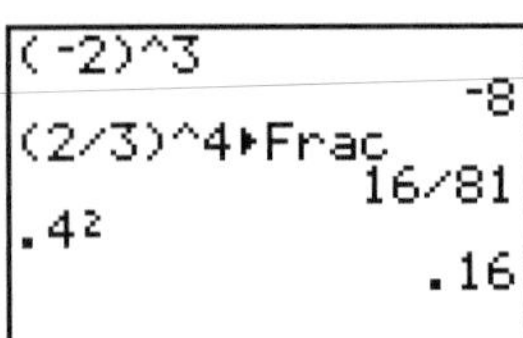

CAUTION Note that $3^3 \neq 9$. We do not multiply the exponent and the base when evaluating an exponential expression.

Be especially careful with exponential expressions involving negative numbers. An exponential expression with a negative base is written with parentheses around the base as in $(-2)^4$:

$$(-2)^4 = (-2)(-2)(-2)(-2) = 16$$

To evaluate $-(2^4)$, use the base 2 as a factor four times, then find the opposite:

$$-(2^4) = -(2 \cdot 2 \cdot 2 \cdot 2) = -(16) = -16$$

We often omit the parentheses in $-(2^4)$ and simply write -2^4. So

$$-2^4 = -(2^4) = -16.$$

To evaluate $-(-2)^4$, use the base -2 as a factor four times, then find the opposite:

$$-(-2)^4 = -(16) = -16$$

EXAMPLE 5 **Evaluating exponential expressions involving negative numbers**

Evaluate.

a) $(-10)^4$ **b)** -10^4

c) $-(-0.5)^2$ **d)** $-(5 - 8)^2$

Solution

a) $(-10)^4 = (-10)(-10)(-10)(-10)$ Use -10 as a factor four times.

$= 10{,}000$

b) $-10^4 = -(10^4)$ Rewrite using parentheses.

$= -(10{,}000)$ Find 10^4.

$= -10{,}000$ Then find the opposite of 10,000.

c) $-(-0.5)^2 = -(-0.5)(-0.5)$ Use -0.5 as a factor two times.

$= -(0.25)$

$= -0.25$

d) $-(5 - 8)^2 = -(-3)^2$ Evaluate within parentheses first.

$= -(9)$ Square -3 to get 9.

$= -9$ Take the opposite of 9 to get -9. ■

The Order of Operations

When we evaluate expressions, operations within grouping symbols are always performed first. For example

$$(3 + 2) \cdot 5 = (5) \cdot 5 = 10 \quad \text{and} \quad (2 \cdot 3)^2 = 6^2 = 36.$$

To make expressions look simpler, we often omit some or all parentheses. In this case we must agree on the order in which to perform the operations. We agree to do multiplication before addition and exponential expressions before multiplication. So

$$3 + 2 \cdot 5 = 3 + 10 = 13 \quad \text{and} \quad 2 \cdot 3^2 = 2 \cdot 9 = 18.$$

We state the complete **order of operations** in the following box.

Helpful Hint

"Please Excuse My Dear Aunt Sally" is often used as a memory aid for the order of operations. Do Parentheses, Exponents, Multiplication, and Division, then Addition and Subtraction. Multiplication and division have equal priority. The same goes for addition and subtraction.

Order of Operations

If an expression contains no grouping symbols, evaluate it using the following order. If an expression contains operations within grouping symbols, evaluate the expressions within grouping symbols first, using the following order.

1. Evaluate each exponential expression (in order from left to right).
2. Perform multiplication and division (in order from left to right).
3. Perform addition and subtraction (in order from left to right).

Multiplication and division have equal priority in the order of operations. If both appear in an expression, they are performed in order from left to right. The same holds for addition and subtraction. For example,

$$8 \div 4 \cdot 3 = 2 \cdot 3 = 6 \quad \text{and} \quad 9 - 3 + 5 = 6 + 5 = 11.$$

EXAMPLE 6 **Using the order of operations**

Evaluate each expression.

a) $2^3 \cdot 3^2$ **b)** $2 \cdot 5 - 3 \cdot 4 + 4^2$ **c)** $2 \cdot 3 \cdot 4 - 3^3 + \frac{8}{2}$

Calculator Close-Up

Most calculators follow the same order of operations shown here. Evaluate these expressions with your calculator.

```
2^3*3²
                72
2*5-3*4+4²
                14
2*3*4-3^3+8/2
                 1
```

Solution

a) $2^3 \cdot 3^2 = 8 \cdot 9$ Evaluate exponential expressions before multiplying.

$= 72$

b) $2 \cdot 5 - 3 \cdot 4 + 4^2 = 2 \cdot 5 - 3 \cdot 4 + 16$ Exponential expressions first

$= 10 - 12 + 16$ Multiplication second

$= 14$ Addition and subtraction from left to right

c) $2 \cdot 3 \cdot 4 - 3^3 + \frac{8}{2} = 2 \cdot 3 \cdot 4 - 27 + \frac{8}{2}$ Exponential expressions first

$= 24 - 27 + 4$ Multiplication and division second

$= 1$ Addition and subtraction from left to right ■

When grouping symbols are used, we perform operations within grouping symbols first. The order of operations is followed within the grouping symbols.

EXAMPLE 7 **Grouping symbols and the order of operations**

Evaluate.

a) $3 - 2(7 - 2^3)$ **b)** $3 - |7 - 3 \cdot 4|$ **c)** $\frac{9 - 5 + 8}{-5^2 - 3(-7)}$

Solution

a) $3 - 2(7 - 2^3) = 3 - 2(7 - 8)$ Evaluate within parentheses first.

$= 3 - 2(-1)$

$= 3 - (-2)$ Multiply.

$= 5$ Subtract.

b) $3 - |7 - 3 \cdot 4| = 3 - |7 - 12|$ Evaluate within the absolute value symbols first.

$= 3 - |-5|$

$= 3 - 5$ Evaluate the absolute value.

$= -2$ Subtract.

c) $\frac{9 - 5 + 8}{-5^2 - 3(-7)} = \frac{12}{-25 + 21} = \frac{12}{-4} = -3$ Numerator and denominator are treated as if in parentheses. ■

When grouping symbols occur within grouping symbols, we evaluate within the innermost grouping symbols first and then work outward. In this case, brackets [ ] can be used as grouping symbols along with parentheses to make the grouping clear.

EXAMPLE 8 **Grouping within grouping**

Evaluate each expression.

a) $6 - 4[5 - (7 - 9)]$ **b)** $-2|3 - (9 - 5)| - |-3|$

Solution

a) $6 - 4[5 - (7 - 9)] = 6 - 4[5 - (-2)]$ Innermost parentheses first

$= 6 - 4[7]$ Next evaluate within the brackets.

$= 6 - 28$ Multiply.

$= -22$ Subtract.

b) $-2|3 - (9 - 5)| - |-3| = -2|3 - 4| - |-3|$ Innermost grouping first

$= -2|-1| - |-3|$ Evaluate within the first absolute value.

$= -2 \cdot 1 - 3$ Evaluate absolute values.

$= -2 - 3$ Multiply.

$= -5$ Subtract. ■

Calculator Close-Up

Graphing calculators can handle grouping symbols within grouping symbols. Since parentheses must occur in pairs, you should have the same number of left parentheses as right parentheses. You might notice other grouping symbols on your calculator, but they may or may not be used for grouping. See your manual.

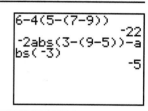

```
6-4(5-(7-9))
                 -22
-2abs(3-(9-5))-a
bs(-3)
                  -5
```

WARM-UPS

True or false? Explain your answer.

1. $(-3)^2 = -6$ False
2. $5 - 3 \cdot 2 = 4$ False
3. $(5 - 3)2 = 4$ True
4. $|5 - 6| = |5| - |6|$ False
5. $5 + 6 \cdot 2 = (5 + 6) \cdot 2$ False
6. $(2 + 3)^2 = 2^2 + 3^2$ False
7. $5 - 3^3 = 8$ False
8. $(5 - 3)^3 = 8$ True
9. $6 - \frac{6}{2} = \frac{0}{2}$ False
10. $\frac{6 - 6}{2} = 0$ True

1.5 EXERCISES

Reading and Writing *After reading this section write out the answers to these questions. Use complete sentences.*

1. What is an arithmetic expression?
An arithmetic expression is the result of writing numbers in a meaningful combination with the ordinary operations of arithmetic.

2. What is the purpose of grouping symbols?
The purpose of grouping symbols is to indicate the order in which to perform operations.

3. What is an exponential expression?
An exponential expression is an expression of the form a^n.

4. What is the difference between -3^6 and $(-3)^6$?
The value of -3^6 is negative while the value of $(-3)^6$ is positive.

5. What is the purpose of the order of operations?
The order of operations tells us the order in which to perform operations when grouping symbols are omitted.

6. What were the different types of grouping symbols used in this section?
Grouping symbols used in this section were parentheses, absolute value bars, and the fraction bar.

Evaluate each expression. See Example 1.

7. $(4 - 3)(5 - 9)$ -4

8. $(5 - 7)(-2 - 3)$ 10

9. $|3 + 4| - |-2 - 4|$ 1

10. $|-4 + 9| + |-3 - 5|$ 13

11. $\dfrac{7 - (-9)}{3 - 5}$ -8

12. $\dfrac{-8 + 2}{-1 - 1}$ 3

13. $(-6 + 5)(7)$ -7

14. $-6 + (5 \cdot 7)$ 29

15. $(-3 - 7) - 6$ -16

16. $-3 - (7 - 6)$ -4

17. $-16 \div (8 \div 2)$ -4

18. $(-16 \div 8) \div 2$ -1

Write each product as an exponential expression. See Example 2.

19. $4 \cdot 4 \cdot 4 \cdot 4$ 4^4

20. $1 \cdot 1 \cdot 1 \cdot 1 \cdot 1$ 1^5

21. $(-5)(-5)(-5)(-5)$ $(-5)^4$

22. $(-7)(-7)(-7)$ $(-7)^3$

23. $(-y)(-y)(-y)$ $(-y)^3$

24. $x \cdot x \cdot x \cdot x \cdot x$ x^5

25. $\frac{3}{7} \cdot \frac{3}{7} \cdot \frac{3}{7} \cdot \frac{3}{7} \cdot \frac{3}{7}$ $\left(\frac{3}{7}\right)^5$

26. $\frac{y}{2} \cdot \frac{y}{2} \cdot \frac{y}{2} \cdot \frac{y}{2}$ $\left(\frac{y}{2}\right)^4$

Write each exponential expression as a product without exponents. See Example 3.

27. 5^3 $5 \cdot 5 \cdot 5$

28. $(-8)^4$ $(-8)(-8)(-8)(-8)$

29. b^2 $b \cdot b$

30. $(-a)^5$ $(-a)(-a)(-a)(-a)(-a)$

31. $\left(-\frac{1}{2}\right)^5$ $\left(-\frac{1}{2}\right)\left(-\frac{1}{2}\right)\left(-\frac{1}{2}\right)\left(-\frac{1}{2}\right)\left(-\frac{1}{2}\right)$

32. $\left(-\frac{13}{12}\right)^3$ $\left(-\frac{13}{12}\right)\left(-\frac{13}{12}\right)\left(-\frac{13}{12}\right)$

33. $(0.22)^4$ $(0.22)(0.22)(0.22)(0.22)$

34. $(1.25)^6$ $(1.25)(1.25)(1.25)(1.25)(1.25)(1.25)$

Evaluate each exponential expression. See Examples 4 and 5.

35. 3^4 81

36. 5^3 125

37. 0^9 0

38. 0^{12} 0

39. $(-5)^4$ 625

40. $(-2)^5$ -32

41. $(-6)^3$ -216

42. $(-12)^2$ 144

43. $(10)^5$ $100,000$

44. $(-10)^6$ $1,000,000$

45. $(-0.1)^3$ -0.001

46. $(-0.2)^2$ 0.04

47. $\left(\frac{1}{2}\right)^3$ $\frac{1}{8}$

48. $\left(\frac{2}{3}\right)^3$ $\frac{8}{27}$

49. $\left(-\frac{1}{2}\right)^2$ $\frac{1}{4}$

50. $\left(-\frac{2}{3}\right)^2$ $\frac{4}{9}$

51. -8^2 -64

52. -7^2 -49

53. -8^4 -4096

54. -7^4 -2401

55. $-(7 - 10)^3$ 27

56. $-(6 - 9)^4$ -81

57. $(-2^2) - (3^2)$ -13

58. $(-3^4) - (-5^2)$ -56

Evaluate each expression. See Example 6.

59. $3^2 \cdot 2^2$ 36

60. $5 \cdot 10^2$ 500

61. $-3 \cdot 2 + 4 \cdot 6$ 18

62. $-5 \cdot 4 - 8 \cdot 3$ -44

63. $(-3)^3 + 2^3$ -19

64. $3^2 - 5(-1)^3$ 14

65. $-21 + 36 \div 3^2$ -17

66. $-18 - 9^2 \div 3^3$ -21

67. $-3 \cdot 2^3 - 5 \cdot 2^2$ -44

68. $2 \cdot 5 - 3^2 + 4 \cdot 0$ 1

69. $\dfrac{-8}{2} + 2 \cdot 3 \cdot 5 - 2^3$ 18

70. $-4 \cdot 2 \cdot 6 - \dfrac{12}{3} + 3^3$ -25

Evaluate each expression. See Example 7.

71. $(-3 + 4^2)(-6)$ -78

72. $-3 \cdot (2^3 + 4) \cdot 5$ -180

73. $(-3 \cdot 2 + 6)^3$ 0

74. $5 - 2(-3 + 2)^3$ 7

75. $2 - 5(3 - 4 \cdot 2)$ 27

76. $(3 - 7)(4 - 6 \cdot 2)$ 32

77. $3 - 2 \cdot |5 - 6|$ 1

78. $3 - |6 - 7 \cdot 3|$ -12

79. $(3^2 - 5) \cdot |3 \cdot 2 - 8|$ 8

80. $|4 - 6 \cdot 3| + |6 - 9|$ 17

81. $\dfrac{3 - 4 \cdot 6}{7 - 10}$ 7

82. $\dfrac{6 - (-8)^2}{-3 - (-1)}$ 29

83. $\dfrac{7 - 9 - 3^2}{9 - 7 - 3}$ 11

84. $\dfrac{3^2 - 2 \cdot 4}{-30 + 2 \cdot 4^2}$ $\frac{1}{2}$

Evaluate each expression. See Example 8.

85. $3 + 4[9 - 6(2 - 5)]$ 111

86. $9 + 3[5 - (3 - 6)^2]$ -3

87. $6^2 - [(2 + 3)^2 - 10]$ 21

88. $3[(2 - 3)^2 + (6 - 4)^2]$ 15

89. $4 - 5 \cdot |3 - (3^2 - 7)|$ -1

90. $2 + 3 \cdot |4 - (7^2 - 6^2)|$ 29

91. $-2|3 - (7 - 3)| - |-9|$ -11

92. $[3 - (2 - 4)][3 + |2 - 4|]$ 25

Evaluate each expression. Use a calculator to check.

93. $1 + 2^3$ 9

94. $(1 + 2)^3$ 27

95. $(-2)^2 - 4(-1)(3)$ 16

96. $(-2)^2 - 4(-2)(-3)$ -20

97. $4^2 - 4(1)(-3)$ 28

98. $3^2 - 4(-2)(3)$ 33

99. $(-11)^2 - 4(5)(0)$ 121

100. $(-12)^2 - 4(3)(0)$ 144

101. $-5^2 - 3 \cdot 4^2$ -73

102. $-6^2 - 5(-3)^2$ -81

103. $[3 + 2(-4)]^2$ 25

104. $[6 - 2(-3)]^2$ 144

105. $|-1| - |-1|$ 0

106. $4 - |1 - 7|$ -2

107. $\dfrac{4 - (-4)}{-2 - 2}$ -2

108. $\dfrac{3 - (-7)}{3 - 5}$ -5

109. $3(-1)^2 - 5(-1) + 4$ 12

110. $-2(1)^2 - 5(1) - 6$ -13

111. $5 - 2^2 + 3^4$ 82

112. $5 + (-2)^2 - 3^2$ 0

113. $-2 \cdot |9 - 6^2|$ -54

114. $8 - 3|5 - 4^2 + 1|$ -22

115. $-3^2 - 5[4 - 2(4 - 9)]$ -79

116. $-2[(3 - 4)^3 - 5] + 7$ 19

117. $1 - 5|5 - (9 + 1)|$ -24

118. $|6 - 3 \cdot 7| + |7 - (5 - 2)|$ 19

Use a calculator to evaluate each expression.

119. $3.2^2 - 4(3.6)(-2.2)$ 41.92

120. $(-4.5)^2 - 4(-2.8)(-4.6)$ -31.27

121. $(5.63)^3 - [4.7 - (-3.3)^2]$ 184.643547

Just as we translated verbal expressions into algebraic expressions, we can translate verbal sentences into algebraic equations. In an algebraic equation we use the equality symbol (=). Equality is indicated in words by phrases such as "is equal to," "is the same as," or simply "is."

EXAMPLE 6 **Writing equations**

Translate each sentence into an equation.

a) The sum of x and 7 is 12.

b) The product of 4 and x is the same as the sum of y and 5.

c) The quotient of $x + 3$ and 5 is equal to -1.

Solution

a) $x + 7 = 12$ **b)** $4x = y + 5$ **c)** $\frac{x + 3}{5} = -1$ ■

Applications

Algebraic expressions are used to describe or **model** real-life situations. We can evaluate an algebraic expression for many values of a variable to get a collection of data. A graph (picture) of this data can give us useful information. For example, a forensic scientist can use a graph to estimate the length of a person's femur from the person's height.

EXAMPLE 7 **Reading a graph**

A forensic scientist uses the expression $69.1 + 2.2F$ as an estimate of the height in centimeters of a male with a femur of length F centimeters (National Space Biomedical Research Institute, www.nsbri.org).

a) If the femur of a male skeleton measures 50.6 cm, then what was the person's height?

b) Use the graph shown in Fig. 1.17 to estimate the length of a femur for a person who is 150 cm tall.

Study Tip

Find a group of students to work with outside of class. Don't just settle for answers. Make sure that everyone in the group understands the solution to a problem. You will find that you really understand a concept when you can explain it to someone else.

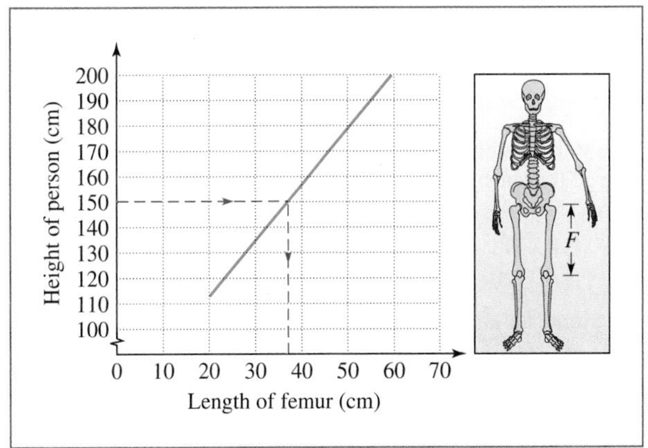

FIGURE 1.17

Solution

a) To find the height of the person, we use $F = 50.6$ in the expression $69.1 + 2.2F$:

$$69.1 + 2.2(50.6) \approx 180.4$$

So the person was approximately 180.4 cm tall.

b) To find the length of a femur for a person who is 150 cm tall, first locate 150 cm on the height scale of the graph in Fig. 1.17. Now draw a horizontal line to the graph and then a vertical line down to the length scale. So the length of a femur for a person who is 150 cm tall is approximately 36 cm. ■

WARM-UPS

True or false? Explain your answer.

1. The expression $2x + 3y$ is referred to as a sum. True
2. The expression $5(y - 9)$ is a difference. False
3. The expression $2(x + 3y)$ is a product. True
4. The expression $\frac{x}{2} + \frac{y}{3}$ is a quotient. False
5. The expression $(a - b)(a + b)$ is a product of a sum and a difference. True
6. If x is -2, then the value of $2x + 4$ is 8. False
7. If $a = -3$, then $a^3 - 5 = 22$. False
8. The number 5 is a solution to the equation $2x - 3 = 13$. False
9. The product of $x + 3$ and 5 is $(x + 3)5$. True
10. The expression $2(x + 7)$ should be read as "the sum of 2 times x plus 7." False

1.6 EXERCISES

Reading and Writing *After reading this section write out the answers to these questions. Use complete sentences.*

1. What is an algebraic expression?
   An algebraic expression is the result of combining numbers and variables with the operations of arithmetic in some meaningful way.
2. What is the difference between an algebraic expression and an arithmetic expression?
   An arithmetic expression involves only numbers.
3. How can you tell whether an algebraic expression should be referred to as a sum, difference, product, quotient, or square?
   An algebraic expression is named according to the last operation to be performed.
4. How do you evaluate an algebraic expression?
   An algebraic expression is evaluated by replacing the variables with numbers and evaluating the resulting arithmetic expression.
5. What is an equation?
   An equation is a sentence that expresses equality between two algebraic expressions.
6. What is a solution to an equation?
   If an equation is true when the variable is replaced by a number, then that number is a solution to the equation.

Identify each expression as a sum, difference, product, quotient, square, or cube. See Example 1.

7. $a^3 - 1$ Difference
8. $b(b - 1)$ Product
9. $(w - 1)^3$ Cube
10. $m^2 + n^2$ Sum
11. $3x + 5y$ Sum
12. $\dfrac{a - b}{b - a}$ Quotient
13. $\dfrac{u}{v} - \dfrac{v}{u}$ Difference
14. $(s - t)^2$ Square
15. $3(x + 5y)$ Product
16. $a - \dfrac{a}{2}$ Difference
17. $\left(\dfrac{2}{z}\right)^2$ Square
18. $(2q - p)^3$ Cube

Use the term sum, difference, product, quotient, square, or cube to translate each algebraic expression into a verbal expression. See Example 2.

19. $x^2 - a^2$ The difference of x^2 and a^2
20. $a^3 + b^3$ The sum of a^3 and b^3
21. $(x - a)^2$ The square of $x - a$
22. $(a + b)^3$ The cube of $a + b$
23. $\dfrac{x - 4}{2}$ The quotient of $x - 4$ and 2
24. $2(x - 3)$ The product of 2 and $x - 3$
25. $\dfrac{x}{2} - 4$ The difference of $\dfrac{x}{2}$ and 4

26. $2x - 3$ The difference of $2x$ and 3

27. $(ab)^3$ The cube of ab

28. a^3b^3 The product of a^3 and b^3

Translate each verbal expression into an algebraic expression. See Example 3.

29. The sum of $2x$ and $3y$ $2x + 3y$

30. The product of $5x$ and z $5xz$

31. The difference of 8 and $7x$ $8 - 7x$

32. The quotient of 6 and $x + 4$ $\frac{6}{x + 4}$

33. The square of $a + b$ $(a + b)^2$

34. The difference of a^3 and b^3 $a^3 - b^3$

35. The product of $x + 9$ and $x + 12$ $(x + 9)(x + 12)$

36. The cube of x x^3

37. The quotient of $x - 7$ and $7 - x$ $\frac{x - 7}{7 - x}$

38. The product of -3 and $x - 1$ $-3(x - 1)$

Evaluate each expression using $a = -1$, $b = 2$, and $c = -3$. See Example 4.

39. $-(a - b)$ 3

40. $b - a$ 3

41. $-b^2 + 7$ 3

42. $-c^2 - b^2$ -13

43. $c^2 - 2c + 1$ 16

44. $b^2 - 2b + 4$ 4

45. $a^3 - b^3$ -9

46. $b^3 - c^3$ 35

47. $(a - b)(a + b)$ -3

48. $(a - c)(a + c)$ -8

49. $b^2 - 4ac$ -8

50. $a^2 - 4bc$ 25

51. $\frac{a - c}{a - b}$ $-\frac{2}{3}$

52. $\frac{b - c}{b + a}$ 5

53. $\frac{2}{a} + \frac{6}{b} - \frac{9}{c}$ 4

54. $\frac{c}{a} + \frac{6}{b} - \frac{b}{a}$ 8

55. $a \div |-a|$ -1

56. $|a| \div a$ -1

57. $|b| - |a|$ 1

58. $|c| + |b|$ 5

59. $-|-a - c|$ -4

60. $-|-a - b|$ -1

61. $(3 - |a - b|)^2$ 0

62. $(|b + c| - 2)^3$ -1

Determine whether the given number is a solution to the equation following it. See Example 5.

63. $2,\ 3x + 7 = 13$ Yes

64. $-1,\ -3x + 7 = 10$ Yes

65. $-2,\ \frac{3x - 4}{2} = 5$ No

66. $-3,\ \frac{-2x + 9}{3} = 5$ Yes

67. $-2,\ -x + 4 = 6$ Yes

68. $-9,\ -x + 3 = 12$ Yes

69. $4,\ 3x - 7 = x + 1$ Yes

70. $5,\ 3x - 7 = 2x + 1$ No

71. $3,\ -2(x - 1) = 2 - 2x$ Yes

72. $-8,\ x - 9 = -(9 - x)$ Yes

73. $1,\ x^2 + 3x - 4 = 0$ Yes

74. $-1,\ x^2 + 5x + 4 = 0$ Yes

75. $8,\ \frac{x}{x - 8} = 0$ No

76. $3,\ \frac{x - 3}{x + 3} = 0$ Yes

77. $-6,\ \frac{x + 6}{x + 6} = 1$ No

78. $9,\ \frac{9}{x - 9} = 0$ No

Translate each sentence into an equation. See Example 6.

79. The sum of $5x$ and $3x$ is $8x$. $5x + 3x = 8x$

80. The sum of $\frac{y}{2}$ and 3 is 7. $\frac{y}{2} + 3 = 7$

81. The product of 3 and $x + 2$ is equal to 12. $3(x + 2) = 12$

82. The product of -6 and $7y$ is equal to 13. $-6(7y) = 13$

83. The quotient of x and 3 is the same as the product of x and 5. $\frac{x}{3} = 5x$

84. The quotient of $x + 3$ and $5y$ is the same as the product of x and y. $\frac{x + 3}{5y} = xy$

85. The square of the sum of a and b is equal to 9. $(a + b)^2 = 9$

86. The sum of the squares of a and b is equal to the square of c. $a^2 + b^2 = c^2$

Fill in the tables with the appropriate values for the given expressions.

87.

x	$2x - 3$
-2	-7
-1	-5
0	-3
1	-1
2	1

88.

x	$-\frac{1}{2}x + 4$
-4	6
-2	5
0	4
2	3
4	2

89.

a	a^2	a^3	a^4
2	4	8	16
$\frac{1}{2}$	$\frac{1}{4}$	$\frac{1}{8}$	$\frac{1}{16}$
10	100	1000	10,000
0.1	0.01	0.001	0.0001

90.

b	$\frac{1}{b}$	$\frac{1}{b^2}$	$\frac{1}{b^3}$
3	$\frac{1}{3}$	$\frac{1}{9}$	$\frac{1}{27}$
$\frac{1}{3}$	3	9	27
10	0.1	0.01	0.001
0.1	10	100	1000

Use a calculator to find the value of $b^2 - 4ac$ for each of the following choices of a, b, and c.

91. $a = 4.2, b = 6.7, c = 1.8$ 14.65

92. $a = -3.5, b = 9.1, c = 3.6$ 133.21

93. $a = -1.2, b = 3.2, c = 5.6$ 37.12

94. $a = 2.4, b = -8.5, c = -5.8$ 127.93

Solve each problem. See Example 7.

95. ***Forensics.*** A forensic scientist uses the expression $81.7 + 2.4T$ to estimate the height in centimeters of a male with a tibia of length T centimeters. If a male skeleton has a tibia of length 36.5 cm, then what was the height of the person? Use the accompanying graph to estimate the length of a tibia for a male with a height of 180 cm. 169.3 cm, 41 cm

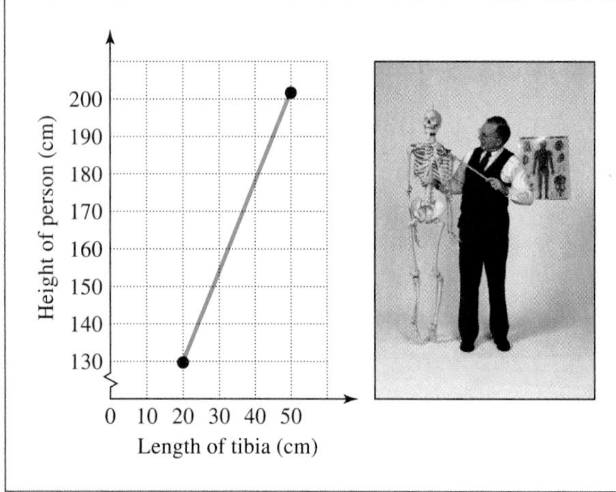

FIGURE FOR EXERCISE 95

96. ***Forensics.*** A forensic scientist uses the expression $72.6 + 2.5T$ to estimate the height in centimeters of a female with a tibia of length T centimeters. If a female skeleton has a tibia of length 32.4 cm, then what was the height of the person? Find the length of your tibia in centimeters, and use the expression from this exercise or the previous exercise to estimate your height. 153.6 cm

97. ***Games behind.*** In baseball a team's standing is measured by its percentage of wins and by the number of games it is behind the leading team in its division. The expression

$$\frac{(X - x) + (y - Y)}{2}$$

gives the number of games behind for a team with x wins and y losses, where the division leader has X wins and Y losses. The table shown here gives the won-lost records for the American League East at the end of 2001 (www.espn.com). Fill in the column for the games behind (GB). 13.5, 16, 32.5, 34

	W	L	Pct	GB
NY Yankees	95	65	0.594	–
Boston	82	79	0.509	?
Toronto	80	82	0.494	?
Baltimore	63	98	0.391	?
Tampa Bay	62	100	0.383	?

TABLE FOR EXERCISE 97

98. ***Fly ball.*** The approximate distance in feet that a baseball travels when hit at an angle of 45° is given by the expression

$$\frac{(v_0)^2}{32}$$

where v_0 is the initial velocity in feet per second. If Barry Bonds of the Giants hits a ball at a 45° angle with an initial velocity of 120 feet per second, then how far will the ball travel? Use the accompanying graph to estimate the initial velocity for a ball that has traveled 370 feet. 450 feet, 109 feet per second

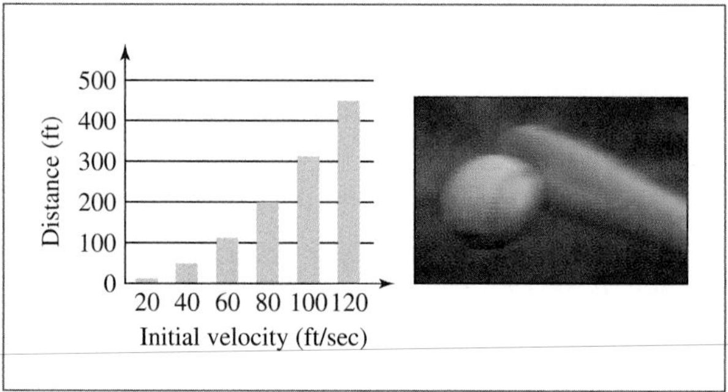

FIGURE FOR EXERCISE 98

99. ***Football field.*** The expression $2L + 2W$ gives the perimeter of a rectangle with length L and width W. What is the perimeter of a football field with length 100 yards and width 160 feet? 920 feet

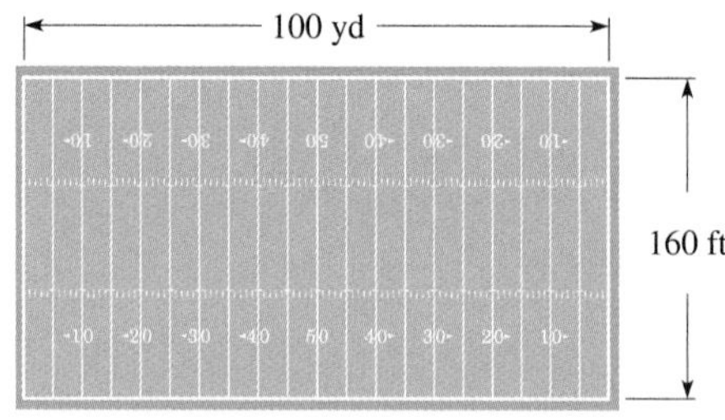

FIGURE FOR EXERCISE 99

100. ***Crop circles.*** The expression πr^2 gives the area of a circle with radius r. How many square meters of wheat were destroyed when an alien ship made a crop circle of

diameter 25 meters in the wheat field at the Southwind Ranch?
490.9 m^2

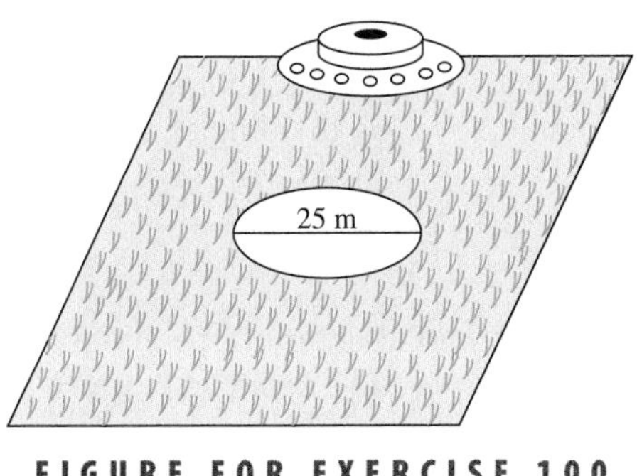

FIGURE FOR EXERCISE 100

GETTING MORE INVOLVED

101. ***Cooperative learning.*** Find some examples of algebraic expressions outside of this class, and explain to the class what they are used for.

102. ***Discussion.*** Why do we use letters to represent numbers? Wouldn't it be simpler to just use numbers?
Using letters, we can make statements that are true for all numbers or statements that are true for only one unknown number.

103. ***Writing.*** Explain why the square of the sum of two numbers is different from the sum of the squares of two numbers.
For the square of the sum consider $(2 + 3)^2 = 5^2 = 25$. For the sum of the squares consider $2^2 + 3^2 = 4 + 9 = 13$. So $(2 + 3)^2 \neq 2^2 + 3^2$.

1.7 PROPERTIES OF THE REAL NUMBERS

In This Section

- The Commutative Properties
- The Associative Properties
- The Distributive Property
- The Identity Properties
- The Inverse Properties
- Multiplication Property of Zero
- Applications

Everyone knows that the price of a hamburger plus the price of a Coke is the same as the price of a Coke plus the price of a hamburger. But do you know that this example illustrates the commutative property of addition? The properties of the real numbers are commonly used by anyone who performs the operations of arithmetic. In algebra we must have a thorough understanding of these properties.

The Commutative Properties

We get the same result whether we evaluate $3 + 5$ or $5 + 3$. This example illustrates the commutative property of addition. The fact that $4 \cdot 6$ and $6 \cdot 4$ are equal illustrates the commutative property of multiplication.

Commutative Properties

For any real numbers a and b,

$$a + b = b + a \qquad \text{and} \qquad ab = ba.$$

EXAMPLE 1 **The commutative property of addition**

Use the commutative property of addition to rewrite each expression.

a) $2 + (-10)$ **b)** $8 + x^2$ **c)** $2y - 4x$

Solution

a) $2 + (-10) = -10 + 2$

b) $8 + x^2 = x^2 + 8$

c) $2y - 4x = 2y + (-4x) = -4x + 2y$ ■

EXAMPLE 2 **The commutative property of multiplication**

Use the commutative property of multiplication to rewrite each expression.

a) $n \cdot 3$ **b)** $(x + 2) \cdot 3$ **c)** $5 - yx$

Solution

a) $n \cdot 3 = 3 \cdot n = 3n$ **b)** $(x + 2) \cdot 3 = 3(x + 2)$

c) $5 - yx = 5 - xy$ ■

Addition and multiplication are commutative operations, but what about subtraction and division? Since $5 - 3 = 2$ and $3 - 5 = -2$, subtraction is not commutative. To see that division is not commutative, try dividing $8 among 4 people and $4 among 8 people.

The Associative Properties

Helpful Hint

In arithmetic we would probably write $(2 + 3) + 7 = 12$ without thinking about the associative property. In algebra, we need the associative property to understand that

$$(x + 3) + 7 = x + (3 + 7) = x + 10.$$

Consider the computation of $2 + 3 + 6$. Using the order of operations, we add 2 and 3 to get 5 and then add 5 and 6 to get 11. If we add 3 and 6 first to get 9 and then add 2 and 9, we also get 11. So

$$(2 + 3) + 6 = 2 + (3 + 6).$$

We get the same result for either order of addition. This property is called the **associative property of addition.** The commutative and associative properties of addition are the reason that a hamburger, a Coke, and French fries cost the same as French fries, a hamburger, and a Coke.

We also have an **associative property of multiplication.** Consider the following two ways to find the product of 2, 3, and 4:

$$(2 \cdot 3)4 = 6 \cdot 4 = 24$$
$$2(3 \cdot 4) = 2 \cdot 12 = 24$$

We get the same result for either arrangement.

Associative Properties

For any real numbers a, b, and c,

$$(a + b) + c = a + (b + c) \quad \text{and} \quad (ab)c = a(bc).$$

EXAMPLE 3 **Using the properties of multiplication**

Use the commutative and associative properties of multiplication and exponential notation to rewrite each product.

a) $(3x)(x)$ **b)** $(xy)(5yx)$

Study Tip

Find out what help is available at your school. Accompanying this text are video tapes, solution manuals, and a computer tutorial. Around most campuses you will find tutors available for hire, but most schools have a math lab where you can get help for free. Some schools even have free one-on-one tutoring available through special programs.

Solution

a) $(3x)(x) = 3(x \cdot x) = 3x^2$

b) The commutative and associative properties of multiplication allow us to rearrange the multiplication in any order. We generally write numbers before variables, and we usually write variables in alphabetical order:

$$(xy)(5yx) = 5xxyy = 5x^2y^2$$ ■

Consider the expression

$$3 - 9 + 7 - 5 - 8 + 4 - 13.$$

According to the accepted order of operations, we could evaluate this by computing from left to right. However, using the definition of subtraction, we can rewrite this expression as addition:

$$3 + (-9) + 7 + (-5) + (-8) + 4 + (-13)$$

The commutative and associative properties of addition allow us to add these numbers in any order we choose. It is usually faster to add the positive numbers, add the negative numbers, and then combine those two totals:

$$3 + 7 + 4 + (-9) + (-5) + (-8) + (-13) = 14 + (-35) = -21$$

Note that by performing the operations in this manner, we must subtract only once. There is no need to rewrite this expression as we have done here. We can sum the positive numbers and the negative numbers from the original expression and then combine their totals.

EXAMPLE 4 **Using the properties of addition**

Evaluate.

a) $3 - 7 + 9 - 5$ **b)** $4 - 5 - 9 + 6 - 2 + 4 - 8$

Solution

a) First add the positive numbers and the negative numbers:

$$\begin{aligned} 3 - 7 + 9 - 5 &= 12 + (-12) \\ &= 0 \end{aligned}$$

b) $\begin{aligned} 4 - 5 - 9 + 6 - 2 + 4 - 8 &= 14 + (-24) \\ &= -10 \end{aligned}$ ■

It is certainly not essential that we evaluate the expressions of Example 4 as shown. We get the same answer by adding and subtracting from left to right. However, in algebra, just getting the answer is not always the most important point. Learning new methods often increases understanding.

Even though addition is associative, subtraction is not an associative operation. For example, $(8 - 4) - 3 = 1$ and $8 - (4 - 3) = 7$. So

$$(8 - 4) - 3 \neq 8 - (4 - 3).$$

We can also use a numerical example to show that division is not associative. For instance, $(16 \div 4) \div 2 = 2$ and $16 \div (4 \div 2) = 8$. So

$$(16 \div 4) \div 2 \neq 16 \div (4 \div 2).$$

Helpful Hint

To visualize the distributive property, we can determine the number of circles shown here in two ways:

o o o o   o o o o o
o o o o   o o o o o
o o o o   o o o o o

There are $3 \cdot 9$ or 27 circles, or there are $3 \cdot 4$ circles in the first group and $3 \cdot 5$ circles in the second group for a total of 27 circles.

The Distributive Property

If four men and five women pay $3 each for a movie, there are two ways to find the total amount spent:

$$3(4 + 5) = 3 \cdot 9 = 27$$
$$3 \cdot 4 + 3 \cdot 5 = 12 + 15 = 27$$

Since we get $27 either way, we can write

$$3(4 + 5) = 3 \cdot 4 + 3 \cdot 5.$$

We say that the multiplication by 3 is *distributed* over the addition. This example illustrates the **distributive property.**

Consider the following expressions involving multiplication and subtraction:

$$5(6 - 4) = 5 \cdot 2 = 10$$
$$5 \cdot 6 - 5 \cdot 4 = 30 - 20 = 10$$

Since both expressions have the same value, we can write

$$5(6 - 4) = 5 \cdot 6 - 5 \cdot 4.$$

Multiplication by 5 is distributed over each number in the parentheses. This example illustrates that multiplication distributes over subtraction.

Distributive Property

For any real numbers a, b, and c,

$$a(b + c) = ab + ac \quad \text{and} \quad a(b - c) = ab - ac.$$

We can use the distributive property to remove parentheses. If we start with $4(x + 3)$ and write

$$4(x + 3) = 4x + 4 \cdot 3 = 4x + 12,$$

we are using it to multiply 4 and $x + 3$ or to remove the parentheses. We wrote the product $4(x + 3)$ as the sum $4x + 12$.

EXAMPLE 5

Writing a product as a sum or difference

Use the distributive property to remove the parentheses.

a) $a(3 - b)$ **b)** $-3(x - 2)$

Solution

a) $a(3 - b) = a3 - ab$ Distributive property

$= 3a - ab$ $a3 = 3a$

b) $-3(x - 2) = -3x - (-3)(2)$ Distributive property

$= -3x - (-6)$ $(-3)(2) = -6$

$= -3x + 6$ Simplify. ■

When we write a number or an expression as a product, we are **factoring.** If we start with $3x + 15$ and write

$$3x + 15 = 3x + 3 \cdot 5 = 3(x + 5),$$

we are using the distributive property to factor $3x + 15$. We factored out the common factor 3.

EXAMPLE 6

Writing a sum or difference as a product

Use the distributive property to factor each expression.

a) $7x - 21$ **b)** $5a + 5$

Solution

a) $7x - 21 = 7x - 7 \cdot 3$ Write 21 as $7 \cdot 3$.

$= 7(x - 3)$ Distributive property

b) $5a + 5 = 5a + 5 \cdot 1$ Write 5 as $5 \cdot 1$.

$= 5(a + 1)$ Factor out the common factor 5. ■

Study Tip

Don't cram for a test. Some students try to cram weeks of work into one "all-nighter." These same students are seen frantically paging through the text up until the moment that the test papers are handed out. These practices create a lot of test anxiety and will only make you sick. Start studying for a test several days in advance, and get a good night's sleep before a test. If you keep up with homework, then there will be no need to cram.

The Identity Properties

The numbers 0 and 1 have special properties. Multiplication of a number by 1 does not change the number, and addition of 0 to a number does not change the number. That is why 1 is called the **multiplicative identity** and 0 is called the **additive identity.**

Identity Properties

For any real number a,

$$a \cdot 1 = 1 \cdot a = a \quad \text{and} \quad a + 0 = 0 + a = a.$$

The Inverse Properties

The idea of additive inverses was introduced in Section 1.3. Every real number a has an **additive inverse** or **opposite,** $-a$, such that $a + (-a) = 0$. Every nonzero real number a also has a **multiplicative inverse** or **reciprocal,** written $\frac{1}{a}$, such that $a \cdot \frac{1}{a} = 1$. Note that the sum of additive inverses is the additive identity and that the product of multiplicative inverses is the multiplicative identity.

Inverse Properties

For any real number a there is a number $-a$, such that

$$a + (-a) = 0.$$

For any nonzero real number a there is a number $\frac{1}{a}$ such that

$$a \cdot \frac{1}{a} = 1.$$

We are already familiar with multiplicative inverses for rational numbers. For example, the multiplicative inverse of $\frac{2}{3}$ is $\frac{3}{2}$ because

$$\frac{2}{3} \cdot \frac{3}{2} = \frac{6}{6} = 1.$$

EXAMPLE 7 **Multiplicative inverses**

Find the multiplicative inverse of each number.

a) 5 **b)** 0.3

c) $-\frac{3}{4}$ **d)** 1.7

Calculator Close-Up

You can find multiplicative inverses with a calculator as shown here.

```
1/.3►Frac
                10/3
1/(-3/4)►Frac
                -4/3
1/1.7►Frac
               10/17
```

When the divisor is a fraction, it must be in parentheses.

Solution

a) The multiplicative inverse of 5 is $\frac{1}{5}$ because

$$5 \cdot \frac{1}{5} = 1.$$

b) To find the reciprocal of 0.3, we first write 0.3 as a ratio of integers:

$$0.3 = \frac{3}{10}$$

The multiplicative inverse of 0.3 is $\frac{10}{3}$ because

$$\frac{3}{10} \cdot \frac{10}{3} = 1.$$

Study Tip

When you get a test back, do not simply file it in your notebook or the waste basket. While the material is fresh in your mind, rework all problems that you missed. Ask questions about anything that you don't understand and save your test for future reference.

c) The reciprocal of $-\frac{3}{4}$ is $-\frac{4}{3}$ because

$$\left(-\frac{3}{4}\right)\left(-\frac{4}{3}\right) = 1.$$

d) First convert 1.7 to a ratio of integers:

$$1.7 = 1\frac{7}{10} = \frac{17}{10}$$

The multiplicative inverse is $\frac{10}{17}$. ■

Multiplication Property of Zero

Zero has a property that no other number has. Multiplication involving zero always results in zero.

Multiplication Property of Zero

For any real number a,

$$0 \cdot a = 0 \quad \text{and} \quad a \cdot 0 = 0.$$

EXAMPLE 8 **Identifying the properties**

Name the property that justifies each equation.

a) $5 \cdot 7 = 7 \cdot 5$

b) $4 \cdot \frac{1}{4} = 1$

c) $1 \cdot 864 = 864$

d) $6 + (5 + x) = (6 + 5) + x$

e) $3x + 5x = (3 + 5)x$

f) $6 + (x + 5) = 6 + (5 + x)$

g) $\pi x^2 + \pi y^2 = \pi(x^2 + y^2)$

h) $325 + 0 = 325$

i) $-3 + 3 = 0$

j) $455 \cdot 0 = 0$

Solution

a) Commutative

b) Multiplicative inverse

c) Multiplicative identity

d) Associative

e) Distributive

f) Commutative

g) Distributive

h) Additive identity

i) Additive inverse

j) Multiplication property of 0 ■

Applications

Reciprocals are important in problems involving work. For example, if you wax one car in 3 hours, then your rate is $\frac{1}{3}$ of a car per hour. If you can wash one car in 12 minutes $\left(\frac{1}{5}\text{ of an hour}\right)$, then you are washing cars at the rate of 5 cars per hour. In general, if you can complete a task in x hours, then your rate is $\frac{1}{x}$ tasks per hour.

EXAMPLE 9 **Washing rates**

A car wash has two machines. The old machine washes one car in 0.1 hour, while the new machine washes one car in 0.08 hour. If both machines are operating, then at what rate (in cars per hour) are the cars being washed?

Helpful Hint

When machines or people are working together, we can add their rates provided they do not interfere with each other's work. If operating both car wash machines causes a traffic jam, then the rate together might not be 22.5 cars per hour.

Solution

The old machine is working at the rate of $\frac{1}{0.1}$ cars per hour, and the new machine is working at the rate of $\frac{1}{0.08}$ cars per hour. Their rate working together is the sum of their individual rates:

$$\frac{1}{0.1} + \frac{1}{0.08} = 10 + 12.5 = 22.5$$

So working together, the machines are washing 22.5 cars per hour. ■

WARM-UPS

True or false? Explain your answer.

1. $24 \div (4 \div 2) = (24 \div 4) \div 2$ False
2. $1 \div 2 = 2 \div 1$ False
3. $6 - 5 = -5 + 6$ True
4. $9 - (4 - 3) = (9 - 4) - 3$ False
5. Multiplication is a commutative operation. True
6. $5x + 5 = 5(x + 1)$ for any value of x. True
7. The multiplicative inverse of 0.02 is 50. True
8. $-3(x - 2) = -3x + 6$ for any value of x. True
9. $3x + 2x = (3 + 2)x$ for any value of x. True
10. The additive inverse of 0 is 0. True

1.7 EXERCISES

Reading and Writing *After reading this section write out the answers to these questions. Use complete sentences.*

1. What is the difference between the commutative property of addition and the associative property of addition?
   The commutative property says that $a + b = b + a$ and the associative property says that $(a + b) + c = a + (b + c)$.
2. Which property involves two different operations?
   The distributive property involves multiplication and addition.
3. What is factoring?
   Factoring is the process of writing an expression or number as a product.
4. Which two numbers play a prominent role in the properties studied here?
   The number 0 is the additive identity and the number 1 is the multiplicative identity.
5. What is the purpose of studying the properties of real numbers?
   The properties help us to understand the operations and how they are related to each other.
6. What is the relationship between rate and time?
   If one task is completed in x hours, then the rate is $1/x$ tasks per hour.

Use the commutative property of addition to rewrite each expression. See Example 1.

7. $9 + r$
   $r + 9$
8. $t + 6$
   $6 + t$
9. $3(2 + x)$
   $3(x + 2)$
10. $P(1 + rt)$
    $P(rt + 1)$
11. $4 - 5x$
    $-5x + 4$
12. $b - 2a$
    $-2a + b$

Use the commutative property of multiplication to rewrite each expression. See Example 2.

13. $x \cdot 6$
    $6x$
14. $y \cdot (-9)$
    $-9y$
15. $(x - 4)(-2)$
    $-2(x - 4)$
16. $a(b + c)$
    $(b + c)a$
17. $4 - y \cdot 8$
    $4 - 8y$
18. $z \cdot 9 - 2$
    $9z - 2$

Use the commutative and associative properties of multiplication and exponential notation to rewrite each product. See Example 3.

19. $(4w)(w)$
    $4w^2$
20. $(y)(2y)$
    $2y^2$
21. $3a(ba)$
    $3a^2b$
22. $(x \cdot x)(7x)$
    $7x^3$
23. $(x)(9x)(xz)$
    $9x^3z$
24. $y(y \cdot 5)(wy)$
    $5y^3w$

Evaluate by finding first the sum of the positive numbers and then the sum of the negative numbers. See Example 4.

25. $8 - 4 + 3 - 10$ $\quad -3$
26. $-3 + 5 - 12 + 10$ $\quad 0$

27. $8 - 10 + 7 - 8 - 7$ -10

28. $6 - 11 + 7 - 9 + 13 - 2$ 4

29. $-4 - 11 + 7 - 8 + 15 - 20$ -21

30. $-8 + 13 - 9 - 15 + 7 - 22 + 5$ -29

31. $-3.2 + 2.4 - 2.8 + 5.8 - 1.6$ 0.6

32. $5.4 - 5.1 + 6.6 - 2.3 + 9.1$ 13.7

33. $3.26 - 13.41 + 5.1 - 12.35 - 5$ -22.4

34. $5.89 - 6.1 + 8.58 - 6.06 - 2.34$ -0.03

Use the distributive property to remove the parentheses. See Example 5.

35. $3(x - 5)$ $3x - 15$ **36.** $4(b - 1)$ $4b - 4$

37. $a(2 + t)$ $2a + at$ **38.** $b(a + w)$ $ab + bw$

39. $-3(w - 6)$ $-3w + 18$ **40.** $-3(m - 5)$ $-3m + 15$

41. $-4(5 - y)$ $-20 + 4y$ **42.** $-3(6 - p)$ $-18 + 3p$

43. $-1(a - 7)$ $-a + 7$ **44.** $-1(c - 8)$ $-c + 8$

45. $-1(t + 4)$ $-t - 4$ **46.** $-1(x + 7)$ $-x - 7$

Use the distributive property to factor each expression. See Example 6.

47. $2m + 12$ $2(m + 6)$ **48.** $3y + 6$ $3(y + 2)$

49. $4x - 4$ $4(x - 1)$ **50.** $6y + 6$ $6(y + 1)$

51. $4y - 16$ $4(y - 4)$ **52.** $5x + 15$ $5(x + 3)$

53. $4a + 8$ $4(a + 2)$ **54.** $7a - 35$ $7(a - 5)$

Find the multiplicative inverse (reciprocal) of each number. See Example 7.

55. $\frac{1}{2}$ 2 **56.** $\frac{1}{3}$ 3 **57.** -5 $-\frac{1}{5}$

58. -6 $-\frac{1}{6}$ **59.** 7 $\frac{1}{7}$ **60.** 8 $\frac{1}{8}$

61. 1 1 **62.** -1 -1 **63.** -0.25 -4

64. 0.75 $\frac{4}{3}$ **65.** 2.5 $\frac{2}{5}$ **66.** 3.5 $\frac{2}{7}$

Name the property that justifies each equation. See Example 8.

67. $3 \cdot x = x \cdot 3$ Commutative property of multiplication

68. $x + 5 = 5 + x$ Commutative property of addition

69. $2(x - 3) = 2x - 6$ Distributive property

70. $a(bc) = (ab)c$ Associative property of multiplication

71. $-3(xy) = (-3x)y$ Associative property of multiplication

72. $3(x + 1) = 3x + 3$ Distributive property

73. $4 + (-4) = 0$ Inverse properties

74. $1.3 + 9 = 9 + 1.3$ Commutative property of addition

75. $x^2 \cdot 5 = 5x^2$ Commutative property of multiplication

76. $0 \cdot \pi = 0$ Multiplication property of 0

77. $1 \cdot 3y = 3y$ Identity property

78. $(0.1)(10) = 1$ Inverse property

79. $2a + 5a = (2 + 5)a$ Distributive property

80. $3 + 0 = 3$ Identity property

81. $-7 + 7 = 0$ Inverse property

82. $1 \cdot b = b$ Identity property

83. $(2346)0 = 0$ Multiplication property of 0

84. $4x + 4 = 4(x + 1)$ Distributive property

85. $ay + y = y(a + 1)$ Distributive property

86. $ab + bc = b(a + c)$ Distributive property

Complete each equation, using the property named.

87. $a + y =$ ____, commutative $y + a$

88. $6x + 6 =$ ____, distributive $6(x + 1)$

89. $5(aw) =$ ____, associative $(5a)w$

90. $x + 3 =$ ____, commutative $3 + x$

91. $\frac{1}{2}x + \frac{1}{2} =$ ____, distributive $\frac{1}{2}(x + 1)$

92. $-3(x - 7) =$ ____, distributive $-3x + 21$

93. $6x + 15 =$ ____, distributive $3(2x + 5)$

94. $(x + 6) + 1 =$ ____, associative $x + (6 + 1)$

95. $4(0.25) =$ ____, inverse property 1

96. $-1(5 - y) =$ ____, distributive $-5 + y$

97. $0 = 96($____$)$, multiplication property of zero 0

98. $3 \cdot ($____$) = 3$, identity property 1

99. $0.33($____$) = 1$, inverse property $\frac{100}{33}$

100. $-8(1) =$ ____, identity property -8

Solve each problem. See Example 9.

101. ***Laying bricks.*** A bricklayer lays one brick in 0.04 hour, while his apprentice lays one brick in 0.05 hour.

a) If both are working, then at what combined rate (in bricks per hour) are they laying bricks? 45 bricks/hour

b) Which person is working faster? Bricklayer

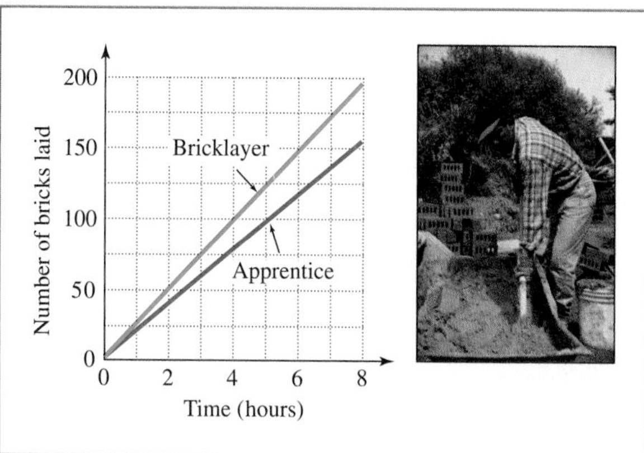

FIGURE FOR EXERCISE 101

102. ***Recovering golf balls.*** Susan and Joan are diving for golf balls in a large water trap. Susan recovers a golf ball every 0.016 hour while Joan recovers a ball every 0.025 hour. If both are working, then at what rate (in golf balls per hour) are they recovering golf balls? 102.5 balls/hour

103. ***Population explosion.*** In 2002 the population of the earth was increasing by one person every 0.4104 second (U.S. Census Bureau, www.census.gov).

a) At what rate in people per second is the population of the earth increasing? 2.4366 people/second

b) At what rate in people per week is the population of the earth increasing? 1,473,684 people/week

104. ***Farmland conversion.*** The amount of farmland in the United States is decreasing by one acre every 0.00876 hours as farmland is being converted to nonfarm use (American Farmland Trust, www.farmland.org). At what rate in acres per day is the farmland decreasing? 2740 acres/day

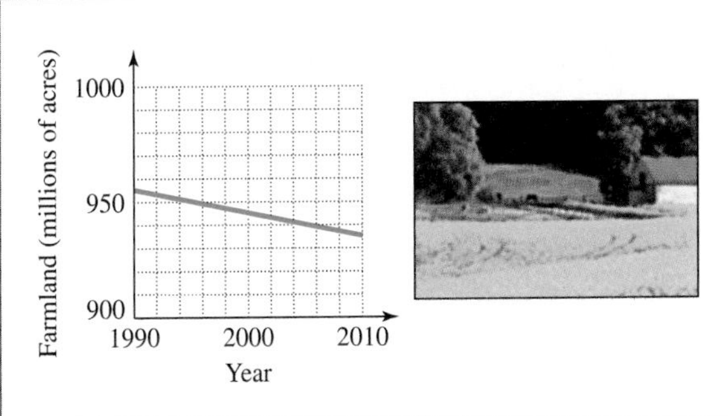

FIGURE FOR EXERCISE 104

GETTING MORE INVOLVED

105. ***Writing.*** The perimeter of a rectangle is the sum of twice the length and twice the width. Write in words another way to find the perimeter that illustrates the distributive property.
The perimeter is twice the sum of the length and width.

106. ***Discussion.*** Eldrid bought a loaf of bread for \$1.69 and a gallon of milk for \$2.29. Using a tax rate of 5%, he correctly figured that the tax on the bread would be 8 cents and the tax on the milk would be 11 cents, for a total of \$4.17. However, at the cash register he was correctly charged \$4.18. How could this happen? Which property of the real numbers is in question in this case?
Due to rounding off, the tax on each item separately does not equal the tax on the total. It looks like the distributive property fails.

107. ***Exploration.*** Determine whether each of the following pairs of tasks are "commutative." That is, does the order in which they are performed produce the same result?

a) Put on your coat; put on your hat.
Commutative

b) Put on your shirt; put on your coat.
Not commutative

Find another pair of "commutative" tasks and another pair of "noncommutative" tasks.

1.8 USING THE PROPERTIES TO SIMPLIFY EXPRESSIONS

In This Section

- Using the Properties in Computation
- Like Terms
- Combining Like Terms
- Products and Quotients
- Removing Parentheses

The properties of the real numbers can be helpful when we are doing computations. In this section we will see how the properties can be applied in arithmetic and algebra.

Using the Properties in Computation

The properties of the real numbers can often be used to simplify computations. For example, to find the product of 26 and 200, we can write

$$\begin{aligned}(26)(200) &= (26)(2 \cdot 100)\\ &= (26 \cdot 2)(100)\\ &= 52 \cdot 100\\ &= 5200\end{aligned}$$

It is the associative property that allows us to multiply 26 by 2 to get 52, then multiply 52 by 100 to get 5200.

EXAMPLE 1 **Using the properties**

Use the appropriate property to aid you in evaluating each expression.

a) $347 + 35 + 65$ **b)** $3 \cdot 435 \cdot \frac{1}{3}$ **c)** $6 \cdot 28 + 4 \cdot 28$

Study Tip

Being a full-time student is a full-time job. A successful student spends from two to four hours studying outside of class for every hour spent in the classroom. It is rare to find a person who can handle two full-time jobs and it is just as rare to find a successful full-time student who also works full time.

Solution

a) Notice that the sum of 35 and 65 is 100. So apply the associative property as follows:

$$\begin{aligned} 347 + (35 + 65) &= 347 + 100 \\ &= 447 \end{aligned}$$

b) Use the commutative and associative properties to rearrange this product. We can then do the multiplication quickly:

$$\begin{aligned} 3 \cdot 435 \cdot \frac{1}{3} &= 435\left(3 \cdot \frac{1}{3}\right) && \text{Commutative and associative properties} \\ &= 435 \cdot 1 && \text{Inverse property} \\ &= 435 && \text{Identity property} \end{aligned}$$

c) Use the distributive property to rewrite this expression.

$$\begin{aligned} 6 \cdot 28 + 4 \cdot 28 &= (6 + 4)28 \\ &= 10 \cdot 28 \\ &= 280 \end{aligned}$$

■

Like Terms

An expression containing a number or the product of a number and one or more variables raised to powers is called a **term.** For example,

$$-3, \qquad 5x, \qquad -3x^2y, \qquad a, \qquad \text{and} \qquad -abc$$

are terms. The number preceding the variables in a term is called the **coefficient.** In the term $5x$, the coefficient of x is 5. In the term $-3x^2y$ the coefficient of x^2y is -3. In the term a, the coefficient of a is 1 because $a = 1 \cdot a$. In the term $-abc$ the coefficient of abc is -1 because $-abc = -1 \cdot abc$. If two terms contain the same variables with the same exponents, they are called **like terms.** For example, $3x^2$ and $-5x^2$ are like terms, but $3x^2$ and $-5x^3$ are not like terms.

Combining Like Terms

Using the distributive property on an expression involving the sum of like terms allows us to combine the like terms as shown in Example 2.

EXAMPLE 2 **Combining like terms**

Use the distributive property to perform the indicated operations.

a) $3x + 5x$ **b)** $-5xy - (-4xy)$

Solution

a)
$$\begin{aligned} 3x + 5x &= (3 + 5)x && \text{Distributive property} \\ &= 8x && \text{Add the coefficients.} \end{aligned}$$

Because the distributive property is valid for any real numbers, we have $3x + 5x = 8x$ no matter what number is used for x.

b)
$$\begin{aligned} -5xy - (-4xy) &= [-5 - (-4)]xy && \text{Distributive property} \\ &= -1xy && -5 - (-4) = -5 + 4 = -1 \\ &= -xy && \text{Multiplying by } -1 \text{ is the same as taking the opposite.} \end{aligned}$$

■

Of course, we do not want to write out all of the steps shown in Example 2 every time we combine like terms. We can combine like terms as easily as we can add or subtract their coefficients.

EXAMPLE 3 **Combining like terms**

Perform the indicated operations.

a) $w + 2w$ **b)** $-3a + (-7a)$ **c)** $-9x + 5x$

d) $7xy - (-12xy)$ **e)** $2x^2 + 4x^2$

Solution

a) $w + 2w = 1w + 2w = 3w$ **b)** $-3a + (-7a) = -10a$

c) $-9x + 5x = -4x$ **d)** $7xy - (-12xy) = 19xy$

e) $2x^2 + 4x^2 = 6x^2$ ■

CAUTION There are no like terms in expressions such as

$$2 + 5x, \quad 3xy + 5y, \quad 3w + 5a, \quad \text{and} \quad 3z^2 + 5z$$

The terms in these expressions cannot be combined.

Products and Quotients

In Example 4 we use the associative property of multiplication to simplify the product of two expressions.

EXAMPLE 4 **Finding products**

Simplify.

a) $3(5x)$ **b)** $2\left(\frac{x}{2}\right)$

c) $(4x)(6x)$ **d)** $(-2a)(4b)$

Solution

a) $3(5x) = (3 \cdot 5)x$ Associative property

$= (15)x$ Multiply.

$= 15x$ Remove unnecessary parentheses.

b) $2\left(\frac{x}{2}\right) = 2\left(\frac{1}{2} \cdot x\right)$ Multiplying by $\frac{1}{2}$ is the same as dividing by 2.

$= \left(2 \cdot \frac{1}{2}\right)x$ Associative property

$= 1 \cdot x$ Multiplicative inverse

$= x$ Multiplicative identity is 1.

c) $(4x)(6x) = 4 \cdot 6 \cdot x \cdot x$ Commutative and associative properties

$= 24x^2$ Definition of exponent

d) $(-2a)(4b) = -2 \cdot 4 \cdot a \cdot b = -8ab$ ■

Study Tip

Note how the exercises are keyed to the examples. This serves two purposes. If you have missed class and are studying on your own, you should study an example and then immediately try to work the corresponding exercises. If you have seen an explanation in class, then you can start the exercises and refer back to the examples as necessary.

CAUTION Be careful with expressions such as $3(5x)$ and $3(5 + x)$. In $3(5x)$ we multiply 5 by 3 to get $3(5x) = 15x$. In $3(5 + x)$, both 5 and x are multiplied by the 3 to get $3(5 + x) = 15 + 3x$.

In Example 4 we showed how the properties are used to simplify products. However, in practice we usually do not write out any steps for these problems—we can write just the answer.

EXAMPLE 5

Finding products quickly

Find each product.

a) $(-3)(4x)$ **b)** $(-4a)(-7a)$ **c)** $(-3a)\left(\frac{b}{3}\right)$ **d)** $6 \cdot \frac{x}{2}$

Solution

a) $-12x$ **b)** $28a^2$ **c)** $-ab$ **d)** $3x$ ■

In Section 1.1 we found the quotient of two numbers by inverting the divisor and then multiplying. Since $a \div b = a \cdot \frac{1}{b}$, any quotient can be written as a product.

EXAMPLE 6

Simplifying quotients

Simplify.

a) $\frac{10x}{5}$ **b)** $\frac{4x + 8}{2}$

Solution

a) Since dividing by 5 is equivalent to multiplying by $\frac{1}{5}$, we have

$$\frac{10x}{5} = \frac{1}{5}(10x) = \left(\frac{1}{5} \cdot 10\right)x = (2)x = 2x.$$

Note that you can simply divide 10 by 5 to get 2.

b) Since dividing by 2 is equivalent to multiplying by $\frac{1}{2}$, we have

$$\frac{4x + 8}{2} = \frac{1}{2}(4x + 8) = 2x + 4.$$

Note that both 4 and 8 are divided by 2. ■

CAUTION It is not correct to divide only one term in the numerator by the denominator. For example,

$$\frac{4 + 7}{2} \neq 2 + 7$$

because $\frac{4 + 7}{2} = \frac{11}{2}$ and $2 + 7 = 9$.

Study Tip

Take notes in class. Write down everything you can. As soon as possible after class, rewrite your notes. Fill in details and make corrections. Make a note of examples and exercises in the text that are similar to examples in your notes. If your instructor takes the time to work an example in class, it is a good bet that your instructor expects you to understand the concepts involved.

Removing Parentheses

Multiplying a number by -1 merely changes the sign of the number. For example,

$$(-1)(7) = -7 \quad \text{and} \quad (-1)(-8) = 8.$$

So -1 times a number is the *opposite* of the number. Using variables, we write

$$(-1)x = -x \quad \text{or} \quad -1(y + 5) = -(y + 5).$$

Calculator Close-Up

A negative sign in front of parentheses changes the sign of every term inside the parentheses.

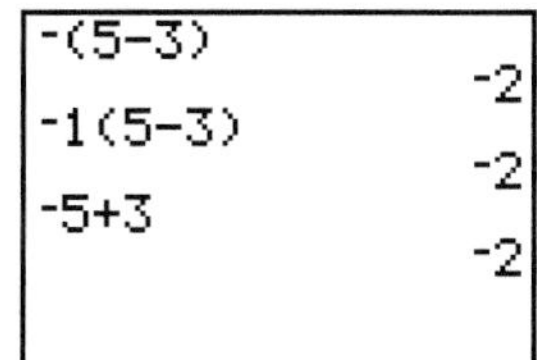

When a minus sign appears in front of a sum, we can change the minus sign to -1 and use the distributive property. For example,

$$\begin{aligned} -(w + 4) &= -1(w + 4) \\ &= (-1)w + (-1)4 && \text{Distributive property} \\ &= -w + (-4) && \textit{Note: } -1 \cdot w = -w,\ -1 \cdot 4 = -4 \\ &= -w - 4 \end{aligned}$$

Note how the minus sign in front of the parentheses caused all of the signs to change: $-(w + 4) = -w - 4$. As another example, consider the following:

$$\begin{aligned} -(x - 3) &= -1(x - 3) \\ &= (-1)x - (-1)3 \\ &= -x + 3 \end{aligned}$$

CAUTION When removing parentheses preceded by a minus sign, you must change the sign of *every* term within the parentheses.

EXAMPLE 7 **Removing parentheses**

Simplify each expression.

a) $5 - (x + 3)$ **b)** $3x - 6 - (2x - 4)$ **c)** $-6x - (-x + 2)$

Solution

a)
$$\begin{aligned} 5 - (x + 3) &= 5 - x - 3 && \text{Change the sign of each term in parentheses.} \\ &= 5 - 3 - x && \text{Commutative property} \\ &= 2 - x && \text{Combine like terms.} \end{aligned}$$

b)
$$\begin{aligned} 3x - 6 - (2x - 4) &= 3x - 6 - 2x + 4 && \text{Remove parentheses and change signs.} \\ &= 3x - 2x - 6 + 4 && \text{Commutative property} \\ &= x - 2 && \text{Combine like terms.} \end{aligned}$$

c)
$$\begin{aligned} -6x - (-x + 2) &= -6x + x - 2 && \text{Remove parentheses and change signs.} \\ &= -5x - 2 && \text{Combine like terms.} \end{aligned}$$

■

The commutative and associative properties of addition allow us to rearrange the terms so that we may combine the like terms. However, it is not necessary to actually write down the rearrangement. We can identify the like terms and combine them without rearranging.

EXAMPLE 8 **Simplifying algebraic expressions**

Simplify.

a) $(-2x + 3) + (5x - 7)$ **b)** $-3x + 6x + 5(4 - 2x)$

c) $-2x(3x - 7) - (x - 6)$ **d)** $x - 0.02(x + 500)$

Solution

a) $(-2x + 3) + (5x - 7) = 3x - 4$ Combine like terms.

b)
$$\begin{aligned} -3x + 6x + 5(4 - 2x) &= -3x + 6x + 20 - 10x && \text{Distributive property} \\ &= -7x + 20 && \text{Combine like terms.} \end{aligned}$$

c) $-2x(3x - 7) - (x - 6) = -6x^2 + 14x - x + 6$ Distributive property

$= -6x^2 + 13x + 6$ Combine like terms.

d) $x - 0.02(x + 500) = 1x - 0.02x - 10$ Distributive property

$= 0.98x - 10$ Combine like terms. ■

WARM-UPS

True or false? Explain your answer.

A statement involving variables should be marked true only if it is true for all values of the variable.

1. $3(x + 6) = 3x + 18$ True

2. $-3x + 9 = -3(x + 9)$ False

3. $-1(x - 4) = -x + 4$ True

4. $3a + 4a = 7a$ True

5. $(3a)(4a) = 12a$ False

6. $3(5 \cdot 2) = 15 \cdot 6$ False

7. $x + x = x^2$ False

8. $x \cdot x = 2x$ False

9. $3 + 2x = 5x$ False

10. $-(5x - 2) = -5x + 2$ True

1.8 EXERCISES

Reading and Writing *After reading this section write out the answers to these questions. Use complete sentences.*

1. What are like terms?
Like terms are terms with the same variables and exponents.

2. What is the coefficient of a term?
The coefficient of a term is the number preceding the variable.

3. What can you do to like terms that you cannot do to unlike terms?
We can add or subtract like terms.

4. What operations can you perform with unlike terms?
Unlike terms can be multiplied and divided.

5. What is the difference between a positive sign preceding a set of parentheses and a negative sign preceding a set of parentheses?
If a negative sign precedes a set of parentheses, then signs for all terms in the parentheses are changed when the parentheses are removed.

6. What happens when a number is multiplied by -1?
Multiplying a number by -1 changes the sign of the number.

Use the appropriate properties to evaluate the expressions. See Example 1.

7. $35(200)$ 7000

8. $15(300)$ 4500

9. $\frac{4}{3}(0.75)$ 1

10. $5(0.2)$ 1

11. $256 + 78 + 22$ 356

12. $12 + 88 + 376$ 476

13. $35 \cdot 3 + 35 \cdot 7$ 350

14. $98 \cdot 478 + 2 \cdot 478$ 47,800

15. $18 \cdot 4 \cdot 2 \cdot \frac{1}{4}$ 36

16. $19 \cdot 3 \cdot 2 \cdot \frac{1}{3}$ 38

17. $(120)(300)$ 36,000

18. $150 \cdot 200$ 30,000

19. $12 \cdot 375(-6 + 6)$ 0

20. $354^2(-2 \cdot 4 + 8)$ 0

21. $78 + 6 + 8 + 4 + 2$ 98

22. $-47 + 12 - 6 - 12 + 6$ -47

Combine like terms where possible. See Examples 2 and 3.

23. $5w + 6w$ $11w$

24. $4a + 10a$ $14a$

25. $4x - x$ $3x$

26. $a - 6a$ $-5a$

27. $2x - (-3x)$ $5x$

28. $2b - (-5b)$ $7b$

29. $-3a - (-2a)$ $-a$

30. $-10m - (-6m)$ $-4m$

31. $-a - a$ $-2a$

32. $a - a$ 0

33. $10 - 6t$ $10 - 6t$

34. $9 - 4w$ $9 - 4w$

35. $3x^2 + 5x^2$ $8x^2$

36. $3r^2 + 4r^2$ $7r^2$

37. $-4x + 2x^2$ $-4x + 2x^2$

38. $6w^2 - w$ $6w^2 - w$

39. $5mw^2 - 12mw^2$ $-7mw^2$

40. $4ab^2 - 19ab^2$ $-15ab^2$

Simplify the following products or quotients. See Examples 4–6.

41. $3(4h)$ $12h$

42. $2(5h)$ $10h$

43. $6b(-3)$ $-18b$

44. $-3m(-1)$ $3m$

45. $(-3m)(3m)$ $-9m^2$

46. $(2x)(-2x)$ $-4x^2$

47. $(-3d)(-4d)$ $12d^2$

48. $(-5t)(-2t)$ $10t^2$

49. $(-y)(-y)$ y^2

50. $y(-y)$ $-y^2$

51. $-3a(5b)$ $-15ab$

52. $-7w(3r)$ $-21rw$

53. $-3a(2 + b)$ $-6a - 3ab$

54. $-2x(3 + y)$ $-6x - 2xy$

55. $-k(1 - k)$ $-k + k^2$

56. $-t(t - 1)$ $-t^2 + t$

57. $\frac{3y}{3}$ y

58. $\frac{-9t}{9}$ $-t$

59. $\frac{-15y}{5}$ $-3y$

60. $\frac{-12b}{2}$ $-6b$

61. $2\left(\frac{y}{2}\right)$ y

62. $6\left(\frac{m}{3}\right)$ $2m$

63. $8y\left(\frac{y}{4}\right)$ $2y^2$

64. $10\left(\frac{2a}{5}\right)$ $4a$

65. $\frac{6a - 3}{3}$ $2a - 1$

66. $\frac{-8x + 6}{2}$ $-4x + 3$

67. $\frac{-9x + 6}{-3}$ $3x - 2$

68. $\frac{10 - 5x}{-5}$ $-2 + x$

Simplify each expression. See Example 7.

69. $x - (3x - 1)$ $-2x + 1$

70. $4x - (2x - 5)$ $2x + 5$

71. $5 - (y - 3)$ $8 - y$

72. $8 - (m - 6)$ $-m + 14$

73. $2m + 3 - (m + 9)$ $m - 6$

74. $7 - 8t - (2t + 6)$ $-10t + 1$

75. $-3 - (-w + 2)$ $w - 5$

76. $-5x - (-2x + 9)$ $-3x - 9$

Simplify the following expressions by combining like terms. See Example 8.

77. $3x + 5x + 6 + 9$ $8x + 15$

78. $2x + 6x + 7 + 15$ $8x + 22$

79. $-2x + 3 + 7x - 4$ $5x - 1$

80. $-3x + 12 + 5x - 9$ $2x + 3$

81. $3a - 7 - (5a - 6)$ $-2a - 1$

82. $4m - 5 - (m - 2)$ $3m - 3$

83. $2(a - 4) - 3(-2 - a)$ $5a - 2$

84. $2(w + 6) - 3(-w - 5)$ $5w + 27$

85. $-5m + 6(m - 3) + 2m$ $3m - 18$

86. $-3a + 2(a - 5) + 7a$ $6a - 10$

87. $5 - 3(x + 2) - 6$ $-3x - 7$

88. $7 + 2(k - 3) - k + 6$ $k + 7$

89. $x - 0.05(x + 10)$ $0.95x - 0.5$

90. $x - 0.02(x + 300)$ $0.98x - 6$

91. $4.5 - 3.2(x - 5.3) - 8.75$ $-3.2x + 12.71$

92. $0.03(4.5x - 3.9) + 0.06(9.8x - 45)$ $0.723x - 2.817$

Simplify each expression.

93. $3x - (4 - x)$ $4x - 4$

94. $2 + 8x - 11x$ $2 - 3x$

95. $y - 5 - (-y - 9)$ $2y + 4$

96. $a - (b - c - a)$ $2a - b + c$

97. $7 - (8 - 2y - m)$ $2y + m - 1$

98. $x - 8 - (-3 - x)$ $2x - 5$

99. $\frac{1}{2}(10 - 2x) + \frac{1}{3}(3x - 6)$ 3

100. $\frac{1}{2}(x - 20) - \frac{1}{5}(x + 15)$ $\frac{3}{10}x - 13$

101. $0.2(x + 3) - 0.05(x + 20)$ $0.15x - 0.4$

102. $0.08x + 0.12(x + 100)$ $0.2x + 12$

103. $2k + 1 - 3(5k - 6) - k + 4$ $-14k + 23$

104. $2w - 3 + 3(w - 4) - 5(w - 6)$ 15

105. $-3m - 3[2m - 3(m + 5)]$ 45

106. $6h + 4[2h - 3(h - 9) - (h - 1)]$ $-2h + 112$

Solve each problem.

107. ***Married filing jointly.*** The value of the expression

$$0.15(45{,}200) + 0.275(x - 45{,}200)$$

is the 2001 federal income tax for a married couple filing jointly with a taxable income of x dollars, where x is over \$45,200 but not over \$109,250 (Internal Revenue Service, www.irs.gov).

a) Simplify the expression. $0.275x - 5650$

b) Use the expression to find the amount of tax for a couple with a taxable income of \$80,000. \$16,350

c) Use the accompanying graph to estimate the 2001 federal income tax for a couple with a taxable income of \$200,000 \$55,000

d) Use the accompanying graph to estimate the taxable income for a couple who paid \$80,000 in federal income tax. \$275,000

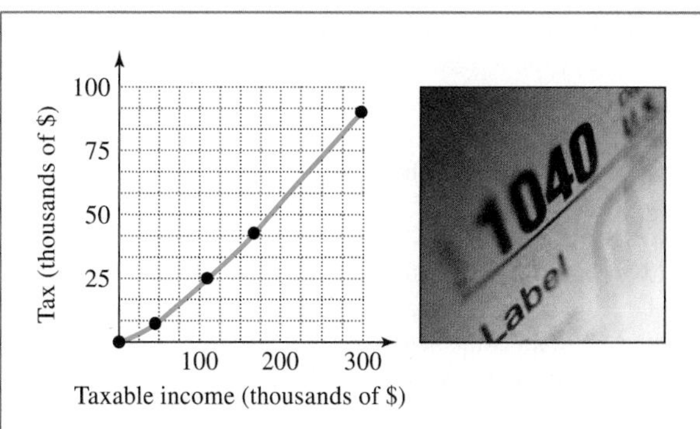

FIGURE FOR EXERCISE 107

108. ***Marriage penalty.*** The value of the expression

$$0.15(27{,}050) + 0.275(x - 27{,}050)$$

is the 2001 federal income tax for a single taxpayer with taxable income of x dollars, where x is over $27,050 but not over $65,550.

a) Simplify the expression. $0.275x - 3381.25$

b) Find the amount of tax for a single taxpayer with taxable income of $40,000. $7619

c) Who pays more, a married couple with a joint taxable income of $80,000 or two single taxpayers with taxable incomes of $40,000 each? See Exercise 107. Married couple pays $1112 more.

109. ***Perimeter of a corral.*** The perimeter of a rectangular corral that has width x feet and length $x + 40$ feet is $2(x) + 2(x + 40)$. Simplify the expression for the perimeter. Find the perimeter if $x = 30$ feet.
$4x + 80$, 200 feet

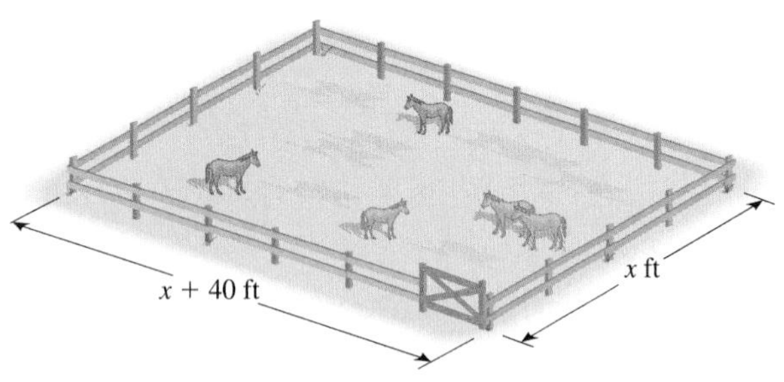

FIGURE FOR EXERCISE 109

GETTING MORE INVOLVED

110. ***Discussion.*** What is wrong with the way in which each of the following expressions is simplified?

a) $4(2 + x) = 8 + x$
$4(2 + x) = 8 + 4x$

b) $4(2x) = 8 \cdot 4x = 32x$
$4(2x) = (4 \cdot 2)x = 8x$

c) $\frac{4 + x}{2} = 2 + x$
$\frac{4 + x}{2} = \frac{1}{2}(4 + x) = 2 + \frac{1}{2}x$

d) $5 - (x - 3) = 5 - x - 3 = 2 - x$
$5 - (x - 3) = 5 - x + 3 = 8 - x$

111. ***Discussion.*** An instructor asked his class to evaluate the expression $1/2x$ for $x = 5$. Some students got 0.1; others got 2.5. Which answer is correct and why?
If $x = 5$, then $1/2 \cdot 5 = \frac{1}{2} \cdot 5 = 2.5$ because we do division and multiplication from left to right.

COLLABORATIVE ACTIVITIES

Walking the Number Line

Grouping: Four students per group
Topic: Signed numbers

This activity will help you understand adding and subtracting integers. There are four roles that will be rotated in your group; positive sign holder, negative sign holder, problem reader, and number-line walker.

Preparation. Using 13 note cards or pieces of paper write the integers from -6 to $+6$, one number to a card. Also make a card with a positive sign on it and another card with a negative sign. Place the 13 number cards in order on the floor about a step apart to create a number line.

Assume Your Position. After choosing your roles, have the positive sign holder stand at the positive end of the 13-card number line after $+6$, facing to the center. The negative sign holder will stand at the negative end of the line after -6, facing the center. The walker will be near the line ready to start. The problem reader is close by ready to read the first problem.

Rules. The number line walker will stand on the first number that the problem reader reads. The walker will face the positive sign holder if the second number is positive or face the negative sign holder if the second number is negative. For addition the walker will walk forward the number of spaces of the second number and for subtraction the walker will walk backward the number of spaces of the second number.

Examples. The reader reads $-1 + (-3)$. The walker begins at -1, faces the negative sign holder, then walks forward three steps. The walker should be on -4, which is the correct result for $-1 + (-3)$.

The reader reads $1 - (-4)$. The walker begins at 1, faces the negative sign holder, then walks backward 4 steps. The walker should be on 5, which is the correct result of $1 - (-4)$.

Exercises. Try the following exercises. After each two exercises, rotate the roles. Be sure to check that the walker is on the correct result.

1. $-2 + 3$ **2.** $-2 + (-3)$

3. $2 + (-3)$ **4.** $2 - (-3)$

5. $2 + 3$ **6.** $2 - 3$

7. $-2 + (-3)$ **8.** $-2 - (-3)$

Extension. Make up your own problems using integers between -6 and $+6$ and walk out the results on the number line.

WRAP-UP CHAPTER 1

SUMMARY

The Real Numbers		Examples
Counting or natural numbers	$\{1, 2, 3, \ldots\}$	
Whole numbers	$\{0, 1, 2, 3, \ldots\}$	
Integers	$\{\ldots, -3, -2, -1, 0, 1, 2, 3, \ldots\}$	
Rational numbers	$\left\{\frac{a}{b} \,\middle|\, a \text{ and } b \text{ are integers with } b \neq 0\right\}$	$\frac{3}{2}, 5, -6, 0$
Irrational numbers	$\{x \mid x \text{ is a real number that is not rational}\}$	$\sqrt{2}, \sqrt{3}, \pi$
Real numbers	The set of real numbers consists of all rational numbers together with all irrational numbers.	

Fractions		Examples
Reducing fractions	$\frac{a \cdot c}{b \cdot c} = \frac{a}{b}$	$\frac{4}{6} = \frac{2 \cdot 2}{2 \cdot 3} = \frac{2}{3}$
Building up fractions	$\frac{a}{b} = \frac{a \cdot c}{b \cdot c}$	$\frac{3}{8} = \frac{3 \cdot 5}{8 \cdot 5} = \frac{15}{40}$
Multiplying fractions	$\frac{a}{b} \cdot \frac{c}{d} = \frac{ac}{bd}$	$\frac{2}{3} \cdot \frac{4}{5} = \frac{8}{15}$
Dividing fractions	$\frac{a}{b} \div \frac{c}{d} = \frac{a}{b} \cdot \frac{d}{c}$	$\frac{2}{3} \div \frac{4}{5} = \frac{2}{3} \cdot \frac{5}{4} = \frac{10}{12} = \frac{5}{6}$
Adding or subtracting fractions	$\frac{a}{b} + \frac{c}{b} = \frac{a+c}{b}$ $\frac{a}{b} - \frac{c}{b} = \frac{a-c}{b}$	$\frac{1}{5} + \frac{2}{5} = \frac{3}{5}$ $\frac{3}{5} - \frac{2}{5} = \frac{1}{5}$
Least common denominator	The smallest number that is a multiple of all denominators.	$\frac{1}{4} + \frac{1}{6} = \frac{3}{12} + \frac{2}{12} = \frac{5}{12}$

Operations with Real Numbers		**Examples**
Absolute value	$\lvert a \rvert = \begin{cases} a & \text{if } a \text{ is positive or zero} \\ -a & \text{if } a \text{ is negative} \end{cases}$	$\lvert 3 \rvert = 3, \lvert 0 \rvert = 0$ $\lvert -3 \rvert = 3$
Sum of two numbers with like signs	Add their absolute values. The sum has the same sign as the given numbers.	$-3 + (-4) = -7$
Sum of two numbers with unlike signs (and different absolute values)	Subtract the absolute values of the numbers. The answer is positive if the number with the larger absolute value is positive. The answer is negative if the number with the larger absolute value is negative.	$-4 + 7 = 3$ $-7 + 4 = -3$
Sum of opposites	The sum of any number and its opposite is 0.	$-6 + 6 = 0$
Subtraction of signed numbers	$a - b = a + (-b)$ Subtract any number by adding its opposite.	$3 - 5 = 3 + (-5) = -2$ $4 - (-3) = 4 + 3 = 7$
Product or quotient	Like signs $\leftrightarrow$ Positive result Unlike signs $\leftrightarrow$ Negative result	$(-3)(-2) = 6$ $(-8) \div 2 = -4$
Definition of exponents	For any counting number n, $a^n = \underbrace{a \cdot a \cdot a \cdot \ldots \cdot a}_{n \text{ factors}}.$	$2^3 = 2 \cdot 2 \cdot 2 = 8$
Order of operations	No parentheses or absolute value present: 1. Exponential expressions 2. Multiplication and division 3. Addition and subtraction With parentheses or absolute value: First evaluate within each set of parentheses or absolute value, using the order of operations.	 $5 + 2^3 = 13$ $2 + 3 \cdot 5 = 17$ $4 + 5 \cdot 3^2 = 49$ $(2 + 3)(5 - 7) = -10$ $2 + 3 \lvert 2 - 5 \rvert = 11$

Properties of the Real Numbers		**Examples**
Commutative properties	$a + b = b + a$ $a \cdot b = b \cdot a$	$5 + 7 = 7 + 5$ $6 \cdot 3 = 3 \cdot 6$
Associative properties	$a + (b + c) = (a + b) + c$ $a \cdot (b \cdot c) = (a \cdot b) \cdot c$	$1 + (2 + 3) = (1 + 2) + 3$ $2(3 \cdot 4) = (2 \cdot 3)4$
Distributive properties	$a(b + c) = ab + ac$ $a(b - c) = ab - ac$	$2(3 + x) = 6 + 2x$ $-2(x - 5) = -2x + 10$

Identity properties	$a + 0 = a$ and $0 + a = a$ Zero is the additive identity. $1 \cdot a = a$ and $a \cdot 1 = a$ One is the multiplicative identity.	$5 + 0 = 0 + 5 = 5$ $7 \cdot 1 = 1 \cdot 7 = 7$
Inverse properties	For any real number a, there is a number $-a$ (additive inverse or opposite) such that $a + (-a) = 0$ and $-a + a = 0$. For any nonzero real number a there is a number $\frac{1}{a}$ (multiplicative inverse or reciprocal) such that $a \cdot \frac{1}{a} = 1$ and $\frac{1}{a} \cdot a = 1$.	$3 + (-3) = 0$ $-3 + 3 = 0$ $3 \cdot \frac{1}{3} = 1$ $\frac{1}{3} \cdot 3 = 1$
Multiplication property of 0	$a \cdot 0 = 0$ and $0 \cdot a = 0$	$5 \cdot 0 = 0$ $0(-7) = 0$

ENRICHING YOUR MATHEMATICAL WORD POWER

For each mathematical term, choose the correct meaning.

1. **like terms**
   a. terms that are identical
   b. the terms of a sum
   c. terms that have the same variables with the same exponents
   d. terms with the same variables c
2. **equivalent fractions**
   a. identical fractions
   b. fractions that represent the same number
   c. fractions with the same denominator
   d. fractions with the same numerator b
3. **variable**
   a. a letter that is used to represent some numbers
   b. the letter x
   c. an equation with a letter in it
   d. not the same a
4. **reducing**
   a. less than
   b. losing weight
   c. making equivalent
   d. dividing out common factors d
5. **lowest terms**
   a. numerator is smaller than the denominator
   b. no common factors
   c. the best interest rate
   d. when the numerator is 1 b
6. **additive inverse**
   a. the number -1
   b. the number 0
   c. the opposite of addition
   d. opposite d
7. **order of operations**
   a. the order in which operations are to be performed in the absence of grouping symbols
   b. the order in which the operations were invented
   c. the order in which operations are written
   d. a list of operations in alphabetical order a
8. **least common denominator**
   a. the smallest divisor of all denominators
   b. the denominator that appears the least
   c. the smallest identical denominator
   d. the least common multiple of the denominators d
9. **absolute value**
   a. definite value
   b. positive number
   c. distance from 0 on the number line
   d. the opposite of a number c
10. **natural numbers**
    a. the counting numbers
    b. numbers that are not irrational
    c. the nonnegative numbers
    d. numbers that we find in nature a

REVIEW EXERCISES

1.1 *Which of the numbers $-\sqrt{5}$, $-2, 0, 1, 2, 3.14, \pi$, and 10 are*

1. whole numbers? 0, 1, 2, 10
2. natural numbers? 1, 2, 10
3. integers? −2, 0, 1, 2, 10
4. rational numbers? −2, 0, 1, 2, 3.14, 10
5. irrational numbers? $-\sqrt{5}$, π
6. real numbers? All of them

Study Tip

Note how the review exercises are arranged according to the sections in this chapter. If you are having trouble with a certain type of problem, refer back to the appropriate section for examples and explanations.

True or false? Explain your answer.

7. Every whole number is a rational number. True
8. Zero is not a rational number. False
9. The counting numbers between −4 and 4 are −3, −2, −1, 0, 1, 2, and 3. False
10. There are infinitely many integers. True
11. The set of counting numbers smaller than the national debt is infinite. False
12. The decimal number 0.25 is a rational number. True
13. Every integer greater than −1 is a whole number. True
14. Zero is the only number that is neither rational nor irrational. False

1.2 *Perform the indicated operations.*

15. $\frac{1}{3} + \frac{3}{8}$ $\frac{17}{24}$
16. $\frac{2}{3} - \frac{1}{4}$ $\frac{5}{12}$
17. $\frac{3}{5} \cdot 10$ 6
18. $\frac{3}{5} \div 10$ $\frac{3}{50}$
19. $\frac{2}{5} \cdot \frac{15}{14}$ $\frac{3}{7}$
20. $7 \div \frac{1}{2}$ 14
21. $4 + \frac{2}{3}$ $\frac{14}{3}$
22. $\frac{7}{12} - \frac{1}{4}$ $\frac{1}{3}$
23. $\frac{1}{2} + \frac{1}{3} + \frac{1}{4}$ $\frac{13}{12}$
24. $\frac{3}{4} \div 9$ $\frac{1}{12}$

1.3 *Evaluate.*

25. $-5 + 7$ 2
26. $-9 + (-4)$ −13
27. $35 - 48$ −13
28. $-3 - 9$ −12
29. $-12 + 5$ −7
30. $-12 - 5$ −17
31. $-12 - (-5)$ −7
32. $-9 - (-9)$ 0
33. $-0.05 + 12$ 11.95
34. $-0.03 + (-2)$ −2.03
35. $-0.1 - (-0.05)$ −0.05
36. $-0.3 + 0.3$ 0
37. $\frac{1}{3} - \frac{1}{2}$ $-\frac{1}{6}$
38. $-\frac{2}{3} + \frac{1}{4}$ $-\frac{5}{12}$
39. $-\frac{1}{3} + \left(-\frac{2}{5}\right)$ $-\frac{11}{15}$
40. $\frac{1}{3} - \left(-\frac{1}{4}\right)$ $\frac{7}{12}$

1.4 *Evaluate.*

41. $(-3)(5)$ −15
42. $(-9)(-4)$ 36
43. $(-8) \div (-2)$ 4
44. $50 \div (-5)$ −10
45. $\frac{-20}{-4}$ 5
46. $\frac{30}{-5}$ −6
47. $\left(-\frac{1}{2}\right)\left(-\frac{1}{3}\right)$ $\frac{1}{6}$
48. $8 \div \left(-\frac{1}{3}\right)$ −24
49. $-0.09 \div 0.3$ −0.3
50. $4.2 \div (-0.3)$ −14
51. $(0.3)(-0.8)$ −0.24
52. $0 \div (-0.0538)$ 0
53. $(-5)(-0.2)$ 1
54. $\frac{1}{2}(-12)$ −6

1.5 *Evaluate.*

55. $3 + 7(9)$ 66
56. $(3 + 7)9$ 90
57. $(3 + 4)^2$ 49
58. $3 + 4^2$ 19
59. $3 + 2 \cdot |5 - 6 \cdot 4|$ 41
60. $3 - (8 - 9)$ 4
61. $(3 - 7) - (4 - 9)$ 1
62. $3 - 7 - 4 - 9$ −17
63. $-2 - 4(2 - 3 \cdot 5)$ 50
64. $3^2 - 7 + 5^2$ 27
65. $3^2 - (7 + 5)^2$ −135
66. $|4 - 6 \cdot 3| - |7 - 9|$ 12
67. $\frac{-3 - 5}{2 - (-2)}$ −2
68. $\frac{1 - 9}{4 - 6}$ 4
69. $\frac{6 + 3}{3} - 5 \cdot 4 + 1$ −16
70. $\frac{2 \cdot 4 + 4}{3} - 3(1 - 2)$ 7

1.6 *Let $a = -1$, $b = -2$, and $c = 3$. Find the value of each algebraic expression.*

71. $b^2 - 4ac$ 16
72. $a^2 - 4b$ 9
73. $(c - b)(c + b)$ 5
74. $(a + b)(a - b)$ −3
75. $a^2 + 2ab + b^2$ 9
76. $a^2 - 2ab + b^2$ 1
77. $a^3 - b^3$ 7
78. $a^3 + b^3$ −9
79. $\frac{b + c}{a + b}$ $-\frac{1}{3}$
80. $\frac{b - c}{2b - a}$ $\frac{5}{3}$
81. $|a - b|$ 1
82. $|b - a|$ 1
83. $(a + b)c$ −9
84. $ac + bc$ −9

Determine whether the given number is a solution to the equation following it.

85. 4, $3x - 2 = 10$ Yes
86. 1, $5(x + 3) = 20$ Yes
87. −6, $\frac{3x}{2} = 9$ No
88. −30, $\frac{x}{3} - 4 = 6$ No
89. 15, $\frac{x + 3}{2} = 9$ Yes
90. 1, $\frac{12}{2x + 1} = 4$ Yes
91. 4, $-x - 3 = 1$ No
92. 7, $-x + 1 = 6$ No

1.7 *Name the property that justifies each statement.*

93. $a(x + y) = ax + ay$ Distributive property
94. $3(4y) = (3 \cdot 4)y$ Associative property of multiplication
95. $(0.001)(1000) = 1$ Inverse property
96. $xy = yx$ Commutative property
97. $0 + y = y$ Identity property

98. $325 \cdot 1 = 325$ Identity property

99. $3 + (2 + x) = (3 + 2) + x$ Associative property of addition

100. $2x - 6 = 2(x - 3)$ Distributive property

101. $5 \cdot 200 = 200 \cdot 5$ Commutative property of multiplication

102. $3 + (x + 2) = (x + 2) + 3$ Commutative property of addition

103. $-50 + 50 = 0$ Inverse property

104. $43 \cdot 59 \cdot 82 \cdot 0 = 0$ Multiplication property of 0

105. $12 \cdot 1 = 12$ Identity property

106. $3x + 1 = 1 + 3x$ Commutative property of addition

1.8 *Simplify by combining like terms.*

107. $3a + 7 - (4a - 5)$ $-a + 12$

108. $2m + 6 - (m - 2)$ $m + 8$

109. $2a(3a - 5) + 4a$ $6a^2 - 6a$

110. $3a(a - 5) + 5a(a + 2)$ $8a^2 - 5a$

111. $3(t - 2) - 5(3t - 9)$ $-12t + 39$

112. $2(m + 3) - 3(3 - m)$ $5m - 3$

113. $0.1(a + 0.3) - (a + 0.6)$ $-0.9a - 0.57$

114. $0.1(x + 0.3) - (x - 0.9)$ $-0.9x + 0.93$

115. $0.05(x - 20) - 0.1(x + 30)$ $-0.05x - 4$

116. $0.02(x - 100) + 0.2(x - 50)$ $0.22x - 12$

117. $5 - 3x(-5x - 2) + 12x^2$ $27x^2 + 6x + 5$

118. $7 - 2x(3x - 7) - x^2$ $-7x^2 + 14x + 7$

119. $-(a - 2) - 2 - a$ $-2a$

120. $-(w - y) - 3(y - w)$ $-2y + 2w$

121. $x(x + 1) + 3(x - 1)$ $x^2 + 4x - 3$

122. $y(y - 2) + 3(y + 1)$ $y^2 + y + 3$

MISCELLANEOUS

Evaluate each expression. Use a calculator to check.

123. $752(-13) + 752(13)$ 0

124. $75 - (-13)$ 88

125. $|15 - 23|$ 8

126. $4^2 - 6^2$ -20

127. $-6^2 + 3(5)$ -21

128. $(0.03)(-200)$ -6

129. $\frac{2}{5} + \frac{1}{10}$ $\frac{1}{2}$

130. $\frac{2 + 1}{5 + 10}$ $\frac{1}{5}$

131. $(0.05) \div (-0.1)$ -0.5

132. $(4 - 9)^2 + (2 \cdot 3 - 1)^2$ 50

133. $2\left(-\frac{1}{2}\right)^2 + \left(-\frac{1}{2}\right) - 1$ -1

134. $\left(-\frac{6}{7}\right)\left(\frac{21}{26}\right)$ $-\frac{9}{13}$

Simplify each expression if possible.

135. $\frac{2x + 4}{2}$ $x + 2$

136. $4(2x)$ $8x$

137. $4 + 2x$ $4 + 2x$

138. $4(2 + x)$ $8 + 4x$

139. $4 \cdot \frac{x}{2}$ $2x$

140. $4 - (x - 2)$ $-x + 6$

141. $-4(x - 2)$ $-4x + 8$

142. $(4x)(2x)$ $8x^2$

143. $4x + 2x$ $6x$

144. $2 + (x + 4)$ $x + 6$

145. $4 \cdot \frac{x}{4}$ x

146. $4 \cdot \frac{3x}{2}$ $6x$

147. $2 \cdot x \cdot 4$ $8x$

148. $4 - 2(2 - x)$ $2x$

Fill in the tables with the appropriate values for the given expressions.

149.

x	$-\frac{1}{3}x + 1$
-6	3
-3	2
0	1
3	0
6	-1

150.

x	$\frac{1}{2}x + 3$
-4	1
-2	2
0	3
2	4
4	5

151.

a	a^2	a^3	a^4
5	25	125	625
-4	16	-64	256

152.

b	$\frac{1}{b}$	$\frac{1}{b^2}$	$\frac{1}{b^3}$
-3	$-\frac{1}{3}$	$\frac{1}{9}$	$-\frac{1}{27}$
$-\frac{1}{2}$	-2	4	-8

Solve each problem.

153. ***Telemarketing.*** Brenda and Nicki sell memberships in an automobile club over the telephone. Brenda sells one membership every 0.125 hour, and Nicki sells one membership every 0.1 hour. At what rate (in memberships per hour) are the memberships being sold when both are working? 18 memberships per hour

154. ***High-income bracket.*** The expression

$$93{,}374 + 0.391(x - 297{,}350)$$

represents the amount for the 2001 federal income tax in dollars for a single taxpayer with x dollars of taxable

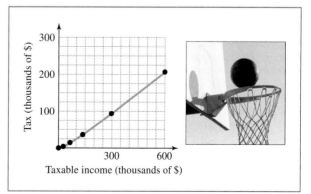

FIGURE FOR EXERCISE 154

income, where x is over \$297,350 (www.irs.gov).

a) Simplify the expression. $0.391x - 22{,}889.85$

b) Use the graph in the accompanying figure to estimate the amount of tax for a single taxpayer with a taxable income of \$300,000. \$100,000.

c) Find the amount of tax for NBA player Kevin Garnett for 2001. At \$22.4 million he was the highest paid player that year (www.usatoday.com). \$8,735,510

CHAPTER 1 TEST

Which of the numbers $-3, -\sqrt{3}, -\frac{1}{4}, 0, \sqrt{5}, \pi$, *and* 8 *are*

1. Whole numbers? 0, 8

2. Integers? $-3, 0, 8$

3. Rational numbers? $-3, -\frac{1}{4}, 0, 8$

4. Irrational numbers? $-\sqrt{3}, \sqrt{5}, \pi$

Evaluate each expression.

5. $6 + 3(-9)$ -21

6. $(-2)^2 - 4(-2)(-1)$ -4

7. $\dfrac{-3^2 - 9}{3 - 5}$ 9

8. $-5 + 6 - 12 + 4$ -7

9. $0.05 - 1$ -0.95

10. $(5 - 9)(5 + 9)$ -56

11. $(878 + 89) + 11$ 978

12. $6 + |3 - 5(2)|$ 13

13. $8 - 3|7 - 10|$ -1

14. $(839 + 974)[3(-4) + 12]$ 0

15. $974(7) + 974(3)$ 9740

16. $-\frac{2}{3} + \frac{3}{8}$ $-\frac{7}{24}$

17. $(-0.05)(400)$ -20

18. $\left(-\frac{3}{4}\right)\left(\frac{2}{9}\right)$ $-\frac{1}{6}$

19. $13 \div \left(-\frac{1}{3}\right)$ -39

Identify the property that justifies each equation.

20. $2(x + 7) = 2x + 14$ Distributive property

21. $48 \cdot 1000 = 1000 \cdot 48$ Commutative property of multiplication

22. $2 + (6 + x) = (2 + 6) + x$ Associative property of addition

23. $-348 + 348 = 0$ Inverse property

24. $1 \cdot (-6) = -6$ Identity property

25. $0 \cdot 388 = 0$ Multiplication property of 0

Use the distributive property to write each sum or difference as a product.

26. $3x + 30$ $3(x + 10)$

27. $7w - 7$ $7(w - 1)$

Simplify each expression.

28. $6 + 4x + 2x$ $6x + 6$

29. $6 + 4(x - 2)$ $4x - 2$

30. $5x - (3 - 2x)$ $7x - 3$

31. $x + 10 - 0.1(x + 25)$ $0.9x + 7.5$

32. $2a(4a - 5) - 3a(-2a - 5)$ $14a^2 + 5a$

33. $\dfrac{6x + 12}{6}$ $x + 2$

34. $8 \cdot \frac{t}{2}$ $4t$

35. $(-9xy)(-6xy)$ $54x^2y^2$

Evaluate each expression if $a = -2$, $b = 3$, *and* $c = 4$.

36. $b^2 - 4ac$ 41

37. $\dfrac{a - b}{b - c}$ 5

38. $(a - c)(a + c)$ -12

Determine whether the given number is a solution to the equation following it.

39. -2, $3x - 4 = 2$ No

40. 13, $\dfrac{x + 3}{8} = 2$ Yes

41. -3, $-x + 5 = 8$ Yes

Solve each problem.

42. Burke and Nora deliver pizzas for Godmother's Pizza. Burke averages one delivery every 0.25 hour, and Nora averages one delivery every 0.2 hour. At what rate (in deliveries per hour) are the deliveries made when both are working? 9 deliveries per hour

43. A forensic scientist uses the expression $80.405 + 3.660R - 0.06(A - 30)$ to estimate the height in centimeters for a male with a radius (bone in the forearm) of length R centimeters and age A in years, where A is over 30. Simplify the expression. Use the expression to estimate the height of an 80-year-old male with a radius of length 25 cm.
$3.66R - 0.06A + 82.205$, 168.905 cm

Study Tip

Before you take an in-class exam on this chapter, work the sample test given here. Set aside one hour to work this test and use the answers in the back of this book to grade yourself. Even though your instructor might not ask exactly the same questions, you will get a good idea of your test readiness.

CHAPTER 2

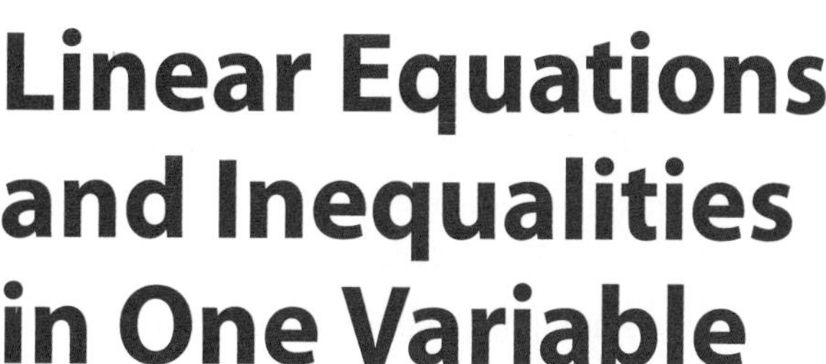

Linear Equations and Inequalities in One Variable

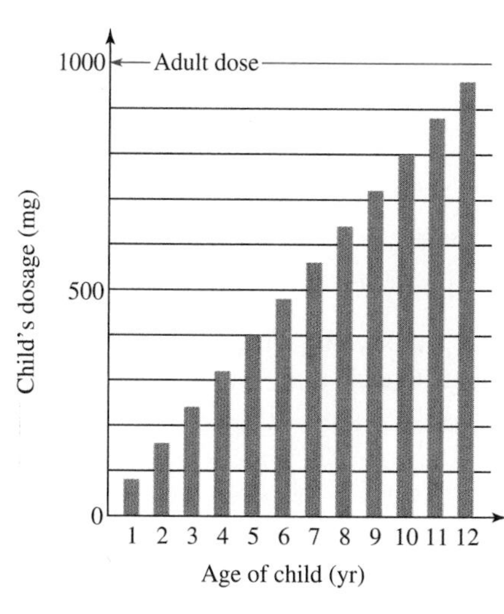

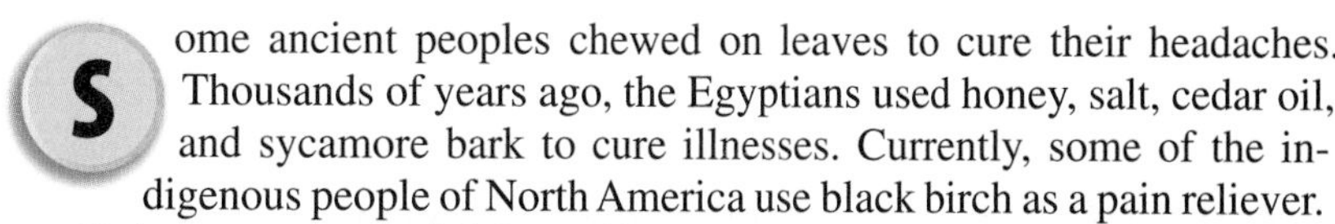

Some ancient peoples chewed on leaves to cure their headaches. Thousands of years ago, the Egyptians used honey, salt, cedar oil, and sycamore bark to cure illnesses. Currently, some of the indigenous people of North America use black birch as a pain reliever.

Today, we are grateful for modern medicine and the seemingly simple cures for illnesses. From our own experiences we know that just the right amount of a drug can work wonders but too much of a drug can do great harm. Even though physicians often prescribe the same drug for children and adults, the amount given must be tailored to the individual. The portion of a drug given to children is usually reduced on the basis of factors such as the weight and height of the child. Likewise, older adults frequently need a lower dosage of medication than what would be prescribed for a younger, more active person.

Various algebraic formulas have been developed for determining the proper dosage for a child and an older adult. In Exercises 95 and 96 of Section 2.4 you will see two formulas that are used to determine a child's dosage by using the adult dosage and the child's age.

2.1 THE ADDITION AND MULTIPLICATION PROPERTIES OF EQUALITY

In This Section

- The Addition Property of Equality
- The Multiplication Property of Equality
- Variables on Both Sides
- Applications

In Section 1.6, you learned that an equation is a statement that two expressions are equal. You also learned how to determine whether a number is a solution to an equation. In this section you will learn systematic procedures for finding solutions to equations.

The Addition Property of Equality

If two workers have equal salaries and each gets a \$1000 raise, then they will have equal salaries after the raise. If two people are the same age now, then in 5 years they will still be the same age. If you add the same number to two equal quantities, the results will be equal. This idea is called the *addition property of equality:*

The Addition Property of Equality

Adding the same number to both sides of an equation does not change the solution to the equation. In symbols, if $a = b$, then

$$a + c = b + c.$$

To **solve** an equation means to find all of the solutions to the equation. The set of all solutions to an equation is the **solution set** to the equation. Equations that have the same solution set are **equivalent equations.** In our first example, we will use the addition property of equality to solve an equation.

EXAMPLE 1 **Adding the same number to both sides**

Solve $x - 3 = -7$.

Helpful Hint

Think of an equation like a balance scale. To keep the scale in balance, what you add to one side you must also add to the other side.

Solution

We can remove the 3 from the left side of the equation by adding 3 to each side of the equation:

$$x - 3 = -7$$
$$x - 3 + 3 = -7 + 3 \quad \text{Add 3 to each side.}$$
$$x + 0 = -4 \quad \text{Simplify each side.}$$
$$x = -4 \quad \text{Zero is the additive identity.}$$

Since -4 satisfies the last equation, it should also satisfy the original equation because all of the previous equations are equivalent. Check that -4 satisfies the original equation by replacing x by -4:

$$x - 3 = -7 \quad \text{Original equation}$$
$$-4 - 3 = -7 \quad \text{Replace } x \text{ by } -4.$$
$$-7 = -7 \quad \text{Simplify.}$$

Since $-4 - 3 = -7$ is correct, $\{-4\}$ is the solution set to the equation. ■

The equations that we work with in this section and the next two are called linear equations.

Linear Equation

A **linear equation in one variable** x is an equation that can be written in the form

$$ax + b = 0,$$

where a and b are real numbers and $a \neq 0$.

An equation such as $2x + 3 = 0$ is a linear equation. We also refer to equations such as

$$x + 8 = 0, \quad 3x = 7, \quad 2x + 5 = 9 - 5x, \quad \text{and} \quad 3 + 5(x - 1) = -7 + x$$

as linear equations, because these equations could be written in the form $ax + b = 0$ using the properties of equality.

In Example 1, we used addition to isolate the variable on the left-hand side of the equation. Once the variable is isolated, we can determine the solution to the equation. Because subtraction is defined in terms of addition, we can also use subtraction to isolate the variable.

EXAMPLE 2 **Subtracting the same number from both sides**

Solve $9 + x = -2$.

Solution

We can remove the 9 from the left side by adding -9 to each side or by subtracting 9 from each side of the equation:

$$9 + x = -2$$
$$9 + x - 9 = -2 - 9 \quad \text{Subtract 9 from each side.}$$
$$x = -11 \quad \text{Simplify each side.}$$

Check that -11 satisfies the original equation by replacing x by -11:

$$9 + x = -2 \quad \text{Original equation}$$
$$9 + (-11) = -2 \quad \text{Replace } x \text{ by } -11.$$

Since $9 + (-11) = -2$ is correct, $\{-11\}$ is the solution set to the equation. ■

Our goal in solving equations is to isolate the variable. In Examples 1 and 2, the variable was isolated on the left side of the equation. In Example 3, we isolate the variable on the right side of the equation.

EXAMPLE 3 **Isolating the variable on the right side**

Solve $\frac{1}{2} = -\frac{1}{4} + y$.

Solution

We can remove $-\frac{1}{4}$ from the right side by adding $\frac{1}{4}$ to both sides of the equation:

$$\frac{1}{2} = -\frac{1}{4} + y$$
$$\frac{1}{2} + \frac{1}{4} = -\frac{1}{4} + y + \frac{1}{4} \quad \text{Add } \tfrac{1}{4} \text{ to each side.}$$
$$\frac{3}{4} = y \quad \text{Simplify each side.}$$

Study Tip

Don't simply work exercises to get answers. Keep reminding yourself of what it is that you are doing. Look for the big picture. What properties are you using? What does a solution mean? Is it reasonable? Does it check?

Check that $\frac{3}{4}$ satisfies the original equation by replacing y by $\frac{3}{4}$:

$$\frac{1}{2} = -\frac{1}{4} + y \quad \text{Original equation}$$

$$\frac{1}{2} = -\frac{1}{4} + \frac{3}{4} \quad \text{Replace } y \text{ by } \tfrac{3}{4}.$$

$$\frac{1}{2} = \frac{2}{4} \quad \text{Simplify.}$$

Since $\frac{1}{2} = \frac{2}{4}$ is correct, $\left\{\frac{3}{4}\right\}$ is the solution set to the equation. ■

The Multiplication Property of Equality

To isolate a variable that is involved in a product or a quotient, we need the multiplication property of equality.

The Multiplication Property of Equality

Multiplying both sides of an equation by the same nonzero number does not change the solution to the equation. In symbols, if $a = b$ and $c \neq 0$, then

$$ac = bc.$$

We specified that $c \neq 0$ in the multiplication property of equality because multiplying by 0 can change the solution to an equation. For example, $x = 4$ is satisfied only by 4, but $0 \cdot x = 0 \cdot 4$ is true for any real number x.

In Example 4 we use the multiplication property of equality to solve an equation.

EXAMPLE 4 **Multiplying both sides by the same number**

Solve $\frac{z}{2} = 6$.

Solution

We isolate the variable z by multiplying each side of the equation by 2.

$$\frac{z}{2} = 6 \quad \text{Original equation}$$

$$2 \cdot \frac{z}{2} = 2 \cdot 6 \quad \text{Multiply each side by 2.}$$

$$1z = 12 \quad \text{Because } 2 \cdot \tfrac{z}{2} = 2 \cdot \tfrac{1}{2}z = 1z$$

$$z = 12 \quad \text{Multiplicative identity}$$

Because $\frac{12}{2} = 6$, $\{12\}$ is the solution set to the equation. ■

Because dividing by a number is the same as multiplying by its reciprocal, the multiplication property of equality allows us to divide each side of the equation by any nonzero number.

EXAMPLE 5 **Dividing both sides by the same number**

Solve $-5w = 30$.

Solution

Since w is multiplied by -5, we can isolate w by multiplying by $-\frac{1}{5}$ or by dividing each side by -5:

$$-5w = 30 \quad \text{Original equation}$$
$$\frac{-5w}{-5} = \frac{30}{-5} \quad \text{Divide each side by } -5.$$
$$1 \cdot w = -6 \quad \text{Because } \frac{-5}{-5} = 1$$
$$w = -6 \quad \text{Multiplicative identity}$$

Because $-5(-6) = 30$, $\{-6\}$ is the solution set to the equation. ■

In Example 6, the coefficient of the variable is a fraction. We could divide each side by the coefficient as we did in Example 5, but it is easier to multiply each side by the reciprocal of the coefficient.

EXAMPLE 6 **Multiplying by the reciprocal**

Solve $\frac{4}{5}p = 40$.

Helpful Hint

You could solve this equation by multiplying each side by 5 to get $4p = 200$, and then dividing each side by 4 to get $p = 50$.

Solution

Multiply each side by $\frac{5}{4}$, the reciprocal of $\frac{4}{5}$, to isolate p on the left side.

$$\frac{4}{5}p = 40$$
$$\frac{5}{4} \cdot \frac{4}{5}p = \frac{5}{4} \cdot 40 \quad \text{Multiply each side by } \frac{5}{4}.$$
$$1 \cdot p = 50 \quad \text{Multiplicative inverses}$$
$$p = 50 \quad \text{Multiplicative identity}$$

Because $\frac{4}{5} \cdot 50 = 40$, we can be sure that the solution set is $\{50\}$. ■

If the coefficient of the variable is an integer, we usually divide each side by that integer, as we did in solving $-5w = 30$ in Example 5. Of course we could also solve that equation by multiplying each side by $-\frac{1}{5}$. If the coefficient of the variable is a fraction, we usually multiply each side by the reciprocal of the fraction as we did in solving $\frac{4}{5}p = 40$ in Example 6. Of course we could also solve that equation by dividing each side by $\frac{4}{5}$. If $-x$ appears in an equation, we can multiply by -1 to get x or divide by -1 to get x, because $-1(-x) = x$ and $\frac{-x}{-1} = x$.

EXAMPLE 7 **Multiplying by −1**

Solve $-h = 12$.

Solution

Multiply each side by -1 to get h on the left side.

$$-h = 12$$
$$-1(-h) = -1 \cdot 12$$
$$h = -12$$

Since $-(-12) = 12$, the solution set is $\{-12\}$. ■

Variables on Both Sides

In the next example, the variable occurs on both sides of the equation. Because the variable represents a real number, we can still isolate the variable by using the addition property of equality. Note that it does not matter whether the variable ends up on the right side or the left side.

EXAMPLE 8 **Subtracting an algebraic expression from both sides**

Solve $-9 + 6y = 7y$.

Helpful Hint

It does not matter whether the variable ends up on the left or right side of the equation. Whether we get $y = -9$ or $-9 = y$ we can still conclude that the solution is -9.

Solution

The expression $6y$ can be removed from the left side of the equation by subtracting $6y$ from both sides.

$$-9 + 6y = 7y$$

$$-9 + 6y - 6y = 7y - 6y \quad \text{Subtract } 6y \text{ from each side.}$$

$$-9 = y \quad \text{Simplify each side.}$$

Check by replacing y by -9 in the original equation:

$$-9 + 6(-9) = 7(-9)$$

$$-63 = -63$$

The solution set to the equation is $\{-9\}$. ■

Applications

In the next example, we use the multiplication property of equality in an applied situation.

EXAMPLE 9 **Comparing populations**

In the 2000 census, Georgia had $\frac{2}{3}$ as many people as Illinois (U.S. Bureau of Census, www.census.gov). If the population of Georgia was 8 million, then what was the population of Illinois?

Solution

If p represents the population of Illinois, then $\frac{2}{3}p$ represents the population of Georgia. Since the population of Georgia was 8 million we can write the equation $\frac{2}{3}p = 8$. To find p, solve the equation:

$$\frac{2}{3}p = 8$$

$$\frac{3}{2} \cdot \frac{2}{3}p = \frac{3}{2} \cdot 8 \quad \text{Multiply each side by } \tfrac{3}{2}.$$

$$p = 12 \quad \text{Simplify.}$$

So the population of Illinois was 12 million in 2000. ■

WARM-UPS

True or false? Explain your answer.

1. The solution to $x - 5 = 5$ is 10. True
2. The equation $\frac{x}{2} = 4$ is equivalent to the equation $x = 8$. True
3. To solve $\frac{3}{4}y = 12$, we should multiply each side by $\frac{3}{4}$. False
4. The equation $\frac{x}{7} = 4$ is equivalent to $\frac{1}{7}x = 4$. True
5. Multiplying each side of an equation by any real number will result in an equation that is equivalent to the original equation. False
6. To isolate t in $2t = 7 + t$, subtract t from each side. True
7. To solve $\frac{2r}{3} = 30$, we should multiply each side by $\frac{3}{2}$. True
8. Adding any real number to both sides of an equation will result in an equation that is equivalent to the original equation. True
9. The equation $5x = 0$ is equivalent to $x = 0$. True
10. The solution to $2x - 3 = x + 1$ is 4. True

2.1 EXERCISES

Reading and Writing *After reading this section, write out the answers to these questions. Use complete sentences.*

1. What does the addition property of equality say?
The addition property of equality says that adding the same number to each side of an equation does not change the solution to the equation.
2. What are equivalent equations?
Equivalent equations are equations that have the same solution set.
3. What is the multiplication property of equality?
The multiplication property of equality says that multiplying both sides of an equation by the same nonzero number does not change the solution to the equation.
4. What is a linear equation in one variable?
A linear equation in one variable is an equation of the form $ax + b = 0$ where $a \neq 0$.
5. How can you tell if your solution to an equation is correct?
Replace the variable in the equation with your solution. If the resulting statement is correct, then the solution is correct.
6. To obtain an equivalent equation, what are you not allowed to do to both sides of the equation?
In solving equations, you are not allowed to multiply or divide both sides by 0.

Solve each equation. Show your work and check your answer. See Example 1.

7. $x - 6 = -5$ $\{1\}$
8. $x - 7 = -2$ $\{5\}$
9. $-13 + x = -4$ $\{9\}$
10. $-8 + x = -12$ $\{-4\}$
11. $y - \frac{1}{2} = \frac{1}{2}$ $\{1\}$
12. $y - \frac{1}{4} = \frac{1}{2}$ $\left\{\frac{3}{4}\right\}$
13. $w - \frac{1}{3} = \frac{1}{3}$ $\left\{\frac{2}{3}\right\}$
14. $w - \frac{1}{3} = \frac{1}{2}$ $\left\{\frac{5}{6}\right\}$

Solve each equation. Show your work and check your answer. See Example 2.

15. $x + 3 = -6$ $\{-9\}$
16. $x + 4 = -3$ $\{-7\}$
17. $12 + x = -7$ $\{-19\}$
18. $19 + x = -11$ $\{-30\}$
19. $t + \frac{1}{2} = \frac{3}{4}$ $\left\{\frac{1}{4}\right\}$
20. $t + \frac{1}{3} = 1$ $\left\{\frac{2}{3}\right\}$
21. $\frac{1}{19} + m = \frac{1}{19}$ $\{0\}$
22. $\frac{1}{3} + n = \frac{1}{2}$ $\left\{\frac{1}{6}\right\}$

Solve each equation. Show your work and check your answer. See Example 3.

23. $2 = x + 7$ $\{-5\}$
24. $3 = x + 5$ $\{-2\}$
25. $-13 = y - 9$ $\{-4\}$
26. $-14 = z - 12$ $\{-2\}$
27. $0.5 = -2.5 + x$ $\{3\}$
28. $0.6 = -1.2 + x$ $\{1.8\}$
29. $\frac{1}{8} = -\frac{1}{8} + r$ $\left\{\frac{1}{4}\right\}$
30. $\frac{1}{6} = -\frac{1}{6} + h$ $\left\{\frac{1}{3}\right\}$

Solve each equation. Show your work and check your answer. See Example 4.

31. $\frac{x}{2} = -4$ $\{-8\}$
32. $\frac{x}{3} = -6$ $\{-18\}$
33. $0.03 = \frac{y}{60}$ $\{1.8\}$
34. $0.05 = \frac{y}{80}$ $\{4\}$

35. $\frac{a}{2} = \frac{1}{3}$ $\left\{\frac{2}{3}\right\}$ **36.** $\frac{b}{2} = \frac{1}{5}$ $\left\{\frac{2}{5}\right\}$

37. $\frac{1}{6} = \frac{c}{3}$ $\left\{\frac{1}{2}\right\}$ **38.** $\frac{1}{12} = \frac{d}{3}$ $\left\{\frac{1}{4}\right\}$

Solve each equation. Show your work and check your answer. See Example 5.

39. $-3x = 15$ $\{-5\}$ **40.** $-5x = -20$ $\{4\}$

41. $20 = 4y$ $\{5\}$ **42.** $18 = -3a$ $\{-6\}$

43. $2w = 2.5$ $\{1.25\}$ **44.** $-2x = -5.6$ $\{2.8\}$

45. $5 = 20x$ $\left\{\frac{1}{4}\right\}$ **46.** $-3 = 27d$ $\left\{-\frac{1}{9}\right\}$

47. $5x = \frac{3}{4}$ $\left\{\frac{3}{20}\right\}$ **48.** $3x = -\frac{2}{3}$ $\left\{-\frac{2}{9}\right\}$

Solve each equation. Show your work and check your answer. See Example 6.

49. $\frac{3}{2}x = -3$ $\{-2\}$ **50.** $\frac{2}{3}x = -8$ $\{-12\}$

51. $90 = \frac{3y}{4}$ $\{120\}$ **52.** $14 = \frac{7y}{8}$ $\{16\}$

53. $-\frac{3}{5}w = -\frac{1}{3}$ $\left\{\frac{5}{9}\right\}$ **54.** $-\frac{5}{2}t = -\frac{3}{5}$ $\left\{\frac{6}{25}\right\}$

55. $\frac{2}{3} = -\frac{4x}{3}$ $\left\{-\frac{1}{2}\right\}$ **56.** $\frac{1}{14} = -\frac{6p}{7}$ $\left\{-\frac{1}{12}\right\}$

Solve each equation. Show your work and check your answer. See Example 7.

57. $-x = 8$ $\{-8\}$ **58.** $-x = 4$ $\{-4\}$

59. $-y = -\frac{1}{3}$ $\left\{\frac{1}{3}\right\}$ **60.** $-y = -\frac{7}{8}$ $\left\{\frac{7}{8}\right\}$

61. $3.4 = -z$ $\{-3.4\}$ **62.** $4.9 = -t$ $\{-4.9\}$

63. $-k = -99$ $\{99\}$ **64.** $-m = -17$ $\{17\}$

Solve each equation. Show your work and check your answer. See Example 8.

65. $4x = 3x - 7$ $\{-7\}$ **66.** $3x = 2x + 9$ $\{9\}$

67. $9 - 6y = -5y$ $\{9\}$ **68.** $12 - 18w = -17w$ $\{12\}$

69. $-6x = 8 - 7x$ $\{8\}$ **70.** $-3x = -6 - 4x$ $\{-6\}$

71. $\frac{1}{2}c = 5 - \frac{1}{2}c$ $\{5\}$ **72.** $-\frac{1}{2}h = 13 - \frac{3}{2}h$ $\{13\}$

Use the appropriate property of equality to solve each equation.

73. $12 = x + 17$ $\{-5\}$ **74.** $-3 = x + 6$ $\{-9\}$

75. $\frac{3}{4}y = -6$ $\{-8\}$ **76.** $\frac{5}{9}z = -10$ $\{-18\}$

77. $-3.2 + x = -1.2$ $\{2\}$ **78.** $t - 3.8 = -2.9$ $\{0.9\}$

79. $2a = \frac{1}{3}$ $\left\{\frac{1}{6}\right\}$ **80.** $-3w = \frac{1}{2}$ $\left\{-\frac{1}{6}\right\}$

81. $-9m = 3$ $\left\{-\frac{1}{3}\right\}$ **82.** $-4h = -2$ $\left\{\frac{1}{2}\right\}$

83. $-b = -44$ $\{44\}$ **84.** $-r = 55$ $\{-55\}$

85. $\frac{2}{3}x = \frac{1}{2}$ $\left\{\frac{3}{4}\right\}$ **86.** $\frac{3}{4}x = \frac{1}{3}$ $\left\{\frac{4}{9}\right\}$

87. $-5x = 7 - 6x$ $\{7\}$ **88.** $-\frac{1}{2} + 3y = 4y$ $\left\{-\frac{1}{2}\right\}$

89. $\frac{5a}{7} = -10$ $\{-14\}$ **90.** $\frac{7r}{12} = -14$ $\{-24\}$

91. $\frac{1}{2}v = -\frac{1}{2}v + \frac{3}{8}$ $\left\{\frac{3}{8}\right\}$ **92.** $\frac{1}{3}s + \frac{7}{9} = \frac{4}{3}s$ $\left\{\frac{7}{9}\right\}$

Solve each problem by writing and solving an equation. See Example 9.

93. ***Births to teenagers.*** In 2000 there were 48.5 births per 1000 females 15 to 19 years of age (National Center for Health Statistics, www.cdc.gov/nchs). This birth rate is $\frac{4}{5}$ of the birth rate for teenagers in 1991.

a) Write an equation and solve it to find the birth rate for teenagers in 1991? 60.6 births per 1000 females

b) Use the accompanying graph to estimate the birth rate to teenagers in 1996. 54 births per 1000 females

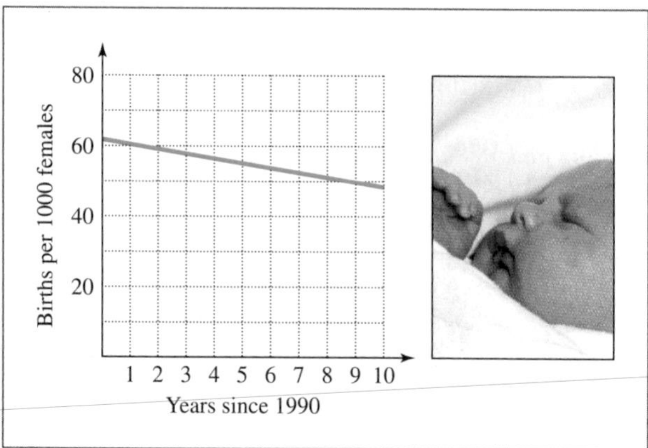

FIGURE FOR EXERCISE 93

94. ***World grain demand.*** Freeport McMoRan projects that in 2010 world grain supply will be 1.8 trillion metric

FIGURE FOR EXERCISE 94

tons and the supply will be only $\frac{3}{4}$ of world grain demand. What will world grain demand be in 2010? 2.4 trillion metric tons

95. ***Advancers and decliners.*** On Thursday, $\frac{2}{3}$ of the stocks traded on the New York Stock Exchange advanced in price. If 1918 stocks advanced, then how many stocks were traded on that day? 2877 stocks

96. ***Births in the United States.*** In 2000, one-third of all births in the United States were to unmarried women (National Center for Health Statistics, www.cdc.gov/nchs). If there were 1,352,938 births to unmarried women, then how many births were there in 2000? 4,058,814

2.2 SOLVING GENERAL LINEAR EQUATIONS

In This Section

- Equations of the Form $ax + b = 0$
- Equations of the Form $ax + b = cx + d$
- Equations with Parentheses
- Applications

All of the equations that we solved in Section 2.1 required only a single application of a property of equality. In this section you will solve equations that require more than one application of a property of equality.

Equations of the Form $ax + b = 0$

To solve an equation of the form $ax + b = 0$ we might need to apply both the addition property of equality and the multiplication property of equality.

EXAMPLE 1 **Using the addition and multiplication properties of equality**

Solve $3r - 5 = 0$.

Helpful Hint

If we divide each side by 3 first, we must divide each term on the left side by 3 to get $r - \frac{5}{3} = 0$. Then add $\frac{5}{3}$ to each side to get $r = \frac{5}{3}$. Although we get the correct answer, we usually save division to the last step so that fractions do not appear until necessary.

Solution

To isolate r, first add 5 to each side, then divide each side by 3.

$3r - 5 = 0$	Original equation
$3r - 5 + 5 = 0 + 5$	Add 5 to each side.
$3r = 5$	Combine like terms.
$\frac{3r}{3} = \frac{5}{3}$	Divide each side by 3.
$r = \frac{5}{3}$	Simplify.

Checking $\frac{5}{3}$ in the original equation gives

$$3 \cdot \frac{5}{3} - 5 = 5 - 5 = 0.$$

So $\left\{\frac{5}{3}\right\}$ is the solution set to the equation. ■

CAUTION In solving $ax + b = 0$ we usually use the addition property of equality first and the multiplication property last. Note that this is the reverse of the order of operations (multiplication before addition), because we are undoing the operations that are done in the expression $ax + b$.

EXAMPLE 2 **Using the addition and multiplication properties of equality**

Solve $-\frac{2}{3}x + 8 = 0$.

Solution

To isolate x, first subtract 8 from each side, then multiply each side by $-\frac{3}{2}$.

$$-\frac{2}{3}x + 8 = 0 \qquad \text{Original equation}$$

$$-\frac{2}{3}x + 8 - 8 = 0 - 8 \qquad \text{Subtract 8 from each side.}$$

$$-\frac{2}{3}x = -8 \qquad \text{Combine like terms.}$$

$$-\frac{3}{2}\left(-\frac{2}{3}x\right) = -\frac{3}{2}(-8) \qquad \text{Multiply each side by } -\frac{3}{2}.$$

$$x = 12 \qquad \text{Simplify.}$$

Checking 12 in the original equation gives

$$-\frac{2}{3}(12) + 8 = -8 + 8 = 0.$$

So $\{12\}$ is the solution set to the equation. ■

Equations of the Form $ax + b = cx + d$

In solving equations our goal is to isolate the variable. We use the addition property of equality to eliminate unwanted terms. Note that it does not matter whether the variable ends up on the right or left side. For some equations we will perform fewer steps if we isolate the variable on the right side.

EXAMPLE 3 **Isolating the variable on the right side**

Solve $3w - 8 = 7w$.

Study Tip

Talk to your classmates. Discuss new terms and ideas. How does this lesson fit in with the last lesson? Form a study group. Does your college have a learning lab where you can study together?

Solution

To eliminate the $3w$ from the left side, we can subtract $3w$ from both sides.

$$3w - 8 = 7w \qquad \text{Original equation}$$

$$3w - 8 - 3w = 7w - 3w \qquad \text{Subtract } 3w \text{ from each side.}$$

$$-8 = 4w \qquad \text{Simplify each side.}$$

$$-\frac{8}{4} = \frac{4w}{4} \qquad \text{Divide each side by 4.}$$

$$-2 = w \qquad \text{Simplify.}$$

To check, replace w with -2 in the original equation:

$$3w - 8 = 7w \qquad \text{Original equation}$$

$$3(-2) - 8 = 7(-2)$$

$$-14 = -14$$

Since -2 satisfies the original equation, the solution set is $\{-2\}$. ■

You should solve the equation in Example 3 by isolating the variable on the left side to see that it takes more steps. In Example 4, it is simplest to isolate the variable on the left side.

EXAMPLE 5 **Solving for y**

Solve $2x - 3y = 9$ for y. Write the answer in the form $y = mx + b$, where m and b are real numbers. (When we study lines in Chapter 3 you will see that $y = mx + b$ is the slope-intercept form of the equation of a line.)

Solution

$$2x - 3y = 9 \quad \text{Original equation}$$
$$-3y = -2x + 9 \quad \text{Subtract } 2x \text{ from each side.}$$
$$\frac{-3y}{-3} = \frac{-2x + 9}{-3} \quad \text{Divide each side by } -3.$$
$$y = \frac{-2x}{-3} + \frac{9}{-3} \quad \text{By the distributive property, each term is divided by } -3.$$
$$y = \frac{2}{3}x - 3 \quad \text{Simplify.}$$

Even though we wrote $y = \frac{2}{3}x - 3$ in Example 5, the equation is still considered to be in the form $y = mx + b$ because we could have written $y = \frac{2}{3}x + (-3)$.

Finding the Value of a Variable

In many situations we know the values of all variables in a formula except one. We use the formula to determine the unknown value.

EXAMPLE 6 **Finding the value of a variable in a formula**

If $2x - 3y = 9$, find y when $x = 6$.

Solution

Method 1: First solve the equation for y. Because we have already solved this equation for y in Example 5 we will not repeat that process in this example. We have

$$y = \frac{2}{3}x - 3.$$

Now replace x by 6 in this equation:

$$y = \frac{2}{3}(6) - 3$$
$$= 4 - 3 = 1$$

So, when $x = 6$, we have $y = 1$.

Method 2: First replace x by 6 in the original equation, then solve for y:

$$2x - 3y = 9 \quad \text{Original equation}$$
$$2 \cdot 6 - 3y = 9 \quad \text{Replace } x \text{ by 6.}$$
$$12 - 3y = 9 \quad \text{Simplify.}$$
$$-3y = -3 \quad \text{Subtract 12 from each side.}$$
$$y = 1 \quad \text{Divide each side by } -3.$$

So when $x = 6$, we have $y = 1$.

If we had to find the value of y for many different values of x, it would be best to solve the equation for y, then insert the various values of x. Method 1 of Example 6 would be the better method. If we must find only one value of y, it does not matter which method we use. When doing the exercises corresponding to this example, you should try both methods.

The next example involves the simple interest formula $I = Prt$, where I is the amount of interest, P is the principal or the amount invested, r is the annual interest rate, and t is the time in years. The interest rate is generally expressed as a percent. When using a rate in computations, you must convert it to a decimal.

EXAMPLE 7 **Using the simple interest formula**

If the simple interest is \$120, the principal is \$400, and the time is 2 years, find the rate.

Helpful Hint

All interest computation is based on simple interest. However, depositors do not like to wait two years to get interest as in Example 7. More often the time is $\frac{1}{12}$ year or $\frac{1}{365}$ year. Simple interest computed every month is said to be compounded monthly. Simple interest computed every day is said to be compounded daily.

Solution

First, solve the formula $I = Prt$ for r, then insert values of P, I, and t:

$Prt = I$ Simple interest formula

$\frac{Prt}{Pt} = \frac{I}{Pt}$ Divide each side by Pt.

$r = \frac{I}{Pt}$ Simplify.

$r = \frac{120}{400 \cdot 2}$ Substitute the values of I, P, and t.

$r = 0.15$ Simplify.

$r = 15\%$ Move the decimal point two places to the right. ■

In solving a geometric problem, it is always helpful to draw a diagram, as we do in Example 8.

EXAMPLE 8 **Using a geometric formula**

The perimeter of a rectangle is 36 feet. If the width is 6 feet, then what is the length?

Solution

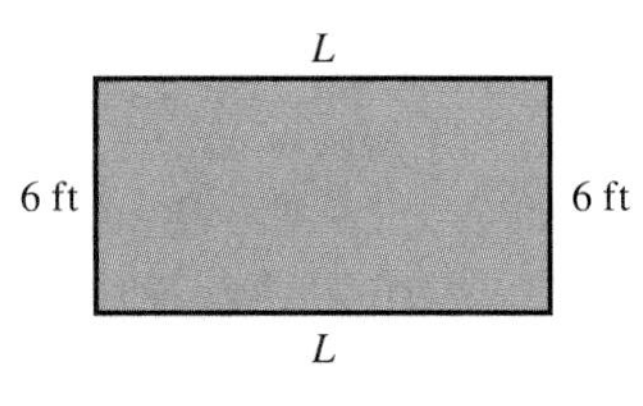

FIGURE 2.1

First, put the given information on a diagram as shown in Fig. 2.1. Substitute the given values into the formula for the perimeter of a rectangle found at the back of the book, and then solve for L. (We could solve for L first and then insert the given values.)

$P = 2L + 2W$ Perimeter of a rectangle

$36 = 2L + 2 \cdot 6$ Substitute 36 for P and 6 for W.

$36 = 2L + 12$ Simplify.

$24 = 2L$ Subtract 12 from each side.

$12 = L$ Divide each side by 2.

Check: If $L = 12$ and $W = 6$, then $P = 2(12) + 2(6) = 36$ feet. So we can be certain that the length is 12 feet. ■

If L is the list price or original price of an item and r is the rate of discount, then the amount of discount is rL, the product of the rate and the list price. The sale price S is the list price minus the amount of discount. So $S = L - rL$. The rate of discount is generally expressed as a percent. In computations, rates must be written as decimals or fractions.

EXAMPLE 9 **Finding the original price**

What was the original price of a stereo that sold for \$560 after a 20% discount.

Study Tip

Don't wait for inspiration to strike, it probably won't. Algebra is learned one tiny step at a time. So do lots of exercises and keep taking those tiny steps.

Solution

Express 20% as the decimal 0.20 or 0.2 and use the formula $S = L - rL$:

$$\text{Selling price} = \text{list price} - \text{amount of discount}$$

$$560 = L - 0.2L$$

$$10(560) = 10(L - 0.2L) \quad \text{Multiply each side by 10.}$$

$$5600 = 10L - 2L \quad \text{Remove the parentheses.}$$

$$5600 = 8L \quad \text{Combine like terms.}$$

$$\frac{5600}{8} = \frac{8L}{8} \quad \text{Divide each side by 8.}$$

$$700 = L$$

Since 20% of \$700 is \$140 and \$700 − \$140 = \$560, we can be sure that the original price was \$700. Note that if the discount is 20%, then the selling price is 80% of the list price. So we could have started with the equation $560 = 0.80L$. ■

WARM-UPS

True or false? Explain your answer.

1. If we solve $D = R \cdot T$ for T, we get $T \cdot R = D$. False
2. If we solve $a - b = 3a - m$ for a, we get $a = 3a - m + b$. False
3. Solving $A = LW$ for L, we get $L = \frac{W}{A}$. False
4. Solving $D = RT$ for R, we get $R = \frac{d}{t}$. False
5. The perimeter of a rectangle is the product of its length and width. False
6. The volume of a shoe box is the product of its length, width, and height. True
7. The sum of the length and width of a rectangle is one-half of its perimeter. True
8. Solving $y - x = 5$ for y gives us $y = x + 5$. True
9. If $x = -1$ and $y = -3x + 6$, then $y = 3$. False
10. The circumference of a circle is the product of its diameter and the number π. True

2.4 EXERCISES

Reading and Writing *After reading this section, write out the answers to these questions. Use complete sentences.*

1. What is a formula?
   A formula is an equation with two or more variables.
2. What is a literal equation?
   A literal equation is a formula.
3. What does it mean to solve a formula for a certain variable?
   To solve for a variable means to find an equivalent equation in which the variable is isolated.
4. How do you solve a formula for a variable that appears on both sides?
   If the variable appears on both sides, then get all terms with the variable onto the same side. Then use the distributive property to get one occurrence of the variable.
5. What are the two methods shown for finding the value of a variable in a formula?
   To find the value of a variable in a formula, we can solve for the variable and then insert values for the other variables, or insert values for the other variables and then solve for the variable.

6. What formula expresses the perimeter of a rectangle in terms of its length and width?
The formula for the perimeter of a rectangle is $P = 2L + 2W$.

Solve each formula for the specified variable. See Examples 1 and 2.

7. $D = RT$ for R $\quad R = \frac{D}{T}$

8. $A = LW$ for W $\quad W = \frac{A}{L}$

9. $C = \pi D$ for D $\quad D = \frac{C}{\pi}$

10. $F = ma$ for a $\quad a = \frac{F}{m}$

11. $I = Prt$ for P $\quad P = \frac{I}{rt}$

12. $I = Prt$ for t $\quad t = \frac{I}{Pr}$

13. $F = \frac{9}{5}C + 32$ for C $\quad C = \frac{5}{9}(F - 32)$

14. $y = \frac{3}{4}x - 7$ for x $\quad x = \frac{4y + 28}{3}$

15. $A = \frac{1}{2}bh$ for h $\quad h = \frac{2A}{b}$

16. $A = \frac{1}{2}bh$ for b $\quad b = \frac{2A}{h}$

17. $P = 2L + 2W$ for L $\quad L = \frac{P - 2W}{2}$

18. $P = 2L + 2W$ for W $\quad W = \frac{P - 2L}{2}$

19. $A = \frac{1}{2}(a + b)$ for a $\quad a = 2A - b$

20. $A = \frac{1}{2}(a + b)$ for b $\quad b = 2A - a$

21. $S = P + Prt$ for r $\quad r = \frac{S - P}{Pt}$

22. $S = P + Prt$ for t $\quad t = \frac{S - P}{Pr}$

23. $A = \frac{1}{2}h(a + b)$ for a $\quad a = \frac{2A - hb}{h}$

24. $A = \frac{1}{2}h(a + b)$ for b $\quad b = \frac{2A - ah}{h}$

Solve each equation for x. See Example 3.

25. $5x + a = 3x + b$ $\quad x = \frac{b - a}{2}$

26. $2c - x = 4x + c - 5b$ $\quad x = \frac{c + 5b}{5}$

27. $4(a + x) - 3(x - a) = 0$ $\quad x = -7a$

28. $-2(x - b) - (5a - x) = a + b$ $\quad x = b - 6a$

29. $3x - 2(a - 3) = 4x - 6 - a$ $\quad x = 12 - a$

30. $2(x - 3w) = -3(x + w)$ $\quad x = \frac{3w}{5}$

31. $3x + 2ab = 4x - 5ab$ $\quad x = 7ab$

32. $x - a = -x + a + 4b$ $\quad x = a + 2b$

Solve each equation for y. See Examples 4 and 5.

33. $x + y = -9$ $\quad y = -x - 9$

34. $3x + y = -5$ $\quad y = -3x - 5$

35. $x + y - 6 = 0$ $\quad y = -x + 6$

36. $4x + y - 2 = 0$ $\quad y = -4x + 2$

37. $2x - y = 2$ $\quad y = 2x - 2$

38. $x - y = -3$ $\quad y = x + 3$

39. $3x - y + 4 = 0$ $\quad y = 3x + 4$

40. $-2x - y + 5 = 0$ $\quad y = -2x + 5$

41. $x + 2y = 4$ $\quad y = -\frac{1}{2}x + 2$

42. $3x + 2y = 6$ $\quad y = -\frac{3}{2}x + 3$

43. $2x - 2y = 1$ $\quad y = x - \frac{1}{2}$

44. $3x - 2y = -6$ $\quad y = \frac{3}{2}x + 3$

45. $y + 2 = 3(x - 4)$ $\quad y = 3x - 14$

46. $y - 3 = -3(x - 1)$ $\quad y = -3x + 6$

47. $y - 1 = \frac{1}{2}(x - 2)$ $\quad y = \frac{1}{2}x$

48. $y - 4 = -\frac{2}{3}(x - 9)$ $\quad y = -\frac{2}{3}x + 10$

49. $\frac{1}{2}x - \frac{1}{3}y = -2$ $\quad y = \frac{3}{2}x + 6$

50. $\frac{x}{2} + \frac{y}{4} = \frac{1}{2}$ $\quad y = -2x + 2$

51. $y - 2 = \frac{3}{2}(x + 3)$ $\quad y = \frac{3}{2}x + \frac{13}{2}$

52. $y + 4 = \frac{2}{3}(x - 2)$ $\quad y = \frac{2}{3}x - \frac{16}{3}$

53. $y - \frac{1}{2} = -\frac{1}{4}\left(x - \frac{1}{2}\right)$ $\quad y = -\frac{1}{4}x + \frac{5}{8}$

54. $y + \frac{1}{2} = -\frac{1}{3}\left(x + \frac{1}{2}\right)$ $\quad y = -\frac{1}{3}x - \frac{2}{3}$

Fill in the tables using the given formulas.

55. $y = -3x + 30$

x	y
−10	60
0	30
10	0
20	−30
30	−60

56. $y = 4x - 20$

x	y
−10	−60
−5	−40
0	−20
5	0
10	20

57. $F = \frac{9}{5}C + 32$

C	F
−10	14
−5	23
0	32
40	104
100	212

58. $C = \frac{5}{9}(F - 32)$

F	C
−40	−40
14	−10
32	0
59	15
86	30

59. $T = \frac{400}{R}$

R (mph)	T (hr)
10	40
20	20
40	10
80	5
100	4

60. $R = \frac{100}{T}$

T (hr)	R (mph)
1	100
5	20
20	5
50	2
100	1

For each equation that follows, find y given that $x = 2$. See Example 6.

61. $y = 3x - 4$ 2

62. $y = -2x + 5$ 1

63. $3x - 2y = -8$ 7

64. $4x + 6y = 8$ 0

65. $\frac{3x}{2} - \frac{5y}{3} = 6$ $-\frac{9}{5}$

66. $\frac{2y}{5} - \frac{3x}{4} = \frac{1}{2}$ 5

67. $y - 3 = \frac{1}{2}(x - 6)$ 1

68. $y - 6 = -\frac{3}{4}(x - 2)$ 6

69. $y - 4.3 = 0.45(x - 8.6)$ 1.33

70. $y + 33.7 = 0.78(x - 45.6)$ -67.708

Solve each of the following problems. Appendix A contains some geometric formulas that may be helpful. See Examples 7–9.

71. ***Finding the rate.*** If the simple interest on \$5000 for 3 years is \$600, then what is the rate? 4%

72. ***Finding the rate.*** Wayne paid \$420 in simple interest on a loan of \$1000 for 7 years. What was the rate? 6%

73. ***Finding the time.*** Kathy paid \$500 in simple interest on a loan of \$2500. If the annual interest rate was 5%, then what was the time? 4 years

74. ***Finding the time.*** Robert paid \$240 in simple interest on a loan of \$1000. If the annual interest rate was 8%, then what was the time? 3 years

75. ***Finding the length.*** The area of a rectangle is 28 square yards. The width is 4 yards. Find the length. 7 yards

76. ***Finding the width.*** The area of a rectangle is 60 square feet. The length is 4 feet. Find the width. 15 feet

77. ***Finding the length.*** If it takes 600 feet of wire fencing to fence a rectangular feed lot that has a width of 75 feet, then what is the length of the lot? 225 feet

78. ***Finding the depth.*** If it takes 500 feet of fencing to enclose a rectangular lot that is 104 feet wide, then how deep is the lot? 146 feet

79. ***Finding MSRP.*** What was the manufacturer's suggested retail price (MSRP) for a Lexus SC 430 that sold for \$54,450 after a 10% discount? \$60,500

80. ***Finding MSRP.*** What was the MSRP for a Hummer H1 that sold for \$107,272 after an 8% discount? \$116,600

81. ***Finding the original price.*** Find the original price if there is a 15% discount and the sale price is \$255. \$300

82. ***Finding the list price.*** Find the list price if there is a 12% discount and the sale price is \$4400. \$5000

83. ***Rate of discount.*** Find the rate of discount if the discount is \$40 and the original price is \$200. 20%

84. ***Rate of discount.*** Find the rate of discount if the discount is \$20 and the original price is \$250. 8%

85. ***Width of a football field.*** The perimeter of a football field in the NFL, excluding the end zones, is 920 feet. How wide is the field? 160 feet

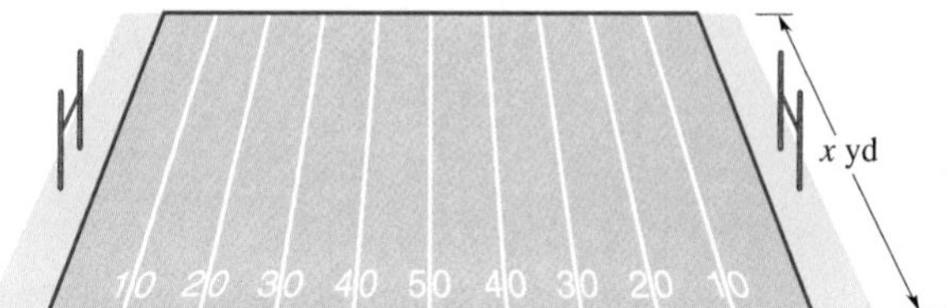

FIGURE FOR EXERCISE 85

86. ***Perimeter of a frame.*** If a picture frame is 16 inches by 20 inches, then what is its perimeter? 72 inches

87. ***Volume of a box.*** A rectangular box measures 2 feet wide, 3 feet long, and 4 feet deep. What is its volume? 24 cubic feet

88. ***Volume of a refrigerator.*** The volume of a rectangular refrigerator is 20 cubic feet. If the top measures 2 feet by 2.5 feet, then what is the height? 4 feet

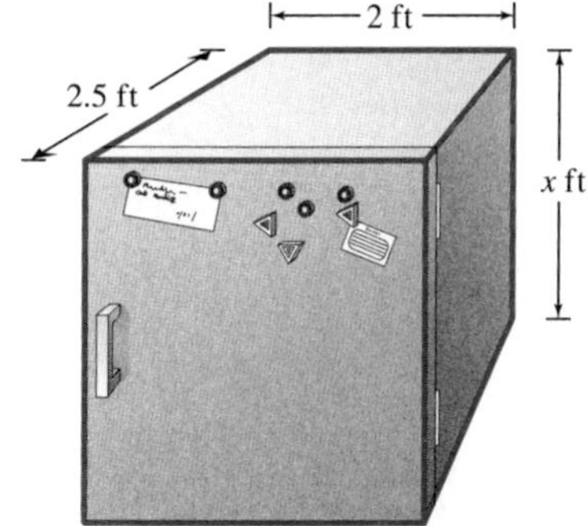

FIGURE FOR EXERCISE 88

89. ***Radius of a pizza.*** If the circumference of a pizza is 8π inches, then what is the radius? 4 inches

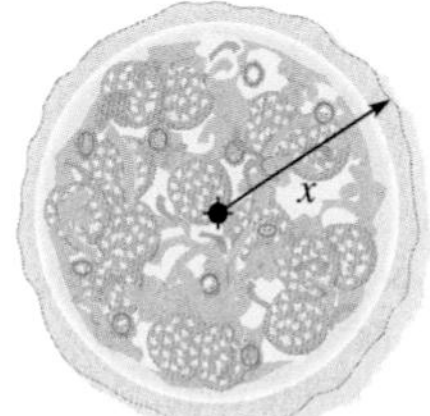

FIGURE FOR EXERCISE 89

90. ***Diameter of a circle.*** If the circumference of a circle is 4π meters, then what is the diameter? 4 meters

91. ***Height of a banner.*** If a banner in the shape of a triangle has an area of 16 square feet with a base of 4 feet, then what is the height of the banner? 8 feet

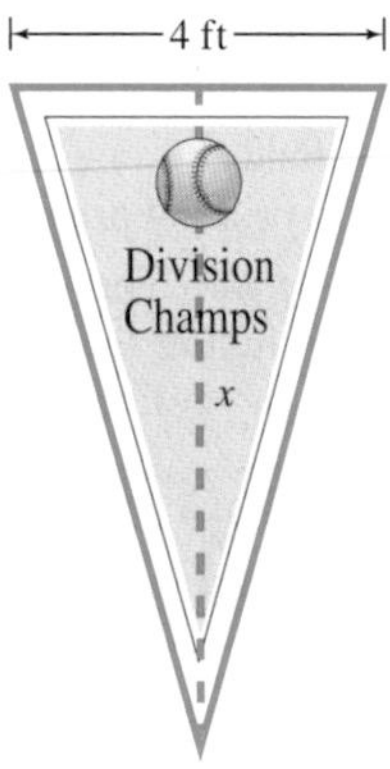

FIGURE FOR EXERCISE 91

92. ***Length of a leg.*** If a right triangle has an area of 14 square meters and one leg is 4 meters in length, then what is the length of the other leg? 7 meters

93. ***Length of the base.*** A trapezoid with height 20 inches and lower base 8 inches has an area of 200 square inches. What is the length of its upper base? 12 inches

94. ***Height of a trapezoid.*** The end of a flower box forms the shape of a trapezoid. The area of the trapezoid is 300 square centimeters. The bases are 16 centimeters and 24 centimeters in length. Find the height. 15 centimeters

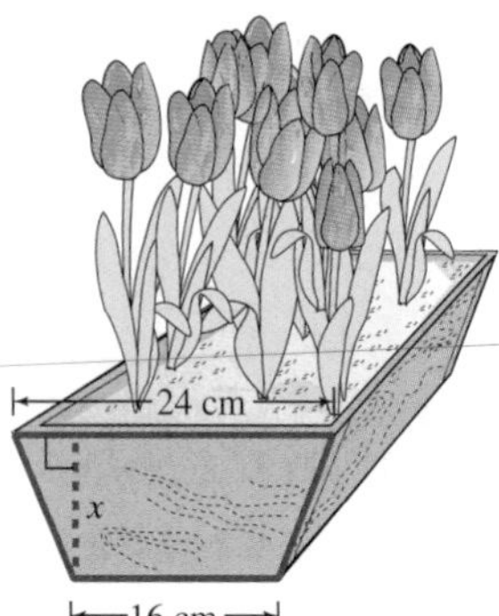

FIGURE FOR EXERCISE 94

95. ***Fried's rule.*** Doctors often prescribe the same drugs for children as they do for adults. The formula $d = 0.08aD$ (Fried's rule) is used to calculate the child's dosage d, where a is the child's age and D is the adult dosage. If a doctor prescribes 1000 milligrams of acetaminophen for an adult, then how many milligrams would the doctor prescribe for an eight-year-old child? Use the bar graph to determine the age at which a child would get the same dosage as an adult. 640 milligrams, age 13

96. ***Cowling's rule.*** Cowling's rule is another method for determining the dosage of a drug to prescribe to a child. For this rule, the formula

$$d = \frac{D(a + 1)}{24}$$

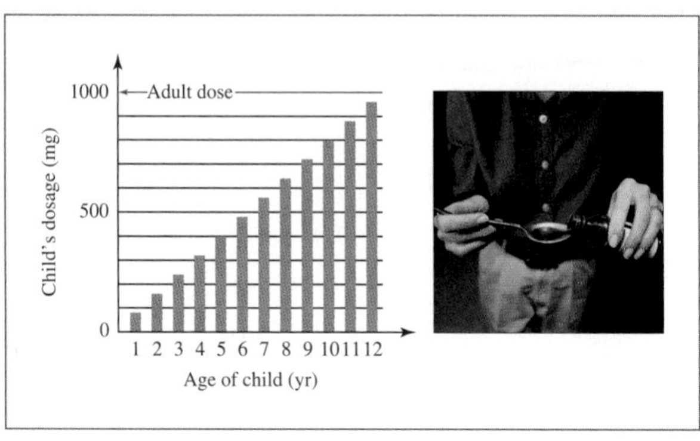

FIGURE FOR EXERCISE 96

gives the child's dosage d, where D is the adult dosage and a is the age of the child in years. If the adult dosage of a drug is 600 milligrams and a doctor uses this formula to determine that a child's dosage is 200 milligrams, then how old is the child? Age 7

97. ***Administering Vancomycin.*** A patient is to receive 750 mg of the antibiotic Vancomycin. However, Vancomycin comes in a solution containing 1 gram (available dose) of Vancomycin per 5 milliliters (quantity) of solution. Use the formula

$$\text{Amount} = \frac{\text{desired dose}}{\text{available dose}} \times \text{quantity}$$

to find the amount of this solution that should be administered to the patient. 3.75 milliliters

98. ***International communications.*** The global investment in telecom infrastructure since 1990 can be modeled by the formula

$$I = 7.5T + 115,$$

where I is in billions of dollars and t is the number of years since 1990 (*Fortune,* www.fortune.com).

a) Use the formula to find the global investment in 2000. \$190 billion

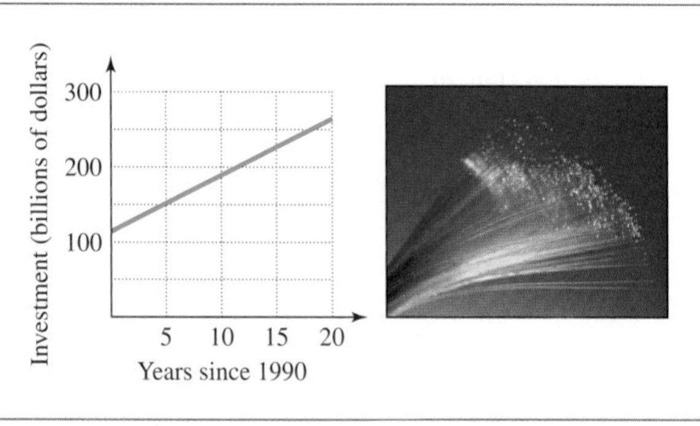

FIGURE FOR EXERCISE 98

b) Use the accompanying graph to estimate the year in which the global investment will reach \$250 billion.
2008

c) Use the formula to find the year in which the global investment will reach \$250 billion.
2008

99. ***The 2.4-meter rule.*** A 2.4-meter sailboat is a one-person boat that is about 13 feet in length, has a displacement of about 550 pounds, and a sail area of about 81 square feet. To compete in the 2.4-meter class, a boat must satisfy the formula

$$2.4 = \frac{L + 2D - F\sqrt{S}}{2.37},$$

where L = length, F = freeboard, D = girth, and S = sail area. Solve the formula for L.
$L = F\sqrt{S} - 2D + 5.688$

PHOTO FOR EXERCISE 99

2.5 TRANSLATING VERBAL EXPRESSIONS INTO ALGEBRAIC EXPRESSIONS

In This Section

- Writing Algebraic Expressions
- Pairs of Numbers
- Consecutive Integers
- Using Formulas
- Writing Equations

You translated some verbal expressions into algebraic expressions in Section 1.6; in this section you will study translating in more detail.

Writing Algebraic Expressions

The following box contains a list of some frequently occurring verbal expressions and their equivalent algebraic expressions.

Translating Words into Algebra

	Verbal Phrase	**Algebraic Expression**
Addition:	The sum of a number and 8	$x + 8$
	Five is added to a number	$x + 5$
	Two more than a number	$x + 2$
	A number increased by 3	$x + 3$
Subtraction:	Four is subtracted from a number	$x - 4$
	Three less than a number	$x - 3$
	The difference between 7 and a number	$7 - x$
	A number decreased by 2	$x - 2$
Multiplication:	The product of 5 and a number	$5x$
	Twice a number	$2x$
	One-half of a number	$\frac{1}{2}x$
	Five percent of a number	$0.05x$
Division:	The ratio of a number to 6	$\frac{x}{6}$
	The quotient of 5 and a number	$\frac{5}{x}$
	Three divided by some number	$\frac{3}{x}$

EXAMPLE 1 **Writing algebraic expressions**

Translate each verbal expression into an algebraic expression.

a) The sum of a number and 9

b) Eighty percent of a number

c) A number divided by 4

d) The result of a number subtracted from 5

e) Three less than a number

Solution

a) If x is the number, then the sum of x and 9 is $x + 9$.

b) If w is the number, then eighty percent of the number is $0.80w$.

c) If y is the number, then the number divided by 4 is $\frac{y}{4}$.

d) If z is the number, then the result of subtracting z from 5 is $5 - z$.

e) If a is the number, then 3 less than a is $a - 3$. ■

Helpful Hint

We know that x and $10 - x$ have a sum of 10 for any value of x. We can easily check that fact by adding:

$$x + 10 - x = 10$$

In general it is not true that x and $x - 10$ have a sum of 10, because

$$x + x - 10 = 2x - 10.$$

For what value of x is the sum of x and $x - 10$ equal to 10?

Pairs of Numbers

There is often more than one unknown quantity in a problem, but a relationship between the unknown quantities is given. For example, if one unknown number is 5 more than another unknown number, we can use

$$x \quad \text{and} \quad x + 5,$$

to represent them. Note that x and $x + 5$ can also be used to represent two unknown numbers that differ by 5, for if two numbers differ by 5, one of the numbers is 5 more than the other.

How would you represent two numbers that have a sum of 10? If one of the numbers is 2, the other is certainly $10 - 2$, or 8. Thus if x is one of the numbers, then $10 - x$ is the other. The expressions

$$x \quad \text{and} \quad 10 - x$$

have a sum of 10 for any value of x.

EXAMPLE 2 **Algebraic expressions for pairs of numbers**

Write algebraic expressions for each pair of numbers.

a) Two numbers that differ by 12 **b)** Two numbers with a sum of -8

Solution

a) The expressions x and $x - 12$ represent two numbers that differ by 12. We can check by subtracting:

$$x - (x - 12) = x - x + 12 = 12$$

Of course, x and $x + 12$ also differ by 12 because $x + 12 - x = 12$.

b) The expressions x and $-8 - x$ have a sum of -8. We can check by addition:

$$x + (-8 - x) = x - 8 - x = -8$$ ■

Pairs of numbers occur in geometry in discussing measures of angles. You will need the following facts about degree measures of angles.

Degree Measures of Angles

Two angles are called **complementary** if the sum of their degree measures is 90°.

Two angles are called **supplementary** if the sum of their degree measures is 180°.

The sum of the degree measures of the three angles of any triangle is 180°.

For complementary angles, we use x and $90 - x$ for their degree measures. For supplementary angles, we use x and $180 - x$. Complementary angles that share a common side form a right angle. Supplementary angles that share a common side form a straight angle or straight line.

EXAMPLE 3 **Degree measures**

Write algebraic expressions for each pair of angles shown.

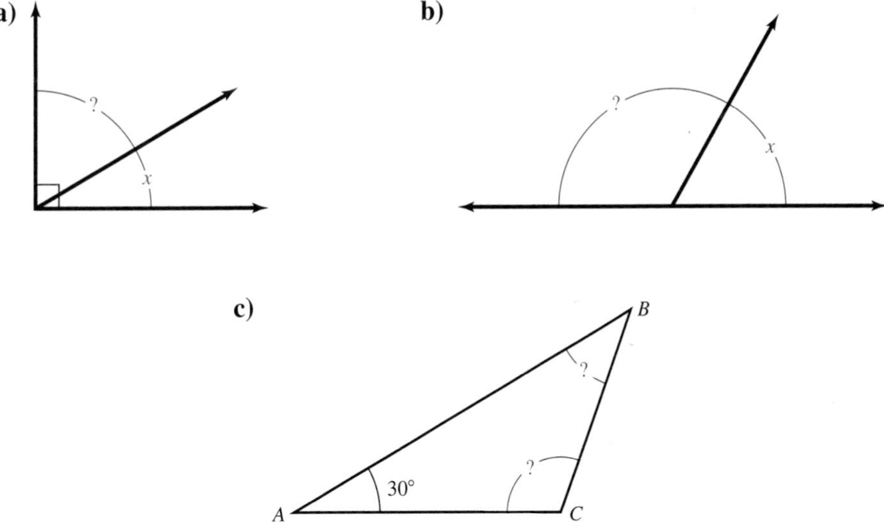

Solution

a) Since the angles shown are complementary, we can use x to represent the degree measure of the smaller angle and $90 - x$ to represent the degree measure of the larger angle.

b) Since the angles shown are supplementary, we can use x to represent the degree measure of the smaller angle and $180 - x$ to represent the degree measure of the larger angle.

c) If we let x represent the degree measure of angle B, then $180 - x - 30$, or $150 - x$, represents the degree measure of angle C. ■

Consecutive Integers

Note that each integer is one larger than the previous integer. For example, if $x = 5$, then $x + 1 = 6$ and $x + 2 = 7$. So if x is an integer, then x, $x + 1$, and $x + 2$ represent three consecutive integers. Each even (or odd) integer is two larger than the previous even (or odd) integer. For example, if $x = 6$, then $x + 2 = 8$, and $x + 4 = 10$. If $x = 7$, then $x + 2 = 9$, and $x + 4 = 11$. So x, $x + 2$, and $x + 4$ represent three consecutive even integers if x is even and three consecutive odd integers if x is odd.

CAUTION The expressions x, $x + 1$, and $x + 3$ do not represent three consecutive odd integers no matter what x represents.

EXAMPLE 4 **Expressions for integers**

Write algebraic expressions for the following unknown integers.

a) Two consecutive integers, the smallest of which is w.

b) Three consecutive even integers, the smallest of which is z.

c) Four consecutive odd integers, the smallest of which is y.

Solution

a) Each integer is 1 larger than the preceding integer. So if w represents the smallest of two consecutive integers, then w and $w + 1$ represent the integers.

b) Each even integer is 2 larger than the preceding even integer. So if z represents the smallest of three consecutive even integers, then z, $z + 2$, and $z + 4$ represent the three consecutive even integers.

c) Each odd integer is 2 larger than the preceding odd integer. So if y represents the smallest of four consecutive odd integers, then y, $y + 2$, $y + 4$, and $y + 6$ represent the four consecutive odd integers. ■

Using Formulas

In writing expressions for unknown quantities, we often use standard formulas such as those given at the back of the book.

EXAMPLE 5 **Writing algebraic expressions using standard formulas**

Find an algebraic expression for

a) the distance if the rate is 30 miles per hour and the time is T hours.

b) the discount if the rate is 40% and the original price is p dollars.

Solution

a) Using the formula $D = RT$, we have $D = 30T$. So $30T$ is an expression that represents the distance in miles.

b) Since the discount is the rate times the original price, an algebraic expression for the discount is $0.40p$ dollars. ■

Writing Equations

To solve a problem using algebra, we describe or **model** the problem with an equation. In this section we write the equations only, and in Section 2.6 we write and solve them. Sometimes we must write an equation from the information given in the problem and sometimes we use a standard model to get the equation. Some standard models are shown in the following box.

Study Tip

It is a good idea to work with others, but don't be misled. Working a problem with help is not the same as working a problem on your own. In the end, mathematics is personal. Make sure that you can do it. Give yourself a weekly quiz by working 10 exercises randomly selected from the material covered since the last test.

Uniform Motion Model

Distance = Rate · Time $\qquad D = R \cdot T$

Percentage Models

What number is 5% of 40? $\qquad x = 0.05 \cdot 40$
Ten is what percent of 80? $\qquad 10 = x \cdot 80$
Twenty is 4% of what number? $\qquad 20 = 0.04 \cdot x$

Selling Price and Discount Model

Discount = Rate of discount · Original price
Selling Price = Original price − Discount

Real Estate Commission Model

Commission = Rate of commission · Selling price
Amount for owner = Selling price − Commission

Geometric Models for Perimeter

Perimeter of any figure = the sum of the lengths of the sides
Rectangle: $P = 2L + 2W$ $\qquad$ Square: $P = 4s$

Geometric Models for Area

Rectangle: $A = LW$ $\qquad$ Square: $A = s^2$
Parallelogram: $A = bh$ $\qquad$ Triangle: $A = \frac{1}{2}bh$

More geometric formulas can be found in Appendix A.

EXAMPLE 6 **Writing equations**

Identify the variable and write an equation that describes each situation.

a) Find two numbers that have a sum of 14 and a product of 45.

b) A coat is on sale for 25% off the list price. If the sale price is \$87, then what is the list price?

c) What percent of 8 is 2?

d) The value of x dimes and $x - 3$ quarters is \$2.05.

Helpful Hint

At this point we are simply learning to write equations that model certain situations. Don't worry about solving these equations now. In Section 2.6 we will solve problems by writing an equation and solving it.

Solution

a) Let x = one of the numbers and $14 - x$ = the other number. Since their product is 45, we have

$$x(14 - x) = 45.$$

b) Let x = the list price and $0.25x$ = the amount of discount. We can write an equation expressing the fact that the selling price is the list price minus the discount:

$$\text{List price} - \text{discount} = \text{selling price}$$
$$x - 0.25x = 87$$

c) If we let x represent the percentage, then the equation is $x \cdot 8 = 2$, or $8x = 2$.

d) The value of x dimes at 10 cents each is $10x$ cents. The value of $x - 3$ quarters at 25 cents each is $25(x - 3)$ cents. We can write an equation expressing the fact that the total value of the coins is 205 cents:

$$\text{Value of dimes} + \text{value of quarters} = \text{total value}$$
$$10x + 25(x - 3) = 205$$

CAUTION The value of the coins in Example 6(d) is either 205 cents or 2.05 dollars. If the total value is expressed in dollars, then all of the values must be expressed in dollars. So we could also write the equation as

$$0.10x + 0.25(x - 3) = 2.05.$$

WARM-UPS

True or false? Explain your answer.

1. For any value of x, the numbers x and $x + 6$ differ by 6. True
2. For any value of a, a and $10 - a$ have a sum of 10. True
3. If Jack ran at x miles per hour for 3 hours, he ran $3x$ miles. True
4. If Jill ran at x miles per hour for 10 miles, she ran for $10x$ hours. False
5. If the realtor gets 6% of the selling price and the house sells for x dollars, the owner gets $x - 0.06x$ dollars. True
6. If the owner got \$50,000 and the realtor got 10% of the selling price, the house sold for \$55,000. False
7. Three consecutive odd integers can be represented by x, $x + 1$, and $x + 3$. False
8. The value in cents of n nickels and d dimes is $0.05n + 0.10d$. False
9. If the sales tax rate is 5% and x represents the price of the goods purchased, then the total bill is $1.05x$. True
10. If the length of a rectangle is 4 feet more than the width w, then the perimeter is $w + (w + 4)$ feet. False

2.5 EXERCISES

Reading and Writing *After reading this section, write out the answers to these questions. Use complete sentences.*

1. What are the different ways of verbally expressing the operation of addition?
   To express addition we use words such as plus, sum, increased by, and more than.
2. How can you algebraically express two numbers using only one variable?
   We can algebraically express two numbers using one variable provided the numbers are related in some known way.
3. What are complementary angles?
   Complementary angles have degree measures with a sum of 90°.
4. What are supplementary angles?
   Supplementary angles have degree measures with a sum of 180°.
5. What is the relationship between distance, rate, and time?
   Distance is the product of rate and time.
6. What is the difference between expressing consecutive even integers and consecutive odd integers algebraically?
   If x is an even integer, then x and $x + 2$ represent consecutive even integers. If x is an odd integer, then x and $x + 2$ represent consecutive odd integers.

Translate each verbal expression into an algebraic expression. See Example 1.

7. The sum of a number and 3 $x + 3$
8. Two more than a number $x + 2$
9. Three less than a number $x - 3$
10. Four subtracted from a number $x - 4$
11. The product of a number and 5 $5x$
12. Five divided by some number $\frac{5}{x}$
13. Ten percent of a number $0.1x$
14. Eight percent of a number $0.08x$
15. The ratio of a number and 3 $\frac{x}{3}$
16. The quotient of 12 and a number $\frac{12}{x}$
17. One-third of a number $\frac{1}{3}x$
18. Three-fourths of a number $\frac{3}{4}x$

Write algebraic expressions for each pair of numbers. See Example 2.

19. Two numbers with a difference of 15 x and $x + 15$
20. Two numbers that differ by 9 x and $x + 9$
21. Two numbers with a sum of 6 x and $6 - x$
22. Two numbers with a sum of 5 x and $5 - x$
23. Two numbers with a sum of -4 x and $-4 - x$
24. Two numbers with a sum of -8 x and $-8 - x$
25. Two numbers such that one is 3 larger than the other x and $x + 3$
26. Two numbers such that one is 8 smaller than the other x and $x + 8$
27. Two numbers such that one is 5% of the other x and $0.05x$
28. Two numbers such that one is 40% of the other x and $0.4x$
29. Two numbers such that one is 30% more than the other x and $1.30x$
30. Two numbers such that one is 20% smaller than the other x and $0.80x$

Each of the following figures shows a pair of angles. Write algebraic expressions for the degree measures of each pair of angles. See Example 3.

31.

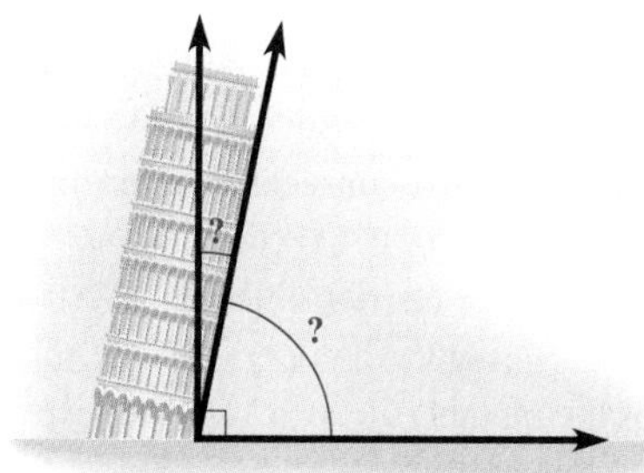

FIGURE FOR EXERCISE 31

x and $90 - x$

32.

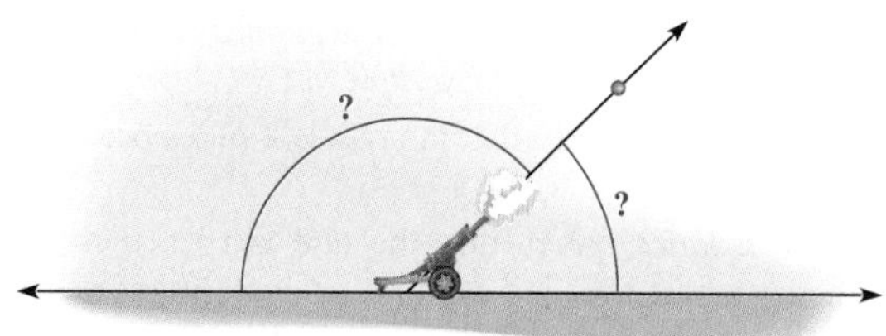

FIGURE FOR EXERCISE 32

x and $180 - x$

33.

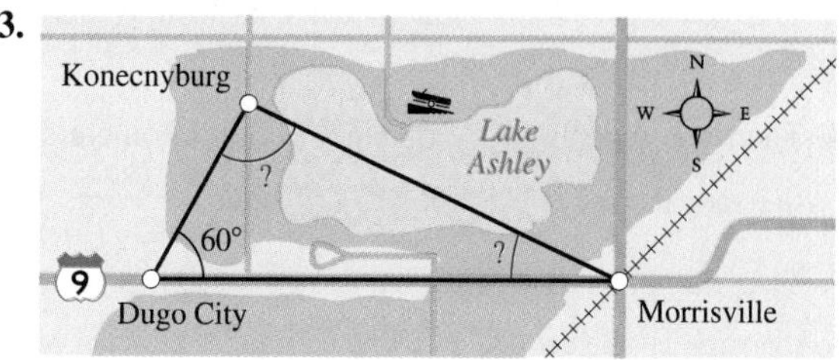

FIGURE FOR EXERCISE 33

x and $120 - x$

34.

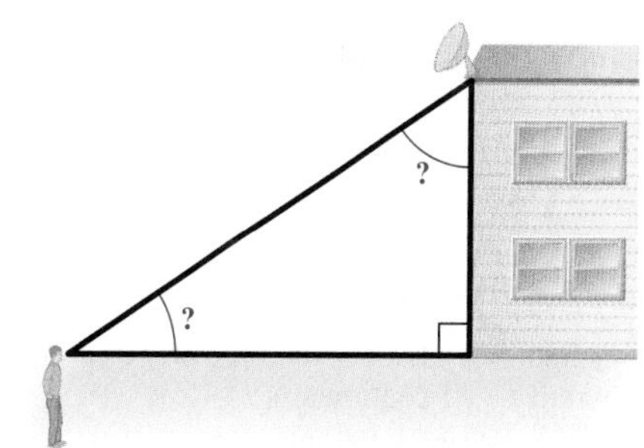

FIGURE FOR EXERCISE 34

x and $90 - x$

Write algebraic expressions for the following unknown integers. See Example 4.

35. Two consecutive even integers, the smallest of which is n n and $n + 2$, where n is an even integer
36. Two consecutive odd integers, the smallest of which is x x and $x + 2$, where x is an odd integer
37. Two consecutive integers x and $x + 1$, where x is an integer
38. Three consecutive even integers x, $x + 2$, and $x + 4$, where x is an even integer
39. Three consecutive odd integers x, $x + 2$, and $x + 4$, where x is an odd integer
40. Three consecutive integers x, $x + 1$, and $x + 2$, where x is an integer
41. Four consecutive even integers x, $x + 2$, $x + 4$, and $x + 6$, where x is an even integer
42. Four consecutive odd integers x, $x + 2$, $x + 4$, and $x + 6$, where x is an odd integer

Find an algebraic expression for the quantity in italics using the given information. See Example 5.

43. The *distance,* given that the rate is x miles per hour and the time is 3 hours $3x$ miles

44. The *distance,* given that the rate is $x + 10$ miles per hour and the time is 5 hours $5x + 50$ miles

45. The *discount,* given that the rate is 25% and the original price is q dollars $0.25q$ dollars

46. The *discount,* given that the rate is 10% and the original price is t yen $0.10t$ yen

47. The *time,* given that the distance is x miles and the rate is 20 miles per hour $\frac{x}{20}$ hour

48. The *time,* given that the distance is 300 kilometers and the rate is $x + 30$ kilometers per hour $\frac{300}{x + 30}$ hour

49. The *rate,* given that the distance is $x - 100$ meters and the time is 12 seconds $\frac{x - 100}{12}$ meters per second

50. The *rate,* given that the distance is 200 feet and the time is $x + 3$ seconds $\frac{200}{x + 3}$ feet per second

51. The *area* of a rectangle with length x meters and width 5 meters $5x$ square meters

52. The *area* of a rectangle with sides b yards and $b - 6$ yards $b(b - 6)$ square yards

53. The *perimeter* of a rectangle with length $w + 3$ inches and width w inches $2w + 2(w + 3)$ inches

54. The *perimeter* of a rectangle with length r centimeters and width $r - 1$ centimeters
$2r + 2(r - 1)$ centimeters

55. The *width* of a rectangle with perimeter 300 feet and length x feet $150 - x$ feet

56. The *length* of a rectangle with area 200 square feet and width w feet $\frac{200}{w}$ feet

57. The *length* of a rectangle, given that its width is x feet and its length is 1 foot longer than twice the width
$2x + 1$ feet

58. The *length* of a rectangle, given that its width is w feet and its length is 3 feet shorter than twice the width
$2w - 3$ feet

59. The *area* of a rectangle, given that the width is x meters and the length is 5 meters longer than the width
$x(x + 5)$ square meters

60. The *perimeter* of a rectangle, given that the length is x yards and the width is 10 yards shorter
$2(x) + 2(x - 10)$ yards

61. The *simple interest,* given that the principal is $x + 1000$, the rate is 18%, and the time is 1 year
$0.18(x + 1000)$

62. The *simple interest,* given that the principal is $3x$, the rate is 6%, and the time is 1 year $0.06(3x)$

63. The *price per pound* of peaches, given that x pounds sold for \$16.50 $\frac{16.50}{x}$ dollars per pound

64. The *rate per hour* of a mechanic who gets \$480 for working x hours $\frac{480}{x}$ dollars per hour

65. The *degree measure* of an angle, given that its complementary angle has measure x degrees $90 - x$ degrees

66. The *degree measure* of an angle, given that its supplementary angle has measure x degrees $180 - x$ degrees

Identify the variable and write an equation that describes each situation. Do not solve the equation. See Example 6.

67. Two numbers differ by 5 and have a product of 8.
x is the smaller number, $x(x + 5) = 8$

68. Two numbers differ by 6 and have a product of -9.
x is the smaller number, $x(x + 6) = -9$

69. Herman's house sold for x dollars. The real estate agent received 7% of the selling price and Herman received \$84,532.
x is the selling price, $x - 0.07x = 84{,}532$

70. Gwen sold her car on consignment for x dollars. The saleswoman's commission was 10% of the selling price and Gwen received \$6570.
x is the selling price, $x - 0.10x = 6570$

71. What percent of 500 is 100?
x is the percent, $500x = 100$

72. What percent of 40 is 120?
x is the percent, $40x = 120$

73. The value of x nickels and $x + 2$ dimes is \$3.80.
x is the number of nickels, $0.05x + 0.10(x + 2) = 3.80$

74. The value of d dimes and $d - 3$ quarters is \$6.75.
d is the number of dimes, $0.10d + 0.25(d - 3) = 6.75$

75. The sum of a number and 5 is 13.
x is the number, $x + 5 = 13$

76. Twelve subtracted from a number is -6.
x is the number, $x - 12 = -6$

77. The sum of three consecutive integers is 42.
x is the smallest integer, $x + (x + 1) + (x + 2) = 42$

78. The sum of three consecutive odd integers is 27.
x is the smallest odd integer, $x + x + 2 + x + 4 = 27$

79. The product of two consecutive integers is 182.
x is the smaller integer, $x(x + 1) = 182$

80. The product of two consecutive even integers is 168.
x is the smaller even integer, $x(x + 2) = 168$

81. Twelve percent of Harriet's income is \$3000.
x is Harriet's income, $0.12x = 3000$

82. If 9% of the members buy tickets, then we will sell 252 tickets to this group.
x is the number of members, $0.09x = 252$

83. Thirteen is 5% of what number?
x is the number, $0.05x = 13$

84. Three hundred is 8% of what number?
x is the number, $0.08x = 300$

85. The length of a rectangle is 5 feet longer than the width, and the area is 126 square feet.
x is the width, $x(x + 5) = 126$

86. The length of a rectangle is 1 yard shorter than twice the width, and the perimeter is 298 yards.
x is the width, $2x + 2(2x - 1) = 298$

87. The value of n nickels and $n - 1$ dimes is 95 cents.
n is the number of nickels, $5n + 10(n - 1) = 95$

88. The value of q quarters, $q + 1$ dimes, and $2q$ nickels is 90 cents.
q is the number of quarters, $25q + 10(q + 1) + 5(2q) = 90$

89. The measure of an angle is 38° smaller than the measure of its supplementary angle.
x is the measure of the larger angle, $x + x - 38 = 180$

90. The measure of an angle is 16° larger than the measure of its complementary angle.
x is the measure of the smaller angle, $x + x + 16 = 90$

91. ***Target heart rate.*** For a cardiovascular work out, fitness experts recommend that you reach your target heart rate and stay at that rate for at least 20 minutes (HealthStatus, www.healthstatus.com). To find your target heart rate, find the sum of your age and your resting heart rate, then subtract that sum from 220. Find 60% of that result and add it to your resting heart rate.

a) Write an equation with variable r expressing the fact that the target heart rate for 30-year-old Bob is 144.

b) Judging from the accompanying graph, does the target heart rate for a 30-year-old increase or decrease as the resting heart rate increases.

a) $r + 0.6(220 - (30 + r)) = 144$, where r is the resting heart rate
b) Target heart rate increases as resting heart rate increases.

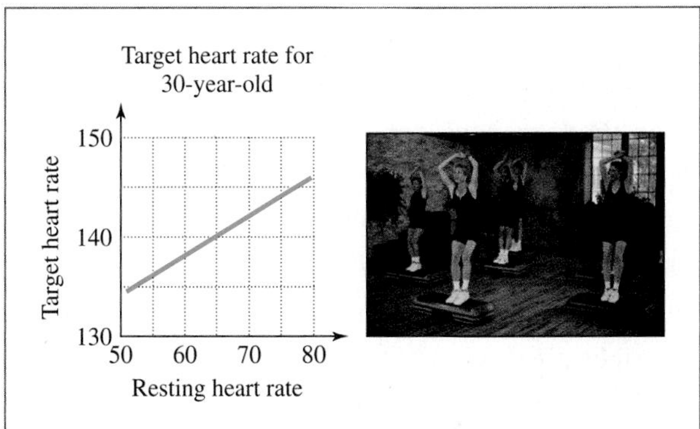

FIGURE FOR EXERCISE 91

92. ***Adjusting the saddle.*** The saddle height on a bicycle should be 109% of the rider's inside leg measurement L (www.harriscyclery.com). See the accompanying figure. Write an equation expressing the fact that the saddle height for Brenda is 36 in.
$1.09L = 36$, where L is the inside leg measurement

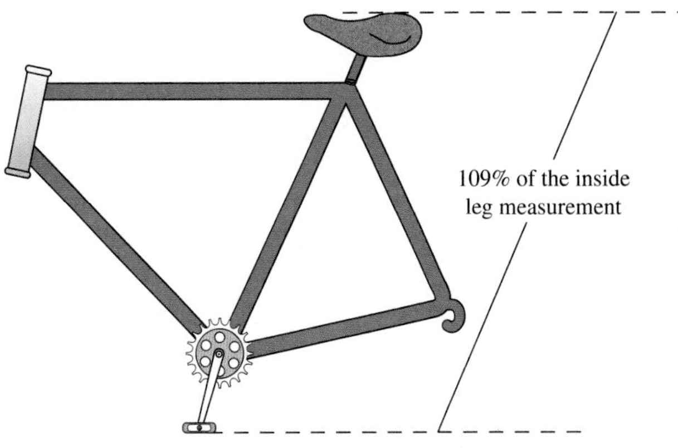

FIGURE FOR EXERCISE 92

Given that the area of each figure is 24 square feet, use the dimensions shown to write an equation expressing this fact. Do not solve the equation.

93.
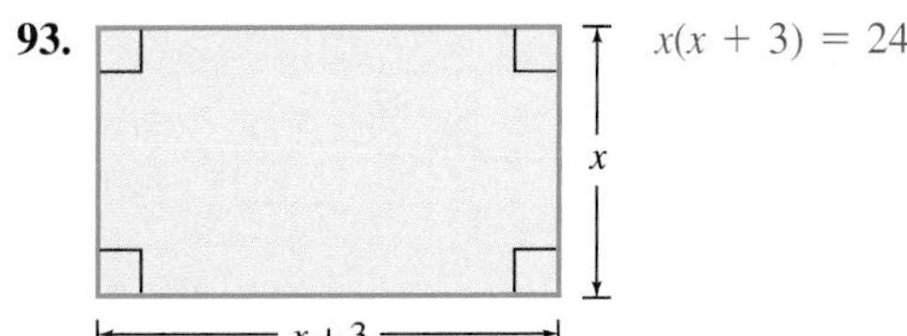

$x(x + 3) = 24$

94.
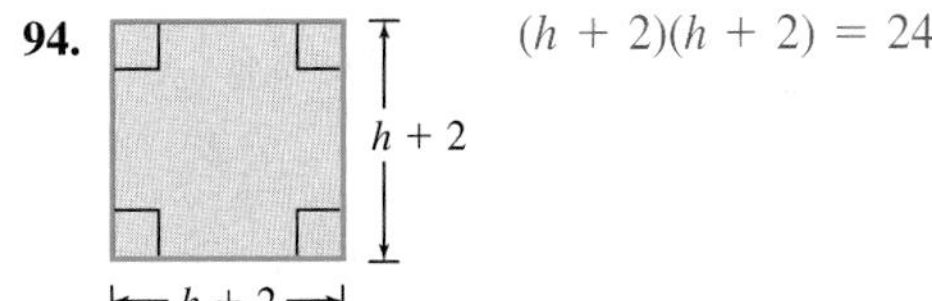

$(h + 2)(h + 2) = 24$

95.
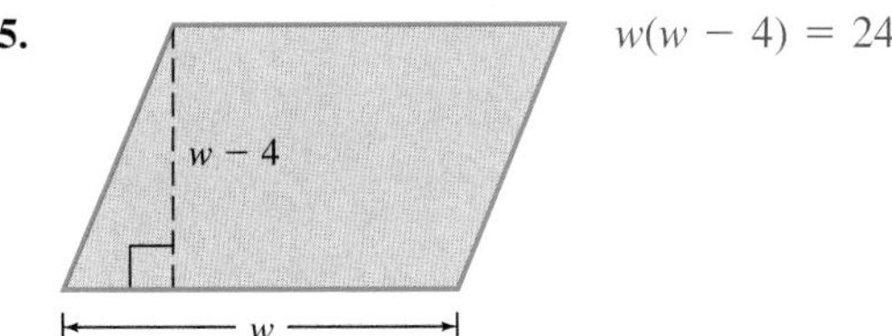

$w(w - 4) = 24$

96.
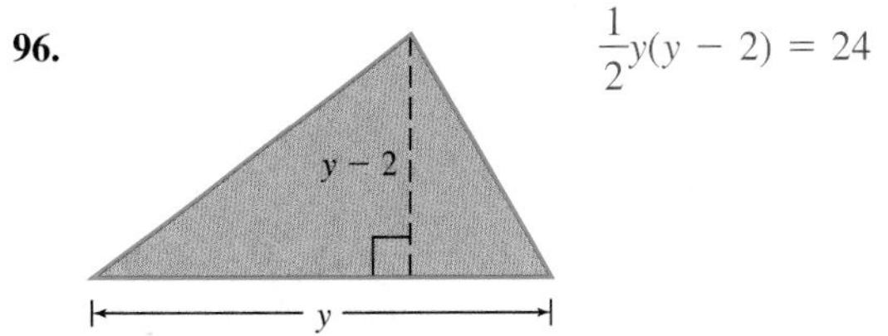

$\frac{1}{2}y(y - 2) = 24$

2.6 NUMBER, GEOMETRIC, AND UNIFORM MOTION APPLICATIONS

In This Section

- Number Problems
- General Strategy for Solving Verbal Problems
- Geometric Problems
- Unit Conversion
- Uniform Motion Problems

In this section, we apply the ideas of Section 2.5 to solving problems. Many of the problems can be solved by using arithmetic only and not algebra. However, remember that we are not just trying to find the answer, we are trying to learn how to apply algebra. So even if the answer is obvious to you, set the problem up and solve it by using algebra as shown in the examples.

Number Problems

Algebra is often applied to problems involving time, rate, distance, interest, or discount. **Number problems** do not involve any physical situation. In number problems we simply find some numbers that satisfy some given conditions. Number problems can provide good practice for solving more complex problems.

EXAMPLE 1 **A consecutive integer problem**

The sum of three consecutive integers is 48. Find the integers.

Solution

If x represents the smallest of the three consecutive integers, then x, $x + 1$, and $x + 2$ represent the three consecutive integers. Since the sum of x, $x + 1$, and $x + 2$ is 48, we write that fact as an equation and solve it:

$$x + (x + 1) + (x + 2) = 48$$

$$3x + 3 = 48 \quad \text{Combine like terms.}$$

$$3x = 45 \quad \text{Subtract 3 from each side.}$$

$$x = 15 \quad \text{Divide each side by 3.}$$

$$x + 1 = 16 \quad \text{If } x \text{ is 15, then } x + 1 \text{ is 16 and } x + 2 \text{ is 17.}$$

$$x + 2 = 17$$

Because $15 + 16 + 17 = 48$, the three consecutive integers that have a sum of 48 are 15, 16, and 17. ■

Helpful Hint

Making a guess can be a good way to get familiar with the problem. For example, let's guess that the answers to Example 1 are 20, 21, and 22. Since $20 + 21 + 22 = 63$, these are not the correct numbers. But now we realize that we should use $x, x + 1$, and $x + 2$ and that the equation should be

$$x + x + 1 + x + 2 = 48.$$

General Strategy for Solving Verbal Problems

You should use the following steps as a guide for solving problems.

Strategy for Solving Problems

1. Read the problem as many times as necessary. Guessing the answer and checking it will help you understand the problem.
2. If possible, draw a diagram to illustrate the problem.
3. Choose a variable and *write* what it represents.
4. Write algebraic expressions for any other unknowns in terms of that variable.
5. Write an equation that describes the situation.
6. Solve the equation.
7. Answer the original question.
8. Check your answer in the original problem (not the equation).

Geometric Problems

Geometric problems involve geometric figures. For these problems you should always draw the figure and label it.

EXAMPLE 2 **A perimeter problem**

The length of a rectangular piece of property is 1 foot less than twice the width. If the perimeter is 748 feet, find the length and width.

Helpful Hint

To get familiar with the problem, guess that the width is 50 ft. Then the length is $2 \cdot 50 - 1$ or 99. The perimeter would be

$$2(50) + 2(99) = 298,$$

which is too small. But now we realize that we should let x be the width, $2x - 1$ be the length, and we should solve

$$2x + 2(2x - 1) = 748.$$

FIGURE 2.2

Solution

Let $x =$ the width. Since the length is 1 foot less than twice the width, $2x - 1 =$ the length. Draw a diagram as in Fig. 2.2. We know that $2L + 2W = P$ is the formula for perimeter of a rectangle. Substituting $2x - 1$ for L and x for W in this formula yields an equation in x:

$$2L + 2W = P$$

$$2(2x - 1) + 2(x) = 748 \quad \text{Replace } L \text{ by } 2x - 1 \text{ and } W \text{ by } x.$$

$$4x - 2 + 2x = 748 \quad \text{Remove the parentheses.}$$

$$6x - 2 = 748 \quad \text{Combine like terms.}$$

$$6x = 750 \quad \text{Add 2 to each side.}$$

$$x = 125 \quad \text{Divide each side by 6.}$$

$$2x - 1 = 249 \quad \text{If } x = 125, \text{ then } 2x - 1 = 2(125) - 1 = 249.$$

Check these answers by computing $2L + 2W$:

$$2(249) + 2(125) = 748$$

So the width is 125 feet, and the length is 249 feet. ■

The next geometric example involves the degree measures of angles. For this problem, the figure is given.

EXAMPLE 3 **Complementary angles**

In Fig. 2.3 the angle formed by the guy wire and the ground is 3.5 times as large as the angle formed by the guy wire and the antenna. Find the degree measure of each of these angles.

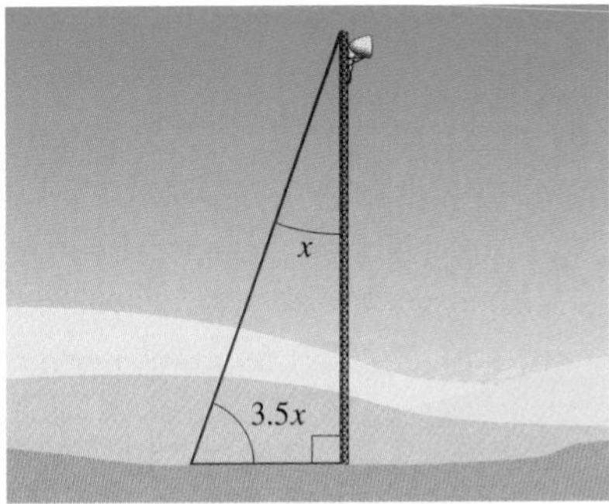

FIGURE 2.3

Solution

Let $x =$ the degree measure of the smaller angle, and let $3.5x =$ the degree measure of the larger angle. Since the antenna meets the ground at a 90° angle, the sum of the degree measures of the other two angles of the right triangle is 90°. (They are complementary angles.) So we have the following equation:

$$x + 3.5x = 90$$
$$4.5x = 90 \quad \text{Combine like terms.}$$
$$x = 20 \quad \text{Divide each side by 4.5.}$$
$$3.5x = 70 \quad \text{Find the other angle.}$$

Check: 70° is $3.5 \cdot 20°$ and $20° + 70° = 90°$. So the smaller angle is 20°, and the larger angle is 70°. ■

Unit Conversion

Most measurements can be expressed in a variety of units. For example, distance could be in miles or kilometers. Converting from one unit of measurement to another can always be done by multiplying by a conversion factor expressed as a fraction. (Some common conversion factors can be found on the inside back cover of this text.) This method is called **cancellation of units,** because the units cancel just like the common factors cancel in multiplication of fractions.

EXAMPLE 4 **Unit conversion**

a) Convert 6 yards to feet.

b) Convert 12 miles to kilometers.

c) Convert 60 miles per hour to feet per second.

Solution

a) Because 3 feet = 1 yard, multiplying by $\frac{3 \text{ feet}}{1 \text{ yard}}$ is equivalent to multiplying by 1.

Notice how yards cancels and the result is feet.

$$6 \text{ yd} = 6 \cancel{\text{yd}} \cdot \frac{3 \text{ ft}}{1 \cancel{\text{yd}}} = 18 \text{ ft}$$

b) There are two ways to convert 12 miles to kilometers using the conversion factors given on the inside back cover:

$$12 \text{ mi} = 12 \cancel{\text{mi}} \cdot \frac{1.609 \text{ km}}{1 \cancel{\text{mi}}} \approx 19.31 \text{ km}$$

$$12 \text{ mi} = 12 \cancel{\text{mi}} \cdot \frac{1 \text{ km}}{0.6215 \cancel{\text{mi}}} \approx 19.31 \text{ km}$$

Notice that in the second method we are also multiplying by a fraction that is equivalent to 1, but we actually divide 12 by 0.6215.

c) Convert 60 miles per hour to feet per second as follows:

$$60 \text{ mi/hr} = \frac{60 \cancel{\text{mi}}}{1 \cancel{\text{hr}}} \cdot \frac{5280 \text{ ft}}{1 \cancel{\text{mi}}} \cdot \frac{1 \cancel{\text{hr}}}{60 \cancel{\text{min}}} \cdot \frac{1 \cancel{\text{min}}}{60 \text{ sec}} = 88 \text{ ft/sec}$$

■

Uniform Motion Problems

Problems involving motion at a constant rate are called **uniform motion problems.** In uniform motion problems we often use an average rate when the actual rate is not

Solve each problem.

39. ***Super Bowl score.*** The 1977 Super Bowl was played in the Rose Bowl in Pasadena. In that football game the Oakland Raiders scored 18 more points than the Minnesota Vikings. If the total number of points scored was 46, then what was the final score for the game?
Raiders 32, Vikings 14

40. ***Top three companies.*** Revenues for the top three companies in 2001, Exxon Mobil, Wal-Mart Stores, and General Motors totaled \$588 billion (Fortune, www.fortune.com). If Exxon Mobil's revenue was \$17 billion greater than Wal-Mart and Wal-Mart's revenue was \$8 billion greater than General Motors, then what was the 2001 revenue for each company?
Exxon \$210 billion, Wal-Mart \$193 billion, GM \$185 billion

41. ***Idabel to Lawton.*** Before lunch, Sally drove from Idabel to Ardmore, averaging 50 mph. After lunch she continued on to Lawton, averaging 53 mph. If her driving time after lunch was 1 hour less than her driving time before lunch and the total trip was 256 miles, then how many hours did she drive before lunch? How far is it from Ardmore to Lawton? 3 hours, 106 miles

42. ***Norfolk to Chadron.*** On Monday, Chuck drove from Norfolk to Valentine, averaging 47 mph. On Tuesday, he continued on to Chadron, averaging 69 mph. His driving time on Monday was 2 hours longer than his driving time on Tuesday. If the total distance from Norfolk to Chadron is 326 miles, then how many hours did he drive on Monday? How far is it from Valentine to Chadron?
4 hours, 138 miles

43. ***Golden oldies.*** Joan Crawford, John Wayne, and James Stewart were born in consecutive years (*Doubleday Almanac*). Joan Crawford was the oldest of the three, and James Stewart was the youngest. In 1950, after all three had their birthdays, the sum of their ages was 129. In what years were they born?
Crawford 1906, Wayne 1907, Stewart 1908

44. ***Leading men.*** Bob Hope was born 2 years after Clark Gable and 2 years before Henry Fonda (*Doubleday Almanac*). In 1951, after all three of them had their birthdays, the sum of their ages was 144. In what years were they born? Hope 1903, Gable 1901, Fonda 1905

45. ***Trimming a garage door.*** A carpenter used 30 ft of molding in three pieces to trim a garage door. If the long piece was 2 ft longer than twice the length of each shorter piece, then how long was each piece? 7 ft, 7 ft, 16 ft

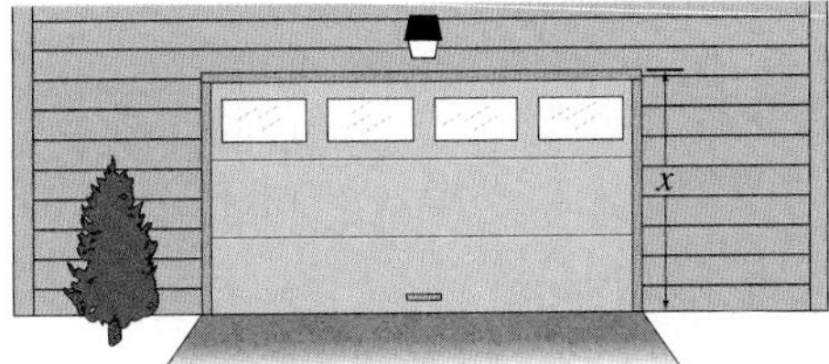

FIGURE FOR EXERCISE 45

46. ***Fencing dog pens.*** Clint is constructing two adjacent rectangular dog pens. Each pen will be three times as long as it is wide, and the pens will share a common long side. If Clint has 65 ft of fencing, what are the dimensions of each pen? 5 ft by 15 ft

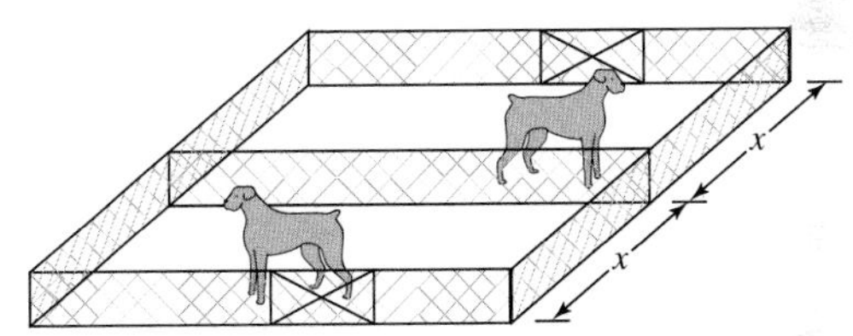

FIGURE FOR EXERCISE 46

2.7 DISCOUNT, INVESTMENT, AND MIXTURE APPLICATIONS

In This Section

- Discount Problems
- Commission Problems
- Investment Problems
- Mixture Problems

In this section, we continue our study of applications of algebra. The problems in this section involve percents.

Discount Problems

When an item is sold at a discount, the amount of the discount is usually described as being a percentage of the original price. The percentage is called the **rate of discount.** Multiplying the rate of discount and the original price gives the amount of the discount.

EXAMPLE 1 **Finding the original price**

Ralph got a 12% discount when he bought his new 2002 Corvette Coupe. If the amount of his discount was \$6606, then what was the original price of the Corvette?

Solution

Let x represent the original price. The discount is found by multiplying the 12% rate of discount and the original price:

$$\text{Rate of discount} \cdot \text{original price} = \text{amount of discount}$$

$$0.12x = 6606$$

$$x = \frac{6606}{0.12} \quad \text{Divide each side by 0.12.}$$

$$x = 55{,}050$$

To check, find 12% of \$55,050. Since $0.12 \cdot 55{,}050 = 6606$, the original price of the Corvette was \$55,050. ■

EXAMPLE 2 **Finding the original price**

When Susan bought her new car, she also got a discount of 12%. She paid \$17,600 for her car. What was the original price of Susan's car?

Helpful Hint

To get familiar with the problem, guess that the original price was \$30,000. Then her discount is 0.12(30,000) or \$3600. The price she paid would be 30,000 − 3600 or \$26,400, which is incorrect.

Solution

Let x represent the original price for Susan's car. The amount of discount is 12% of x, or $0.12x$. We can write an equation expressing the fact that the original price minus the discount is the price Susan paid.

$$\text{Original price} - \text{discount} = \text{sale price}$$

$$x - 0.12x = 17{,}600$$

$$0.88x = 17{,}600 \quad 1.00x - 0.12x = 0.88x$$

$$x = \frac{17{,}600}{0.88} \quad \text{Divide each side by 0.88.}$$

$$x = \$20{,}000$$

Check: 12% of \$20,000 is \$2400, and \$20,000 − \$2400 = \$17,600. The original price of Susan's car was \$20,000. ■

Commission Problems

A salesperson's commission for making a sale is often a percentage of the selling price. **Commission problems** are very similar to other problems involving percents. The commission is found by multiplying the rate of commission and the selling price.

EXAMPLE 3 **Real estate commission**

Sarah is selling her house through a real estate agent whose commission rate is 7%. What should the selling price be so that Sarah can get the \$83,700 she needs to pay off the mortgage?

Solution

Let x be the selling price. The commission is 7% of x (not 7% of \$83,700). Sarah receives the selling price less the sales commission:

$$\text{Selling price} - \text{commission} = \text{Sarah's share}$$

$$x - 0.07x = 83{,}700$$

$$0.93x = 83{,}700 \quad 1.00x - 0.07x = 0.93x$$

$$x = \frac{83{,}700}{0.93}$$

$$x = 90{,}000$$

Check: 7% of \$90,000 is \$6300, and \$90,000 − \$6300 = \$83,700. So the house should sell for \$90,000. ■

Investment Problems

The interest on an investment is a percentage of the investment, just as the sales commission is a percentage of the sale amount. However, in **investment problems** we must often account for more than one investment at different rates. So it is a good idea to make a table, as in Example 4.

EXAMPLE 4 **Diversified investing**

Ruth Ann invested some money in a certificate of deposit with an annual yield of 9%. She invested twice as much in a mutual fund with an annual yield of 10%. Her interest from the two investments at the end of the year was \$232. How much was invested at each rate?

Helpful Hint

To get familiar with the problem, guess that she invested \$1000 at 9% and \$2000 at 10%. Then her interest in one year would be

$$0.09(1000) + 0.10(2000)$$

or \$290, which is close but incorrect.

Solution

When there are many unknown quantities, it is often helpful to identify them in a table. Since the time is 1 year, the amount of interest is the product of the interest rate and the amount invested.

	Interest rate	Amount invested	Interest for 1 year
CD	9%	x	$0.09x$
Mutual fund	10%	$2x$	$0.10(2x)$

Since the total interest from the investments was \$232, we can write the following equation:

$$\begin{aligned} \text{CD interest} + \text{mutual fund interest} &= \text{total interest} \\ 0.09x + 0.10(2x) &= 232 \\ 0.09x + 0.20x &= 232 \\ 0.29x &= 232 \\ x &= \frac{232}{0.29} \\ x &= \$800 \\ 2x &= \$1600 \end{aligned}$$

To check, we find the total interest:

$$\begin{aligned} 0.09(800) + 0.10(1600) &= 72 + 160 \\ &= 232 \end{aligned}$$

So Ruth Ann invested \$800 at 9% and \$1600 at 10%. ■

Study Tip

Finding out what happened in class and attending class are not the same. Attend every class and use class as a learning time. Take notes, ask questions, and make sure that you are learning in the classroom.

Mixture Problems

Mixture problems are concerned with the result of mixing two quantities, each of which contains another substance. Notice how similar the following mixture problem is to the last investment problem.

EXAMPLE 5 **Mixing milk**

How many gallons of milk containing 4% butterfat must be mixed with 80 gallons of 1% milk to obtain 2% milk?

Helpful Hint

To get familiar with the problem, guess that we need 100 gal of 4% milk. Mixing that with 80 gal of 1% milk would produce 180 gal of 2% milk. Now the two milks separately have

$$0.04(100) + 0.01(80)$$

or 4.8 gal of fat. Together the amount of fat is 0.02(180) or 3.6 gal. Since these amounts are not equal, our guess is incorrect.

Solution

It is helpful to draw a diagram and then make a table to classify the given information.

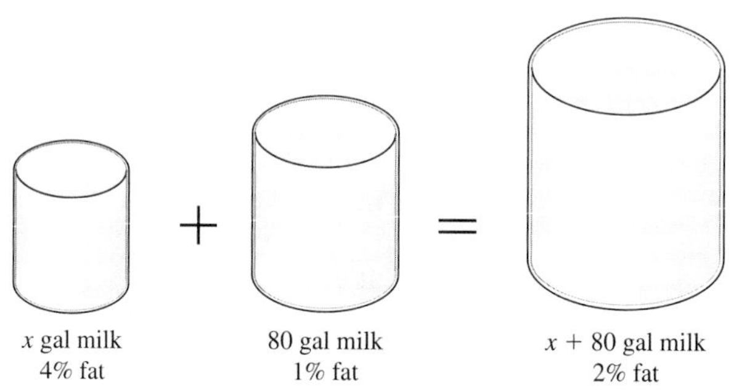

	Percentage of fat	Amount of milk	Amount of fat
4% milk	4%	x	$0.04x$
1% milk	1%	80	$0.01(80)$
2% milk	2%	$x + 80$	$0.02(x + 80)$

The equation expresses the fact that the total fat from the first two types of milk is the same as the fat in the mixture:

$$\text{Fat in 4\% milk} + \text{fat in 1\% milk} = \text{fat in 2\% milk}$$

$$0.04x + 0.01(80) = 0.02(x + 80)$$

$$0.04x + 0.8 = 0.02x + 1.6 \quad \text{Simplify.}$$

$$100(0.04x + 0.8) = 100(0.02x + 1.6) \quad \text{Multiply each side by 100.}$$

$$4x + 80 = 2x + 160 \quad \text{Distributive property.}$$

$$2x + 80 = 160 \quad \text{Subtract } 2x \text{ from each side.}$$

$$2x = 80 \quad \text{Subtract 80 from each side.}$$

$$x = 40 \quad \text{Divide each side by 2.}$$

To check, calculate the total fat:

$$2\% \text{ of 120 gallons} = 0.02(120) = 2.4 \text{ gallons of fat}$$

$$0.04(40) + 0.01(80) = 1.6 + 0.8 = 2.4 \text{ gallons of fat}$$

So we mix 40 gallons of 4% milk with 80 gallons of 1% milk to get 120 gallons of 2% milk. ■

Study Tip

Don't expect to understand a new topic the first time that you see it. Learning mathematics takes time, patience, and repetition. Keep reading the text, asking questions, and working problems. Someone once said, "All mathematics is easy once you understand it."

In mixture problems, the solutions might contain fat, alcohol, salt, or some other substance. We always assume that the substance neither appears nor disappears in the process. For example, if there are 3 grams of salt in one glass of water and 2 grams in another, then there are exactly 5 grams in a mixture of the two.

WARM-UPS

True or false? Explain your answer.

1. If Jim gets a 12% commission for selling a \$1000 Wonder Vac, then his commission is \$120. True
2. If Bob earns a 5% commission on an \$80,000 motorhome sale, then Bob earns \$400. False
3. If Sue gets a 20% discount on a TV with a list price of x dollars, then Sue pays $0.8x$ dollars. True
4. If you get a 6% discount on a car that has an MSRP of x dollars, then your discount is $0.6x$ dollars. False
5. If the original price is w and the discount is 8%, then the selling price is $w - 0.08w$. True
6. If x is the selling price and the commission is 8% of the selling price, then the commission is $0.08x$. True
7. If you need \$40,000 for your house and the agent gets 10% of the selling price, then the agent gets \$4000, and the house sells for \$44,000. False
8. If you mix 10 liters of a 20% acid solution with x liters of a 30% acid solution, then the total amount of acid is $2 + 0.3x$ liters. True
9. A 10% acid solution mixed with a 14% acid solution results in a 24% acid solution. False
10. If a TV costs x dollars and sales tax is 5%, then the total bill is $1.05x$ dollars. True

2.7 EXERCISES

Reading and Writing *After reading this section, write out the answers to these questions. Use complete sentences.*

1. What types of problems are discussed in this section? We studied discount, investment, and mixture problems in this section.
2. What is the difference between discount and rate of discount? The rate of discount is a percentage and the discount is the actual amount that the price is reduced.
3. What is the relationship between discount, original price, rate of discount, and sale price? The product of the rate and the original price gives the amount of discount. The original price minus the discount is the sale price.
4. What do mixture problems and investment problems have in common? Both mixture problems and investment problems involve rates.
5. Why do we make a table when solving certain problems. A table helps us to organize the information given in a problem.
6. What is the relationship between amount of interest, amount invested, and interest rate? The product of the interest rate and the amount invested gives the amount of interest.

Show a complete solution to each problem. See Examples 1 and 2.

7. ***Close-out sale.*** At a 25% off sale, Jose saved \$80 on a 19-inch Panasonic TV. What was the original price of the television. \$320
8. ***Nice tent.*** A 12% discount on a Walrus tent saved Melanie \$75. What was the original price of the tent? \$625
9. ***Circuit city.*** After getting a 20% discount, Robert paid \$320 for a Pioneer CD player for his car. What was the original price of the CD player? \$400
10. ***Chrysler Sebring.*** After getting a 15% discount on the price of a new Chrysler Sebring convertible, Helen paid \$27,000. What was the original price of the convertible? \$31,765

Show a complete solution to each problem. See Example 3.

11. ***Selling price of a home.*** Kirk wants to get $115,000 for his house. The real estate agent gets a commission equal to 8% of the selling price for selling the house. What should the selling price be? $125,000

FIGURE FOR EXERCISE 11

12. ***Horse trading.*** Gene is selling his palomino at an auction. The auctioneer's commission is 10% of the selling price. If Gene still owes $810 on the horse, then what must the horse sell for so that Gene can pay off his loan? $900

13. ***Sales tax collection.*** Merilee sells tomatoes at a roadside stand. Her total receipts including the 7% sales tax were $462.24. What amount of sales tax did she collect? $30.24

14. ***Toyota Corolla.*** Gwen bought a new Toyota Corolla. The selling price plus the 8% state sales tax was $15,714. What was the selling price? $14,550

Show a complete solution to each problem. See Example 4.

15. ***Wise investments.*** Wiley invested some money in the Berger 100 Fund and $3000 more than that amount in the Berger 101 Fund. For the year he was in the fund, the 100 Fund paid 18% simple interest and the 101 Fund paid 15% simple interest. If the income from the two investments totaled $3750 for one year, then how much did he invest in each fund? 100 Fund $10,000, 101 Fund $13,000

16. ***Loan shark.*** Becky lent her brother some money at 8% simple interest, and she lent her sister twice as much at twice the interest rate. If she received a total of 20 cents interest, then how much did she lend to each of them? Brother $0.50, sister $1.00

17. ***Investing in bonds.*** David split his $25,000 inheritance between Fidelity Short-Term Bond Fund with an annual yield of 5% and T. Rowe Price Tax-Free Short-Intermediate Fund with an annual yield of 4%. If his total income for one year on the two investments was $1140, then how much did he invest in each fund? Fidelity $14,000, Price $11,000

18. ***High-risk funds.*** Of the $50,000 that Natasha pocketed on her last real estate deal, $20,000 went to charity. She invested part of the remainder in Dreyfus New Leaders Fund with an annual yield of 16% and the rest in Templeton Growth Fund with an annual yield of 25%. If she made $6060 on these investments in one year, then how much did she invest in each fund? Dreyfus $16,000, Templeton $14,000

Show a complete solution to each problem. See Example 5.

19. ***Mixing milk.*** How many gallons of milk containing 1% butterfat must be mixed with 30 gallons of milk containing 3% butterfat to obtain a mixture containing 2% butterfat? 30 gallons

x gal 1% fat + 30 gal 3% fat = $x + 30$ gal 2% fat

FIGURE FOR EXERCISE 19

20. ***Acid solutions.*** How many gallons of a 5% acid solution should be mixed with 30 gallons of a 10% acid solution to obtain a mixture that is 8% acid? 20 gallons

21. ***Alcohol solutions.*** Gus has on hand a 5% alcohol solution and a 20% alcohol solution. He needs 30 liters of a 10% alcohol solution. How many liters of each solution should he mix together to obtain the 30 liters? 20 liters of 5% alcohol, 10 liters of 20% alcohol

22. ***Adjusting antifreeze.*** Angela needs 20 quarts of 50% antifreeze solution in her radiator. She plans to obtain this by mixing some pure antifreeze with an appropriate amount of a 40% antifreeze solution. How many quarts of each should she use? $\frac{10}{3}$ quarts of pure antifreeze, $\frac{50}{3}$ quarts of 40% solution

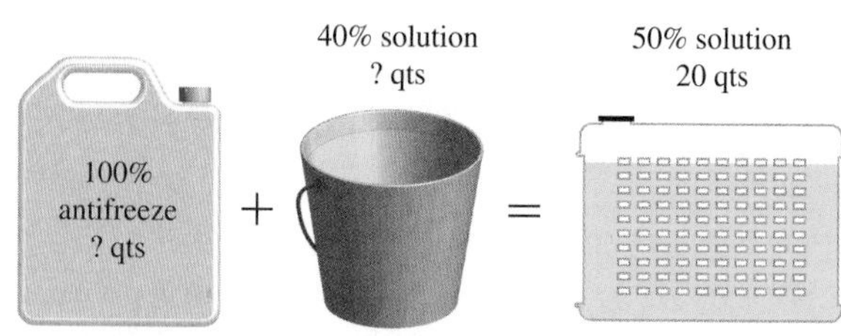

FIGURE FOR EXERCISE 22

b) The length of a certain rectangle must be 4 meters longer than the width, and the perimeter must be at least 120 meters.

c) Fred made a 76 on the midterm exam. To get a B, the average of his mid-term and his final exam must be between 80 and 90.

Study Tip

No two students learn in the same way or at the same speed. No one can tell you exactly how to study and learn. Learning is personal. You must discover what it takes for you to learn mathematics and then do whatever it takes.

Solution

a) If x is the price of the washing machine, then $0.09x$ is the amount of sales tax. Since the total must be less than or equal to \$500, the inequality is

$$x + 0.09x \leq 500.$$

b) If W represents the width of the rectangle, then $W + 4$ represents the length. Since the perimeter $(2W + 2L)$ must be greater than or equal to 120, the inequality is

$$2(W) + 2(W + 4) \geq 120.$$

c) If we let x represent Fred's final exam score, then his average is $\frac{x + 76}{2}$. To indicate that the average is between 80 and 90, we use the compound inequality

$$80 < \frac{x + 76}{2} < 90.$$

■

CAUTION In Example 4(b) you are given that L is 4 meters longer than W. So $L = W + 4$, and you can use $W + 4$ in place of L. If you knew only that L was longer than W, then you would know only that $L > W$.

WARM-UPS

True or false? Explain your answer.

1. $-2 \leq -2$ True **2.** $-5 < 4 < 6$ True **3.** $-3 < 0 < -1$ False

4. The inequalities $7 < x$ and $x > 7$ have the same graph. True

5. The graph of $x < -3$ includes the point at -3. False

6. The number 5 satisfies the inequality $x > 2$. True

7. The number -3 is a solution to $-2 < x$. False

8. The number 4 satisfies the inequality $2x - 1 < 4$. False

9. The number 0 is a solution to the inequality $2x - 3 \leq 5x - 3$. True

10. The inequalities $2x - 1 < x$ and $x < 2x - 1$ have the same solutions. False

2.8 EXERCISES

Reading and Writing *After reading this section, write out the answers to these questions. Use complete sentences.*

1. What are the inequality symbols used in this section?
The inequality symbols are $<$, $\leq$, $>$, and $\geq$.

2. What different looking inequality means the same as $a < b$?
The inequalities $a < b$ and $b > a$ have the same meaning.

3. How do you know when to use an open circle and when to use a solid circle when graphing an inequality on a number line?
For $\leq$ and $\geq$ use the solid circle and for $<$ and $>$ use the open circle.

4. What is a compound inequality?
A compound inequality is a statement involving more than one inequality.

5. What is the meaning of the compound inequality $a < b < c$?
The compound inequality $a < b < c$ means $b > a$ and $b < c$, or b is between a and c.

6. What is the difference between "at most" and "at least?"
"At most" means less than or equal to and "at least" means greater than or equal to.

Determine whether each of the following statements is correct. See Example 1.

7. $-3 < 5$ True

8. $-6 < 0$ True

9. $4 \le 4$ True

10. $-3 \ge -3$ True

11. $-6 > -5$ False

12. $-2 < -9$ False

13. $-4 \le -3$ True

14. $-5 \ge -10$ True

15. $(-3)(4) - 1 < 0 - 3$ True

16. $2(4) - 6 \le -3(5) + 1$ False

17. $-4(5) - 6 \ge 5(-6)$ True

18. $4(8) - 30 > 7(5) - 2(17)$ True

19. $7(4) - 12 \le 3(9) - 2$ True

20. $-3(4) + 12 \le 2(3) - 6$ True

Sketch the graph of each inequality on the number line. See Example 2.

21. $x \le 3$
−1 0 1 2 3 4 5

22. $x \le -7$ 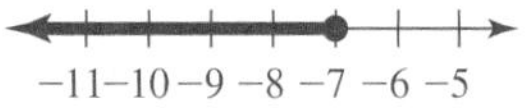
−11 −10 −9 −8 −7 −6 −5

23. $x > -2$
−4 −3 −2 −1 0 1 2

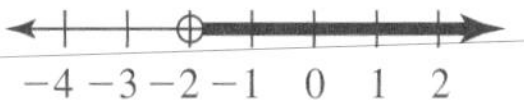

24. $x > 4$
2 3 4 5 6 7 8

25. $-1 > x$
−5 −4 −3 −2 −1 0 1

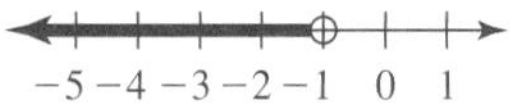

26. $0 > x$
−4 −3 −2 −1 0 1 2

27. $-2 \le x$
−4 −3 −2 −1 0 1 2

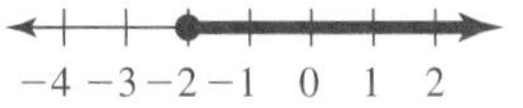

28. $-5 \ge x$
−9 −8 −7 −6 −5 −4 −3

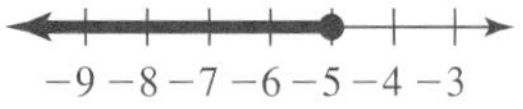

29. $x \ge \frac{1}{2}$
$\frac{1}{2}$
−1 0 1 2 3

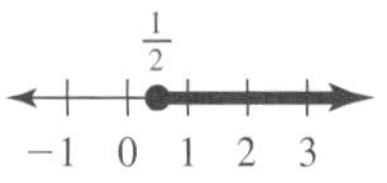

30. $x \ge -\frac{2}{3}$
$-\frac{2}{3}$
−3 −2 −1 0 1 2

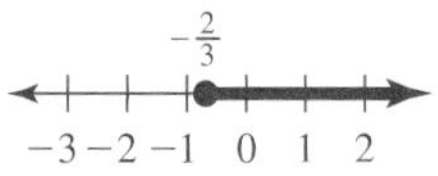

31. $x \le 5.3$
5.3
1 2 3 4 5 6 7

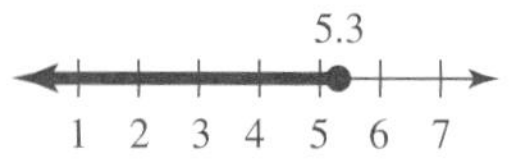

32. $x \le -3.4$
−3.4
−7 −6 −5 −4 −3 −2

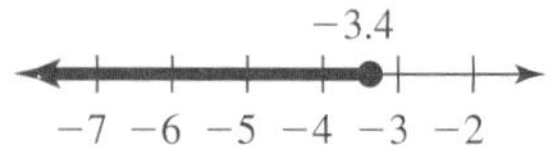

33. $-3 < x < 1$
−4 −3 −2 −1 0 1 2

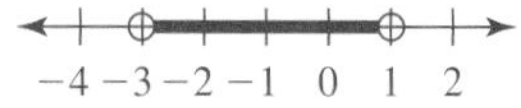

34. $0 < x < 5$
0 1 2 3 4 5

35. $3 \le x \le 7$
2 3 4 5 6 7 8

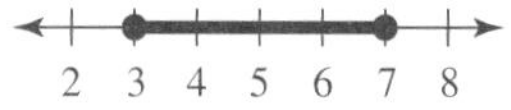

36. $-3 \le x \le -1$
−5 −4 −3 −2 −1 0 1

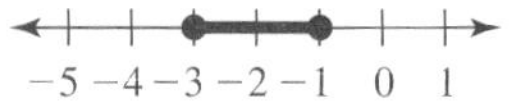

37. $-5 \le x < 0$
−5 −4 −3 −2 −1 0 1

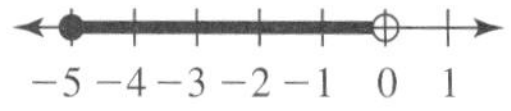

38. $-2 < x \le 2$
−3 −2 −1 0 1 2 3

39. $40 < x \le 100$
20 40 60 80 100 120

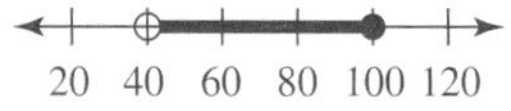

40. $0 \le x < 600$
−200 0 200 400 600 800

For each graph, write an inequality that describes the graph.

41. −3 −2 −1 0 1 2 3 4 5 6 7 — $x > 3$

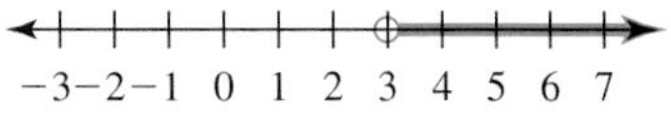

42. −4 −3 −2 −1 0 1 2 3 4 5 6 — $x \le 4$

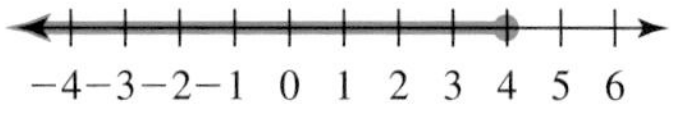

43. −6 −5 −4 −3 −2 −1 0 1 2 3 4 — $x \le 2$

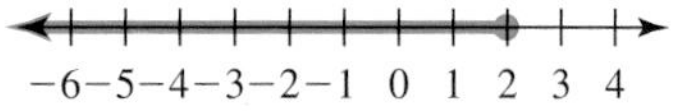

44. −5 −4 −3 −2 −1 0 1 2 3 4 5 — $0 < x \le 3$

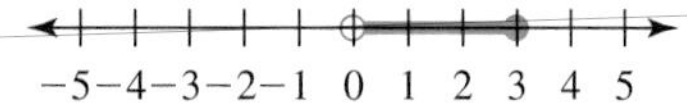

45. −5 −4 −3 −2 −1 0 1 2 3 4 5 — $0 < x < 2$

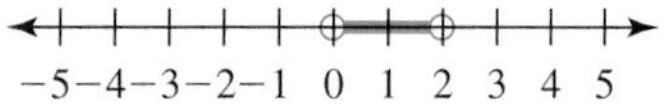

46. −5 −4 −3 −2 −1 0 1 2 3 4 5 — $-1 \le x < 3$

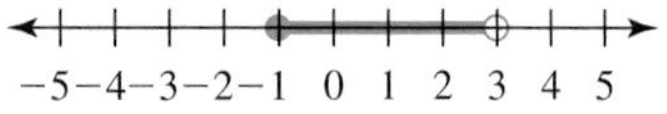

47. −6 −4 −2 0 2 4 6 8 — $-5 < x \le 7$

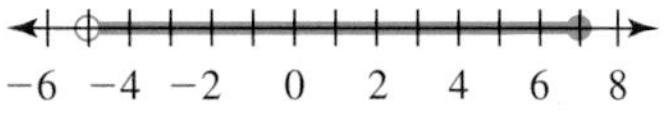

48. −5 −4 −3 −2 −1 0 1 2 3 4 5 — $x < 4$

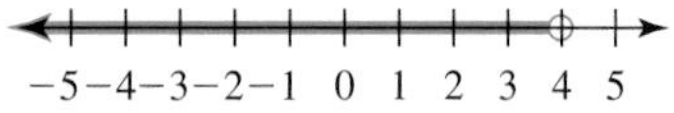

49. −5 −4 −3 −2 −1 0 1 2 3 4 5 — $x > -4$

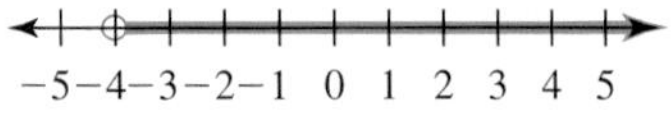

50. −5 −4 −3 −2 −1 0 1 2 3 4 5 — $0 < x \le 2$

Determine whether the given number satisfies the inequality following it. See Example 3.

51. -9, $-x > 3$ Yes

52. 5, $-3 < -x$ No

53. -2, $5 \le x$ No

54. $4, 4 \ge x$ Yes

55. $-6, 2x - 3 > -11$ No

56. $4, 3x - 5 < 7$ No

57. $3, -3x + 4 > -7$ Yes

58. $-4, -5x + 1 > -5$ Yes

59. $0, 3x - 7 \le 5x - 7$ Yes

60. $0, 2x + 6 \ge 4x - 9$ Yes

61. $2.5, -10x + 9 \le 3(x + 3)$ Yes

62. $1.5, 2x - 3 \le 4(x - 1)$ Yes

63. $-7, -5 < x < 9$ No

64. $-9, -6 \le x \le 40$ No

65. $-2, -3 \le 2x + 5 \le 9$ Yes

66. $-5, -3 < -3x - 7 \le 8$ Yes

67. $-3.4, -4.25x - 13.29 < 0.89$ No

68. $4.8, 3.25x - 14.78 \le 1.3$ Yes

For each inequality, determine which of the numbers -5.1, 0, *and* 5.1 *satisfies the inequality.*

69. $x > -5$ 0, 5.1

70. $x \le 0$ −5.1, 0

71. $5 < x$ 5.1

72. $-5 > x$ −5.1

73. $5 < x < 7$ 5.1

74. $5 < -x < 7$ −5.1

75. $-6 < -x < 6$ −5.1, 0, 5.1

76. $-5 \le x - 0.1 \le 5$ 5.1, 0

Write an inequality to describe each situation. See Example 4.

77. ***Sales tax.*** At an 8% sales tax rate, Susan paid more than \$1500 sales tax when she purchased her new Camaro. Let p represent the price of the Camaro.
$0.08p > 1500$

78. ***Fine dining.*** At Burger Brothers the price of a hamburger is twice the price of an order of French fries, and the price of a Coke is \$0.25 more than the price of the fries. Burger Brothers advertises that you can get a complete meal (burger, fries, and Coke) for under \$2.00. Let p represent the price of an order of fries.
$p + 2p + p + 0.25 < 2.00$

79. ***Barely passing.*** Travis made 44 and 72 on the first two tests in algebra and has one test remaining. The average on the three tests must be at least 60 for Travis to pass the course. Let s represent his score on the last test.
$\frac{44 + 72 + s}{3} \ge 60$

80. ***Ace the course.*** Florence made 87 on her midterm exam in psychology. The average of her midterm and her final must be at least 90 to get an A in the course. Let s represent her score on the final. $\frac{87 + s}{2} \ge 90$

81. ***Coast to coast.*** On Howard's recent trip from Bangor to San Diego, he drove for 8 hours each day and traveled between 396 and 453 miles each day. Let R represent his average speed for each day. $396 < 8R < 453$

82. ***Mother's Day present.*** Bart and Betty are looking at color televisions that range in price from \$399.99 to \$579.99. Bart can afford more than Betty and has agreed to spend \$100 more than Betty when they purchase this gift for their mother. Let b represent Betty's portion of the gift. $399.99 < b + b + 100 < 579.99$

83. ***Positioning a ladder.*** Write an inequality in the variable x for the degree measure of the angle at the base of the ladder shown in the figure, given that the angle at the base must be between 60° and 70°.
$60 < 90 - x < 70$

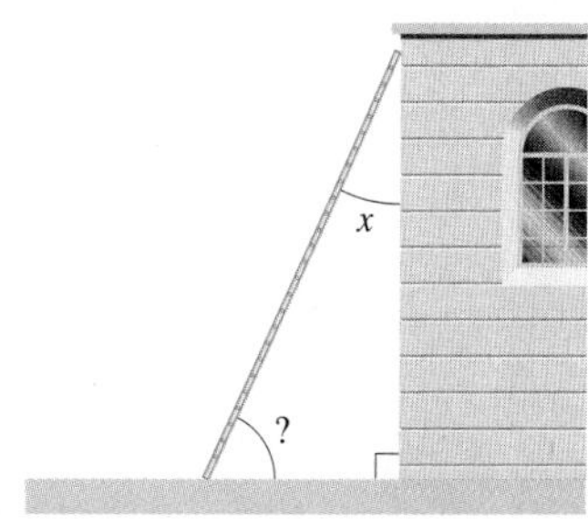

FIGURE FOR EXERCISE 83

84. ***Building a ski ramp.*** Write an inequality in the variable x for the degree measure of the smallest angle of the triangle shown in the figure, given that the degree measure of the smallest angle is at most 30°.
$180 - x - (x + 8) \le 30$

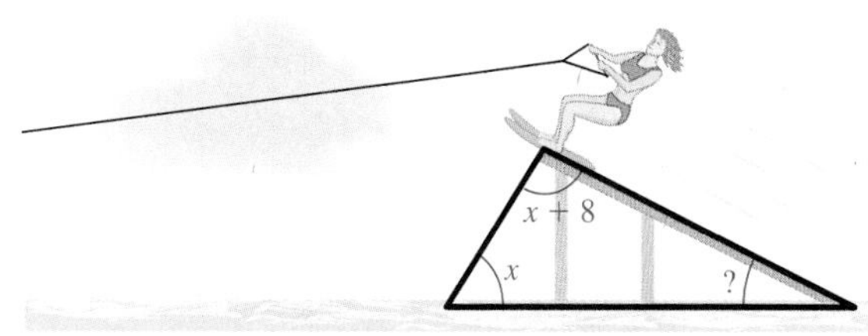

FIGURE FOR EXERCISE 84

85. ***Maximum girth.*** United Parcel Service defines the girth of a box as the sum of the length, twice the width, and twice the height. The maximum girth that UPS will ship is 130 in. If a box has a length of 45 in. and a width of 30 in., then what inequality must be satisfied by the height? $45 + 2(30) + 2h \le 130$

86. ***Batting average.*** Near the end of the season a professional baseball player has 93 hits in 317 times at bat for an average of 93/317 or 0.293. He gets \$1 million bonus if his season average is over 0.300. Let x represent the number of hits in the last 20 at bats of the season. Write the inequality that must be satisfied for him to get the bonus. $\frac{93 + x}{317 + 20} > 0.300$

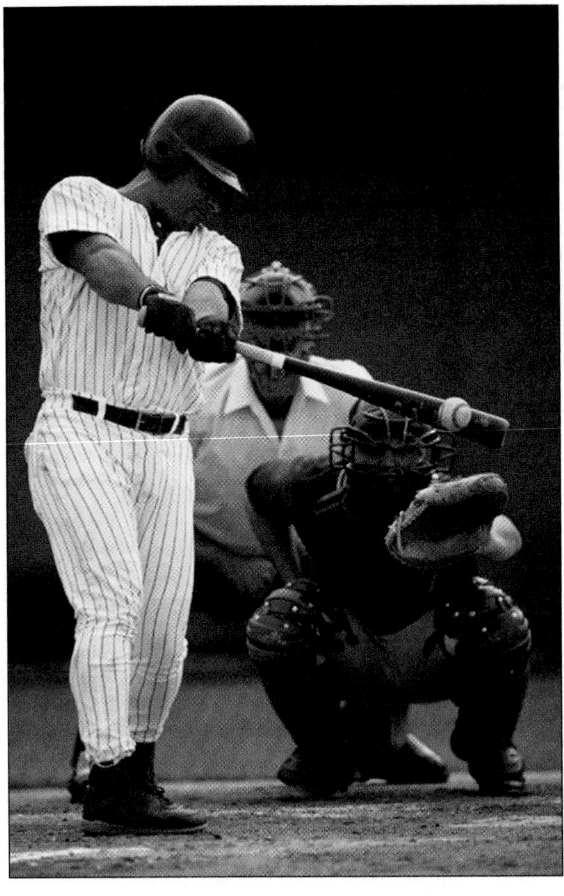

FIGURE FOR EXERCISE 86

Solve.

87. ***Bicycle gear ratios.*** The gear ratio r for a bicycle is defined by the formula

$$r = \frac{Nw}{n},$$

where N is the number of teeth on the chainring (by the pedal), n is the number of teeth on the cog (by the wheel), and w is the wheel diameter in inches (*Cycling*, Burkett and Darst). The following chart gives uses for the various gear ratios.

Ratio	Use
$r > 90$	hard pedaling on level ground
$70 < r \leq 90$	moderate effort on level ground
$50 < r \leq 70$	mild hill climbing
$35 < r \leq 50$	long hill climbing with load

A bicycle with a 27-inch diameter wheel has 50 teeth on the chainring and 17 teeth on the cog. Find the gear ratio and indicate what this gear ratio is good for.
79, moderate effort on level ground

2.9 SOLVING INEQUALITIES AND APPLICATIONS

In This Section

- Rules for Inequalities
- Solving Inequalities
- Applications of Inequalities

To solve equations, we write a sequence of equivalent equations that ends in a very simple equation whose solution is obvious. In this section you will learn that the procedure for solving inequalities is the same. However, the rules for performing operations on each side of an inequality are slightly different from the rules for equations.

Rules for Inequalities

Equivalent inequalities are inequalities that have exactly the same solutions. Inequalities such as $x > 3$ and $x + 2 > 5$ are equivalent because any number that is larger than 3 certainly satisfies $x + 2 > 5$ and any number that satisfies $x + 2 > 5$ must certainly be larger than 3.

We can get equivalent inequalities by performing operations on each side of an inequality just as we do for solving equations. If we start with the inequality $6 < 10$ and add 2 to each side, we get the true statement $8 < 12$. Examine the results of performing the same operation on each side of $6 < 10$.

Helpful Hint

You can think of an inequality like a seesaw that is out of balance.

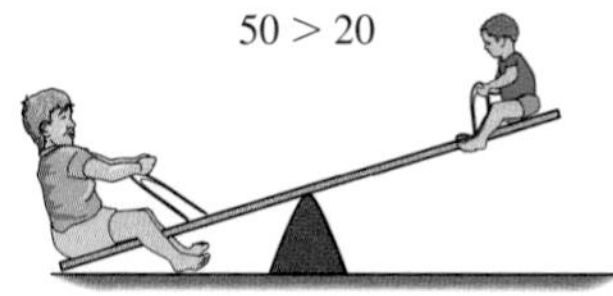

If the same weight is added to or subtracted from each side, it will remain in the same state of imbalance.

Study Tip

Get in the habit of checking your work and having confidence in your answers. The answers to the odd-numbered exercises are in the back of this book, but you should look in the answer section only after you have checked on your own. You will not always have an answer section available.

Helpful Hint

Changing the signs of numbers, changes their relative position on the number line. For example, 3 lies to the left of 5 on the number line, but -3 lies to the right of -5. So $3 < 5$, but $-3 > -5$. Since multiplying and dividing by a negative cause sign changes, these operations reverse the inequality.

Perform these operations on each side:

	Add 2	Subtract 2	Multiply by 2	Divide by 2
Start with $6 < 10$	$8 < 12$	$4 < 8$	$12 < 20$	$3 < 5$

All of the resulting inequalities are correct. Now if we repeat these operations using -2, we get the following results.

Perform these operations on each side:

	Add -2	Subtract -2	Multiply by -2	Divide by -2
Start with $6 < 10$	$4 < 8$	$8 < 12$	$-12 > -20$	$-3 > -5$

Notice that the direction of the inequality symbol is the same for all of the results except the last two. When we multiplied each side by -2 and when we divided each side by -2, we had to reverse the inequality symbol to get a correct result. These tables illustrate the rules for solving inequalities.

Addition Property of Inequality

If we add the same number to each side of an inequality we get an equivalent inequality. If $a < b$, then $a + c < b + c$.

The addition property of inequality also allows us to subtract the same number from each side of an inequality because subtraction is defined in terms of addition.

Multiplication Property of Inequality

If we multiply each side of an inequality by the same *positive* number, we get an equivalent inequality. If $a < b$ and $c > 0$, then $ac < bc$. If we multiply each side of an inequality by the same *negative* number and *reverse the inequality symbol,* we get an equivalent inequality. If $a < b$ and $c < 0$, then $ac > bc$.

The multiplication property of inequality also allows us to divide each side of an inequality by a nonzero number because division is defined in terms of multiplication. So if we multiply or divide each side by a negative number, the inequality symbol is reversed.

EXAMPLE 1 **Writing equivalent inequalities**

Write the appropriate inequality symbol in the blank so that the two inequalities are equivalent.

a) $x + 3 > 9$, x _____ 6 **b)** $-2x \le 6$, x _____ -3

Solution

a) If we subtract 3 from each side of $x + 3 > 9$, we get the equivalent inequality $x > 6$.

b) If we divide each side of $-2x \le 6$ by -2, we get the equivalent inequality $x \ge -3$. ■

CAUTION We use the properties of inequality just as we use the properties of equality. However, when we multiply or divide each side by a negative number, we must reverse the inequality symbol.

Solving Inequalities

To solve inequalities, we use the properties of inequality to isolate x on one side.

EXAMPLE 2 **Isolating the variable on the left side**

Solve and graph the inequality $4x - 5 > 19$.

Solution

$$
\begin{aligned}
4x - 5 &> 19 && \text{Original inequality} \\
4x - 5 + 5 &> 19 + 5 && \text{Add 5 to each side.} \\
4x &> 24 && \text{Simplify.} \\
x &> 6 && \text{Divide each side by 4.}
\end{aligned}
$$

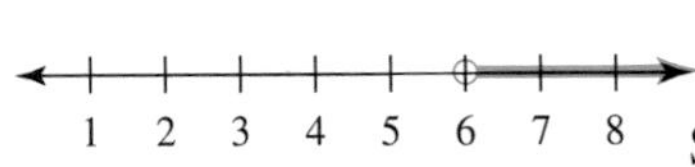

FIGURE 2.13

Since the last inequality is equivalent to the first, they have the same solutions and the same graph, which is shown in Fig. 2.13. ■

Calculator Close-Up

You can use the TABLE feature of a graphing calculator to numerically support the solution to the inequality $4x - 5 > 19$ in Example 2. Use the Y = key to enter the equation $y_1 = 4x - 5$.

```
Plot1 Plot2 Plot3
\Y1=4X-5
\Y2=
\Y3=
\Y4=
\Y5=
\Y6=
\Y7=
```

Next, use TBLSET to set the table so that the values of x start at 4.5 and the change in x is 0.5.

Finally, press TABLE to see lists of x-values and the corresponding y-values.

X	Y1
4.5	13
5	15
5.5	17
6	19
6.5	21
7	23
7.5	25

X=6

Notice that when x is larger than 6, y_1 (or $4x - 5$) is larger than 19. The table verifies or supports the algebraic solution, but it should not replace the algebraic method.

Remember that $5 < x$ is equivalent to $x > 5$. So the variable can be isolated on the right side of an inequality as shown in Example 3.

EXAMPLE 3 **Isolating the variable on the right side**

Solve and graph the inequality $5x - 2 \le 7x - 5$.

Solution

$$
\begin{aligned}
5x - 2 &\le 7x - 5 && \text{Original inequality} \\
5x - 2 - 5x &\le 7x - 5 - 5x && \text{Subtract } 5x \text{ from each side.} \\
-2 &\le 2x - 5 && \text{Simplify.} \\
3 &\le 2x && \text{Add 5 to each side.} \\
\frac{3}{2} &\le x && \text{Divide each side by 2.}
\end{aligned}
$$

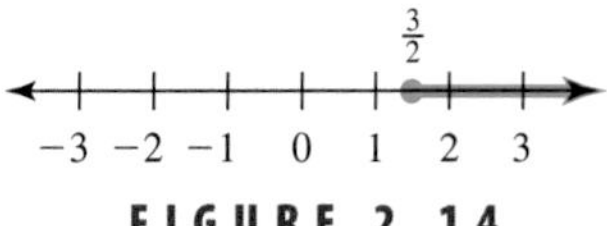

FIGURE 2.14

Note that $\frac{3}{2} \le x$ is equivalent to $x \ge \frac{3}{2}$. The graph is shown in Fig. 2.14. Notice that $\frac{3}{2}$ is half way between 1 and 2 on the number line. ■

Rewriting $\frac{3}{2} \le x$ as $x \ge \frac{3}{2}$ in Example 3 is not "reversing the inequality." Multiplying or dividing each side of $\frac{3}{2} \le x$ by a negative number would reverse the inequality. For example, multiplying by -1 yields $-\frac{3}{2} \ge -x$. In Example 4, we divide each side of an inequality by a negative number and reverse the inequality symbol.

EXAMPLE 4 **Reversing the inequality symbol**

Solve and graph the inequality $5 - 5x \le 1 + 2(5 - x)$.

Solution

$$\begin{aligned} 5 - 5x &\le 1 + 2(5 - x) && \text{Original inequality} \\ 5 - 5x &\le 11 - 2x && \text{Simplify the right side.} \\ 5 - 3x &\le 11 && \text{Add } 2x \text{ to each side.} \\ -3x &\le 6 && \text{Subtract 5 from each side.} \\ x &\ge -2 && \text{Divide each side by } -3 \text{, and reverse the inequality.} \end{aligned}$$

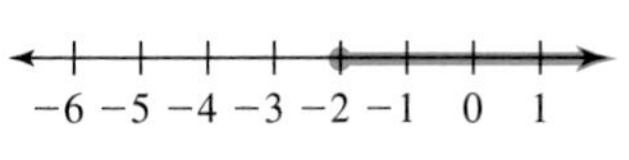

FIGURE 2.15

The inequalities $5 - 5x \le 1 + 2(5 - x)$ and $x \ge -2$ have the same graph, which is shown in Fig. 2.15. ■

We can use the rules for solving inequalities on the compound inequalities that we studied in Section 2.8.

EXAMPLE 5 **Solving a compound inequality**

Solve and graph the inequality $-9 \le \frac{2x}{3} - 7 < 5$.

Solution

$$\begin{aligned} -9 \le \frac{2x}{3} - 7 &< 5 && \text{Original inequality} \\ -9 + 7 \le \frac{2x}{3} - 7 + 7 &< 5 + 7 && \text{Add 7 to each part.} \\ -2 \le \frac{2x}{3} &< 12 && \text{Simplify.} \\ \frac{3}{2}(-2) \le \frac{3}{2} \cdot \frac{2x}{3} &< \frac{3}{2} \cdot 12 && \text{Multiply each part by } \tfrac{3}{2}. \\ -3 \le x &< 18 && \text{Simplify.} \end{aligned}$$

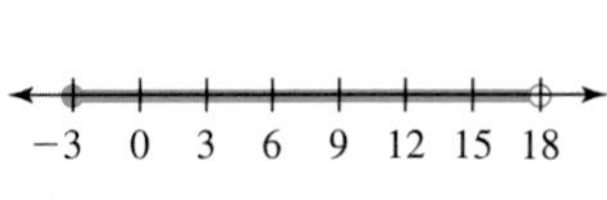

FIGURE 2.16

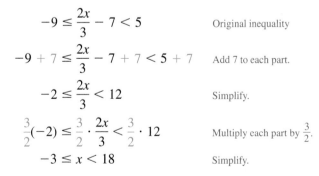

Any number that satisfies $-3 \le x < 18$ also satisfies the original inequality. Figure 2.16 shows all of the solutions to the original inequality. ■

CAUTION There are many negative numbers in Example 5, but the inequality was not reversed, since we did not multiply or divide by a negative number. An inequality is reversed only if you multiply or divide by a negative number.

EXAMPLE 6 **Reversing inequality symbols in a compound inequality**

Solve and graph the inequality $-3 \leq 5 - x \leq 5$.

Solution

$$-3 \leq 5 - x \leq 5 \quad \text{Original inequality}$$

$$-3 - 5 \leq 5 - x - 5 \leq 5 - 5 \quad \text{Subtract 5 from each part.}$$

$$-8 \leq -x \leq 0 \quad \text{Simplify.}$$

$$(-1)(-8) \geq (-1)(-x) \geq (-1)(0) \quad \text{Multiply each part by } -1 \text{, reversing the inequality symbols.}$$

$$8 \geq x \geq 0$$

It is customary to write $8 \geq x \geq 0$ with the smallest number on the left:

$$0 \leq x \leq 8$$

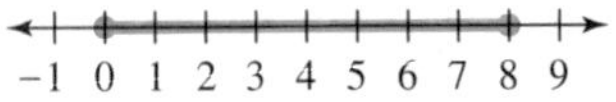

-1 0 1 2 3 4 5 6 7 8 9

FIGURE 2.17

Figure 2.17 shows all numbers that satisfy $-3 \leq 5 - x \leq 5$. ■

Applications of Inequalities

Example 7 shows how inequalities can be used in applications.

EXAMPLE 7 **Averaging test scores**

Mei Lin made a 76 on the midterm exam in history. To get a B, the average of her midterm and her final exam must be between 80 and 90. For what range of scores on the final exam will she get a B?

Helpful Hint

Remember that all inequality symbols in a compound inequality must point in the same direction. We usually have them all point to the left so that the numbers are increasing in size as you go from left to right in the inequality.

Solution

Let x represent the final exam score. Her average is then $\frac{x + 76}{2}$. The inequality expresses the fact that the average must be between 80 and 90:

$$80 < \frac{x + 76}{2} < 90$$

$$2(80) < 2\left(\frac{x + 76}{2}\right) < 2(90) \quad \text{Multiply each part by 2.}$$

$$160 < x + 76 < 180 \quad \text{Simplify.}$$

$$160 - 76 < x + 76 - 76 < 180 - 76 \quad \text{Subtract 76 from each part.}$$

$$84 < x < 104 \quad \text{Simplify.}$$

The last inequality indicates that Mei Lin's final exam score must be between 84 and 104. ■

WARM-UPS

True or false? Explain your answer.

1. The inequality $2x > 18$ is equivalent to $x > 9$. True
2. The inequality $x - 5 > 0$ is equivalent to $x < 5$. False
3. We can divide each side of an inequality by any real number. False
4. The inequality $-2x \leq 6$ is equivalent to $-x \leq 3$. True

WARM-UPS

(continued)

5. The statement "x is at most 7" is written as $x < 7$. False
6. "The sum of x and $0.05x$ is at least 76" is written as $x + 0.05x \geq 76$. True
7. The statement "x is not more than 85" is written as $x < 85$. False
8. The inequality $-3 > x > -9$ is equivalent to $-9 < x < -3$. True
9. If x is the sale price of Glen's truck, the sales tax rate is 8%, and the title fee is \$50, then the total that he pays is $1.08x + 50$ dollars. True
10. If the selling price of the house, x, less the sales commission of 6% must be at least \$60,000, then $x - 0.06x \leq 60{,}000$. False

2.9 EXERCISES

Reading and Writing *After reading this section, write out the answers to these questions. Use complete sentences.*

1. What are equivalent inequalities?
   Equivalent inequalities are inequalities that have the same solutions.
2. What is the addition property of inequality?
   The addition property of inequality says that adding any real number to each side of an inequality produces an equivalent inequality.
3. What is the multiplication property of inequality?
   According to the multiplication property of inequality, the inequality symbol is reversed when multiplying (or dividing) by a negative number and not reversed when multiplying (or dividing) by a positive number.
4. What similarities are there between solving equations and solving inequalities?
   For equations or inequalities we try to isolate the variable. The properties of equality and inequality are similar.
5. How do we solve compound inequalities?
   We solve compound inequalities using the properties of inequality as we do for simple inequalities.
6. How do you know when to reverse the direction of an inequality symbol?
   The direction of the inequality symbol is reversed when we multiply or divide by a negative number.

Write the appropriate inequality symbol in the blank so that the two inequalities are equivalent. See Example 1.

7. $x + 7 > 0$
   x ___ -7
   $x > -7$
8. $x - 6 < 0$
   x ___ 6
   $x < 6$
9. $9 \leq 3w$
   w ___ 3
   $w \geq 3$
10. $10 \geq 5z$
    z ___ 2
    $z \leq 2$
11. $-4k < -4$
    k ___ 1
    $k > 1$
12. $-9t > 27$
    t ___ -3
    $t < -3$
13. $-\frac{1}{2}y \geq 4$
    y ___ -8
    $y \leq -8$
14. $-\frac{1}{3}x \leq 4$
    x ___ -12
    $x \geq -12$

Solve and graph each of the following inequalities. See Examples 2–4.

15. $x + 3 > 0$ $x > -3$ 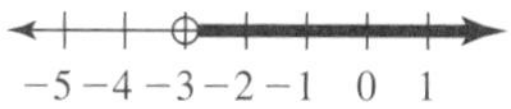
    −5 −4 −3 −2 −1 0 1
16. $x + 9 \leq -8$ $x \leq -17$ 
    −19 −18 −17 −16 −15
17. $-3 < w - 1$ $w > -2$ 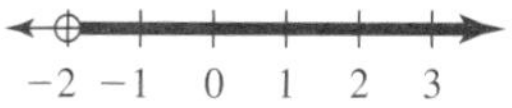
    −2 −1 0 1 2 3
18. $9 > w - 12$ $w < 21$
    19 20 21 22 23
19. $8 > 2b$ $b < 4$
    −1 0 1 2 3 4 5
20. $35 < 7b$ $b > 5$ 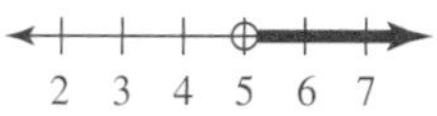
    2 3 4 5 6 7
21. $-8z \leq 4$ $z \geq -\frac{1}{2}$
    $-\frac{1}{2}$
    −2 −1 0 1 2 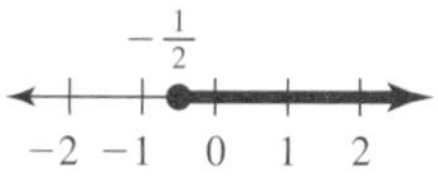
22. $-4y \geq -10$ $y \leq \frac{5}{2}$ 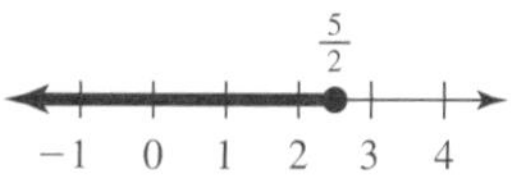
    $\frac{5}{2}$
    −1 0 1 2 3 4
23. $3y - 2 < 7$ $y < 3$
    −1 0 1 2 3 4 5
24. $2y - 5 > -9$ $y > -2$
    −4 −3 −2 −1 0 1 2

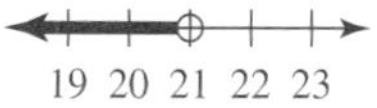

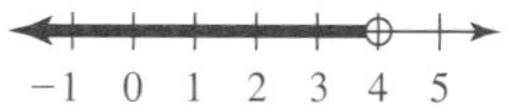

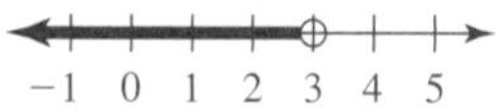

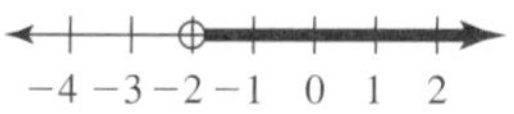

25. $3 - 9z \leq 6 \quad z \geq -\frac{1}{3}$

26. $5 - 6z \geq 13 \quad z \leq -\frac{4}{3}$

27. $6 > -r + 3 \quad r > -3$

28. $6 \leq 12 - r \quad r \leq 6$

29. $5 - 4p > -8 - 3p \quad p < 13$

30. $7 - 9p > 11 - 8p \quad p < -4$

31. $-\frac{5}{6}q \geq -20 \quad q \leq 24$

32. $-\frac{2}{3}q \geq -4 \quad q \leq 6$

33. $1 - \frac{1}{4}t \geq \frac{1}{8} \quad t \leq \frac{7}{2}$

34. $\frac{1}{6} - \frac{1}{3}t > 0 \quad t < \frac{1}{2}$

35. $2x + 5 < x - 6 \quad x < -11$

36. $3x - 4 < 2x + 9 \quad x < 13$

37. $x - 4 < 2(x + 3) \quad x > -10$

38. $2x + 3 < 3(x - 5) \quad x > 18$

39. $0.52x - 35 < 0.45x + 8 \quad x < 614.3$

40. $8455(x - 3.4) > 4320 \quad x > 3.91$

Solve and graph each compound inequality. See Examples 5 and 6.

41. $5 < x - 3 < 7 \quad 8 < x < 10$

42. $2 < x - 5 < 6 \quad 7 < x < 11$

43. $3 < 2v + 1 < 10 \quad 1 < v < \frac{9}{2}$

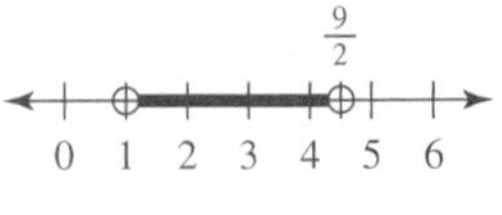

44. $-3 < 3v + 4 < 7 \quad -\frac{7}{3} < v < 1$

45. $-4 \leq 5 - k \leq 7 \quad -2 \leq k \leq 9$

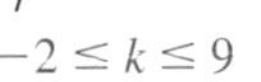

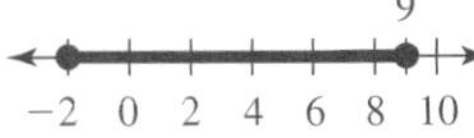

46. $2 \leq 3 - k \leq 8 \quad -5 \leq k \leq 1$

47. $-2 < 7 - 3y \leq 22 \quad -5 \leq y < 3$

48. $-1 \leq 1 - 2y < 3 \quad -1 < y \leq 1$

49. $5 < \frac{2u}{3} - 3 < 17 \quad 12 < u < 30$

50. $-4 < \frac{3u}{4} - 1 < 11 \quad -4 < u < 16$

51. $-2 < \frac{4m - 4}{3} \leq \frac{2}{3} \quad -\frac{1}{2} < m \leq \frac{3}{2}$

52. $0 \leq \frac{3 - 2m}{2} < 9 \quad -\frac{15}{2} < m \leq \frac{3}{2}$

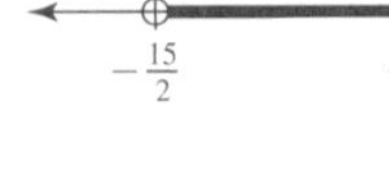

53. $0.02 < 0.54 - 0.0048x < 0.05 \quad 102.1 < x < 108.3$

54. $0.44 < \frac{34.55 - 22.3x}{124.5} < 0.76 \quad -2.69 < x < -0.91$

Solve and graph each inequality.

55. $\frac{1}{2}x - 1 \le 4 - \frac{1}{3}x$ $x \le 6$

56. $\frac{y}{4} - \frac{5}{12} \ge \frac{y}{3} + \frac{1}{4}$ $y \le -8$

57. $\frac{1}{2}\left(x - \frac{1}{4}\right) > \frac{1}{4}\left(6x - \frac{1}{2}\right)$ $x < 0$

58. $-\frac{1}{2}\left(z - \frac{2}{5}\right) < \frac{2}{3}\left(\frac{3}{4}z - \frac{6}{5}\right)$ $z > 1$

59. $\frac{1}{3} < \frac{1}{4}x - \frac{1}{6} < \frac{7}{12}$ $2 < x < 3$

60. $-\frac{3}{5} < \frac{1}{5} - \frac{2}{15}w < -\frac{1}{3}$ $4 < w < 6$

Solve each of the following problems by using an inequality. See Example 7.

61. ***Boat storage.*** The length of a rectangular boat storage shed must be 4 meters more than the width, and the perimeter must be at least 120 meters. What is the range of values for the width? At least 28 meters

62. ***Fencing a garden.*** Elka is planning a rectangular garden that is to be twice as long as it is wide. If she can afford to buy at most 180 feet of fencing, then what are the possible values for the width? At most 30 feet

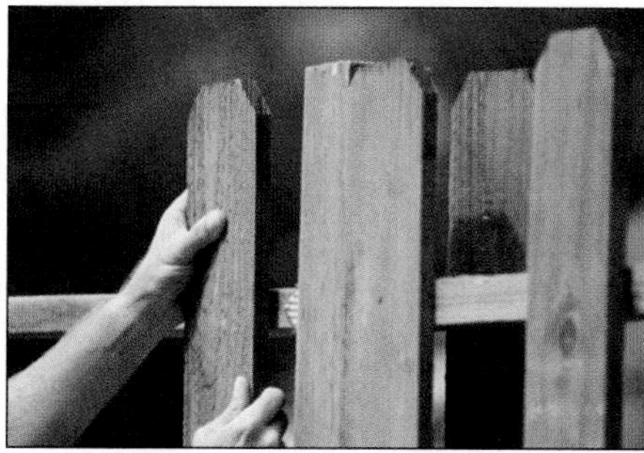

FIGURE FOR EXERCISE 62

63. ***Car shopping.*** Harold Ivan is shopping for a new car. In addition to the price of the car, there is a 5% sales tax and a \$144 title and license fee. If Harold Ivan decides that he will spend less than \$9970 total, then what is the price range for the car? Less than \$9358

64. ***Car selling.*** Ronald wants to sell his car through a broker who charges a commission of 10% of the selling price. Ronald still owes \$11,025 on the car. Ronald must get enough to at least pay off the loan. What is the range of the selling price?
At least \$12,250

65. ***Microwave oven.*** Sherie is going to buy a microwave in a city with an 8% sales tax. She has at most \$594 to spend. In what price range should she look?
At most \$550

66. ***Dining out.*** At Burger Brothers the price of a hamburger is twice the price of an order of French fries, and the price of a Coke is \$0.40 more than the price of the fries. Burger Brothers advertises that you can get a complete meal (burger, fries, and Coke) for under \$4.00. What is the price range of an order of fries?
Less than 90 cents

67. ***Averaging test scores.*** Tilak made 44 and 72 on the first two tests in algebra and has one test remaining. For Tilak to pass the course, the average on the three tests must be at least 60. For what range of scores on his last test will Tilak pass the course?
At least 64

68. ***Averaging income.*** Helen earned \$400 in January, \$450 in February, and \$380 in March. To pay all of her bills, she must average at least \$430 per month. For what income in April would her average for the four months be at least \$430?
At least \$490

69. ***Going for a C.*** Professor Williams gives only a midterm exam and a final exam. The semester average is computed by taking $\frac{1}{3}$ of the midterm exam score plus $\frac{2}{3}$ of the final exam score. To get a C, Stacy must have a semester average between 70 and 79 inclusive. If Stacy scored only 48 on the midterm, then for what range of scores on the final exam will Stacy get a C?
Between 81 and 94.5 inclusive

70. ***Different weights.*** Professor Williamson counts his midterm as $\frac{2}{3}$ of the grade and his final as $\frac{1}{3}$ of the grade. Wendy scored only 48 on the midterm. What range of scores on the final exam would put Wendy's average between 70 and 79 inclusive? Compare to the previous exercise.
Between 114 and 141 inclusive

71. ***Average driving speed.*** On Halley's recent trip from Bangor to San Diego, she drove for 8 hours each day and traveled between 396 and 453 miles each day. In what range was her average speed for each day of the trip? Between 49.5 and 56.625 miles per hour

72. ***Driving time.*** On Halley's trip back to Bangor, she drove at an average speed of 55 mph every day and traveled between 330 and 495 miles per day. In what range was her daily driving time?
Between 6 and 9 hours

73. ***Sailboat navigation.*** As the sloop sailed north along the coast, the captain sighted the lighthouse at points A and B as shown in the figure on the next page. If the degree measure of the angle at the lighthouse is less than 30°, then what are the possible values for x?
Between 55° and 85°

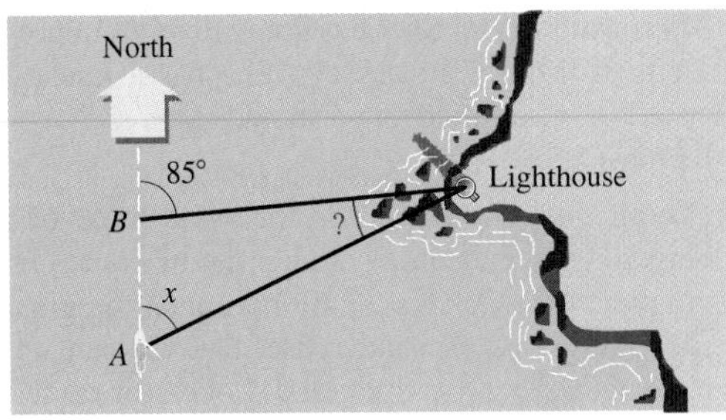

FIGURE FOR EXERCISE 73

74. ***Flight plan.*** A pilot started at point A and flew in the direction shown in the diagram for some time. At point B she made a 110° turn to end up at point C, due east of where she started. If the measure of angle C is less than 85°, then what are the possible values for x?
Between 0° and 65°

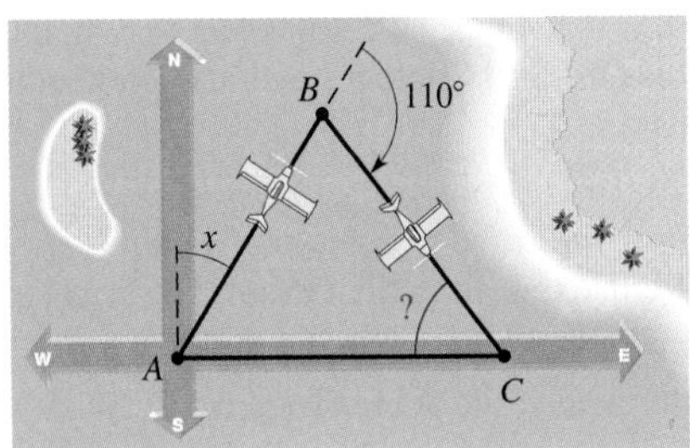

FIGURE FOR EXERCISE 74

75. ***Bicycle gear ratios.*** The gear ratio r for a bicycle is defined by the formula

$$r = \frac{Nw}{n},$$

where N is the number of teeth on the chainring (by the pedal), n is the number of teeth on the cog (by the wheel), and w is the wheel diameter in inches (www.sheldonbrown.com/gears).

a) If the wheel has a diameter of 27 in. and there are 12 teeth on the cog, then for what number of teeth on the chainring is the gear ratio between 60 and 80?
Between 27 and 35 teeth inclusive

b) If a bicycle has 48 teeth on the chainring and 17 teeth on the cog, then for what diameter wheel is the gear ratio between 65 and 70?
Between 23.02 in. and 24.79 in.

c) If a bicycle has a 26-in. diameter wheel and 40 teeth on the chainring, then for what number of teeth on the cog is the gear ratio less than 75?
At least 14 teeth

76. ***Virtual demand.*** The weekly demand (the number bought by consumers) for the Acme Virtual Pet is given by the formula

$$d = 9000 - 60p$$

where p is the price each in dollars.

a) What is the demand when the price is \$30 each?
7200

b) In what price range will the demand be above 6000?
Less than \$50

COLLABORATIVE ACTIVITIES

Expression-Equation-Inequality

In this activity you will practice your skills with simplifying expressions, solving equations, and solving and graphing inequalities. The roles are simplifier, solver, and grapher. The simplifier will simplify all expressions, the solver will solve all equations, and the grapher will first solve and then graph all inequalities.

Sorting Out the Work. Each of the following exercises contains one expression to be simplified, one equation to be solved, and one inequality to be graphed, not necessarily in that order. For each exercise your group will first sort out the three parts: give the expression to the simplifier, the equation to the solver, and the inequality to the grapher. After each person has completed his or her assigned task, pass the paper to the left for checking. Take on the role of the paper that you now have and proceed to the next exercise.

Grouping: Three students per group

Topic: Expressions, Equations, and Inequalities

1. a) $-2x \geq 20$
b) $2(x - 4) + x - 10$
c) $3x - 5 = 4x + 1$

2. a) $\frac{1}{2}x + 3 + \frac{7}{2}x - 5$
b) $4(x + 3) = 3(1 + x)$
c) $5x - 7 > 8x + 5$

3. a) $4(2x - 5) \leq 5x - 1$
b) $3(x + 2) - (12 - x)$
c) $\frac{5}{3}x + 4 = \frac{2}{3}x$

4. a) $5x - 6 + x = 9 + 4x - 1$
b) $\frac{2(5x + 1)}{3} > 2$
c) $5x - x + 2x$

Creating Your Own. Each student now creates and solves the type of problem he or she worked last, using the terms given here. For example, given the terms $4y$, $2y$, and -4 you could make up the expression $4y - 2y + (-4)$, the equation $4y - 2y = -4$, and the inequality $4y + 2y < -4$. Pass your work to the person on your left to check. Then proceed to the next exercise as before.

5. $6x, 2x, 3$

6. $-10x, 5x, 20, -15$

7. $3y, -7y, 5, -4$

8. $-4w, 7, -6w, 22$

WRAP-UP CHAPTER 2

SUMMARY

Equations		**Examples**
Linear equation	An equation of the form $ax + b = 0$ with $a \neq 0$	$3x + 7 = 0$
Identity	An equation that is satisfied by every number for which both sides are defined	$x + x = 2x$
Conditional equation	An equation that has at least one solution but is not an identity	$5x - 10 = 0$
Inconsistent equation	An equation that has no solution	$x = x + 1$
Equivalent equations	Equations that have exactly the same solutions	$2x + 1 = 5$ $2x = 4$
Properties of equality	If the same number is added to or subtracted from each side of an equation, the resulting equation is equivalent to the original equation.	$x - 5 = -9$ $x = -4$
	If each side of an equation is multiplied or divided by the same nonzero number, the resulting equation is equivalent to the original equation.	$9x = 27$ $x = 3$
Solving equations	**1.** Remove parentheses by using the distributive property and then combine like terms to simplify each side as much as possible. **2.** Use the addition property of equality to get like terms from opposite sides onto the same side so that they may be combined. **3.** The multiplication property of equality is generally used last. **4.** Check that the solution satisfies the original equation.	$2(x - 3) = -7 + 3(x - 1)$ $2x - 6 = -10 + 3x$ $-x - 6 = -10$ $-x = -4$ $x = 4$ *Check:* $2(4 - 3) = -7 + 3(4 - 1)$ $2 = 2$

Applications

Steps in solving applied problems

1. Read the problem.
2. If possible, draw a diagram to illustrate the problem.
3. Choose a variable and write down what it represents.
4. Represent any other unknowns in terms of that variable.
5. Write an equation that describes the situation.
6. Solve the equation.
7. Answer the original question.
8. Check your answer by using it to solve the original problem (not the equation).

Inequalities

Properties of inequality

Addition, subtraction, multiplication, and division may be performed on each side of an inequality, just as we do in solving equations, with one exception. When multiplying or dividing by a negative number, the inequality symbol is reversed.

Examples

$$-3x + 1 > 7$$
$$-3x > 6$$
$$x < -2$$

ENRICHING YOUR MATHEMATICAL WORD POWER

For each mathematical term, choose the correct meaning.

1. **linear equation**
   a. an equation in which the terms are in line
   b. an equation of the form $ax + b = 0$ where $a \neq 0$
   c. the equation $a = b$
   d. an equation of the form $a^2 + b^2 = c^2$ b

2. **identity**
   a. an equation that is satisfied by all real numbers
   b. an equation that is satisfied by every real number
   c. an equation that is identical
   d. an equation that is satisfied by every real number for which both sides are defined d

3. **conditional equation**
   a. an equation that has at least one real solution
   b. an equation that is correct
   c. an equation that is satisfied by at least one real number but is not an identity
   d. an equation that we are not sure how to solve c

4. **inconsistent equation**
   a. an equation that is wrong
   b. an equation that is only sometimes consistent
   c. an equation that has no solution
   d. an equation with two variables c

5. **equivalent equations**
   a. equations that are identical
   b. equations that are correct
   c. equations that are equal
   d. equations that have the same solution d

6. **formula**
   a. an equation
   b. a type of race car
   c. a process
   d. an equation involving two or more variables d

7. **literal equation**
   a. a formula
   b. an equation with words
   c. a false equation
   d. a fact a

8. **complementary angles**
   a. angles that compliment each other
   b. angles whose degree measures total 90°
   c. angles whose degree measures total 180°
   d. angles with the same vertex b

9. **supplementary angles**
   a. angles with soft flexible sides
   b. angles whose degree measures total 90°
   c. angles whose degree measures total 180°
   d. angles that form a square c

10. **uniform motion**
    a. movement of an army
    b. movement in a straight line
    c. consistent motion
    d. motion at a constant rate d

b) The y-coordinate of (, −5) is −5. Let $y = -5$ in the equation $y = -3x + 4$:

$$\begin{aligned} -5 &= -3x + 4 \\ -9 &= -3x \\ 3 &= x \end{aligned}$$

The ordered pair (3, −5) satisfies the equation.

c) Replace x by 0 in the equation $y = -3x + 4$:

$$y = -3 \cdot 0 + 4 = 4$$

So (0, 4) satisfies the equation. ■

The Rectangular Coordinate System

We use the **rectangular** (or **Cartesian**) **coordinate system** to get a visual image of ordered pairs of real numbers. The rectangular coordinate system consists of two number lines drawn at a right angle to one another, intersecting at zero on each number line, as shown in Fig. 3.1. The plane containing these number lines is called the **coordinate plane.** On the horizontal number line the positive numbers are to the right of zero, and on the vertical number line the positive numbers are above zero.

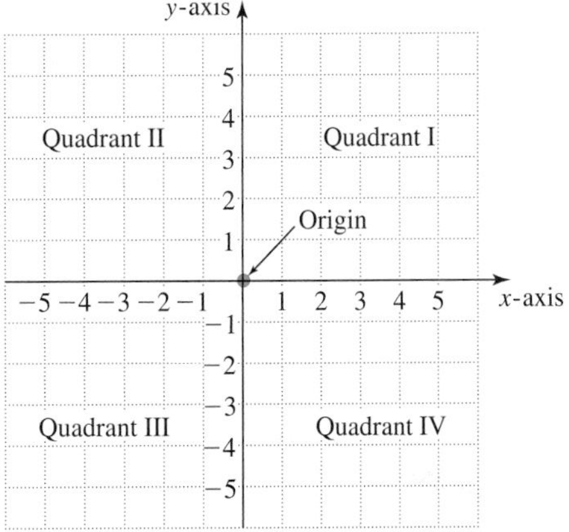

FIGURE 3.1

The horizontal number line is called the ***x*-axis,** and the vertical number line is called the ***y*-axis.** The point at which they intersect is called the **origin.** The two number lines divide the plane into four regions called **quadrants.** They are numbered as shown in Fig. 3.1. The quadrants do not include any points on the axes.

Plotting Points

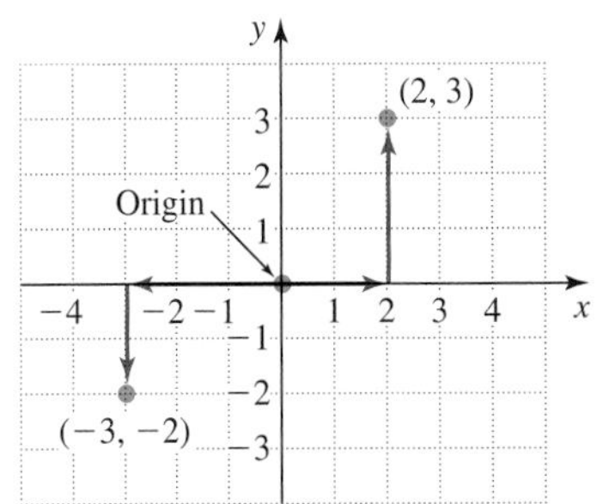

FIGURE 3.2

Just as every real number corresponds to a point on the number line, *every pair of real numbers corresponds to a point in the rectangular coordinate system.* For example, the point corresponding to the pair (2, 3) is found by starting at the origin and moving two units to the right and then three units up. The point corresponding to the pair (−3, −2) is found by starting at the origin and moving three units to the left and then two units down. Both of these points are shown in Fig. 3.2.

When we locate a point in the rectangular coordinate system, we are **plotting** or **graphing** the point. Because ordered pairs of numbers correspond to points in the coordinate plane, we frequently refer to an ordered pair as a point.

EXAMPLE 2 **Plotting points**

Plot the points (2, 5), (−1, 4), (−3, −4), and (3, −2).

Solution

To locate (2, 5), start at the origin, move two units to the right, and then move up five units. To locate (−1, 4), start at the origin, move one unit to the left, and then move up four units. All four points are shown in Fig. 3.3. ■

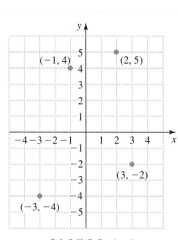

FIGURE 3.3

Calculator Close-Up

You can make a table of values for x and y with a graphing calculator. Enter the equation $y = 2x - 1$ using Y= and then press TABLE.

Graphing a Linear Equation

The **graph** of an equation is an illustration in the coordinate plane that shows all of the ordered pairs that satisfy the equation. When we draw the graph, we are **graphing the equation.**

Consider again the equation $y = 2x - 1$. To find ordered pairs that satisfy $y = 2x - 1$ we can arbitrarily select some x-coordinates and calculate the corresponding y-coordinates:

$$\begin{array}{ll}
\text{If } x = -3, & \text{then } y = 2(-3) - 1 = -7. \\
\text{If } x = -2, & \text{then } y = 2(-2) - 1 = -5. \\
\text{If } x = -1, & \text{then } y = 2(-1) - 1 = -3. \\
\text{If } x = 0, & \text{then } y = 2(0) - 1 = -1. \\
\text{If } x = 1, & \text{then } y = 2(1) - 1 = 1. \\
\text{If } x = 2, & \text{then } y = 2(2) - 1 = 3. \\
\text{If } x = 3, & \text{then } y = 2(3) - 1 = 5.
\end{array}$$

We can make a table for these results as follows:

x	−3	−2	−1	0	1	2	3
$y = 2x - 1$	−7	−5	−3	−1	1	3	5

The ordered pairs (−3, −7), (−2, −5), (−1, −3), (0, −1), (1, 1), (2, 3), and (3, 5) are graphed in Fig. 3.4. Notice that the points lie in a straight line. If we choose any real number for x and find the point that satisfies $y = 2x - 1$, we get another point on this line. Likewise, any point on this line satisfies the equation. So the graph of $y = 2x - 1$ is the straight line in Fig. 3.5. Because it is not possible to show all of the

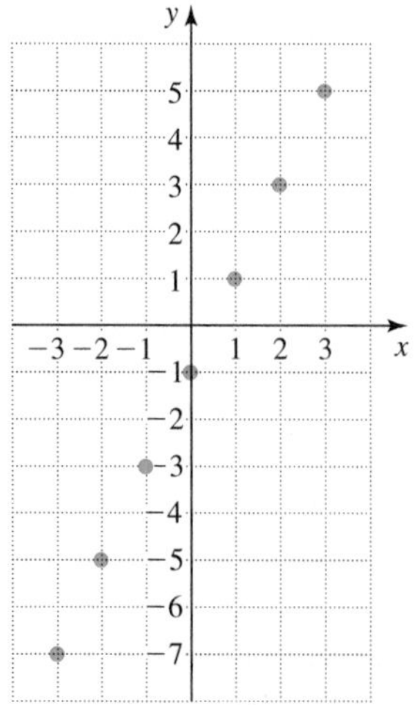

FIGURE 3.4

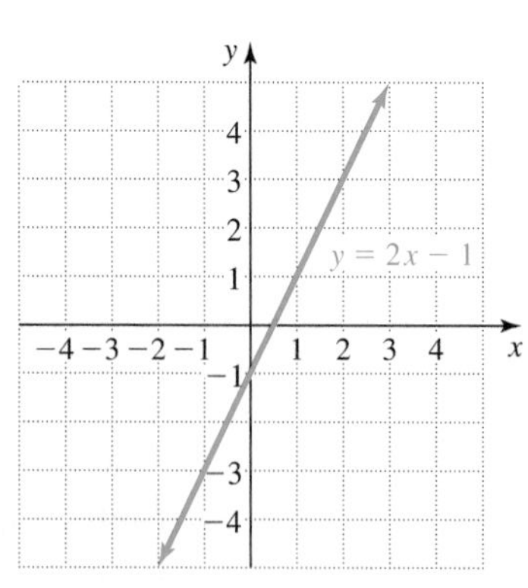

FIGURE 3.5

line, the arrows on the ends of the line indicate that it goes indefinitely in both directions. The equation $y = 2x - 1$ is an example of a *linear equation in two variables.*

Linear Equation in Two Variables

A **linear equation in two variables** is an equation that can be written in the form

$$Ax + By = C,$$

where A, B, and C are real numbers, with A and B not both equal to zero.

Equations such as

$$x - y = 5, \quad y = 2x + 3, \quad 2x - 5y - 9 = 0, \quad \text{and} \quad x = 8$$

are linear equations because they could all be rewritten in the form $Ax + By = C$. The graph of any linear equation is a straight line.

EXAMPLE 3 **Graphing an equation**

Graph the equation $3x + y = 2$. Plot at least five points.

Solution

It is easier to make a table of ordered pairs if the equation is solved for y. So subtract $3x$ from each side to get $y = -3x + 2$. Now select some values for x and then calculate the corresponding y-coordinates:

If $x = -2$, then $y = -3(-2) + 2 = 8$.
If $x = -1$, then $y = -3(-1) + 2 = 5$.
If $x = 0$, then $y = -3(0) + 2 = 2$.
If $x = 1$, then $y = -3(1) + 2 = -1$.
If $x = 2$, then $y = -3(2) + 2 = -4$.

The following table shows these five ordered pairs:

x	-2	-1	0	1	2
$y = -3x + 2$	8	5	2	-1	-4

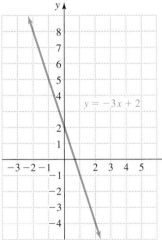

The graph of the line through these points is shown in Fig. 3.6. ■

FIGURE 3.6

Calculator Close-Up

To graph $y = -3x + 2$, enter the equation using the Y= key:

```
Plot1 Plot2 Plot3
\Y1=-3X+2
\Y2=
\Y3=
\Y4=
\Y5=
\Y6=
\Y7=
```

Next, set the viewing window (WINDOW) to get the desired view of the graph. Xmin and Xmax indicate the minimum and maximum x-values used for the graph; likewise for Ymin and Ymax. Xscl and Yscl (scale) give the distance between tick marks on the respective axes.

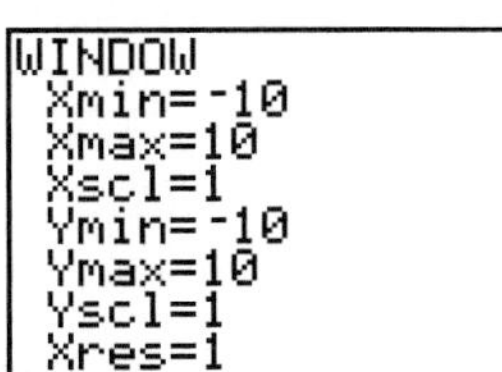

Press GRAPH to get the graph:

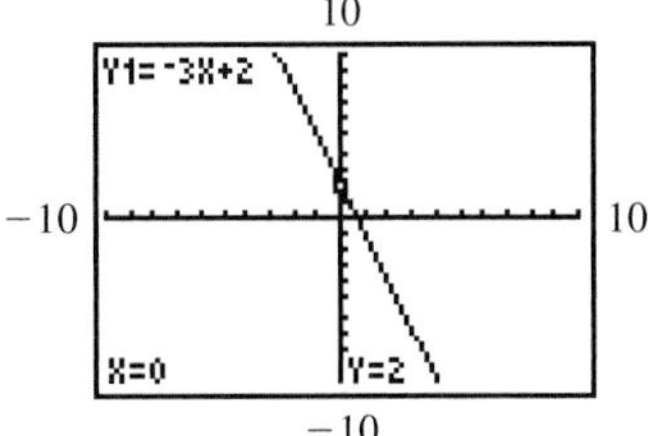

Even though the graph is not really "straight," it is consistent with the graph of $y = -3x + 2$ in Fig. 3.6.

In the linear equation $Ax + By = C$ either A or B could be zero. For example, $0 \cdot x + y = 4$ is a linear equation. Since x is multiplied by 0 this equation is usually written simply as $y = 4$. To graph $y = 4$ in the coordinate plane, we must understand that it comes from $0 \cdot x + y = 4$.

EXAMPLE 4 **Horizontal and vertical lines**

Graph each linear equation.

a) $y = 4$ **b)** $x = 3$

Solution

a) The equation $y = 4$ is a simplification of $0 \cdot x + y = 4$. So if y is replaced with 4, then we can use any real number for x. For example, $(-1, 4)$ satisfies $0 \cdot x + y = 4$ because $0(-1) + 4 = 4$ is correct. The following table shows five ordered pairs that satisfy $y = 4$.

x	-2	-1	0	1	2
$y = 4$	4	4	4	4	4

Figure 3.7 shows a horizontal line through these points.

b) The equation $x = 3$ is a simplification of $x + 0 \cdot y = 3$. So if x is replaced with 3, then we can use any real number for y. For example, $(3, -2)$ satisfies $x + 0 \cdot y = 3$ because $3 + 0(-2) = 3$ is correct. The following table shows five ordered pairs that satisfy $x = 3$.

$x = 3$	3	3	3	3	3
y	-2	-1	0	1	2

Figure 3.8 shows a vertical line through these points.

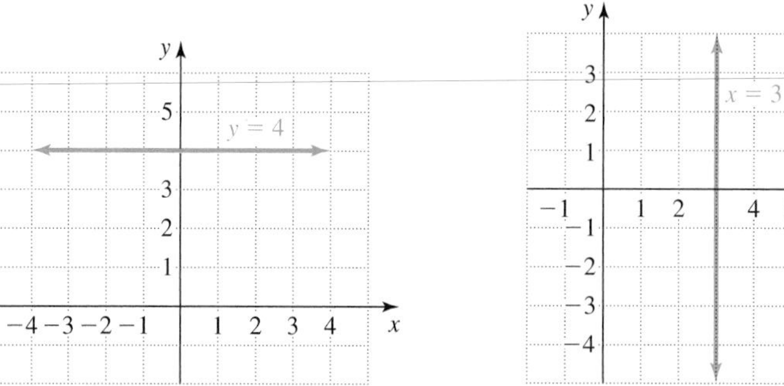

FIGURE 3.7 **FIGURE 3.8** ■

Calculator Close-Up

You cannot graph the vertical line $x = 3$ on most graphing calculators. The only equations that can be graphed are ones in which y is written in terms of x.

CAUTION If an equation such as $x = 3$ is discussed in the context of equations in two variables, then we assume that it is a simplified form of $x + 0 \cdot y = 3$, and there are infinitely many ordered pairs that satisfy the equation. If the equation $x = 3$ is discussed in the context of equations in a single variable, then $x = 3$ has only one solution, 3.

All of the equations we have considered so far have involved single-digit numbers. If an equation involves large numbers, then we must change the scale on the x-axis, the y-axis, or both to accommodate the numbers involved. The change of scale is arbitrary, and the graph will look different for different scales.

EXAMPLE 5

Adjusting the scale

Graph the equation $y = 20x + 500$. Plot at least five points.

Solution

The following table shows five ordered pairs that satisfy the equation.

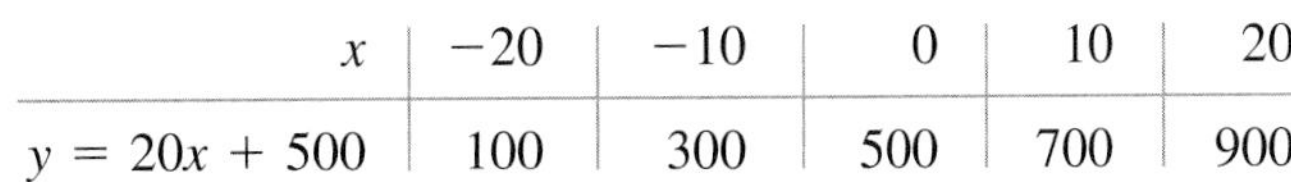

x	-20	-10	0	10	20
$y = 20x + 500$	100	300	500	700	900

To fit these points onto a graph, we change the scale on the x-axis to let each division represent 10 units and change the scale on the y-axis to let each division represent 200 units. The graph is shown in Fig. 3.9. ■

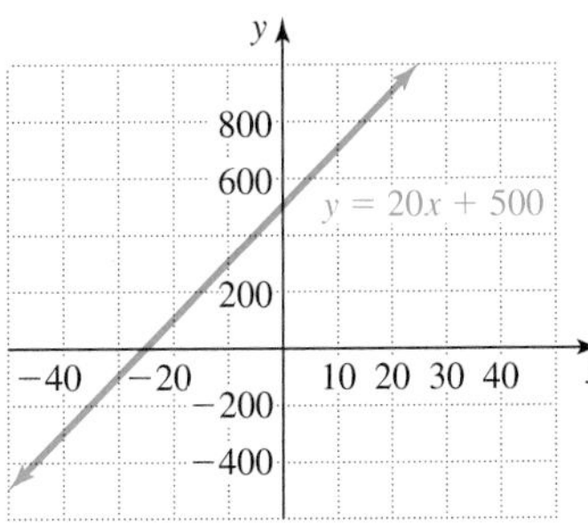

FIGURE 3.9

Graphing a Line Using Intercepts

We know that the graph of a linear equation is a straight line. Because it takes only two points to determine a line, we can graph a linear equation using only two points. The two points that are the easiest to locate are usually the points where the line crosses the axes. The point where the graph crosses the x-axis is the ***x*-intercept,** and the point where the graph crosses the y-axis is the ***y*-intercept.** The x-coordinate of the y-intercept is zero and the y-coordinate of the x-intercept is zero.

EXAMPLE 6

Graphing a line using intercepts

Graph the equation $2x - 3y = 6$ by using the x- and y-intercepts.

Solution

To find the x-intercept, let $y = 0$ in the equation $2x - 3y = 6$:

$$2x - 3 \cdot 0 = 6$$
$$2x = 6$$
$$x = 3$$

The x-intercept is $(3, 0)$. To find the y-intercept, let $x = 0$ in $2x - 3y = 6$:

$$2 \cdot 0 - 3y = 6$$
$$-3y = 6$$
$$y = -2$$

The y-intercept is $(0, -2)$. Locate the intercepts and draw a line through them as shown in Fig. 3.10. To check, find one additional point that satisfies the equation, say $(6, 2)$, and see whether the line goes through that point.

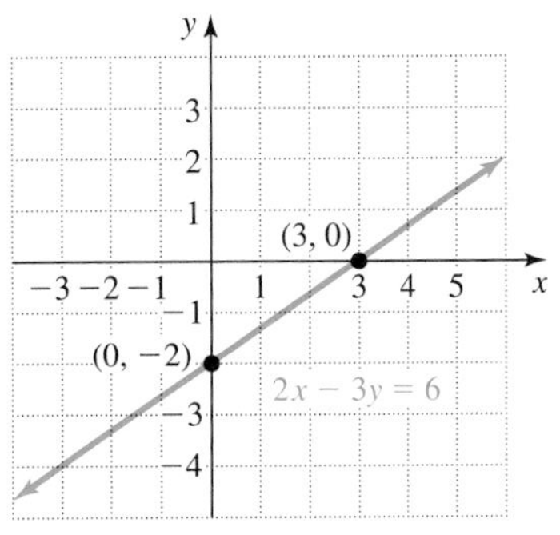

FIGURE 3.10 ■

Helpful Hint

You can find the intercepts for $2x - 3y = 6$ using the *cover-up method.* Cover up $-3y$ with your pencil, then solve $2x = 6$ mentally to get $x = 3$ and an x-intercept of $(3, 0)$. Now cover up $2x$ and solve $-3y = 6$ to get $y = -2$ and a y-intercept of $(0, -2)$.

Calculator Close-Up

To check the result in Example 6, graph $y = (2/3)x - 2$:

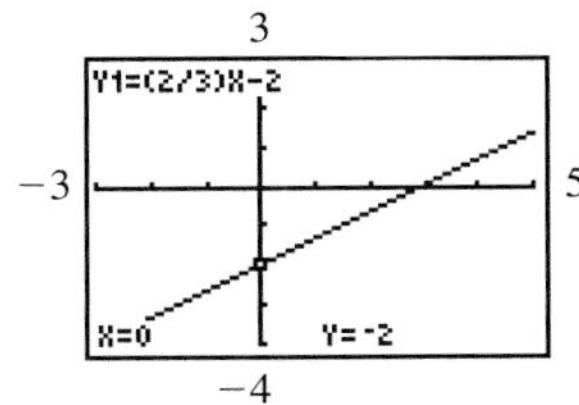

Since the calculator graph appears to be the same as the graph in Fig. 3.10, it supports the conclusion that Fig. 3.10 is correct.

Applications

Linear equations occur in many real-life situations. If the cost of plans for a house is \$475 for one copy plus \$30 for each additional copy, then $C = 475 + 30x$, where x is the number of additional copies. If you have \$1000 budgeted for landscaping with trees at \$50 each and bushes at \$20 each, then $50t + 20b = 1000$, where t is the number of trees and b is the number of bushes. In the next example we see a linear equation that models ticket demand.

EXAMPLE 7 **Ticket demand**

The demand for tickets to see the Ice Gators play hockey can be modeled by the equation $d = 8000 - 100p$, where d is the number of tickets sold and p is the price per ticket in dollars.

a) How many tickets will be sold at \$20 per ticket?

b) Find the intercepts and interpret them.

c) Graph the linear equation.

d) What happens to the demand as the price increases?

Solution

a) If tickets are \$20 each, then $d = 8000 - 100 \cdot 20 = 6000$. So at \$20 per ticket, the demand will be 6000 tickets.

MATH AT WORK

BIOMECHANIST

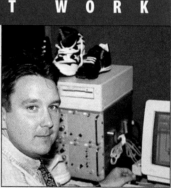

Christopher J. Edington, manager of the Biomechanics Laboratory at Converse, Inc., is a specialist in studying human movements from the hip down. In the past he has worked with diabetics, helping to educate them about the role of shoes and stress points in the shoes and their relationship to preventing foot injuries. More recently, he has helped to design and run tests for Converse's new athletic and leisure shoes. The latest development is a new basketball shoe that combines the lightweight characteristic of a running shoe with the support and durability of a standard basketball sneaker.

Information on how the foot strikes the ground, the length of contact time, and movements of the foot, knee, and hip can be recorded by using high-speed video equipment. This information is then used to evaluate the performance and design requirements of a lightweight, flexible, and well-fitting shoe. To meet the requirements of a good basketball shoe, Mr. Edington helped design and test the "React" shock-absorbing technology that is in Converse's latest sneakers.

In Exercise 88 of this section you will see the motion of a runner's heel as Mr. Edington does.

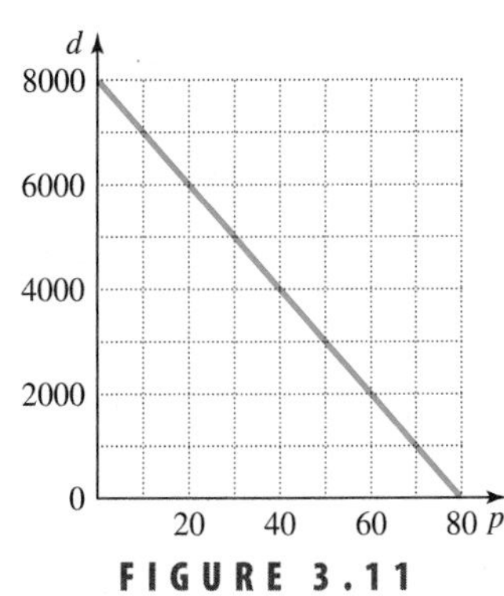

FIGURE 3.11

b) Replace d with 0 in the equation $d = 8000 - 100p$ and solve for p:

$$0 = 8000 - 100p$$
$$100p = 8000 \quad \text{Add } 100p \text{ to each side.}$$
$$p = 80 \quad \text{Divide each side by 100.}$$

If $p = 0$, then $d = 8000 - 100 \cdot 0 = 8000$. So the intercepts are (0, 8000) and (80, 0). If the tickets are free, the demand will be 8000 tickets. At \$80 per ticket, no tickets will be sold.

c) Graph the line using the intercepts (0, 8000) and (80, 0) as shown in Fig. 3.11.

d) When the tickets are free, the demand is high. As the price increases, the demand goes down. At \$80 per ticket, there will be no demand. ■

WARM-UPS

True or false? Explain your answer.

1. The point (2, 4) satisfies the equation $2y - 3x = -8$. False
2. If (1, 5) satisfies an equation, then (5, 1) also satisfies the equation. False
3. The origin is in quadrant I. False
4. The point (4, 0) is on the y-axis. False
5. The graph of $x + 0 \cdot y = 9$ is the same as the graph of $x = 9$. True
6. The graph of $x = -5$ is a vertical line. True
7. The graph of $0 \cdot x + y = 6$ is a horizontal line. True
8. The y-intercept for the line $x + 2y = 5$ is (5, 0). False
9. The point $(5, -3)$ is in quadrant II. False
10. The point $(-349, 0)$ is on the x-axis. True

3.1 EXERCISES

Reading and Writing *After reading this section, write out the answers to these questions. Use complete sentences.*

1. What is an ordered pair?
   An ordered pair is a pair of numbers in which there is a first number and a second number, usually written as (a, b).
2. What is the rectangular coordinate system?
   The rectangular coordinate system is a means of dividing up the plane with two number lines in order to picture all ordered pairs of real numbers.
3. What name is given to the point of intersection of the x-axis and the y-axis?
   The origin is the point of intersection of the x-axis and y-axis.
4. What is the graph of an equation?
   The graph of an equation is a picture of all ordered pairs that satisfy the equation drawn in the rectangular coordinate system.
5. What is a linear equation in two variables?
   A linear equation in two variables is an equation of the form $Ax + By = C$, where A and B are not both zero.
6. What are intercepts?
   Intercepts are the points at which a graph crosses the axes.

Complete each ordered pair so that it satisfies the given equation. See Example 1.

7. $y = 3x + 9$: (0,), (, 24), (2,)
   (0, 9), (5, 24), (2, 15)
8. $y = 2x + 5$: (8,), $(-1,$), (, $-1)$
   (8, 21), $(-1, 3)$, $(-3, -1)$
9. $y = -3x - 7$: (0,), $\left(\frac{1}{3}, \quad\right)$, (, $-5)$
   $(0, -7)$, $\left(\frac{1}{3}, -8\right)$, $\left(-\frac{2}{3}, -5\right)$
10. $y = -5x - 3$: $(-1,$), $\left(-\frac{1}{2}, \quad\right)$, (, $-2)$
    $(-1, 2)$, $\left(-\frac{1}{2}, -\frac{1}{2}\right)$, $\left(-\frac{1}{5}, -2\right)$
11. $y = -12x + 5$: (0,), (10,), (, 17)
    (0, 5), $(10, -115)$, $(-1, 17)$
12. $y = 18x + 200$: (1,), $(-10,$), (, 200)
    (1, 218), $(-10, 20)$, (0, 200)

13. $2x - 3y = 6$: $(3, \quad)$, $(\quad , -2)$, $(12, \quad)$
$(3, 0)$, $(0, -2)$, $(12, 6)$

14. $3x + 5y = 0$: $(-5, \quad)$, $(\quad , -3)$, $(10, \quad)$
$(-5, 3)$, $(5, -3)$, $(10, -6)$

15. $0 \cdot y + x = 5$: $(\quad , -3)$, $(\quad , 5)$, $(\quad , 0)$
$(5, -3)$, $(5, 5)$, $(5, 0)$

16. $0 \cdot x + y = -6$: $(3, \quad)$, $(-1, \quad)$, $(4, \quad)$
$(3, -6)$, $(-1, -6)$, $(4, -6)$

Plot the points on a rectangular coordinate system. See Example 2.

17. $(1, 5)$ **18.** $(4, 3)$

19. $(-2, 1)$ **20.** $(-3, 5)$

21. $\left(3, -\frac{1}{2}\right)$ **22.** $\left(2, -\frac{1}{3}\right)$

23. $(-2, -4)$ **24.** $(-3, -5)$

25. $(0, 3)$ **26.** $(0, -2)$

27. $(-3, 0)$ **28.** $(5, 0)$

29. $(\pi, 1)$ **30.** $(-2, \pi)$

31. $(1.4, 4)$ **32.** $(-3, 0.4)$

17–31 odd

18–32 even

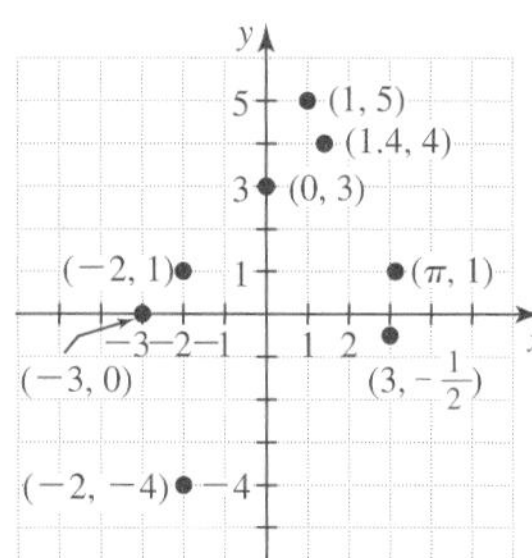

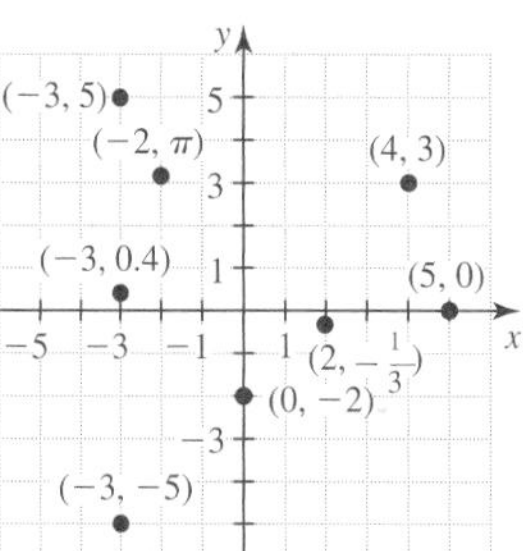

Use the given equations to find the missing coordinates in the following tables.

33. $y = -2x + 5$

x	y
−2	9
0	5
2	1
4	−3
6	−7

34. $y = -x + 4$

x	y
−2	6
0	4
2	2
4	0
6	−2

35. $y = \frac{1}{3}x + 2$

x	y
−6	0
−3	1
0	2
3	3

36. $y = -\frac{1}{2}x + 1$

x	y
−2	2
−1	$\frac{3}{2}$
0	1
1	$\frac{1}{2}$

Graph each equation. Plot at least five points for each equation. Use graph paper. See Examples 3 and 4. If you have a graphing calculator, use it to check your graphs when possible.

37. $y = x + 1$

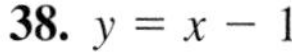
38. $y = x - 1$

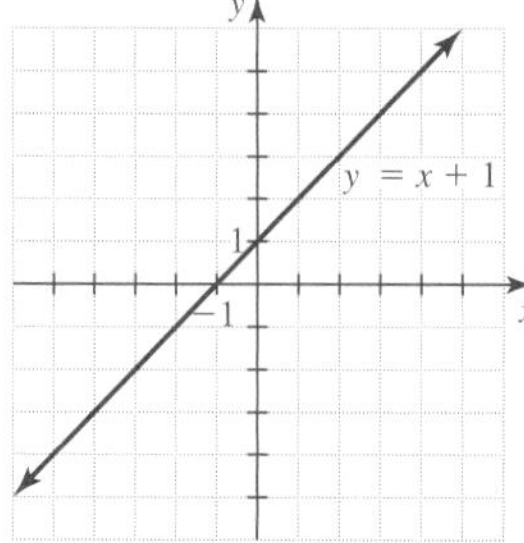

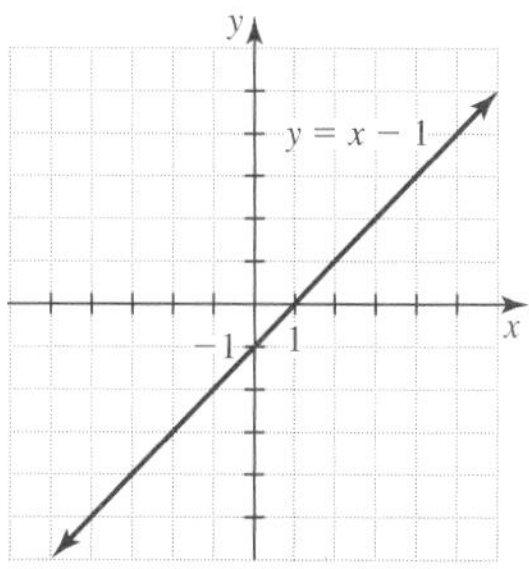

39. $y = 2x + 1$

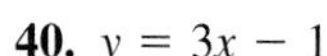
40. $y = 3x - 1$

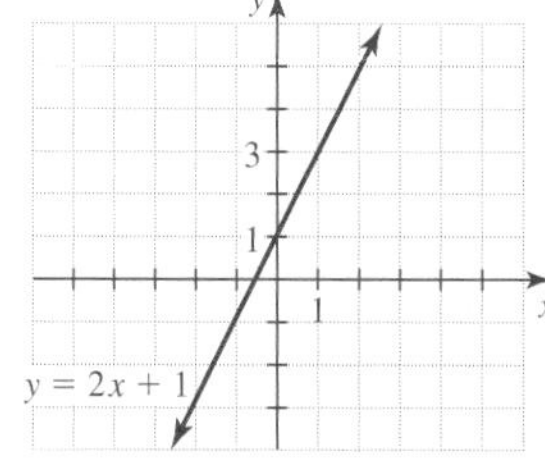

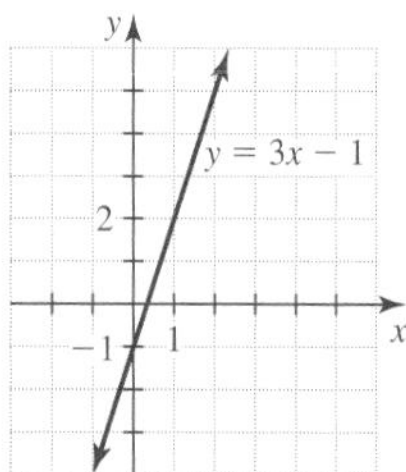

41. $y = 3x - 2$ **42.** $y = 2x + 3$

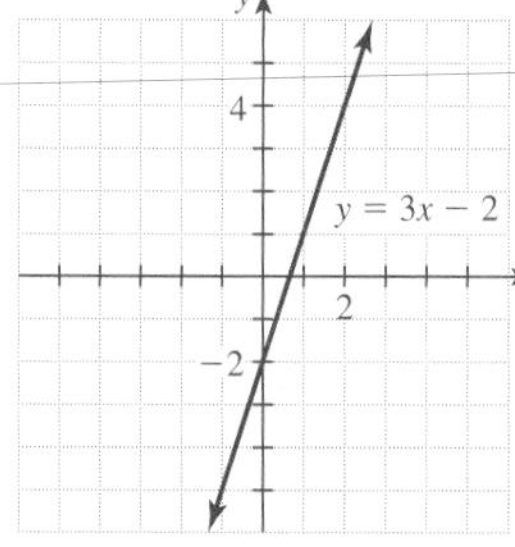

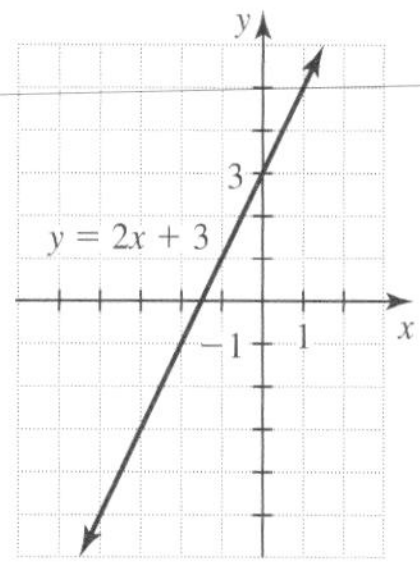

43. $y = x$ **44.** $y = -x$

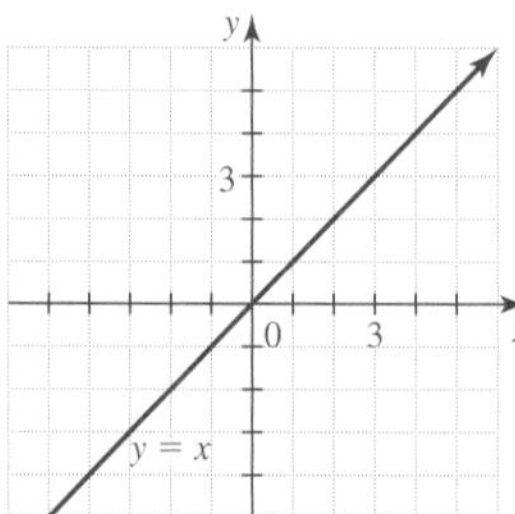

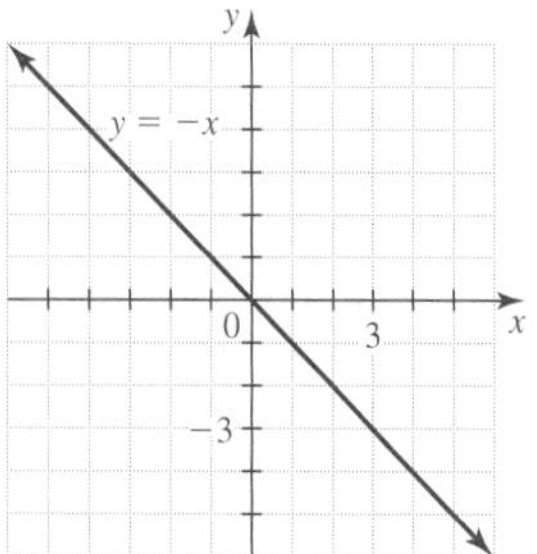

45. $y = 1 - x$

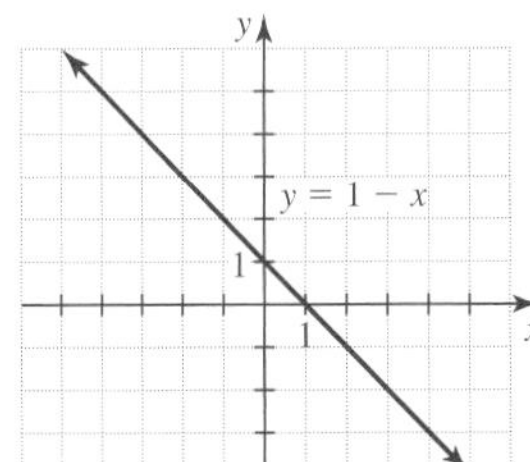

46. $y = 2 - x$

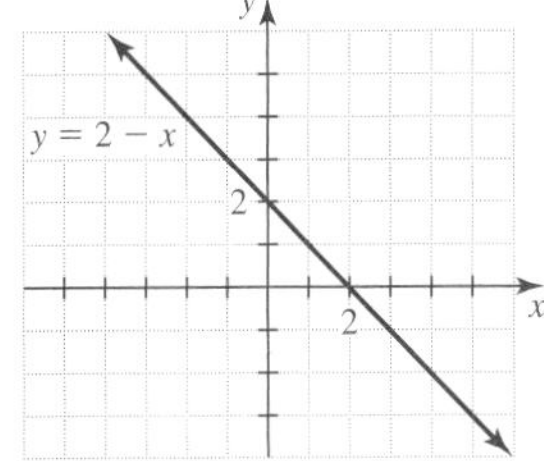

47. $y = -2x + 3$

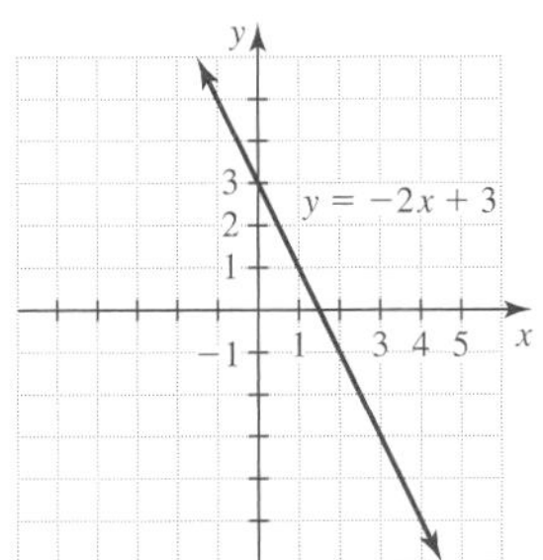

48. $y = -3x + 2$

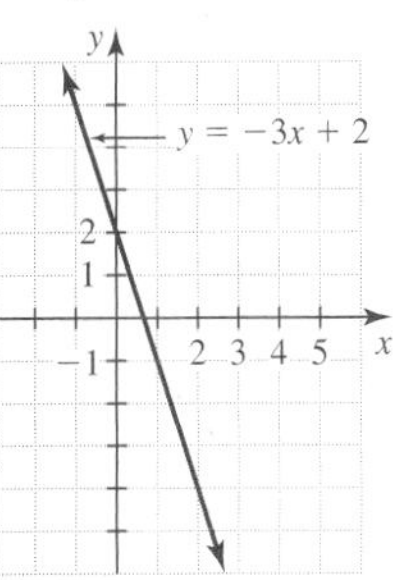

49. $y = -3$

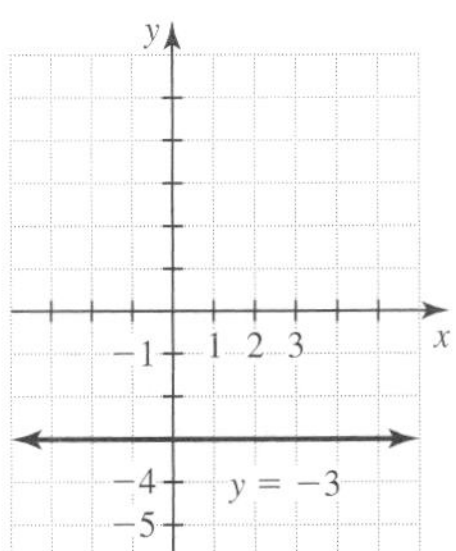

50. $y = 2$

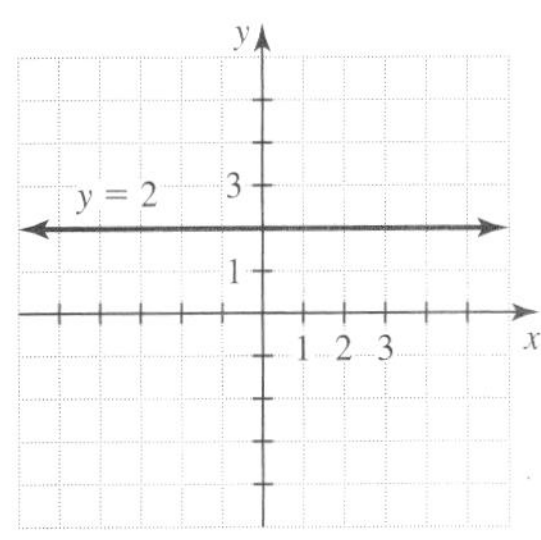

51. $x = 2$

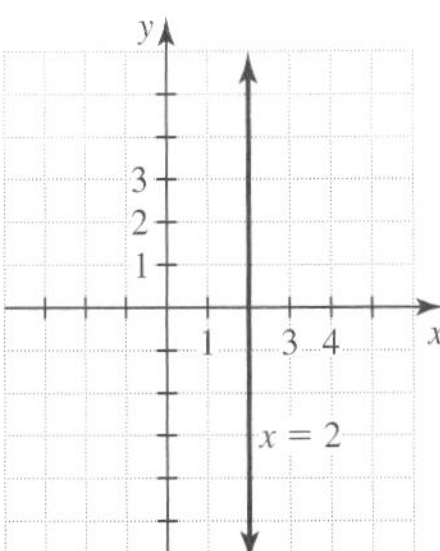

52. $x = -4$

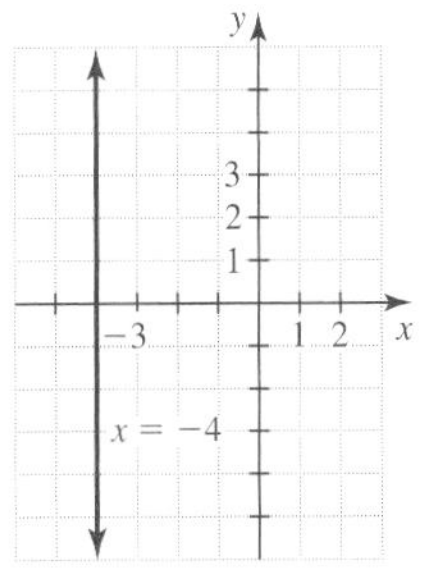

53. $2x + y = 5$

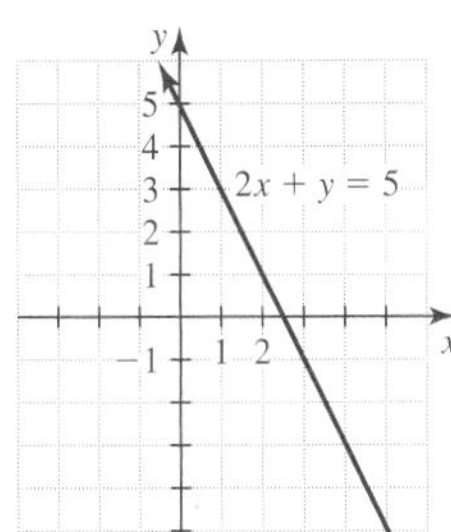

54. $3x + y = 5$

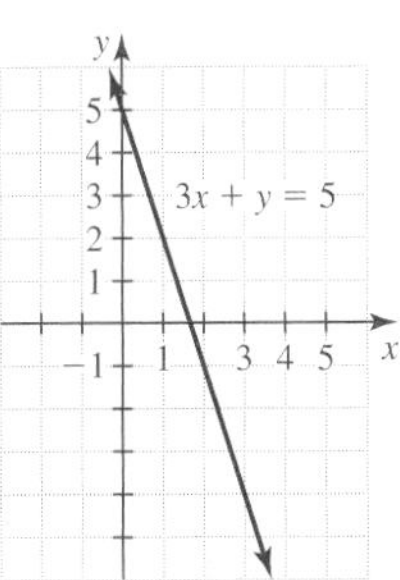

55. $x + 2y = 4$

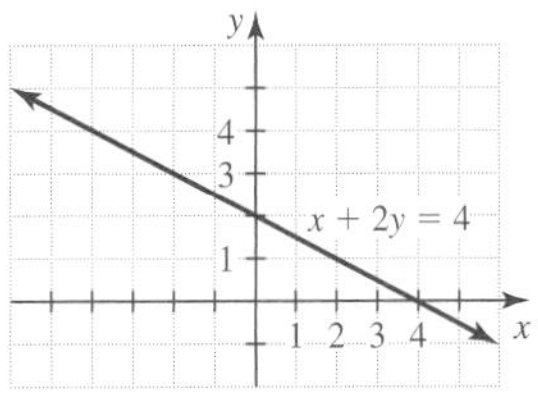

56. $x - 2y = 6$

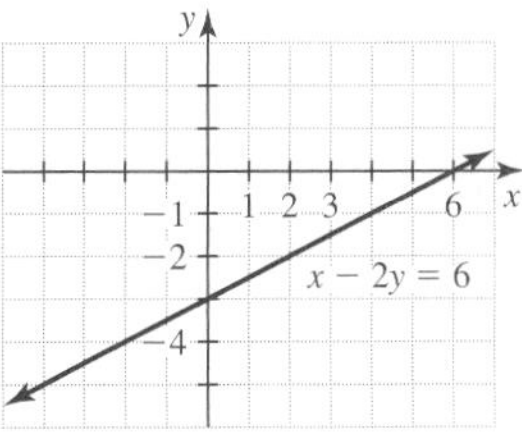

57. $x - 3y = 6$

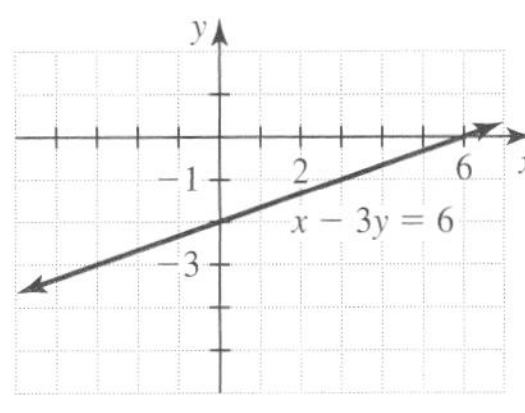

58. $x + 4y = 5$

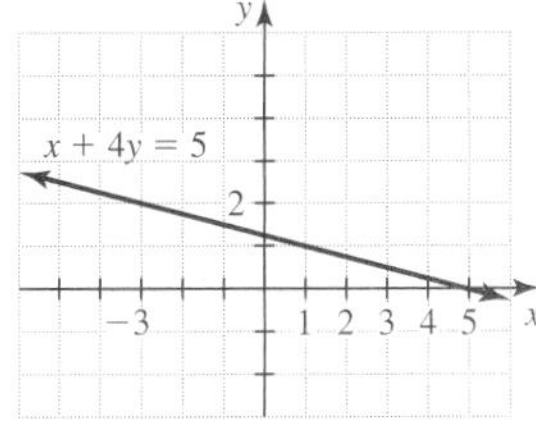

59. $y = 0.36x + 0.4$

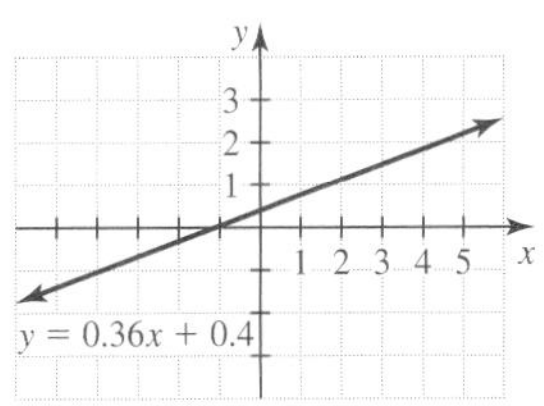

60. $y = 0.27x - 0.42$

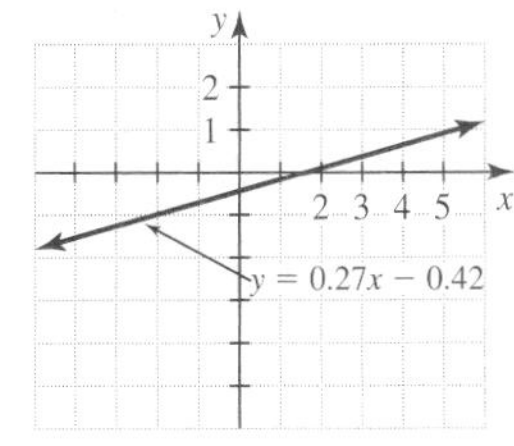

For each point, name the quadrant in which it lies or the axis on which it lies.

61. $(-3, 45)$
Quadrant II

62. $(-33, 47)$
Quadrant II

63. $(-3, 0)$
x-axis

64. $(0, -9)$
y-axis

65. $(-2.36, -5)$
Quadrant III

66. $(89.6, 0)$
x-axis

67. $(3.4, 8.8)$
Quadrant I

68. $(4.1, 44)$
Quadrant I

69. $\left(-\frac{1}{2}, 50\right)$
Quadrant II

70. $\left(-6, -\frac{1}{2}\right)$
Quadrant III

71. $(0, -99)$
y-axis

72. $(\pi, 0)$
x-axis

Graph each equation. Plot at least five points for each equation. Use graph paper. See Example 5. If you have a graphing calculator, use it to check your graphs.

73. $y = x + 1200$

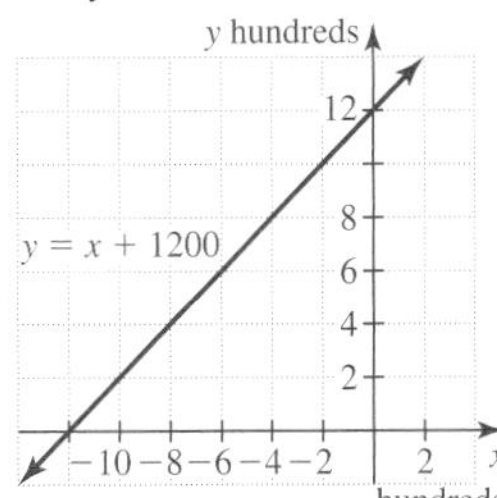

74. $y = 2x - 3000$

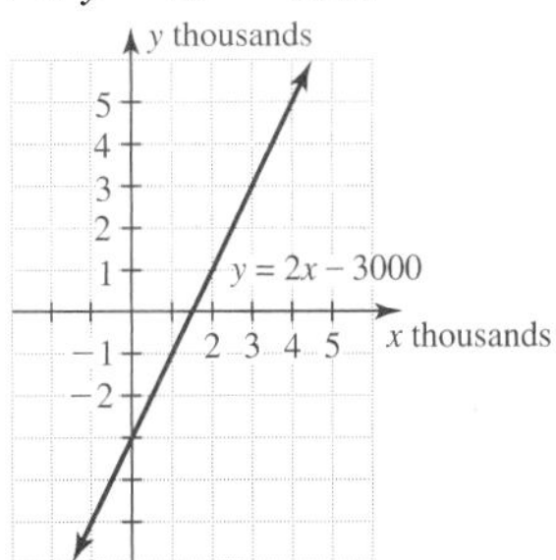

75. $y = 50x - 2000$

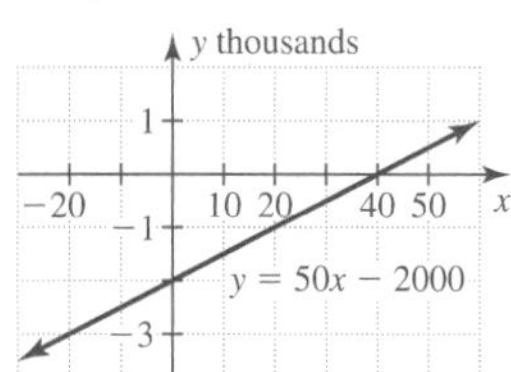

76. $y = -300x + 4500$

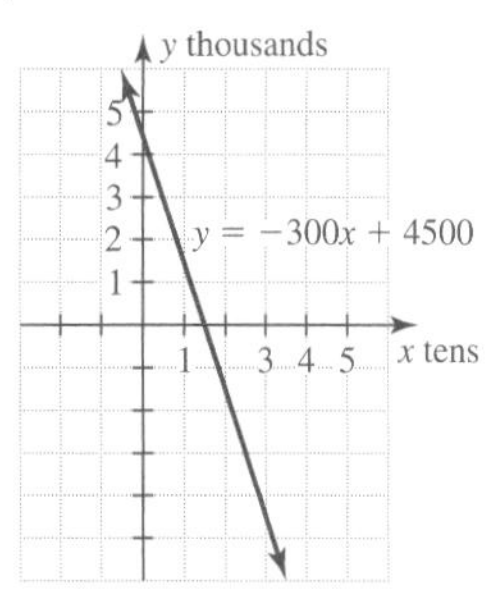

77. $y = -400x + 2000$

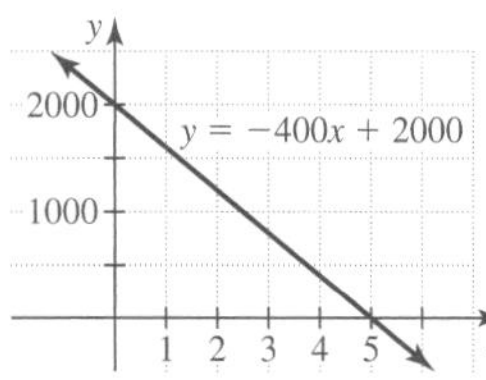

78. $y = 500x + 3$

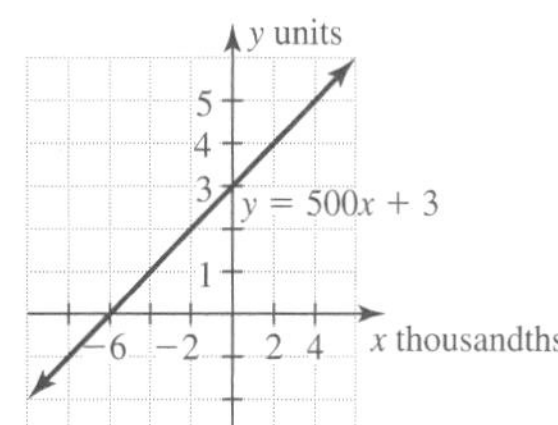

Graph each equation using the x- and y-intercepts. See Example 6. Use a third point to check.

79. $3x + 2y = 6$

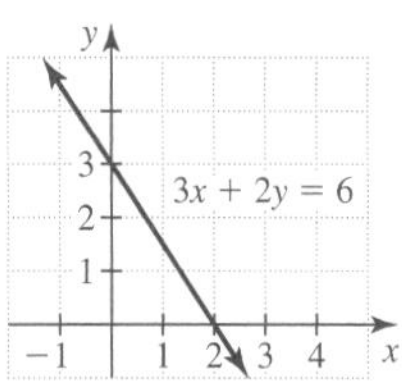

80. $2x + y = 6$

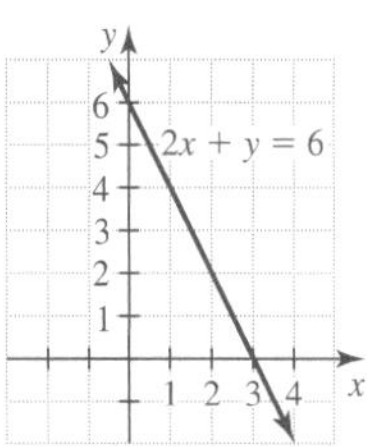

81. $x - 4y = 4$

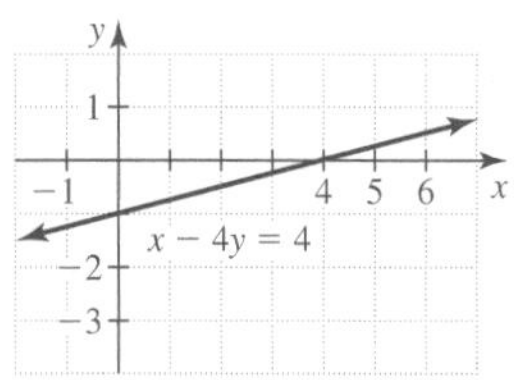

82. $-2x + y = 4$

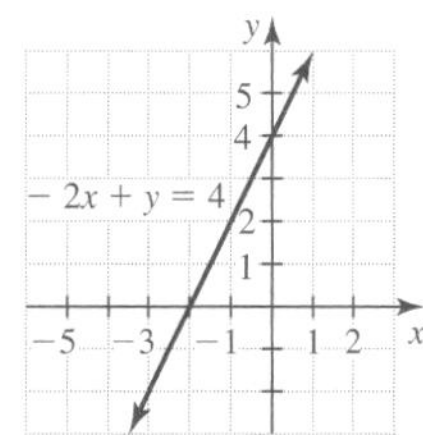

83. $y = \frac{3}{4}x - 9$

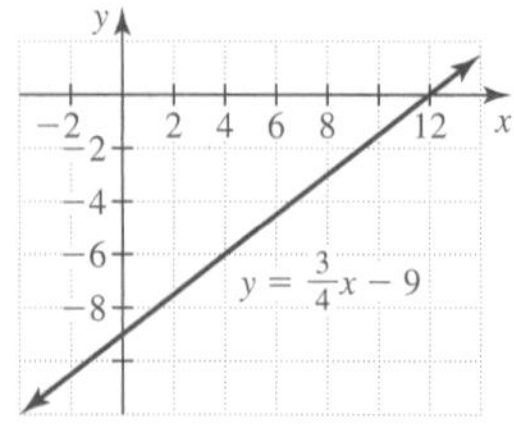

84. $y = -\frac{1}{2}x + 5$

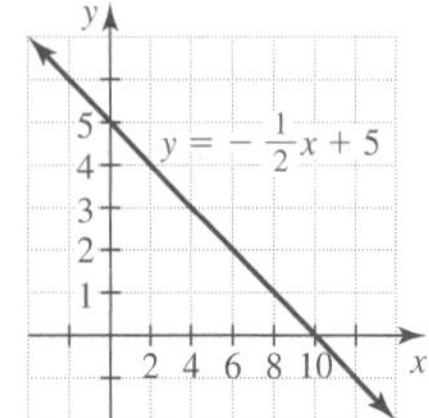

85. $\frac{1}{2}x + \frac{1}{4}y = 1$

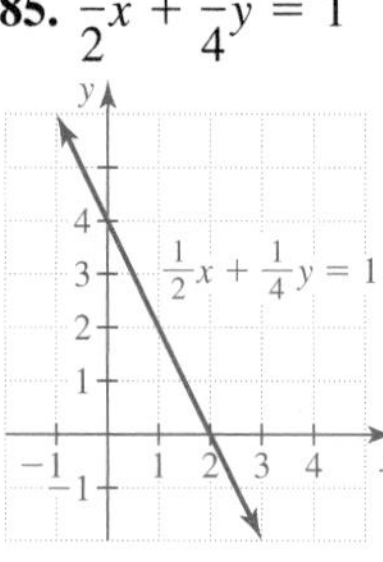

86. $\frac{1}{3}x - \frac{1}{2}y = 3$

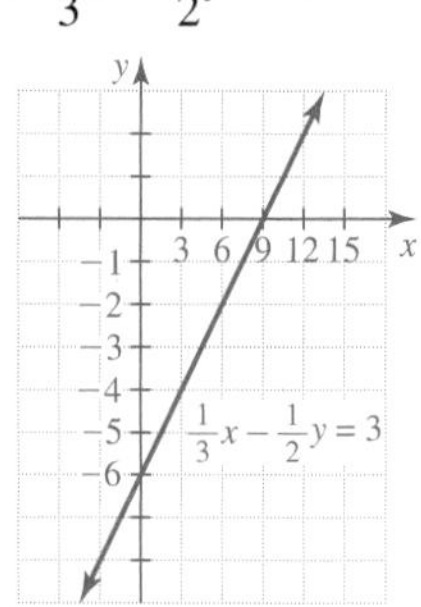

Solve each problem. See Example 7.

87. ***Percentage of full benefit.*** The age at which you retire affects your Social Security benefits. The accompanying graph gives the percentage of full benefit for each age from 62 through 70, based on current legislation and retirement after the year 2005 (Source: Social Security Administration). What percentage of full benefit does a person receive if that person retires at age 63? At what age will a retiree receive the full benefit? For what ages do you receive more than the full benefit?
75%, 67, 68 and up

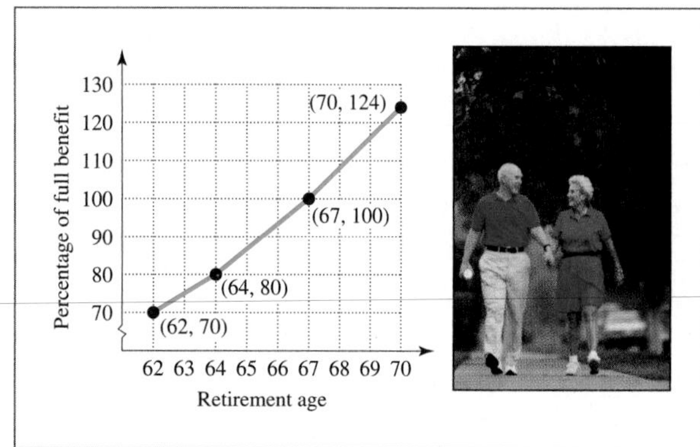

FIGURE FOR EXERCISE 87

88. ***Heel motion.*** When designing running shoes, Chris Edington studies the motion of a runner's foot. The following data gives the coordinates of the heel (in centimeters) at intervals of 0.05 millisecond during one cycle of level treadmill running at 3.8 meters per second (*Sagittal Plane Kinematics, Milliron and Cavanagh*):

(31.7, 5.7), (48.0, 5.7), (68.3, 5.8), (88.9, 6.9),
(107.2, 13.3), (119.4, 24.7), (127.2, 37.8),
(125.7, 52.0), (116.1, 60.2), (102.2, 59.5)
(88.7, 50.2), (73.9, 35.8), (52.6, 20.6),
(29.6, 10.7), (22.4, 5.9).

Graph these ordered pairs to see the heel motion.

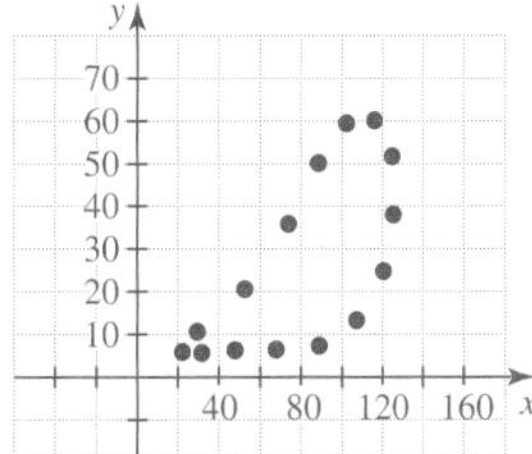

89. ***Medicaid spending.*** The cost in billions of dollars for federal Medicaid (health care for the poor) can be modeled by the equation

$$C = 3.2n + 65.3,$$

where n is the number of years since 1990 (Health Care Financing Administration, www.hcfa.gov).

a) What was the cost of federal Medicaid in 2000?

b) In what year will the cost reach \$150 billion?

c) Graph the equation for n ranging from 0 through 20.

a) \$97.3 billion **b)** 2016

c)

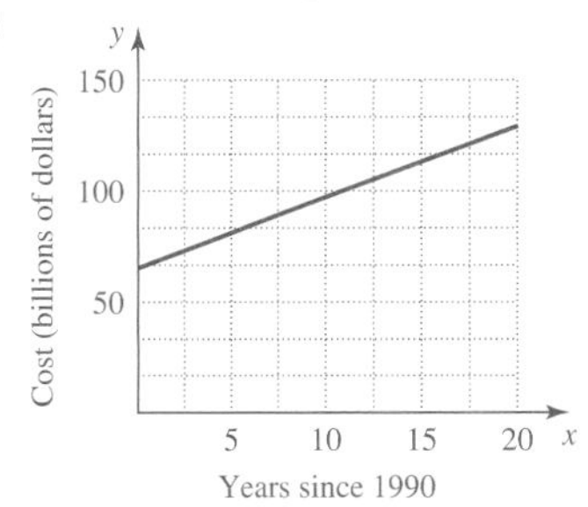

90. ***Dental services.*** The national cost C in billions of dollars for dental services can be modeled by the linear equation

$$C = 2.85n + 30.52,$$

where n is the number of years since 1990 (Health Care Financing Administration, www.hcfa.gov).

a) Find and interpret the C-intercept for the line.

b) Find and interpret the n-intercept for the line.

c) Graph the line for n ranging from 0 through 20.

d) If this trend continues, then in what year will the cost of dental services reach 100 billion?

a) (0, 30.52); The cost was \$30.52 billion in 1990.
b) (−10.71, 0); The cost was zero dollars in 1979.
c) **d)** 2014

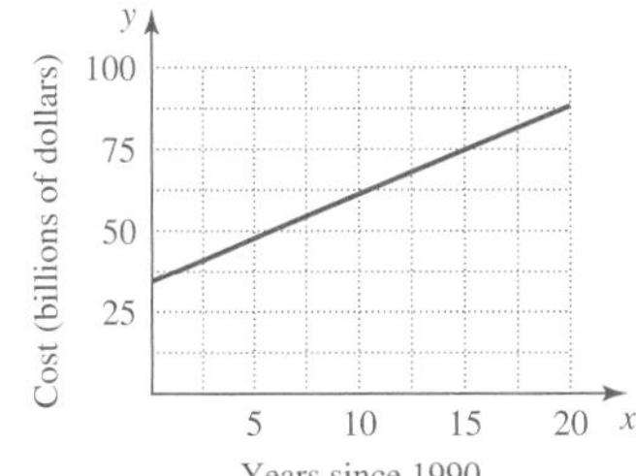

91. ***Hazards of depth.*** The accompanying table shows the depth below sea level and atmospheric pressure (*Encyclopedia of Sports Science,* 1997). The equation

$$A = 0.03d + 1$$

expresses the atmospheric pressure in terms of the depth d.

a) Find the atmospheric pressure at the depth where nitrogen narcosis begins. 4 atm

b) Find the maximum depth for intermediate divers. 130 ft

c) Graph the equation for d ranging from 0 to 250 feet.

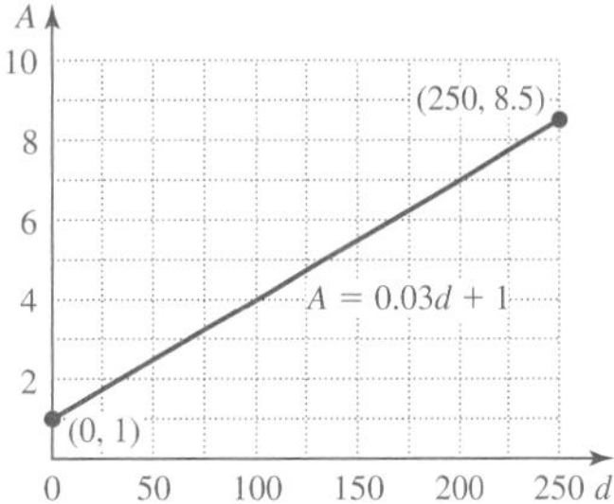

Depth (ft)	Atmospheric pressure (atm)	Comments
21	1.63	Bends are a danger
60	2.8	Maximum for beginners
100		Nitrogen narcosis begins
	4.9	Maximum for intermediate
200	7.0	Severe nitrogen narcosis
250	8.5	Extremely dangerous depth

FIGURE FOR EXERCISE 91

92. ***Demand equation.*** Helen's Health Foods usually sells 400 cans of ProPac Muscle Punch per week when the price is \$5 per can. After experimenting with prices for some time, Helen has determined that the weekly demand can be found by using the equation

$$d = 600 - 40p,$$

where d is the number of cans and p is the price per can.

a) Will Helen sell more or less Muscle Punch if she raises her price from \$5? Less

b) What happens to her sales every time she raises her price by \$1? Goes down by 40 cans

c) Graph the equation.

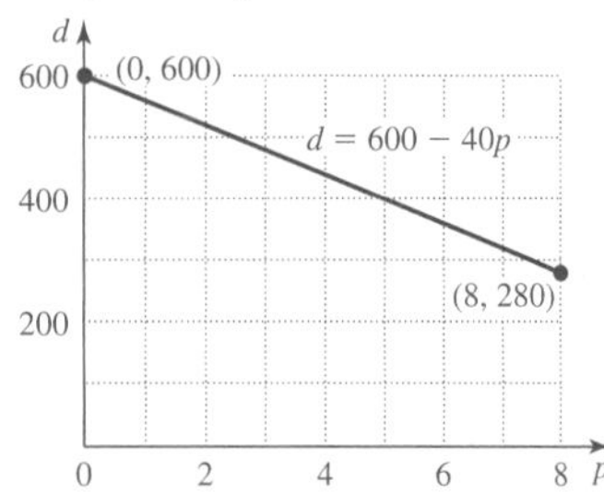

d) What is the maximum price that she can charge and still sell at least one can? $14.97

93. ***Advertising blitz.*** Furniture City in Toronto had \$24,000 to spend on advertising a year-end clearance sale. A 30-second radio ad costs \$300, and a 30-second local television ad costs \$400. To model this situation, the advertising manager wrote the equation $300x + 400y = 24{,}000$. What do x and y represent? Graph the equation. How many solutions are there to the equation, given that the number of ads of each type must be a whole number?
$x =$ the number of radio ads,
$y =$ the number of TV ads, 21 solutions

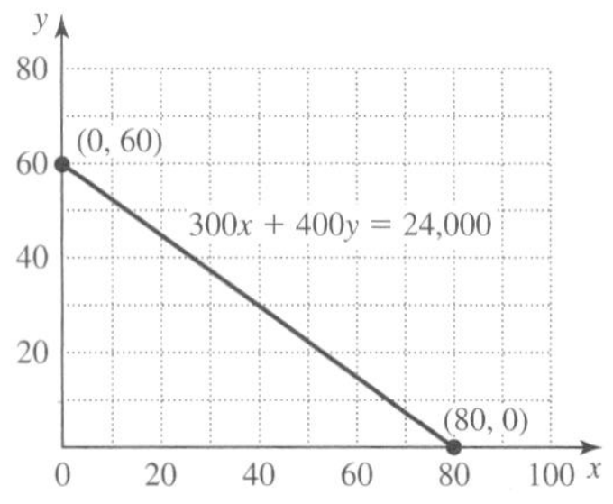

94. ***Material allocation.*** A tent maker had 4500 square yards of nylon tent material available. It takes 45 square yards of nylon to make an 8×10 tent and 50 square yards to make a 9×12 tent. To model this situation, the manager wrote the equation $45x + 50y = 4500$. What do x and y represent? Graph the equation. How many solutions are there to the equation, given that the number of tents of each type must be a whole number?
$x =$ the number of 8×10 tents,
$y =$ the number of 9×12 tents, 11 solutions

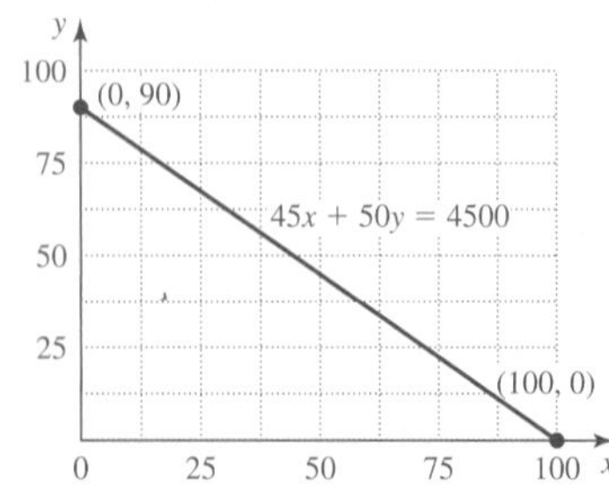

GRAPHING CALCULATOR EXERCISES

Graph each straight line on your graphing calculator using a viewing window that shows both intercepts. Answers may vary.

95. $2x + 3y = 1200$

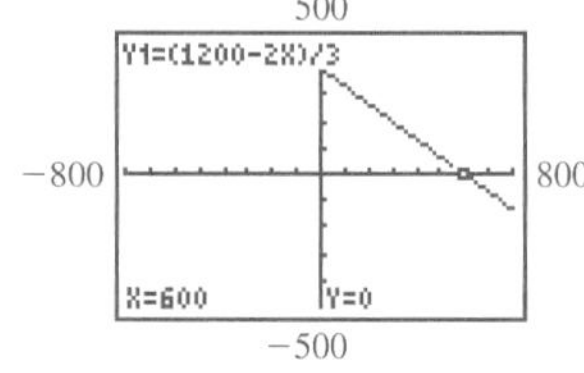

96. $3x - 700y = 2100$

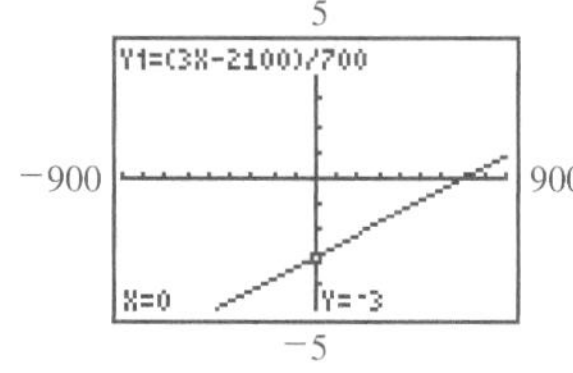

97. $200x - 300y = 6$

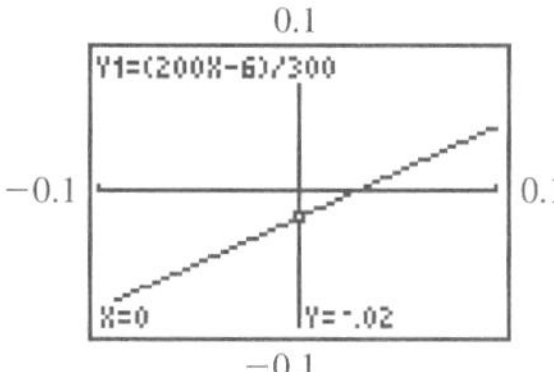

98. $300x + 5y = 20$

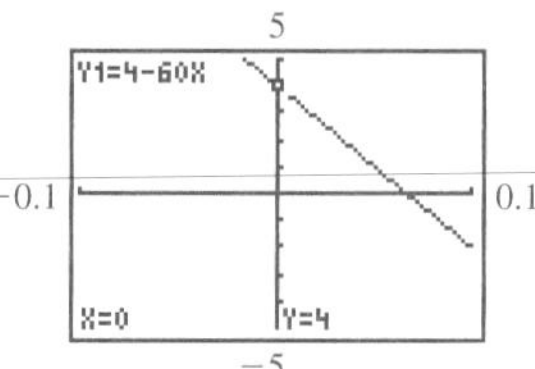

99. $y = 300x - 1$

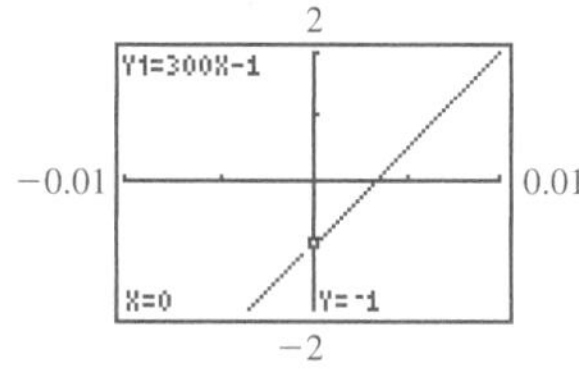

100. $y = 300x - 6000$

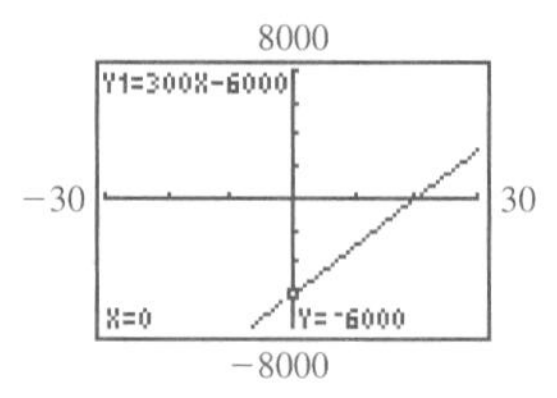

3.2 SLOPE

In This Section

- Slope Concepts
- Slope Using Coordinates
- Graphing a Line Given a Point and Its Slope
- Parallel Lines
- Perpendicular Lines
- Interpreting Slope

In Section 3.1 you learned that the graph of a linear equation is a straight line. In this section, we will continue our study of lines in the coordinate plane.

Slope Concepts

If a highway rises 6 feet in a horizontal run of 100 feet, then the grade is $\frac{6}{100}$ or 6%. See Fig. 3.12. The grade of a road is a measurement of the steepness of the road. It is the rate at which the road is going upward.

FIGURE 3.12

The steepness of a line is called the **slope** of the line and it is measured like the grade of a road. As you move from (1, 1) to (4, 3) in Fig. 3.13 the x-coordinate increases by 3 and the y-coordinate increases by 2. The line rises 2 units in a horizontal run of 3 units. So the slope of the line is $\frac{2}{3}$. The slope is the rate at which the y-coordinate is increasing. It increases 2 units for every 3-unit increase in x or it increases $\frac{2}{3}$ of a unit for every 1-unit increase in x. In general, we have the following definition of slope.

Slope

$$\text{Slope} = \frac{\text{change in } y\text{-coordinate}}{\text{change in } x\text{-coordinate}}$$

If we move from the point (4, 3) to the point (1, 1), there is a change of -2 in the y-coordinate and a change of -3 in the x-coordinate. See Fig. 3.14. In this case we get

$$\text{Slope} = \frac{-2}{-3} = \frac{2}{3}.$$

Note that going from (4, 3) to (1, 1) gives the same slope as going from (1, 1) to (4, 3).

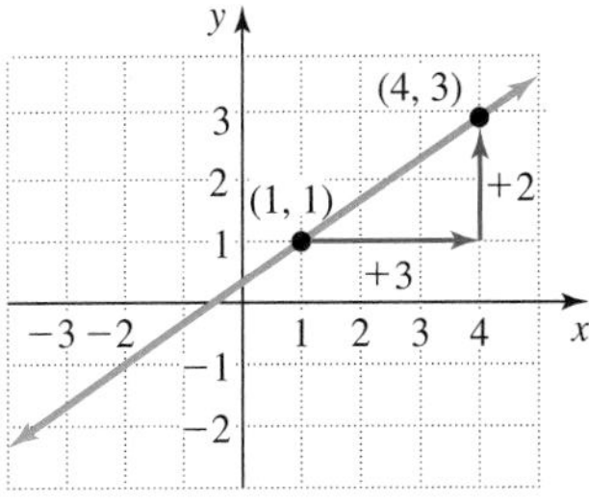

FIGURE 3.13

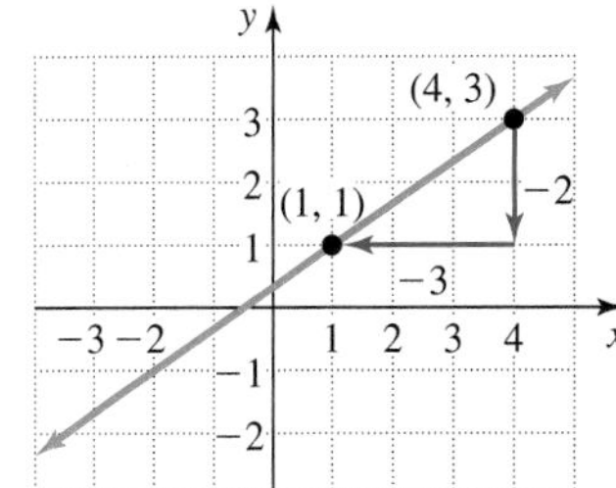

FIGURE 3.14

We call the change in y-coordinate the **rise** and the change in x-coordinate the **run.** Moving up is a positive rise, and moving down is a negative rise. Moving to the right is a positive run, and moving to the left is a negative run. We usually use the letter m to stand for slope. So we have

$$m = \frac{\text{change in } y}{\text{change in } x} = \frac{\text{rise}}{\text{run}}.$$

EXAMPLE 1 **Finding the slope of a line**

Find the slopes of the given lines by going from point A to point B.

a)

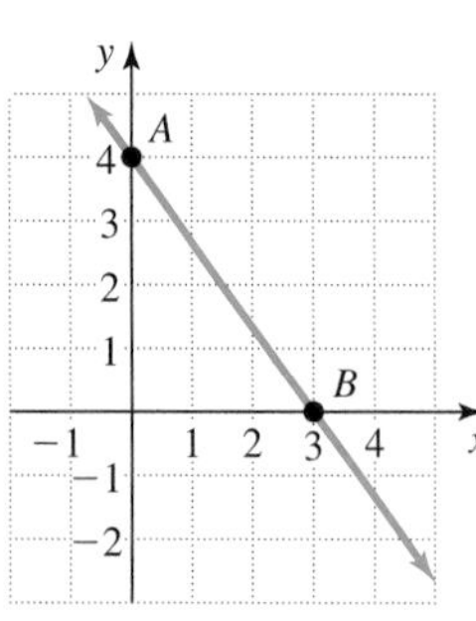

b)

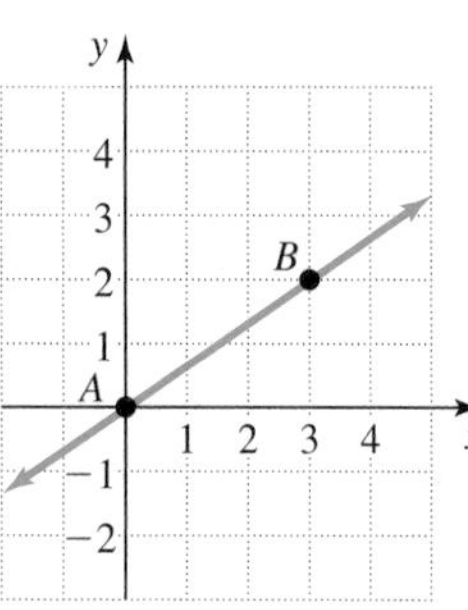

c)

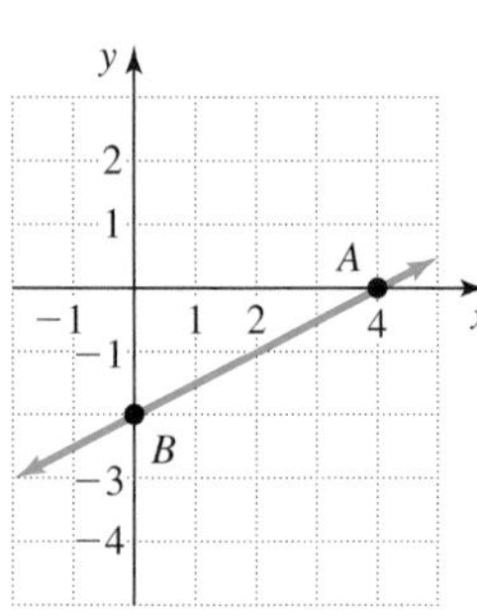

Solution

a) The coordinates of point A are (0, 4), and the coordinates of point B are (3, 0). Going from A to B, the change in y is -4, and the change in x is $+3$. So

$$m = \frac{-4}{3} = -\frac{4}{3}.$$

b) Going from A to B, the rise is 2, and the run is 3. So

$$m = \frac{2}{3}.$$

c) Going from A to B, the rise is -2, and the run is -4. So

$$m = \frac{-2}{-4} = \frac{1}{2}.$$

■

Study Tip

Working problems 1 hour per day every day of the week is better than working problems for 7 hours on one day of the week. It is usually better to spread out your study time than to try and learn everything in one big session.

CAUTION The change in y is always in the numerator, and the change in x is always in the denominator.

The ratio of rise to run is the ratio of the lengths of the two legs of any right triangle whose hypotenuse is on the line. As long as one leg is vertical and the other is horizontal, all such triangles for a certain line have the same shape. These triangles are similar triangles. The ratio of the length of the vertical side to the length of the horizontal side for any two such triangles is the same number. So we get the same value for the slope no matter which two points of the line are used to calculate it or in which order the points are used.

EXAMPLE 2 **Finding slope**

Find the slope of the line shown here using points A and B, points A and C, and points B and C.

Solution

Using A and B, we get

$$m = \frac{\text{rise}}{\text{run}} = \frac{1}{4}.$$

Using A and C, we get

$$m = \frac{\text{rise}}{\text{run}} = \frac{2}{8} = \frac{1}{4}.$$

Using B and C, we get

$$m = \frac{\text{rise}}{\text{run}} = \frac{1}{4}.$$

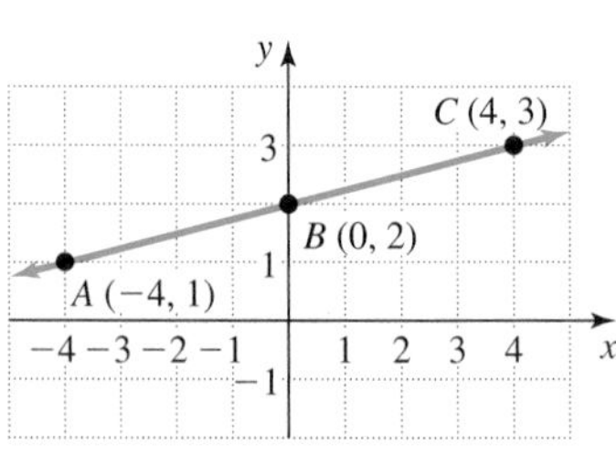

■

Helpful Hint

It is good to think of what the slope represents when x and y are measured quantities rather than just numbers. For example, if the change in y is 50 miles and the change in x is 2 hours, then the slope is 25 mph (or 25 miles per 1 hour). So the slope is the amount of change in y for a change of one in x.

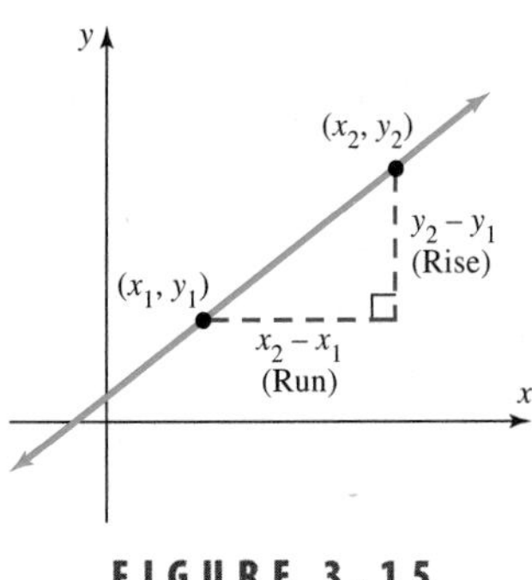

FIGURE 3.15

Slope Using Coordinates

One way to obtain the rise and run is from a graph. The rise and run can also be found by using the coordinates of two points on the line as shown in Fig. 3.15.

Coordinate Formula for Slope

The slope of the line containing the points (x_1, y_1) and (x_2, y_2) is given by

$$m = \frac{y_2 - y_1}{x_2 - x_1},$$

provided that $x_2 - x_1 \neq 0$.

EXAMPLE 3 **Using coordinates to find slope**

Find the slope of each of the following lines.

a) The line through (0, 5) and (6, 3)

b) The line through (−3, 4) and (−5, −2)

c) The line through (−4, 2) and the origin

Study Tip

Students who have difficulty with algebra often schedule it in a class that meets one day per week so they do not have to see it as often. However, many students do better in classes that meet more often for shorter time periods. So schedule your classes to maximize your chances of success.

Solution

a) If $(x_1, y_1) = (0, 5)$ and $(x_2, y_2) = (6, 3)$ then

$$m = \frac{y_2 - y_1}{x_2 - x_1}$$
$$= \frac{3 - 5}{6 - 0} = \frac{-2}{6} = -\frac{1}{3}.$$

If $(x_1, y_1) = (6, 3)$ and $(x_2, y_2) = (0, 5)$ then

$$m = \frac{y_2 - y_1}{x_2 - x_1}$$
$$= \frac{5 - 3}{0 - 6} = \frac{2}{-6} = -\frac{1}{3}.$$

Note that it does not matter which point is called (x_1, y_1) and which is called (x_2, y_2). In either case the slope is $-\frac{1}{3}$.

b) Let $(x_1, y_1) = (-3, 4)$ and $(x_2, y_2) = (-5, -2)$:

$$m = \frac{y_2 - y_1}{x_2 - x_1}$$
$$= \frac{-2 - 4}{-5 - (-3)}$$
$$= \frac{-6}{-2} = 3$$

c) Let $(x_1, y_1) = (0, 0)$ and $(x_2, y_2) = (-4, 2)$:

$$m = \frac{2 - 0}{-4 - 0} = \frac{2}{-4} = -\frac{1}{2}$$

■

CAUTION It does not matter which point is called (x_1, y_1) and which is called (x_2, y_2), but if you divide $y_2 - y_1$ by $x_1 - x_2$, the slope will have the wrong sign.

Because division by zero is undefined, slope is undefined if $x_2 - x_1 = 0$ or $x_2 = x_1$. The x-coordinates of two distinct points on a line are equal only if the points are on a vertical line. *So slope is undefined for vertical lines.* The concept of slope does not exist for a vertical line.

Any two points on a horizontal line have equal y-coordinates. So for points on a horizontal line we have $y_2 - y_1 = 0$. Since $y_2 - y_1$ is in the numerator of the slope formula, *the slope for any horizontal line is zero.* We never refer to a line as having "no slope" because in English no can mean zero or does not exist.

EXAMPLE 4 **Slope for vertical and horizontal lines**

Find the slope of the line through each pair of points.

a) (2, 1) and (2, −3)

b) (−2, 2) and (4, 2)

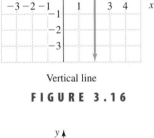

Vertical line

FIGURE 3.16

Solution

a) The points (2, 1) and (2, −3) are on the vertical line shown in Fig. 3.16. Since slope is undefined for vertical lines, this line does not have a slope. Using the slope formula we get

$$m = \frac{-3 - 1}{2 - 2} = \frac{-4}{0}.$$

Since division by zero is undefined, we can again conclude that slope is undefined for the vertical line through the given points.

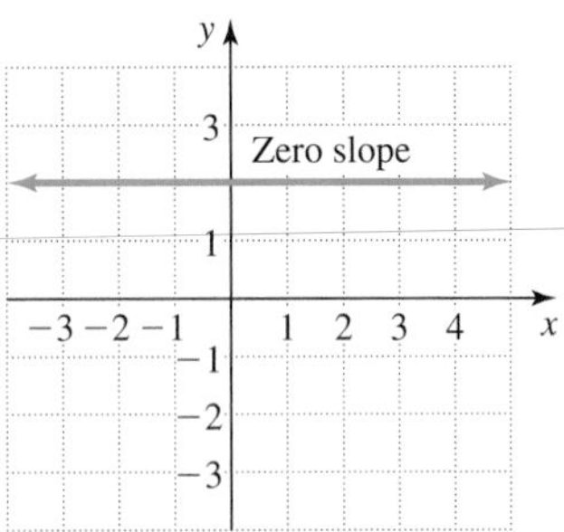

Horizontal line

FIGURE 3.17

b) The points (−2, 2) and (4, 2) are on the horizontal line shown in Fig. 3.17. Using the slope formula we get

$$m = \frac{2 - 2}{-2 - 4} = \frac{0}{-6} = 0.$$

So the slope of the horizontal line through these points is 0. ■

Note that for a line with *positive slope,* the y-values increase as the x-values increase. For a line with *negative slope,* the y-values decrease as the x-values increase. See Fig. 3.18.

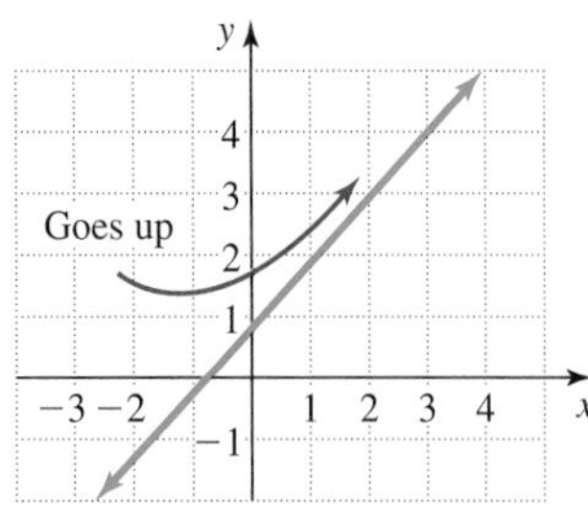

Positive slope

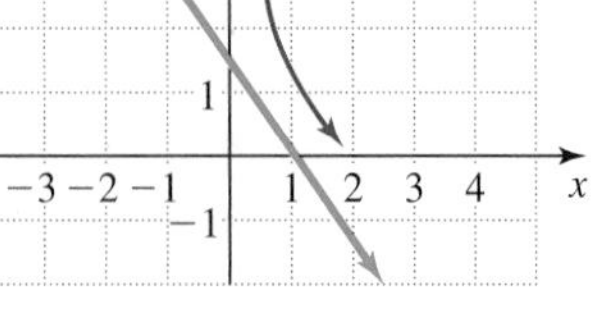

Negative slope

FIGURE 3.18

Graphing a Line Given a Point and Its Slope

To graph a line from its equation we usually make a table of ordered pairs and then draw a line through the points or we use the intercepts. In Example 5 we will graph a line using one point and the slope. From the slope we find additional points by using the rise and the run.

EXAMPLE 5 **Graphing a line given a point and its slope**

Graph each line.

a) The line through (2, 1) with slope $\frac{3}{4}$

b) The line through $(-2, 4)$ with slope -3

Calculator Close-Up

When we graph a line we usually draw a graph that shows both intercepts, because they are important features of the graph. If the intercepts are not between -10 and 10, you will have to adjust the window to get a good graph. The viewing window that has x- and y-values ranging from a minimum of -10 to a maximum of 10 is called the **standard viewing window.**

Solution

a) First locate the point (2, 1). Because the slope is $\frac{3}{4}$, we can find another point on the line by going up three units and to the right four units to get the point (6, 4), as shown in Fig. 3.19. Now draw a line through (2, 1) and (6, 4). Since $\frac{3}{4} = \frac{-3}{-4}$ we could have obtained the second point by starting at (1, 2) and going down 3 units and to the left 4 units.

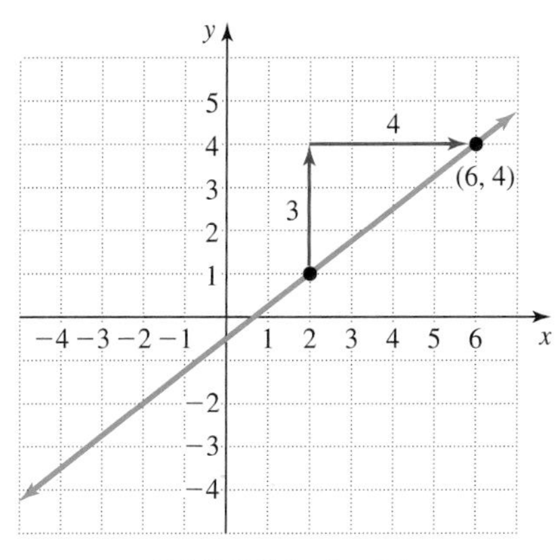

FIGURE 3.19

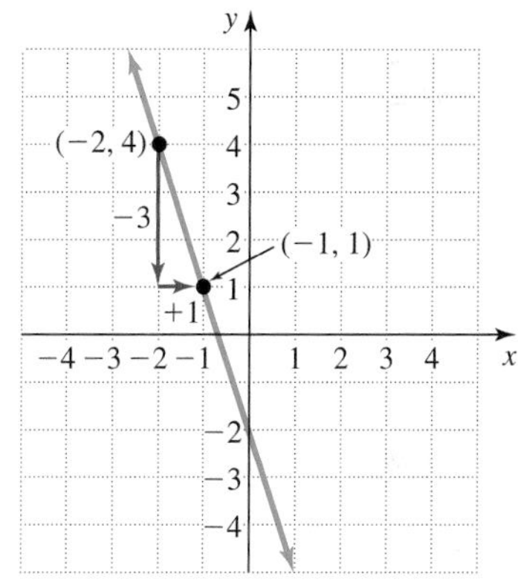

FIGURE 3.20

b) First locate the point $(-2, 4)$. Because the slope is -3, or $\frac{-3}{1}$, we can locate another point on the line by starting at $(-2, 4)$ and moving down three units and then one unit to the right to get the point $(-1, 1)$. Now draw a line through $(-2, 4)$ and $(-1, 1)$ as shown in Fig. 3.20. Since $\frac{-3}{1} = \frac{3}{-1}$ we could have obtained the second point by starting at $(-2, 4)$ and going up 3 units and to the left 1 unit. ■

Parallel Lines

Every nonvertical line has a unique slope, but there are infinitely many lines with a given slope. All lines that have a given slope are parallel.

Parallel Lines

Nonvertical lines are parallel if and only if they have equal slopes. Any two vertical lines are parallel to each other.

EXAMPLE 6 **Graphing parallel lines**

Draw a line through the point $(-2, 1)$ with slope $\frac{1}{2}$ and a line through $(3, 0)$ with slope $\frac{1}{2}$.

Solution

Because slope is the ratio of rise to run, a slope of $\frac{1}{2}$ means that we can locate a second point of the line by starting at $(-2, 1)$ and going up one unit and to the right two units. For the line through $(3, 0)$ we start at $(3, 0)$ and go up one unit and to the right two units. See Fig. 3.21.

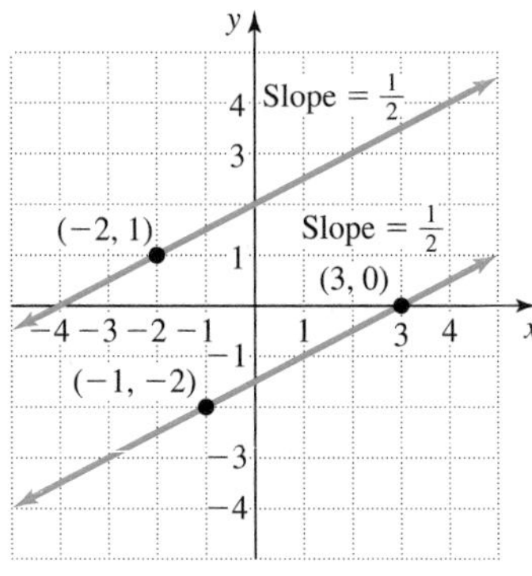

FIGURE 3.21

Perpendicular Lines

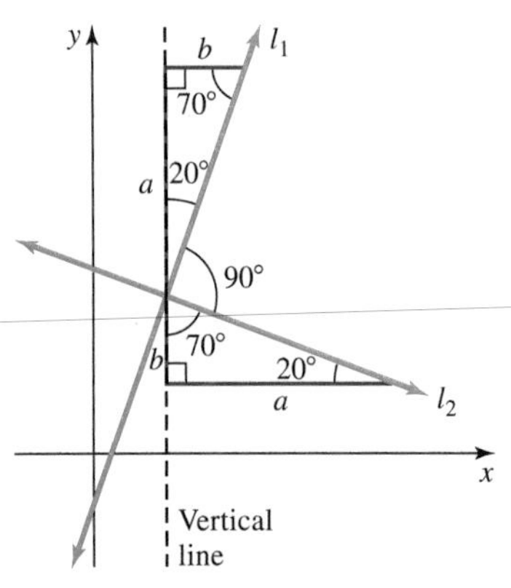

FIGURE 3.22

Figure 3.22 shows two right triangles with acute angles of 20° and 70° and legs with lengths a and b ($a > 0$, $b > 0$) positioned along a vertical line. The angle between lines l_1 and l_2 in Fig. 3.22 must be 90° because that angle along with 20° and 70° together form the vertical line. Now the slope of l_1 is $\frac{a}{b}$ and the slope of l_2 is $\frac{-b}{a}$. That is, the slope of one line is the opposite of the reciprocal of the slope of the other. This example illustrates the following rule.

Perpendicular Lines

Two lines with slopes m_1 and m_2 are perpendicular if and only if

$$m_1 = -\frac{1}{m_2}.$$

Any vertical line is perpendicular to any horizontal line.

Notice that we cannot compare slopes of horizontal and vertical lines to see if they are perpendicular because slope is not defined for vertical lines.

EXAMPLE 7 **Graphing perpendicular lines**

Draw two lines through the point $(-1, 2)$, one with slope $-\frac{1}{3}$ and the other with slope 3.

Solution

Because slope is the ratio of rise to run, a slope of $-\frac{1}{3}$ means that we can locate a second point on the line by starting at $(-1, 2)$ and going down one unit and to the right three units. For the line with slope 3, we start at $(-1, 2)$ and go up three units and to the right one unit. See Fig. 3.23.

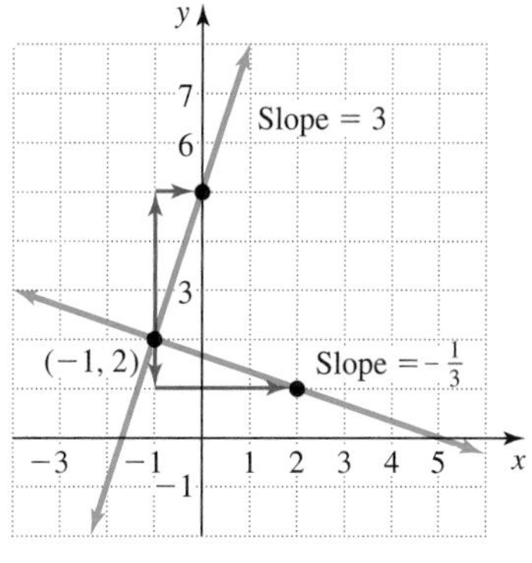

FIGURE 3.23

Helpful Hint

The relationship between the slopes of perpendicular lines can also be remembered as

$$m_1 \cdot m_2 = -1.$$

For example, lines with slopes -3 and $\frac{1}{3}$ are perpendicular because $-3 \cdot \frac{1}{3} = -1$.

Interpreting Slope

Slope of a line is the ratio of the rise and the run. If the rise is measured in dollars and the run in days, then the slope is measured in dollars per day or dollars/day. The slope of a line is the rate at which the dependent variable is increasing or decreasing.

EXAMPLE 8 **Interpreting slope**

A car goes from 60 mph to 0 mph in 120 feet after applying the brakes.

a) Find and interpret the slope of the line shown here.

b) What is the velocity at a distance of 80 feet?

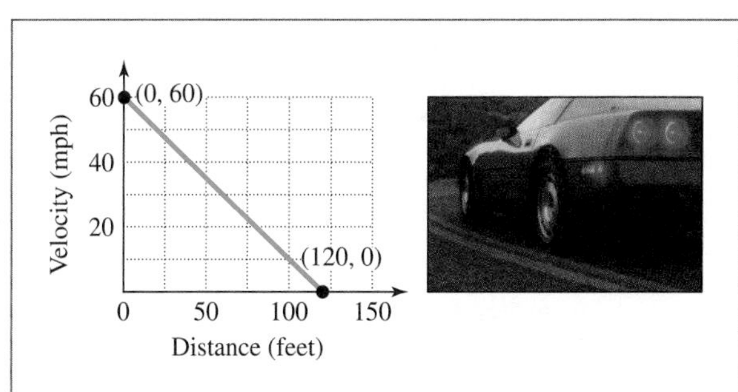

Solution

a) Find the slope of the line through $(0, 60)$ and $(120, 0)$:

$$m = \frac{60 - 0}{0 - 120} = -0.5$$

Because the vertical axis is miles per hour and the horizontal axis is feet, the slope is -0.5 mph/ft, which means the car is losing 0.5 mph of velocity for every foot it travels after the brakes are applied.

b) If the velocity is decreasing 0.5 mph for every foot the car travels, then in 80 feet the velocity goes down 0.5(80) or 40 mph. So the velocity at 80 feet is $60 - 40$ or 20 mph.

EXAMPLE 9 **Finding points when given the slope**

Assume that the base price of a new Jeep Wrangler is increasing $300 per year. Find the data that is missing from the table.

Year	Price (dollars)
2001	15,600
2002	
2003	
	18,300
	20,100

Solution

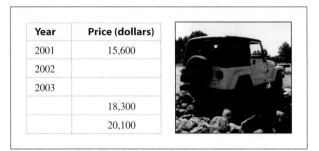

The price in 2002 is $15,900 and in 2003 it is $16,200 because the slope is $300 per year. The rise in price from $16,200 to $18,300 is $2100, which takes 7 years at $300 per year. So in 2010 the price is $18,300. The rise from $18,300 to $20,100 is $1800, which takes 6 years at $300 per year. So in 2016 the price is $20,100. ■

WARM-UPS

True or false? Explain your answer.

1. Slope is a measurement of the steepness of a line. True
2. Slope is rise divided by run. True
3. Every line in the coordinate plane has a slope. False
4. The line through the point (1, 1) and the origin has slope 1. True
5. Slope can never be negative. False
6. A line with slope 2 is perpendicular to any line with slope −2. False
7. The slope of the line through (0, 3) and (4, 0) is $\frac{3}{4}$. False
8. Two different lines cannot have the same slope. False
9. The line through (1, 3) and (−5, 3) has zero slope. True
10. Slope can have units such as feet per second. True

3.2 EXERCISES

Reading and Writing *After reading this section, write out the answers to these questions. Use complete sentences.*

1. What is the slope of a line?
   The slope of a line is the ratio of its rise and run.
2. What is the difference between rise and run?
   Rise is the amount of vertical change and run is the amount of horizontal change.
3. For which lines is slope undefined?
   Slope is undefined for vertical lines.
4. Which lines have zero slope?
   Horizontal lines have zero slope.
5. What is the difference between lines with positive slope and lines with negative slope?
   Lines with positive slope are rising as you go from left to right, while lines with negative slope are falling as you go from left to right.
6. What is the relationship between the slopes of perpendicular lines?
   If m_1 and m_2 are slopes of perpendicular lines, then $m_1 = \frac{-1}{m_2}$.

Solve each problem. See Example 7.

77. ***Marginal cost.*** A manufacturer plans to spend \$150,000 on research and development for a new lawn mower and then \$200 to manufacture each mower. The formula $C = 200n + 150{,}000$ gives the cost in dollars of n mowers. What is the cost of 5000 mowers? What is the cost of 5001 mowers? By how much did the one extra lawn mower increase the cost? (The increase in cost is called the *marginal cost* of the 5001st lawn mower.) \$1,150,000, \$1,150,200, \$200

78. ***Marginal revenue.*** A defense attorney charges her client \$4000 plus \$120 per hour. The formula $R = 120n + 4000$ gives her revenue in dollars for n hours of work. What is her revenue for 100 hours of work? What is her revenue for 101 hours of work? By how much did the one extra hour of work increase the revenue? (The increase in revenue is called the *marginal revenue* for the 101st hour.) \$16,000, \$16,120, \$120

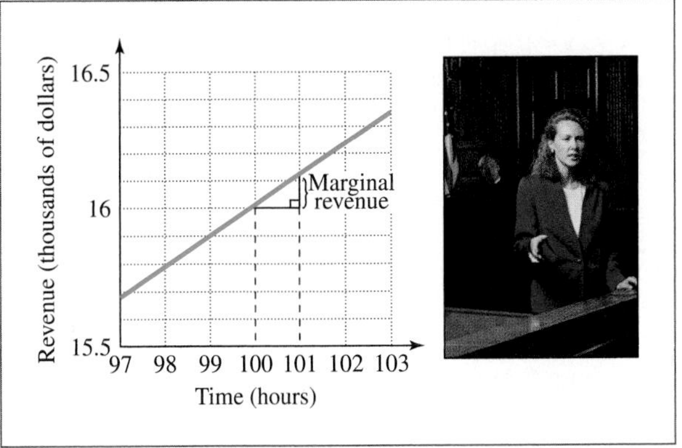

FIGURE FOR EXERCISE 78

79. ***In-house training.*** The accompanying graph shows the percentage of U.S. workers receiving training by their employers (Department of Labor, www.dol.gov). The percentage went from 5% in 1982 to 25% in 2002.

a) Find and interpret the slope of the line.

b) Write the equation of the line in slope-intercept form.

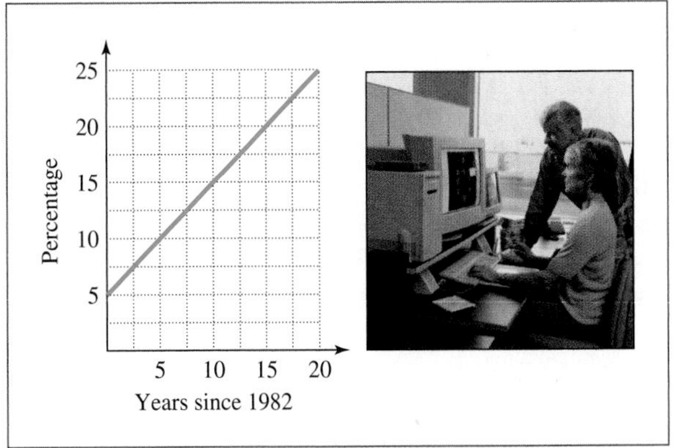

FIGURE FOR EXERCISE 79

c) What is the meaning of the y-intercept?

d) Use your equation to predict the percentage that will be receiving training in 2010.

a) A slope of 1 means that the percentage of workers receiving training is going up 1% per year.

b) $y = x + 5$ where x is the number of years since 1982

c) The y-intercept $(0, 5)$ means that 5% of the workers received training in 1982.

d) 33%

80. ***Single women.*** The percentage of women in the 20–24 age group who have never married went from 55% in 1970 to 73% in 2000 (Census Bureau, www.census.gov). Let 1970 be year 0 and 2000 be year 30.

a) Find and interpret the slope of the line through the points $(0, 55)$ and $(30, 73)$.

b) Find the equation of the line in part (a).

c) What is the meaning of the y-intercept?

d) Use the equation to predict the percentage in 2010.

e) If this trend continues, then in what year will the percentage of women in the 20–24 age group who have never married reach 100%?

a) A slope of 0.6 means that the percentage is increasing by 0.6% per year.

b) $y = 0.6x + 55$

c) The y-intercept $(0, 55)$ means that in 1970 55% of the women between 20 and 24 had never been married.

d) 79%

e) 2045

81. ***Pansies and snapdragons.*** A nursery manager plans to spend \$100 on 6-packs of pansies at 50 cents per pack and snapdragons at 25 cents per pack. The equation $0.50x + 0.25y = 100$ can be used to model this situation.

a) What do x and y represent?

x = the number of packs of pansies, y = the number of packs of snapdragons

b) Graph the equation.

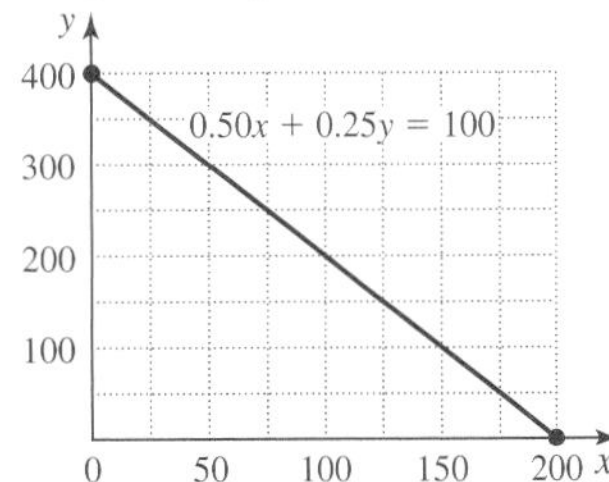

c) Write the equation in slope-intercept form.

$y = -2x + 400$

d) What is the slope of the line? -2

e) What does the slope tell you?

If the number of packs of pansies goes up by 1, then the number of packs of snapdragons goes down by 2.

82. ***Pens and pencils.*** A bookstore manager plans to spend \$60 on pens at 30 cents each and pencils at 10 cents each. The equation $0.10x + 0.30y = 60$ can be used to model this situation.

a) What do x and y represent?

$x =$ the number of pencils, $y =$ the number of pens

b) Graph the equation.

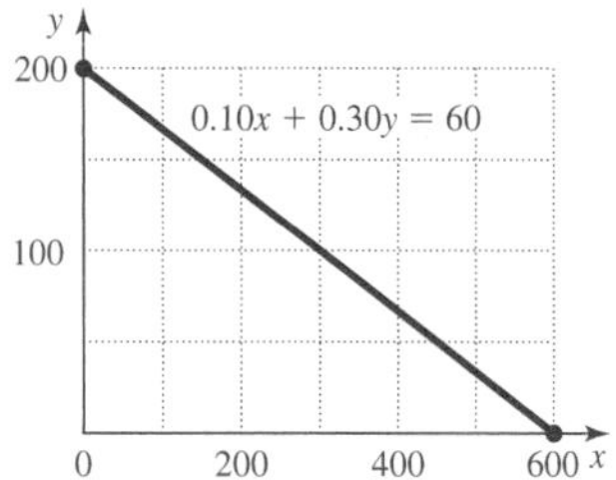

c) Write the equation in slope-intercept form.

$y = -\frac{1}{3}x + 200$

d) What is the slope of the line?

$-\frac{1}{3}$

e) What does the slope tell you?

If the number of pencils increases by 3, then the number of pens goes down by 1.

GRAPHING CALCULATOR EXERCISES

Graph each pair of straight lines on your graphing calculator using a viewing window that makes the lines look perpendicular. Answers may vary.

83. $y = 12x - 100$, $y = -\frac{1}{12}x + 50$

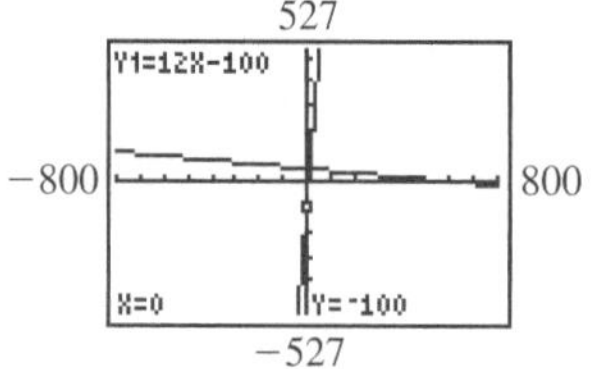

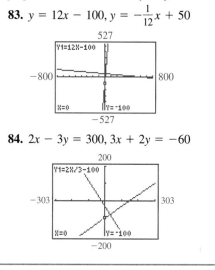

84. $2x - 3y = 300$, $3x + 2y = -60$

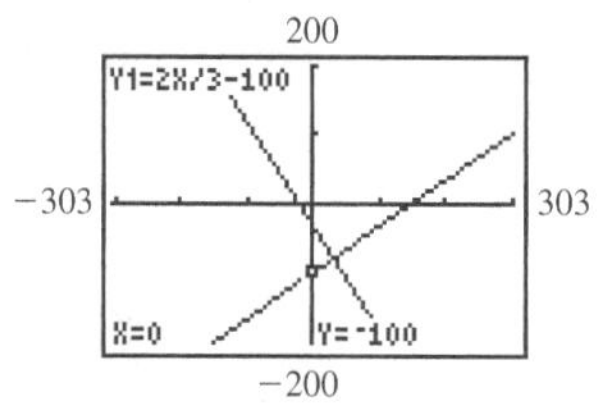

In This Section

- Point-Slope Form
- Parallel Lines
- Perpendicular Lines
- Applications

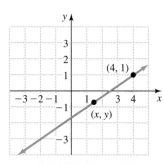

3.4 THE POINT-SLOPE FORM

In Section 3.3 we wrote the equation of a line given its slope and y-intercept. In this section you will learn to write the equation of a line given the slope and *any* other point on the line.

Point-Slope Form

Consider a line through the point (4, 1) with slope $\frac{2}{3}$ as shown in Fig. 3.27. Because the slope can be found by using any two points on the line, we use (4, 1) and an arbitrary point (x, y) in the formula for slope:

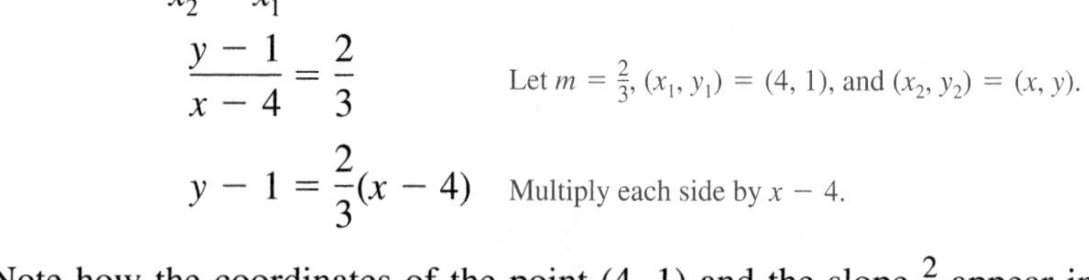

$$\frac{y_2 - y_1}{x_2 - x_1} = m \quad \text{Slope formula}$$

$$\frac{y - 1}{x - 4} = \frac{2}{3} \quad \text{Let } m = \tfrac{2}{3},\ (x_1, y_1) = (4, 1), \text{ and } (x_2, y_2) = (x, y).$$

$$y - 1 = \frac{2}{3}(x - 4) \quad \text{Multiply each side by } x - 4.$$

FIGURE 3.27

Note how the coordinates of the point (4, 1) and the slope $\frac{2}{3}$ appear in the above equation. We can use the same procedure to get the equation of any line given one point on the line and the slope. The resulting equation is called the **point-slope form** of the equation of the line.

Helpful Hint

If a point (x, y) is on a line with slope m through (x_1, y_1), then

$$\frac{y - y_1}{x - x_1} = m.$$

Multiplying each side of this equation by $x - x_1$ gives us the point-slope form.

Point-Slope Form

The equation of the line through the point (x_1, y_1) with slope m is

$$y - y_1 = m(x - x_1).$$

EXAMPLE 1 **Writing an equation given a point and a slope**

Find the equation of the line through $(-2, 3)$ with slope $\frac{1}{2}$, and write it in slope-intercept form.

Solution

Because we know a point and the slope, we can use the point-slope form:

$$y - y_1 = m(x - x_1) \quad \text{Point-slope form}$$

$$y - 3 = \frac{1}{2}[x - (-2)] \quad \text{Substitute } m = \tfrac{1}{2} \text{ and } (x_1, y_1) = (-2, 3).$$

$$y - 3 = \frac{1}{2}(x + 2) \quad \text{Simplify.}$$

$$y - 3 = \frac{1}{2}x + 1 \quad \text{Distributive property}$$

$$y = \frac{1}{2}x + 4 \quad \text{Slope-intercept form}$$

Alternate Solution

Replace m by $\frac{1}{2}$, x by -2, and y by 3 in the slope-intercept form:

$$y = mx + b \quad \text{Slope-intercept form}$$

$$3 = \frac{1}{2}(-2) + b \quad \text{Substitute } m = \tfrac{1}{2} \text{ and } (x, y) = (-2, 3).$$

$$3 = -1 + b \quad \text{Simplify.}$$

$$4 = b$$

Since $b = 4$, we can write $y = \frac{1}{2}x + 4$. ■

The alternate solution to Example 1 is shown because many students have seen that method in the past. This does not mean that you should ignore the point-slope form. It is always good to know more than one method to accomplish a task. The good thing about using the point-slope form is that you immediately write down the equation and then you simplify it. In the alternate solution, the last thing you do is to write the equation.

The point-slope form can be used to find the equation of a line for *any* given point and slope. However, if the given point is the y-intercept, then it is simpler to use the slope-intercept form. Note that it is not necessary that the slope be given, because the slope can be found from any two points. So if we know two points on a line, then we can find the slope and use the slope with either one of the points in the point-slope form.

EXAMPLE 2 **Writing an equation given two points**

Find the equation of the line that contains the points $(-3, -2)$ and $(4, -1)$, and write it in standard form.

Solution

First find the slope using the two given points:

$$m = \frac{-2 - (-1)}{-3 - 4} = \frac{-1}{-7} = \frac{1}{7}$$

Calculator Close-Up

Graph $y = (x + 3)/7 - 2$ to see that the line goes through $(-3, -2)$ and $(4, -1)$.

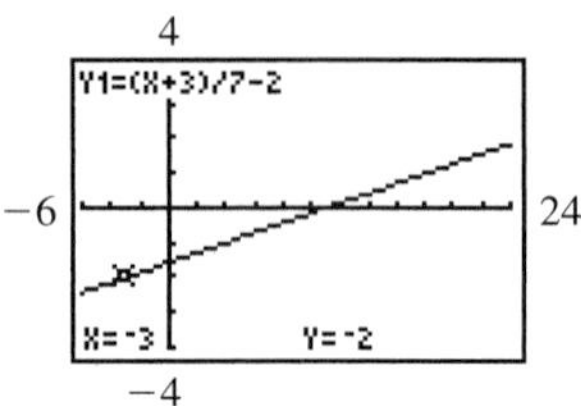

Note that the form of the equation does not matter on the calculator as long as it is solved for y.

Now use one of the points, say $(-3, -2)$, and slope $\frac{1}{7}$ in the point-slope form:

$$y - y_1 = m(x - x_1) \quad \text{Point-slope form}$$

$$y - (-2) = \frac{1}{7}[x - (-3)] \quad \text{Substitute.}$$

$$y + 2 = \frac{1}{7}(x + 3) \quad \text{Simplify.}$$

$$7(y + 2) = 7 \cdot \frac{1}{7}(x + 3) \quad \text{Multiply each side by 7.}$$

$$7y + 14 = x + 3$$

$$7y = x - 11 \quad \text{Subtract 14 from each side.}$$

$$-x + 7y = -11 \quad \text{Subtract } x \text{ from each side.}$$

$$x - 7y = 11 \quad \text{Multiply each side by } -1.$$

The equation in standard form is $x - 7y = 11$. Using the other given point, $(4, -1)$, would give the same final equation in standard form. Try it. ■

Parallel Lines

In Section 3.2 you learned that parallel lines have the same slope. We will use this fact in Example 3.

EXAMPLE 3 **Using point-slope form with parallel lines**

Find the equation of each line. Write the answer in slope-intercept form.

a) The line through $(2, -1)$ that is parallel to $y = -3x + 9$

b) The line through $(3, 4)$ that is parallel to $2x - 3y = 6$

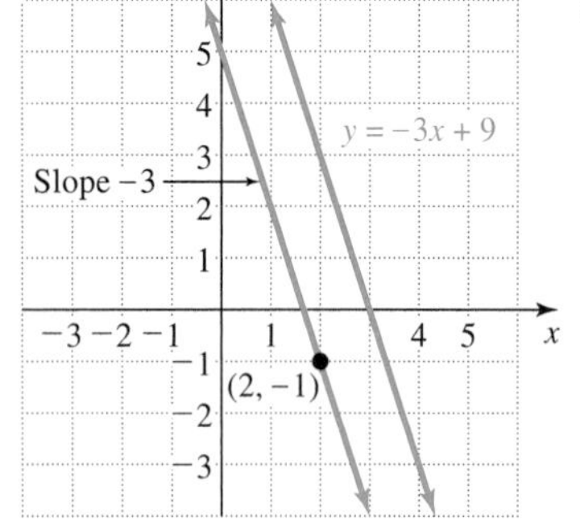

FIGURE 3.28

Solution

a) The slope of $y = -3x + 9$ and any line parallel to it is -3. See Fig. 3.28. Now use the point $(2, -1)$ and slope -3 in point-slope form:

$$y - y_1 = m(x - x_1) \quad \text{Point-slope form}$$

$$y - (-1) = -3(x - 2) \quad \text{Substitute.}$$

$$y + 1 = -3x + 6 \quad \text{Simplify.}$$

$$y = -3x + 5 \quad \text{Slope-intercept form}$$

Since $-1 = -3(2) + 5$ is correct, the line $y = -3x + 5$ goes through $(2, -1)$. It is certainly parallel to $y = -3x + 9$. So $y = -3x + 5$ is the desired equation.

b) Solve $2x - 3y = 6$ for y to determine its slope:

$$2x - 3y = 6$$

$$-3y = -2x + 6$$

$$y = \frac{2}{3}x - 2$$

So the slope of $2x - 3y = 6$ and any line parallel to it is $\frac{2}{3}$. Now use the point $(3, 4)$ and slope $\frac{2}{3}$ in the point-slope form:

$$y - y_1 = m(x - x_1) \quad \text{Point-slope form}$$
$$y - 4 = \frac{2}{3}(x - 3) \quad \text{Substitute.}$$
$$y - 4 = \frac{2}{3}x - 2 \quad \text{Simplify.}$$
$$y = \frac{2}{3}x + 2 \quad \text{Slope-intercept form}$$

Since $4 = \frac{2}{3}(3) + 2$ is correct, the line $y = \frac{2}{3}x + 2$ contains the point $(3, 4)$. Since $y = \frac{2}{3}x + 2$ and $y = \frac{2}{3}x - 2$ have the same slope, they are parallel. So the equation is $y = \frac{2}{3}x + 2$. ■

Perpendicular Lines

In Section 3.2 you learned that lines with slopes m and $-\frac{1}{m}$ (for $m \neq 0$) are perpendicular to each other. For example, the lines

$$y = -2x + 7 \quad \text{and} \quad y = \frac{1}{2}x - 8$$

are perpendicular to each other. In the next example we will write the equation of a line that is perpendicular to a given line and contains a given point.

EXAMPLE 4 **Writing an equation given a point and a perpendicular line**

Write the equation of the line that is perpendicular to $3x + 2y = 8$ and contains the point $(1, -3)$. Write the answer in slope-intercept form.

Calculator Close-Up

Graph $y_1 = (2/3)x - 11/3$ and $y_2 = (-3/2)x + 4$ as shown:

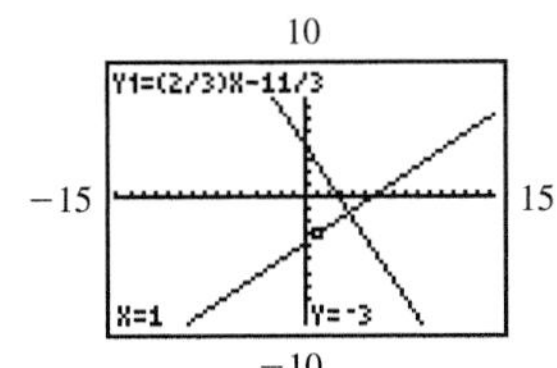

Because the lines look perpendicular and y_1 goes through $(1, -3)$, the graph supports the answer to Example 4.

Solution

First graph $3x + 2y = 8$ and a line through $(1, -3)$ that is perpendicular to $3x + 2y = 8$ as shown in Fig. 3.29. The right angle symbol is used in the figure to indicate that the lines are perpendicular. Now write $3x + 2y = 8$ in slope-intercept form to determine its slope:

$$3x + 2y = 8$$
$$2y = -3x + 8$$
$$y = -\frac{3}{2}x + 4 \quad \text{Slope-intercept form}$$

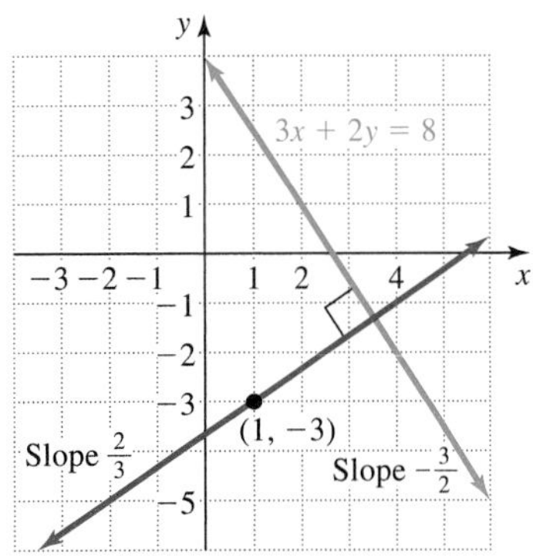

FIGURE 3.29

The slope of the given line is $-\frac{3}{2}$. The slope of any line perpendicular to it is $\frac{2}{3}$. Now we use the point-slope form with the point $(1, -3)$ and the slope $\frac{2}{3}$:

$$y - y_1 = m(x - x_1) \quad \text{Point-slope form}$$

$$y - (-3) = \frac{2}{3}(x - 1)$$

$$y + 3 = \frac{2}{3}x - \frac{2}{3}$$

$$y = \frac{2}{3}x - \frac{2}{3} - 3 \quad \text{Subtract 3 from each side.}$$

$$y = \frac{2}{3}x - \frac{11}{3} \quad \text{Slope-intercept form}$$

So $y = \frac{2}{3}x - \frac{11}{3}$ is the equation of the line that contains $(1, -3)$ and is perpendicular to $3x + 2y = 8$. Check that $(1, -3)$ satisfies $y = \frac{2}{3}x - \frac{11}{3}$. ■

Applications

We use the point-slope form to find the equation of a line given two points on the line. In Example 5 we use that same procedure to find a linear equation that relates two variables in an applied situation.

EXAMPLE 5 **Writing a formula given two points**

A contractor charges \$30 for installing 100 feet of pipe and \$120 for installing 500 feet of pipe. To determine the charge he uses a linear equation that gives the charge C in terms of the length L. Find the equation and find the charge for installing 240 feet of pipe.

Solution

Because C is determined from L, we let C take the place of the dependent variable y and let L take the place of the independent variable x. So the ordered pairs are in the form (L, C). We can use the slope formula to find the slope of the line through the two points (100, 30) and (500, 120) shown in Fig. 3.30.

$$m = \frac{120 - 30}{500 - 100} = \frac{90}{400} = \frac{9}{40}$$

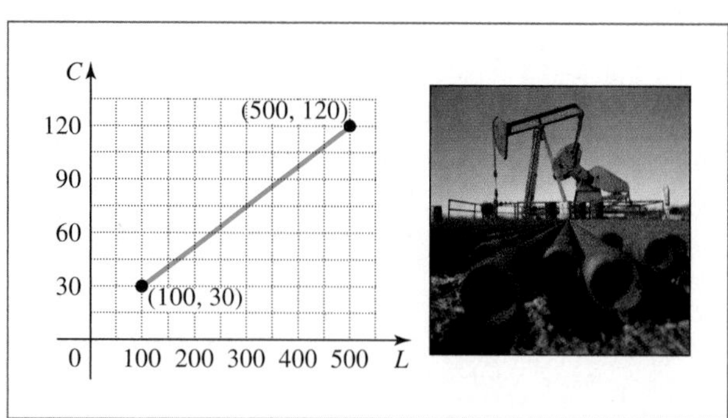

FIGURE 3.30

Now we use the point-slope form with the point (100, 30) and slope $\frac{9}{40}$:

$$y - y_1 = m(x - x_1)$$
$$C - 30 = \frac{9}{40}(L - 100)$$
$$C - 30 = \frac{9}{40}L - \frac{45}{2}$$
$$C = \frac{9}{40}L - \frac{45}{2} + 30$$
$$C = \frac{9}{40}L + \frac{15}{2}$$

Study Tip

When working a test, scan the problems and pick out the ones that are the easiest for you. Do them first. Save the harder problems till last.

Note that $C = \frac{9}{40}L + \frac{15}{2}$ means that the charge is $\frac{9}{40}$ dollars/foot plus a fixed charge of $\frac{15}{2}$ dollars (or \$7.50). We can now find C when $L = 240$:

$$C = \frac{9}{40} \cdot 240 + \frac{15}{2}$$
$$C = 54 + 7.5$$
$$C = 61.5$$

The charge for installing 240 feet of pipe is \$61.50. ■

WARM-UPS

True or false? Explain your answer.

1. The formula $y = m(x - x_1)$ is the point-slope form for a line. False
2. It is impossible to find the equation of a line through (2, 5) and (−3, 1). False
3. The point-slope form will not work for the line through (3, 4) and (3, 6). True
4. The equation of the line through the origin with slope 1 is $y = x$. True
5. The slope of the line $5x + y = 4$ is 5. False
6. The slope of any line perpendicular to the line $y = 4x - 3$ is $-\frac{1}{4}$. True
7. The slope of any line parallel to the line $x + y = 1$ is −1. True
8. The line $2x - y = -1$ goes through the point (−2, −3). True
9. The lines $2x + y = 4$ and $y = -2x + 7$ are parallel. True
10. The equation of the line through (0, 0) perpendicular to $y = x$ is $y = -x$. True

3.4 EXERCISES

Reading and Writing *After reading this section, write out the answers to these questions. Use complete sentences.*

1. What is the point-slope form for the equation of a line?
   Point-slope form is $y - y_1 = m(x - x_1)$.
2. For what is the point-slope form used?
   If we know any point and the slope of a line we can use point-slope form to write the equation.
3. What is the procedure for finding the equation of a line when given two points on the line?
   If you know two points on a line, find the slope. Then use it along with a point in point-slope form to write the equation of the line.
4. How can you find the slope of a line when given the equation of the line?
   Rewrite any equation in slope-intercept form to find the slope of the line.

5. What is the relationship between the slopes of parallel lines?
Nonvertical parallel lines have equal slopes.

6. What is the relationship between the slopes of perpendicular lines?
If lines with slopes m_1 and m_2 are perpendicular, then $m_1 = \frac{-1}{m_2}$.

Write each equation in slope-intercept form. See Example 1.

7. $y - 1 = 5(x + 2)$
$y = 5x + 11$

8. $y + 3 = -3(x - 6)$
$y = -3x + 15$

9. $3x - 4y = 80$
$y = \frac{3}{4}x - 20$

10. $2x + 3y = 90$
$y = -\frac{2}{3}x + 30$

11. $y - \frac{1}{2} = \frac{2}{3}\left(x - \frac{1}{4}\right)$
$y = \frac{2}{3}x + \frac{1}{3}$

12. $y + \frac{2}{3} = -\frac{1}{2}\left(x - \frac{2}{5}\right)$
$y = -\frac{1}{2}x - \frac{7}{15}$

Find the equation of each line. Write each answer in slope-intercept form. See Example 1.

13. The line through (1, 2) with slope 3
$y = 3x - 1$

14. The line through (2, 5) with slope 4
$y = 4x - 3$

15. The line through (2, 4) with slope $\frac{1}{2}$
$y = \frac{1}{2}x + 3$

16. The line through (4, 6) with slope $\frac{1}{2}$
$y = \frac{1}{2}x + 4$

17. The line through (2, 3) with slope $\frac{1}{3}$
$y = \frac{1}{3}x + \frac{7}{3}$

18. The line through (1, 4) with slope $\frac{1}{4}$
$y = \frac{1}{4}x + \frac{15}{4}$

19. The line through $(-2, 5)$ with slope $-\frac{1}{2}$
$y = -\frac{1}{2}x + 4$

20. The line through $(-3, 1)$ with slope $-\frac{1}{3}$
$y = -\frac{1}{3}x$

21. The line with slope -6 that goes through $(-1, -7)$
$y = -6x - 13$

22. The line with slope -8 that goes through $(-1, -5)$
$y = -8x - 13$

Write each equation in standard form using only integers. See Example 2.

23. $y - 3 = 2(x - 5)$
$2x - y = 7$

24. $y + 2 = -3(x - 1)$
$3x + y = 1$

25. $y = \frac{1}{2}x - 3$
$x - 2y = 6$

26. $y = \frac{1}{3}x + 5$
$x - 3y = -15$

27. $y - 2 = \frac{2}{3}(x - 4)$
$2x - 3y = 2$

28. $y + 1 = \frac{3}{2}(x + 4)$
$3x - 2y = -10$

Find the equation of each line. Write each answer in standard form using only integers. See Example 2.

29. The line through the points (1, 3) and (2, 5)
$2x - y = -1$

30. The line through the points (2, 5) and (3, 9)
$4x - y = 3$

31. The line through the points (1, 1) and (2, 2)
$x - y = 0$

32. The line through $(-1, 1)$ and $(1, -1)$
$x + y = 0$

33. The line through the points (1, 2) and (5, 8)
$3x - 2y = -1$

34. The line through the points (3, 5) and (8, 15)
$2x - y = 1$

35. The line through the points $(-2, -1)$ and $(3, -4)$
$3x + 5y = -11$

36. The line through the points $(-1, -3)$ and $(2, -1)$
$2x - 3y = 7$

37. The line through the points $(-2, 0)$ and $(0, 2)$
$x - y = -2$

38. The line through the points (0, 3) and (5, 0)
$3x + 5y = 15$

The lines in each figure are perpendicular. Find the equation (in slope-intercept form) for the solid line.

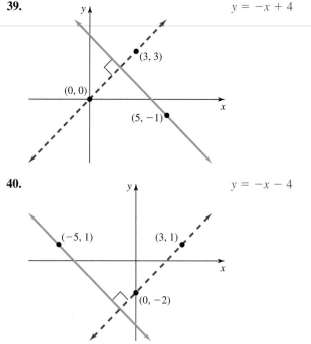

39. $y = -x + 4$

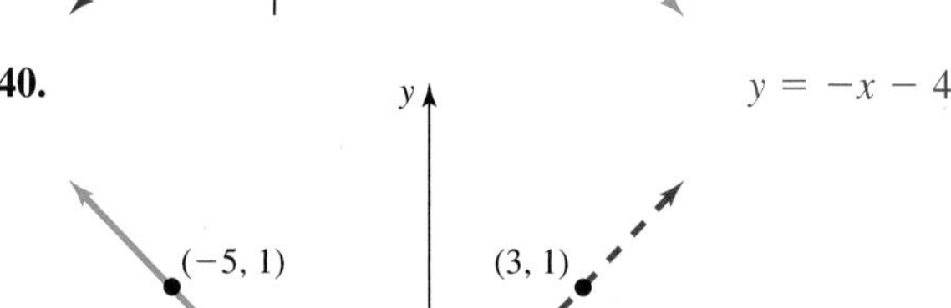

40. $y = -x - 4$

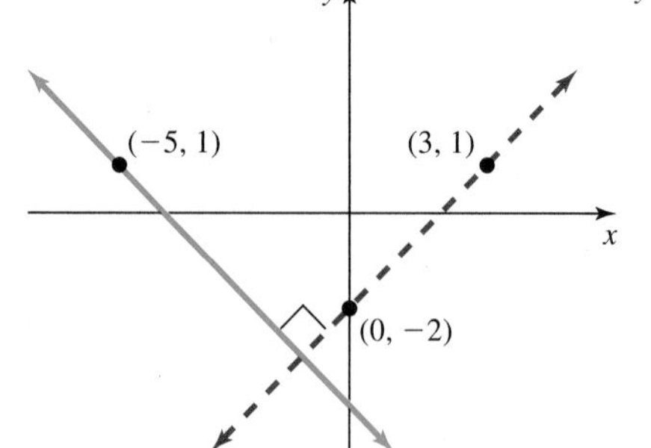

41.

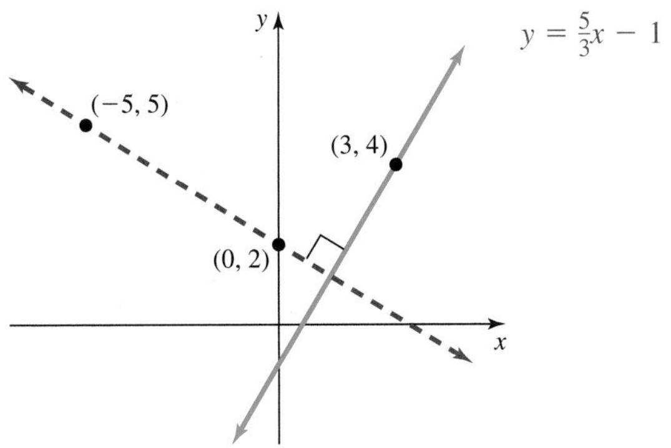

$y = \frac{5}{3}x - 1$

42.

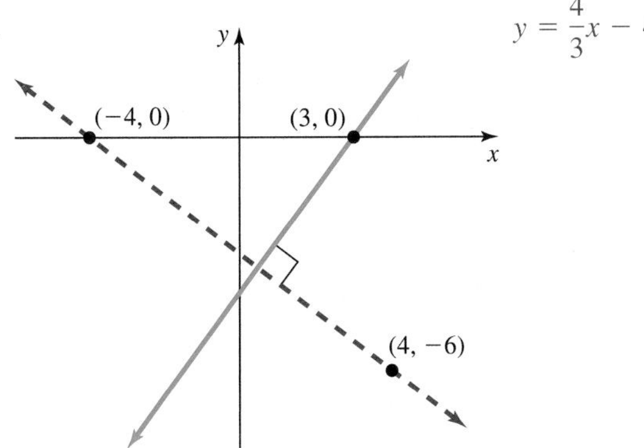

$y = \frac{4}{3}x - 4$

Find the equation of each line. Write each answer in slope-intercept form. See Examples 3 and 4.

43. The line contains the point (3, 4) and is perpendicular to $y = 3x - 1$. $y = -\frac{1}{3}x + 5$

44. The line contains the point $(-2, 3)$ and is perpendicular to $y = 2x + 7$. $y = -\frac{1}{2}x + 2$

45. The line is parallel to $y = x - 9$ and goes through the point (7, 10). $y = x + 3$

46. The line is parallel to $y = -x + 5$ and goes through the point $(-3, 6)$. $y = -x + 3$

47. The line is perpendicular to $3x - 2y = 10$ and passes through the point (1, 1). $y = -\frac{2}{3}x + \frac{5}{3}$

48. The line is perpendicular to $x - 5y = 4$ and passes through the point $(-1, 1)$. $y = -5x - 4$

49. The line is parallel to $2x + y = 8$ and contains the point $(-1, -3)$. $y = -2x - 5$

50. The line is parallel to $-3x + 2y = 9$ and contains the point $(-2, 1)$. $y = \frac{3}{2}x + 4$

51. The line goes through $(-1, 2)$ and is perpendicular to $3x + y = 5$. $y = \frac{1}{3}x + \frac{7}{3}$

52. The line goes through (1, 2) and is perpendicular to $y = \frac{1}{2}x - 3$. $y = -2x + 4$

53. The line goes through (2, 3) and is parallel to $-2x + y = 6$. $y = 2x - 1$

54. The line goes through (1, 4) and is parallel to $x - 2y = 6$. $y = \frac{1}{2}x + \frac{7}{2}$

Find the equation of each line in the form $y = mx + b$ if possible.

55. The line through (3, 2) with slope 0 $y = 2$

56. The line through (3, 2) with undefined slope $x = 3$

57. The line through (3, 2) and the origin $y = \frac{2}{3}x$

58. The line through the origin that is perpendicular to $y = \frac{2}{3}x$ $y = -\frac{3}{2}x$

59. The line through the origin that is parallel to the line through (5, 0) and (0, 5) $y = -x$

60. The line through the origin that is perpendicular to the line through $(-3, 0)$ and $(0, -3)$ $y = x$

61. The line through $(-30, 50)$ that is perpendicular to the line $x = 400$ $y = 50$

62. The line through $(20, -40)$ that is parallel to the line $y = 6000$ $y = -40$

63. The line through $(-5, -1)$ that is perpendicular to the line through (0, 0) and (3, 5) $y = -\frac{3}{5}x - 4$

64. The line through (3, 1) that is parallel to the line through $(-3, -2)$ and (0, 0) $y = \frac{2}{3}x - 1$

Solve each problem.

65. ***Automated tellers.*** ATM volume reached 10.6 billion transactions in 1996 and 14.2 billion transactions in 2000 as shown in the accompanying graph. If 1996 is year 0 and 2000 is year 4, then the line goes through the points (0, 10.6) and (4, 14.2).

a) Find and interpret the slope of the line.

b) Write the equation of the line in slope-intercept form.

c) Use your equation from part (b) to predict the number of transactions at automated teller machines in 2010.

a) Slope 0.9 means that the number of ATM transactions is increasing by 0.9 billion per year.
b) $y = 0.9x + 10.6$ **c)** 23.2 billion

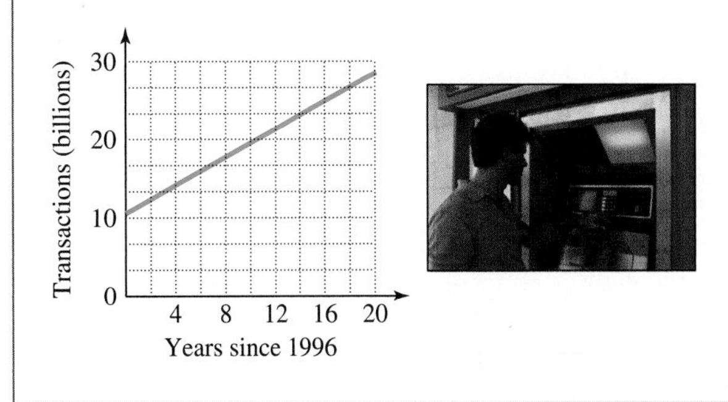

FIGURE FOR EXERCISE 65

66. ***Direct deposit.*** The percentage of workers receiving direct deposit of their paychecks went from 32% in 1994 to 50% in 2000 (www.directdeposit.com). Let 1994 be year 4 and 2000 be year 10.

a) Write the equation of the line through (4, 32) and (10, 50) to model the growth of direct deposit. $y = 3x + 20$

b) Use the accompanying graph to predict the year in which 100% of all workers will receive direct deposit of their paychecks. About 2017

c) Use the equation from part (a) to predict the year in which 100% of all workers will receive direct deposit. 2017

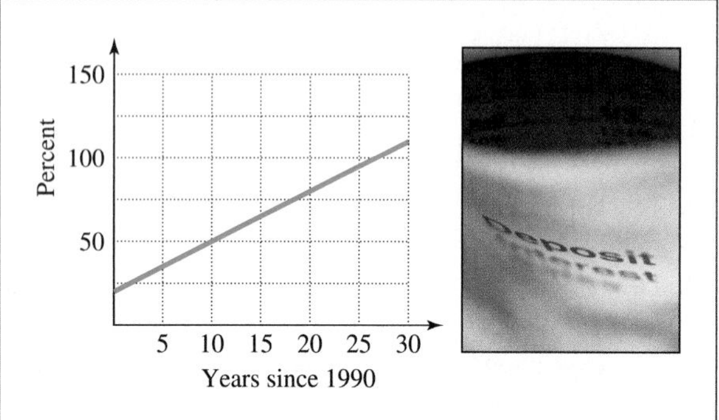

FIGURE FOR EXERCISE 66

67. ***Gross domestic product.*** The U.S. gross domestic product (GDP) per employed person increased from \$62.7 thousand in 1996 to \$65.9 thousand in 1998 (Bureau of Labor Statistics, www.bls.gov). Let 1996 be year 6 and 1998 be year 8.

a) Find the equation of the line through (6, 62.7) and (8, 65.9) to model the gross domestic product.
$y = 1.6x + 53.1$

b) What do x and y represent in your equation?
x = years since 1990, y = GDP in thousands of dollars

c) Use the equation to predict the GDP in 2005.
\$77,100

d) Graph the equation.

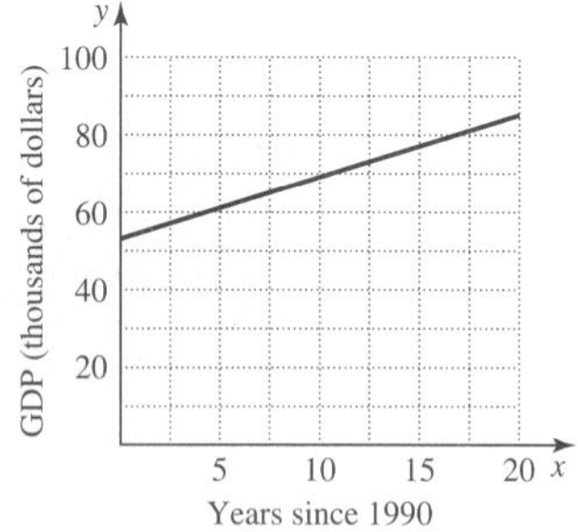

68. ***Age at first marriage.*** The median age at first marriage for females increased from 24.5 years in 1995 to 25.1 years in 2000 (U.S. Census Bureau, www.census.gov). Let 1995 be year 5 and 2000 be year 10.

a) Find the equation of the line through (5, 24.5) and (10, 25.1).
$y = 0.12x + 23.9$

b) What do x and y represent in your equation?
x = the number of years since 1990, y = median age at first marriage

c) Interpret the slope of this line.
Median age increases 0.12 year each year or approximately 1 year in 8 years.

d) In what year will the median age be 30.
2041

e) Graph the equation.

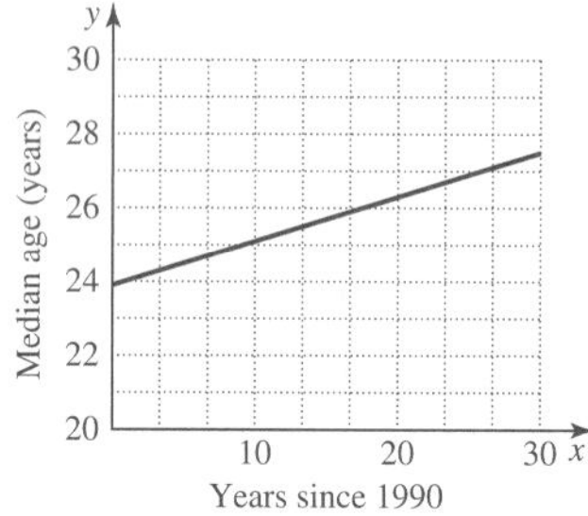

69. ***Plumbing charges.*** Pete the plumber worked 2 hours at Millie's house and charged her \$70. He then worked 4 hours at Rosalee's house and charged her \$110. To determine the amount he charges Pete uses a linear equation that gives the charge C in terms of the number of hours worked n. Find the equation and find the charge for 7 hours at Fred's house. $C = 20n + 30$, \$170

70. ***Interior angles.*** The sum of the measures of the interior angles of a triangle is 180°. The sum of the measures of the interior angles of a square is 360°. Let S represent the sum of the measures of the interior angles of a polygon and n represent the number of sides of the polygon. There is a linear equation that gives S in terms of n. Find the equation and find the sum of the measures of the interior angles of the stop sign shown in the accompanying figure. $S = 180n - 360$, 1080°

FIGURE FOR EXERCISE 70

71. ***Shoe sizes.*** If a child's foot is 7.75 inches long, then the child wears a size 13 shoe. If a child's foot is 5.75 inches long, then the child wears a size 7 shoe. Let S represent the shoe size and L represent the length of the foot in inches. There is a linear equation that gives S in terms of L. Find the equation and find the shoe size for a child with a 6.25-inch foot.
$S = 3L - \frac{41}{4}$, 8.5

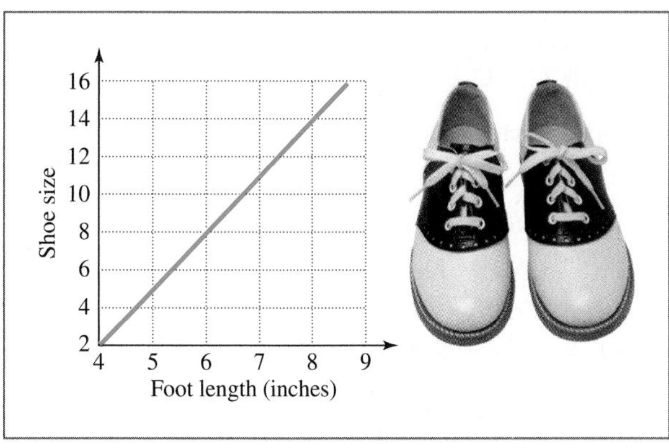

FIGURE FOR EXERCISE 71

72. ***Celsius to Fahrenheit.*** Water freezes at 0°C or 32°F and boils at 100°C or 212°F. There is a linear equation that expresses the number of degrees Fahrenheit (F) in terms of the number of degrees Celsius (C). Find the equation and find the Fahrenheit temperature when the Celsius temperature is 45°.
$F = \frac{9}{5}C + 32$, 113°F

73. ***Velocity of a projectile.*** A ball is thrown downward from the top of a tall building. Its velocity is 42 feet per second after 1 second and 74 feet per second after 2 seconds. There is a linear equation that expresses the velocity v in terms of the time t. Find the equation and find the velocity after 3.5 seconds.
$v = 32t + 10$, 122 ft/sec

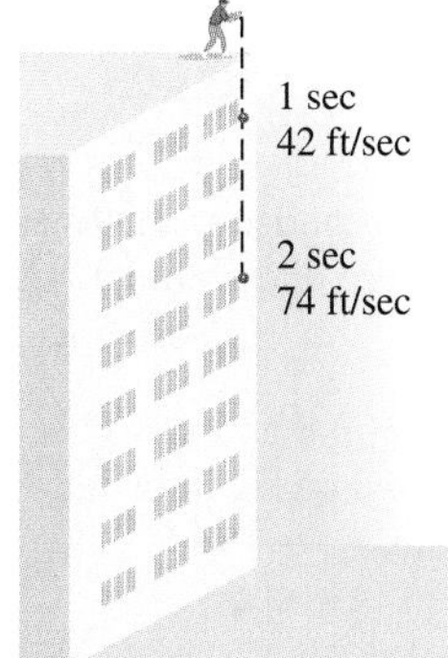

FIGURE FOR EXERCISE 73

74. ***Natural gas.*** The cost of 1000 cubic feet of natural gas is \$39 and the cost of 3000 cubic feet is \$99. There is a linear equation that expresses the cost C in terms of the number of cubic feet n. Find the equation and find the cost of 2400 cubic feet of natural gas.
$C = 0.03n + 9$, \$81

75. ***Expansion joint.*** When the temperature is 90°F the width of an expansion joint on a bridge is 0.75 inch. When the temperature is 30°F the width is 1.25 inches. There is a linear equation that expresses the width w in terms of the temperature t.
   a) Find the equation.
   b) What is the width when the temperature is 80°F?
   c) What is the temperature when the width is 1 inch?
   a) $w = -\frac{1}{120}t + \frac{3}{2}$ **b)** $\frac{5}{6}$ inch **c)** 60°F

76. ***Perimeter of a rectangle.*** A rectangle has a fixed width and a variable length. Let P represent the perimeter and L represent the length. $P = 28$ inches when $L = 6.5$ inches and $P = 36$ inches when $L = 10.5$ inches. There is a linear equation that expresses P in terms of L.
   a) Find the equation.
   b) What is the perimeter when the $L = 40$ inches?
   c) What is the length when $P = 215$ inches?
   d) What is the width of the rectangle?
   a) $P = 2L + 15$ **b)** 95 in. **c)** 100 in. **d)** 7.5 in.

77. ***Stretching a spring.*** A weight of 3 pounds stretches a spring 1.8 inches beyond its natural length and weight of 5 pounds stretches the same spring 3 inches beyond its natural length. Let A represent the amount of stretch and w the weight. There is a linear equation that expresses A in terms of w. Find the equation and find the amount that the spring will stretch with a weight of 6 pounds.
$A = 0.6w$, 3.6 in.

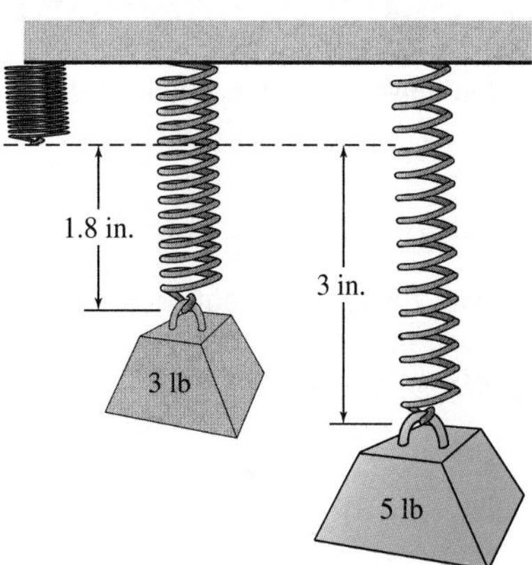

FIGURE FOR EXERCISE 77

78. ***Velocity of a bullet.*** A gun is fired straight upward. The bullet leaves the gun at 100 feet per second (time $t = 0$). After 2 seconds the velocity of the bullet is 36 feet per second. There is a linear equation that gives the velocity v in terms of the time t. Find the equation and find the velocity after 3 seconds.
$v = -32t + 100$, 4 ft/sec

79. ***Enzyme concentration.*** The amount of light absorbed by a certain liquid depends on the concentration of an enzyme in the liquid. A concentration of 2 milligrams per milliliter (mg/ml) produces an absorption of 0.16 and a concentration of 5 mg/ml produces an absorption of 0.40. There is a linear equation that expresses the absorption a in terms of the concentration c.
   a) Find the equation.
   b) What is the absorption when the concentration is 3 mg/ml?

c) Use the accompanying graph to estimate the concentration when the absorption is 0.50.
a) $a = 0.08c$ **b)** 0.24 **c)** 6.25 mg/ml

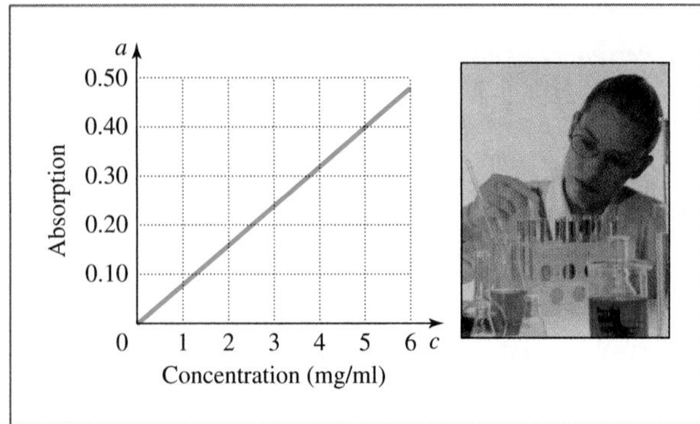

FIGURE FOR EXERCISE 79

80. ***Basal energy requirement.*** The basal energy requirement B is the number of calories that a person needs to maintain the life process. For a 28-year-old female with a height of 160 centimeters and a weight of 45 kilograms (kg), B is 1300 calories. If her weight increases to 50 kg, then B is 1365 calories. There is a linear equation that expresses B in terms of her weight w. Find the equation and find the basal energy requirement if her weight is 53.2 kg. $B = 13w + 715$, 1406.6 calories

GETTING MORE INVOLVED

81. ***Exploration.*** Each linear equation in the following table is given in standard form $Ax + By = C$. In each case identify A, B, and the slope of the line.

Equation	A	B	Slope
$2x + 3y = 9$	2	3	$-\frac{2}{3}$
$4x - 5y = 6$	4	-5	$\frac{4}{5}$
$\frac{1}{2}x + 3y = 1$	$\frac{1}{2}$	3	$-\frac{1}{6}$
$2x - \frac{1}{3}y = 7$	2	$-\frac{1}{3}$	6

82. ***Exploration.*** Find a pattern in the table of Exercise 81 and write a formula for the slope of $Ax + By = C$, where $B \neq 0$. $m = -\frac{A}{B}$

GRAPHING CALCULATOR EXERCISES

83. Graph each equation on a graphing calculator. Choose a viewing window that includes both the x- and y-intercepts. Use the calculator output to help you draw the graph on paper.

a) $y = 20x - 300$
b) $y = -30x + 500$
c) $2x - 3y = 6000$

a)

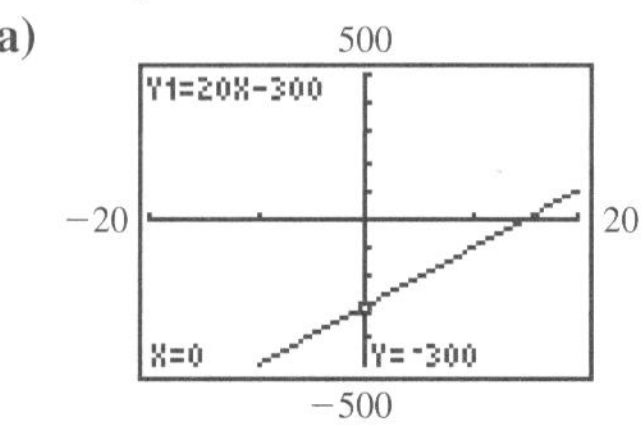

b)

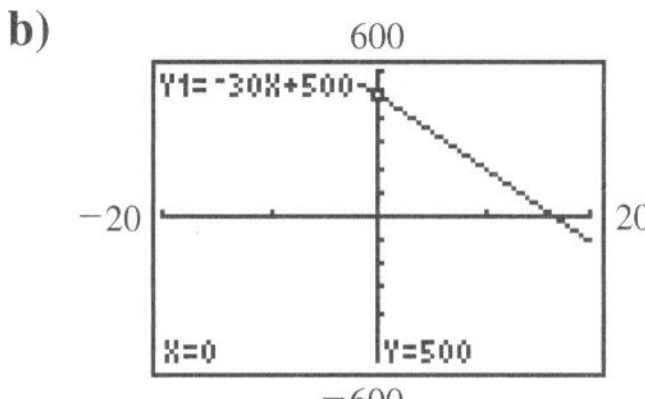

c)

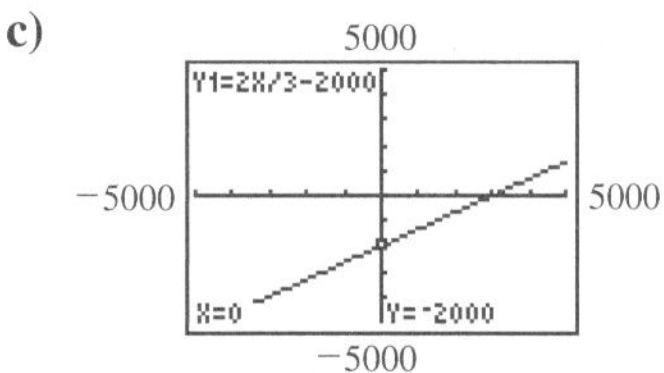

84. Graph $y = 2x + 1$ and $y = 1.99x - 1$ on a graphing calculator. Are these lines parallel? Explain your answer. They look parallel, but they are not.

85. Graph $y = 0.5x + 0.8$ and $y = 0.5x + 0.7$ on a graphing calculator. Find a viewing window in which the two lines are separate.
$-1 \leq x \leq 1, -1 \leq y \leq 1$

86. Graph $y = 3x + 1$ and $y = -\frac{1}{3}x + 2$ on a graphing calculator. Do the lines look perpendicular? Explain. They will look perpendicular in the right window.

In This Section

- Direct Variation
- Finding the Constant
- Inverse Variation
- Joint Variation

3.5 VARIATION

If $y = 5x$, then the value of y depends on the value of x. As x varies, so does y. Simple relationships like $y = 5x$ are customarily expressed in terms of variation. In this section you will learn the language of variation and learn to write formulas from verbal descriptions.

Direct Variation

Suppose you average 60 miles per hour on the freeway. The distance D that you travel depends on the amount of time T that you travel. Using the formula $D = R \cdot T$, we can write

$$D = 60T.$$

Consider the possible values for T and D given in the following table.

T (hours)	1	2	3	4	5	6
D (miles)	60	120	180	240	300	360

The graph of $D = 60T$ is shown in Fig. 3.31. Note that as T gets larger, so does D. In this situation we say that *D varies directly with T,* or D is *directly proportional* to T. The constant rate of 60 miles per hour is called the **variation constant** or **proportionality constant.** Notice that $D = 60T$ is simply a linear equation. We are just introducing some new terms to express an old idea.

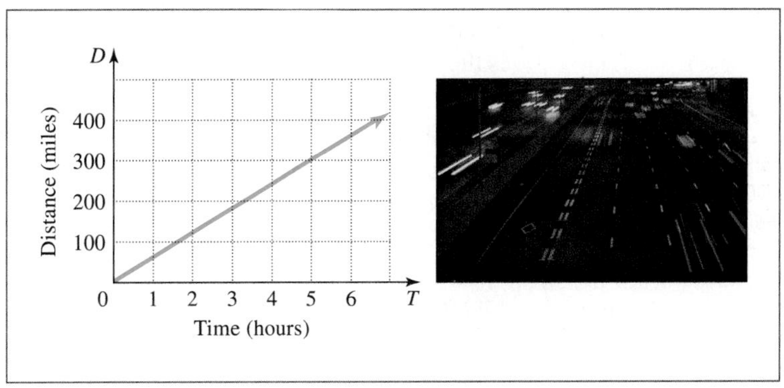

FIGURE 3.31

Direct Variation

The statement **"y varies directly as x"** or **"y is directly proportional to x"** means that

$$y = kx$$

for some constant k. The constant of variation k is a fixed nonzero real number.

CAUTION Direct variation refers only to equations of the form $y = kx$ (lines through the origin). We do *not* refer to $y = 3x + 5$ as a direct variation.

Finding the Constant

If we know one ordered pair in a direct variation, then we can find the constant of variation.

EXAMPLE 1 **Finding a constant of variation**

Natasha is traveling by car, and the distance D that she travels varies directly as the rate R at which she drives. At 45 miles per hour, Natasha travels 135 miles. Find the constant of variation, and write a formula for D in terms of R.

Helpful Hint

In any variation problem you must first determine the general form of the relationship. Because this problem involves direct variation, the general form is $y = kx$.

Solution

Because D varies directly as R, there is a constant k such that

$$D = kR.$$

Because $D = 135$ when $R = 45$, we can write

$$135 = k \cdot 45$$

or

$$3 = k.$$

So $D = 3R$. ■

In Example 2 we find the constant of variation and use it to solve a variation problem.

EXAMPLE 2 **A direct variation problem**

Your electric bill at Middle States Electric Co-op varies directly with the amount of electricity that you use. If the bill for 2800 kilowatts of electricity is \$196, then what is the bill for 4000 kilowatts of electricity?

Study Tip

If your grades are not what you would like and you are doing all of your work, then see your teacher for advice. Your grade varies directly with the amount of effort that you put forth.

Solution

Because the amount A of the electric bill varies directly as the amount E of electricity used, we have

$$A = kE$$

for some constant k. Because 2800 kilowatts cost \$196, we have

$$196 = k2800$$

or

$$0.07 = k.$$

So $A = 0.07E$. Now if $E = 4000$ we get

$$A = 0.07(4000) = 280.$$

The bill for 4000 kilowatts would be \$280. ■

Inverse Variation

If you plan to make a 400-mile trip by car, the time it will take depends on your rate of speed. Using the formula $D = RT$, we can write

$$T = \frac{400}{R}.$$

Consider the possible values for R and T given in the following table:

R (mph)	10	20	40	50	80	100
T (hours)	40	20	10	8	5	4

The graph of $T = \frac{400}{R}$ is shown in Fig. 3.32. As your rate increases, the time for the trip decreases. In this situation we say that the time is *inversely proportional* to the speed. Note that the graph of $T = \frac{400}{R}$ is not a straight line because $T = \frac{400}{R}$ is not a linear equation.

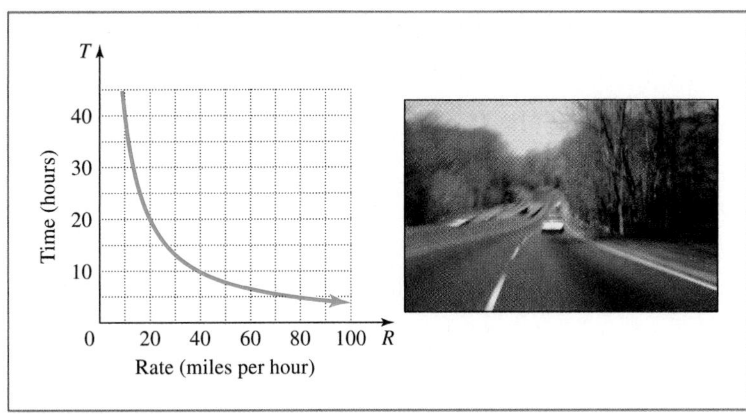

FIGURE 3.32

Inverse Variation

The statement **"y varies inversely as x"**, or **"y is inversely proportional to x"** means that

$$y = \frac{k}{x}$$

for some nonzero constant of variation k.

CAUTION The constant of variation is usually positive because most physical examples involve positive quantities. However, the definitions of direct and inverse variation do not rule out a negative constant.

EXAMPLE 3 An inverse variation problem

The volume of a gas in a cylinder is inversely proportional to the pressure on the gas. If the volume is 12 cubic centimeters when the pressure on the gas is 200 kilograms per square centimeter, then what is the volume when the pressure is 150 kilograms per square centimeter? See Fig. 3.33.

$P = 200$ kg/cm^2 $P = 150$ kg/cm^2

$V = 12$ cm^3 $V = ?$

FIGURE 3.33

Solution

Because the volume V is inversely proportional to the pressure P, we have

$$V = \frac{k}{P}$$

for some constant k. Because $V = 12$ when $P = 200$, we can find k:

$$12 = \frac{k}{200}$$

$$200 \cdot 12 = 200 \cdot \frac{k}{200} \quad \text{Multiply each side by 200.}$$

$$2400 = k$$

Now to find V when $P = 150$, we can use the formula $V = \frac{2400}{P}$:

$$V = \frac{2400}{150} = 16$$

So the volume is 16 cubic centimeters when the pressure is 150 kilograms per square centimeter. ■

Joint Variation

If the price of carpet is \$30 per square yard, then the cost C of carpeting a rectangular room depends on the width W (in yards) and the length L (in yards). As the width or length of the room increases, so does the cost. We can write the cost in terms of the two variables L and W:

$$C = 30LW$$

We say that C *varies jointly* as L and W.

Joint Variation

The statement **"y varies jointly as x and z"** or **"y is jointly proportional to x and z"** means that

$$y = kxz$$

for some nonzero constant of variation k.

EXAMPLE 4 **A joint variation problem**

The cost of shipping a piece of machinery by truck varies jointly with the weight of the machinery and the distance that it is shipped. It costs \$3000 to ship a 2500-lb milling machine a distance of 600 miles. Find the cost for shipping a 1500-lb lathe a distance of 800 miles.

Helpful Hint

Because the variation in this problem is joint, we know the general form is $y = kxz$, where k is the constant of variation.

Solution

Because the cost C varies jointly with the weight w and the distance d, we have

$$C = kwd$$

where k is the constant of variation. To find k, we use $C = 3000$, $w = 2500$, and $d = 600$:

$$3000 = k \cdot 2500 \cdot 600$$

$$\frac{3000}{2500 \cdot 600} = k \quad \text{Divide each side by } 2500 \cdot 600.$$

$$0.002 = k$$

Now use $w = 1500$ and $d = 800$ in the formula $C = 0.002wd$:

$$C = 0.002 \cdot 1500 \cdot 800$$
$$= 2400$$

So the cost of shipping the lathe is \$2400. ■

CAUTION The variation words (directly, inversely, or jointly) are never used to indicate addition or subtraction. We use multiplication in the formula unless we see the word "inversely." We use division for inverse variation.

WARM-UPS

True or false? Explain your answer.

1. If y varies directly as z, then $y = kz$ for some constant k. True
2. If a varies inversely as b, then $a = \frac{b}{k}$ for some constant k. False
3. If y varies directly as x and $y = 8$ when $x = 2$, then the variation constant is 4. True
4. If y varies inversely as x and $y = 8$ when $x = 2$, then the variation constant is $\frac{1}{4}$. False
5. If C varies jointly as h and t, then $C = ht$. False
6. The amount of sales tax on a new car varies directly with the purchase price of the car. True
7. If z varies inversely as w and $z = 10$ when $w = 2$, then $z = \frac{20}{w}$. True
8. The time that it takes to travel a fixed distance varies inversely with the rate. True
9. If m varies directly as w, then $m = w + k$ for some constant k. False
10. If y varies jointly as x and z, then $y = k(x + z)$ for some constant k. False

3.5 EXERCISES

Reading and Writing *After reading this section, write out the answers to these questions. Use complete sentences.*

1. What does it mean to say that y varies directly as x?
   If y varies directly as x, then there is a constant k such that $y = kx$.
2. What is a variation constant?
   A variation constant is the constant k in the formulas $y = kx$ or $y = \frac{k}{x}$.
3. What does it mean to say that y is inversely proportional to x?
   If y is inversely proportional to x, then there is a constant k such that $y = \frac{k}{x}$.
4. What does it mean to say that y varies jointly as x and z?
   If y varies jointly as x and z, then there is a constant k such that $y = kxz$.

Write a formula that expresses the relationship described by each statement. Use k for the constant in each case. See Examples 1–4.

5. T varies directly as h.
   $T = kh$
6. m varies directly as p.
   $m = kp$
7. y varies inversely as r.
   $y = \frac{k}{r}$
8. u varies inversely as n.
   $u = \frac{k}{n}$
9. R is jointly proportional to t and s.
   $R = kts$
10. W varies jointly as u and v.
    $W = kuv$
11. i is directly proportional to b.
    $i = kb$
12. p is directly proportional to x.
    $p = kx$
13. A is jointly proportional to y and m.
    $A = kym$
14. t is inversely proportional to e.
    $t = \frac{k}{e}$

Find the variation constant, and write a formula that expresses the indicated variation. See Example 1.

15. y varies directly as x, and $y = 5$ when $x = 3$.
    $y = \frac{5}{3}x$
16. m varies directly as w, and $m = \frac{1}{2}$ when $w = \frac{1}{4}$.
    $m = 2w$
17. A varies inversely as B, and $A = 3$ when $B = 2$.
    $A = \frac{6}{B}$
18. c varies inversely as d, and $c = 5$ when $d = 2$.
    $c = \frac{10}{d}$

19. m varies inversely as p, and $m = 22$ when $p = 9$. $m = \frac{198}{p}$

20. s varies inversely as v, and $s = 3$ when $v = 4$. $s = \frac{12}{v}$

21. A varies jointly as t and u, and $A = 24$ when $t = 6$ and $u = 2$. $A = 2tu$

22. N varies jointly as p and q, and $N = 720$ when $p = 3$ and $q = 2$. $N = 120pq$

23. T varies directly as u, and $T = 9$ when $u = 2$. $T = \frac{9}{2}u$

24. R varies directly as p, and $R = 30$ when $p = 6$. $R = 5p$

Solve each variation problem. See Examples 2–4.

25. Y varies directly as x, and $Y = 100$ when $x = 20$. Find Y when $x = 5$. 25

26. n varies directly as q, and $n = 39$ when $q = 3$. Find n when $q = 8$. 104

27. a varies inversely as b, and $a = 3$, when $b = 4$. Find a when $b = 12$. 1

28. y varies inversely as w, and $y = 9$ when $w = 2$. Find y when $w = 6$. 3

29. P varies jointly as s and t, and $P = 56$ when $s = 2$ and $t = 4$. Find P when $s = 5$ and $t = 3$. 105

30. B varies jointly as u and v, and $B = 12$ when $u = 4$ and $v = 6$. Find B when $u = 5$ and $v = 8$. 20

Use the given formula to fill in the missing entries in each table and determine whether b varies directly or inversely as a.

31. $b = \frac{300}{a}$

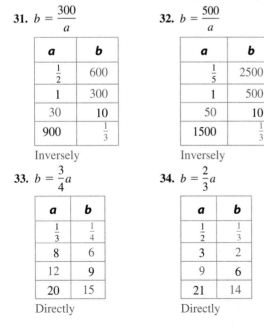

a	b
$\frac{1}{2}$	600
1	300
30	10
900	$\frac{1}{3}$

Inversely

32. $b = \frac{500}{a}$

a	b
$\frac{1}{5}$	2500
1	500
50	10
1500	$\frac{1}{3}$

Inversely

33. $b = \frac{3}{4}a$

a	b
$\frac{1}{3}$	$\frac{1}{4}$
8	6
12	9
20	15

Directly

34. $b = \frac{2}{3}a$

a	b
$\frac{1}{2}$	$\frac{1}{3}$
3	2
9	6
21	14

Directly

For each table, determine whether y varies directly or inversely as x and find a formula for y in terms of x.

35.

x	y
2	7
3	10.5
4	14
5	17.5

Directly, $y = 3.5x$

36.

x	y
10	5
15	7.5
20	10
25	12.5

Directly, $y = 0.5x$

37.

x	y
2	10
4	5
10	2
20	1

Inversely, $y = \frac{20}{x}$

38.

x	y
5	100
10	50
50	10
250	2

Inversely, $y = \frac{500}{x}$

Solve each problem.

39. ***Distance.*** With the cruise control set at 65 mph, the distance traveled varies directly with the time spent traveling. Fill in the missing entries in the following table.

Time (hours)	1	2	3	4
Distance (miles)	65	130	195	260

40. ***Cost.*** With gas selling for \$1.60 per gallon, the cost of filling your tank varies directly with the amount of gas that you pump. Fill in the missing entries in the following table.

Amount (gallons)	5	10	15	20
Cost (dollars)	8	16	24	32

41. ***Time.*** The time that it takes to complete a 400-mile trip varies inversely with your average speed. Fill in the missing entries in the following table.

Speed (mph)	20	40	50	200
Time (hours)	20	10	8	2

42. ***Amount.*** The amount of gasoline that you can buy for \$20 varies inversely with the price per gallon. Fill in the missing entries in the following table.

Price per gallon (dollars)	1	2	4	10
Amount (gallons)	20	10	5	2

43. ***Carpeting.*** The cost C of carpeting a rectangular living room with \$20 per square yard carpet varies jointly with the length L and the width W. Fill in the missing entries in the following table.

Length (yd)	Width (yd)	Cost ($)
8	10	1600
10	12	2400
12	14	3360

44. ***Waterfront property.*** At $50 per square foot, the price of a rectangular waterfront lot varies jointly with the length and width. Fill in the missing entries in the following table.

Length (ft)	Width (ft)	Cost ($)
60	100	300,000
80	90	360,000
100	150	750,000

45. ***Aluminum flatboat.*** The weight of an aluminum flatboat varies directly with the length of the boat. If a 12-foot boat weighs 86 pounds, then what is the weight of a 14-foot boat?
100.3 pounds

46. ***Christmas tree.*** The price of a Christmas tree varies directly with the height. If a 5-foot tree costs $20, then what is the price of a 6-foot tree?
$24

47. ***Sharing the work.*** The time it takes to erect the big circus tent varies inversely as the number of elephants working on the job. If it takes four elephants 75 minutes, then how long would it take six elephants?
50 minutes

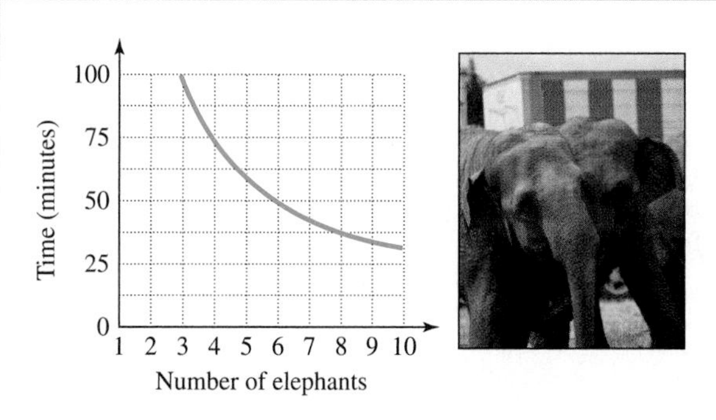

FIGURE FOR EXERCISE 47

48. ***Gas laws.*** The volume of a gas is inversely proportional to the pressure on the gas. If the volume is 6 cubic centimeters when the pressure on the gas is 8 kilograms per square centimeter, then what is the volume when the pressure is 12 kilograms per square centimeter?
4 cm^3

49. ***Steel tubing.*** The cost of steel tubing is jointly proportional to its length and diameter. If a 10-foot tube with a 1-inch diameter costs $5.80, then what is the cost of a 15-foot tube with a 2-inch diameter?
$17.40

50. ***Sales tax.*** The amount of sales tax varies jointly with the number of Cokes purchased and the price per Coke. If the sales tax on eight Cokes at 65 cents each is 26 cents, then what is the sales tax on six Cokes at 90 cents each?
27 cents

51. ***Approach speed.*** The approach speed of an airplane is directly proportional to its landing speed. If the approach speed for a Piper Cheyenne is 90 mph with a landing speed of 75 mph, then what is the landing speed for an airplane with an approach speed of 96 mph?
80 mph

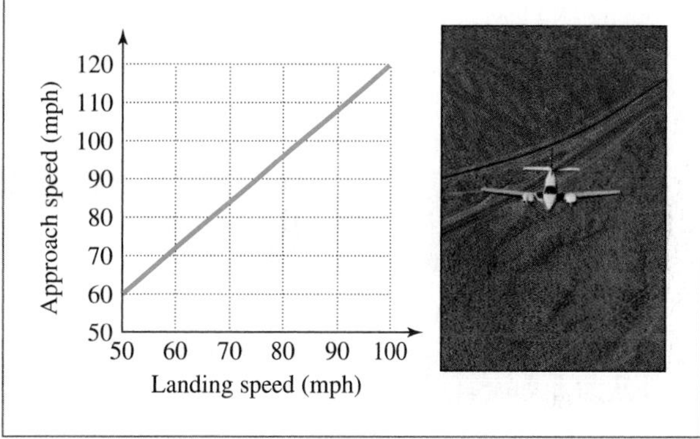

FIGURE FOR EXERCISE 51

52. ***Ideal waist size.*** According to Dr. Aaron R. Folsom of the University of Minnesota School of Public Health, your maximum ideal waist size is directly proportional to your hip size. For a woman with 40-inch hips, the maximum ideal waist size is 32 inches. What is the maximum ideal waist size for a woman with 35-inch hips?
28 inches

GETTING MORE INVOLVED

53. ***Discussion.*** If y varies directly as x, then the graph of the equation is a straight line. What is its slope? What is the y-intercept? If $y = 3x + 2$, then does y vary directly as x? Which straight lines correspond to direct variations?
k, (0, 0), no, $y = kx$

54. ***Writing.*** Write a summary of the three types of variation. Include an example of each type that is not found in this text.

COLLABORATIVE ACTIVITIES

Inches or Centimeters?

In this activity you will generate data by measuring in both inches and centimeters the height of each member of your group. Then you will plot the points on a graph and use any two of your points to find the conversion formula for converting inches to centimeters.

Part I: Measure the height of each person in your group and fill out a table like the one shown here:

Name	Height in inches	Height in centimeters

Grouping: Three to four students

Topic: Plotting points, graphing lines

Part II: The numbers for inches and centimeters from the table will give you three or four ordered pairs to graph. Plot these points on a graph. Let inches be the horizontal x-axis and centimeters be the vertical y-axis. Let each mark on the axes represent 10 units. When graphing, you will need to estimate the place to plot fractional values.

Part III: Use any two of your points to find an equation of the line you have graphed. What is the slope of your line? Where does it cross the horizontal axis?

Extension: Look up the conversion formula for converting inches to centimeters. Is it the same as the one you found by measuring? If it is different, what could account for the difference?

WRAP-UP CHAPTER 3

SUMMARY

Slope of a Line

Examples

Slope

The slope of the line through (x_1, y_1) and (x_2, y_2) is given by

$$m = \frac{y_2 - y_1}{x_2 - x_1}, \text{ provided that } x_2 - x_1 \neq 0.$$

Example: $(0, 1), (3, 5)$

$$m = \frac{5 - 1}{3 - 0} = \frac{4}{3}$$

Slope is the ratio of the rise to the run for any two points on the line:

$$m = \frac{\text{change in } y}{\text{change in } x} = \frac{\text{rise}}{\text{run}}$$

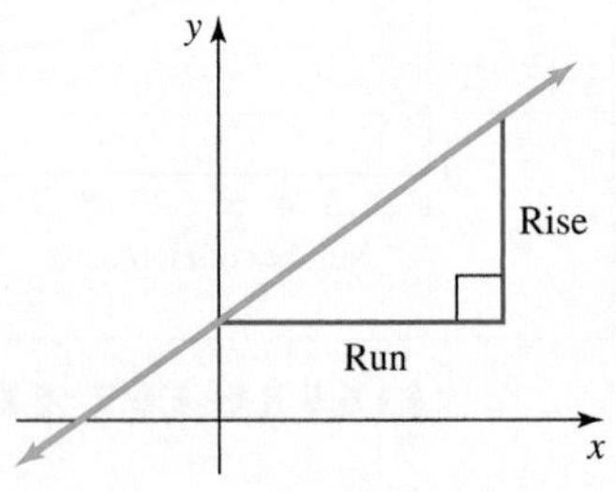

Types of slope

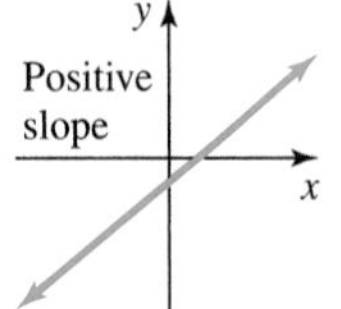

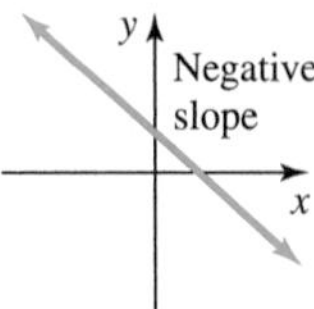

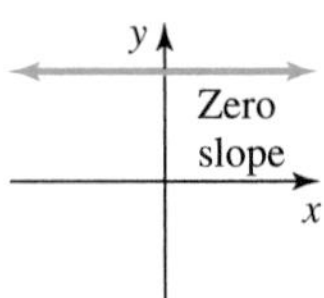

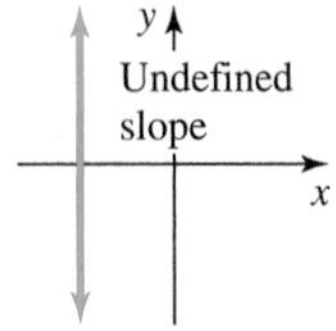

Parallel lines	Nonvertical parallel lines have equal slopes. Two vertical lines are parallel.	The lines $y = 3x - 9$ and $y = 3x + 7$ are parallel lines.
Perpendicular lines	Lines with slopes m and $-\frac{1}{m}$ are perpendicular. Any vertical line is perpendicular to any horizontal line.	The lines $y = -5x + 7$ and $y = \frac{1}{5}x$ are perpendicular.
Equations of Lines		**Examples**
Slope-intercept form	The equation of the line with y-intercept $(0, b)$ and slope m is $y = mx + b$.	$y = 3x - 1$ has slope 3 and y-intercept $(0, -1)$.
Point-slope form	The equation of the line with slope m that contains the point (x_1, y_1) is $y - y_1 = m(x - x_1)$.	The line through $(2, -1)$ with slope -5 is $y + 1 = -5(x - 2)$.
Standard form	Every line has an equation of the form $Ax + By = C$, where A, B, and C are real numbers with A and B not both equal to zero.	$4x - 9y = 15$ $x = 5$ (vertical line) $y = -7$ (horizontal line)
Graphing a line using y-intercept and slope	1. Write the equation in slope-intercept form. 2. Plot the y-intercept. 3. Use the rise and run to locate a second point. 4. Draw a line through the two points.	
Variation		**Examples**
Direct	If $y = kx$, then y varies directly as x.	$D = 50T$
Inverse	If $y = \frac{k}{x}$, then y varies inversely as x.	$R = \frac{400}{T}$
Joint	If $y = kxz$, then y varies jointly as x and z.	$V = 6LW$

ENRICHING YOUR MATHEMATICAL WORD POWER

For each mathematical term, choose the correct meaning.

1. **graph of an equation**
   a. the Cartesian coordinate system
   b. two number lines that intersect at a right angle
   c. the x-axis and y-axis
   d. an illustration in the coordinate plane that shows all ordered pairs that satisfy an equation
   d

2. **x-coordinate**
   a. the first number in an ordered pair
   b. the second number in an ordered pair
   c. a point on the x-axis
   d. a point where a graph crosses the x-axis
   a

3. **y-intercept**
   a. the second number in an ordered pair
   b. a point at which a graph intersects the y-axis
   c. any point on the y-axis
   d. the point where the y-axis intersects the x-axis
   b

4. **coordinate plane**
   a. a matching plane
   b. when the x-axis is coordinated with the y-axis
   c. a plane with a rectangular coordinate system
   d. a coordinated system for graphs
   c

5. **slope**
   a. the change in x divided by the change in y
   b. a measure of the steepness of a line
   c. the run divided by the rise
   d. the slope of a line
   b

6. **slope-intercept form**
   a. $y = mx + b$
   b. rise over run
   c. the point at which a line crosses the y-axis
   d. $y - y_1 = m(x - x_1)$
   a

7. **point-slope form**
   a. $Ax + By = C$
   b. rise over run
   c. $y - y_1 = m(x - x_1)$
   d. the slope of a line at a single point
   c

8. **independent variable**
   a. a rational constant
   b. an irrational constant
   c. the first variable of an ordered pair
   d. the second variable of an ordered pair
   c

9. **dependent variable**
   a. an irrational variable
   b. a rational variable
   c. the first variable of an ordered pair
   d. the second variable of an ordered pair
   d

10. **direct variation**
    a. $y = \pi$
    b. $y = kx$
    c. $y = k/x$
    d. $y = kxz$
    b

11. **inverse variation**
    a. $y = \pi$
    b. $y = kx$
    c. $y = k/x$
    d. $y = kxz$
    c

12. **joint variation**
    a. $y = \pi$
    b. $y = kx$
    c. $y = k/x$
    d. $y = kxz$
    d

REVIEW EXERCISES

3.1 *For each point, name the quadrant in which it lies or the axis on which it lies.*

1. $(-2, 5)$
   Quadrant II

2. $(-3, -5)$
   Quadrant III

3. $(3, 0)$
   x-axis

4. $(9, 10)$
   Quadrant I

5. $(0, -6)$
   y-axis

6. $(0, \pi)$
   y-axis

7. $(1.414, -3)$
   Quadrant IV

8. $(-4, 1.732)$
   Quadrant II

Complete the given ordered pairs so that each ordered pair satisfies the given equation.

9. $y = 3x - 5$: $(0,\ ), (-3,\ ), (4,\ )$
   $(0, -5), (-3, -14), (4, 7)$

10. $y = -2x + 1$: $(9,\ ), (3,\ ), (-1,\ )$
    $(9, -17), (3, -5), (-1, 3)$

11. $2x - 3y = 8$: $(0,\ ), (3,\ ), (-6,\ )$
    $\left(0, -\frac{8}{3}\right), \left(3, -\frac{2}{3}\right), \left(-6, -\frac{20}{3}\right)$

Study Tip

Note how the review exercises are arranged according to the sections in this chapter. If you are having trouble with a certain type of problem, refer back to the appropriate section for examples and explanations.

12. $x + 2y = 1$: $(0,\ ), (-2,\ ), (2,\ )$
    $\left(0, \frac{1}{2}\right), \left(-2, \frac{3}{2}\right), \left(2, -\frac{1}{2}\right)$

Sketch the graph of each equation by finding three ordered pairs that satisfy each equation.

13. $y = -3x + 4$

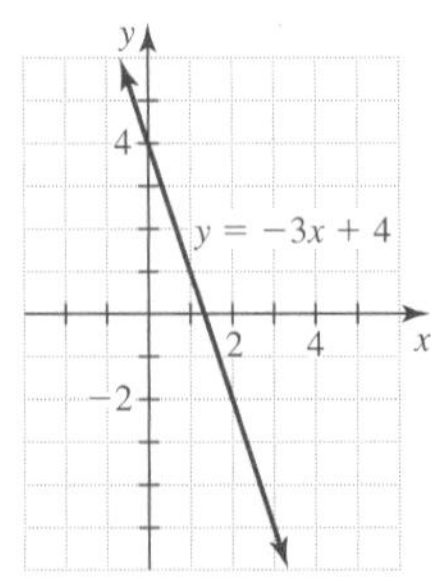

14. $y = 2x - 6$

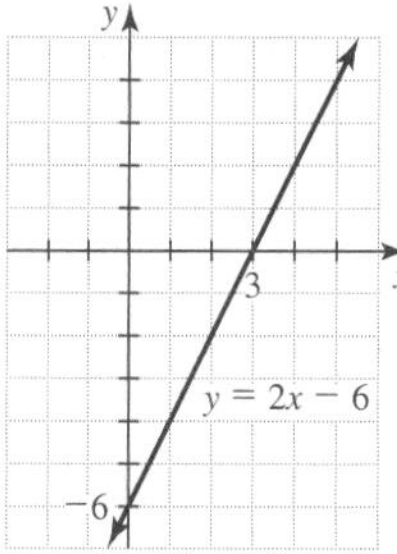

15. $x + y = 7$

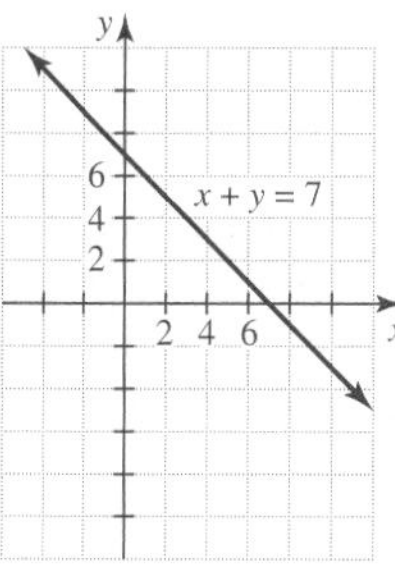

16. $x - y = 4$

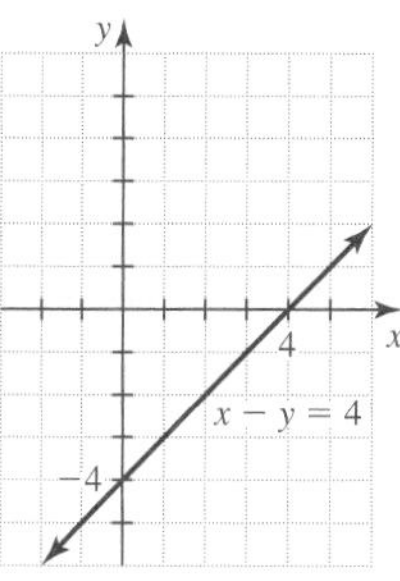

3.2 *Determine the slope of the line that goes through each pair of points.*

17. (0, 0) and (1, 1) 1

18. (−1, 1) and (2, −2) −1

19. (−2, −3) and (0, 0) $\frac{3}{2}$

20. (−1, −2) and (4, −1) $\frac{1}{5}$

21. (−4, −2) and (3, 1) $\frac{3}{7}$

22. (0, 4) and (5, 0) $-\frac{4}{5}$

3.3 *Find the slope and y-intercept for each line.*

23. $y = 3x - 18$ 3, (0, −18)

24. $y = -x + 5$ −1, (0, 5)

25. $2x - y = 3$ 2, (0, −3)

26. $x - 2y = 1$ $\frac{1}{2}, \left(0, -\frac{1}{2}\right)$

27. $4x - 2y - 8 = 0$ 2, (0, −4)

28. $3x + 5y + 10 = 0$ $-\frac{3}{5}$, (0, −2)

Sketch the graph of each equation.

29. $y = \frac{2}{3}x - 5$

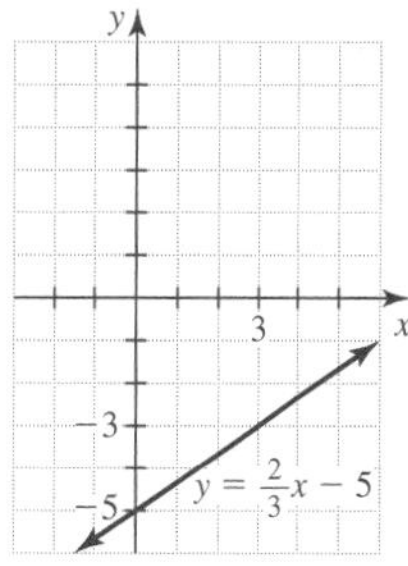

30. $y = \frac{3}{2}x + 1$

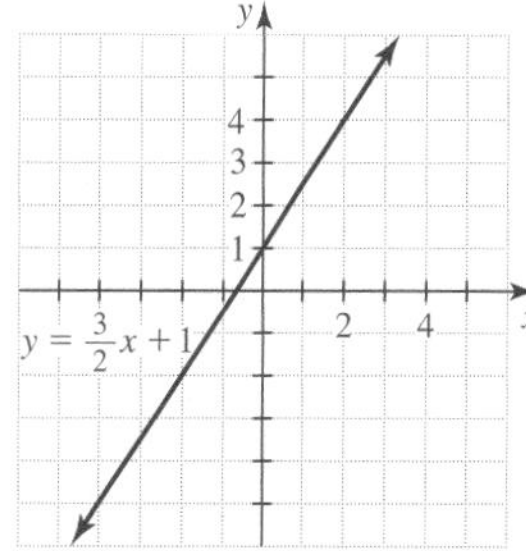

31. $2x + y = -6$

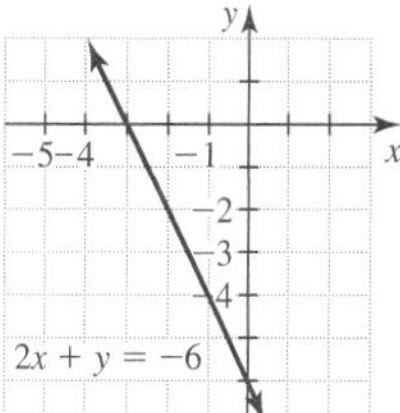

32. $-3x - y = 2$

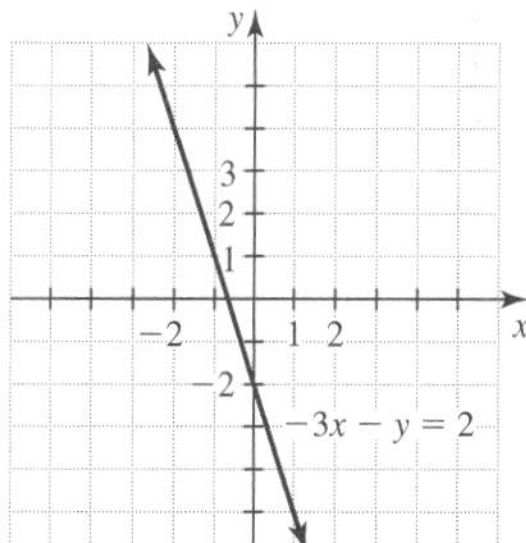

33. $y = -4$

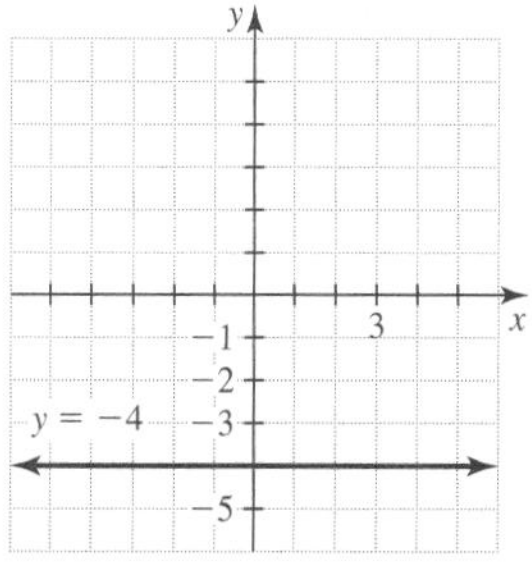

34. $x = 9$

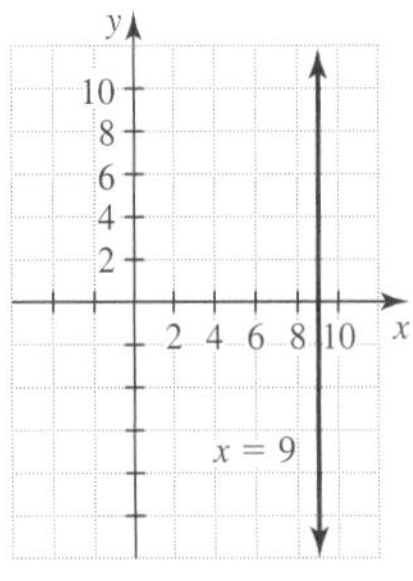

Determine the equation of each line. Write the answer in standard form using only integers as the coefficients.

35. The line through (0, 4) with slope $\frac{1}{3}$ $x - 3y = -12$

36. The line through (−2, 0) with slope $-\frac{3}{4}$ $3x + 4y = -6$

37. The line through the origin that is perpendicular to the line $y = 2x - 1$ $x + 2y = 0$

38. The line through (0, 9) that is parallel to the line $3x + 5y = 15$ $3x + 5y = 45$

39. The line through (3, 5) that is parallel to the x-axis $y = 5$

40. The line through (−2, 4) that is perpendicular to the x-axis $x = -2$

3.4 *Write each equation in slope-intercept form.*

41. $y - 3 = \frac{2}{3}(x + 6)$
$y = \frac{2}{3}x + 7$

42. $y + 2 = -6(x - 1)$
$y = -6x + 4$

43. $3x - 7y - 14 = 0$
$y = \frac{3}{7}x - 2$

44. $1 - x - y = 0$
$y = -x + 1$

45. $y - 5 = -\frac{3}{4}(x + 1)$
$y = -\frac{3}{4}x + \frac{17}{4}$

46. $y + 8 = -\frac{2}{5}(x - 2)$
$y = -\frac{2}{5}x - \frac{36}{5}$

Determine the equation of each line. Write the answer in slope-intercept form.

47. The line through (−4, 7) with slope −2
$y = -2x - 1$

48. The line through (9, 0) with slope $\frac{1}{2}$
$y = \frac{1}{2}x - \frac{9}{2}$

49. The line through the two points (−2, 1) and (3, 7)
$y = \frac{6}{5}x + \frac{17}{5}$

50. The line through the two points (4, 0) and (−3, −5)
$y = \frac{5}{7}x - \frac{20}{7}$

51. The line through (3, −5) that is parallel to the line $y = 3x - 1$ $y = 3x - 14$

52. The line through (4, 0) that is perpendicular to the line $x + y = 3$ $y = x - 4$

Solve each problem.

53. ***Rental charge.*** The charge for renting an air hammer for two days is \$113 and the charge for five days is \$209. The charge C is determined by the number of days n using a linear equation. Find the equation and find the charge for a four-day rental.
$C = 32n + 49$, \$177

54. ***Time on a treadmill.*** After 2 minutes on a treadmill, Jenny has a heart rate of 82. After 3 minutes she has a heart rate of 86. Assume that there is a linear equation that gives her heart rate h in terms of time on the treadmill t. Find the equation and use it to predict her heart rate after 10 minutes on the treadmill.
$h = 4t + 74$, 114

55. ***Probability of rain.*** If the probability p of rain is 90%, the probability q that it does not rain is 10%. If the probability of rain is 80%, then the probability that it does not rain is 20%. There is a linear equation that gives q in terms of p.

a) Find the equation.

b) Use the accompanying graph to determine the probability of rain if the probability that it does not rain is 0.

a) $q = 1 - p$
b) 1

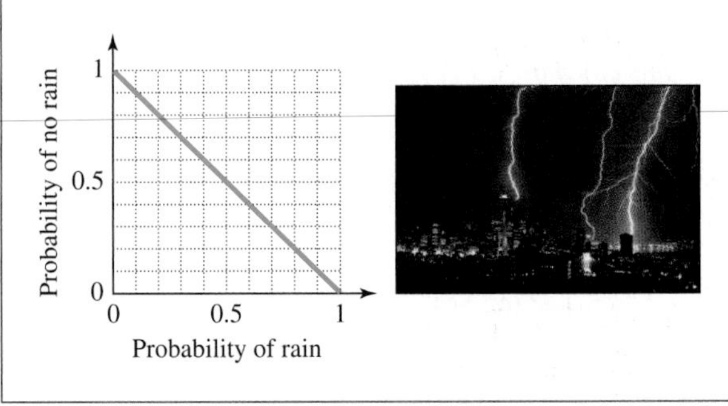

FIGURE FOR EXERCISE 55

56. ***Social Security benefits.*** If you earned an average of \$25,000 over your working life and you retire after 2005 at age 62, 63, or 64, then your annual Social Security benefit will be \$7000, \$7500, or \$8000, respectively (Social Security Administration, www.ssa.gov). There is a linear equation that gives the annual benefit b in terms of age a for these three years. Find the equation.
$b = 500a - 24{,}000$

57. ***Predicting freshman GPA.*** A researcher who is studying the relationship between ACT score and grade point average for freshman gathered the data shown in the

ACT score (x)	GPA (y)
4	1.0
14	2.0
24	3.0
34	4.0

TABLE FOR EXERCISE 57

accompanying table. Find the equation of the line in slope-intercept form that goes through these points.
$y = 0.1x + 0.6$

58. ***Interest rates.*** A credit manager rates each applicant for a car loan on a scale of 1 through 5 and then determines the interest rate from the accompanying table. Find the equation of the line in slope-intercept form that goes through these points.
$y = -4x + 28$

Credit rating	Interest rate (%)
1	24
2	20
3	16
4	12
5	8

TABLE FOR EXERCISE 58

3.5 *Solve each variation problem.*

59. Suppose y varies directly as w. If $y = 48$ when $w = 4$, then what is y when $w = 11$?
132

60. Suppose m varies directly as t. If $m = 13$ when $t = 2$, then what is m when $t = 6$?
39

61. If y varies inversely as v and $y = 8$ when $v = 6$, then what is y when $v = 24$?
2

62. If y varies inversely as r and $y = 9$ when $r = 3$, then what is y when $r = 9$?
3

63. Suppose y varies jointly as u and v, and $y = 72$ when $u = 3$ and $v = 4$. Find y when $u = 5$ and $v = 2$.
60

64. Suppose q varies jointly as s and t, and $q = 10$ when $s = 4$ and $t = 3$. Find q when $s = 25$ and $t = 6$.
125

65. ***Taxi fare.*** The cost of a taxi ride varies directly with the length of the ride in minutes. A 12-minute ride costs \$9.00.

a) Write the cost in terms of the length of the ride.

b) What is the cost of a 20-minute ride?

c) Is the cost increasing or decreasing as the length of the ride increases?

a) $C = 0.75T$ **b)** \$15 **c)** Increasing

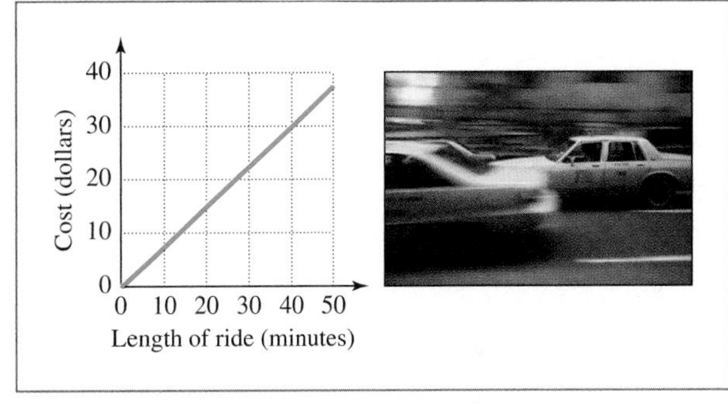

FIGURE FOR EXERCISE 65

66. ***Applying shingles.*** The number of hours that it takes to apply 296 bundles of shingles varies inversely with the number of roofers working on the job. Three workers can complete the job in 40 hours.

a) Write the number of hours in terms of the number of roofers on the job.

b) How long would it take five roofers to complete the job?

c) Is the time to complete the job increasing or decreasing as the number of workers increases?

a) $h = \frac{120}{n}$ **b)** 24 hours **c)** Decreasing

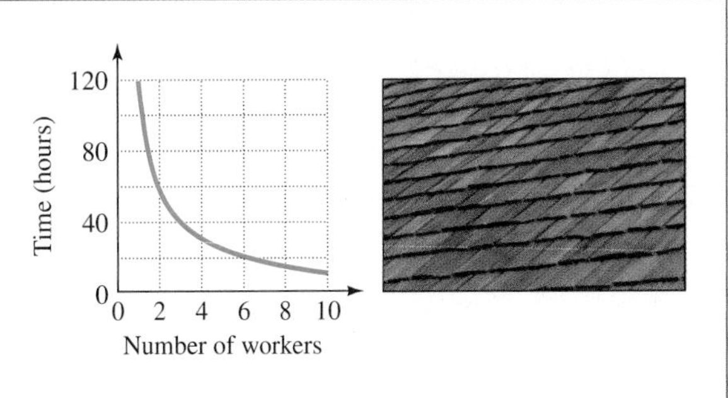

FIGURE FOR EXERCISE 66

CHAPTER 3 TEST

For each point, name the quadrant in which it lies or the axis on which it lies.

1. $(-2, 7)$ Quadrant II
2. $(-\pi, 0)$ x-axis
3. $(3, -6)$ Quadrant IV
4. $(0, 1785)$ y-axis

Find the slope of the line through each pair of points.

5. $(3, 3)$ and $(4, 4)$ 1
6. $(-2, -3)$ and $(4, -8)$ $-\frac{5}{6}$

Find the slope of each line.

7. The line $y = 3x - 5$ 3
8. The line $y = 3$ 0
9. The line $x = 5$ Undefined
10. The line $2x - 3y = 4$ $\frac{2}{3}$

Write the equation of each line. Give the answer in slope-intercept form.

11. The line through $(0, 3)$ with slope $-\frac{1}{2}$ $y = -\frac{1}{2}x + 3$
12. The line through $(-1, -2)$ with slope $\frac{3}{7}$ $y = \frac{3}{7}x - \frac{11}{7}$

Write the equation of each line. Give the answer in standard form using only integers as the coefficients.

13. The line through $(2, -3)$ that is perpendicular to the line $y = -3x + 12$ $x - 3y = 11$
14. The line through $(3, 4)$ that is parallel to the line $5x + 3y = 9$
 $5x + 3y = 27$

Sketch the graph of each equation.

15. $y = \frac{1}{2}x - 3$

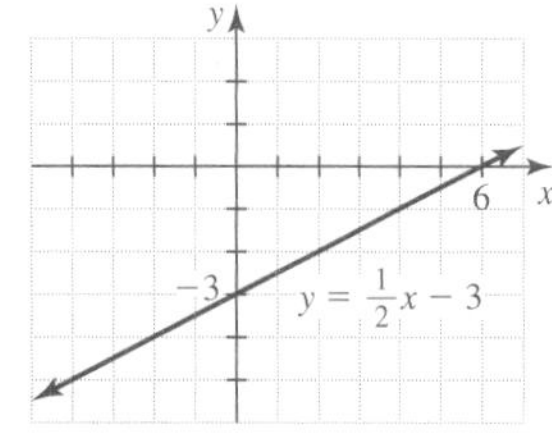

16. $2x - 3y = 6$

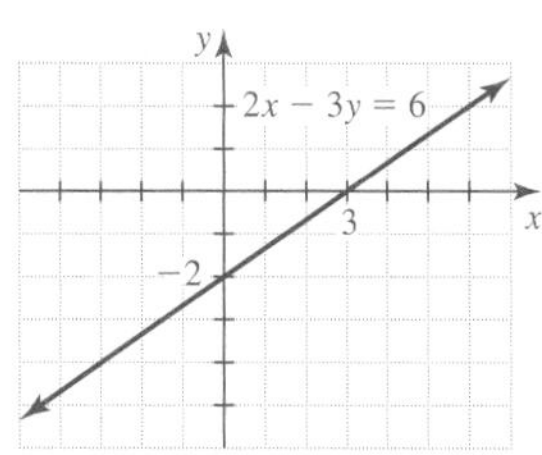

Study Tip

Before you take an in-class exam on this chapter, work the sample test given here. Set aside one hour to work this test and use the answers in the back of this book to grade yourself. Even though your instructor might not ask exactly the same questions, you will get a good idea of your test readiness.

17. $y = 4$

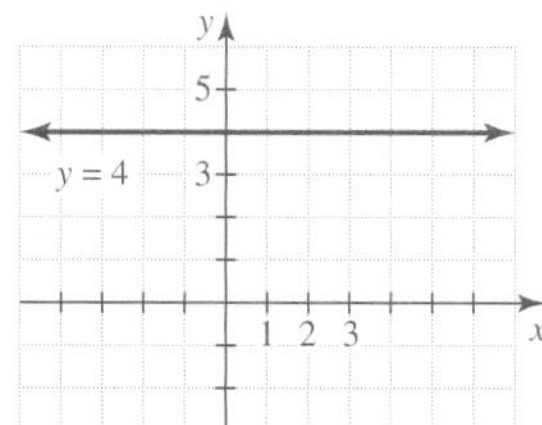

18. $x = -2$

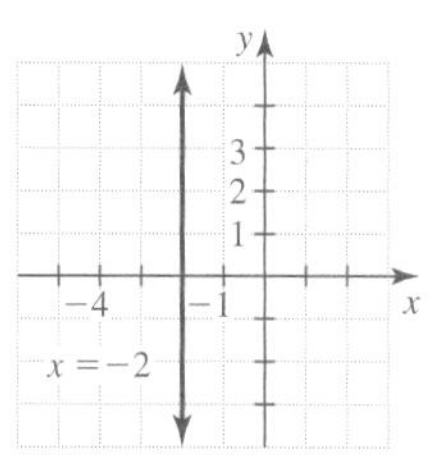

Solve each problem.

19. Julie's mail-order CD club charges a shipping and handling fee of \$2.50 plus \$0.75 per CD for each order shipped. Write the shipping and handling fee S in terms of the number n of CDs in the order.
 $S = 0.75n + 2.50$
20. A 10-ounce soft drink sells for 50 cents, and a 16-ounce soft drink sells for 68 cents. The price P is determined from the volume of the cup v by a linear equation. Find the equation and find the price for a 20-ounce soft drink.
 $P = 3v + 20$, 80 cents
21. The price of a watermelon varies directly with its weight. If the price of a 30-pound watermelon is \$4.20, then what is the price of a 20-pound watermelon?
 \$2.80
22. The number of days that Jerry spends on the road is inversely proportional to his sales for the previous month. If Jerry spent 15 days on the road when his previous month's sales were \$75,000, then how many days would he spend on the road when his previous month's sales were \$60,000? Does his road time increase or decrease as his sales increase?
 18.75 days, decreases
23. The labor cost for installing ceramic floor tile in a rectangular room varies jointly with the length and width. For a room that is 8 feet by 10 feet the cost is \$400. For a room that is 9 feet by 12 feet the cost is \$540. What is the cost for a room that is 11 feet by 14 feet?
 \$770

Graph Paper

Use these grids for graphing. Make as many copies of this page as you need.

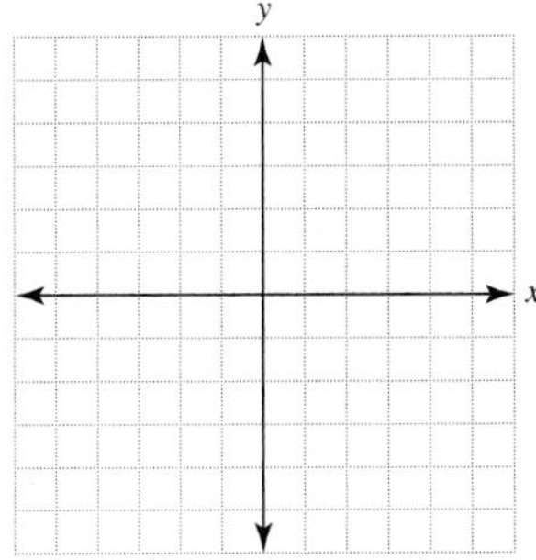

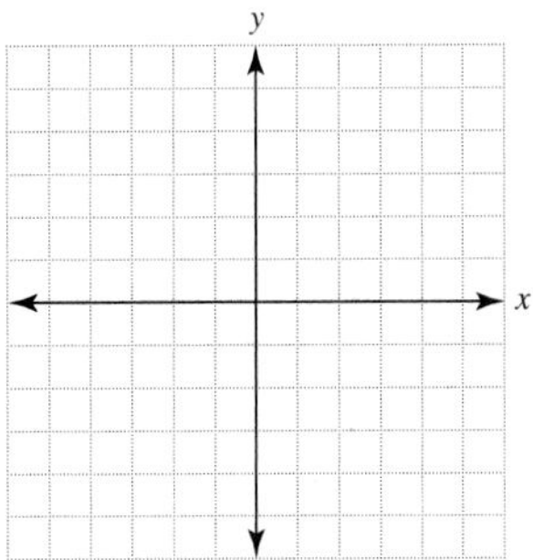

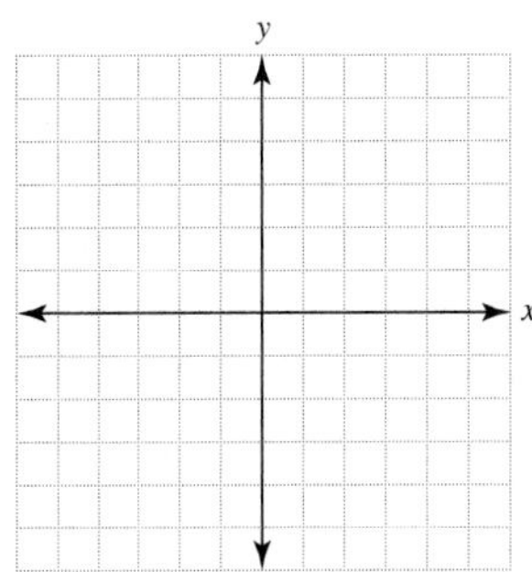

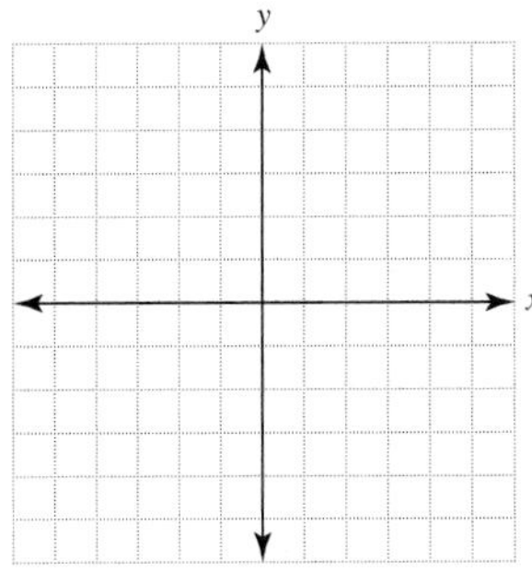

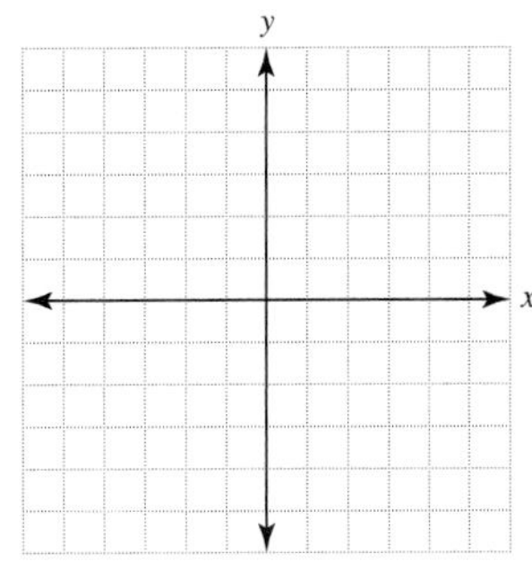

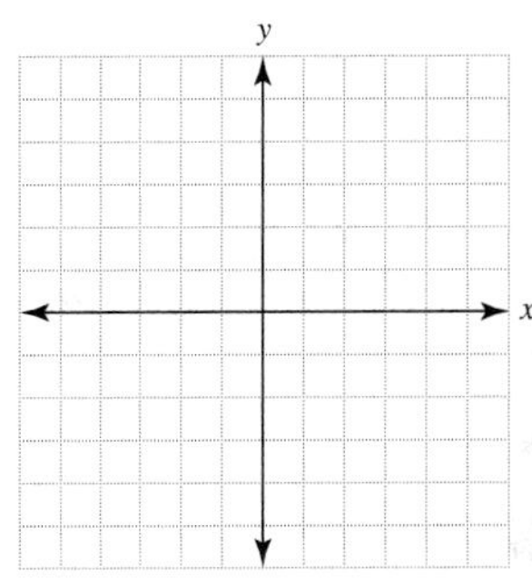

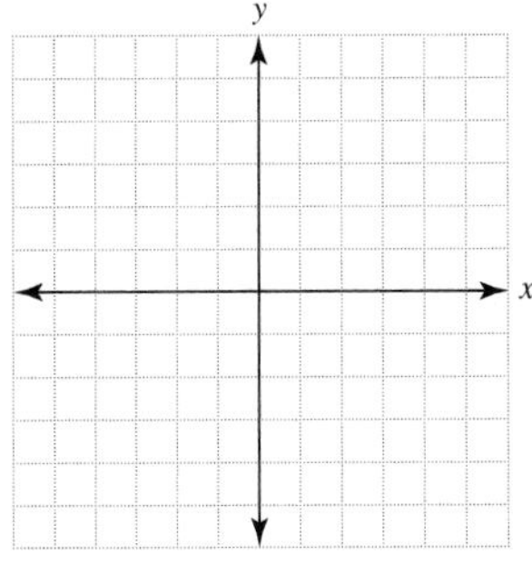

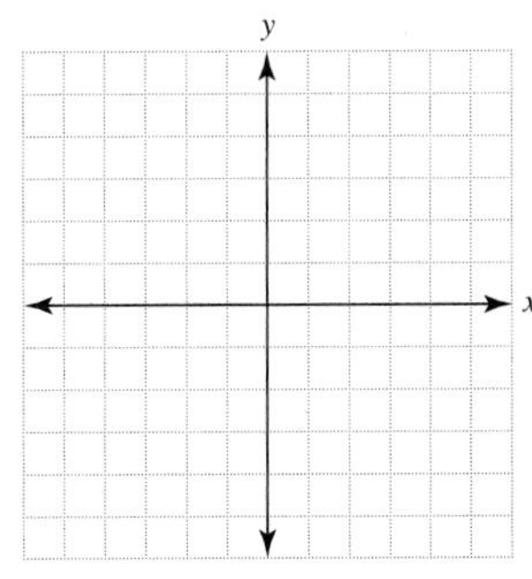

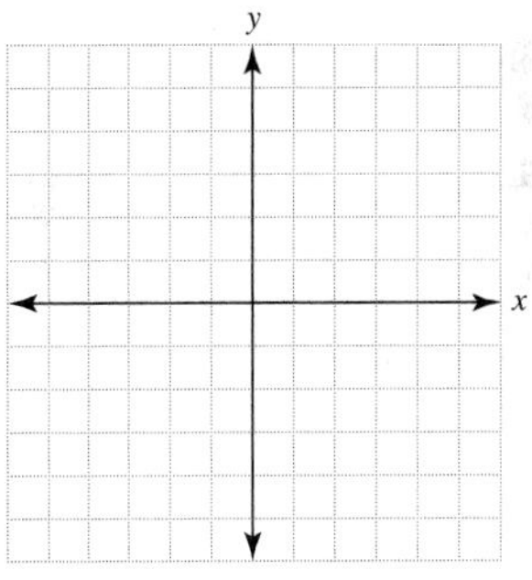

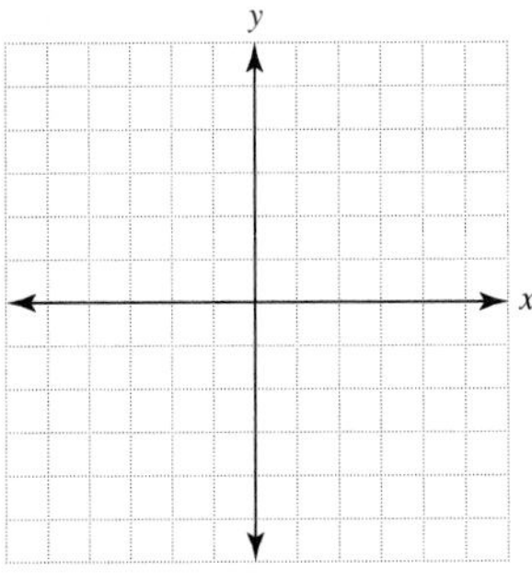

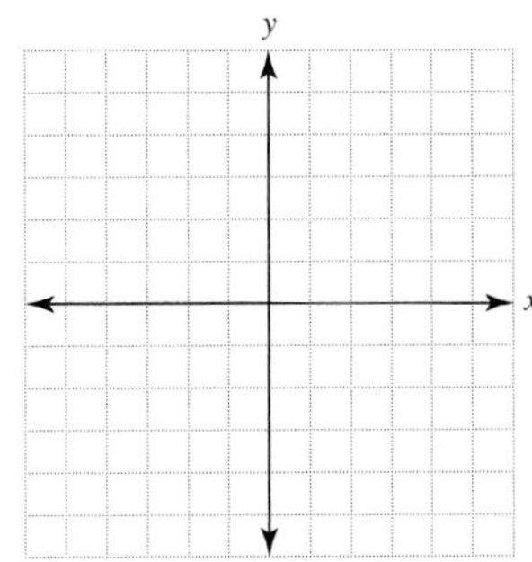

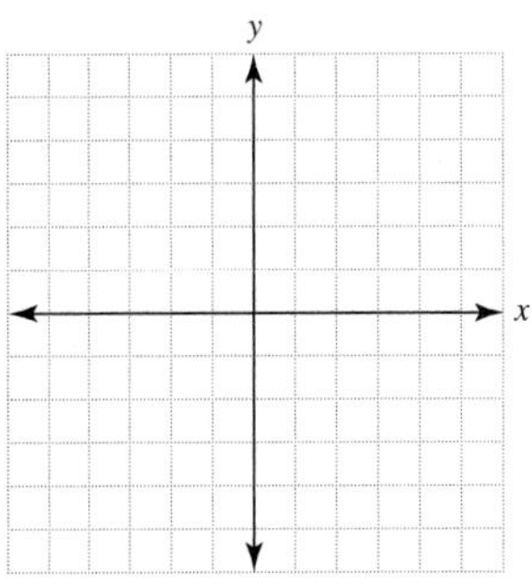

Simplify each arithmetic expression.

1. $9 - 5 \cdot 2$ -1

2. $-4 \cdot 5 - 7 \cdot 2$ -34

3. $3^2 - 2^3$ 1

4. $3^2 \cdot 2^3$ 72

5. $(-4)^2 - 4(1)(5)$ -4

6. $-4^2 - 4 \cdot 3$ -28

7. $\dfrac{-5 - 9}{2 - (-2)}$ $-\dfrac{7}{2}$

8. $\dfrac{6 - 3.6}{6}$ 0.4

9. $\dfrac{1 - \frac{1}{2}}{4 - (-1)}$ $\dfrac{1}{10}$

10. $\dfrac{4 - (-6)}{1 - \frac{1}{3}}$ 15

Simplify the given expression or solve the given equation, whichever is appropriate.

11. $4x - (-9x)$ $13x$

12. $4(x - 9) - x$ $3x - 36$

13. $5(x - 3) + x = 0$ $\left\{\dfrac{5}{2}\right\}$

14. $5 - 2(x - 1) = x$ $\left\{\dfrac{7}{3}\right\}$

15. $\dfrac{1}{2} - \dfrac{1}{3}$ $\dfrac{1}{6}$

16. $\dfrac{1}{4} + \dfrac{1}{6}$ $\dfrac{5}{12}$

17. $\dfrac{1}{2}x - \dfrac{1}{3} = \dfrac{1}{4}x + \dfrac{1}{6}$ $\{2\}$

18. $\dfrac{2}{3}x + \dfrac{1}{5} = \dfrac{3}{5}x - \dfrac{1}{15}$ $\{-4\}$

19. $\dfrac{4x - 8}{2}$ $2x - 4$

20. $\dfrac{-5x - 10}{-5}$ $x + 2$

21. $\dfrac{6 - 2(x - 3)}{2} = 1$ $\{5\}$

22. $\dfrac{20 - 5(x - 5)}{5} = 6$ $\{3\}$

23. $-4(x - 9) - 4 = -4x$ $\varnothing$

24. $4(x - 6) = -4(6 - x)$ All real numbers

Study Tip

Don't wait until the final exam to review material. Do some review on a regular basis. The Making Connections exercises on this page can be used to review, compare, and contrast different concepts that you have studied. A good time to work these exercises is between a test and the start of new material.

Solve each equation for y.

25. $3\pi y + 2 = t$ $y = \dfrac{t - 2}{3\pi}$

26. $x = \dfrac{y - b}{m}$ $y = mx + b$

27. $3x - 3y - 12 = 0$ $y = x - 4$

28. $2y - 3 = 9$ $y = 6$

29. $\dfrac{y}{2} - \dfrac{y}{4} = \dfrac{1}{5}$ $y = \dfrac{4}{5}$

30. $0.6y - 0.06y = 108$ $y = 200$

Solve.

31. ***Financial planning.*** Financial advisors at Fidelity Investments use the information in the accompanying graph as a guide for retirement investing.

a) What is the slope of the line segment for ages 35 through 50?

b) What is the slope of the line segment for ages 50 through 65?

c) If a 38-year-old man is making \$40,000 per year, then what percent of his income should he be saving?

d) If a 58-year-old woman has an annual salary of \$60,000, then how much should she have saved and how much should she be saving per year?

a) $\dfrac{2}{15}$ **b)** $\dfrac{1}{5}$ **c)** About 13% per year

d) \$276,000 saved, \$12,000 per year

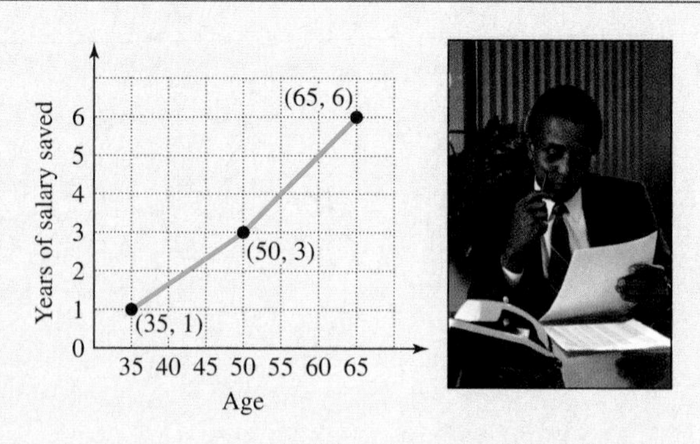

FIGURE FOR EXERCISE 31

CHAPTER 4

Systems of Linear Equations and Inequalities

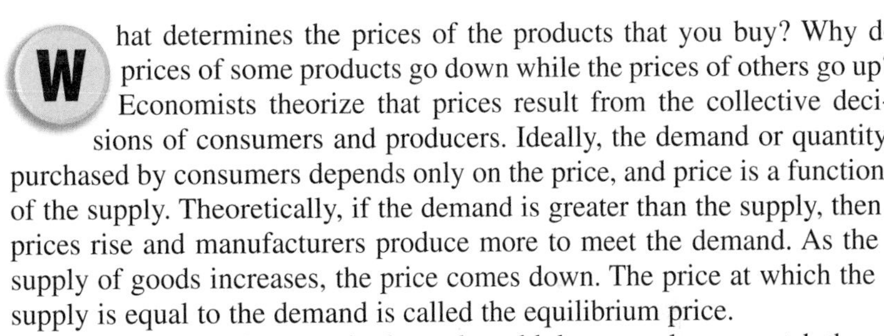

What determines the prices of the products that you buy? Why do prices of some products go down while the prices of others go up? Economists theorize that prices result from the collective decisions of consumers and producers. Ideally, the demand or quantity purchased by consumers depends only on the price, and price is a function of the supply. Theoretically, if the demand is greater than the supply, then prices rise and manufacturers produce more to meet the demand. As the supply of goods increases, the price comes down. The price at which the supply is equal to the demand is called the equilibrium price.

However, what happens in the real world does not always match the theory. Manufacturers cannot always control the supply, and factors other than price can affect a consumer's decision to buy. For example, droughts in Brazil decreased the supply of coffee and drove coffee prices up. Floods in California did the same to the prices of produce. With one of the most abundant wheat crops ever in 1994, cattle gained weight more quickly, increasing the supply of cattle ready for market. With supply going up, prices went down. Decreased demand for beef in Japan and Mexico drove the price of beef down further. With lower prices, consumers should be buying more beef, but increased competition from chicken and pork products, as well as health concerns, have kept consumer demand low.

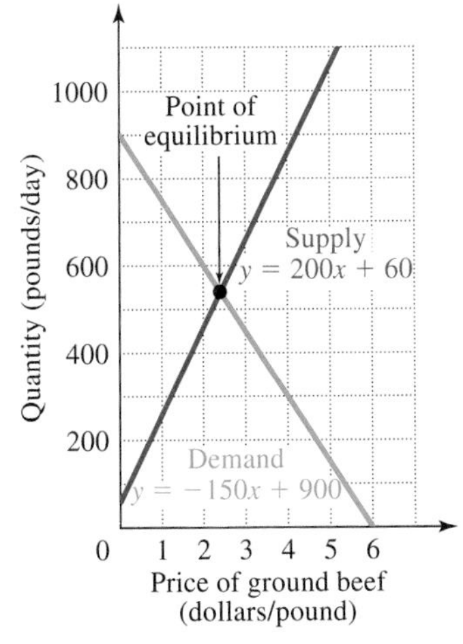

The two functions that govern supply and demand form a system of equations. In this chapter you will learn how to solve systems of equations. In Exercise 43 of Section 4.2 you will see an example of supply and demand equations for ground beef.

In This Section

- Solving a System of Linear Equations by Graphing
- Independent, Inconsistent, and Dependent Equations
- Applications

4.1 THE GRAPHING METHOD

You studied linear equations in two variables in Chapter 3. In this section you will learn to solve systems of linear equations in two variables and use systems to solve problems.

Solving a System of Linear Equations by Graphing

Consider the linear equation $y = 2x - 1$. The graph of this equation is a straight line, and every point on the line is a solution to the equation. Now consider a second linear equation, $x + y = 2$. The graph of this equation is also a straight line, and every point on the line is a solution to this equation. The pair of equations

$$y = 2x - 1$$
$$x + y = 2$$

is called a **system of equations.** A point that satisfies both equations is called a **solution to the system.**

EXAMPLE 1 **A solution to a system**

Determine whether the point $(-1, 3)$ is a solution to each system of equations.

a) $3x - y = -6$
$x + 2y = 5$

b) $y = 2x - 1$
$x + y = 2$

Calculator Close-Up

Graph $y_1 = 3x + 6$ and $y_2 = (5 - x)/2$ to see that $(-1, 3)$ is on both lines.

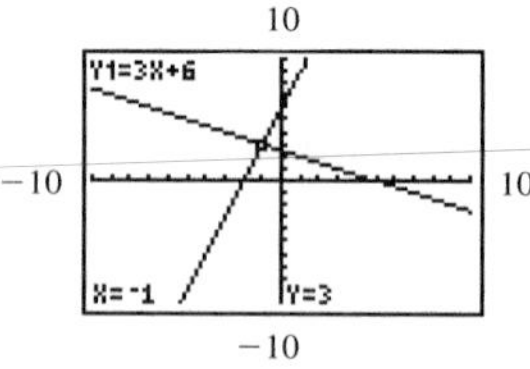

Graph $y_1 = 2x - 1$ and $y_2 = 2 - x$ to see that $(-1, 3)$ is on one line but not the other.

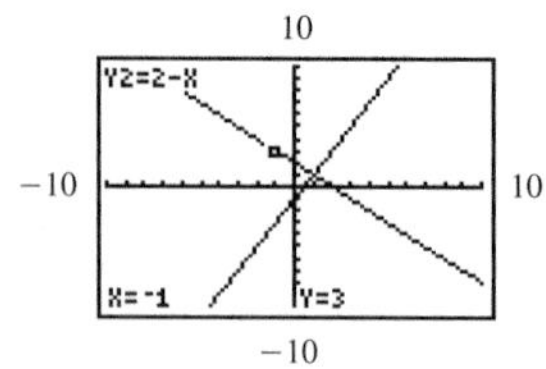

Solution

a) If we let $x = -1$ and $y = 3$ in both equations of the system, we get the following equations:

$$3(-1) - 3 = -6 \quad \text{Correct}$$
$$-1 + 2(3) = 5 \quad \text{Correct}$$

Because both of these equations are correct, $(-1, 3)$ is a solution to the system.

b) If we let $x = -1$ and $y = 3$ in both equations of the system, we get the following equations:

$$3 = 2(-1) - 1 \quad \text{Incorrect}$$
$$-1 + 3 = 2 \quad \text{Correct}$$

Because the first equation is not satisfied by $(-1, 3)$, the point $(-1, 3)$ is not a solution to the system. ■

If we graph each equation of a system on the same coordinate plane, then we may be able to see the points that they have in common. Any point that is on both graphs is a solution to the system.

EXAMPLE 2 **Using graphing to solve a system**

Solve the system by graphing:

$$y = 2x - 1$$
$$x + y = 2$$

4.2 THE SUBSTITUTION METHOD

In This Section

- Solving a System of Linear Equations by Substitution
- Inconsistent and Dependent Systems
- Applications

Solving a system by graphing is certainly limited by the accuracy of the graph. If the lines intersect at a point whose coordinates are not integers, then it is difficult to identify the solution from a graph. In this section we introduce a method for solving systems of linear equations in two variables that does not depend on a graph and is totally accurate.

Solving a System of Linear Equations by Substitution

The next example shows how to solve a system without graphing. The method is called **substitution.**

EXAMPLE 1 **Solving a system by substitution**

Solve:

$$2x - 3y = 9$$
$$y - 4x = -8$$

Calculator Close-Up

To check Example 1, graph

$$y_1 = (2x - 9)/3$$

and

$$y_2 = 4x - 8.$$

Use the intersect feature of your calculator to find the point of intersection.

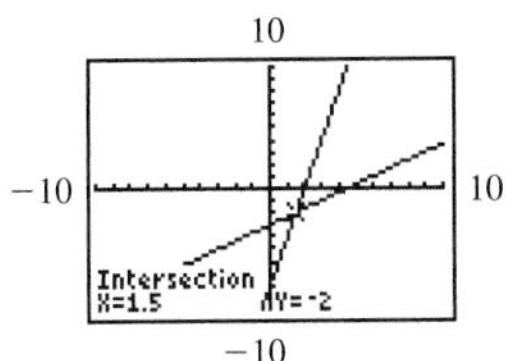

Solution

First solve the second equation for y:

$$y - 4x = -8$$
$$y = 4x - 8$$

Now substitute $4x - 8$ for y in the first equation:

$$2x - 3y = 9$$
$$2x - 3(4x - 8) = 9 \quad \text{Substitute } 4x - 8 \text{ for y.}$$
$$2x - 12x + 24 = 9 \quad \text{Simplify.}$$
$$-10x + 24 = 9$$
$$-10x = -15$$
$$x = \frac{-15}{-10}$$
$$= \frac{3}{2}$$

Use the value $x = \frac{3}{2}$ in $y = 4x - 8$ to find y:

$$y = 4 \cdot \frac{3}{2} - 8$$
$$= -2$$

Check that $\left(\frac{3}{2}, -2\right)$ satisfies both of the original equations. The solution to the system is $\left(\frac{3}{2}, -2\right)$. ■

EXAMPLE 2 **Solving a system by substitution**

Solve:

$$3x + 4y = 5$$
$$x = y - 1$$

Calculator Close-Up

To check Example 2, graph

$$y_1 = (5 - 3x)/4$$

and

$$y_2 = x + 1.$$

Use the intersect feature of your calculator to find the point of intersection.

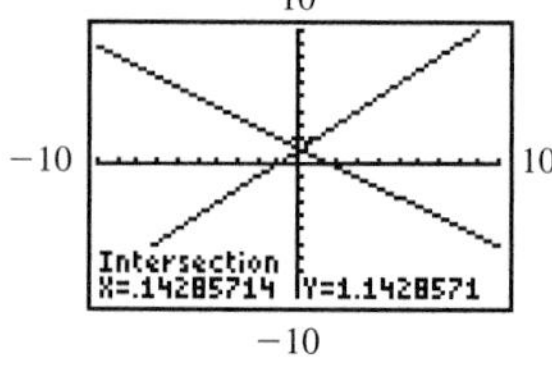

Solution

Because the second equation is already solved for x in terms of y, we can substitute $y - 1$ for x in the first equation:

$$
\begin{aligned}
3x + 4y &= 5 \\
3(y - 1) + 4y &= 5 && \text{Replace } x \text{ with } y - 1. \\
3y - 3 + 4y &= 5 && \text{Simplify.} \\
7y - 3 &= 5 \\
7y &= 8 \\
y &= \frac{8}{7}
\end{aligned}
$$

Now use the value $y = \frac{8}{7}$ in one of the original equations to find x. The simplest one to use is $x = y - 1$:

$$
\begin{aligned}
x &= \frac{8}{7} - 1 \\
x &= \frac{1}{7}
\end{aligned}
$$

Check that $\left(\frac{1}{7}, \frac{8}{7}\right)$ satisfies both equations. The solution to the system is $\left(\frac{1}{7}, \frac{8}{7}\right)$. ■

Use the following strategy for solving by substitution.

Strategy for Solving a System by Substitution

1. Solve one of the equations for one variable in terms of the other.
2. Substitute this value into the other equation to eliminate one of the variables.
3. Solve for the remaining variable.
4. Insert this value into one of the original equations to find the value of the other variable.
5. Check your solution in both equations.

Inconsistent and Dependent Systems

Examples 3 and 4 illustrate how the inconsistent and dependent cases appear when we use substitution to solve the system.

EXAMPLE 3 **An inconsistent system**

Solve by substitution:

$$
\begin{aligned}
3x - 6y &= 9 \\
x &= 2y + 5
\end{aligned}
$$

Calculator Close-Up

To check Example 3, graph $y_1 = (3x - 9)/6$ and $y_2 = (x - 5)/2$. Since the lines appear to be parallel, there is no solution to the system.

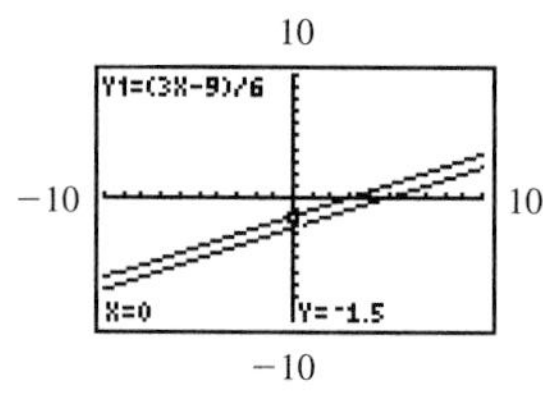

Solution

Use $x = 2y + 5$ to replace x in the first equation:

$$
\begin{aligned}
3x - 6y &= 9 \\
3(2y + 5) - 6y &= 9 && \text{Replace } x \text{ by } 2y + 5. \\
6y + 15 - 6y &= 9 && \text{Simplify.} \\
15 &= 9
\end{aligned}
$$

No values for x and y will make 15 equal to 9. So there is no ordered pair that satisfies both equations. This system is inconsistent. It has no solution. The equations are the equations of parallel lines. ■

EXAMPLE 4 **A dependent system**

Solve:

$$2(y - x) = x + y - 1$$
$$y = 3x - 1$$

Helpful Hint

The purpose of Examples 3 and 4 is to show what happens when you try to solve an inconsistent or dependent system by substitution. If we had first written the equations in slope-intercept form, we would have seen that the lines in Example 3 are parallel and the lines in Example 4 are the same.

Solution

Because the second equation is solved for y, we will eliminate the variable y in the substitution. Substitute $y = 3x - 1$ into the first equation:

$$2(3x - 1 - x) = x + (3x - 1) - 1$$
$$2(2x - 1) = 4x - 2$$
$$4x - 2 = 4x - 2$$

Any value for x makes the last equation true because both sides are identical. So any value for x can be used as a solution to the original system as long as we choose $y = 3x - 1$. The system is dependent. The two equations are equations for the same straight line. The solution to the system is the set of all points on that line,

$$\{(x, y) \mid y = 3x - 1\}.$$ ■

When solving a system by substitution, we can recognize an inconsistent system or dependent system as follows:

Inconsistent and Dependent Systems

An inconsistent system leads to a false statement.
A dependent system leads to a statement that is always true.

Applications

Many of the problems that we solved in previous chapters had two unknown quantities, but we wrote only one equation to solve the problem. For problems with two unknown quantities we can use two variables and a system of equations.

EXAMPLE 5 **Two investments**

Mrs. Robinson invested a total of \$25,000 in two investments, one paying 6% and the other paying 8%. If her total income from these investments was \$1790, then how much money did she invest in each?

Helpful Hint

In Chapter 2, we would have done Example 5 with one variable by letting x represent the amount invested at 6% and $25{,}000 - x$ represent the amount invested at 8%.

Solution

Let x represent the amount invested at 6%, and let y represent the amount invested at 8%. The following table organizes the given information.

	Interest rate	Amount invested	Amount of interest
First investment	6%	x	$0.06x$
Second investment	8%	y	$0.08y$

Write one equation describing the total of the investments, and the other equation describing the total interest:

$$x + y = 25{,}000 \quad \text{Total investments}$$
$$0.06x + 0.08y = 1790 \quad \text{Total interest}$$

To solve the system, we solve the first equation for y:

$$y = 25{,}000 - x$$

Substitute $25{,}000 - x$ for y in the second equation:

$$\begin{aligned} 0.06x + 0.08(25{,}000 - x) &= 1790 \\ 0.06x + 2000 - 0.08x &= 1790 \\ -0.02x + 2000 &= 1790 \\ -0.02x &= -210 \\ x &= \frac{-210}{-0.02} \\ &= 10{,}500 \end{aligned}$$

Let $x = 10{,}500$ in the equation $y = 25{,}000 - x$ to find y:

$$\begin{aligned} y &= 25{,}000 - 10{,}500 \\ &= 14{,}500 \end{aligned}$$

Check these values for x and y in the original problem. Mrs. Robinson invested \$10,500 at 6% and \$14,500 at 8%. ■

Calculator Close-Up

You can use a calculator to check the answers in Example 5:

```
10500+14500
            25000
.06*10500+.08*14
500
             1790
```

MATH AT WORK

RACE CAR DRIVER

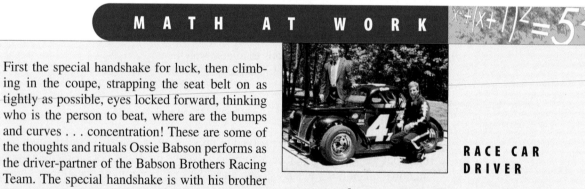

First the special handshake for luck, then climbing in the coupe, strapping the seat belt on as tightly as possible, eyes locked forward, thinking who is the person to beat, where are the bumps and curves . . . concentration! These are some of the thoughts and rituals Ossie Babson performs as the driver-partner of the Babson Brothers Racing Team. The special handshake is with his brother Dave Babson, who is the crew chief of the team. The car is a $\frac{5}{8}$-scale model of a 1940 Ford Coupe, specially built for the Legends of Nascar Racing Series.

There are strict rules on the weight distribution, frame height, tire size, and engine size for these cars, but for best results the car should be set up to lean to the left and front. The challenge is to meet all these criteria for a successful race. Before the race, Ossie drives on the track and makes observations and recommendations on how the car handles going into the turns and what adjustments to make for better performance under specific track conditions. Dave then supervises the changes to the car. This process continues until both driver and crew chief are satisfied. Ultimately, a combination of art, how the car feels, and science makes the car perform to its ultimate capabilities.

In Exercises 41 and 42 of this section you will see how the Babsons use a system of equations to determine the proper weight distribution for their car.

WARM-UPS

True or false? Explain your answer.

For Exercises 1–7, use the following systems:

a) $y = x - 7$
$2x + 3y = 4$

b) $x + 2y = 1$
$2x - 4y = 0$

1. If we substitute $x - 7$ for y in system (a), we get $2x + 3(x - 7) = 4$. True
2. The x-coordinate of the solution to system (a) is 5. True
3. The solution to system (a) is $(5, -2)$. True
4. The point $\left(\frac{1}{2}, \frac{1}{4}\right)$ satisfies system (b). True
5. It would be difficult to solve system (b) by graphing. True
6. Either x or y could be eliminated by substitution in system (b). True
7. System (b) is a dependent system. False
8. Solving an inconsistent system by substitution will result in a false statement. True
9. Solving a dependent system by substitution results in an equation that is always true. True
10. Any system of two linear equations can be solved by substitution. True

4.2 EXERCISES

Reading and Writing *After reading this section, write out the answers to these questions. Use complete sentences.*

1. What method is used in this section to solve systems of equations?
   In this section we learned the substitution method.
2. What is wrong with the graphing method for solving systems?
   Graphing is not accurate enough.
3. What is a dependent system?
   A dependent system is one in which the equations are equivalent.
4. What is an inconsistent system?
   An inconsistent system is one with no solution.
5. What happens when you try to solve a dependent system by substitution?
   Using substitution on a dependent system results in an equation that is always true.
6. What happens when you try to solve an inconsistent system by substitution?
   Using substitution on an inconsistent system results in a false equation.

Solve each system by the substitution method. See Examples 1 and 2.

7. $y = x + 3$
   $2x - 3y = -11$ $(2, 5)$
8. $y = x - 5$
   $x + 2y = 8$ $(6, 1)$
9. $x = 2y - 4$
   $2x + y = 7$ $(2, 3)$
10. $x = y - 2$
   $-2x + y = -1$ $(3, 5)$
11. $2x + y = 5$
   $5x + 2y = 8$ $(-2, 9)$
12. $5y - x = 0$
   $6x - y = 2$ $\left(\frac{10}{29}, \frac{2}{29}\right)$
13. $x + y = 0$
   $3x + 2y = -5$ $(-5, 5)$
14. $x - y = 6$
   $3x + 4y = -3$ $(3, -3)$
15. $x + y = 1$
   $4x - 8y = -4$ $\left(\frac{1}{3}, \frac{2}{3}\right)$
16. $x - y = 2$
   $3x - 6y = 8$ $\left(\frac{4}{3}, -\frac{2}{3}\right)$
17. $2x + 3y = 2$
   $4x - 9y = -1$ $\left(\frac{1}{2}, \frac{1}{3}\right)$
18. $x - 2y = 1$
   $3x + 10y = -1$ $\left(\frac{1}{2}, -\frac{1}{4}\right)$

Solve each system by substitution, and identify each system as independent, dependent, or inconsistent. See Examples 3 and 4.

19. $x - 2y = -2$
   $x + 2y = 8$ $\left(3, \frac{5}{2}\right)$, independent
20. $y = -3x + 1$
   $y = 2x + 4$ $\left(-\frac{3}{5}, \frac{14}{5}\right)$, independent

21. $x = 4 - 2y$
$4y + 2x = -8$ No solution, inconsistent

22. $21x - 35 = 7y$
$3x - y = 5$ $\{(x, y) \mid 3x - y = 5\}$, dependent

23. $y - 3 = 2(x - 1)$
$y = 2x + 3$ No solution, inconsistent

24. $y + 1 = 5(x + 1)$
$y = 5x - 1$ No solution, inconsistent

25. $3x - 2y = 7$
$3x + 2y = 7$ $\left(\frac{7}{3}, 0\right)$, independent

26. $2x + 5y = 5$
$3x - 5y = 6$ $\left(\frac{11}{5}, \frac{3}{25}\right)$, independent

27. $x + 5y = 4$
$x + 5y = 4y$ $(-1, 1)$, independent

28. $2x + y = 3x$
$3x - y = 2y$ $\{(x, y) \mid y = x\}$, dependent

Solve each system by the graphing method shown in Section 4.1 and by substitution.

29. $x + y = 5$
$x - y = 1$ (3, 2)

30. $x + y = 6$
$2x - y = 3$ (3, 3)

31. $y = x - 2$
$y = 4 - x$ (3, 1)

32. $y = 2x - 3$
$y = -x + 3$ (2, 1)

33. $y = 3x - 2$
$y - 3x = 1$ No solution

34. $x + y = 5$
$y = 2 - x$ No solution

Write a system of two equations in two unknowns for each problem. Solve each system by substitution. See Example 5.

35. ***Investing in the future.*** Mrs. Miller invested \$20,000 and received a total of \$1600 in interest. If she invested part of the money at 10% and the remainder at 5%, then how much did she invest at each rate?
\$12,000 at 10%, \$8000 at 5%

36. ***Stocks and bonds.*** Mr. Walker invested \$30,000 in stocks and bonds and had a total return of \$2880 in one year. If his stock investment returned 10% and his bond investment returned 9%, then how much did he invest in each? \$18,000 at 10%, \$12,000 at 9%

37. ***Gross receipts.*** As of March of 2002, the two highest grossing movies of all time were *Titanic* and *Star Wars* with total receipts of \$1062 million (www.movieweb.com). If the gross receipts for *Titanic* exceeded the gross receipts for *Star Wars* by \$140 million, then what were the gross receipts for each movie?
Titanic \$601 million, *Star Wars* \$461 million

38. ***Tennis court dimensions.*** The singles court in tennis is four yards longer than it is wide. If its perimeter is 44 yards, then what are the length and width?
Length 13 yd, width 9 yd

39. ***Mowing and shoveling.*** When Mr. Wilson came back from his vacation, he paid Frank \$50 for mowing his lawn three times and shoveling his sidewalk two times. During Mr. Wilson's vacation last year, Frank earned \$45 for mowing the lawn two times and shoveling the sidewalk three times. How much does Frank make for mowing the lawn once? How much does Frank make for shoveling the sidewalk once?
Lawn \$12, sidewalk \$7

40. ***Burgers and fries.*** Donna ordered four burgers and one order of fries at the Hamburger Palace. However, the waiter put three burgers and two orders of fries in the bag and charged Donna the correct price for three burgers and two orders of fries, \$3.15. When Donna discovered the mistake, she went back to complain. She found out that the price for four burgers and one order of fries is \$3.45 and decided to keep what she had. What is the price of one burger, and what is the price of one order of fries?
Burger \$0.75, fries \$0.45

41. ***Racing rules.*** According to Nascar rules, no more than 52% of a car's total weight can be on any pair of tires. For optimal performance a driver of a 1150-pound car wants to have 50% of its weight on the left rear and left front tires and 48% of its weight on the left rear and right front tires. If the right front weight is determined to be 264 pounds, then what amount of weight should be on the left rear and left front? Are the Nascar rules satisfied with this weight distribution?
Left rear 288 pounds, left front 287 pounds, no

42. ***Weight distribution.*** A driver of a 1200-pound car wants to have 50% of the car's weight on the left front and left rear tires, 48% on the left rear and right front tires, and 51% on the left rear and right rear tires. How much weight should be on each of these tires?
Left front 306, left rear 294, right front 282, right rear 318 pounds

43. ***Price of hamburger.*** A grocer will supply y pounds of ground beef per day when the retail price is x dollars per pound, where $y = 200x + 60$. Consumer studies show that consumer demand for ground beef is y pounds per day, where $y = -150x + 900$. What is the price at

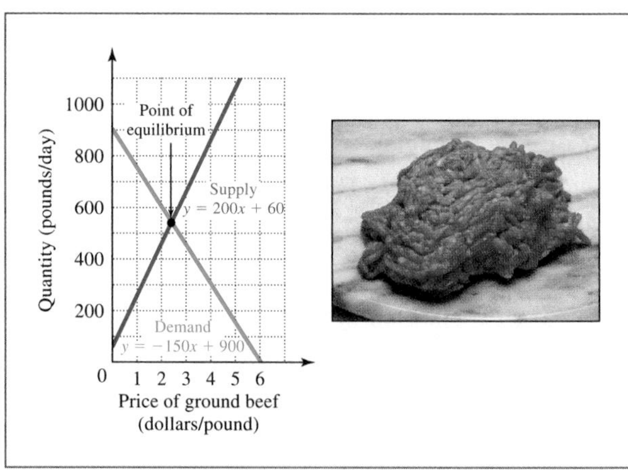

FIGURE FOR EXERCISE 43

which the supply is equal to the demand, the equilibrium price? See the accompanying figure.
\$2.40 per pound

44. ***Tweedle Dum and Dee.*** Tweedle Dum said to Tweedle Dee "The sum of my weight and twice yours is 361 pounds." Tweedle Dee said to Tweedle Dum "Contrariwise the sum of my weight and twice yours is 362 pounds." Find the weight of each.
Dum 121 lb, Dee 120 lb

GRAPHING CALCULATOR EXERCISE

45. ***Life expectancy.*** Since 1950, the life expectancy of a U.S. male born in year x is modeled by the formula

$$y = 0.165x - 256.7,$$

and the life expectancy of a U.S. female born in year x is modeled by

$$y = 0.186x - 290.6$$

(National Center for Health Statistics, www.cdc.gov).

a) Find the life expectancy of a U.S. male born in 1975 and a U.S. female born in 1975.

b) Graph both equations on your graphing calculator for $1950 < x < 2050$.

c) Will U.S. males ever catch up with U.S. females in life expectancy?

d) Assuming that these equations were valid before 1950, solve the system to find the year of birth for which U.S. males and females had the same life expectancy.

a) 69.2 years, 76.8 years

b)

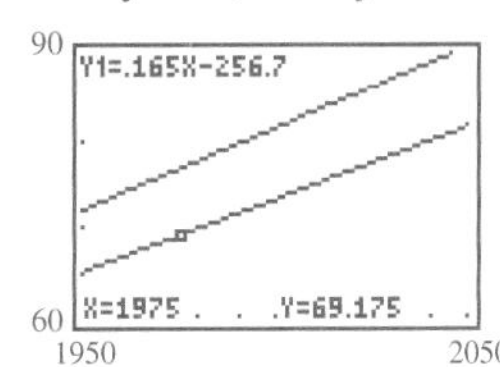

c) No d) 1614

4.3 THE ADDITION METHOD

In This Section

- Solving a System of Linear Equations by Addition
- Inconsistent and Dependent Systems
- Applications

In Section 4.2 we solved systems of equations by using substitution. We substituted one equation into the other to eliminate a variable. The addition method of this section is another method for eliminating a variable to solve a system of equations.

Solving a System of Linear Equations by Addition

In the substitution method we solve for one variable in terms of the other variable. When doing this, we may get an expression involving fractions, which must be substituted into the other equation. The addition method avoids fractions and is easier to use on certain systems.

EXAMPLE 1 **Solving a system by addition**

Solve: $3x - y = 5$
$2x + y = 10$

Calculator Close-Up

To check Example 1, graph $y_1 = 3x - 5$ and $y_2 = 10 - 2x$. The lines appear to intersect at (3, 4).

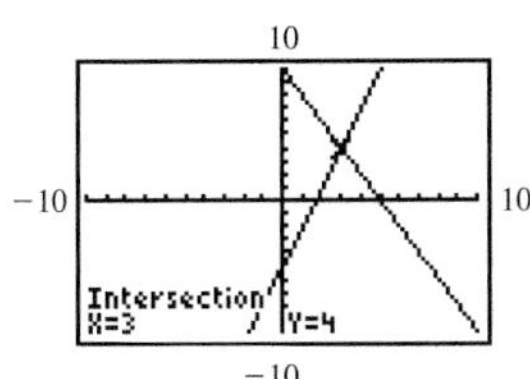

Solution

The addition property of equality allows us to add the same number to each side of an equation. If we assume that x and y are numbers that satisfy $3x - y = 5$, then adding these equations is equivalent to adding 5 to each side of $2x + y = 10$:

$$\begin{aligned} 3x - y &= 5 \\ 2x + y &= 10 \quad \text{Add.} \\ \hline 5x \quad &= 15 \quad -y + y = 0 \\ x &= 3 \end{aligned}$$

Note that the y-term was eliminated when we added the equations because the coefficients of y in the two equations were opposites. Now use $x = 3$ in either one of the original equations to find y:

$$2x + y = 10$$
$$2(3) + y = 10 \quad \text{Let } x = 3.$$
$$y = 4$$

Check that (3, 4) satisfies both equations. The solution to the system is (3, 4). ■

The addition method is based on the addition property of equality. We are adding equal quantities to each side of an equation. The form of the equations does not matter as long as the equal signs and the like terms are in line.

In Example 1, y was eliminated by the addition because the coefficients of y in the two equations were opposites. If no variable will be eliminated by addition, we can use the multiplication property of equality to change the coefficients of the variables. In the next example the coefficient of x in one equation is a multiple of the coefficient of x in the other equation. We use the multiplication property of equality to get opposite coefficients for x.

EXAMPLE 2 **Solving a system by addition**

Solve: $-x + 4y = -14$
$2x - 3y = 18$

Solution

If we add these equations as they are written, we will not eliminate any variables. However, if we multiply each side of the first equation by 2, then we will be adding $-2x$ and $2x$, and x will be eliminated:

$$2(-x + 4y) = 2(-14) \quad \text{Multiply each side by 2.}$$
$$2x - 3y = 18$$

$$-2x + 8y = -28$$
$$\underline{2x - 3y = 18} \quad \text{Add.}$$
$$5y = -10$$
$$y = -2$$

Now replace y by -2 in one of the original equations:

$$-x + 4(-2) = -14$$
$$-x - 8 = -14$$
$$-x = -6$$
$$x = 6$$

Check $x = 6$ and $y = -2$ in the original equations.

$$-6 + 4(-2) = -14$$
$$2(6) - 3(-2) = 18$$

The solution to the system is $(6, -2)$. ■

In Example 3 we need to use a multiple of each equation to eliminate a variable by addition.

EXAMPLE 3 **Solving a system by addition**

Solve: $2x + 3y = 7$

$3x + 4y = 10$

Study Tip

Read carefully. Ask yourself questions and look for the answers. Every sentence says something pertaining to the subject. You cannot read a mathematics textbook like you read a novel. You can read a novel passively, but a textbook requires more concentration and retention.

Solution

To eliminate x by addition, the coefficients of x in the two equations must be opposites. So we multiply the first equation by -3 and the second by 2:

$$-3(2x + 3y) = -3(7)$$
$$2(3x + 4y) = 2(10)$$

$$-6x - 9y = -21$$
$$6x + 8y = 20 \quad \text{Add.}$$
$$-y = -1$$
$$y = 1$$

Replace y with 1 in one of the original equations:

$$2x + 3y = 7$$
$$2x + 3(1) = 7$$
$$2x + 3 = 7$$
$$2x = 4$$
$$x = 2$$

Check $x = 2$ and $y = 1$ in the original equations.

$$2(2) + 3(1) = 7$$
$$3(2) + 4(1) = 10$$

The solution to the system is (2, 1). ■

If the equations have fractions, you can multiply each equation by the LCD to eliminate the fractions. Once the fractions are cleared, it is easier to see how to eliminate a variable by addition.

EXAMPLE 4 **A system involving fractions**

Solve: $\frac{1}{2}x - \frac{2}{3}y = 2$

$\frac{1}{4}x + \frac{1}{2}y = 6$

Solution

Multiply the first equation by 6 and the second by 4 to eliminate the fractions:

$$6\left(\frac{1}{2}x - \frac{2}{3}y\right) = 6 \cdot 2$$
$$4\left(\frac{1}{4}x + \frac{1}{2}y\right) = 4 \cdot 6$$
$$3x - 4y = 12$$
$$x + 2y = 24$$

Now multiply $x + 2y = 24$ by 2 to get $2x + 4y = 48$, and then add:

$$\begin{aligned} 3x - 4y &= 12 \\ 2x + 4y &= 48 \\ \hline 5x \qquad &= 60 \\ x &= 12 \end{aligned}$$

Let $x = 12$ in $x + 2y = 24$:

$$\begin{aligned} 12 + 2y &= 24 \\ 2y &= 12 \\ y &= 6 \end{aligned}$$

Check $x = 12$ and $y = 6$ in the original equations. The solution is (12, 6). ■

Use the following strategy to solve a system by addition.

Strategy for Solving a System by Addition

1. Write both equations in standard form.
2. If a variable will be eliminated by adding, then add the equations.
3. If necessary, obtain multiples of one or both equations so that a variable will be eliminated by adding the equations.
4. After one variable is eliminated, solve for the remaining variable.
5. Use the value of the remaining variable to find the value of the eliminated variable.
6. Check the solution in the original system.

Inconsistent and Dependent Systems

When the addition method is used, an inconsistent system will be indicated by a false statement. A dependent system will be indicated by an equation that is always true.

EXAMPLE 5 **Inconsistent and dependent systems**

Use the addition method to solve each system.

a) $-2x + 3y = 9$
$2x - 3y = 18$

b) $2x - y = 1$
$4x - 2y = 2$

Solution

a) Add the equations:

$$\begin{aligned} -2x + 3y &= 9 \\ 2x - 3y &= 18 \\ \hline 0 &= 27 \quad \text{False} \end{aligned}$$

There is no solution to the system. The system is inconsistent.

b) Multiply the first equation by -2, and then add the equations:

$$-2(2x - y) = -2(1)$$
$$4x - 2y = 2$$

$$\begin{array}{r} -4x + 2y = -2 \\ 4x - 2y = 2 \\ \hline 0 = 0 \quad \text{True} \end{array}$$

Because the equation $0 = 0$ is correct for any value of x, the system is dependent. The set of points satisfying the system is $\{(x, y) \mid 2x - y = 1\}$. ■

Applications

In Example 6 we solve a problem using a system of equations and the addition method.

EXAMPLE 6 **Milk and bread**

Lea purchased two gallons of milk and three loaves of bread for \$8.25. Yesterday she purchased five gallons of milk and two loaves of bread for \$13.75. What is the price of a single gallon of milk? What is the price of a single loaf of bread?

Helpful Hint

You can see from Example 6, that the standard form $Ax + By = C$ occurs naturally in accounting. This form will occur whenever we have the price each and quantity of two items and we want to express the total cost.

Solution

Let x represent the price of one gallon of milk. Let y represent the price of one loaf of bread. We can write two equations about the milk and bread:

$$2x + 3y = 8.25 \quad \text{Today's purchase}$$
$$5x + 2y = 13.75 \quad \text{Yesterday's purchase}$$

To eliminate x, multiply the first equation by -5 and the second by 2:

$$-5(2x + 3y) = -5(8.25)$$
$$2(5x + 2y) = 2(13.75)$$

$$\begin{array}{r} -10x - 15y = -41.25 \\ 10x + 4y = 27.50 \quad \text{Add.} \\ \hline -11y = -13.75 \\ y = 1.25 \end{array}$$

Replace y by 1.25 in the first equation:

$$2x + 3(1.25) = 8.25$$
$$2x + 3.75 = 8.25$$
$$2x = 4.50$$
$$x = 2.25$$

A gallon of milk costs \$2.25, and a loaf of bread costs \$1.25. ■

WARM-UPS

True or false? Explain your answer.

Use the following systems for these exercises:

a) $3x + 2y = 7$
$4x - 5y = -6$

b) $y = -3x + 2$
$2y + 6x - 4 = 0$

c) $y = x - 5$
$x = y + 6$

1. To eliminate x by addition in system (a), we multiply the first equation by 4 and the second equation by 3. False

2. Either variable in system (a) can be eliminated by the addition method. True

3. The ordered pair (1, 2) is a solution to system (a). True

4. The addition method can be used to eliminate a variable in system (b). True

5. Both (0, 2) and (1, −1) satisfy system (b). True

6. The solution to system (c) is $\{(x, y) \mid y = x - 5\}$. False

7. System (c) is independent. False

8. System (b) is inconsistent. False

9. System (a) is dependent. False

10. The graphs of the equations in system (c) are parallel lines. True

4.3 EXERCISES

Reading and Writing *After reading this section, write out the answers to these questions. Use complete sentences.*

1. What method is used in this section to solve systems of equations?
In this section we learned to solve systems by the addition method.

2. What three methods have now been presented for solving a system of linear equations?
The three methods presented are graphing, substitution, and addition.

3. What do the addition method and the substitution method have in common?
In addition and substitution we eliminate a variable and solve for the remaining variable.

4. What do we sometimes do before we add the equations?
It is sometimes necessary to use the multiplication property of equality before adding the equations.

5. How do you decide which variable to eliminate when using the addition method?
Eliminate the variable that is easiest to eliminate.

6. How do you identify inconsistent and dependent systems when using the addition method?
In the addition method inconsistent systems result in a false equation and dependent systems result in an equation that is always true.

Solve each system by the addition method. See Examples 1–4.

7. $2x + y = 5$
$3x - y = 10$ (3, −1)

8. $3x - y = 3$
$4x + y = 11$ (2, 3)

9. $x + 2y = 7$
$-x + 3y = 18$ (−3, 5)

10. $x + 2y = 7$
$-x + 4y = 5$ (3, 2)

11. $x + 2y = 2$
$-4x + 3y = 25$ (−4, 3)

12. $2x - 3y = -7$
$5x + y = -9$ (−2, 1)

13. $x + 3y = 4$
$2x - y = -1$ $\left(\frac{1}{7}, \frac{9}{7}\right)$

14. $x - y = 0$
$x - 2y = 0$ (0, 0)

15. $y = 4x - 1$
$y = 3x + 7$ (8, 31)

16. $x = 3y + 45$
$x = 2y + 40$ (30, −5)

17. $4x = 3y + 1$
$2x = y - 1$ (−2, −3)

18. $2x = y - 9$
$x = -1 - 3y$ (−4, 1)

19. $2x - 5y = -22$
$-6x + 3y = 18$ (−1, 4)

20. $4x - 3y = 7$
$5x + 6y = -1$ (1, −1)

21. $2x + 3y = 4$
$-3x + 5y = 13$ (−1, 2)

22. $-5x + 3y = 1$
$2x - 7y = 17$ (−2, −3)

23. $2x - 5y = 11$
$3x - 2y = 11$ (3, −1)

24. $4x - 3y = 17$
$3x - 5y = 21$ (2, −3)

25. $5x + 4y = 13$
$2x + 3y = 8$ (1, 2)

26. $4x + 3y = 8$
$6x + 5y = 14$ (−1, 4)

Use either the addition method or substitution to solve each system. State whether the system is independent, inconsistent, or dependent. See Example 5.

27. $x + y = 5$
$x + y = 6$ No solution, inconsistent

28. $x + y = 5$
$x + 2y = 6$ (4, 1), independent

29. $x + y = 5$
$2x + 2y = 10$ $\{(x, y) \mid x + y = 5\}$, dependent

30. $2x + 3y = 4$
$2x - 3y = 4$ (2, 0), independent

31. $2x = y + 3$
$2y = 4x - 6$ $\{(x, y) \mid 2x = y + 3\}$, dependent

32. $y = 2x - 1$
$2x - y + 5 = 0$ No solution, inconsistent

33. $x + 3y = 3$
$5x = 15 - 15y$ $\{(x, y) \mid x + 3y = 3\}$, dependent

34. $y - 3x = 2$
$5y = -15x + 10$ (0, 2), independent

35. $6x - 2y = -2$
$\frac{1}{3}y = x + \frac{4}{3}$ No solution, inconsistent

36. $x + y = 8$
$\frac{1}{3}x - \frac{1}{2}y = 1$ (6, 2), independent

37. $\frac{1}{2}x - \frac{2}{3}y = -6$
$-\frac{3}{4}x - \frac{1}{2}y = -18$ (12, 18), independent

38. $\frac{1}{2}x - y = 3$
$\frac{1}{5}x + 2y = 6$ (10, 2), independent

39. $0.04x + 0.09y = 7$
$x + y = 100$ (40, 60), independent

40. $0.08x - 0.05y = 0.2$
$2x + y = 140$ (40, 60), independent

41. $0.1x - 0.2y = -0.01$
$0.3x + 0.5y = 0.08$ (0.1, 0.1), independent

42. $0.5y = 0.2x - 0.25$
$0.1y = 0.8x - 1.57$ (2, 0.3), independent

Use a calculator to assist you in finding the exact solution to each system.

43. $2.33x - 4.58y = 16.319$
$4.98x + 3.44y = -2.162$ (1.5, −2.8)

44. $234x - 499y = 1337$
$282x + 312y = 51{,}846$ (123, 55)

Solve each system by graphing (Section 4.1), by substitution (Section 4.2), and by addition.

45. $x + y = 7$
$x - y = 1$ (4, 3)

46. $x + y = 8$
$2x - y = 4$ (4, 4)

47. $y = x - 3$
$y = 5 - x$ (4, 1)

48. $y = 2x - 5$
$y = -x + 4$ (3, 1)

49. $y = 2x - 1$
$y - 2x = 3$ No solution

50. $x + y = 3$
$y = 4 - x$ No solution

Use two variables and a system of equations to solve each problem. See Example 6.

51. ***Cars and trucks.*** An automobile dealer had 250 vehicles on his lot during the month of June. He must pay a monthly inventory tax of \$3 per car and \$4 per truck. If his tax bill for June was \$850, then how many cars and how many trucks did he have on his lot during June?
150 cars, 100 trucks

FIGURE FOR EXERCISE 51

52. ***Dimes and nickels.*** Kimberly opened a parking meter and removed 30 coins consisting of dimes and nickels. If the value of these coins is \$2.30, then how many of each type does she have?
16 dimes, 14 nickels

53. ***Adults and children.*** The Audubon Zoo charges \$5.50 for each adult admission and \$2.75 for each child. The total bill for the 30 people on the Spring Creek Elementary School kindergarten field trip was \$99. How many adults and how many children went on the field trip?
6 adults, 24 children

FIGURE FOR EXERCISE 53

54. ***Coffee and doughnuts.*** Jorge has worked at Dandy Doughnuts so long that he has memorized the amounts for many of the common orders. For example, six doughnuts and five coffees cost \$4.35, while four doughnuts and three coffees cost \$2.75. What are the prices of one cup of coffee and one doughnut?
Coffee \$0.45, doughnut \$0.35

55. ***Marketing research.*** The Independent Marketing Research Corporation found 130 smokers among 300 adults surveyed. If one-half of the men and one-third of the women were smokers, then how many men and how many women were in the survey?
180 men, 120 women

56. ***Time and a half.*** In one month, Shelly earned \$1800 for 210 hours of work. If she earns \$8 per hour for regular time and \$12 per hour for overtime, then how many hours of each type did she work?
180 hours regular time, 30 hours overtime

GETTING MORE INVOLVED

57. ***Discussion.*** Compare and contrast the three methods for solving systems of linear equations in two variables that were presented in this chapter. What are the advantages and disadvantages of each method? How do you choose which method to use?

58. ***Exploration.*** Consider the following system:

$$a_1x + b_1y = c_1$$
$$a_2x + b_2y = c_2$$

a) Multiply the first equation by a_2 and the second equation by $-a_1$. Add the resulting equations and solve for y to get a formula for y in terms of the a's, b's, and c's.
$y = \dfrac{a_2c_1 - a_1c_2}{a_2b_1 - a_1b_2}$

b) Multiply the first equation by b_2 and the second by $-b_1$. Add the resulting equations and solve for x to get a formula for x in terms of the a's, b's, and c's.
$x = \dfrac{b_2c_1 - b_1c_2}{b_2a_1 - b_1a_2}$

c) Use the formulas that you found in (a) and (b) to find the solution to the following system:

$$2x + 3y = 7$$
$$5x + 4y = 14$$

(2, 1)

4.4 GRAPHING LINEAR INEQUALITIES IN TWO VARIABLES

In This Section

- Definition
- Graph of a Linear Inequality
- Using a Test Point to Graph an Inequality
- Applications

You studied linear equations and inequalities in one variable in Chapter 2. In this section we extend the ideas of linear equations in two variables to study linear inequalities in two variables.

Definition

Linear inequalities in two variables have the same form as linear equations in two variables. An inequality symbol is used in place of the equal sign.

Linear Inequality in Two Variables

If A, B, and C are real numbers with A and B not both zero, then

$$Ax + By < C$$

is called a **linear inequality in two variables.** In place of $<$, we can also use $\leq$, $>$, or $\geq$.

The inequalities

$$3x - 4y \leq 8, \qquad y > 2x - 3, \qquad \text{and} \qquad x - y + 9 < 0$$

are linear inequalities. Not all of these are in the form of the definition, but they could all be rewritten in that form.

An ordered pair is a solution to an inequality in two variables if the ordered pair satisfies the inequality.

EXAMPLE 1 **Satisfying a linear inequality**

Determine whether each point satisfies the inequality $2x - 3y \geq 6$.

a) (4, 1) **b)** (3, 0) **c)** (3, −2)

Study Tip

Write about what you read in the text. Sum things up in your own words. Write out important facts on note cards. When you have a few spare minutes in between classes review your note cards. Try to get the information on the cards into your memory.

Solution

a) To determine whether (4, 1) is a solution to the inequality, we replace x by 4 and y by 1 in the inequality $2x - 3y \geq 6$:

$$2(4) - 3(1) \geq 6$$
$$8 - 3 \geq 6$$
$$5 \geq 6 \quad \text{Incorrect}$$

So (4, 1) does not satisfy the inequality $2x - 3y \geq 6$.

b) Replace x by 3 and y by 0:

$$2(3) - 3(0) \geq 6$$
$$6 \geq 6 \quad \text{Correct}$$

So the point (3, 0) satisfies the inequality $2x - 3y \geq 6$.

c) Replace x by 3 and y by -2:

$$2(3) - 3(-2) \geq 6$$
$$6 + 6 \geq 6$$
$$12 \geq 6 \quad \text{Correct}$$

So the point (3, −2) satisfies the inequality $2x - 3y \geq 6$. ■

Graph of a Linear Inequality

The graph of a linear inequality in two variables consists of all points in the rectangular coordinate system that satisfy the inequality. For example, the graph of the inequality

$$y > x + 2$$

consists of all points where the y-coordinate is larger than the x-coordinate plus 2. Consider the point (3, 5) on the line

$$y = x + 2.$$

The y-coordinate of (3, 5) is equal to the x-coordinate plus 2. If we choose a point with a larger y-coordinate, such as (3, 6), it satisfies the inequality and it is above the line $y = x + 2$. In fact, any point above the line $y = x + 2$ satisfies $y > x + 2$. Likewise, all points below the line $y = x + 2$ satisfy the inequality $y < x + 2$. See Fig. 4.6.

Helpful Hint

Why do we keep drawing graphs? When we solve $2x + 1 = 7$, we don't bother to draw a graph showing 3, because the solution set is so simple. However, the solution set to a linear inequality is a very large set of ordered pairs. Graphing gives us a way to visualize the solution set.

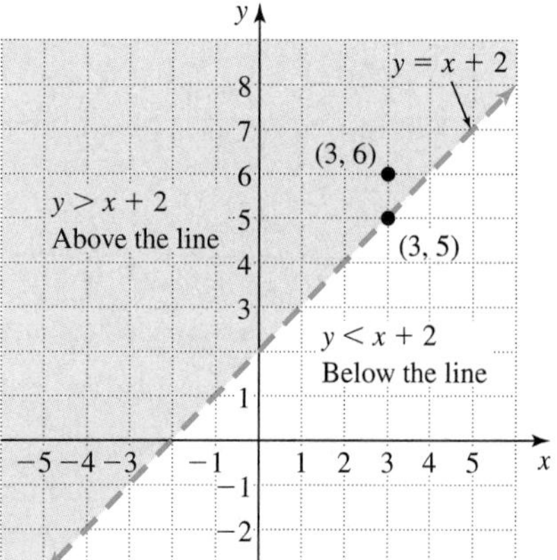

FIGURE 4.6

To graph the inequality, we shade all points above the line $y = x + 2$. To indicate that the line is not included in the graph of $y > x + 2$, we use a dashed line.

The procedure for graphing linear inequalities is summarized as follows.

Strategy for Graphing a Linear Inequality in Two Variables

1. Solve the inequality for y, then graph $y = mx + b$.
   $y > mx + b$ is the region above the line.
   $y = mx + b$ is the line itself.
   $y < mx + b$ is the region below the line.
2. If the inequality involves only x, then graph the vertical line $x = k$.
   $x > k$ is the region to the right of the line.
   $x = k$ is the line itself.
   $x < k$ is the region to the left of the line.

EXAMPLE 2 **Graphing a linear inequality**

Graph each inequality.

a) $y < \frac{1}{3}x + 1$ **b)** $y \geq -2x + 3$

c) $2x - 3y < 6$

Solution

a) The set of points satisfying this inequality is the region below the line

$$y = \frac{1}{3}x + 1.$$

To show this region, we first graph the boundary line. The slope of the line is $\frac{1}{3}$, and the y-intercept is $(0, 1)$. We draw the line dashed because it is not part of the graph of $y < \frac{1}{3}x + 1$. In Fig. 4.7 the graph is the shaded region.

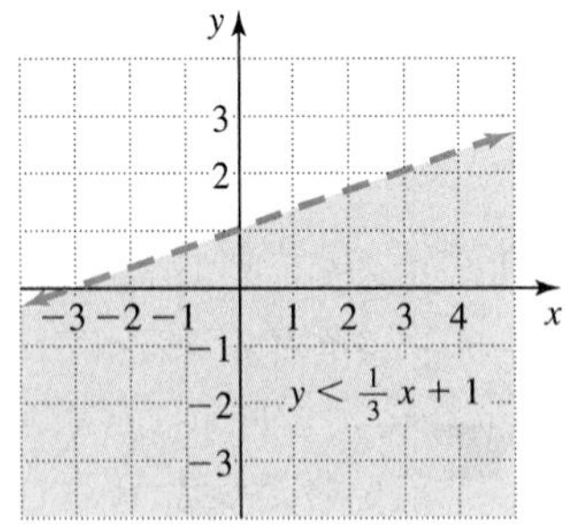

FIGURE 4.7

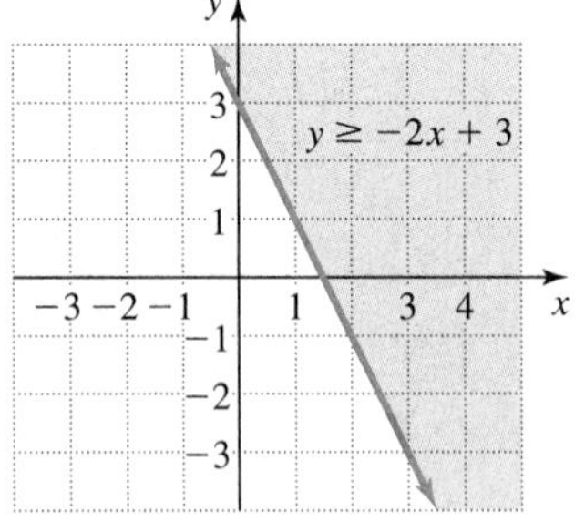

FIGURE 4.8

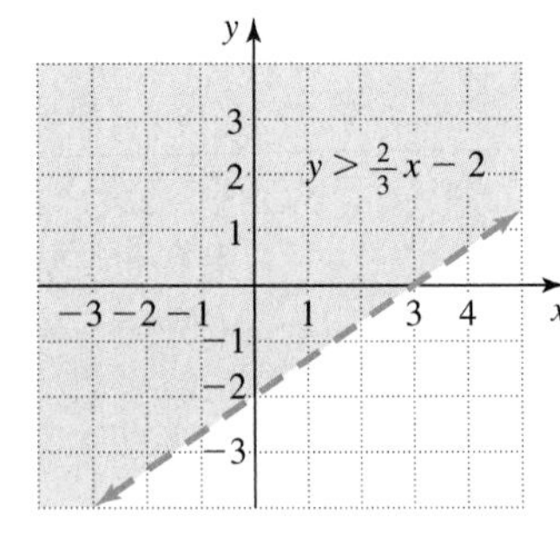

FIGURE 4.9

b) Because the inequality symbol is $\geq$, every point on or above the line satisfies this inequality. We use the fact that the slope of this line is -2 and the y-intercept is $(0, 3)$ to draw the graph of the line. To show that the line $y = -2x + 3$ is included in the graph, we make it a solid line and shade the region above. See Fig. 4.8.

c) First solve for y:

$$2x - 3y < 6$$
$$-3y < -2x + 6$$
$$y > \frac{2}{3}x - 2 \quad \text{Divide by } -3 \text{ and reverse the inequality.}$$

To graph this inequality, we first graph the line with slope $\frac{2}{3}$ and y-intercept $(0, -2)$. We use a dashed line for the boundary because it is not included, and we shade the region above the line. Remember, "less than" means below the line and "greater than" means above the line only when the inequality is solved for y. See Fig. 4.9 for the graph. ■

EXAMPLE 3 **Horizontal and vertical boundary lines**

Graph each inequality.

a) $y \le 4$ **b)** $x > 3$

Solution

a) The line $y = 4$ is the horizontal line with y-intercept $(0, 4)$. We draw a solid horizontal line and shade below it as in Fig. 4.10.

b) The line $x = 3$ is a vertical line through $(3, 0)$. Any point to the right of this line has an x-coordinate larger than 3. The graph is shown in Fig. 4.11.

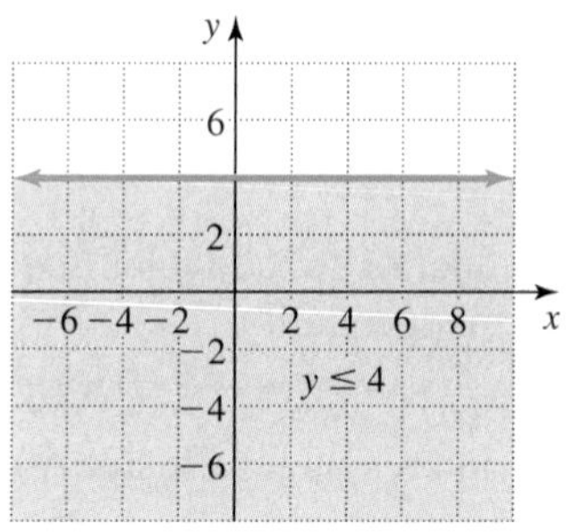

FIGURE 4.10

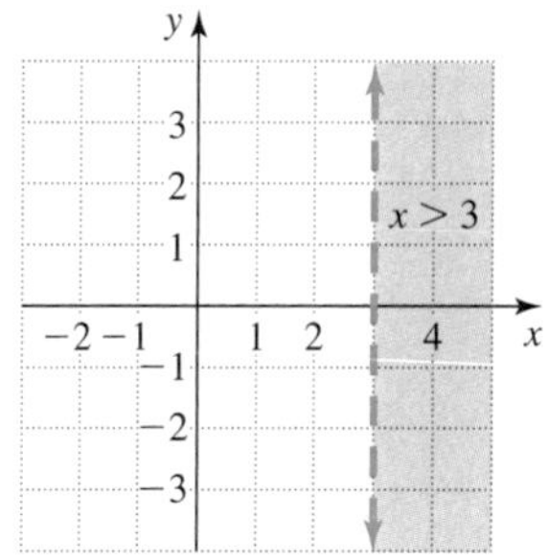

FIGURE 4.11 ■

Using a Test Point to Graph an Inequality

The graph of a linear equation such as $2x - 3y = 6$ separates the coordinate plane into two regions. One region satisfies the inequality $2x - 3y > 6$, and the other region satisfies the inequality $2x - 3y < 6$. We can tell which region satisfies which inequality by testing a point in one region. With this method it is not necessary to solve the inequality for y.

EXAMPLE 4 **Using a test point**

Graph the inequality $2x - 3y > 6$.

Helpful Hint

Some people always like to choose (0, 0) as the test point for lines that do not go through (0, 0). The arithmetic for testing (0, 0) is generally easier than for any other point.

Solution

First graph the equation $2x - 3y = 6$ using the x-intercept $(3, 0)$ and the y-intercept $(0, -2)$ as shown in Fig. 4.12. Select a point on one side of the line, say $(0, 1)$, to test in the inequality. Because

$$2(0) - 3(1) > 6$$

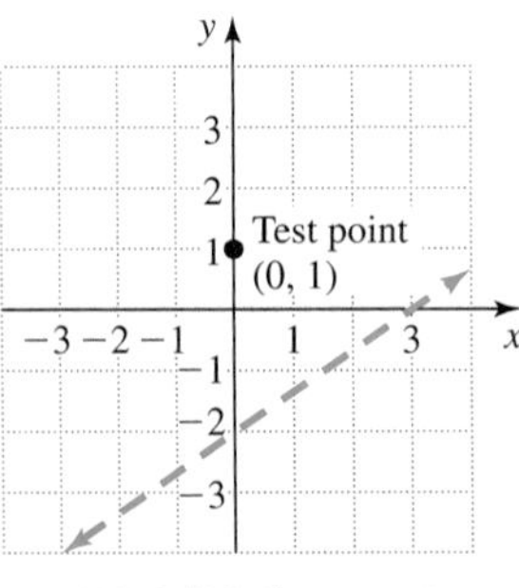

FIGURE 4.12

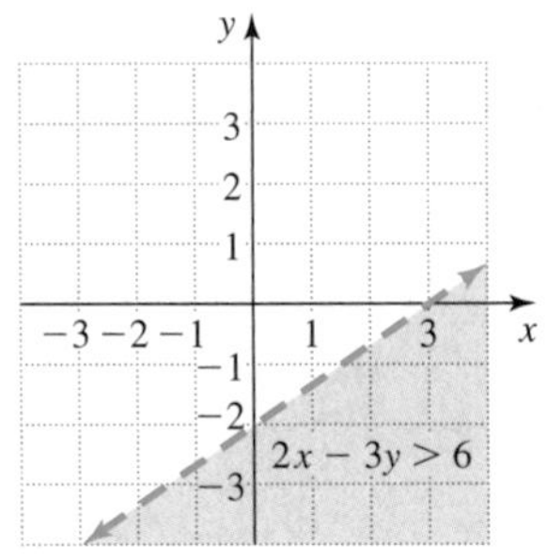

FIGURE 4.13

is false, the region on the other side of the line satisfies the inequality. The graph of $2x - 3y > 6$ is shown in Fig. 4.13. ■

Applications

The values of variables used in applications are often restricted to nonnegative numbers. So solutions to inequalities in these applications are graphed in the first quadrant only.

EXAMPLE 5 **Manufacturing tables**

The Ozark Furniture Company can obtain at most 8000 board feet of oak lumber for making two types of tables. It takes 50 board feet to make a round table and 80 board feet to make a rectangular table. Write an inequality that limits the possible number of tables of each type that can be made. Draw a graph showing all possibilities for the number of tables that can be made.

Solution

If x is the number of round tables and y is the number of rectangular tables, then x and y satisfy the inequality

$$50x + 80y \leq 8000.$$

Now find the intercepts for the line $50x + 80y = 8000$:

$$\begin{aligned} 50 \cdot 0 + 80y &= 8000 \\ 80y &= 8000 \\ y &= 100 \end{aligned} \qquad \begin{aligned} 50x + 80 \cdot 0 &= 8000 \\ 50x &= 8000 \\ x &= 160 \end{aligned}$$

Draw the line through (0, 100) and (160, 0). Because (0, 0) satisfies the inequality, the number of tables must be below the line. Since the number of tables cannot be negative, the number of tables made must be below the line and in the first quadrant as shown in Fig. 4.14. Assuming that Ozark will not make a fraction of a table, only points in Fig. 4.14 with whole-number coordinates are practical.

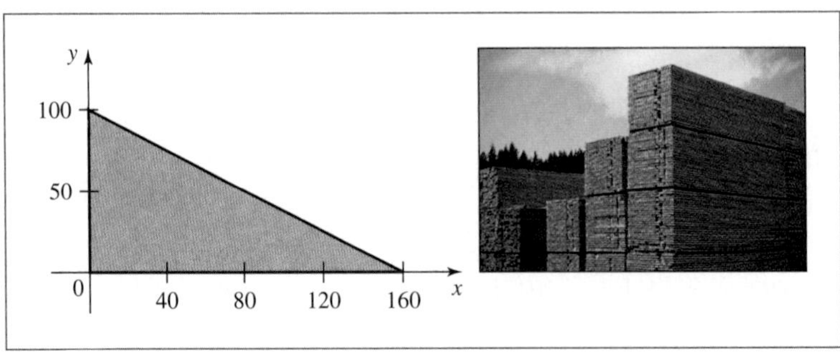

FIGURE 4.14

■

WARM-UPS

True or false? Explain your answer.

1. The point $(-1, 4)$ satisfies the inequality $y > 3x + 1$. True
2. The point $(2, -3)$ satisfies the inequality $3x - 2y \geq 12$. True
3. The graph of the inequality $y > x + 9$ is the region above the line $y = x + 9$. True
4. The graph of the inequality $x < y + 2$ is the region below the line $x = y + 2$. False
5. The graph of $x = 3$ is a single point on the x-axis. False
6. The graph of $y \leq 5$ is the region below the horizontal line $y = 5$. False
7. The graph of $x < 3$ is the region to the left of the vertical line $x = 3$. True
8. In graphing the inequality $y \geq x$ we use a dashed boundary line. False
9. The point $(0, 0)$ is on the graph of the inequality $y \geq x$. True
10. The point $(0, 0)$ lies above the line $y = 2x + 1$. False

4.4 EXERCISES

Reading and Writing *After reading this section, write out the answers to these questions. Use complete sentences.*

1. What is a linear inequality in two variables?
   A linear inequality has the same form as a linear equation except that an inequality symbol is used.
2. How can you tell if an ordered pair satisfies a linear inequality in two variables?
   An ordered pair satisfies a linear inequality if the inequality is correct when the variables are replaced by the coordinates of the ordered pair.
3. How do you determine whether to draw the boundary line of the graph of a linear inequality dashed or solid?
   If the inequality symbol includes equality, then the boundary line is solid; otherwise it is dashed.
4. How do you decide which side of the boundary line to shade?
   We shade the side that satisfies the inequality.
5. What is the test point method?
   In the test point method we test a point to see which side of the boundary line satisfies the inequality.
6. What is the advantage of the test point method?
   With the test point method you can use the inequality in any form.

Determine which of the points following each inequality satisfy that inequality. See Example 1.

7. $x - y > 5$ $(2, 3), (-3, -9), (8, 3)$ $(-3, -9)$
8. $2x + y < 3$ $(-2, 6), (0, 3), (3, 0)$ $(-2, 6)$
9. $y \geq -2x + 5$ $(3, 0), (1, 3), (-2, 5)$ $(3, 0), (1, 3)$
10. $y \leq -x + 6$ $(2, 0), (-3, 9), (-4, 12)$ $(2, 0), (-3, 9)$
11. $x > -3y + 4$ $(2, 3), (7, -1), (0, 5)$ $(2, 3), (0, 5)$
12. $x < -y - 3$ $(1, 2), (-3, -4), (0, -3)$ $(-3, -4)$

Graph each inequality. See Examples 2 and 3.

13. $y < x + 4$

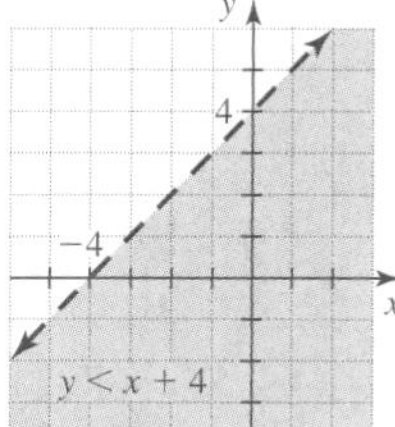

14. $y < 2x + 2$

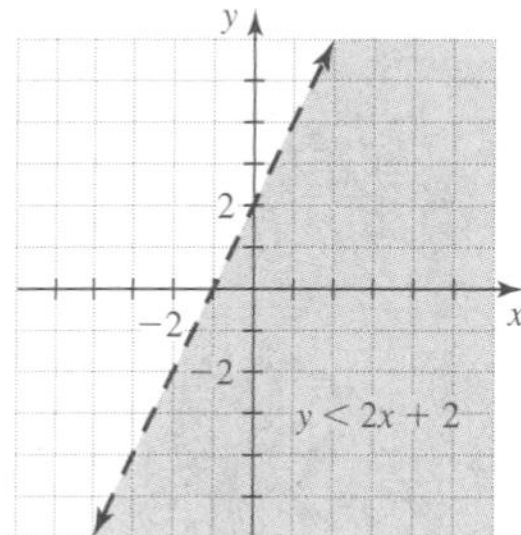

15. $y > -x + 3$

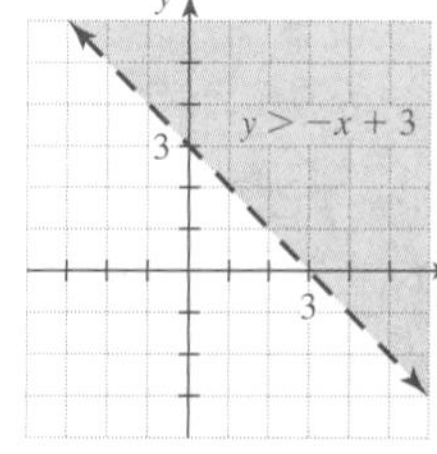

16. $y < -2x + 1$

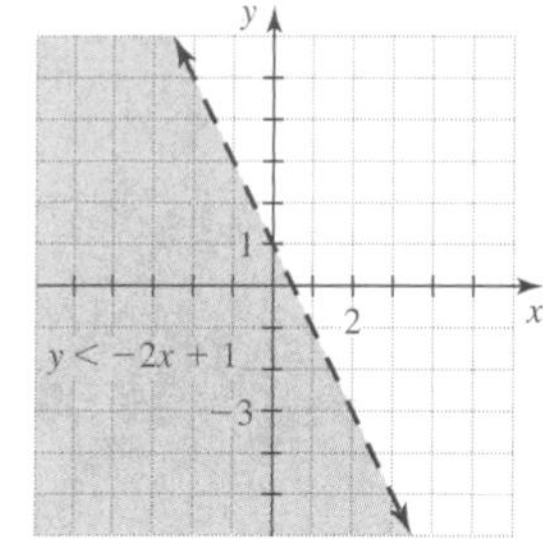

17. $y > \frac{2}{3}x - 3$

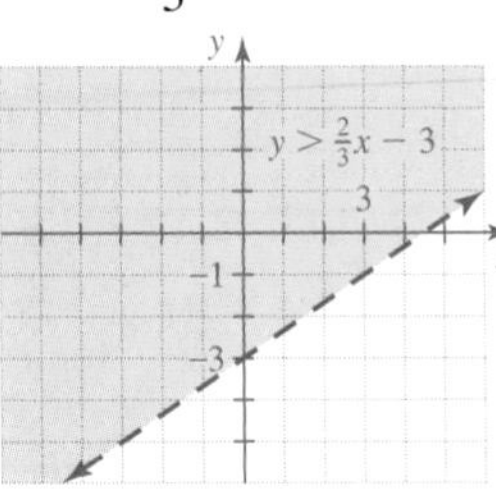

18. $y < \frac{1}{2}x + 1$

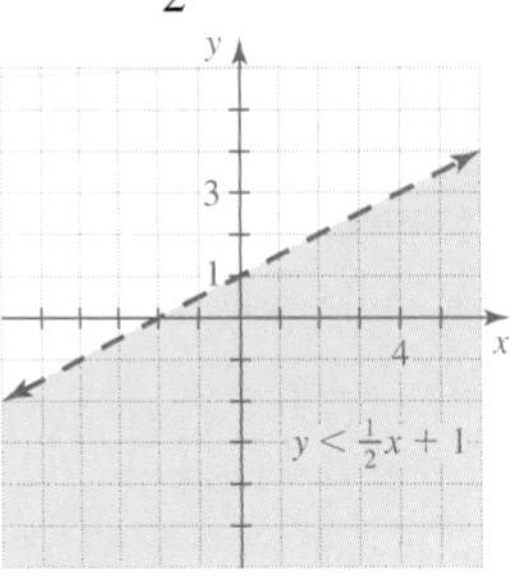

27. $y \geq 2$

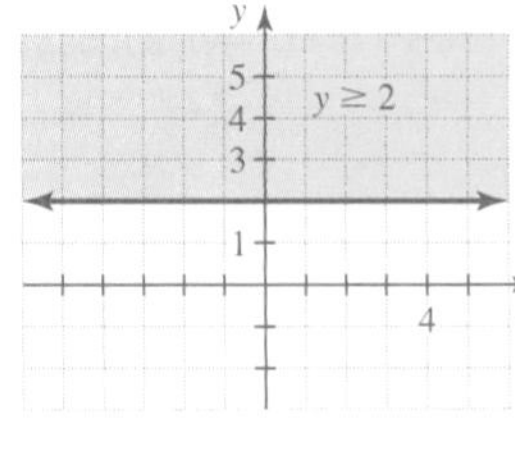

28. $y < 7$

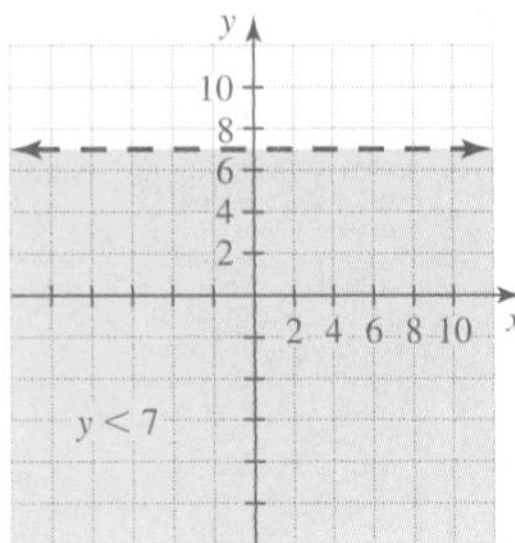

19. $y \leq -\frac{2}{5}x + 2$

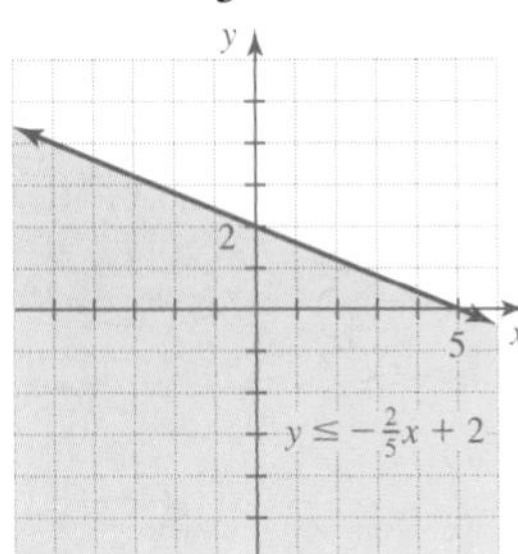

20. $y \geq -\frac{1}{2}x + 3$

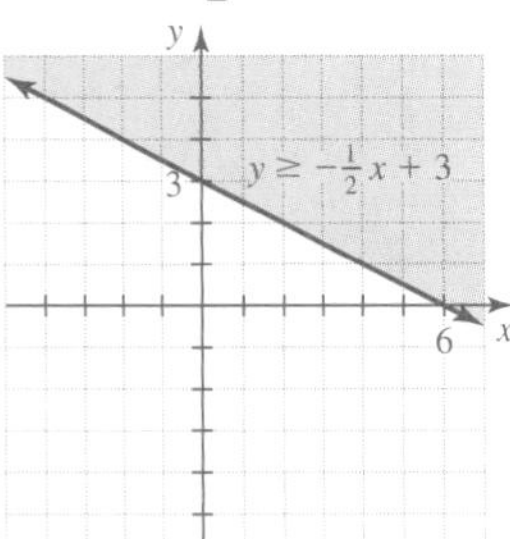

29. $x > 9$

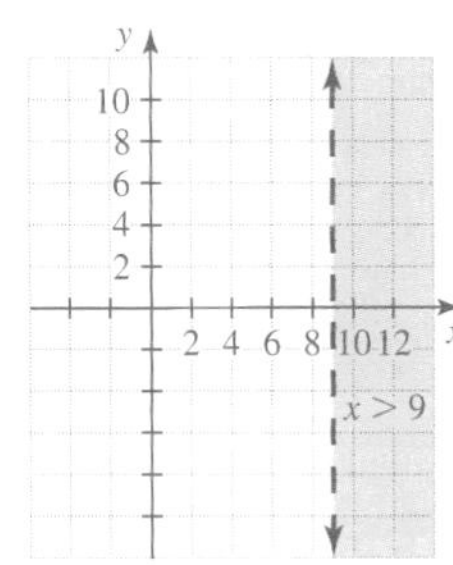

30. $x \leq 1$

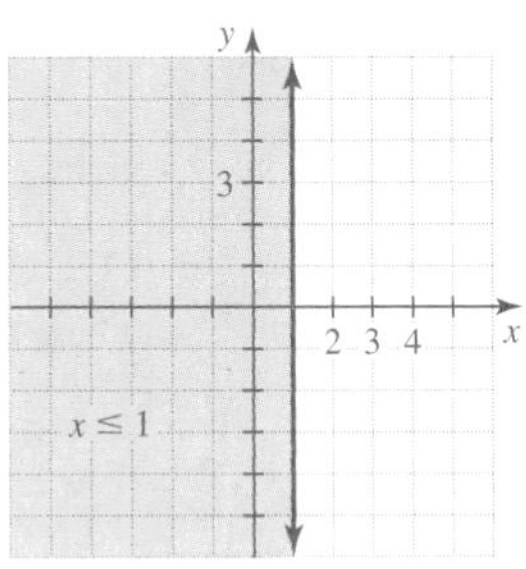

21. $y - x \geq 0$

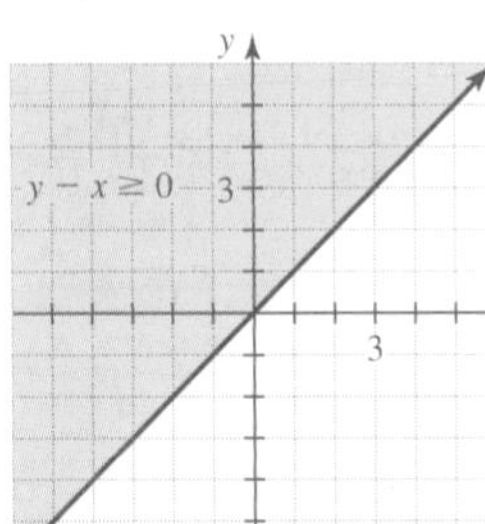

22. $x - 2y \leq 0$

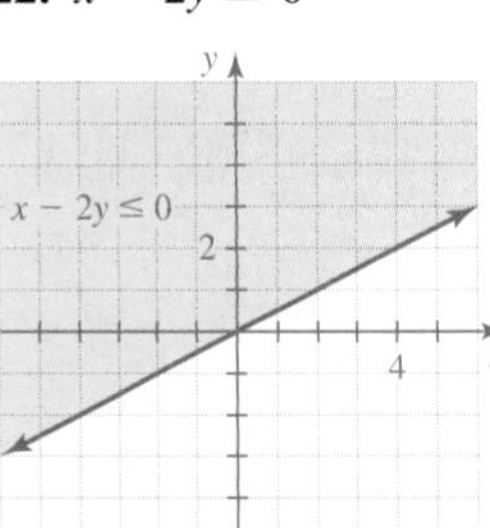

31. $x + y \leq 60$

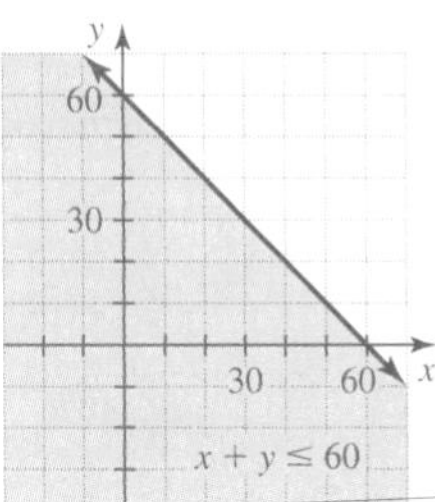

32. $x - y \leq 90$

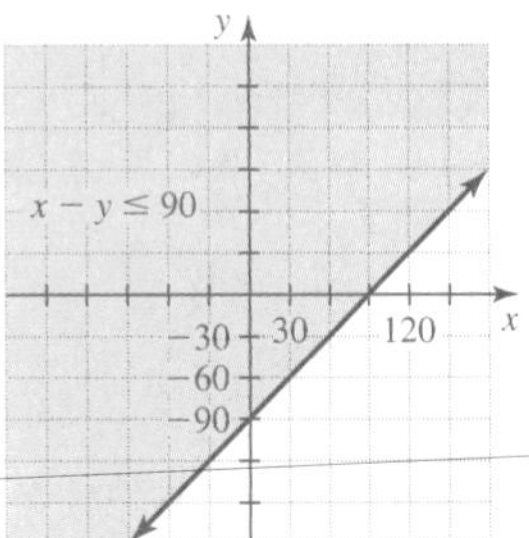

23. $x > y - 5$

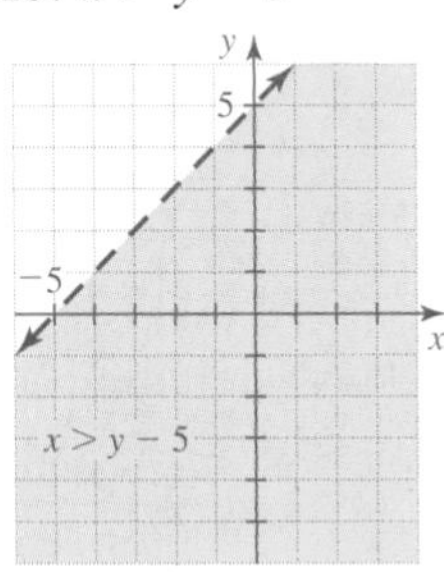

24. $2x < 3y + 6$

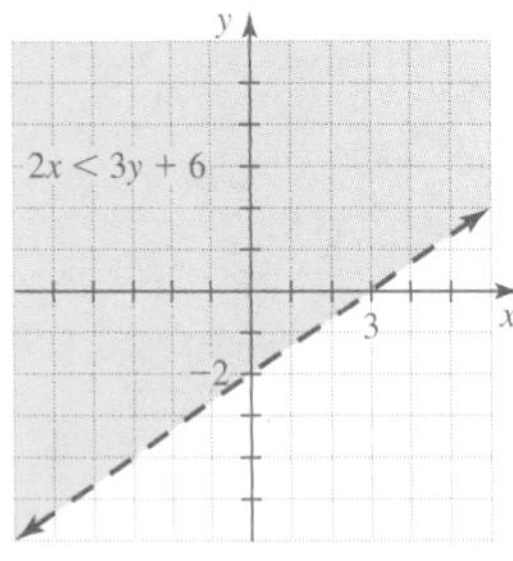

33. $x \leq 100y$

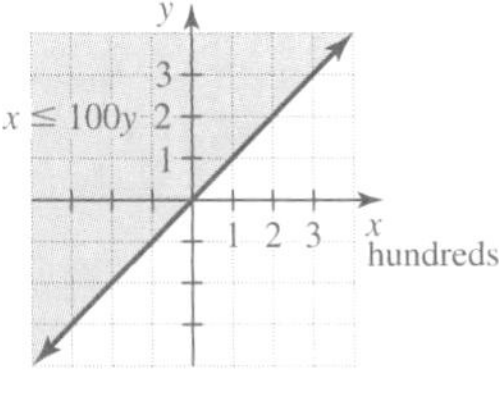

34. $y \geq 600x$

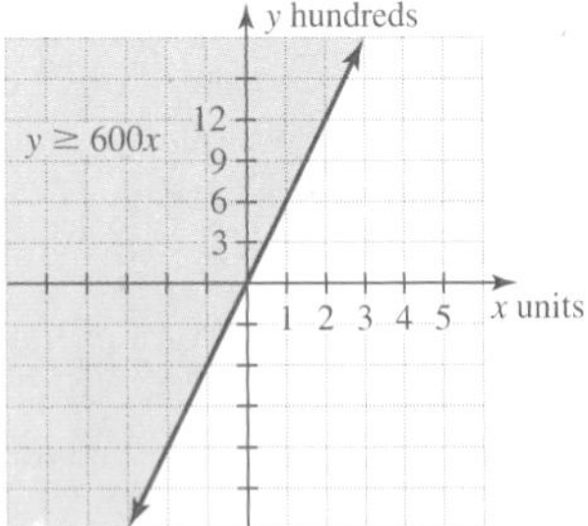

25. $x - 2y + 4 \leq 0$

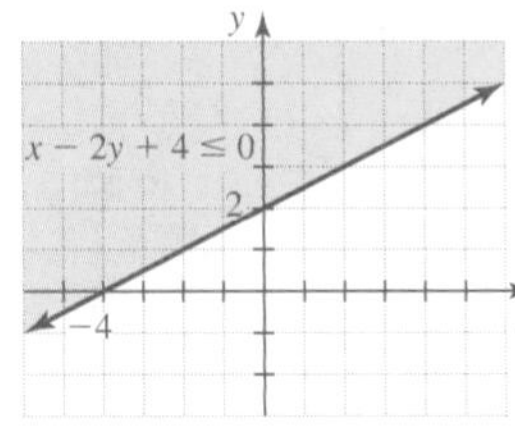

26. $2x - y + 3 \geq 0$

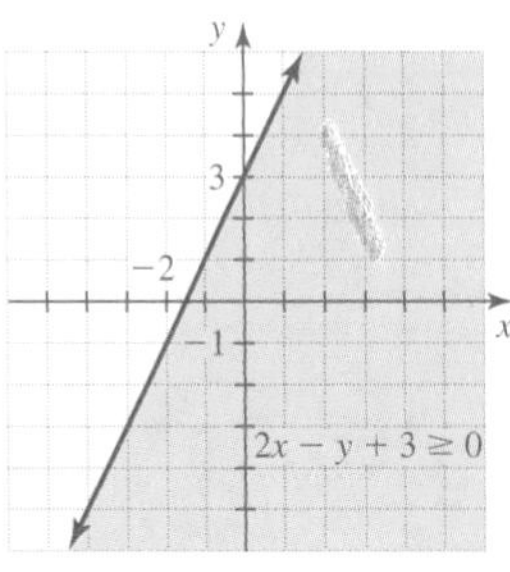

35. $3x - 4y \leq 8$

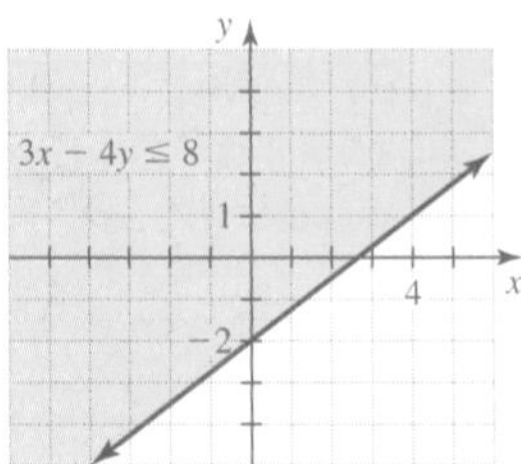

36. $2x + 5y \geq 10$

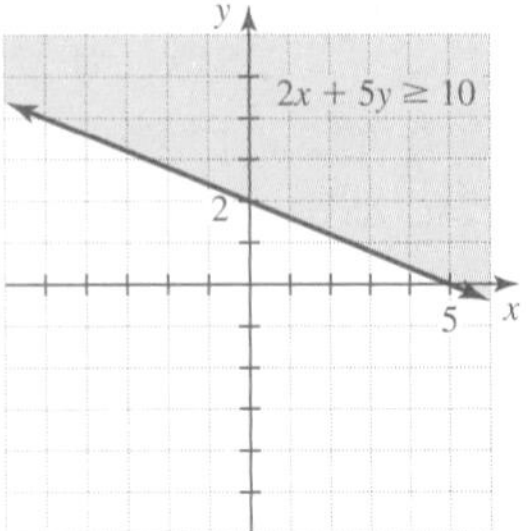

Graph each inequality. Use the test point method of Example 4.

37. $2x - 3y < 6$

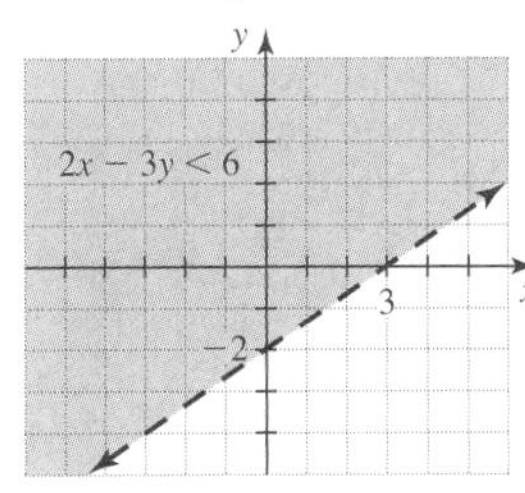

38. $x - 4y > 4$

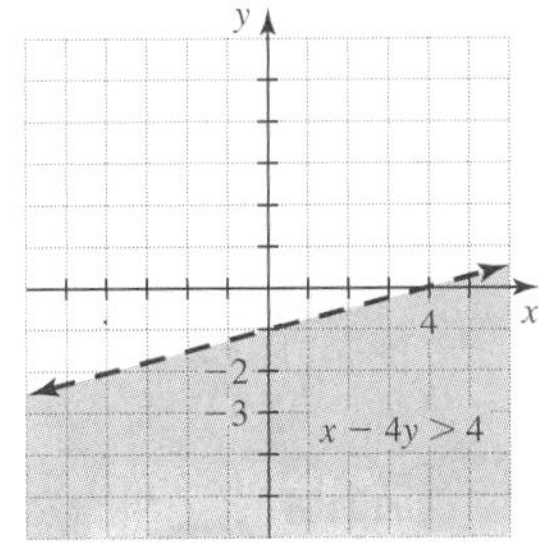

39. $x - 4y \leq 8$

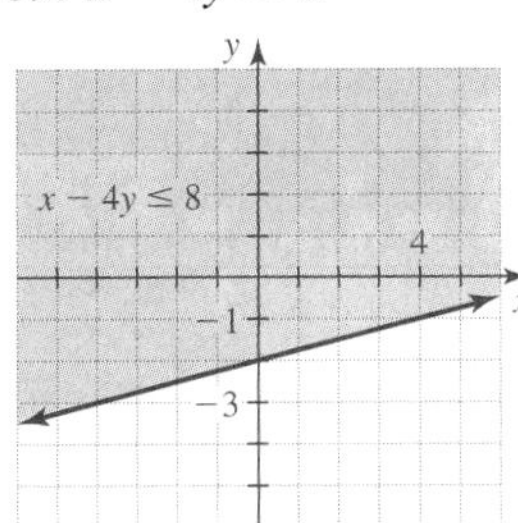

40. $3y - 5x \geq 15$

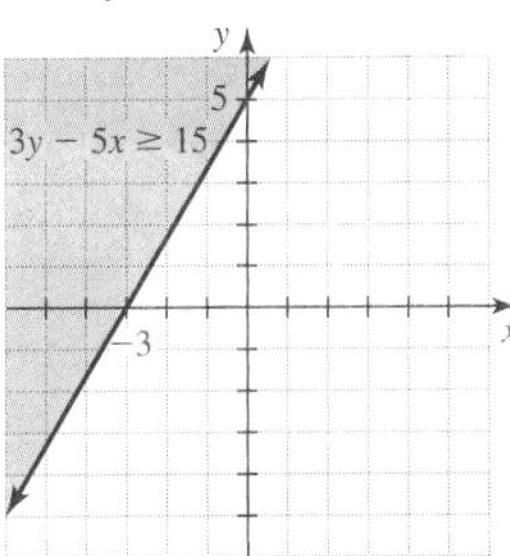

41. $y - \frac{7}{2}x \leq 7$

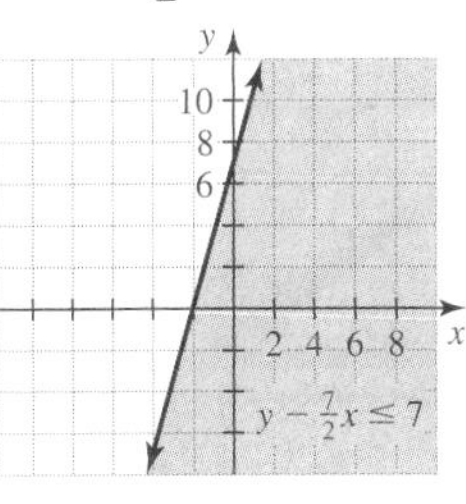

42. $\frac{2}{3}x + 3y \leq 12$

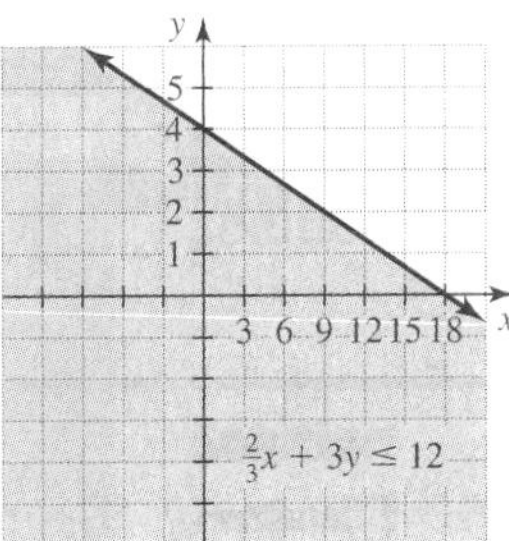

43. $x - y < 5$

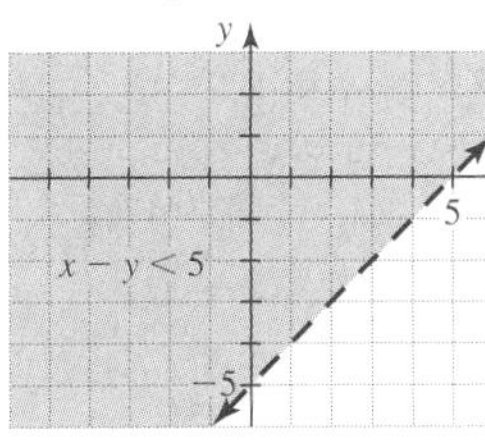

44. $y - x > -3$

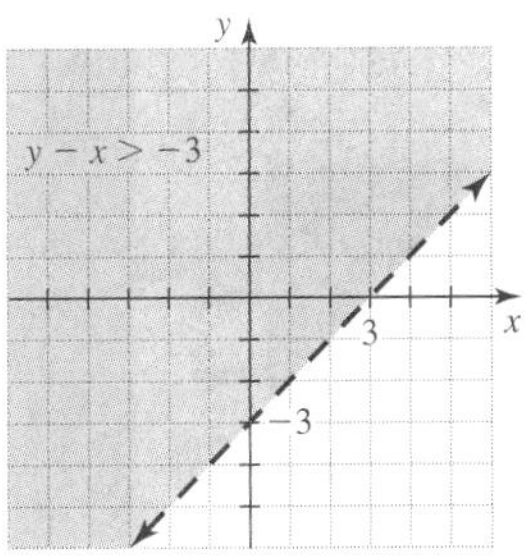

45. $3x - 4y < -12$

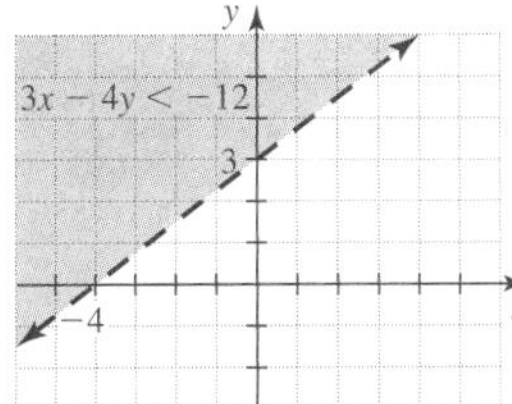

46. $4x + 3y > 24$

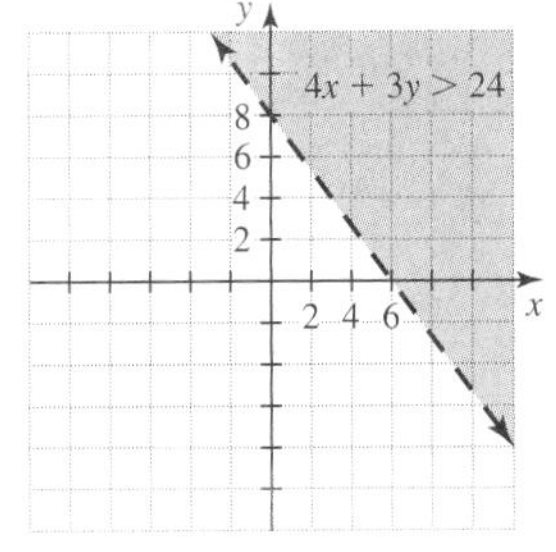

47. $x < 5y - 100$

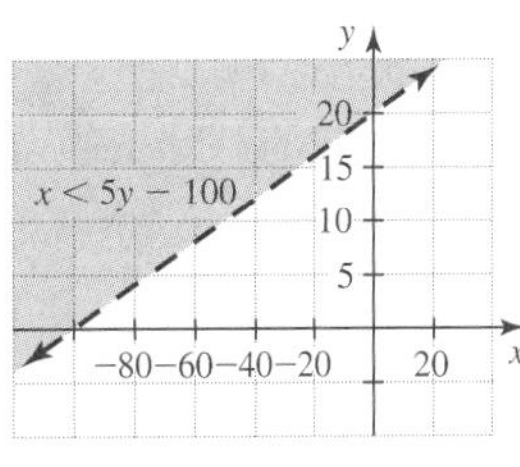

48. $-x > 70 - y$

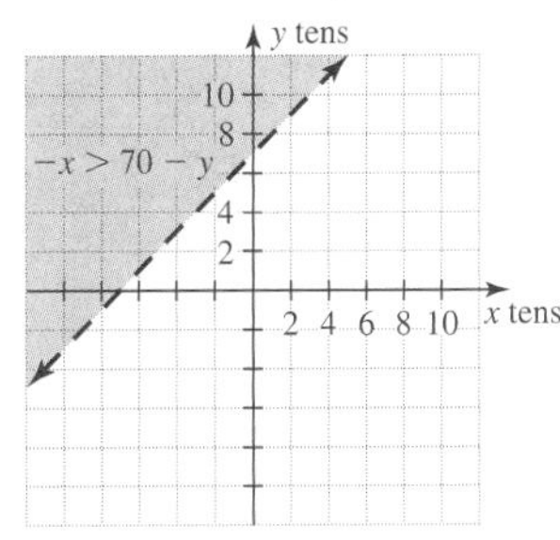

Solve each problem. See Example 5.

49. ***Storing the tables.*** Ozark Furniture Company must store its oak tables before shipping. A round table is packaged in a carton with a volume of 25 cubic feet (ft^3), and a rectangular table is packaged in a carton with a volume of 35 ft^3. The warehouse has at most 3850 ft^3 of space available for these tables. Write an inequality that limits the possible number of tables of each type that can be stored, and graph the inequality in the first quadrant.
$5x + 7y \leq 770$

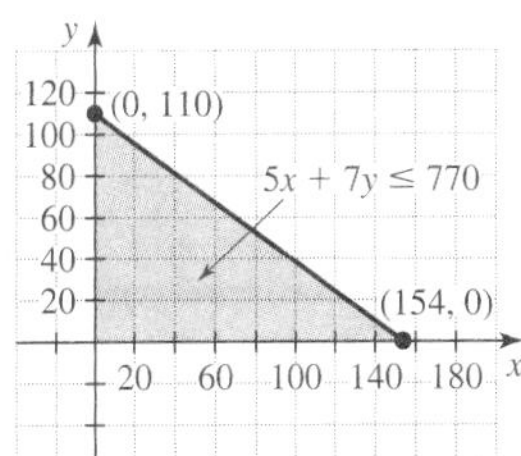

50. ***Maple rockers.*** Ozark Furniture Company can obtain at most 3000 board feet of maple lumber for making its classic and modern maple rocking chairs. A classic

FIGURE FOR EXERCISE 50

maple rocker requires 15 board feet of maple, and a modern rocker requires 12 board feet of maple. Write an inequality that limits the possible number of maple rockers of each type that can be made, and graph the inequality in the first quadrant.
$5x + 4y \leq 1000$

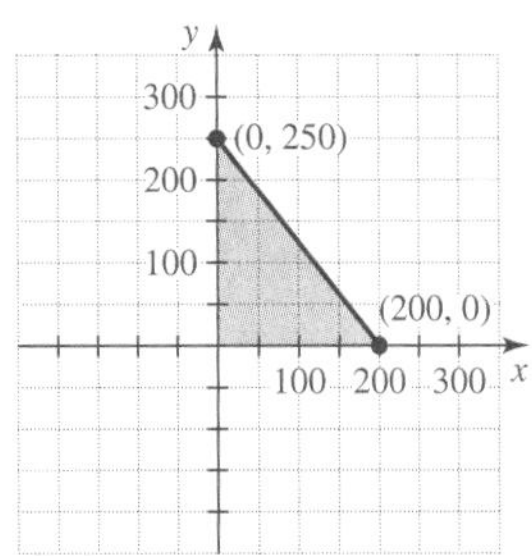

51. ***Enzyme concentration.*** A food chemist tests enzymes for their ability to break down pectin in fruit juices (Dennis Callas, *Snapshots of Applications in Mathematics*). Excess pectin makes juice cloudy. In one test, the chemist measures the concentration of the enzyme, c, in milligrams per milliliter and the fraction of light absorbed by the liquid, a. If $a > 0.07c + 0.02$, then the enzyme is working as it should. Graph the inequality for $0 < c < 5$.

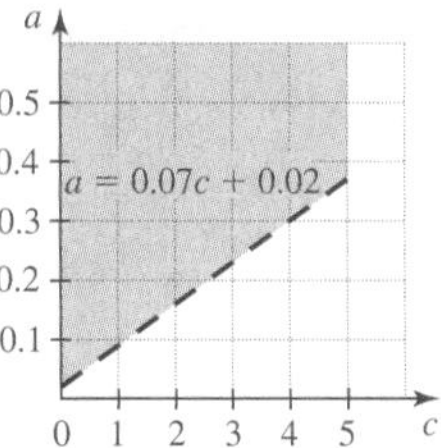

GETTING MORE INVOLVED

52. ***Discussion.*** When asked to graph the inequality $x + 2y < 12$, a student found that (0, 5) and (8, 0) both satisfied $x + 2y < 12$. The student then drew a dashed line through these two points and shaded the region below the line. What is wrong with this method? Do all of the points graphed by this student satisfy the inequality?

53. ***Writing.*** Compare and contrast the two methods presented in this section for graphing linear inequalities. What are the advantages and disadvantages of each method? How do you choose which method to use?

4.5 GRAPHING SYSTEMS OF LINEAR INEQUALITIES

In This Section

- The Solution to a System of Inequalities
- Graphing a System of Inequalities

In Section 4.4 you learned how to solve a linear inequality. In this section you will solve systems of linear inequalities.

The Solution to a System of Inequalities

A point is a solution to a system of equations if it satisfies both equations. Similarly, a point is a solution to a system of inequalities if it satisfies both inequalities.

EXAMPLE 1 **Satisfying a system of inequalities**

Determine whether each point is a solution to the system of inequalities:

$$2x + 3y < 6$$
$$y > 2x - 1$$

a) $(-3, 2)$ **b)** $(4, -3)$ **c)** $(5, 1)$

Solution

a) The point $(-3, 2)$ is a solution to the system if it satisfies both inequalities. Let $x = -3$ and $y = 2$ in each inequality:

$2x + 3y < 6$	$y > 2x - 1$
$2(-3) + 3(2) < 6$	$2 > 2(-3) - 1$
$0 < 6$	$2 > -7$

Study Tip

Read the text and recite to yourself what you have read. Ask questions and answer them out loud. Listen to your answers to see if they are complete and correct. Would other students understand your answers?

Because both inequalities are satisfied, the point $(-3, 2)$ is a solution to the system.

b) Let $x = 4$ and $y = -3$ in each inequality:

$$\begin{aligned} 2x + 3y &< 6 \\ 2(4) + 3(-3) &< 6 \\ -1 &< 6 \end{aligned} \qquad \begin{aligned} y &> 2x - 1 \\ -3 &> 2(4) - 1 \\ -3 &> 7 \end{aligned}$$

Because only one inequality is satisfied, the point $(4, -3)$ is not a solution to the system.

c) Let $x = 5$ and $y = 1$ in each inequality:

$$\begin{aligned} 2x + 3y &< 6 \\ 2(5) + 3(1) &< 6 \\ 13 &< 6 \end{aligned} \qquad \begin{aligned} y &> 2x - 1 \\ 1 &> 2(5) - 1 \\ 1 &> 9 \end{aligned}$$

Because neither inequality is satisfied, the point $(5, 1)$ is not a solution to the system. ■

Graphing a System of Inequalities

There are infinitely many points that satisfy a typical system of inequalities. The best way to describe the solution to a system of inequalities is with a graph showing all points that satisfy the system. When we graph the points that satisfy a system, we say that we are graphing the system.

EXAMPLE 2 **Graphing a system of inequalities**

Graph all ordered pairs that satisfy the following system of inequalities:

$$y > x - 2$$
$$y < -2x + 3$$

Solution

We want a graph showing all points that satisfy both inequalities. The lines $y = x - 2$ and $y = -2x + 3$ divide the coordinate plane into four regions as shown in Fig. 4.15 on the next page. To determine which of the four regions contains points that satisfy the system, we check one point in each region to see whether it satisfies both inequalities. The points are shown in Fig. 4.15.

Check (0, 0):

$0 > 0 - 2$ Correct

$0 < -2(0) + 3$ Correct

Check (0, 5):

$5 > 0 - 2$ Correct

$5 < -2(0) + 3$ Incorrect

Check (0, −5):

$-5 > 0 - 2$ Incorrect

$-5 < -2(0) + 3$ Correct

Check (4, 0):

$0 > 4 - 2$ Incorrect

$0 < -2(4) + 3$ Incorrect

The only point that satisfies both inequalities of the system is (0, 0). So every point in the region containing (0, 0) also satisfies both inequalities. The points that satisfy the system are graphed in Fig. 4.16 on the next page.

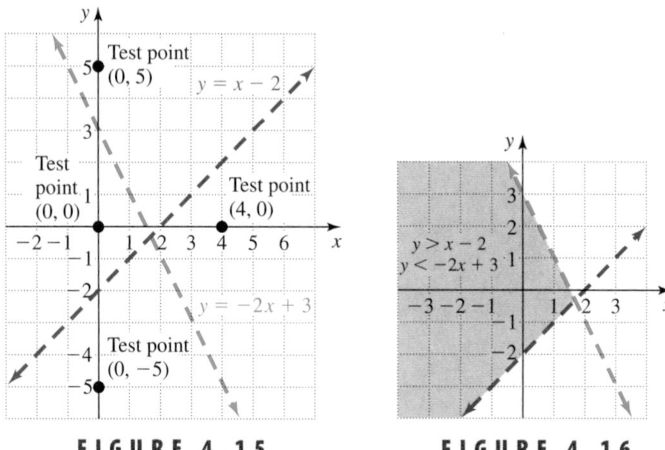

FIGURE 4.15 FIGURE 4.16

■

EXAMPLE 3 **Graphing a system of inequalities**

Graph all ordered pairs that satisfy the following system of inequalities:

$$y > -3x + 4$$
$$2y - x > 2$$

Solution

First graph the equations $y = -3x + 4$ and $2y - x = 2$. Now we select the points (0, 0), (0, 2), (0, 6), and (5, 0). We leave it to you to check each point in the system of inequalities. You will find that only (0, 6) satisfies the system. So only the region containing (0, 6) is shaded in Fig. 4.17.

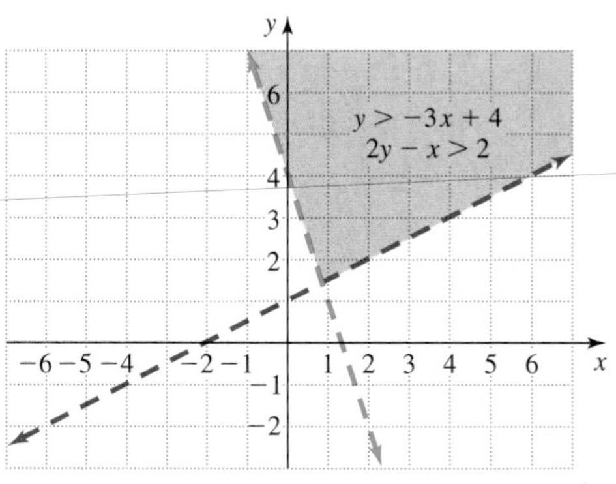

FIGURE 4.17

■

EXAMPLE 4 **Horizontal and vertical boundary lines**

Graph the system of inequalities:

$$x > 4$$
$$y < 3$$

Solution

We first graph the vertical line $x = 4$ and the horizontal line $y = 3$. The points that satisfy both inequalities are those points that lie to the right of the vertical line $x = 4$ and below the horizontal line $y = 3$. See Fig. 4.18 for the graph of the system. ■

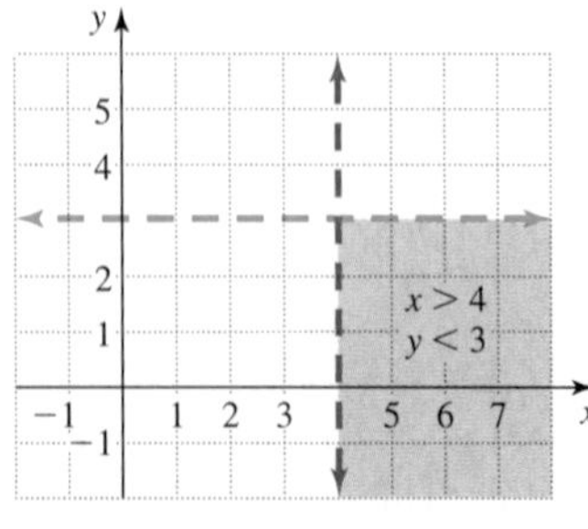

FIGURE 4.18

COLLABORATIVE ACTIVITIES

Which Cider?

Grouping: Three to four students per group

Topic: Systems of equations

For the Fall Harvest Festival your math club decides to sell apple cider. You can get a good deal on bulk apple cider by buying 30 gallons for \$120. At the club meeting, one member suggests buying the apple cider from her uncle, who has a nearby organic apple orchard. He will lower his price for the club to \$6 per gallon. You find that you can only get paper cups in batches of 100 at \$3 per 100. To decide which cider to buy, analyze the two options using the profit equation: Profit = Sales − Cost.

1. If you want to sell all 30 gallons in 8-ounce cups, how many cups will you need to buy?
2. What will your total costs be for paper cups and bulk cider? For paper cups and local cider?
3. The club wants to sell the cider for \$1 per cup. Write a profit equation (Profit = Sales − Cost) for each type of cider in terms of number of cups sold. Let c = number of cups sold and P = profit.
4. Graph both equations using a graphing calculator or on graph paper. Let the vertical (y-axis) be profit and the horizontal (x-axis) be the number of cups sold. Decide on a scale for your graph. If using a graphing calculator set your x- and y-max to at least 300.
5. How many cups do you need to sell to make a profit for local cider? For bulk cider?
6. Is there a point at which the number of cups and the profit are the same for both types?

The member who wants the local cider points out that your club could sell it for more, since it will be fresher and of higher quality. She suggests selling it for \$1.75 per cup.

7. Write an equation for local cider at \$1.75 per cup. Graph this equation with the one you had for bulk cider.
8. When are the profits greater for the local cider? Estimate this from your graph. Find this answer algebraically, using elimination or substitution.
9. Decide which cider your club should sell and at which price.

WRAP-UP CHAPTER 4

SUMMARY

Systems of Linear Equations in Two Variables

Examples

Graphing method — Sketch each graph and identify the points they have in common.

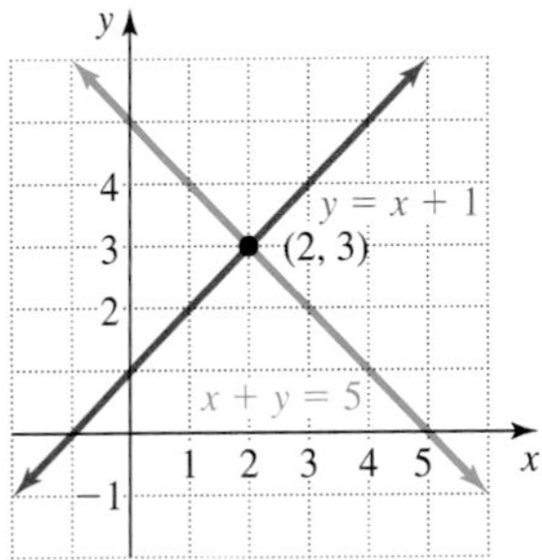

Substitution method — Solve one equation for one variable in terms of the other, then substitute into the other equation.

$$y = x - 4$$
$$x + y = 9$$
$$x + (x - 4) = 9$$

Addition method — Multiply each equation as necessary to eliminate a variable upon addition of the equations.

$$\begin{array}{rl} 5x - 3y &= 4 \\ x + 3y &= 1 \\ \hline 6x \quad\quad &= 5 \end{array}$$

Independent	Only one point satisfies both equations. The graphs cross at one point.	$y = x - 4$ $y = 2x + 5$
Inconsistent	No solution The graphs are parallel lines.	$y = 5x - 3$ $y = 5x + 1$
Dependent	Infinitely many solutions One equation is a multiple of the other. The graphs coincide.	$5x + 3y = 2$ $10x + 6y = 4$

Linear Inequalities in Two Variables		Examples
Graphing the solution to an inequality in two variables	1. Solve the inequality for y, then graph $y = mx + b$. $y > mx + b$ is the region above the line. $y = mx + b$ is the line itself. $y < mx + b$ is the region below the line.	 $y > x + 3$ $y = x + 3$ $y < x + 3$
	Remember that "less than" means below the line and "greater than" means above the line only when the inequality is solved for y.	
	2. If the inequality involves only x, then graph the vertical line $x = k$. $x > k$ is the region to the right of the line. $x = k$ is the line itself. $x < k$ is the region to the left of the line.	$x > 5$ Region to right of vertical line $x = 5$
Test points	A linear inequality may also be graphed by graphing the equation and then testing a point to determine which region satisfies the inequality.	$x + y > 4$ (0, 6) satisfies the inequality.
Graphing a system of inequalities	Graph the equations and use test points to see which regions satisfy both inequalities.	$x + y > 4$ $x - y < 1$ (0, 6) satisfies the system.

ENRICHING YOUR MATHEMATICAL WORD POWER

For each mathematical term, choose the correct meaning.

1. system of equations
- a. a systematic method for classifying equations
- b. a method for solving an equation
- c. two or more equations
- d. the properties of equality c

2. independent linear system
- a. a system with exactly one solution
- b. an equation that is satisfied by every real number
- c. equations that are identical
- d. a system of lines a

3. inconsistent system
- a. a system with no solution
- b. a system of inconsistent equations
- c. a system that is incorrect
- d. a system that we are not sure how to solve a

4. dependent system
- a. a system that is independent
- b. a system that depends on a variable
- c. a system that has no solution
- d. a system for which the graphs coincide d

5. **substitution method**
   a. replacing the variables by the correct answer
   b. a method of eliminating a variable by substituting one equation into the other
   c. the replacement method
   d. any method of solving a system b

6. **linear inequality in two variables**
   a. when two lines are not equal
   b. line segments that are unequal in length
   c. an inequality of the form $Ax + By \geq C$ or with another symbol of inequality
   d. an inequality of the form $Ax^2 + By^2 < C^2$ c

7. **rational numbers**
   a. the numbers 1, 2, 3, and so on
   b. the integers
   c. numbers that make sense
   d. numbers of the form a/b where a and b are integers with $b \neq 0$ d

8. **irrational numbers**
   a. the cube roots
   b. numbers that cannot be expressed as a ratio of integers
   c. numbers that do not make sense
   d. the integers b

9. **additive identity**
   a. the number 0
   b. the number 1
   c. the opposite of a number
   d. when two sums are identical a

10. **multiplicative identity**
    a. the number 0
    b. the number 1
    c. the reciprocal
    d. when two products are identical b

REVIEW EXERCISES

4.1 *Solve each system by graphing.*

1. $y = 2x + 1$
   $x + y = 4$ (1, 3)
2. $y = -x + 1$
   $y = -x + 3$ No solution
3. $y = 2x + 3$
   $y = -2x - 1$ (−1, 1)
4. $x + y = 6$
   $x - y = -10$ (−2, 8)

4.2 *Solve each system by the substitution method.*

5. $y = 3x$
   $2x + 3y = 22$ (2, 6)
6. $x + y = 3$
   $3x - 2y = -11$ (−1, 4)
7. $x = y - 5$
   $2x - 3y = -7$ (−8, −3)
8. $2x + y = 5$
   $6x - 9 = 3y$ (2, 1)

4.3 *In Exercises 9–20, solve each system by the addition method. Indicate whether each system is independent, inconsistent, or dependent.*

9. $x - y = 4$
   $2x + y = 5$ (3, −1), independent
10. $x + 2y = -5$
    $x - 3y = 10$ (1, −3), independent
11. $2x - 4y = 8$
    $x - 2y = 4$ $\{(x, y) \mid x - 2y = 4\}$, dependent
12. $x + 3y = 7$
    $2x + 6y = 5$ No solution, inconsistent
13. $y = 3x - 5$
    $2y = -x - 3$ (1, −2), independent
14. $3x + 4y = 6$
    $4x + 3y = 1$ (−2, 3), independent
15. $2x + 7y = 0$
    $7x + 2y = 0$ (0, 0), independent
16. $3x - 5y = 1$
    $10y = 6x - 1$ No solution, inconsistent
17. $x - y = 6$
    $2x - 12 = 2y$ $\{(x, y) \mid x - y = 6\}$, dependent
18. $y = 4x$
    $y = 3x$ (0, 0), independent
19. $y = 4x$
    $y = 4x + 3$ No solution, inconsistent
20. $3x - 5y = 21$
    $4x + 7y = -13$ (2, −3), independent

4.4 *Graph each inequality.*

21. $y > \frac{1}{3}x - 5$

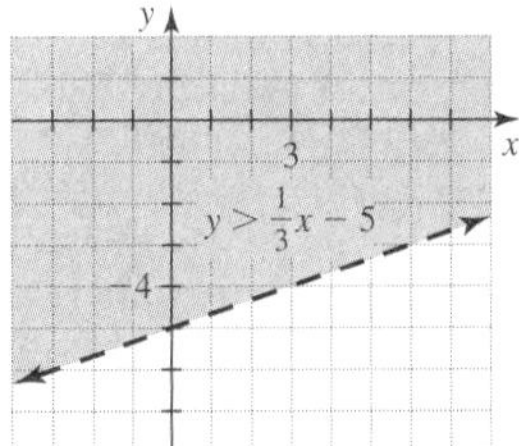

22. $y < \frac{1}{2}x + 2$

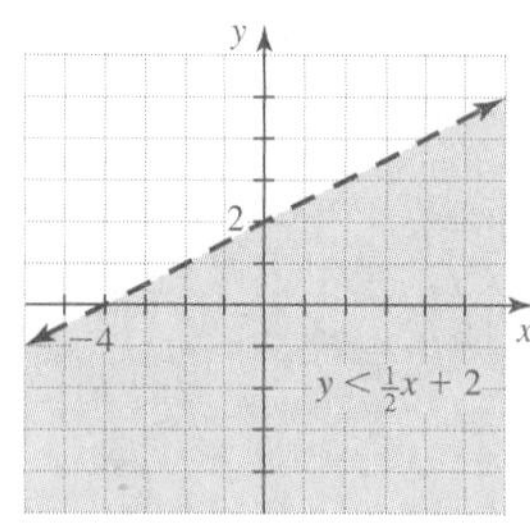

23. $y \le -2x + 7$

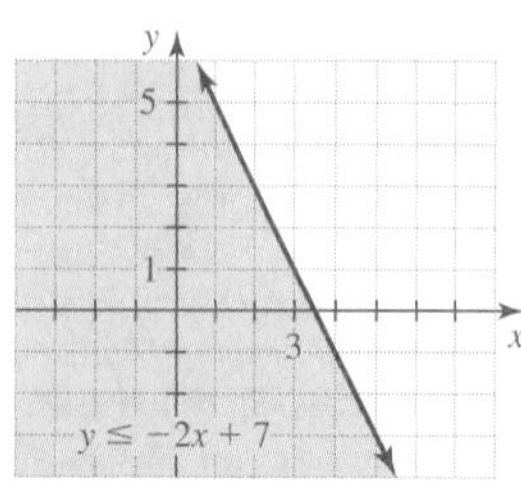

24. $y \ge x - 6$

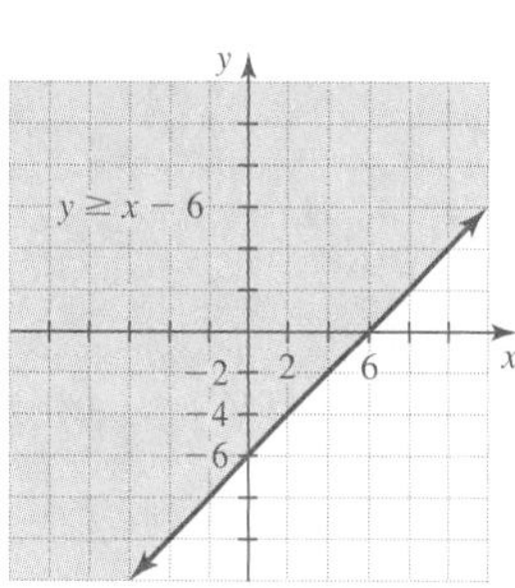

25. $y \le 8$

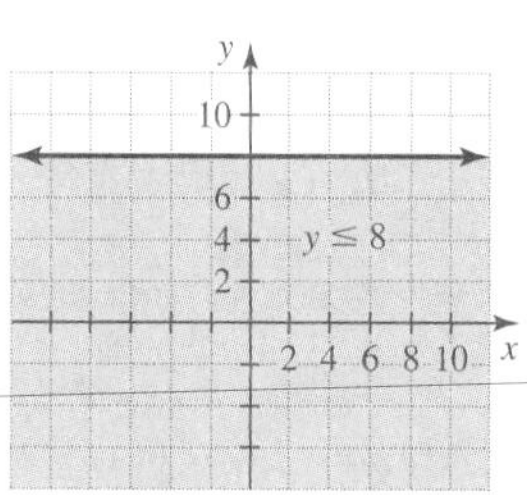

26. $x \ge -6$

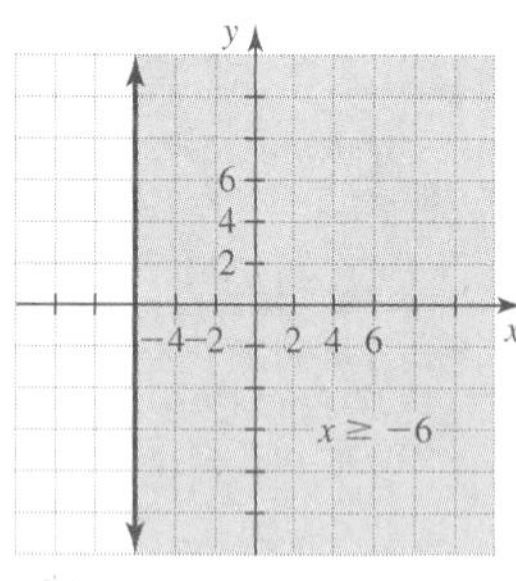

27. $2x + 3y \le -12$

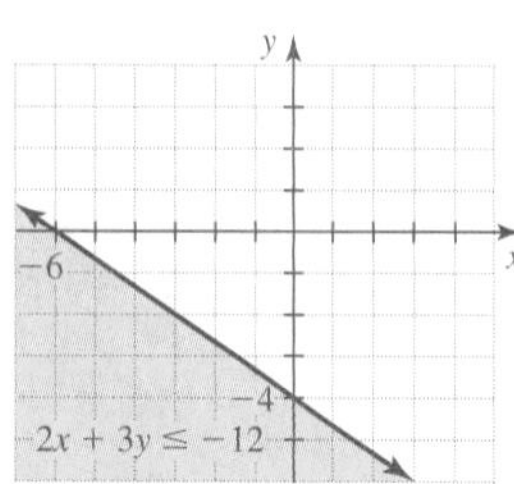

28. $x - 3y < 9$

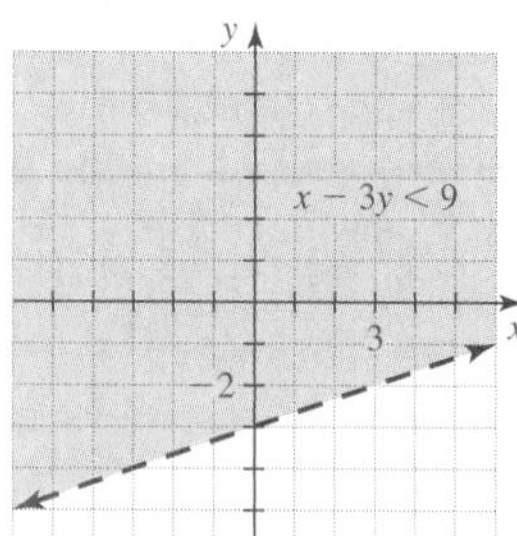

4.5 *Graph each system of inequalities.*

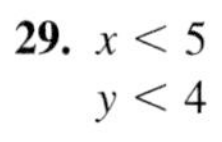

29. $x < 5$
$y < 4$

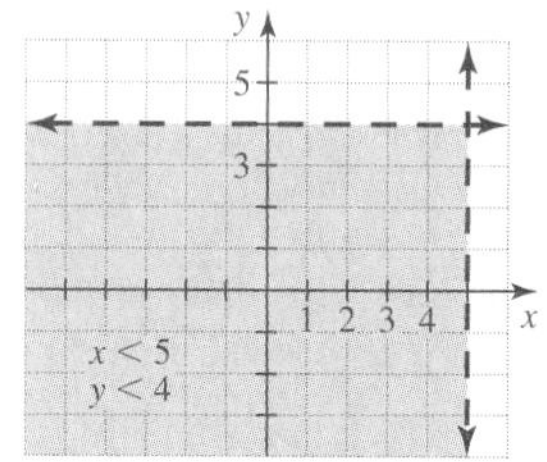

30. $y > -2$
$x < 1$

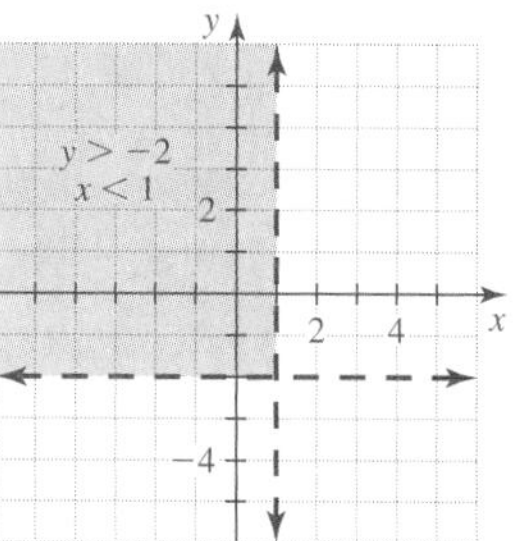

31. $x + y < 2$
$y > 2x - 3$

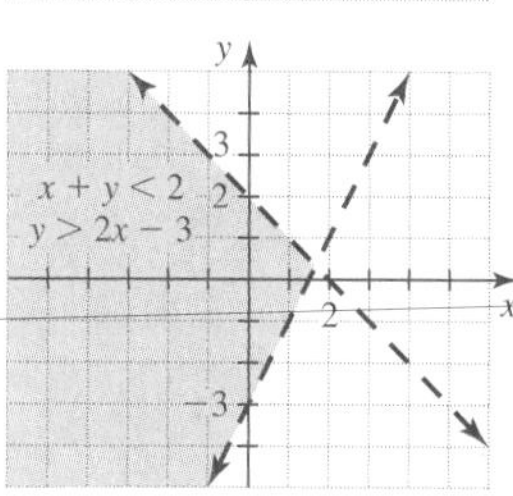

32. $x - y > 4$
$2y > x - 4$

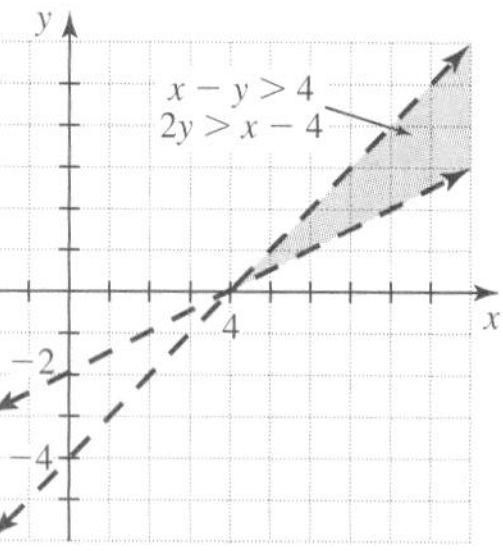

33. $y > 5x - 7$
$y < 5x + 1$

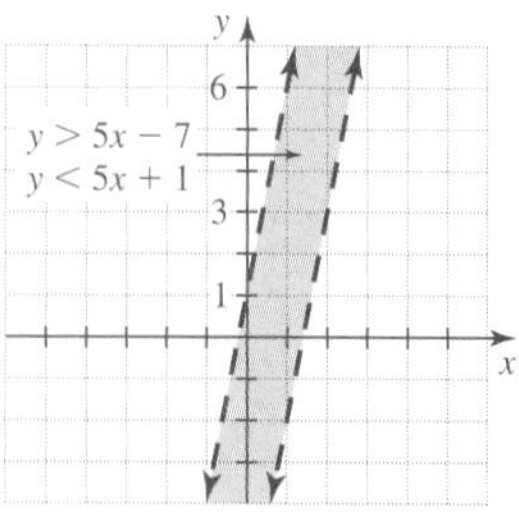

34. $y > x - 6$
$y < x - 5$

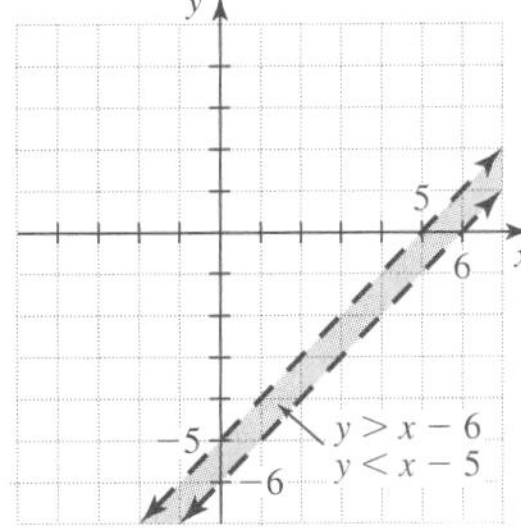

35. $y < 3x + 5$
$y < 3x$

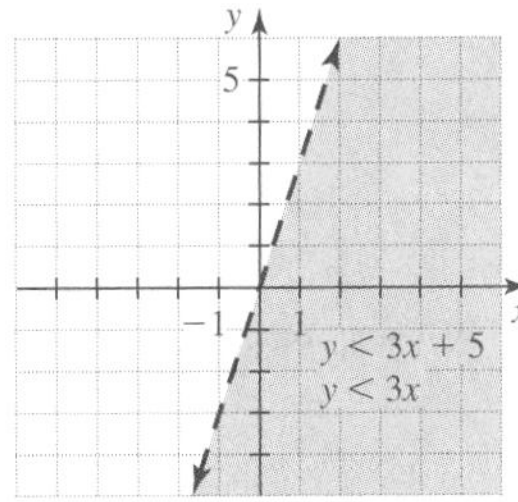

36. $y > -2x$
$y > -3x$

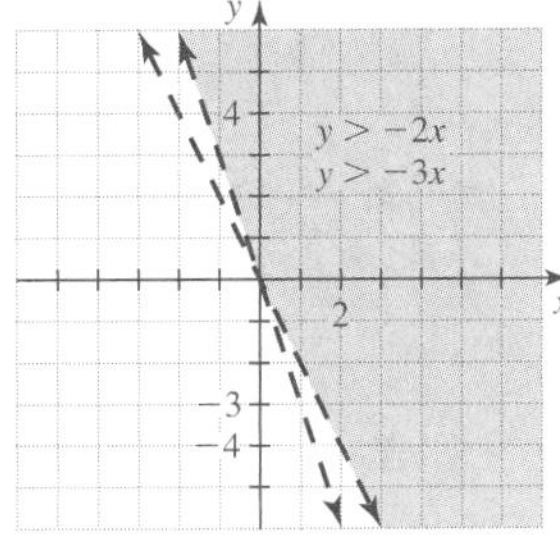

MISCELLANEOUS

Use a system of equations in two variables to solve each problem. Solve the system by the method of your choice.

37. ***Apples and oranges.*** Two apples and three oranges cost \$1.95, and three apples and two oranges cost \$2.05. What are the costs of one apple and one orange?
Apple \$0.45, orange \$0.35

FIGURE FOR EXERCISE 37

38. ***Small or medium.*** Three small drinks and one medium drink cost \$2.30, and two small drinks and four medium drinks cost \$3.70. What is the cost of one small drink? What is the cost of one medium drink?
Small \$0.55, medium \$0.65

39. ***Gambling fever.*** After a long day at the casinos in Biloxi, Louis returned home and told his wife Lois that he had won \$430 in \$5 bills and \$10 bills. On counting them again, he realized that he had mixed up the number of bills of each denomination, and he had really won only \$380. How many bills of each denomination does Louis have?
32 fives, 22 tens

40. ***Diversifying investments.*** Diane invested her \$10,000 bonus in a municipal bond fund and an emerging market fund. In one year the amount invested in the bond fund earned 8%, and the amount invested in the emerging market fund earned 10%. If the total income from these two investments for one year was \$880, then how much did she invest in each fund?
\$6000 at 8%, \$4000 at 10%

41. ***Protein and carbohydrates.*** One serving of green beans contains 1 gram of protein and 4 grams of carbohydrates. One serving of chicken soup contains 3 grams of protein and 9 grams of carbohydrates. The Westdale Diet recommends a lunch of 13 grams of protein and 43 grams of carbohydrates. How many servings of each are necessary to obtain the recommended amounts?
4 servings green beans, 3 servings chicken soup

FIGURE FOR EXERCISE 41

42. ***Advertising revenue.*** A television station aired four 30-second commercials and three 60-second commercials during the first hour of the midnight movie. During the second hour, it aired six 30-second commercials and five 60-second commercials. The advertising revenue for the first hour was \$7700, and that for the second hour was \$12,300. What is the cost of each type of commercial?
30-second \$800, 60-second \$1500

CHAPTER 4 TEST

Solve the system by graphing.

1. $x + y = 2$
$y = 2x + 5$ $(-1, 3)$

Solve each system by substitution.

2. $y = 2x - 3$
$2x + 3y = 7$ $(2, 1)$

3. $x - y = 4$
$3x - 2y = 11$ $(3, -1)$

Solve each system by the addition method.

4. $2x + 5y = 19$
$4x - 3y = -1$ $(2, 3)$

5. $3x - 2y = 10$
$2x + 5y = 13$ $(4, 1)$

Determine whether each system is independent, inconsistent, or dependent.

6. $y = 4x - 9$
$y = 4x + 8$ Inconsistent

7. $3x - 3y = 12$
$y = x - 4$ Dependent

8. $y = 2x$
$y = 5x$ Independent

Graph each inequality.

9. $y > 3x - 5$

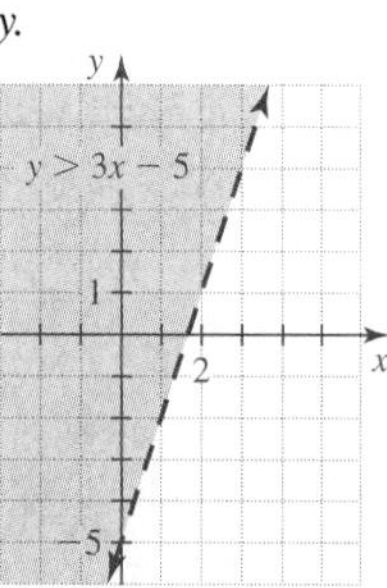

10. $x - y < 3$

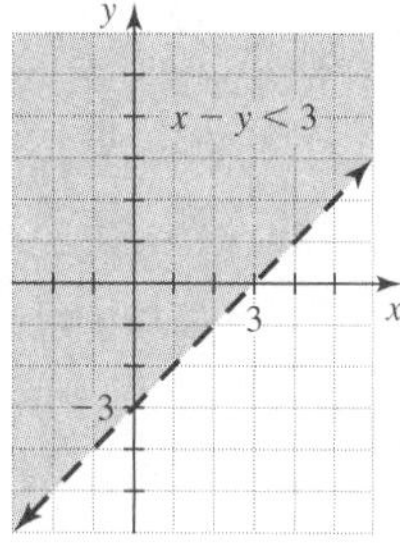

11. $x - 2y \geq 4$

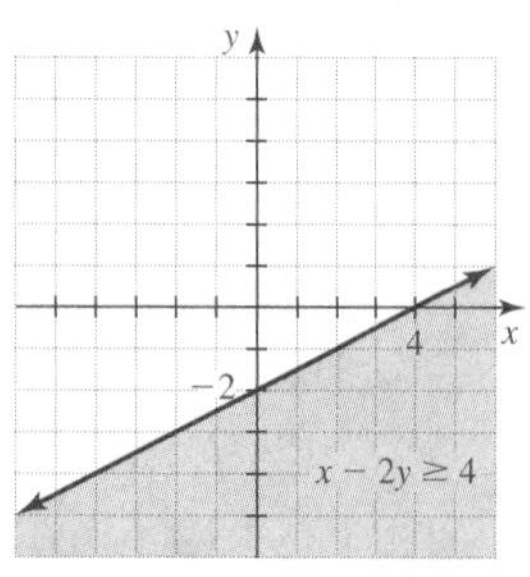

Graph each system of inequalities.

12. $x < 6$
$y > -1$

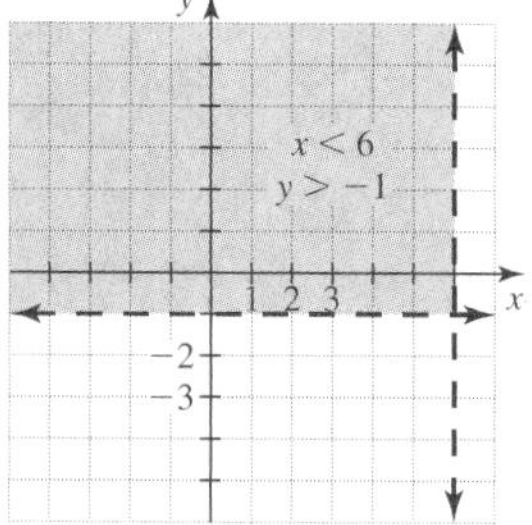

13. $2x + 3y > 6$
$3x - y < 3$

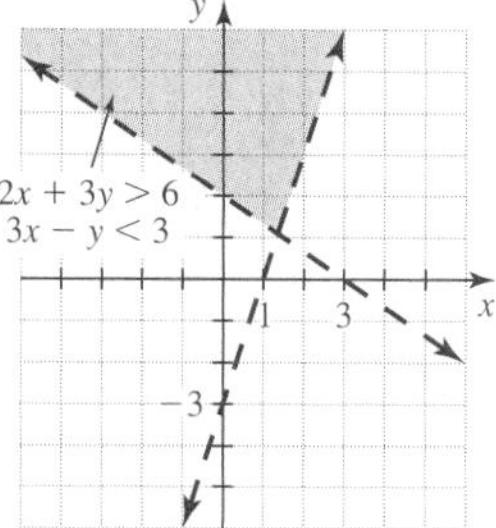

14. $y > 3x - 4$
$3x - y > 3$

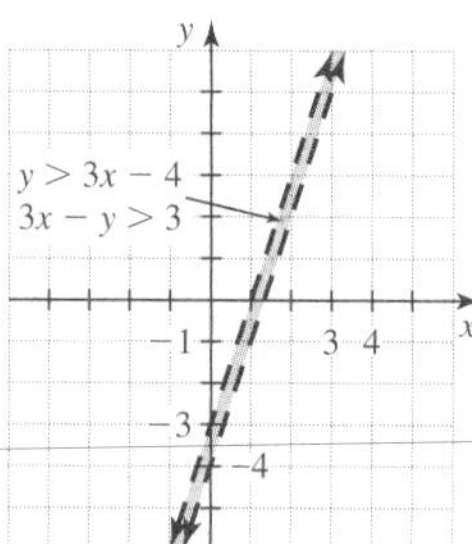

For each problem, write a system of equations in two variables. Use the method of your choice to solve each system.

15. Kathy and Chris studied a total of 54 hours for the CPA exam. If Chris studied only one-half as many hours as Kathy, then how many hours did each of them study?
Kathy 36 hours, Chris 18 hours

16. The Rest-Is-Easy Motel just outside Amarillo rented five singles and three doubles on Monday night for a total of \$188. On Tuesday night it rented three singles and four doubles for a total of \$170. On Wednesday night it rented only one single and one double. How much rent did the motel receive on Wednesday night? \$48

MAKING CONNECTIONS CHAPTERS 1–4

Solve each equation.

1. $2(x - 5) + 3x = 25$ $\{7\}$
2. $3x - 5 = 0$ $\left\{\frac{5}{3}\right\}$
3. $\frac{x}{3} - \frac{2}{5} = \frac{x}{2} - \frac{12}{5}$ $\{12\}$
4. $x - 0.05x = 950$ $\{1000\}$
5. $3(x - 5) - 5x = 5 - 2(x - 4)$ $\varnothing$
6. $7x - 4(5 - x) = 5(2x - 4) + x$ All real numbers

Solve each inequality in one variable, and sketch the graph on the number line.

7. $3(2 - x) < -6$
   $x > 4$

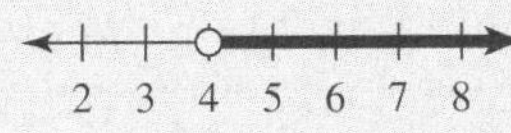

8. $-3 \le 2x - 4 \le 6$
   $\frac{1}{2} \le x \le 5$

9. $4 \ge 5 - x$
   $x \ge 1$

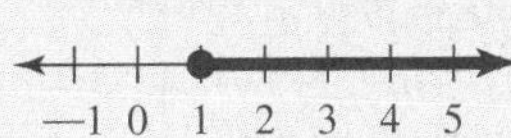

Sketch the graph of each equation.

10. $y = 3x - 7$

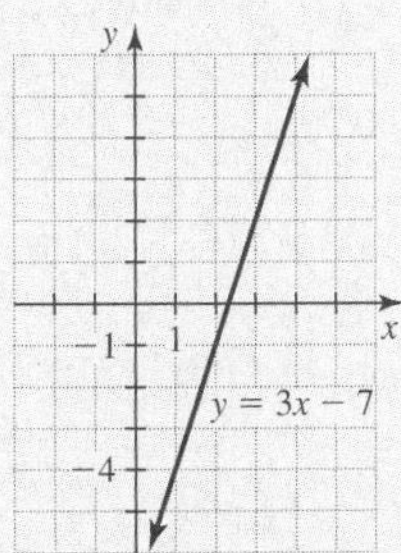

11. $y = 5 - x$

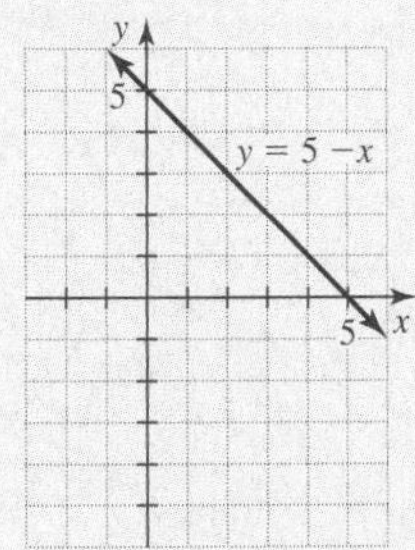

12. $y = x - 1$

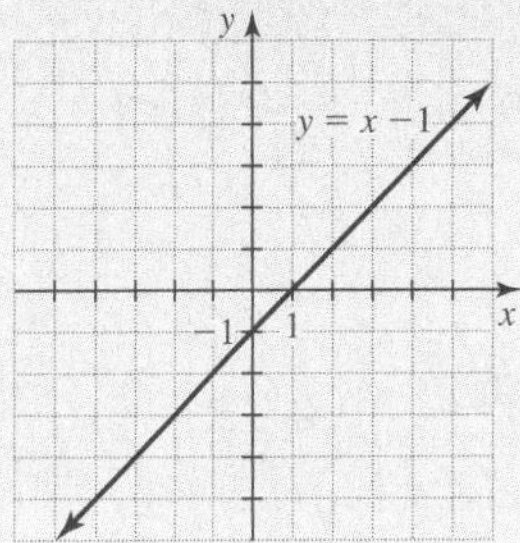

13. $y = x + 1$

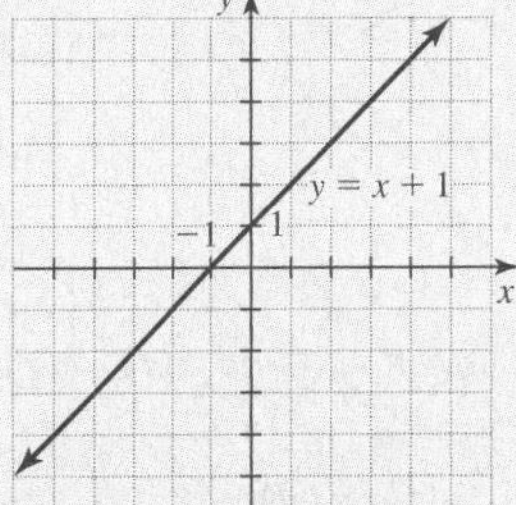

14. $y = -2x + 4$

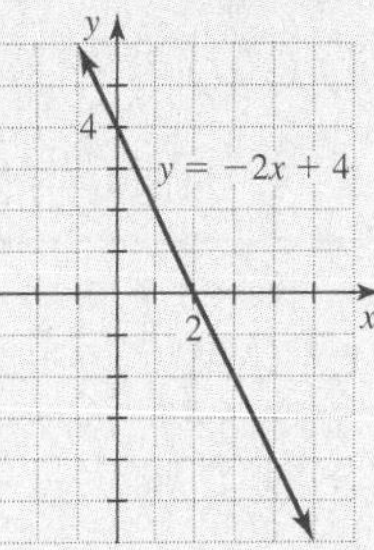

15. $y = -4x - 1$

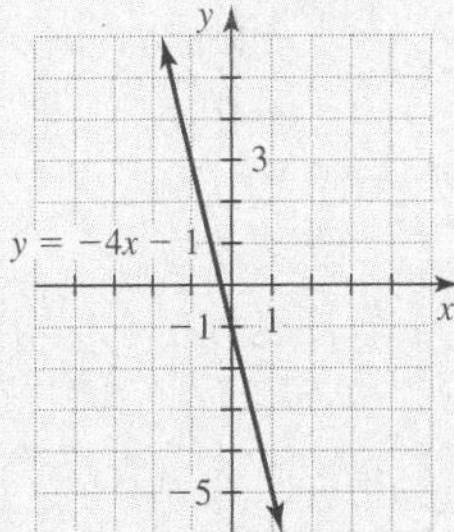

Graph each inequality in two variables.

16. $y \ge 3x - 7$

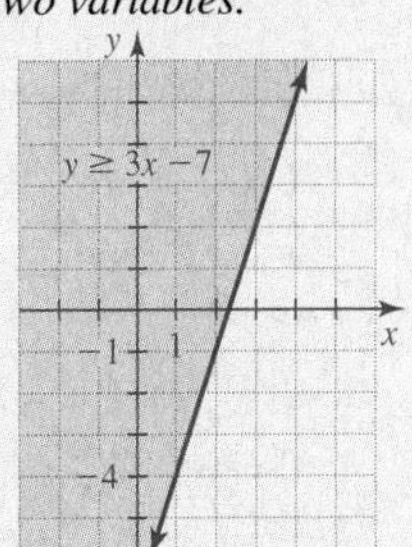

17. $x - 2y < 6$

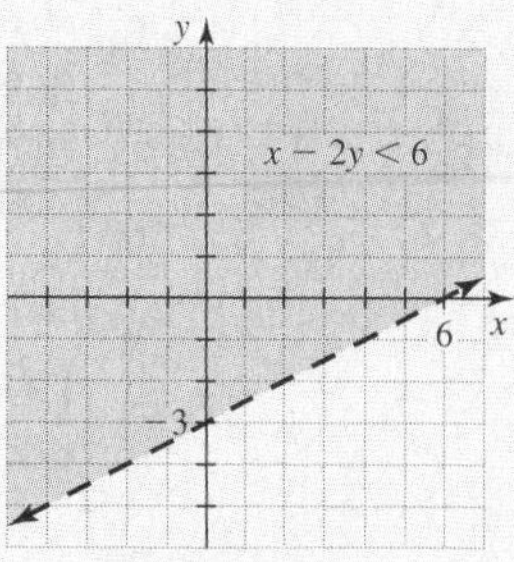

18. $x > 1$

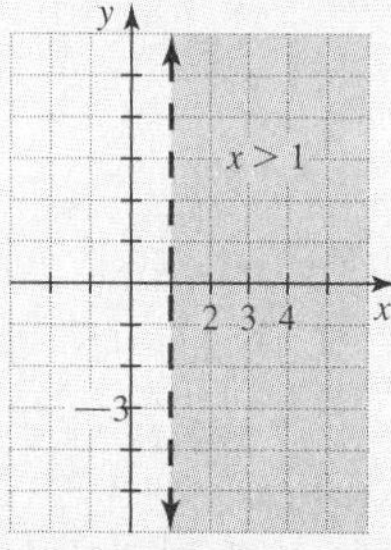

Write the equation of the line going through each pair of points.

19. (0, 36) and (8, 84) $y = 6x + 36$

20. (1, 88) and (12, 11) $y = -7x + 95$

Study Tip

Don't wait until the final exam to review material. Do some review on a regular basis. The Making Connections exercises on this page can be used to review, compare, and contrast different concepts that you have studied. A good time to work these exercises is between a test and the start of new material.

Solve the problem.

21. ***Decreasing market share.*** The market share for Toys "R" Us has gone from 25% of the toy market in 1990 to 16% in 2000 as shown in the accompanying graph (*Forbes,* www.forbes.com).

a) Write the market share p in terms of x, where x is the number of years since 1990.

$p = -\frac{9}{10}x + 25$

b) Wal-Mart's market share of the toy market went from 10% in 1990 to 19% in 2000. Write Wal-Mart's market share p in terms of x, where x is the number of years since 1990.

$p = \frac{9}{10}x + 10$

c) Solve the system of equations that you found in parts (a) and (b) to find the year in which Wal-Mart passed up Toys "R" Us in market share for toys.

$8\frac{1}{3}$ years after 1990 or 1998.

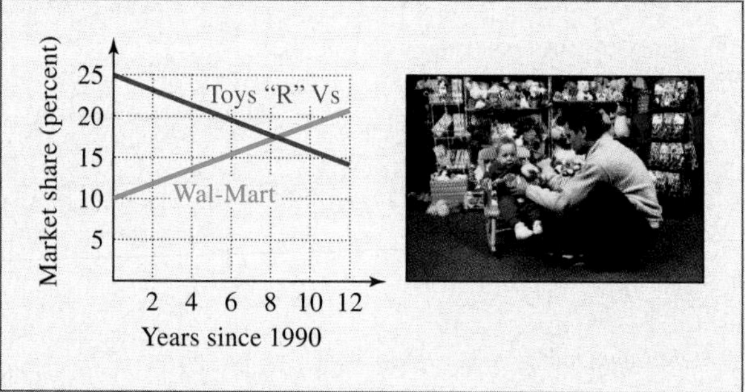

FIGURE FOR EXERCISE 21

CHAPTER 5

Polynomials and Exponents

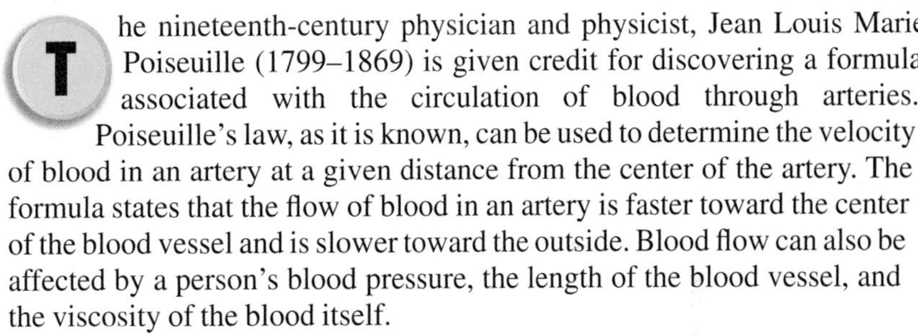

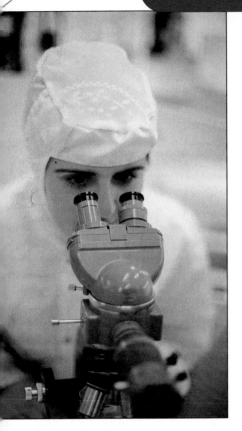

The nineteenth-century physician and physicist, Jean Louis Marie Poiseuille (1799–1869) is given credit for discovering a formula associated with the circulation of blood through arteries. Poiseuille's law, as it is known, can be used to determine the velocity of blood in an artery at a given distance from the center of the artery. The formula states that the flow of blood in an artery is faster toward the center of the blood vessel and is slower toward the outside. Blood flow can also be affected by a person's blood pressure, the length of the blood vessel, and the viscosity of the blood itself.

In later years, Poiseuille's continued interest in blood circulation led him to experiments to show that blood pressure rises and falls when a person exhales and inhales. In modern medicine, physicians can use Poiseuille's law to determine how much the radius of a blocked blood vessel must be widened to create a healthy flow of blood.

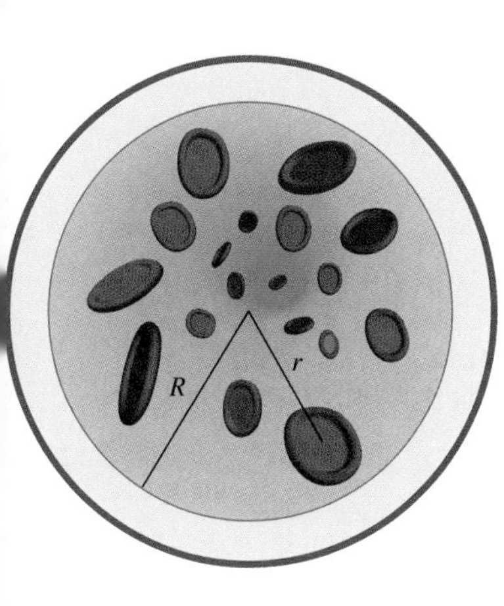

In this chapter you will study polynomials, the fundamental expressions of algebra. Polynomials are to algebra what integers are to arithmetic. We use polynomials to represent quantities in general, such as perimeter, area, revenue, and the volume of blood flowing through an artery. In Exercise 85 of Section 5.4, you will see Poiseuille's law represented by a polynomial.

5.1 ADDITION AND SUBTRACTION OF POLYNOMIALS

In This Section

- Polynomials
- Naming and Evaluating Polynomials
- Addition of Polynomials
- Subtraction of Polynomials
- Applications

We first used polynomials in Chapter 1 but did not identify them as polynomials. Polynomials also occurred in the equations and inequalities of Chapter 2. In this section we will define polynomials and begin a thorough study of polynomials.

Polynomials

In Chapter 1 we defined a **term** as an expression containing a number or the product of a number and one or more variables raised to powers. Some examples of terms are

$$4x^3,\ -x^2y^3,\ 6ab, \text{ and } -2.$$

A **polynomial** is a single term or a finite sum of terms. The powers of the variables in a polynomial must be positive integers. For example,

$$4x^3 + (-15x^2) + x + (-2)$$

is a polynomial. Because it is simpler to write addition of a negative as subtraction, this polynomial is usually written as

$$4x^3 - 15x^2 + x - 2.$$

Study Tip

Many commuting students find it difficult to get help. Students are often stuck on a minor point that can be easily resolved over the telephone. So ask your instructor if he or she will answer questions over the telephone and during what hours.

The **degree of a polynomial** in one variable is the highest power of the variable in the polynomial. So $4x^3 - 15x^2 + x - 2$ has degree 3 and $7w - w^2$ has degree 2. The **degree of a term** is the power of the variable in the term. Because the last term has no variable, its degree is 0.

$$4x^3 - 15x^2 + x - 2$$

$4x^3$	$15x^2$	x	2
Third-degree term	Second-degree term	First-degree term	Zero-degree term

A single number is called a **constant** and so the last term is the **constant term.** The degree of a polynomial consisting of a single number such as 8 is 0.

The number preceding the variable in each term is called the **coefficient** of that variable or the coefficient of that term. In $4x^3 - 15x^2 + x - 2$ the coefficient of x^3 is 4, the coefficient of x^2 is -15, and the coefficient of x is 1 because $x = 1 \cdot x$.

EXAMPLE 1 **Identifying coefficients**

Determine the coefficients of x^3 and x^2 in each polynomial:

a) $x^3 + 5x^2 - 6$ **b)** $4x^6 - x^3 + x$

Solution

a) Write the polynomial as $1 \cdot x^3 + 5x^2 - 6$ to see that the coefficient of x^3 is 1 and the coefficient of x^2 is 5.

b) The x^2-term is missing in $4x^6 - x^3 + x$. Because $4x^6 - x^3 + x$ can be written as

$$4x^6 - 1 \cdot x^3 + 0 \cdot x^2 + x,$$

the coefficient of x^3 is -1 and the coefficient of x^2 is 0. ■

For simplicity we generally write polynomials with the exponents decreasing from left to right and the constant term last. So we write

$$x^3 - 4x^2 + 5x + 1 \quad \text{rather than} \quad -4x^2 + 1 + 5x + x^3.$$

When a polynomial is written with decreasing exponents, the coefficient of the first term is called the **leading coefficient.**

Certain polynomials are given special names. A **monomial** is a polynomial that has one term, a **binomial** is a polynomial that has two terms, and a **trinomial** is a polynomial that has three terms. For example, $3x^5$ is a monomial, $2x - 1$ is a binomial, and $4x^6 - 3x + 2$ is a trinomial.

EXAMPLE 2

Types of polynomials

Identify each polynomial as a monomial, binomial, or trinomial and state its degree.

a) $5x^2 - 7x^3 + 2$ **b)** $x^{43} - x^2$ **c)** $5x$ **d)** -12

Solution

a) The polynomial $5x^2 - 7x^3 + 2$ is a third-degree trinomial.

b) The polynomial $x^{43} - x^2$ is a binomial with degree 43.

c) Because $5x = 5x^1$, this polynomial is a monomial with degree 1.

d) The polynomial -12 is a monomial with degree 0. ■

Study Tip

Be active in class. Don't be embarrassed to ask questions or answer questions. You can often learn more from a wrong answer than a right one. Your instructor knows that you are not yet an expert in algebra. Instructors love active classrooms and they will not think less of you for speaking out.

Naming and Evaluating Polynomials

A polynomial such as $x^2 - x + 3$ has a value if a number is chosen for x. For example, if $x = 2$, then

$$x^2 - x + 3 = 2^2 - 2 + 3 = 5.$$

So the value of the polynomial $x^2 - x + 3$ is 5 when $x = 2$. To make it easier to discuss polynomials and their values, we often name them with capital letters. For example, if $P = x^2 - x + 3$, then $P = 5$ when $x = 2$.

Another notation that is commonly used in mathematics, computer science, and on graphing calculators is to follow the letter that names the polynomial with the number used for x and write the result of the evaluation as a single equation. So $P = 5$ when $x = 2$ is written as $P(2) = 5$ and read as "P evaluated at 2 is 5" or simply "P of 2 is 5." This notation is called **function notation.** (See Section 9.7 for more information on functions.) Using function notation we write the polynomial as $P(x) = x^2 - x + 3$ rather than $P = x^2 - x + 3$. (Read $P(x)$ as "P of x.")

Function notation is very useful when we are evaluating a polynomial at several values of x. For example, if $P(x) = x^2 - x + 3$, then $P(0) = 3$, $P(1) = 3$, and $P(2) = 5$. Note that in function notation $P(x)$ does *not* mean P times x.

EXAMPLE 3

Evaluating polynomials

a) Find the value of $-3x^4 - x^3 + 20x + 3$ when $x = 1$.

b) Find the value of $-3x^4 - x^3 + 20x + 3$ when $x = -2$.

c) If $P(x) = -3x^4 - x^3 + 20x + 3$, find $P(1)$.

Solution

a) Replace x by 1 in the polynomial:

$$\begin{aligned} -3x^4 - x^3 + 20x + 3 &= -3(1)^4 - (1)^3 + 20(1) + 3 \\ &= -3 - 1 + 20 + 3 \\ &= 19 \end{aligned}$$

So the value of the polynomial is 19 when $x = 1$.

Calculator Close-Up

To evaluate the polynomial in Example 3 with a calculator, first use Y = to define the polynomial.

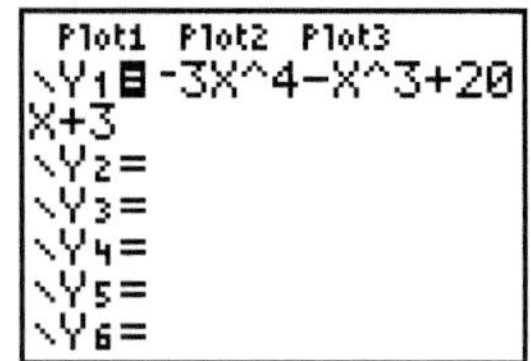

Then find $y_1(-2)$ and $y_1(1)$.

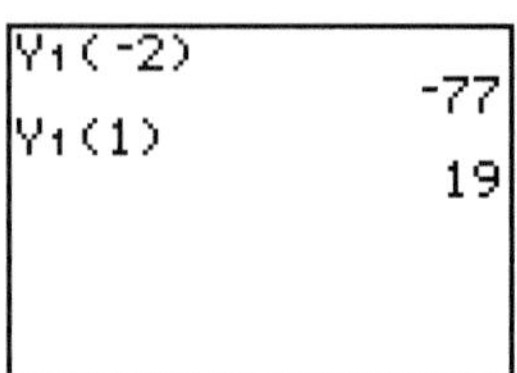

b) Replace x by -2 in the polynomial:

$$\begin{aligned}-3x^4 - x^3 + 20x + 3 &= -3(-2)^4 - (-2)^3 + 20(-2) + 3\\ &= -3(16) - (-8) - 40 + 3\\ &= -48 + 8 - 40 + 3\\ &= -77\end{aligned}$$

So the value of the polynomial is -77 when $x = -2$.

c) To find $P(1)$, replace x by 1 in the formula for $P(x)$:

$$\begin{aligned}P(x) &= -3x^4 - x^3 + 20x + 3\\ P(1) &= -3(1)^4 - (1)^3 + 20(1) + 3\\ &= 19\end{aligned}$$

So $P(1) = 19$. The value of the polynomial when $x = 1$ is 19. ■

Addition of Polynomials

You learned how to combine like terms in Chapter 1. Also, you combined like terms when solving equations in Chapter 2. Addition of polynomials is done simply by adding the like terms.

Addition of Polynomials

To add two polynomials, add the like terms.

Polynomials can be added horizontally or vertically, as shown in Example 4.

EXAMPLE 4 **Adding polynomials**

Perform the indicated operation.

a) $(x^2 - 6x + 5) + (-3x^2 + 5x - 9)$

b) $(-5a^3 + 3a - 7) + (4a^2 - 3a + 7)$

Helpful Hint

When we perform operations with polynomials and write the results as equations, those equations are identities. For example,

$$(x + 1) + (3x + 5) = 4x + 6$$

is an identity. This equation is satisfied by every real number.

Solution

a) We can use the commutative and associative properties to get the like terms next to each other and then combine them:

$$\begin{aligned}(x^2 - 6x + 5) + (-3x^2 + 5x - 9) &= x^2 - 3x^2 - 6x + 5x + 5 - 9\\ &= -2x^2 - x - 4\end{aligned}$$

b) When adding vertically, we line up the like terms:

$$\begin{array}{rrrr} -5a^3 & & +\,3a & -\,7\\ & 4a^2 & -\,3a & +\,7\\ \hline -5a^3 & +\,4a^2 & & \end{array} \quad \text{Add.}$$

■

Subtraction of Polynomials

When we subtract polynomials, we subtract the like terms. Because $a - b = a + (-b)$, we can subtract by adding the opposite of the second polynomial to the first polynomial. Remember that a negative sign in front of parentheses changes the

Applications

EXAMPLE 6 **Multiplying polynomials**

A parking lot is 20 yards wide and 30 yards long. If the college increases the length and width by the same amount to handle an increasing number of cars, then what polynomial represents the area of the new lot? What is the new area if the increase is 15 yards?

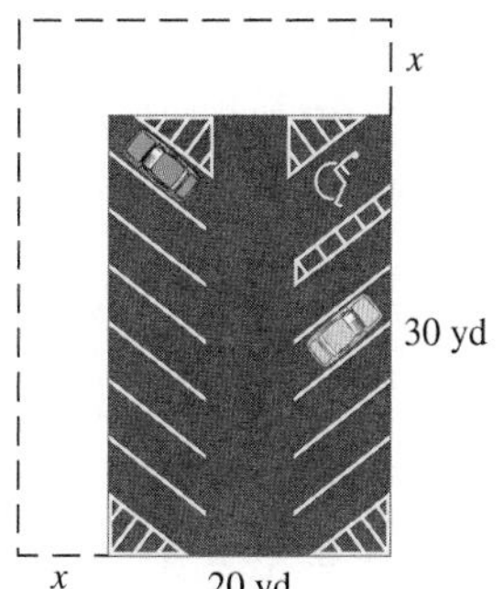

FIGURE 5.1

Solution

If x is the amount of increase, then the new lot will be $x + 20$ yards wide and $x + 30$ yards long as shown in Fig. 5.1. Multiply the length and width to get the area:

$$(x + 20)(x + 30) = (x + 20)x + (x + 20)30$$
$$= x^2 + 20x + 30x + 600$$
$$= x^2 + 50x + 600$$

The polynomial $x^2 + 50x + 600$ represents the area of the new lot. If $x = 15$, then

$$x^2 + 50x + 600 = (15)^2 + 50(15) + 600 = 1575.$$

If the increase is 15 yards, then the area of the lot will be 1575 square yards. ■

WARM-UPS

True or false? Explain your answer.

1. $3x^3 \cdot 5x^4 = 15x^{12}$ for any value of x. False
2. $3x^2 \cdot 2x^7 = 5x^9$ for any value of x. False
3. $(3y^3)^2 = 9y^6$ for any value of y. True
4. $-3x(5x - 7x^2) = -15x^3 + 21x^2$ for any value of x. False
5. $2x(x^2 - 3x + 4) = 2x^3 - 6x^2 + 8x$ for any number x. True
6. $-2(3 - x) = 2x - 6$ for any number x. True
7. $(a + b)(c + d) = ac + ad + bc + bd$ for any values of a, b, c, and d. True
8. $-(x - 7) = 7 - x$ for any value of x. True
9. $83 - 37 = -(37 - 83)$ True
10. The opposite of $x + 3$ is $x - 3$ for any number x. False

5.2 EXERCISES

Reading and Writing *After reading this section, write out the answers to these questions. Use complete sentences.*

1. What is the product rule for exponents?
   The product rule for exponents says that $a^m \cdot a^n = a^{m+n}$.
2. Why is the sum of two monomials not necessarily a monomial?
   The sum of two monomials can be a binomial if the terms are not like terms.
3. What property of the real numbers is used when multiplying a monomial and a polynomial?
   To multiply a monomial and a polynomial we use the distributive property.
4. What property of the real numbers is used when multiplying two binomials?
   To multiply two binomials we use the distributive property twice.

5. How do we multiply any two polynomials?
To multiply any two polynomials we multiply each term of the first polynomial by every term of the second polynomial.

6. How do we find the opposite of a polynomial?
To find the opposite of a polynomial, change the sign of each term in the polynomial.

Find each product. See Example 1.

7. $3x^2 \cdot 9x^3$ $27x^5$
8. $5x^7 \cdot 3x^5$ $15x^{12}$
9. $2a^3 \cdot 7a^8$ $14a^{11}$
10. $3y^{12} \cdot 5y^{15}$ $15y^{27}$
11. $-6x^2 \cdot 5x^2$ $-30x^4$
12. $-2x^2 \cdot 8x^5$ $-16x^7$
13. $(-9x^{10})(-3x^7)$ $27x^{17}$
14. $(-2x^2)(-8x^9)$ $16x^{11}$
15. $-6st \cdot 9st$ $-54s^2t^2$
16. $-12sq \cdot 3s$ $-36qs^2$
17. $3wt \cdot 8w^7t^6$ $24t^7w^8$
18. $h^8k^3 \cdot 5h$ $5h^9k^3$
19. $(5y)^2$ $25y^2$
20. $(6x)^2$ $36x^2$
21. $(2x^3)^2$ $4x^6$
22. $(3y^5)^2$ $9y^{10}$

Find each product. See Example 2.

23. $4y^2(y^5 - 2y)$ $4y^7 - 8y^3$
24. $6t^3(t^5 + 3t^2)$ $6t^8 + 18t^5$
25. $-3y(6y - 4)$ $-18y^2 + 12y$
26. $-9y(y^2 - 1)$ $-9y^3 + 9y$
27. $(y^2 - 5y + 6)(-3y)$ $-3y^3 + 15y^2 - 18y$
28. $(x^3 - 5x^2 - 1)7x^2$ $7x^5 - 35x^4 - 7x^2$
29. $-x(y^2 - x^2)$ $-xy^2 + x^3$
30. $-ab(a^2 - b^2)$ $ab^3 - a^3b$
31. $(3ab^3 - a^2b^2 - 2a^3b)5a^3$ $15a^4b^3 - 5a^5b^2 - 10a^6b$
32. $(3c^2d - d^3 + 1)8cd^2$ $24c^3d^3 - 8cd^5 + 8cd^2$
33. $-\frac{1}{2}t^2v(4t^3v^2 - 6tv - 4v)$ $-2t^5v^3 + 3t^3v^2 + 2t^2v^2$
34. $-\frac{1}{3}m^2n^3(-6mn^2 + 3mn - 12)$
$2m^3n^5 - m^3n^4 + 4m^2n^3$

Use the distributive property to find each product. See Example 3.

35. $(x + 1)(x + 2)$ $x^2 + 3x + 2$
36. $(x + 6)(x + 3)$ $x^2 + 9x + 18$
37. $(x - 3)(x + 5)$ $x^2 + 2x - 15$
38. $(y - 2)(y + 4)$ $y^2 + 2y - 8$
39. $(t - 4)(t - 9)$ $t^2 - 13t + 36$
40. $(w - 3)(w - 5)$ $w^2 - 8w + 15$
41. $(x + 1)(x^2 + 2x + 2)$ $x^3 + 3x^2 + 4x + 2$
42. $(x - 1)(x^2 + x + 1)$ $x^3 - 1$
43. $(3y + 2)(2y^2 - y + 3)$ $6y^3 + y^2 + 7y + 6$
44. $(4y + 3)(y^2 + 3y + 1)$ $4y^3 + 15y^2 + 13y + 3$
45. $(y^2z - 2y^4)(y^2z + 3z^2 - y^4)$
$2y^8 - 3y^6z - 5y^4z^2 + 3y^2z^3$
46. $(m^3 - 4mn^2)(6m^4n^2 - 3m^6 + m^2n^4)$
$18m^7n^2 - 23m^5n^4 - 3m^9 - 4m^3n^6$

Find each product vertically. See Example 4.

47. $2a - 3$ times $a + 5$: $2a^2 + 7a - 15$
48. $2w - 6$ times $w + 5$: $2w^2 + 4w - 30$
49. $7x + 30$ times $2x + 5$: $14x^2 + 95x + 150$
50. $5x + 7$ times $3x + 6$: $15x^2 + 51x + 42$
51. $5x + 2$ times $4x - 3$: $20x^2 - 7x - 6$
52. $4x + 3$ times $2x - 6$: $8x^2 - 18x - 18$
53. $m - 3n$ times $2a + b$: $2am - 6an + mb - 3nb$
54. $3x + 7$ times $a - 2b$: $3ax + 7a - 6xb - 14b$
55. $x^2 + 3x - 2$ times $x + 6$: $x^3 + 9x^2 + 16x - 12$
56. $-x^2 + 3x - 5$ times $x - 7$: $-x^3 + 10x^2 - 26x + 35$
57. $2a^3 - 3a^2 + 4$ times $-2a - 3$: $-4a^4 + 9a^2 - 8a - 12$
58. $-3x^2 + 5x - 2$ times $-5x - 6$: $15x^3 - 7x^2 - 20x + 12$
59. $x - y$ times $x + y$: $x^2 - y^2$
60. $a^2 + b^2$ times $a^2 - b^2$: $a^4 - b^4$
61. $x^2 - xy + y^2$ times $x + y$: $x^3 + y^3$
62. $4w^2 + 2wv + v^2$ times $2w - v$: $8w^3 - v^3$

Find the opposite of each polynomial. See Example 5.

63. $3t - u$ $u - 3t$
64. $-3t - u$ $3t + u$
65. $3x + y$ $-3x - y$
66. $x - 3y$ $3y - x$
67. $-3a^2 - a + 6$ $3a^2 + a - 6$
68. $3b^2 - b - 6$ $-3b^2 + b + 6$
69. $3v^2 + v - 6$ $-3v^2 - v + 6$
70. $-3t^2 + t - 6$ $3t^2 - t + 6$

Perform the indicated operation.

71. $-3x(2x - 9)$ $-6x^2 + 27x$
72. $-1(2 - 3x)$ $3x - 2$
73. $2 - 3x(2x - 9)$ $-6x^2 + 27x + 2$
74. $6 - 3(4x - 8)$ $30 - 12x$
75. $(2 - 3x) + (2x - 9)$ $-x - 7$
76. $(2 - 3x) - (2x - 9)$ $-5x + 11$
77. $(6x^6)^2$ $36x^{12}$
78. $(-3a^3b)^2$ $9a^6b^2$
79. $3ab^3(-2a^2b^7)$ $-6a^3b^{10}$
80. $-4xst \cdot 8xs$ $-32s^2tx^2$
81. $(5x + 6)(5x + 6)$ $25x^2 + 60x + 36$
82. $(5x - 6)(5x - 6)$ $25x^2 - 60x + 36$
83. $(5x - 6)(5x + 6)$ $25x^2 - 36$
84. $(2x - 9)(2x + 9)$ $4x^2 - 81$

85. $2x^2(3x^5 - 4x^2)$
$6x^7 - 8x^4$

86. $4a^3(3ab^3 - 2ab^3)$
$4a^4b^3$

87. $(m - 1)(m^2 + m + 1)$
$m^3 - 1$

88. $(a + b)(a^2 - ab + b^2)$
$a^3 + b^3$

89. $(3x - 2)(x^2 - x - 9)$ $3x^3 - 5x^2 - 25x + 18$

90. $(5 - 6y)(3y^2 - y - 7)$ $-18y^3 + 21y^2 + 37y - 35$

Solve each problem. See Example 6.

91. ***Office space.*** The length of a professor's office is x feet, and the width is $x + 4$ feet. Write a polynomial that represents the area. Find the area if $x = 10$ ft.
$x^2 + 4x$ square feet, 140 square feet

92. ***Swimming space.*** The length of a rectangular swimming pool is $2x - 1$ meters, and the width is $x + 2$ meters. Write a polynomial that represents the area. Find the area if x is 5 meters.
$2x^2 + 3x - 2$ square meters, 63 square meters

93. ***Area.*** A roof truss is in the shape of a triangle with height of x feet and a base of $2x + 1$ feet. Write a polynomial $A(x)$ that represents the area of the triangle. Find $A(5)$.
$A(x) = x^2 + \frac{1}{2}x$, $A(5) = 27.5$ square feet

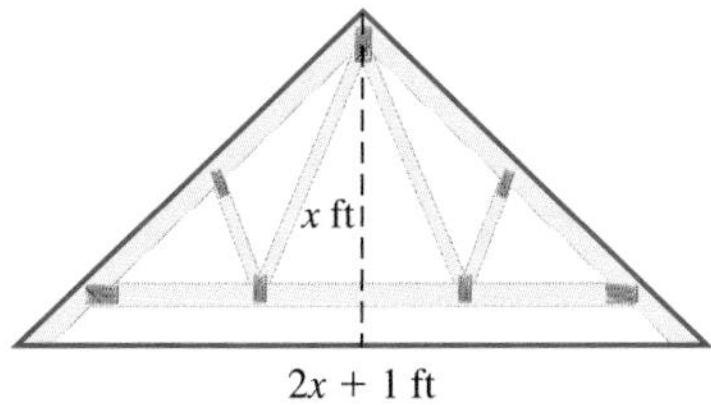

FIGURE FOR EXERCISE 93

94. ***Volume.*** The length, width, and height of a box are x, $2x$, and $3x - 5$ inches, respectively. Write a polynomial $V(x)$ that represents its volume. Find $V(3)$.
$V(x) = 6x^3 - 10x^2$, $V(3) = 72$ cubic inches

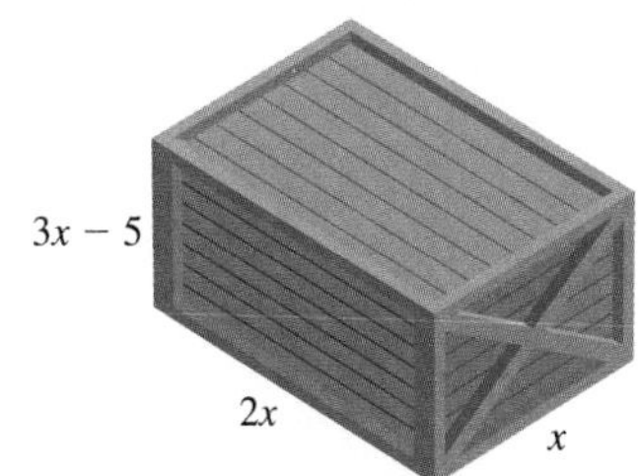

FIGURE FOR EXERCISE 94

95. ***Number pairs.*** If two numbers differ by 5, then what polynomial represents their product? $x^2 + 5x$

96. ***Number pairs.*** If two numbers have a sum of 9, then what polynomial represents their product? $9x - x^2$

97. ***Area of a rectangle.*** The length of a rectangle is $2.3x + 1.2$ meters, and its width is $3.5x + 5.1$ meters. What polynomial represents its area?
$8.05x^2 + 15.93x + 6.12$ square meters

98. ***Patchwork.*** A quilt patch cut in the shape of a triangle has a base of $5x$ inches and a height of $1.732x$ inches. What polynomial represents its area?
$4.33x^2$ square inches

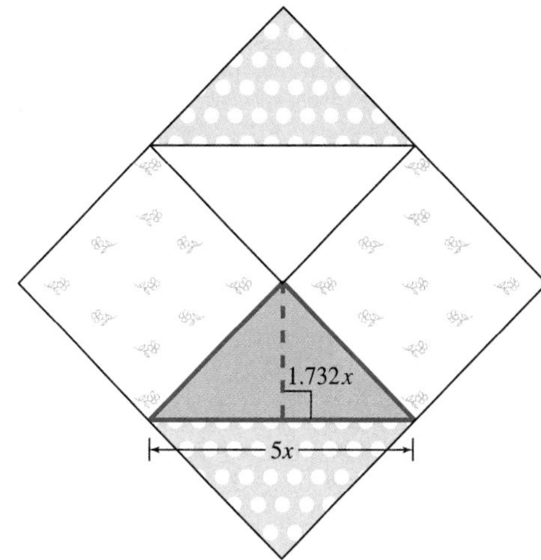

FIGURE FOR EXERCISE 98

99. ***Total revenue.*** If a promoter charges p dollars per ticket for a concert in Tulsa, then she expects to sell $40{,}000 - 1000p$ tickets to the concert. How many tickets will she sell if the tickets are \$10 each? Find the total revenue when the tickets are \$10 each. What polynomial represents the total revenue expected for the concert when the tickets are p dollars each?
30,000, \$300,000, $40{,}000p - 1000p^2$

100. ***Manufacturing shirts.*** If a manufacturer charges p dollars each for rugby shirts, then he expects to sell $2000 - 100p$ shirts per week. What polynomial represents the total revenue expected for a week? How many shirts will be sold if the manufacturer charges \$20 each for the shirts? Find the total revenue when the shirts are sold for \$20 each. Use the bar graph to determine the price that will give the maximum total revenue. $2000p - 100p^2$, 0, \$0, \$10

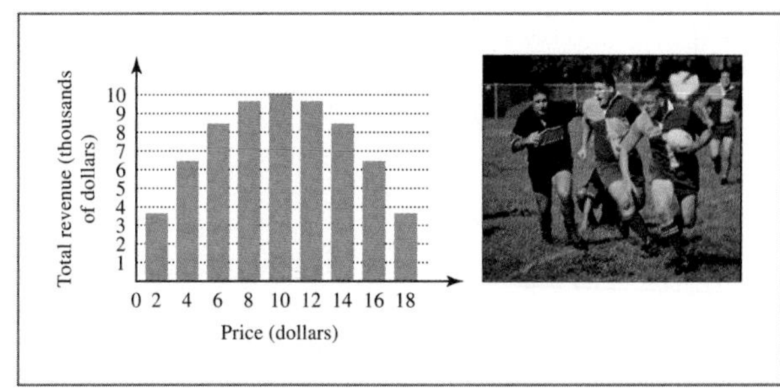

FIGURE FOR EXERCISE 100

101. ***Periodic deposits.*** At the beginning of each year for 5 years, an investor invests \$10 in a mutual fund with

an average annual return of r. If we let $x = 1 + r$, then at the end of the first year (just before the next investment) the value is $10x$ dollars. Because \$10 is then added to the $10x$ dollars, the amount at the end of the second year is $(10x + 10)x$ dollars. Find a polynomial that represents the value of the investment at the end of the fifth year. Evaluate this polynomial if $r = 10\%$.
$10x^5 + 10x^4 + 10x^3 + 10x^2 + 10x$, \$67.16

102. ***Increasing deposits.*** At the beginning of each year for 5 years, an investor invests in a mutual fund with an average annual return of r. The first year, she invests \$10; the second year, she invests \$20; the third year, she invests \$30; the fourth year, she invests \$40; the fifth year, she invests \$50. Let $x = 1 + r$ as in Exercise 101 and write a polynomial in x that represents the value of the investment at the end of the fifth year. Evaluate this polynomial for $r = 8\%$.
$10x^5 + 20x^4 + 30x^3 + 40x^2 + 50x$, \$180.35

GETTING MORE INVOLVED

103. ***Discussion.*** Name all properties of the real numbers that are used in finding the following products:
a) $-2ab^3c^2 \cdot 5a^2bc$ **b)** $(x^2 + 3)(x^2 - 8x - 6)$

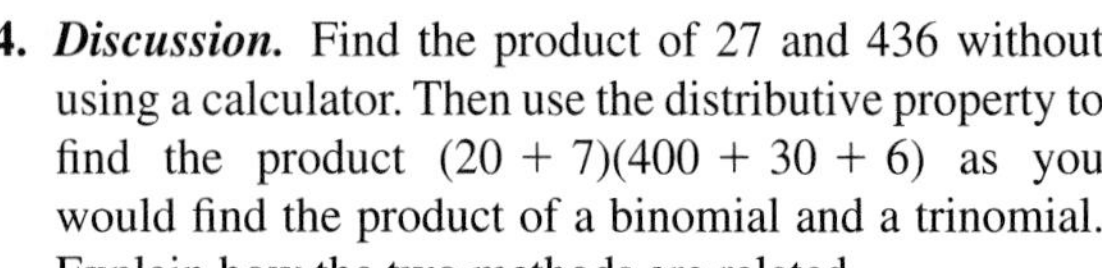

104. ***Discussion.*** Find the product of 27 and 436 without using a calculator. Then use the distributive property to find the product $(20 + 7)(400 + 30 + 6)$ as you would find the product of a binomial and a trinomial. Explain how the two methods are related.

5.3 MULTIPLICATION OF BINOMIALS

In This Section

- The FOIL Method

- Multiplying Binomials Quickly

In Section 5.2 you learned to multiply polynomials. In this section you will learn a rule that makes multiplication of binomials simpler.

The FOIL Method

We can use the distributive property to find the product of two binomials. For example,

$$\begin{aligned}(x + 2)(x + 3) &= (x + 2)x + (x + 2)3 && \text{Distributive property}\\ &= x^2 + 2x + 3x + 6 && \text{Distributive property}\\ &= x^2 + 5x + 6 && \text{Combine like terms.}\end{aligned}$$

There are four terms in $x^2 + 2x + 3x + 6$. The term x^2 is the product of the *first* term of each binomial, x and x. The term $3x$ is the product of the two *outer* terms, 3 and x. The term $2x$ is the product of the two *inner* terms, 2 and x. The term 6 is the product of the last term of each binomial, 2 and 3. We can connect the terms multiplied by lines as follows:

$$(x + 2)(x + 3)$$

F = First terms
O = Outer terms
I = Inner terms
L = Last terms

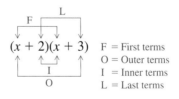

If you remember the word FOIL, you can get the product of the two binomials much faster than writing out all of the steps above. This method is called the **FOIL method.** The name should make it easier to remember.

EXAMPLE 1 **Using the FOIL method**

Find each product.

a) $(x + 2)(x - 4)$ **b)** $(2x + 5)(3x - 4)$

c) $(a - b)(2a - b)$ **d)** $(x + 3)(y + 5)$

Helpful Hint

You may have to practice FOIL a while to get good at it. However, the better you are at FOIL, the easier you will find factoring in Chapter 6.

Solution

a) $(x + 2)(x - 4) = x^2 - 4x + 2x - 8$ (F O I L)
$= x^2 - 2x - 8$ Combine the like terms.

b) $(2x + 5)(3x - 4) = 6x^2 - 8x + 15x - 20$
$= 6x^2 + 7x - 20$ Combine the like terms.

c) $(a - b)(2a - b) = 2a^2 - ab - 2ab + b^2$
$= 2a^2 - 3ab + b^2$

d) $(x + 3)(y + 5) = xy + 5x + 3y + 15$ There are no like terms to combine. ■

FOIL can be used to multiply any two binomials. The binomials in Example 2 have higher powers than those of Example 1.

EXAMPLE 2 **Using the FOIL method**

Find each product.

a) $(x^3 - 3)(x^3 + 6)$ **b)** $(2a^2 + 1)(a^2 + 5)$

Study Tip

Remember that everything we do in solving problems is based on principles (which are also called rules, theorems, and definitions). These principles justify the steps we take. Be sure that you understand the reasons. If you just memorize procedures without understanding, you will soon forget the procedures.

Solution

a) $(x^3 - 3)(x^3 + 6) = x^6 + 6x^3 - 3x^3 - 18$
$= x^6 + 3x^3 - 18$

b) $(2a^2 + 1)(a^2 + 5) = 2a^4 + 10a^2 + a^2 + 5$
$= 2a^4 + 11a^2 + 5$ ■

Multiplying Binomials Quickly

The outer and inner products in the FOIL method are often like terms, and we can combine them without writing them down. Once you become proficient at using FOIL, you can find the product of two binomials without writing anything except the answer.

EXAMPLE 3 **Using FOIL to find a product quickly**

Find each product. Write down only the answer.

a) $(x + 3)(x + 4)$ **b)** $(2x - 1)(x + 5)$ **c)** $(a - 6)(a + 6)$

Solution

a) $(x + 3)(x + 4) = x^2 + 7x + 12$ Combine like terms: $3x + 4x = 7x$.

b) $(2x - 1)(x + 5) = 2x^2 + 9x - 5$ Combine like terms: $10x - x = 9x$.

c) $(a - 6)(a + 6) = a^2 - 36$ Combine like terms: $6a - 6a = 0$. ■

EXAMPLE 4 **Area of a garden**

Sheila has a square garden with sides of length x feet. If she increases the length by 7 feet and decreases the width by 2 feet, then what trinomial represents the area of the new rectangular garden?

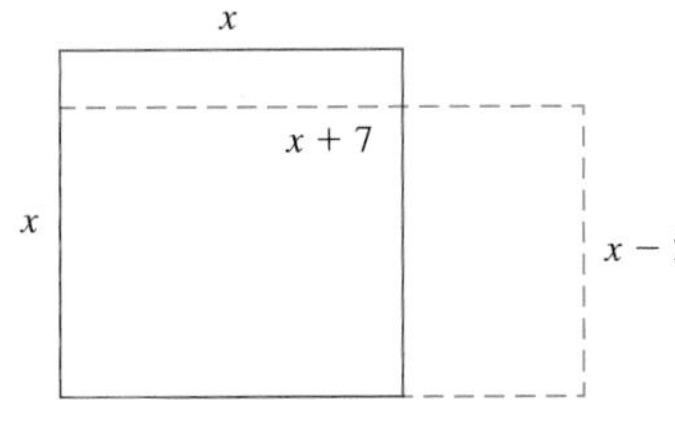

FIGURE 5.2

Solution

The length of the new garden is $x + 7$ and the width is $x - 2$ as shown in Fig. 5.2. The area is $(x + 7)(x - 2)$ or $x^2 + 5x - 14$ square feet. ■

WARM-UPS

True or false? Answer true only if the equation is true for all values of the variable or variables. Explain your answer.

1. $(x + 3)(x + 2) = x^2 + 6$ False
2. $(x + 2)(y + 1) = xy + x + 2y + 2$ True
3. $(3a - 5)(2a + 1) = 6a^2 + 3a - 10a - 5$ True
4. $(y + 3)(y - 2) = y^2 + y - 6$ True
5. $(x^2 + 2)(x^2 + 3) = x^4 + 5x^2 + 6$ True
6. $(3a^2 - 2)(3a^2 + 2) = 9a^2 - 4$ False
7. $(t + 3)(t + 5) = t^2 + 8t + 15$ True
8. $(y - 9)(y - 2) = y^2 - 11y - 18$ False
9. $(x + 4)(x - 7) = x^2 + 4x - 28$ False
10. It is not necessary to learn FOIL as long as you can get the answer. False

5.3 EXERCISES

Reading and Writing *After reading this section, write out the answers to these questions. Use complete sentences.*

1. What property of the real numbers do we usually use to find the product of two binomials?
We use the distributive property to find the product of two binomials.

2. What does FOIL stand for?
FOIL stands for first, outer, inner, and last.

3. What is the purpose of FOIL?
The purpose of FOIL is to provide a faster method for finding the product of two binomials.

4. What is the maximum number of terms that can be obtained when two binomials are multiplied?
The maximum number of terms obtained in multiplying binomials is four.

Use FOIL to find each product. See Example 1.

5. $(x + 2)(x + 4)$ $x^2 + 6x + 8$
6. $(x + 3)(x + 5)$ $x^2 + 8x + 15$
7. $(a - 3)(a + 2)$ $a^2 - a - 6$
8. $(b - 1)(b + 2)$ $b^2 + b - 2$
9. $(2x - 1)(x - 2)$ $2x^2 - 5x + 2$
10. $(2y - 5)(y - 2)$ $2y^2 - 9y + 10$
11. $(2a - 3)(a + 1)$ $2a^2 - a - 3$
12. $(3x - 5)(x + 4)$ $3x^2 + 7x - 20$
13. $(w - 50)(w - 10)$ $w^2 - 60w + 500$
14. $(w - 30)(w - 20)$ $w^2 - 50w + 600$
15. $(y - a)(y + 5)$ $y^2 - ay + 5y - 5a$
16. $(a + t)(3 - y)$ $3a + 3t - ay - ty$
17. $(5 - w)(w + m)$ $5w - w^2 + 5m - mw$
18. $(a - h)(b + t)$ $ab - hb + at - ht$
19. $(2m - 3t)(5m + 3t)$ $10m^2 - 9mt - 9t^2$
20. $(2x - 5y)(x + y)$ $2x^2 - 3xy - 5y^2$
21. $(5a + 2b)(9a + 7b)$ $45a^2 + 53ab + 14b^2$
22. $(11x + 3y)(x + 4y)$ $11x^2 + 47xy + 12y^2$

Use FOIL to find each product. See Example 2.

23. $(x^2 - 5)(x^2 + 2)$ $x^4 - 3x^2 - 10$
24. $(y^2 + 1)(y^2 - 2)$ $y^4 - y^2 - 2$
25. $(h^3 + 5)(h^3 + 5)$ $h^6 + 10h^3 + 25$
26. $(y^6 + 1)(y^6 - 4)$ $y^{12} - 3y^6 - 4$
27. $(3b^3 + 2)(b^3 + 4)$ $3b^6 + 14b^3 + 8$
28. $(5n^4 - 1)(n^4 + 3)$ $5n^8 + 14n^4 - 3$
29. $(y^2 - 3)(y - 2)$ $y^3 - 2y^2 - 3y + 6$
30. $(x - 1)(x^2 - 1)$ $x^3 - x^2 - x + 1$
31. $(3m^3 - n^2)(2m^3 + 3n^2)$ $6m^6 + 7m^3n^2 - 3n^4$
32. $(6y^4 - 2z^2)(6y^4 - 3z^2)$ $36y^8 - 30y^4z^2 + 6z^4$
33. $(3u^2v - 2)(4u^2v + 6)$ $12u^4v^2 + 10u^2v - 12$
34. $(5y^3w^2 + z)(2y^3w^2 + 3z)$ $10y^6w^4 + 17y^3w^2z + 3z^2$

Find each product. Try to write only the answer. See Example 3.

35. $(b + 4)(b + 5)$ $b^2 + 9b + 20$
36. $(y + 8)(y + 4)$ $y^2 + 12y + 32$
37. $(x - 3)(x + 9)$ $x^2 + 6x - 27$
38. $(m + 7)(m - 8)$ $m^2 - m - 56$
39. $(a + 5)(a + 5)$ $a^2 + 10a + 25$

40. $(t - 4)(t - 4)$ $t^2 - 8t + 16$
41. $(2x - 1)(2x - 1)$ $4x^2 - 4x + 1$
42. $(3y + 4)(3y + 4)$ $9y^2 + 24y + 16$
43. $(z - 10)(z + 10)$ $z^2 - 100$
44. $(3h - 5)(3h + 5)$ $9h^2 - 25$
45. $(a + b)(a + b)$ $a^2 + 2ab + b^2$
46. $(x - y)(x - y)$ $x^2 - 2xy + y^2$
47. $(a - 1)(a - 2)$ $a^2 - 3a + 2$
48. $(b - 8)(b - 1)$ $b^2 - 9b + 8$
49. $(2x - 1)(x + 3)$ $2x^2 + 5x - 3$
50. $(3y + 5)(y - 3)$ $3y^2 - 4y - 15$
51. $(5t - 2)(t - 1)$ $5t^2 - 7t + 2$
52. $(2t - 3)(2t - 1)$ $4t^2 - 8t + 3$
53. $(h - 7)(h - 9)$ $h^2 - 16h + 63$
54. $(h - 7w)(h - 7w)$ $h^2 - 14hw + 49w^2$
55. $(h + 7w)(h + 7w)$ $h^2 + 14hw + 49w^2$
56. $(h - 7q)(h + 7q)$ $h^2 - 49q^2$
57. $(2h^2 - 1)(2h^2 - 1)$ $4h^4 - 4h^2 + 1$
58. $(3h^2 + 1)(3h^2 + 1)$ $9h^4 + 6h^2 + 1$

Perform the indicated operations.

59. $\left(2a + \frac{1}{2}\right)\left(4a - \frac{1}{2}\right)$ $8a^2 + a - \frac{1}{4}$
60. $\left(3b + \frac{2}{3}\right)\left(6b - \frac{1}{3}\right)$ $18b^2 + 3b - \frac{2}{9}$
61. $\left(\frac{1}{2}x - \frac{1}{3}\right)\left(\frac{1}{4}x + \frac{1}{2}\right)$ $\frac{1}{8}x^2 + \frac{1}{6}x - \frac{1}{6}$
62. $\left(\frac{2}{3}t - \frac{1}{4}\right)\left(\frac{1}{2}t - \frac{1}{2}\right)$ $\frac{1}{3}t^2 - \frac{11}{24}t + \frac{1}{8}$
63. $-2x^4(3x - 1)(2x + 5)$ $-12x^6 - 26x^5 + 10x^4$
64. $4xy^3(2x - y)(3x + y)$ $24x^3y^3 - 4x^2y^4 - 4xy^5$
65. $(x - 1)(x + 1)(x + 3)$ $x^3 + 3x^2 - x - 3$
66. $(a - 3)(a + 4)(a - 5)$ $a^3 - 4a^2 - 17a + 60$
67. $(3x - 2)(3x + 2)(x + 5)$ $9x^3 + 45x^2 - 4x - 20$
68. $(x - 6)(9x + 4)(9x - 4)$ $81x^3 - 486x^2 - 16x + 96$
69. $(x - 1)(x + 2) - (x + 3)(x - 4)$ $2x + 10$
70. $(k - 4)(k + 9) - (k - 3)(k + 7)$ $k - 15$

Solve each problem.

71. ***Area of a rug.*** Find a trinomial that represents the area of a rectangular rug whose sides are $x + 3$ feet and $2x - 1$ feet. $2x^2 + 5x - 3$ square feet

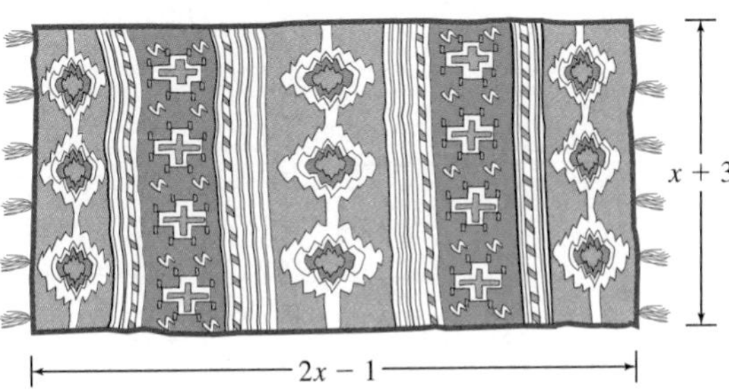

FIGURE FOR EXERCISE 71

72. ***Area of a parallelogram.*** Find a trinomial that represents the area of a parallelogram whose base is $3x + 2$ meters and whose height is $2x + 3$ meters. $6x^2 + 13x + 6$ square meters

73. ***Area of a sail.*** The sail of a tall ship is triangular in shape with a base of $4.57x + 3$ meters and a height of $2.3x - 1.33$ meters. Find a polynomial that represents the area of the triangle. $5.2555x^2 + 0.41095x - 1.995$ square meters

74. ***Area of a square.*** A square has a side of length $1.732x + 1.414$ meters. Find a polynomial that represents its area. $2.9998x^2 + 4.8981x + 1.9994$ square meters

GETTING MORE INVOLVED

75. ***Exploration.*** Find the area of each of the four regions shown in the figure. What is the total area of the four regions? What does this exercise illustrate? 12 ft^2, $3h$ ft^2, $4h$ ft^2, h^2 ft^2, $h^2 + 7h + 12$ ft^2, $(h + 3)(h + 4) = h^2 + 7h + 12$

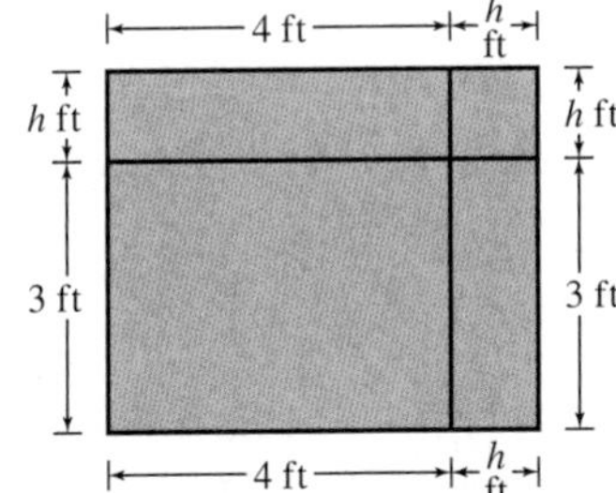

FIGURE FOR EXERCISE 75

76. ***Exploration.*** Find the area of each of the four regions shown in the figure. What is the total area of the four regions? What does this exercise illustrate? a^2, ab, ab, b^2, $a^2 + 2ab + b^2$, $(a + b)(a + b) = a^2 + 2ab + b^2$

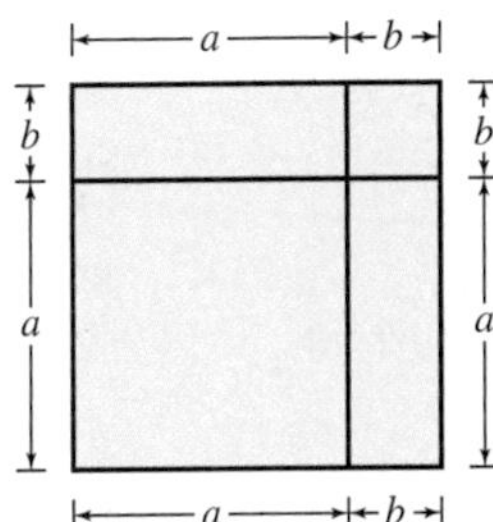

FIGURE FOR EXERCISE 76

5.4 SPECIAL PRODUCTS

In This Section

- The Square of a Binomial
- Product of a Sum and a Difference
- Higher Powers of Binomials
- Applications to Area

In Section 5.3 you learned the FOIL method to make multiplying binomials simpler. In this section you will learn rules for squaring binomials and for finding the product of a sum and a difference. These products are called **special products.**

Helpful Hint

To visualize the square of a sum, draw a square with sides of length $a + b$ as shown.

	a	b
a	a^2	ab
b	ab	b^2

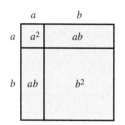

The area of the large square is $(a + b)^2$. It comes from four terms as stated in the rule for the square of a sum.

The Square of a Binomial

To compute $(a + b)^2$, the square of a binomial, we can write it as $(a + b)(a + b)$ and use FOIL:

$$\begin{aligned}(a + b)^2 &= (a + b)(a + b)\\ &= a^2 + ab + ab + b^2\\ &= a^2 + 2ab + b^2\end{aligned}$$

So to square $a + b$, *we square the first term* (a^2), *add twice the product of the two terms* $(2ab)$, *then add the square of the last term* (b^2). The square of a binomial occurs so frequently that it is helpful to learn this new rule to find it. The rule for squaring a sum is given symbolically as follows.

The Square of a Sum

$$(a + b)^2 = a^2 + 2ab + b^2$$

EXAMPLE 1 **Using the rule for squaring a sum**

Find the square of each sum.

a) $(x + 3)^2$ **b)** $(2a + 5)^2$

Helpful Hint

You can mentally find the square of a number that ends in 5. To find 25^2 take the 2 (from 25), multiply it by 3 (the next integer) to get 6, and then annex 25 for 625. To find 35^2, find $3 \cdot 4 = 12$ and annex 25 for 1225. To see why this works, write

$$\begin{aligned}&(30 + 5)^2\\ &= 30^2 + 2 \cdot 30 \cdot 5 + 5^2\\ &= 30(30 + 2 \cdot 5) + 25\\ &= 30 \cdot 40 + 25\\ &= 1200 + 25\\ &= 1225.\end{aligned}$$

Now find 45^2 and 55^2 mentally.

Solution

a) $(x + 3)^2 = x^2 + 2(x)(3) + 3^2 = x^2 + 6x + 9$

(x^2: Square of first; $2(x)(3)$: Twice the product; 3^2: Square of last)

b) $(2a + 5)^2 = (2a)^2 + 2(2a)(5) + 5^2$
$= 4a^2 + 20a + 25$ ■

CAUTION Do not forget the middle term when squaring a sum. The equation $(x + 3)^2 = x^2 + 6x + 9$ is an identity, but $(x + 3)^2 = x^2 + 9$ is not an identity. For example, if $x = 1$ in $(x + 3)^2 = x^2 + 9$, then we get $4^2 = 1^2 + 9$, which is false.

When we use FOIL to find $(a - b)^2$, we see that

$$\begin{aligned}(a - b)^2 &= (a - b)(a - b)\\ &= a^2 - ab - ab + b^2\\ &= a^2 - 2ab + b^2.\end{aligned}$$

So to square $a - b$, *we square the first term* (a^2), *subtract twice the product of the two terms* ($-2ab$), *and add the square of the last term* (b^2). The rule for squaring a difference is given symbolically as follows.

The Square of a Difference

$$(a - b)^2 = a^2 - 2ab + b^2$$

EXAMPLE 2 **Using the rule for squaring a difference**

Find the square of each difference.

a) $(x - 4)^2$

b) $(4b - 5y)^2$

Helpful Hint

Many students keep using FOIL to find the square of a sum or difference. However, learning the new rules for these special cases will pay off in the future.

Solution

a) $(x - 4)^2 = x^2 - 2(x)(4) + 4^2$
$= x^2 - 8x + 16$

b) $(4b - 5y)^2 = (4b)^2 - 2(4b)(5y) + (5y)^2$
$= 16b^2 - 40by + 25y^2$ ■

Product of a Sum and a Difference

If we multiply the sum $a + b$ and the difference $a - b$ by using FOIL, we get

$$\begin{aligned}(a + b)(a - b) &= a^2 - ab + ab - b^2\\ &= a^2 - b^2.\end{aligned}$$

The inner and outer products have a sum of 0. So *the product of a sum and a difference of the same two terms is equal to the difference of two squares.*

The Product of a Sum and a Difference

$$(a + b)(a - b) = a^2 - b^2$$

EXAMPLE 3 **Product of a sum and a difference**

Find each product.

a) $(x + 2)(x - 2)$ **b)** $(b + 7)(b - 7)$ **c)** $(3x - 5)(3x + 5)$

Helpful Hint

You can use
$(a + b)(a - b) = a^2 - b^2$
to perform mental arithmetic tricks like
$19 \cdot 21 = (20 - 1)(20 + 1)$
$= 400 - 1$
$= 399$
What is $29 \cdot 31$? $28 \cdot 32$?

Solution

a) $(x + 2)(x - 2) = x^2 - 4$

b) $(b + 7)(b - 7) = b^2 - 49$

c) $(3x - 5)(3x + 5) = 9x^2 - 25$ ■

Higher Powers of Binomials

To find a power of a binomial that is higher than 2, we can use the rule for squaring a binomial along with the method of multiplying binomials using the distributive property. Finding the second or higher power of a binomial is called **expanding the binomial** because the result has more terms than the original.

EXAMPLE 4

Higher powers of a binomial

Expand each binomial.

a) $(x + 4)^3$ **b)** $(y - 2)^4$

Study Tip

Correct answers often have more than one form. If your answer to an exercise doesn't agree with the one in the back of this text, try to determine if it is simply a different form of the answer. For example, $\frac{1}{2}x$ and $\frac{x}{2}$ look different but they are equivalent expressions.

Solution

a) $(x + 4)^3 = (x + 4)^2(x + 4)$
$= (x^2 + 8x + 16)(x + 4)$
$= (x^2 + 8x + 16)x + (x^2 + 8x + 16)4$
$= x^3 + 8x^2 + 16x + 4x^2 + 32x + 64$
$= x^3 + 12x^2 + 48x + 64$

b) $(y - 2)^4 = (y - 2)^2(y - 2)^2$
$= (y^2 - 4y + 4)(y^2 - 4y + 4)$
$= (y^2 - 4y + 4)(y^2) + (y^2 - 4y + 4)(-4y) + (y^2 - 4y + 4)(4)$
$= y^4 - 4y^3 + 4y^2 - 4y^3 + 16y^2 - 16y + 4y^2 - 16y + 16$
$= y^4 - 8y^3 + 24y^2 - 32y + 16$ ■

Applications to Area

EXAMPLE 5

Area of a pizza

A pizza parlor saves money by making all of its round pizzas one inch smaller in radius than advertised. Write a trinomial for the actual area of a pizza with an advertised radius of r inches.

Solution

A pizza advertised as r inches has an actual radius of $r - 1$ inches. The actual area is $\pi(r - 1)^2$:

$$\pi(r - 1)^2 = \pi(r^2 - 2r + 1) = \pi r^2 - 2\pi r + \pi.$$

So $\pi r^2 - 2\pi r + \pi$ is a trinomial representing the actual area. ■

WARM-UPS

True or false? Explain your answer.

1. $(2 + 3)^2 = 2^2 + 3^2$ False
2. $(x + 3)^2 = x^2 + 6x + 9$ for any value of x. True
3. $(3 + 5)^2 = 9 + 30 + 25$ True
4. $(2x + 7)^2 = 4x^2 + 28x + 49$ for any value of x. True
5. $(y + 8)^2 = y^2 + 64$ for any value of y. False
6. The product of a sum and a difference of the same two terms is equal to the difference of two squares. True
7. $(40 - 1)(40 + 1) = 1599$ True
8. $49 \cdot 51 = 2499$ True
9. $(x - 3)^2 = x^2 - 3x + 9$ for any value of x. False
10. The square of a sum is equal to a sum of two squares. False

5.4 EXERCISES

Reading and Writing *After reading this section, write out the answers to these questions. Use complete sentences.*

1. What are the special products?
   The special products are $(a + b)^2$, $(a - b)^2$, and $(a + b)(a - b)$.
2. What is the rule for squaring a sum?
   $(a + b)^2 = a^2 + 2ab + b^2$
3. Why do we need a new rule to find the square of a sum when we already have FOIL?
   It is faster to do by the new rule than with FOIL.
4. What happens to the inner and outer products in the product of a sum and a difference?
   In $(a + b)(a - b)$ the inner and outer products have a sum of zero.
5. What is the rule for finding the product of a sum and a difference?
   $(a + b)(a - b) = a^2 - b^2$
6. How can you find higher powers of binomials?
   Higher powers of binomials are found by using the distributive property.

Square each binomial. See Example 1.

7. $(x + 1)^2$
   $x^2 + 2x + 1$
8. $(y + 2)^2$
   $y^2 + 4y + 4$
9. $(y + 4)^2$
   $y^2 + 8y + 16$
10. $(z + 3)^2$
    $z^2 + 6z + 9$
11. $(3x + 8)^2$
    $9x^2 + 48x + 64$
12. $(2m + 7)^2$
    $4m^2 + 28m + 49$
13. $(s + t)^2$
    $s^2 + 2st + t^2$
14. $(x + z)^2$
    $x^2 + 2xz + z^2$
15. $(2x + y)^2$
    $4x^2 + 4xy + y^2$
16. $(3t + v)^2$
    $9t^2 + 6tv + v^2$
17. $(2t + 3h)^2$
    $4t^2 + 12ht + 9h^2$
18. $(3z + 5k)^2$
    $9z^2 + 30kz + 25k^2$

Square each binomial. See Example 2.

19. $(a - 3)^2$
    $a^2 - 6a + 9$
20. $(w - 4)^2$
    $w^2 - 8w + 16$
21. $(t - 1)^2$
    $t^2 - 2t + 1$
22. $(t - 6)^2$
    $t^2 - 12t + 36$
23. $(3t - 2)^2$
    $9t^2 - 12t + 4$
24. $(5a - 6)^2$
    $25a^2 - 60a + 36$
25. $(s - t)^2$
    $s^2 - 2st + t^2$
26. $(r - w)^2$
    $r^2 - 2rw + w^2$
27. $(3a - b)^2$
    $9a^2 - 6ab + b^2$
28. $(4w - 7)^2$
    $16w^2 - 56w + 49$
29. $(3z - 5y)^2$
    $9z^2 - 30yz + 25y^2$
30. $(2z - 3w)^2$
    $4z^2 - 12wz + 9w^2$

Find each product. See Example 3.

31. $(a - 5)(a + 5)$
    $a^2 - 25$
32. $(x - 6)(x + 6)$
    $x^2 - 36$
33. $(y - 1)(y + 1)$
    $y^2 - 1$
34. $(p + 2)(p - 2)$
    $p^2 - 4$
35. $(3x - 8)(3x + 8)$
    $9x^2 - 64$
36. $(6x + 1)(6x - 1)$
    $36x^2 - 1$
37. $(r + s)(r - s)$
    $r^2 - s^2$
38. $(b - y)(b + y)$
    $b^2 - y^2$
39. $(8y - 3a)(8y + 3a)$
    $64y^2 - 9a^2$
40. $(4u - 9v)(4u + 9v)$
    $16u^2 - 81v^2$
41. $(5x^2 - 2)(5x^2 + 2)$
    $25x^4 - 4$
42. $(3y^2 + 1)(3y^2 - 1)$
    $9y^4 - 1$

Expand each binomial. See Example 4.

43. $(x + 1)^3$ $x^3 + 3x^2 + 3x + 1$
44. $(y - 1)^3$ $y^3 - 3y^2 + 3y - 1$
45. $(2a - 3)^3$ $8a^3 - 36a^2 + 54a - 27$
46. $(3w - 1)^3$ $27w^3 - 27w^2 + 9w - 1$
47. $(a - 3)^4$ $a^4 - 12a^3 + 54a^2 - 108a + 81$
48. $(2b + 1)^4$ $16b^4 + 32b^3 + 24b^2 + 8b + 1$
49. $(a + b)^4$ $a^4 + 4a^3b + 6a^2b^2 + 4ab^3 + b^4$
50. $(2a - 3b)^4$ $16a^4 - 96a^3b + 216a^2b^2 - 216ab^3 + 81b^4$

Find each product.

51. $(a - 20)(a + 20)$
    $a^2 - 400$
52. $(1 - x)(1 + x)$
    $1 - x^2$
53. $(x + 8)(x + 7)$
    $x^2 + 15x + 56$
54. $(x - 9)(x + 5)$
    $x^2 - 4x - 45$
55. $(4x - 1)(4x + 1)$
    $16x^2 - 1$
56. $(9y - 1)(9y + 1)$
    $81y^2 - 1$
57. $(9y - 1)^2$
    $81y^2 - 18y + 1$
58. $(4x - 1)^2$
    $16x^2 - 8x + 1$
59. $(2t - 5)(3t + 4)$
    $6t^2 - 7t - 20$
60. $(2t + 5)(3t - 4)$
    $6t^2 + 7t - 20$
61. $(2t - 5)^2$
    $4t^2 - 20t + 25$
62. $(2t + 5)^2$
    $4t^2 + 20t + 25$
63. $(2t + 5)(2t - 5)$
    $4t^2 - 25$
64. $(3t - 4)(3t + 4)$
    $9t^2 - 16$
65. $(x^2 - 1)(x^2 + 1)$
    $x^4 - 1$
66. $(y^3 - 1)(y^3 + 1)$
    $y^6 - 1$
67. $(2y^3 - 9)^2$
    $4y^6 - 36y^3 + 81$
68. $(3z^4 - 8)^2$
    $9z^8 - 48z^4 + 64$
69. $(2x^3 + 3y^2)^2$
    $4x^6 + 12x^3y^2 + 9y^4$
70. $(4y^5 + 2w^3)^2$
    $16y^{10} + 16y^5w^3 + 4w^6$
71. $\left(\frac{1}{2}x + \frac{1}{3}\right)^2$
    $\frac{1}{4}x^2 + \frac{1}{3}x + \frac{1}{9}$
72. $\left(\frac{2}{3}y - \frac{1}{2}\right)^2$
    $\frac{4}{9}y^2 - \frac{2}{3}y + \frac{1}{4}$
73. $(0.2x - 0.1)^2$ $0.04x^2 - 0.04x + 0.01$
74. $(0.1y + 0.5)^2$ $0.01y^2 + 0.1y + 0.25$

75. $(a + b)^3$ $a^3 + 3a^2b + 3ab^2 + b^3$

76. $(2a - 3b)^3$ $8a^3 - 36a^2b + 54ab^2 - 27b^3$

77. $(1.5x + 3.8)^2$
$2.25x^2 + 11.4x + 14.44$

78. $(3.45a - 2.3)^2$
$11.9025a^2 - 15.87a + 5.29$

79. $(3.5t - 2.5)(3.5t + 2.5)$
$12.25t^2 - 6.25$

80. $(4.5h + 5.7)(4.5h - 5.7)$
$20.25h^2 - 32.49$

In Exercises 81–90, solve each problem.

81. ***Shrinking garden.*** Rose's garden is a square with sides of length x feet. Next spring she plans to make it rectangular by lengthening one side 5 feet and shortening the other side by 5 feet. What polynomial represents the new area? By how much will the area of the new garden differ from that of the old garden?
$x^2 - 25$ square feet, 25 square feet smaller

82. ***Square lot.*** Sam lives on a lot that he thought was a square, 157 feet by 157 feet. When he had it surveyed, he discovered that one side was actually 2 feet longer than he thought and the other was actually 2 feet shorter than he thought. How much less area does he have than he thought he had? 4 square feet

83. ***Area of a circle.*** Find a polynomial that represents the area of a circle whose radius is $b + 1$ meters. Use the value 3.14 for π.
$3.14b^2 + 6.28b + 3.14$ square meters

84. ***Comparing dart boards.*** A toy store sells two sizes of circular dartboards. The larger of the two has a radius that is 3 inches greater than that of the other. The radius of the smaller dartboard is t inches. Find a polynomial that represents the difference in area between the two dartboards. $6\pi t + 9\pi$ square centimeters

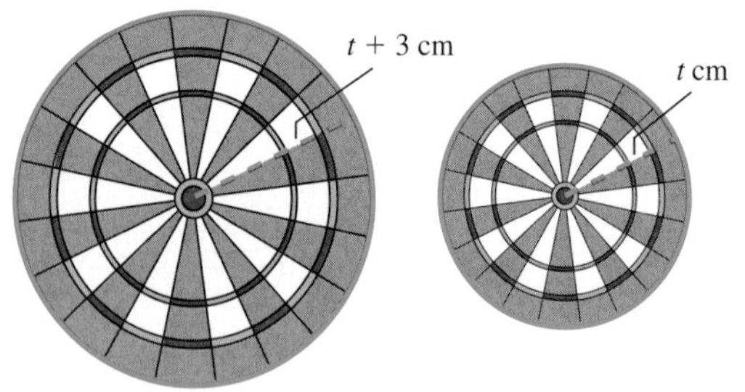

FIGURE FOR EXERCISE 84

85. ***Poiseuille's law.*** According to the nineteenth-century physician Poiseuille, the velocity (in centimeters per second) of blood r centimeters from the center of an artery of radius R centimeters is given by

$$v = k(R - r)(R + r),$$

where k is a constant. Rewrite the formula using a special product rule. $v = k(R^2 - r^2)$

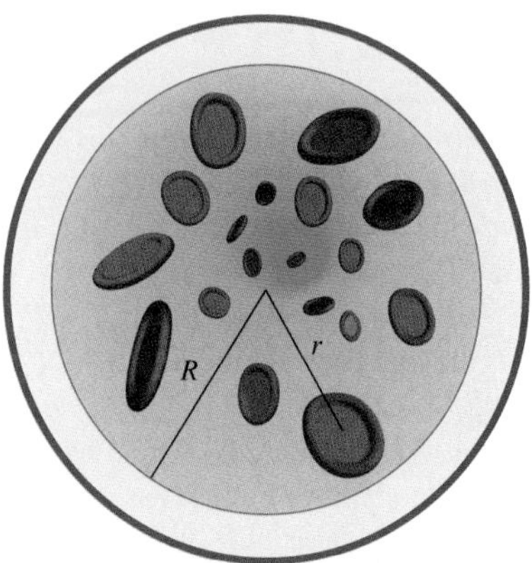

FIGURE FOR EXERCISE 85

86. ***Going in circles.*** A promoter is planning a circular race track with an inside radius of r feet and a width of w feet. The cost in dollars for paving the track is given by the formula

$$C = 1.2\pi[(r + w)^2 - r^2].$$

Use a special product rule to simplify this formula. What is the cost of paving the track if the inside radius is 1000 feet and the width of the track is 40 feet?
$C = 1.2\pi[2rw + w^2]$, \$307,624.75

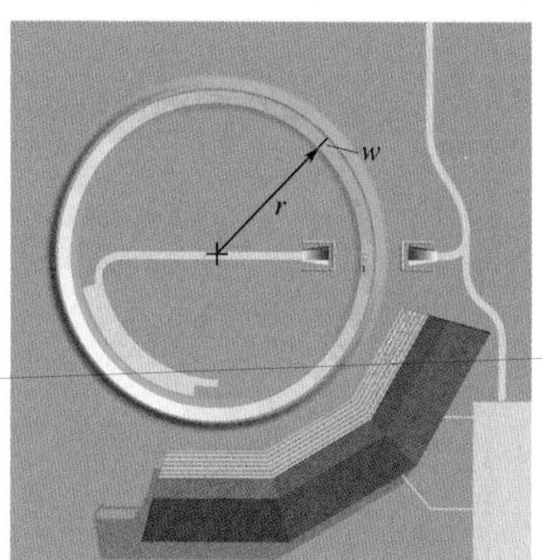

FIGURE FOR EXERCISE 86

87. ***Compounded annually.*** P dollars is invested at annual interest rate r for 2 years. If the interest is compounded annually, then the polynomial $P(1 + r)^2$ represents the value of the investment after 2 years. Rewrite this expression without parentheses. Evaluate the polynomial if $P = \$200$ and $r = 10\%$.
$P + 2Pr + Pr^2$, \$242

88. ***Compounded semiannually.*** P dollars is invested at annual interest rate r for 1 year. If the interest is compounded semiannually, then the polynomial $P\left(1 + \frac{r}{2}\right)^2$ represents the value of the investment after 1 year. Rewrite this expression without parentheses. Evaluate the polynomial if $P = \$200$ and $r = 10\%$.
$P + Pr + \frac{Pr^2}{4}$, \$220.50

89. ***Investing in treasury bills.*** An investment advisor uses the polynomial $P(1 + r)^{10}$ to predict the value in 10 years of a client's investment of P dollars with an average annual return r. The accompanying graph shows historic average annual returns for the last 20 years for various asset classes (T. Rowe Price, www.troweprice.com). Use the historical average return to predict the value in 10 years of an investment of \$10,000 in U.S. treasury bills? \$20,230.06

90. ***Comparing investments.*** How much more would the investment in Exercise 89 be worth in 10 years if the client invests in large company stocks rather than U.S. treasury bills? \$26,619.83

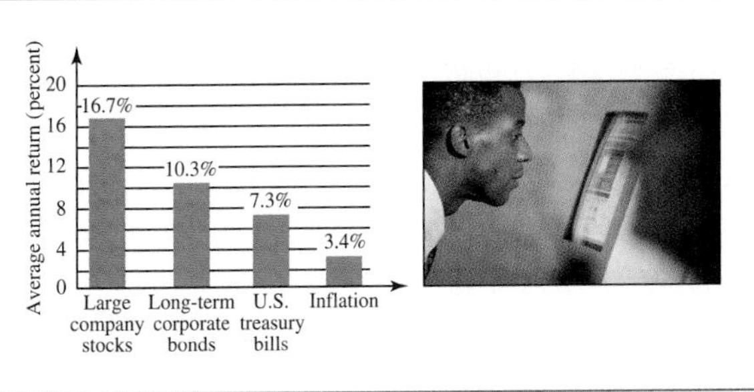

FIGURE FOR EXERCISES 89 AND 90

GETTING MORE INVOLVED

91. ***Writing.*** What is the difference between the equations $(x + 5)^2 = x^2 + 10x + 25$ and $(x + 5)^2 = x^2 + 25$?
The first is an identity and the second is a conditional equation.

92. ***Writing.*** Is it possible to square a sum or a difference without using the rules presented in this section? Why should you learn the rules given in this section?
A sum or difference can be squared with the distributive property, FOIL, or the special product rules. It is easier with the special product rules.

5.5 DIVISION OF POLYNOMIALS

In This Section

- Dividing Monomials Using the Quotient Rule
- Dividing a Polynomial by a Monomial
- Dividing a Polynomial by a Binomial

You multiplied polynomials in Section 5.2. In this section you will learn to divide polynomials.

Dividing Monomials Using the Quotient Rule

In Chapter 1 we used the definition of division to divide signed numbers. Because the definition of division applies to any division, we restate it here.

Division of Real Numbers

If a, b, and c are any numbers with $b \neq 0$, then

$$a \div b = c \qquad \text{provided that} \qquad c \cdot b = a.$$

Study Tip

Establish a regular routine of eating, sleeping, and exercise. The ability to concentrate depends on adequate sleep, decent nutrition, and the physical well-being that comes with exercise.

If $a \div b = c$, we call a the **dividend,** b the **divisor,** and c (or $a \div b$) the **quotient.**

You can find the quotient of two monomials by writing the quotient as a fraction and then reducing the fraction. For example,

$$x^5 \div x^2 = \frac{x^5}{x^2} = \frac{x \cdot x \cdot x \cdot \cancel{x} \cdot \cancel{x}}{\cancel{x} \cdot \cancel{x}} = x^3.$$

You can be sure that x^3 is correct by checking that $x^3 \cdot x^2 = x^5$. You can also divide x^2 by x^5, but the result is not a monomial:

$$x^2 \div x^5 = \frac{x^2}{x^5} = \frac{1 \cdot \cancel{x} \cdot \cancel{x}}{x \cdot x \cdot x \cdot \cancel{x} \cdot \cancel{x}} = \frac{1}{x^3}$$

Note that the exponent 3 can be obtained in either case by subtracting 5 and 2. These examples illustrate the quotient rule for exponents.

Quotient Rule

Suppose $a \neq 0$, and m and n are positive integers.

$$\text{If } m \geq n, \text{ then } \frac{a^m}{a^n} = a^{m-n}.$$

$$\text{If } n > m, \text{ then } \frac{a^m}{a^n} = \frac{1}{a^{n-m}}.$$

Note that if you use the quotient rule to subtract the exponents in $x^4 \div x^4$, you get the expression x^{4-4}, or x^0, which has not been defined yet. Because we must have $x^4 \div x^4 = 1$ if $x \neq 0$, we define the zero power of a nonzero real number to be 1. We do not define the expression 0^0.

Zero Exponent

For any nonzero real number a,

$$a^0 = 1.$$

EXAMPLE 1 **Using the definition of zero exponent**

Simplify each expression. Assume that all variables are nonzero real numbers.

a) 5^0 b) $(3xy)^0$ c) $a^0 + b^0$

Solution

a) $5^0 = 1$ b) $(3xy)^0 = 1$ c) $a^0 + b^0 = 1 + 1 = 2$ ■

With the definition of zero exponent the quotient rule is valid for all positive integers as stated.

EXAMPLE 2 **Using the quotient rule in dividing monomials**

Find each quotient.

a) $\dfrac{y^9}{y^5}$ b) $\dfrac{12b^2}{3b^7}$

c) $-6x^3 \div (2x^9)$ d) $\dfrac{x^8y^2}{x^2y^2}$

Solution

a) $\dfrac{y^9}{y^5} = y^{9-5} = y^4$

Use the definition of division to check that $y^4 \cdot y^5 = y^9$.

b) $\dfrac{12b^2}{3b^7} = \dfrac{12}{3} \cdot \dfrac{b^2}{b^7} = 4 \cdot \dfrac{1}{b^{7-2}} = \dfrac{4}{b^5}$

Use the definition of division to check that

$$\frac{4}{b^5} \cdot 3b^7 = \frac{12b^7}{b^5} = 12b^2.$$

c) $-6x^3 \div (2x^9) = \dfrac{-6x^3}{2x^9} = \dfrac{-3}{x^6}$

Use the definition of division to check that

$$\frac{-3}{x^6} \cdot 2x^9 = \frac{-6x^9}{x^6} = -6x^3.$$

d) $\dfrac{x^8y^2}{x^2y^2} = \dfrac{x^8}{x^2} \cdot \dfrac{y^2}{y^2} = x^6 \cdot y^0 = x^6$

Use the definition of division to check that $x^6 \cdot x^2y^2 = x^8y^2$. ■

Note that the parentheses in Example 2(c) are important. According to the order of operations, multiplication and division are performed from left to right in the absence of parentheses. So without parentheses, $-6x^3 \div 2x^9 = \frac{-6x^3}{2} \cdot x^9$.

Dividing a Polynomial by a Monomial

We divided some simple polynomials by monomials in Chapter 1. For example,

$$\frac{6x + 8}{2} = \frac{1}{2}(6x + 8) = 3x + 4.$$

We use the distributive property to take one-half of $6x$ and one-half of 8 to get $3x + 4$. So both $6x$ and 8 are divided by 2. To divide any polynomial by a monomial, we divide each term of the polynomial by the monomial.

EXAMPLE 3 **Dividing a polynomial by a monomial**

Find the quotient for $(-8x^6 + 12x^4 - 4x^2) \div (4x^2)$.

Solution

$$\frac{-8x^6 + 12x^4 - 4x^2}{4x^2} = \frac{-8x^6}{4x^2} + \frac{12x^4}{4x^2} - \frac{4x^2}{4x^2}$$
$$= -2x^4 + 3x^2 - 1$$

The quotient is $-2x^4 + 3x^2 - 1$. We can check by multiplying.

$$4x^2(-2x^4 + 3x^2 - 1) = -8x^6 + 12x^4 - 4x^2.$$ ■

Study Tip

Play offensive math, not defensive math. A student who says, "Give me a question and I'll see if I can answer it," is playing defensive math. The student is taking a passive approach to learning. A student who takes an active approach and knows the usual questions and answers for each topic is playing offensive math.

Because division by zero is undefined, we will always assume that the divisor is nonzero in any quotient involving variables. For example, the division in Example 3 is valid only if $4x^2 \neq 0$, or $x \neq 0$.

Dividing a Polynomial by a Binomial

Division of whole numbers is often done with a procedure called **long division.** For example, 253 is divided by 7 as follows:

```
              36    ← Quotient
Divisor →  7)253    ← Dividend
             21
              43
              42
               1    ← Remainder
```

Note that $36 \cdot 7 + 1 = 253$. It is always true that

$$(\text{quotient})(\text{divisor}) + (\text{remainder}) = \text{dividend}.$$

To divide a polynomial by a binomial, we perform the division like long division of whole numbers. For example, to divide $x^2 - 3x - 10$ by $x + 2$, we get the first term of the quotient by dividing the first term of $x + 2$ into the first term of $x^2 - 3x - 10$. So divide x^2 by x to get x, then multiply and subtract as follows:

$$\begin{array}{lll}
1 \text{ Divide:} & x & x^2 \div x = x \\
2 \text{ Multiply:} & x + 2\overline{)x^2 - 3x - 10} & \\
 & \underline{x^2 + 2x} & x \cdot (x + 2) = x^2 + 2x \\
3 \text{ Subtract:} & -5x & -3x - 2x = -5x
\end{array}$$

Now bring down -10 and continue the process. We get the second term of the quotient (below) by dividing the first term of $x + 2$ into the first term of $-5x - 10$. So divide $-5x$ by x to get -5:

$$\begin{array}{lll}
1 \text{ Divide:} & x - 5 & -5x \div x = -5 \\
2 \text{ Multiply:} & x + 2\overline{)x^2 - 3x - 10} & \\
 & \underline{x^2 + 2x} \quad \downarrow & \text{Bring down } -10. \\
 & -5x - 10 & \\
 & \underline{-5x - 10} & -5(x + 2) = -5x - 10 \\
3 \text{ Subtract:} & 0 & -10 - (-10) = 0
\end{array}$$

So the quotient is $x - 5$, and the remainder is 0.

In Example 4 there is a term missing in the dividend. To account for the missing term we insert a term with a zero coefficient.

EXAMPLE 4 **Dividing a polynomial by a binomial**

Determine the quotient and remainder when $x^3 - 5x - 1$ is divided by $x - 4$.

Solution

Because the x^2-term in the dividend $x^3 - 5x - 1$ is missing, we write $0 \cdot x^2$ for it:

$$\begin{array}{ll}
x^2 + 4x + 11 & x^3 \div x = x^2 \\
x - 4\overline{)x^3 + 0 \cdot x^2 - 5x - 1} & \\
\underline{x^3 - 4x^2} & x^2(x - 4) = x^3 - 4x^2 \\
4x^2 - 5x & 0 \cdot x^2 - (-4x^2) = 4x^2 \\
\underline{4x^2 - 16x} & 4x(x - 4) = 4x^2 - 16x \\
11x - 1 & -5x - (-16x) = 11x \\
\underline{11x - 44} & 11(x - 4) = 11x - 44 \\
43 & -1 - (-44) = 43
\end{array}$$

The quotient is $x^2 + 4x + 11$ and the remainder is 43. ■

In Example 5 the terms of the dividend are not in order of decreasing exponents and there is a missing term.

EXAMPLE 5 **Dividing a polynomial by a binomial**

Divide $2x^3 - 4 - 7x^2$ by $2x - 3$, and identify the quotient and the remainder.

Power of a Quotient Rule

If a and b are real numbers, $b \neq 0$, and n is a positive integer, then

$$\left(\frac{a}{b}\right)^n = \frac{a^n}{b^n}.$$

EXAMPLE 4 **Using the power of a quotient rule**

Simplify. Assume that the variables are nonzero.

a) $\left(\frac{2}{5x^3}\right)^2$ b) $\left(\frac{3x^4}{2y^3}\right)^3$ c) $\left(\frac{-12a^5b}{4a^2b^7}\right)^3$

Solution

a) $\left(\frac{2}{5x^3}\right)^2 = \frac{2^2}{(5x^3)^2}$ Power of a quotient rule

$= \frac{4}{25x^6}$ $(5x^3)^2 = 5^2(x^3)^2 = 25x^6$

b) $\left(\frac{3x^4}{2y^3}\right)^3 = \frac{3^3x^{12}}{2^3y^9}$ Power of a quotient and power of a product rules

$= \frac{27x^{12}}{8y^9}$ Simplify.

c) Use the quotient rule to simplify the expression inside the parentheses before using the power of a quotient rule.

$\left(\frac{-12a^5b}{4a^2b^7}\right)^3 = \left(\frac{-3a^3}{b^6}\right)^3$ Use the quotient rule first.

$= \frac{-27a^9}{b^{18}}$ Power of a quotient rule ■

Summary of Rules

The rules for exponents are summarized in the following box.

Helpful Hint

Note that the rules of exponents show how exponents behave with respect to multiplication and division only. We studied the more complicated problem of using exponents with addition and subtraction in Section 5.4 when we learned rules for $(a + b)^2$ and $(a - b)^2$.

Rules for Nonnegative Integral Exponents

The following rules hold for nonzero real numbers a and b and nonnegative integers m and n.

1. $a^0 = 1$ Definition of zero exponent
2. $a^m \cdot a^n = a^{m+n}$ Product rule
3. $\frac{a^m}{a^n} = a^{m-n}$ for $m \geq n$,

   $\frac{a^m}{a^n} = \frac{1}{a^{n-m}}$ for $n > m$ Quotient rule
4. $(a^m)^n = a^{mn}$ Power rule
5. $(ab)^n = a^n \cdot b^n$ Power of a product rule
6. $\left(\frac{a}{b}\right)^n = \frac{a^n}{b^n}$ Power of a quotient rule

WARM-UPS

True or false? Assume that all variables represent nonzero real numbers. A statement involving variables is to be marked true only if it is an identity. Explain your answer.

1. $-3^0 = 1$ False **2.** $2^5 \cdot 2^8 = 4^{13}$ False **3.** $2^3 \cdot 3^2 = 6^5$ False
4. $(2x)^4 = 2x^4$ False **5.** $(q^3)^5 = q^8$ False **6.** $(-3x^2)^3 = 27x^6$ False
7. $(ab^3)^4 = a^4b^{12}$ True **8.** $\dfrac{a^{12}}{a^4} = a^3$ False **9.** $\dfrac{6w^4}{3w^9} = 2w^5$ False
10. $\left(\dfrac{2y^3}{9}\right)^2 = \dfrac{4y^6}{81}$ True

5.6 EXERCISES

Reading and Writing *After reading this section, write out the answers to these questions. Use complete sentences.*

1. What is the product rule for exponents?
The product rule says that $a^m a^n = a^{m+n}$.

2. What is the quotient rule for exponents?
The quotient rule says that $a^m/a^n = a^{m-n}$ if $m \geq n$ and $a^m/a^n = \dfrac{1}{a^{n-m}}$ if $n > m$.

3. Why must the bases be the same in these rules?
These rules do not make sense without identical bases.

4. What is the power rule for exponents?
The power rule for exponents says that $(a^m)^n = a^{mn}$.

5. What is the power of a product rule?
The power of a product rule says that $(ab)^n = a^n b^n$.

6. What is the power of a quotient rule?
The power of a quotient rule says that $\left(\dfrac{a}{b}\right)^n = \dfrac{a^n}{b^n}$.

For all exercises in this section, assume that the variables represent nonzero real numbers.

Simplify the exponential expressions. See Example 1.

7. $2^2 \cdot 2^5$ 128
8. $x^6 \cdot x^7$ x^{13}
9. $(-3u^8)(-2u^2)$ $6u^{10}$
10. $(3r^4)(-6r^2)$ $-18r^6$
11. $a^3b^4 \cdot ab^6(ab)^0$ a^4b^{10}
12. $x^2y \cdot x^3y^6(x+y)^0$ x^5y^7
13. $\dfrac{-2a^3}{4a^7}$ $\dfrac{-1}{2a^4}$
14. $\dfrac{-3t^9}{6t^{18}}$ $\dfrac{-1}{2t^9}$
15. $\dfrac{2a^5b \cdot 3a^7b^3}{15a^6b^8}$ $\dfrac{2a^6}{5b^4}$
16. $\dfrac{3xy^8 \cdot 5xy^9}{20x^3y^{14}}$ $\dfrac{3y^3}{4x}$
17. $2^3 \cdot 5^2$ 200
18. $2^2 \cdot 10^3$ 4000

Simplify. See Example 2.

19. $(x^2)^3$ x^6
20. $(y^2)^4$ y^8
21. $2x^2 \cdot (x^2)^5$ $2x^{12}$
22. $(y^2)^6 \cdot 3y^5$ $3y^{17}$
23. $\dfrac{(t^2)^5}{(t^3)^4}$ $\dfrac{1}{t^2}$
24. $\dfrac{(r^4)^2}{(r^5)^3}$ $\dfrac{1}{r^7}$
25. $\dfrac{3x(x^5)^2}{6x^3(x^2)^4}$ $\dfrac{1}{2}$
26. $\dfrac{5y^3(y^5)^2}{10y^5(y^2)^6}$ $\dfrac{1}{2y^4}$

Simplify. See Example 3.

27. $(xy^2)^3$ x^3y^6
28. $(wy^2)^6$ w^6y^{12}
29. $(-2t^5)^3$ $-8t^{15}$
30. $(-3r^3)^3$ $-27r^9$
31. $(-2x^2y^5)^3$ $-8x^6y^{15}$
32. $(-3y^2z^3)^3$ $-27y^6z^9$
33. $\dfrac{(a^4b^2c^5)^3}{a^3b^4c}$ $a^9b^2c^{14}$
34. $\dfrac{(2ab^2c^3)^5}{(2a^3bc)^4}$ $\dfrac{2b^6c^{11}}{a^7}$

Simplify. See Example 4.

35. $\left(\dfrac{x^4}{4}\right)^3$ $\dfrac{x^{12}}{64}$
36. $\left(\dfrac{y^2}{2}\right)^3$ $\dfrac{y^6}{8}$
37. $\left(\dfrac{-2a^2}{b^3}\right)^4$ $\dfrac{16a^8}{b^{12}}$
38. $\left(\dfrac{-9r^3}{t^5}\right)^2$ $\dfrac{81r^6}{t^{10}}$
39. $\left(\dfrac{2x^2y}{-4y^2}\right)^3$ $-\dfrac{x^6}{8y^3}$
40. $\left(\dfrac{3y^8}{2zy^2}\right)^4$ $\dfrac{81y^{24}}{16z^4}$
41. $\left(\dfrac{-6x^2y^4z^9}{3x^6y^4z^3}\right)^2$ $\dfrac{4z^{12}}{x^8}$
42. $\left(\dfrac{-10rs^9t^4}{2rs^2t^7}\right)^3$ $\dfrac{-125s^{21}}{t^9}$

Simplify each expression. Your answer should be an integer or a fraction. Do not use a calculator.

43. $3^2 + 6^2$ 45
44. $(5-3)^2$ 4
45. $(3+6)^2$ 81
46. $5^2 - 3^2$ 16
47. $2^3 - 3^3$ -19
48. $3^3 + 4^3$ 91
49. $(2-3)^3$ -1
50. $(3+4)^3$ 343
51. $\left(\dfrac{2}{5}\right)^3$ $\dfrac{8}{125}$
52. $\left(\dfrac{3}{4}\right)^3$ $\dfrac{27}{64}$
53. $5^2 \cdot 2^3$ 200
54. $10^3 \cdot 3^3$ 27,000
55. $2^3 \cdot 2^4$ 128
56. $10^2 \cdot 10^4$ 1,000,000
57. $\left(\dfrac{2^3}{2^5}\right)^2$ $\dfrac{1}{16}$
58. $\left(\dfrac{3}{3^3}\right)^2$ $\dfrac{1}{81}$

Simplify each expression.

59. $3x^4 \cdot 5x^7$ $15x^{11}$

60. $-2y^3(3y)$ $-6y^4$

61. $(-5x^4)^3$ $-125x^{12}$

62. $(4z^3)^3$ $64z^9$

63. $-3y^5z^{12} \cdot 9yz^7$ $-27y^6z^{19}$

64. $2a^4b^5 \cdot 2a^9b^2$ $4a^{13}b^7$

65. $\dfrac{-9u^4v^9}{-3u^5v^8}$ $\dfrac{3v}{u}$

66. $\dfrac{-20a^5b^{13}}{5a^4b^{13}}$ $-4a$

67. $(-xt^2)(-2x^2t))^4$ $-16x^9t^6$

68. $(-ab)^3(-3ba^2)^4$ $-81a^{11}b^7$

69. $\left(\dfrac{2x^2}{x^4}\right)^3$ $\dfrac{8}{x^6}$

70. $\left(\dfrac{3y^8}{y^5}\right)^2$ $9y^6$

71. $\left(\dfrac{-8a^3b^4}{4c^5}\right)^5$ $\dfrac{-32a^{15}b^{20}}{c^{25}}$

72. $\left(\dfrac{-10a^5c}{5a^5b^4}\right)^5$ $\dfrac{-32c^5}{b^{20}}$

73. $\left(\dfrac{-8x^4y^7}{-16x^5y^6}\right)^5$ $\dfrac{y^5}{32x^5}$

74. $\left(\dfrac{-5x^2yz^3}{-5x^2yz}\right)^5$ z^{10}

Solve each problem.

75. ***Long-term investing.*** Sheila invested P dollars at annual rate r for 10 years. At the end of 10 years her investment was worth $P(1 + r)^{10}$ dollars. She then reinvested this money for another 5 years at annual rate r. At the end of the second time period her investment was worth $P(1 + r)^{10}(1 + r)^5$ dollars. Which law of exponents can be used to simplify the last expression? Simplify it.
$P(1 + r)^{15}$

76. ***CD rollover.*** Ronnie invested P dollars in a 2-year CD with an annual rate of return of r. After the CD rolled over two times, its value was $P((1 + r)^2)^3$. Which law of exponents can be used to simplify the expression? Simplify it. $P(1 + r)^6$

GETTING MORE INVOLVED

77. ***Writing.*** When we square a product, we square each factor in the product. For example, $(3b)^2 = 9b^2$. Explain why we cannot square a sum by simply squaring each term of the sum.

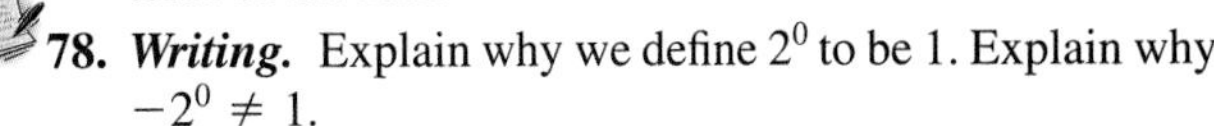

78. ***Writing.*** Explain why we define 2^0 to be 1. Explain why $-2^0 \neq 1$.

5.7 NEGATIVE EXPONENTS AND SCIENTIFIC NOTATION

In This Section

- Negative Integral Exponents
- Rules for Integral Exponents
- Converting from Scientific Notation
- Converting to Scientific Notation
- Computations with Scientific Notation

We defined exponential expressions with positive integral exponents in Chapter 1 and learned the rules for positive integral exponents in Section 5.6. In this section you will first study negative exponents and then see how positive and negative integral exponents are used in scientific notation.

Negative Integral Exponents

If x is nonzero, the reciprocal of x is written as $\frac{1}{x}$. For example, the reciprocal of 2^3 is written as $\frac{1}{2^3}$. To write the reciprocal of an exponential expression in a simpler way, we use a negative exponent. So $2^{-3} = \frac{1}{2^3}$. In general we have the following definition.

Negative Integral Exponents

If a is a nonzero real number and n is a positive integer, then

$$a^{-n} = \frac{1}{a^n}. \quad \text{(If } n \text{ is positive, } -n \text{ is negative.)}$$

EXAMPLE 1 **Simplifying expressions with negative exponents**

Simplify.

a) 2^{-5} **b)** $(-2)^{-5}$ **c)** $\dfrac{2^{-3}}{3^{-2}}$

Calculator Close-Up

You can evaluate expressions with negative exponents on a calculator as shown here.

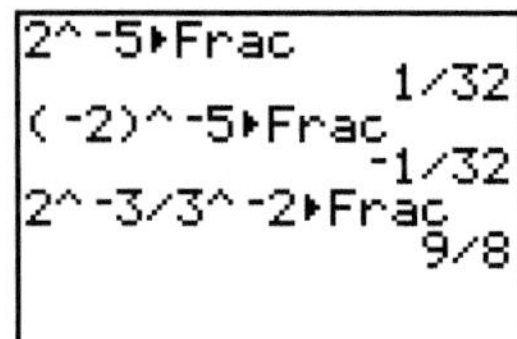

Solution

a) $2^{-5} = \frac{1}{2^5} = \frac{1}{32}$

b) $(-2)^{-5} = \frac{1}{(-2)^5}$ Definition of negative exponent

$= \frac{1}{-32} = -\frac{1}{32}$

c) $\frac{2^{-3}}{3^{-2}} = 2^{-3} \div 3^{-2}$

$= \frac{1}{2^3} \div \frac{1}{3^2}$

$= \frac{1}{8} \div \frac{1}{9} = \frac{1}{8} \cdot \frac{9}{1} = \frac{9}{8}$ ■

CAUTION In simplifying -5^{-2}, the negative sign preceding the 5 is used after 5 is squared and the reciprocal is found. So $-5^{-2} = -(5^{-2}) = -\frac{1}{25}$.

To evaluate a^{-n}, you can first find the nth power of a and then find the reciprocal. However, the result is the same if you first find the reciprocal of a and then find the nth power of the reciprocal. For example,

$$3^{-2} = \frac{1}{3^2} = \frac{1}{9} \quad \text{or} \quad 3^{-2} = \left(\frac{1}{3}\right)^2 = \frac{1}{3} \cdot \frac{1}{3} = \frac{1}{9}.$$

Helpful Hint

Just because the exponent is negative, it doesn't mean the expression is negative. Note that $(-2)^{-3} = -\frac{1}{8}$ while $(-2)^{-4} = \frac{1}{16}$.

So the power and the reciprocal can be found in either order. If the exponent is -1, we simply find the reciprocal. For example,

$$5^{-1} = \frac{1}{5}, \quad \left(\frac{1}{4}\right)^{-1} = 4, \quad \text{and} \quad \left(-\frac{3}{5}\right)^{-1} = -\frac{5}{3}.$$

Because $3^{-2} \cdot 3^2 = 1$, the reciprocal of 3^{-2} is 3^2, and we have

$$\frac{1}{3^{-2}} = 3^2.$$

These examples illustrate the following rules.

Rules for Negative Exponents

If a is a nonzero real number and n is a positive integer, then

$$a^{-n} = \left(\frac{1}{a}\right)^n, \quad a^{-1} = \frac{1}{a}, \quad \frac{1}{a^{-n}} = a^n, \quad \text{and} \quad \left(\frac{a}{b}\right)^{-n} = \left(\frac{b}{a}\right)^n.$$

EXAMPLE 2 **Using the rules for negative exponents**

Simplify.

a) $\left(\frac{3}{4}\right)^{-3}$ **b)** $10^{-1} + 10^{-1}$ **c)** $\frac{2}{10^{-3}}$

Solution

a) We can find the third power and the reciprocal in either order:

$$\left(\frac{3}{4}\right)^{-3} = \left(\frac{4}{3}\right)^3 = \frac{64}{27} \qquad \left(\frac{3}{4}\right)^{-3} = \left(\frac{27}{64}\right)^{-1} = \frac{64}{27}$$

Calculator Close-Up

You can use a calculator to demonstrate that the product rule for exponents holds when the exponents are negative numbers.

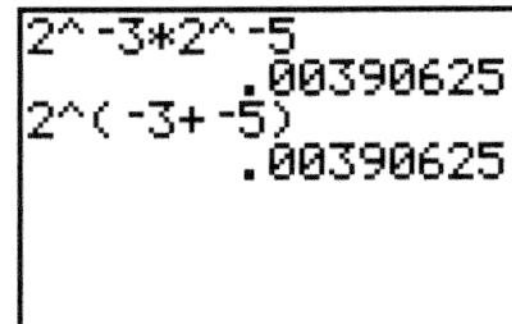

b) $10^{-1} + 10^{-1} = \frac{1}{10} + \frac{1}{10} = \frac{2}{10} = \frac{1}{5}$

c) $\frac{2}{10^{-3}} = 2 \cdot \frac{1}{10^{-3}} = 2 \cdot 10^3 = 2 \cdot 1000 = 2000$ ■

Rules for Integral Exponents

Negative exponents are used to make expressions involving reciprocals simpler looking and easier to write. Negative exponents have the added benefit of working in conjunction with all of the rules of exponents that you learned in Section 5.6. For example, we can use the product rule to get

$$x^{-2} \cdot x^{-3} = x^{-2+(-3)} = x^{-5}$$

and the quotient rule to get

$$\frac{y^3}{y^5} = y^{3-5} = y^{-2}.$$

With negative exponents there is no need to state the quotient rule in two parts as we did in Section 5.6. It can be stated simply as

$$\frac{a^m}{a^n} = a^{m-n}$$

for any integers m and n. We list the rules of exponents here for easy reference.

Helpful Hint

The definitions of the different types of exponents are a really clever mathematical invention. The fact that we have rules for performing arithmetic with those exponents makes the notation of exponents even more amazing.

Rules for Integral Exponents

The following rules hold for nonzero real numbers a and b and any integers m and n.

1. $a^0 = 1$ — Definition of zero exponent
2. $a^m \cdot a^n = a^{m+n}$ — Product rule
3. $\frac{a^m}{a^n} = a^{m-n}$ — Quotient rule
4. $(a^m)^n = a^{mn}$ — Power rule
5. $(ab)^n = a^n \cdot b^n$ — Power of a product rule
6. $\left(\frac{a}{b}\right)^n = \frac{a^n}{b^n}$ — Power of a quotient rule

EXAMPLE 3 **The product and quotient rules for integral exponents**

Simplify. Write your answers without negative exponents. Assume that the variables represent nonzero real numbers.

a) $b^{-3}b^5$

b) $-3x^{-3} \cdot 5x^2$

c) $\frac{m^{-6}}{m^{-2}}$

d) $\frac{4y^5}{-12y^{-3}}$

Solution

a) $b^{-3}b^5 = b^{-3+5}$ Product rule

$= b^2$ Simplify.

b) $-3x^{-3} \cdot 5x^2 = -15x^{-1}$ Product rule

$= -\frac{15}{x}$ Definition of negative exponent

c) $\frac{m^{-6}}{m^{-2}} = m^{-6-(-2)}$ Quotient rule

$= m^{-4}$ Simplify.

$= \frac{1}{m^4}$ Definition of negative exponent

Note that we could use the rules for negative exponents and the old quotient rule:

$$\frac{m^{-6}}{m^{-2}} = \frac{m^2}{m^6} = \frac{1}{m^4}$$

d) $\frac{4y^5}{-12y^{-3}} = \frac{y^{5-(-3)}}{-3} = \frac{-y^8}{3}$ ■

In Example 4 we use the power rules with negative exponents.

EXAMPLE 4 **The power rules for integral exponents**

Simplify each expression. Write your answers with positive exponents only. Assume that all variables represent nonzero real numbers.

a) $(a^{-3})^2$ **b)** $(10x^{-3})^{-2}$ **c)** $\left(\frac{4x^{-5}}{y^2}\right)^{-2}$

Calculator Close-Up

You can use a calculator to demonstrate that the power rule for exponents holds when the exponents are negative integers.

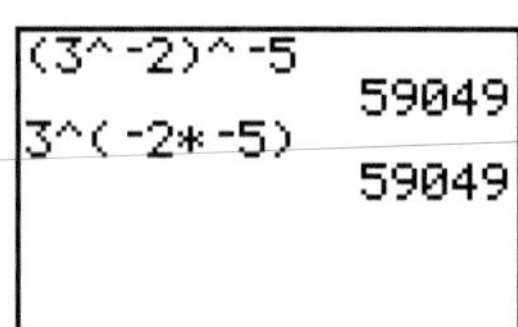

Solution

a) $(a^{-3})^2 = a^{-3 \cdot 2}$ Power rule

$= a^{-6}$

$= \frac{1}{a^6}$ Definition of negative exponent

b) $(10x^{-3})^{-2} = 10^{-2}(x^{-3})^{-2}$ Power of a product rule

$= 10^{-2}x^{(-3)(-2)}$ Power rule

$= \frac{x^6}{10^2}$ Definition of negative exponent

$= \frac{x^6}{100}$

c) $\left(\frac{4x^{-5}}{y^2}\right)^{-2} = \frac{(4x^{-5})^{-2}}{(y^2)^{-2}}$ Power of a quotient rule

$= \frac{4^{-2}x^{10}}{y^{-4}}$ Power of a product rule and power rule

$= 4^{-2} \cdot x^{10} \cdot \frac{1}{y^{-4}}$ Because $\frac{a}{b} = a \cdot \frac{1}{b}$.

$= \frac{1}{4^2} \cdot x^{10} \cdot y^4$ Definition of negative exponent

$= \frac{x^{10}y^4}{16}$ Simplify. ■

Helpful Hint

The exponent rules in this section apply to expressions that involve only multiplication and division. This is not too surprising since exponents, multiplication, and division are closely related. Recall that $a^3 = a \cdot a \cdot a$ and $a \div b = a \cdot b^{-1}$.

Converting from Scientific Notation

Many of the numbers occurring in science are either very large or very small. The speed of light is 983,569,000 feet per second. One millimeter is equal to 0.000001 kilometer. In scientific notation, numbers larger than 10 or smaller than 1 are written by using positive or negative exponents.

Scientific notation is based on multiplication by integral powers of 10. Multiplying a number by a positive power of 10 moves the decimal point to the right:

$$10(5.32) = 53.2$$
$$10^2(5.32) = 100(5.32) = 532$$
$$10^3(5.32) = 1000(5.32) = 5320$$

Multiplying by a negative power of 10 moves the decimal point to the left:

$$10^{-1}(5.32) = \frac{1}{10}(5.32) = 0.532$$
$$10^{-2}(5.32) = \frac{1}{100}(5.32) = 0.0532$$
$$10^{-3}(5.32) = \frac{1}{1000}(5.32) = 0.00532$$

Calculator Close-Up

On a graphing calculator you can write scientific notation by actually using the power of 10 or press EE to get the letter E, which indicates that the following number is the power of 10.

```
3.27*10^9
          3270000000
3.27E9
          3270000000
```

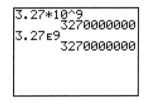

Note that if the exponent is not too large, scientific notation is converted to standard notation when you press ENTER.

So if n is a positive integer, multiplying by 10^n moves the decimal point n places to the right and multiplying by 10^{-n} moves it n places to the left.

A number in scientific notation is written as a product of a number between 1 and 10 and a power of 10. The times symbol $\times$ indicates multiplication. For example, 3.27×10^9 and 2.5×10^{-4} are numbers in scientific notation. In scientific notation there is one digit to the left of the decimal point.

To convert 3.27×10^9 to standard notation, move the decimal point nine places to the right:

$$3.27 \times 10^9 = 3{,}270{,}000{,}000$$

9 places to the right

Of course, it is not necessary to put the decimal point in when writing a whole number.

To convert 2.5×10^{-4} to standard notation, the decimal point is moved four places to the left:

$$2.5 \times 10^{-4} = 0.00025$$

4 places to the left

In general, we use the following strategy to convert from scientific notation to standard notation.

Strategy for Converting from Scientific Notation to Standard Notation

1. Determine the number of places to move the decimal point by examining the exponent on the 10.
2. Move to the right for a positive exponent and to the left for a negative exponent.

EXAMPLE 5 **Converting scientific notation to standard notation**

Write in standard notation.

a) 7.02×10^6 **b)** 8.13×10^{-5}

Solution

a) Because the exponent is positive, move the decimal point six places to the right:

$$7.02 \times 10^6 = 7020000. = 7,020,000$$

b) Because the exponent is negative, move the decimal point five places to the left.

$$8.13 \times 10^{-5} = 0.0000813$$

■

Converting to Scientific Notation

To convert a positive number to scientific notation, we just reverse the strategy for converting from scientific notation.

Strategy for Converting to Scientific Notation

1. Count the number of places (n) that the decimal must be moved so that it will follow the first nonzero digit of the number.
2. If the original number was larger than 10, use 10^n.
3. If the original number was smaller than 1, use 10^{-n}.

Remember that the scientific notation for a number larger than 10 will have a positive power of 10 and the scientific notation for a number between 0 and 1 will have a negative power of 10.

EXAMPLE 6 **Converting numbers to scientific notation**

Write in scientific notation.

a) 7,346,200 **b)** 0.0000348 **c)** 135×10^{-12}

Calculator Close-Up

To convert to scientific notation, set the mode to scientific. In scientific mode all results are given in scientific notation.

```
7346200
             7.3462E6
.0000348
              3.48E-5
135E-12
             1.35E-10
```

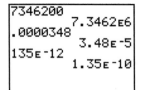

Solution

a) Because 7,346,200 is larger than 10, the exponent on the 10 will be positive:

$$7,346,200 = 7.3462 \times 10^6$$

b) Because 0.0000348 is smaller than 1, the exponent on the 10 will be negative:

$$0.0000348 = 3.48 \times 10^{-5}$$

c) There should be only one nonzero digit to the left of the decimal point:

$$135 \times 10^{-12} = 1.35 \times 10^2 \times 10^{-12} \quad \text{Convert 135 to scientific notation.}$$
$$= 1.35 \times 10^{-10} \quad \text{Product rule}$$

■

Computations with Scientific Notation

An important feature of scientific notation is its use in computations. Numbers in scientific notation are nothing more than exponential expressions, and you have already studied operations with exponential expressions in this section. We use the same rules of exponents on numbers in scientific notation that we use on any other exponential expressions.

EXAMPLE 7 **Using the rules of exponents with scientific notation**

Perform the indicated computations. Write the answers in scientific notation.

a) $(3 \times 10^6)(2 \times 10^8)$ b) $\dfrac{4 \times 10^5}{8 \times 10^{-2}}$ c) $(5 \times 10^{-7})^3$

Calculator Close-Up

With a calculator's built-in scientific notation, some parentheses can be omitted as shown below. Writing out the powers of 10 can lead to errors.

```
4E5/8E-2
                 5E6
4*10^5/8*10^-2
                 5E2
```

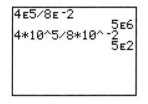

Try these computations with your calculator.

Solution

a) $(3 \times 10^6)(2 \times 10^8) = 3 \cdot 2 \cdot 10^6 \cdot 10^8 = 6 \times 10^{14}$

b) $\dfrac{4 \times 10^5}{8 \times 10^{-2}} = \dfrac{4}{8} \cdot \dfrac{10^5}{10^{-2}} = \dfrac{1}{2} \cdot 10^{5-(-2)}$ Quotient rule

$= (0.5)10^7$ $\frac{1}{2} = 0.5$

$= 5 \times 10^{-1} \cdot 10^7$ Write 0.5 in scientific notation.

$= 5 \times 10^6$ Product rule

c) $(5 \times 10^{-7})^3 = 5^3(10^{-7})^3$ Power of a product rule

$= 125 \cdot 10^{-21}$ Power rule

$= 1.25 \times 10^2 \times 10^{-21}$ $125 = 1.25 \times 10^2$

$= 1.25 \times 10^{-19}$ Product rule ■

EXAMPLE 8 **Converting to scientific notation for computations**

Perform these computations by first converting each number into scientific notation. Give your answer in scientific notation.

a) (3,000,000)(0.0002) b) $(20{,}000{,}000)^3(0.0000003)$

Solution

a) $(3{,}000{,}000)(0.0002) = 3 \times 10^6 \cdot 2 \times 10^{-4}$ Scientific notation

$= 6 \times 10^2$ Product rule

b) $(20{,}000{,}000)^3(0.0000003) = (2 \times 10^7)^3(3 \times 10^{-7})$ Scientific notation

$= 8 \times 10^{21} \cdot 3 \times 10^{-7}$ Power of a product rule

$= 24 \times 10^{14}$

$= 2.4 \times 10^1 \times 10^{14}$ $24 = 2.4 \times 10^1$

$= 2.4 \times 10^{15}$ Product rule ■

WARM-UPS

True or false? Explain your answer.

1. $10^{-2} = \dfrac{1}{100}$ True
2. $\left(-\dfrac{1}{5}\right)^{-1} = 5$ False
3. $3^{-2} \cdot 2^{-1} = 6^{-3}$ False
4. $\dfrac{3^{-2}}{3^{-1}} = \dfrac{1}{3}$ True
5. $23.7 = 2.37 \times 10^{-1}$ False
6. $0.000036 = 3.6 \times 10^{-5}$ True
7. $25 \cdot 10^7 = 2.5 \times 10^8$ True
8. $0.442 \times 10^{-3} = 4.42 \times 10^{-4}$ True
9. $(3 \times 10^{-9})^2 = 9 \times 10^{-18}$ True
10. $(2 \times 10^{-5})(4 \times 10^4) = 8 \times 10^{-20}$ False

5.7 EXERCISES

Reading and Writing *After reading this section, write out the answers to these questions. Use complete sentences.*

1. What does a negative exponent mean?
   A negative exponent means "reciprocal," as in $a^{-n} = \frac{1}{a^n}$.
2. What is the correct order for evaluating the operations indicated by a negative exponent?
   The operations can be evaluated in any order.
3. What is the new quotient rule for exponents?
   The new quotient rule is $a^m/a^n = a^{m-n}$ for any integers m and n.
4. How do you convert a number from scientific notation to standard notation?
   Convert from scientific notation by multiplying by the appropriate power of 10.
5. How do you convert a number from standard notation to scientific notation?
   Convert from standard notation by counting the number of places the decimal must move so that there is one nonzero digit to the left of the decimal point.
6. Which numbers are not usually written in scientific notation?
   Numbers between 1 and 10 are not written in scientific notation.

Variables in all exercises represent positive real numbers. Evaluate each expression. See Example 1.

7. 3^{-1} $\frac{1}{3}$ **8.** 3^{-3} $\frac{1}{27}$ **9.** $(-2)^{-4}$ $\frac{1}{16}$

10. $(-3)^{-4}$ $\frac{1}{81}$ **11.** -4^{-2} $-\frac{1}{16}$ **12.** -2^{-4} $-\frac{1}{16}$

13. $\frac{5^{-2}}{10^{-2}}$ 4 **14.** $\frac{3^{-4}}{6^{-2}}$ $\frac{4}{9}$

Simplify. See Example 2.

15. $\left(\frac{5}{2}\right)^{-3}$ $\frac{8}{125}$ **16.** $\left(\frac{4}{3}\right)^{-2}$ $\frac{9}{16}$ **17.** $6^{-1} + 6^{-1}$ $\frac{1}{3}$

18. $2^{-1} + 4^{-1}$ $\frac{3}{4}$ **19.** $\frac{10}{5^{-3}}$ 1250 **20.** $\frac{1}{25 \cdot 10^{-4}}$ 400

21. $\frac{1}{4^{-3}} + \frac{3^2}{2^{-1}}$ 82 **22.** $\frac{2^3}{10^{-2}} - \frac{2}{7^{-2}}$ 702

Simplify. Write answers without negative exponents. See Example 3.

23. $x^{-1}x^2$ x **24.** $y^{-3}y^5$ y^2

25. $-2x^2 \cdot 8x^{-6}$ $-\frac{16}{x^4}$ **26.** $5y^5(-6y^{-7})$ $-\frac{30}{y^2}$

27. $-3a^{-2}(-2a^{-3})$ $\frac{6}{a^5}$ **28.** $(-b^{-3})(-b^{-5})$ $\frac{1}{b^8}$

29. $\frac{u^{-5}}{u^3}$ $\frac{1}{u^8}$ **30.** $\frac{w^{-4}}{w^6}$ $\frac{1}{w^{10}}$

31. $\frac{8t^{-3}}{-2t^{-5}}$ $-4t^2$ **32.** $\frac{-22w^{-4}}{-11w^{-3}}$ $\frac{2}{w}$

33. $\frac{-6x^5}{-3x^{-6}}$ $2x^{11}$ **34.** $\frac{-51y^6}{17y^{-9}}$ $-3y^{15}$

Simplify each expression. Write answers without negative exponents. See Example 4.

35. $(x^2)^{-5}$ $\frac{1}{x^{10}}$ **36.** $(y^{-2})^4$ $\frac{1}{y^8}$ **37.** $(a^{-3})^{-3}$ a^9

38. $(b^{-5})^{-2}$ b^{10} **39.** $(2x^{-3})^{-4}$ $\frac{x^{12}}{16}$ **40.** $(3y^{-1})^{-2}$ $\frac{y^2}{9}$

41. $(4x^2y^{-3})^{-2}$ $\frac{y^6}{16x^4}$ **42.** $(6s^{-2}t^4)^{-1}$ $\frac{s^2}{6t^4}$

43. $\left(\frac{2x^{-1}}{y^{-3}}\right)^{-2}$ $\frac{x^2}{4y^6}$ **44.** $\left(\frac{a^{-2}}{3b^3}\right)^{-3}$ $27a^6b^9$

45. $\left(\frac{2a^{-3}}{ac^{-2}}\right)^{-4}$ $\frac{a^{16}}{16c^8}$ **46.** $\left(\frac{3w^2}{w^4x^3}\right)^{-2}$ $\frac{w^4x^6}{9}$

Simplify. Write answers without negative exponents.

47. $2^{-1} \cdot 3^{-1}$ $\frac{1}{6}$ **48.** $2^{-1} + 3^{-1}$ $\frac{5}{6}$

49. $(2 \cdot 3^{-1})^{-1}$ $\frac{3}{2}$ **50.** $(2^{-1} + 3)^{-1}$ $\frac{2}{7}$

51. $(x^{-2})^{-3} + 3x^7(-5x^{-1})$ $-14x^6$

52. $(ab^{-1})^2 - ab(-ab^{-3})$ $\frac{2a^2}{b^2}$

53. $\frac{a^3b^{-2}}{a^{-1}} + \left(\frac{b^6a^{-2}}{b^5}\right)^{-2}$ $\frac{2a^4}{b^2}$

54. $\left(\frac{x^{-3}y^{-1}}{2x}\right)^{-3} + \frac{6x^9y^3}{-3x^{-3}}$ $6x^{12}y^3$

Write each number in standard notation. See Example 5.

55. 9.86×10^9 9,860,000,000 **56.** 4.007×10^4 40,070

57. 1.37×10^{-3} 0.00137 **58.** 9.3×10^{-5} 0.000093

59. 1×10^{-6} 0.000001 **60.** 3×10^{-1} 0.3

61. 6×10^5 600,000 **62.** 8×10^6 8,000,000

Write each number in scientific notation. See Example 6.

63. 9000 9×10^3

64. 5,298,000 5.298×10^6

65. 0.00078 7.8×10^{-4}

66. 0.000214 2.14×10^{-4}

67. 0.0000085 8.5×10^{-6}

68. 5,670,000,000 5.67×10^9

69. 525×10^9 5.25×10^{11}

70. 0.0034×10^{-8} 3.4×10^{-11}

Perform the computations. Write answers in scientific notation. See Example 7.

71. $(3 \times 10^5)(2 \times 10^{-15})$ 6×10^{-10}

72. $(2 \times 10^{-9})(4 \times 10^{23})$ 8×10^{14}

73. $\dfrac{4 \times 10^{-8}}{2 \times 10^{30}}$ 2×10^{-38}

74. $\dfrac{9 \times 10^{-4}}{3 \times 10^{-6}}$ 3×10^{2}

75. $\dfrac{3 \times 10^{20}}{6 \times 10^{-8}}$ 5×10^{27}

76. $\dfrac{1 \times 10^{-8}}{4 \times 10^{7}}$ 2.5×10^{-16}

77. $(3 \times 10^{12})^2$ 9×10^{24}

78. $(2 \times 10^{-5})^3$ 8×10^{-15}

79. $(5 \times 10^{4})^3$ 1.25×10^{14}

80. $(5 \times 10^{14})^{-1}$ 2×10^{-15}

81. $(4 \times 10^{32})^{-1}$ 2.5×10^{-33}

82. $(6 \times 10^{11})^2$ 3.6×10^{23}

Perform the following computations by first converting each number into scientific notation. Write answers in scientific notation. See Example 8.

83. (4300)(2,000,000) 8.6×10^{9}

84. (40,000)(4,000,000,000) 1.6×10^{14}

85. (4,200,000)(0.00005) 2.1×10^{2}

86. (0.00075)(4,000,000) 3×10^{3}

87. $(300)^3(0.000001)^5$ 2.7×10^{-23}

88. $(200)^4(0.0005)^3$ 2×10^{-1}

89. $\dfrac{(4000)(90{,}000)}{0.00000012}$ 3×10^{15}

90. $\dfrac{(30{,}000)(80{,}000)}{(0.000006)(0.002)}$ 2×10^{17}

Perform the following computations with the aid of a calculator. Write answers in scientific notation. Round to three decimal places.

91. $(6.3 \times 10^{6})(1.45 \times 10^{-4})$ 9.135×10^{2}

92. $(8.35 \times 10^{9})(4.5 \times 10^{3})$ 3.758×10^{13}

93. $(5.36 \times 10^{-4}) + (3.55 \times 10^{-5})$ 5.715×10^{-4}

94. $(8.79 \times 10^{8}) + (6.48 \times 10^{9})$ 7.359×10^{9}

95. $\dfrac{(3.5 \times 10^{5})(4.3 \times 10^{-6})}{3.4 \times 10^{-8}}$ 4.426×10^{7}

96. $\dfrac{(3.5 \times 10^{-8})(4.4 \times 10^{-4})}{2.43 \times 10^{45}}$ 6.337×10^{-57}

97. $(3.56 \times 10^{85})(4.43 \times 10^{96})$ 1.577×10^{182}

98. $(8 \times 10^{99}) + (3 \times 10^{99})$ 1.1×10^{100}

Solve each problem.

99. ***Distance to the sun.*** The distance from the earth to the sun is 93 million miles. Express this distance in feet. (1 mile = 5280 feet.) 4.910×10^{11} feet

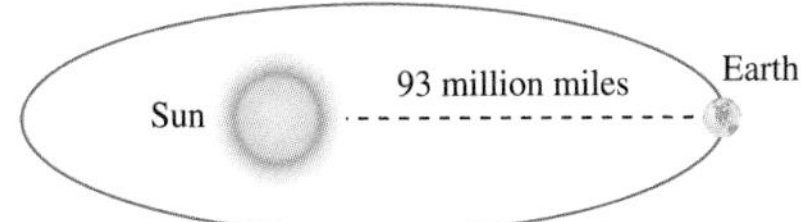

FIGURE FOR EXERCISE 99

100. ***Speed of light.*** The speed of light is 9.83569×10^{8} feet per second. How long does it take light to travel from the sun to the earth? See Exercise 99. 8.3 minutes

101. ***Warp drive, Scotty.*** How long does it take a spacecraft traveling at 2×10^{35} miles per hour (warp factor 4) to travel 93 million miles. 4.65×10^{-28} hours

102. ***Area of a dot.*** If the radius of a very small circle is 2.35×10^{-8} centimeters, then what is the circle's area? 1.735×10^{-15} cm^2

103. ***Circumference of a circle.*** If the circumference of a circle is 5.68×10^{9} feet, then what is its radius? 9.040×10^{8} feet

104. ***Diameter of a circle.*** If the diameter of a circle is 1.3×10^{-12} meters, then what is its radius? 6.5×10^{-13} meters

105. ***Extracting metals from ore.*** Thomas Sherwood studied the relationship between the concentration of a metal in commercial ore and the price of the metal. The accompanying graph shows the Sherwood plot with the locations of several metals marked. Even though the scales on this graph are not typical, the graph can be read in the same manner as other graphs. Note also that a concentration of 100 is 100%.

a) Use the figure to estimate the price of copper (Cu) and its concentration in commercial ore. \$1 per pound and 1%

b) Use the figure to estimate the price of a metal that has a concentration of 10^{-6} percent in commercial ore. \$1,000,000 per pound.

c) Would the four points shown in the graph lie along a straight line if they were plotted in our usual coordinate system? No

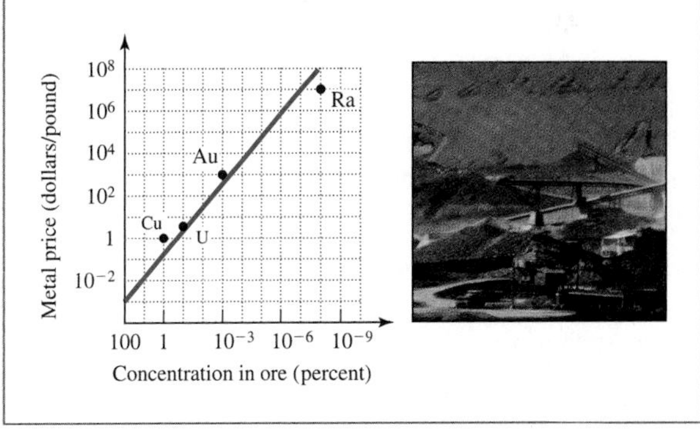

FIGURE FOR EXERCISE 105

106. ***Recycling metals.*** The graph on the next page shows the prices of various metals that are being recycled and the minimum concentration in waste required for recycling. The straight line is the line from the figure for Exercise 105. Points above the line correspond to metals for which it is economically feasible to increase recycling efforts.

a) Use the figure to estimate the price of mercury (Hg) and the minimum concentration in waste required for recycling mercury. $1 per pound and 1%

b) Use the figure to estimate the price of silver (Ag) and the minimum concentration in waste required for recycling silver. $100 per pound and 0.1%

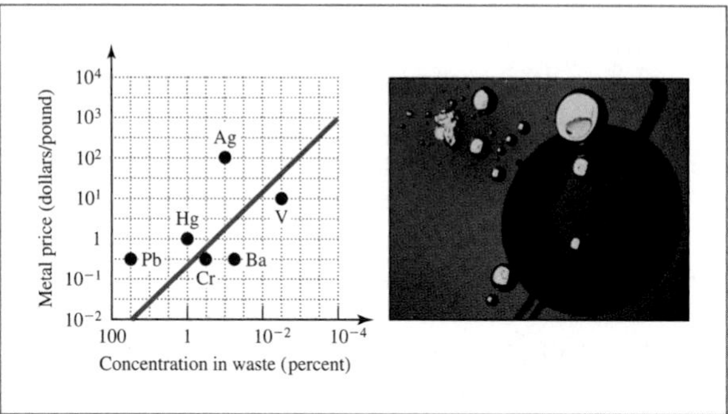

FIGURE FOR EXERCISE 106

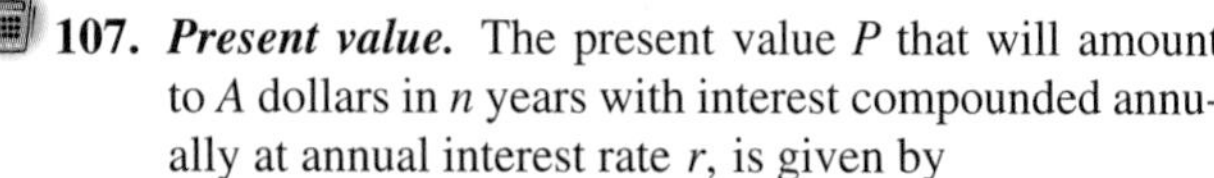

107. ***Present value.*** The present value P that will amount to A dollars in n years with interest compounded annually at annual interest rate r, is given by

$$P = A(1 + r)^{-n}.$$

Find the present value that will amount to $50,000 in 20 years at 8% compounded annually. $10,727.41

108. ***Investing in stocks.*** U.S. small company stocks have returned an average of 14.9% annually for the last 50 years (T. Rowe Price, www.troweprice.com). Use the present value formula from the previous exercise to find the amount invested today in small company stocks that would be worth $1 million in 50 years, assuming that small company stocks continue to return 14.9% annually for the next 50 years. $963.83

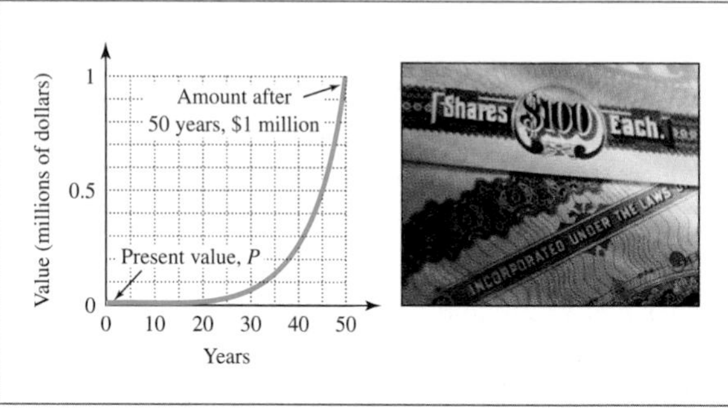

FIGURE FOR EXERCISE 108

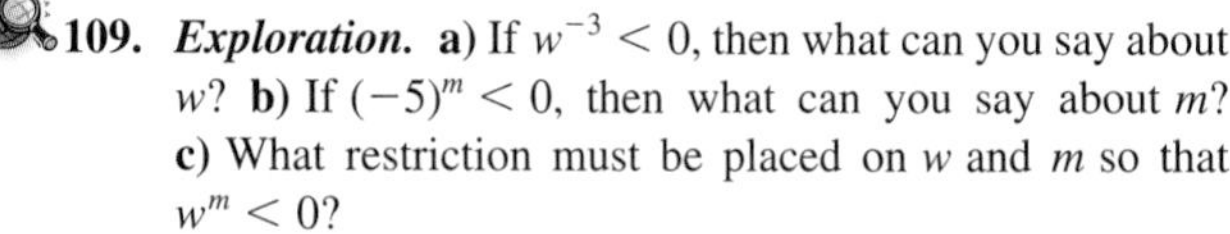

GETTING MORE INVOLVED

109. ***Exploration.*** **a)** If $w^{-3} < 0$, then what can you say about w? **b)** If $(-5)^m < 0$, then what can you say about m? **c)** What restriction must be placed on w and m so that $w^m < 0$?

a) $w < 0$ b) m is odd c) $w < 0$ and m odd

110. ***Discussion.*** Which of the following expressions is not equal to -1? Explain your answer.

a) -1^{-1} **b)** -1^{-2}

c) $(-1^{-1})^{-1}$ **d)** $(-1)^{-1}$

e) $(-1)^{-2}$ e

COLLABORATIVE ACTIVITIES

Area as a Model of Binomial Multiplication

Drawings and diagrams are often used to illustrate mathematical ideas. In this activity we use areas of rectangles to illustrate multiplication of binomials.

Example. The product $15 \cdot 13$ is the area of a 15 by 13 rectangle. Rewrite the product as $(10 + 5)(10 + 3)$ and make the following drawing.

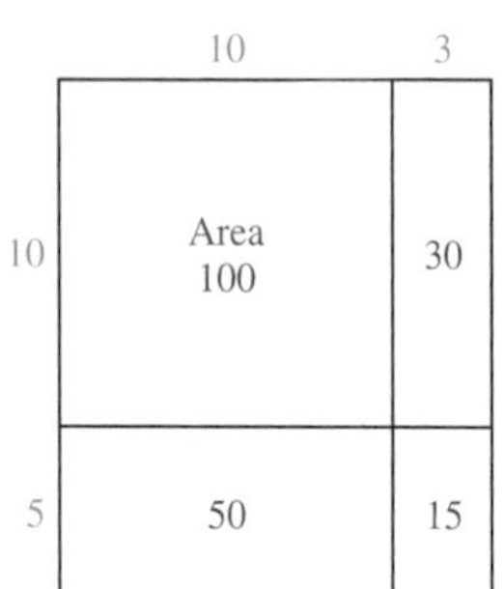

Grouping: Two students per group

Topic: Multiplying Polynomials

Note that the areas of the four regions are the four parts of FOIL:

$$\begin{aligned}(10 + 5)(10 + 3) &= \overset{\text{First}}{10^2} + \overset{\text{Outer}}{10 \cdot 3} + \overset{\text{Inner}}{5 \cdot 10} + \overset{\text{Last}}{5 \cdot 3}\\ &= 100 + 30 + 50 + 15\\ &= 195\end{aligned}$$

Exercises

1. Make a drawing using graph paper to illustrate the product $12 \cdot 13$ as the product of two binomials $(10 + 2)(10 + 3)$. Have each square of your graph paper represent one unit. One partner should draw the rectangle and find the total of the areas of the four regions that correspond to FOIL while the other should find $12 \cdot 13$ in the "usual" way. Check your answers.

2. We have been using 10 for the first term of our binomials, but now we will use x. Choose any positive integer for x and draw an $x + 2$ by $x + 4$ rectangle on your graph paper. Have one partner draw the rectangle and find the total of the areas of the four regions that correspond to FOIL while the other will find $(x + 2)(x + 4)$ in the "usual" way. Check your answers.
3. Suppose that the area of the $x + 2$ by $x + 4$ rectangle drawn in the last exercise is actually 120. Does your rectangle have an area of 120? If not, then draw an $x + 2$ by $x + 4$ rectangle with an area of 120. What is x? Is x an integer?
4. Is there an $x + 3$ by $x + 6$ rectangle for which the area is 120 and x is a positive integer?
5. Is there an $x + 7$ by $x + 14$ rectangle for which the area is 120 and x is a positive integer?
6. How many ways are there to partition a 10 by 12 rectangle into four regions using sides $x + a$ and $x + b$, where x, a, and b are positive integers?

WRAP-UP CHAPTER 5

SUMMARY

Polynomials		**Examples**
Term	A number or the product of a number and one or more variables raised to powers	$5x^3, -4x, 7$
Polynomial	A single term or a finite sum of terms	$2x^5 - 9x^2 + 11$
Degree of a polynomial	The highest degree of any of the terms	Degree of $2x - 9$ is 1. Degree of $5x^3 - x^2$ is 3.
Naming a Polynomial	A polynomial can be named with a letter such as P or $P(x)$ (function notation).	$P = x^2 - 1$ $P(x) = x^2 - 1$
Evaluating a polynomial	The value of a polynomial is the real number that is obtained when the variable (x) is replaced with a real number.	If $x = 3$ then $P = 8$, or $P(3) = 8$.

Adding, Subtracting, and Multiplying Polynomials		**Examples**
Add or subtract polynomials	Add or subtract the like terms.	$(x + 1) + (x - 4) = 2x - 3$ $(x^2 - 3x) - (4x^2 - x)$ $= -3x^2 - 2x$
Multiply monomials	Use the product rule for exponents.	$-2x^5 \cdot 6x^8 = -12x^{13}$
Multiply polynomials	Multiply each term of one polynomial by every term of the other polynomial, then combine like terms.	$\begin{array}{r} x^2 + 2x + 5 \\ x - 1 \\ \hline -x^2 - 2x - 5 \\ x^3 + 2x^2 + 5x \quad \\ \hline x^3 + x^2 + 3x - 5 \end{array}$

Binomials		**Examples**
FOIL	A method for multiplying two binomials quickly	$(x - 2)(x + 3) = x^2 + x - 6$
Square of a sum	$(a + b)^2 = a^2 + 2ab + b^2$	$(x + 3)^2 = x^2 + 6x + 9$
Square of a difference	$(a - b)^2 = a^2 - 2ab + b^2$	$(m - 5)^2 = m^2 - 10m + 25$
Product of a sum and a difference	$(a - b)(a + b) = a^2 - b^2$	$(x + 2)(x - 2) = x^2 - 4$

Dividing Polynomials		**Examples**
Dividing monomials	Use the quotient rule for exponents	$8x^5 \div (2x^2) = 4x^3$
Divide a polynomial by a monomial	Divide each term of the polynomial by the monomial.	$\dfrac{3x^5 + 9x}{3x} = x^4 + 3$
Divide a polynomial by a binomial	If the divisor is a binomial, use long division. (divisor)(quotient) + (remainder) = dividend	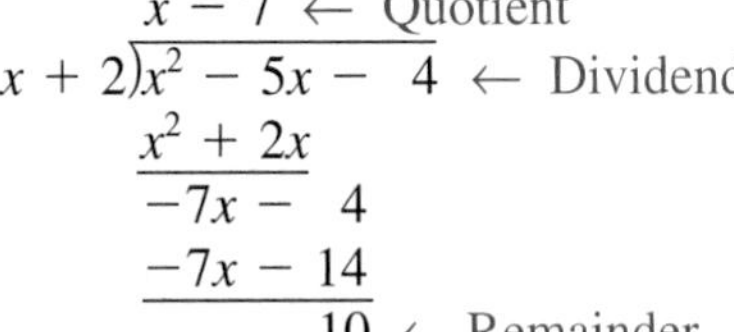

Rules of Exponents		**Examples**
The following rules hold for any integers m and n, and nonzero real numbers a and b.		
Zero exponent	$a^0 = 1$	$2^0 = 1, (-34)^0 = 1$
Product rule	$a^m \cdot a^n = a^{m+n}$	$a^2 \cdot a^3 = a^5$ $3x^6 \cdot 4x^9 = 12x^{15}$
Quotient rule	$\dfrac{a^m}{a^n} = a^{m-n}$	$x^8 \div x^2 = x^6, \dfrac{y^3}{y^3} = y^0 = 1$ $\dfrac{c^7}{c^9} = c^{-2} = \dfrac{1}{c^2}$
Power rule	$(a^m)^n = a^{mn}$	$(2^2)^3 = 2^6, (w^5)^3 = w^{15}$
Power of a product rule	$(ab)^n = a^n b^n$	$(2t)^3 = 8t^3$
Power of a quotient rule	$\left(\dfrac{a}{b}\right)^n = \dfrac{a^n}{b^n}$	$\left(\dfrac{x}{3}\right)^3 = \dfrac{x^3}{27}$

Negative Exponents		**Examples**
Negative integral exponents	If n is a positive integer and a is a nonzero real number, then $a^{-n} = \dfrac{1}{a^n}$	$3^{-2} = \dfrac{1}{3^2}, x^{-5} = \dfrac{1}{x^5}$

Rules for negative exponents	If a is a nonzero real number and n is a positive integer, then $a^{-n} = \left(\frac{1}{a}\right)^n$, $a^{-1} = \frac{1}{a}$, and $\frac{1}{a^{-n}} = a^n$.	$\left(\frac{2}{3}\right)^{-3} = \left(\frac{3}{2}\right)^3$, $5^{-1} = \frac{1}{5}$ $\frac{1}{w^{-8}} = w^8$

Scientific Notation		**Examples**
Converting from scientific notation	1. Find the number of places to move the decimal point by examining the exponent on the 10. 2. Move to the right for a positive exponent and to the left for a negative exponent.	$5.6 \times 10^3 = 5600$ $9 \times 10^{-4} = 0.0009$
Converting into scientific notation (positive numbers)	1. Count the number of places (n) that the decimal point must be moved so that it will follow the first nonzero digit of the number. 2. If the original number was larger than 10, use 10^n. 3. If the original number was smaller than 1, use 10^{-n}.	$304.6 = 3.046 \times 10^2$ $0.0035 = 3.5 \times 10^{-3}$

ENRICHING YOUR MATHEMATICAL WORD POWER

For each mathematical term, choose the correct meaning.

1. **term**
   a. an expression containing a number or the product of a number and one or more variables
   b. the amount of time spent in this course
   c. a word that describes a number
   d. a variable a

2. **polynomial**
   a. four or more terms
   b. many numbers
   c. a sum of four or more numbers
   d. a single term or a finite sum of terms d

3. **degree of a polynomial**
   a. the number of terms in a polynomial
   b. the highest degree of any of the terms of a polynomial
   c. the value of a polynomial when $x = 0$
   d. the largest coefficient of any of the terms of a polynomial b

4. **leading coefficient**
   a. the first coefficient
   b. the largest coefficient
   c. the coefficient of the first term when a polynomial is written with decreasing exponents
   d. the most important coefficient c

5. **monomial**
   a. a single polynomial
   b. one number
   c. an equation that has only one solution
   d. a polynomial that has one term d

6. **FOIL**
   a. a method for adding polynomials
   b. first, outer, inner, last
   c. an equation with no solution
   d. a polynomial with five terms b

7. **dividend**
   a. a in a/b
   b. b in a/b
   c. the result of a/b
   d. what a bank pays on deposits a

8. **divisor**
   a. a in a/b
   b. b in a/b
   c. the result of a/b
   d. two visors b

9. **quotient**
   a. a in a/b
   b. b in a/b
   c. a/b
   d. the divisor plus the remainder c

10. **binomial**
   a. a polynomial with two terms
   b. any two numbers
   c. the two coordinates in an ordered pair
   d. an equation with two variables a

11. **integral exponent**
   a. an exponent that is an integer
   b. a positive exponent
   c. a rational exponent
   d. a fractional exponent a

12. **scientific notation**
   a. the notation of rational exponents
   b. the notation of algebra
   c. a notation for expressing large or small numbers with powers of 10
   d. radical notation c

REVIEW EXERCISES

5.1 *Perform the indicated operations.*

1. $(2w - 6) + (3w + 4)$ $5w - 2$
2. $(1 - 3y) + (4y - 6)$ $y - 5$
3. $(x^2 - 2x - 5) - (x^2 + 4x - 9)$ $-6x + 4$
4. $(3 - 5x - x^2) - (x^2 - 7x + 8)$ $-2x^2 + 2x - 5$
5. $(5 - 3w + w^2) + (w^2 - 4w - 9)$ $2w^2 - 7w - 4$
6. $(-2t^2 + 3t - 4) + (t^2 - 7t + 2)$ $-t^2 - 4t - 2$
7. $(4 - 3m - m^2) - (m^2 - 6m + 5)$ $-2m^2 + 3m - 1$
8. $(n^3 - n^2 + 9) - (n^4 - n^3 + 5)$ $-n^4 + 2n^3 - n^2 + 4$

5.2 *Perform the indicated operations.*

9. $5x^2 \cdot (-10x^9)$ $-50x^{11}$
10. $3h^3t^2 \cdot 2h^2t^5$ $6h^5t^7$
11. $(-11a^7)^2$ $121a^{14}$
12. $(12b^3)^2$ $144b^6$
13. $x - 5(x - 3)$ $-4x + 15$
14. $x - 4(x - 9)$ $-3x + 36$
15. $5x + 3(x^2 - 5x + 4)$ $3x^2 - 10x + 12$
16. $5 + 4x^2(x - 5)$ $4x^3 - 20x^2 + 5$
17. $3m^2(5m^3 - m + 2)$ $15m^5 - 3m^3 + 6m^2$
18. $-4a^4(a^2 + 2a + 4)$ $-4a^6 - 8a^5 - 16a^4$
19. $(x - 5)(x^2 - 2x + 10)$ $x^3 - 7x^2 + 20x - 50$
20. $(x + 2)(x^2 - 2x + 4)$ $x^3 + 8$
21. $(x^2 - 2x + 4)(3x - 2)$ $3x^3 - 8x^2 + 16x - 8$
22. $(5x + 3)(x^2 - 5x + 4)$ $5x^3 - 22x^2 + 5x + 12$

5.3 *Perform the indicated operations.*

23. $(q - 6)(q + 8)$ $q^2 + 2q - 48$
24. $(w + 5)(w + 12)$ $w^2 + 17w + 60$
25. $(2t - 3)(t - 9)$ $2t^2 - 21t + 27$
26. $(5r + 1)(5r + 2)$ $25r^2 + 15r + 2$
27. $(4y - 3)(5y + 2)$ $20y^2 - 7y - 6$
28. $(11y + 1)(y + 2)$ $11y^2 + 23y + 2$
29. $(3x^2 + 5)(2x^2 + 1)$ $6x^4 + 13x^2 + 5$
30. $(x^3 - 7)(2x^3 + 7)$ $2x^6 - 7x^3 - 49$

5.4 *Perform the indicated operations. Try to write only the answers.*

31. $(z - 7)(z + 7)$ $z^2 - 49$
32. $(a - 4)(a + 4)$ $a^2 - 16$
33. $(y + 7)^2$ $y^2 + 14y + 49$
34. $(a + 5)^2$ $a^2 + 10a + 25$
35. $(w - 3)^2$ $w^2 - 6w + 9$
36. $(a - 6)^2$ $a^2 - 12a + 36$
37. $(x^2 - 3)(x^2 + 3)$ $x^4 - 9$
38. $(2b^2 - 1)(2b^2 + 1)$ $4b^4 - 1$
39. $(3a + 1)^2$ $9a^2 + 6a + 1$
40. $(1 - 3c)^2$ $1 - 6c + 9c^2$
41. $(4 - y)^2$ $16 - 8y + y^2$
42. $(9 - t)^2$ $81 - 18t + t^2$

Study Tip

Note how the review exercises are arranged according to the sections in this chapter. If you are having trouble with a certain type of problem, refer back to the appropriate section for examples and explanations.

5.5 *In Exercises 43–54, find each quotient.*

43. $-10x^5 \div (2x^3)$ $-5x^2$
44. $-6x^4y^2 \div (-2x^2y^2)$ $3x^2$
45. $\dfrac{6a^5b^7c^6}{-3a^3b^9c^6}$ $\dfrac{-2a^2}{b^2}$
46. $\dfrac{-9h^5t^9r^2}{3h^7t^6r^2}$ $\dfrac{-3t^3}{h^2}$
47. $\dfrac{3x - 9}{-3}$ $-x + 3$
48. $\dfrac{7 - y}{-1}$ $y - 7$
49. $\dfrac{9x^3 - 6x^2 + 3x}{-3x}$ $-3x^2 + 2x - 1$
50. $\dfrac{-8x^3y^5 + 4x^2y^4 - 2xy^3}{2xy^2}$ $-4x^2y^3 + 2xy^2 - y$
51. $(a - 1) \div (1 - a)$ -1
52. $(t - 3) \div (3 - t)$ -1
53. $(m^4 - 16) \div (m - 2)$ $m^3 + 2m^2 + 4m + 8$
54. $(x^4 - 1) \div (x - 1)$ $x^3 + x^2 + x + 1$

Find the quotient and remainder.

55. $(3m^3 - 9m^2 + 18m) \div (3m)$ $m^2 - 3m + 6, 0$

56. $(8x^3 - 4x^2 - 18x) \div (2x)$ $4x^2 - 2x - 9, 0$

57. $(b^2 - 3b + 5) \div (b + 2)$ $b - 5, 15$

58. $(r^2 - 5r + 9) \div (r - 3)$ $r - 2, 3$

59. $(4x^2 - 9) \div (2x + 1)$ $2x - 1, -8$

60. $(9y^3 + 2y) \div (3y + 2)$ $3y^2 - 2y + 2, -4$

61. $(x^3 + x^2 - 11x + 10) \div (x - 1)$ $x^2 + 2x - 9, 1$

62. $(y^3 - 9y^2 + 3y - 6) \div (y + 1)$
$y^2 - 10y + 13, -19$

Write each expression in the form

$$quotient + \frac{remainder}{divisor}.$$

63. $\dfrac{2x}{x-3}$ $2 + \dfrac{6}{x-3}$

64. $\dfrac{3x}{x-4}$ $3 + \dfrac{12}{x-4}$

65. $\dfrac{2x}{1-x}$ $-2 + \dfrac{2}{1-x}$

66. $\dfrac{3x}{5-x}$ $-3 + \dfrac{15}{5-x}$

67. $\dfrac{x^2-3}{x+1}$ $x - 1 - \dfrac{2}{x+1}$

68. $\dfrac{x^2+3x+1}{x-3}$ $x + 6 + \dfrac{19}{x-3}$

69. $\dfrac{x^2}{x+1}$ $x - 1 + \dfrac{1}{x+1}$

70. $\dfrac{-2x^2}{x-3}$ $-2x - 6 + \dfrac{-18}{x-3}$

5.6 *Simplify each expression.*

71. $2y^{10} \cdot 3y^{20}$ $6y^{30}$

72. $(-3a^5)(5a^3)$ $-15a^8$

73. $\dfrac{-10b^5c^3}{2b^5c^9}$ $\dfrac{-5}{c^6}$

74. $\dfrac{-30k^3y^9}{15k^3y^2}$ $-2y^7$

75. $(b^5)^6$ b^{30}

76. $(y^5)^8$ y^{40}

77. $(-2x^3y^2)^3$ $-8x^9y^6$

78. $(-3a^4b^6)^4$ $81a^{16}b^{24}$

79. $\left(\dfrac{2a}{b}\right)^3$ $\dfrac{8a^3}{b^3}$

80. $\left(\dfrac{3y}{2}\right)^3$ $\dfrac{27y^3}{8}$

81. $\left(\dfrac{-6x^2y^5}{-3z^6}\right)^3$ $\dfrac{8x^6y^{15}}{z^{18}}$

82. $\left(\dfrac{-3a^4b^8}{6a^3b^{12}}\right)^4$ $\dfrac{a^4}{16b^{16}}$

For the following exercises, assume that all of the variables represent positive real numbers.

5.7 *Simplify each expression. Use only positive exponents in answers.*

83. 2^{-5} $\dfrac{1}{32}$

84. -2^{-4} $-\dfrac{1}{16}$

85. 10^{-3} $\dfrac{1}{1000}$

86. $5^{-1} \cdot 5^0$ $\dfrac{1}{5}$

87. x^5x^{-8} $\dfrac{1}{x^3}$

88. $a^{-3}a^{-9}$ $\dfrac{1}{a^{12}}$

89. $\dfrac{a^{-8}}{a^{-12}}$ a^4

90. $\dfrac{a^{10}}{a^{-4}}$ a^{14}

91. $\dfrac{a^3}{a^{-7}}$ a^{10}

92. $\dfrac{b^{-2}}{b^{-6}}$ b^4

93. $(x^{-3})^4$ $\dfrac{1}{x^{12}}$

94. $(x^5)^{-10}$ $\dfrac{1}{x^{50}}$

95. $(2x^{-3})^{-3}$ $\dfrac{x^9}{8}$

96. $(3y^{-5})^2$ $\dfrac{9}{y^{10}}$

97. $\left(\dfrac{a}{3b^{-3}}\right)^{-2}$ $\dfrac{9}{a^2b^6}$

98. $\left(\dfrac{a^{-2}}{5b}\right)^{-3}$ $125a^6b^3$

Convert each number in scientific notation to a number in standard notation, and convert each number in standard notation to a number in scientific notation.

99. 5000
5×10^3

100. 0.00009
9×10^{-5}

101. 3.4×10^5
340,000

102. 5.7×10^{-8}
0.000000057

103. 0.0000461
4.61×10^{-5}

104. 44,000
4.4×10^4

105. 5.69×10^{-6}
0.00000569

106. 5.5×10^9
5,500,000,000

Perform each computation without using a calculator. Write answers in scientific notation.

107. $(3.5 \times 10^8)(2.0 \times 10^{-12})$ 7×10^{-4}

108. $(9 \times 10^{12})(2 \times 10^{17})$ 1.8×10^{30}

109. $(2 \times 10^{-4})^4$ 1.6×10^{-15}

110. $(-3 \times 10^5)^3$ -2.7×10^{16}

111. (0.00000004)(2,000,000,000) 8×10^1

112. (3,000,000,000) ÷ (0.000002) 1.5×10^{15}

113. $(0.0000002)^5$ 3.2×10^{-34}

114. $(50{,}000{,}000{,}000)^3$ 1.25×10^{32}

MISCELLANEOUS

Perform the indicated operations.

115. $(x + 3)(x + 7)$ $x^2 + 10x + 21$

116. $(k + 5)(k + 4)$ $k^2 + 9k + 20$

117. $(t - 3y)(t - 4y)$ $t^2 - 7ty + 12y^2$

118. $(t + 7z)(t + 6z)$ $t^2 + 13tz + 42z^2$

119. $(2x^3)^0 + (2y)^0$ 2

120. $(4y^2 - 9)^0$ 1

121. $(-3ht^6)^3$ $-27h^3t^{18}$

122. $(-9y^3c^4)^2$ $81y^6c^8$

123. $(2w + 3)(w - 6)$ $2w^2 - 9w - 18$

124. $(3x + 5)(2x - 6)$ $6x^2 - 8x - 30$

125. $(3u - 5v)(3u + 5v)$ $9u^2 - 25v^2$

126. $(9x^2 - 2)(9x^2 + 2)$ $81x^4 - 4$

127. $(3h + 5)^2$ $9h^2 + 30h + 25$

128. $(4v - 3)^2$ $16v^2 - 24v + 9$

129. $(x + 3)^3$ $x^3 + 9x^2 + 27x + 27$

130. $(k - 10)^3$ $k^3 - 30k^2 + 300k - 1000$

131. $(-7s^2t)(-2s^3t^5)$ $14s^5t^6$

132. $-5w^3r^2 \cdot 2w^4r^8$ $-10w^7r^{10}$

133. $\left(\dfrac{k^4m^2}{2k^2m^2}\right)^4$ $\dfrac{k^8}{16}$

134. $\left(\dfrac{-6h^3y^5}{2h^7y^2}\right)^4$ $\dfrac{81y^{12}}{h^{16}}$

135. $(5x^2 - 8x - 8) - (4x^2 + x - 3)$ $x^2 - 9x - 5$

136. $(4x^2 - 6x - 8) - (9x^2 - 5x + 7)$ $-5x^2 - x - 15$

137. $(2x^2 - 2x - 3) + (3x^2 + x - 9)$ $5x^2 - x - 12$

138. $(x^2 - 3x - 1) + (x^2 - 2x + 1)$ $2x^2 - 5x$

139. $(x + 4)(x^2 - 5x + 1)$ $x^3 - x^2 - 19x + 4$

140. $(2x^2 - 7x + 4)(x + 3)$ $2x^3 - x^2 - 17x + 12$

141. $(x^2 + 4x - 12) \div (x - 2)$ $x + 6$

142. $(a^2 - 3a - 10) \div (a - 5)$ $a + 2$

Solve each problem.

143. ***Roundball court.*** The length of a basketball court is 44 feet more than its width w. Find polynomials $P(w)$ and $A(w)$ that represent its perimeter and area. Find $P(50)$ and $A(50)$.
$P(w) = 4w + 88$, $A(w) = w^2 + 44w$, $P(50) = 288$ ft, $A(50) = 4700$ ft^2

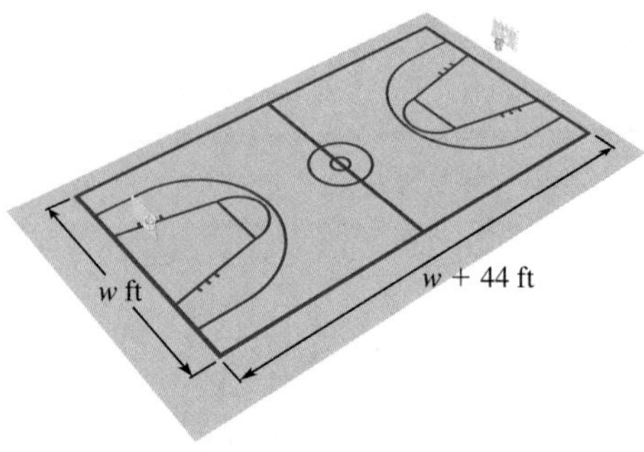

FIGURE FOR EXERCISE 143

144. ***Badminton court.*** The width of a badminton court is 24 feet less than its length x. Find polynomials $P(x)$ and $A(x)$ that represent its perimeter and area. Find $P(44)$ and $A(44)$.
$P(x) = 4x - 48$, $A(x) = x^2 - 24x$, $P(44) = 128$ ft, $A(44) = 880$ ft^2

145. ***Smoke alert.*** A retailer of smoke alarms knows that at a price of p dollars each, she can sell $600 - 15p$ smoke alarms per week. Find a polynomial that represents the weekly revenue for the smoke alarms. Find the revenue for a week in which the price is \$12 per smoke alarm. Use the bar graph to find the price per smoke alarm that gives the maximum weekly revenue.
$R = -15p^2 + 600p$, \$5040, \$20

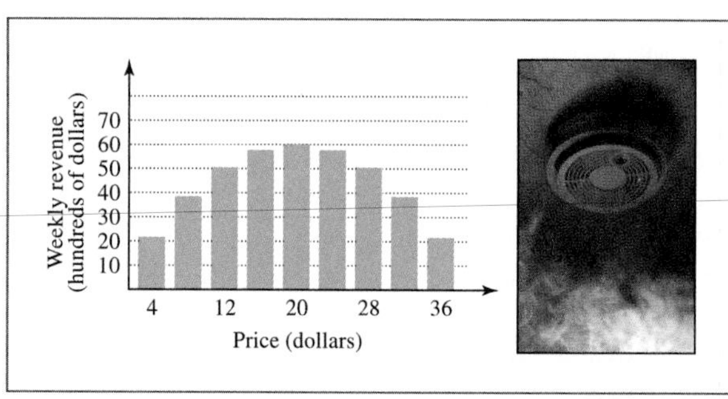

FIGURE FOR EXERCISE 145

146. ***Boom box sales.*** A retailer of boom boxes knows that at a price of q dollars each, he can sell $900 - 3q$ boom boxes per month. Find a polynomial that represents the monthly revenue for the boom boxes? How many boom boxes will he sell if the price is \$300 each?
$R = -3q^2 + 900q$, 0

CHAPTER 5 TEST

Perform the indicated operations.

1. $(7x^3 - x^2 - 6) + (5x^2 + 2x - 5)$ $7x^3 + 4x^2 + 2x - 11$

2. $(x^2 - 3x - 5) - (2x^2 + 6x - 7)$ $-x^2 - 9x + 2$

3. $\dfrac{6y^3 - 9y^2}{-3y}$ $-2y^2 + 3y$

4. $(x - 2) \div (2 - x)$ -1

5. $(x^3 - 2x^2 - 4x + 3) \div (x - 3)$ $x^2 + x - 1$

6. $3x^2(5x^3 - 7x^2 + 4x - 1)$ $15x^5 - 21x^4 + 12x^3 - 3x^2$

Find the products.

7. $(x + 5)(x - 2)$
$x^2 + 3x - 10$

8. $(3a - 7)(2a + 5)$
$6a^2 + a - 35$

9. $(a - 7)^2$
$a^2 - 14a + 49$

10. $(4x + 3y)^2$
$16x^2 + 24xy + 9y^2$

11. $(b - 3)(b + 3)$
$b^2 - 9$

12. $(3t^2 - 7)(3t^2 + 7)$
$9t^4 - 49$

13. $(4x^2 - 3)(x^2 + 2)$
$4x^4 + 5x^2 - 6$

14. $(x - 2)(x + 3)(x - 4)$
$x^3 - 3x^2 - 10x + 24$

Write each expression in the form

$$\text{quotient} + \frac{\text{remainder}}{\text{divisor}}.$$

15. $\dfrac{2x}{x - 3}$
$2 + \dfrac{6}{x - 3}$

16. $\dfrac{x^2 - 3x + 5}{x + 2}$
$x - 5 + \dfrac{15}{x + 2}$

Use the rules of exponents to simplify each expression. Write answers without negative exponents.

17. $-5x^3 \cdot 7x^5$ $-35x^8$

18. $3x^3y \cdot (2xy^4)^2$ $12x^5y^9$

19. $-4a^6b^5 \div (2a^5b)$ $-2ab^4$

20. $3x^{-2} \cdot 5x^7$ $15x^5$

21. $\left(\dfrac{-2a}{b^2}\right)^5$ $\dfrac{-32a^5}{b^{10}}$

22. $\dfrac{-6a^7b^6c^2}{-2a^3b^8c^2}$ $\dfrac{3a^4}{b^2}$

23. $\dfrac{6t^{-7}}{2t^9}$ $\dfrac{3}{t^{16}}$

24. $\dfrac{w^{-6}}{w^{-4}}$ $\dfrac{1}{w^2}$

25. $(-3s^{-3}t^2)^{-2}$ $\dfrac{s^6}{9t^4}$

26. $(-2x^{-6}y)^3$ $\dfrac{-8y^3}{x^{18}}$

Study Tip

Before you take an in-class exam on this chapter, work the sample test given here. Set aside one hour to work this test and use the answers in the back of this book to grade yourself. Even though your instructor might not ask exactly the same questions, you will get a good idea of your test readiness.

Convert to scientific notation.

27. 5,433,000 5.433×10^6

28. 0.0000065 6.5×10^{-6}

Perform each computation by converting to scientific notation. Give answers in scientific notation.

29. (80,000)(0.000006) 4.8×10^{-1}

30. $(0.0000003)^4$ 8.1×10^{-27}

Solve each problem.

31. Find the quotient and remainder when $x^2 - 5x + 9$ is divided by $x - 3$.
$x - 2$, 3

32. Subtract $3x^2 - 4x - 9$ from $x^2 - 3x + 6$.
$-2x^2 + x + 15$

33. The width of a pool table is x feet, and the length is 4 feet longer than the width. Find polynomials $A(x)$ and $P(x)$ that represent the area and perimeter of the pool table. Find $A(4)$ and $P(4)$.
$A(x) = x^2 + 4x$, $P(x) = 4x + 8$, $A(4) = 32$ ft^2, $P(4) = 24$ ft

34. If a manufacturer charges q dollars each for footballs, then he can sell $3000 - 150q$ footballs per week. Find a polynomial that represents the revenue for one week. Find the weekly revenue if the price is $8 for each football.
$R = -150q^2 + 3000q$, $14,400

MAKING CONNECTIONS CHAPTERS 1–5

Evaluate each arithmetic expression.

1. $-16 \div (-2)$ 8

2. $-16 \div \left(-\frac{1}{2}\right)$ 32

3. $(-5)^2 - 3(-5) + 1$ 41

4. $-5^2 - 4(-5) + 3$ -2

5. $2^{15} \div 2^{10}$ 32

6. $2^6 - 2^5$ 32

7. $-3^2 \cdot 4^2$ -144

8. $(-3 \cdot 4)^2$ 144

9. $\left(\frac{1}{2}\right)^3 + \frac{1}{2}$ $\frac{5}{8}$

10. $\left(\frac{2}{3}\right)^2 - \frac{1}{3}$ $\frac{1}{9}$

11. $(5 + 3)^2$ 64

12. $5^2 + 3^2$ 34

13. $3^{-1} + 2^{-1}$ $\frac{5}{6}$

14. $2^{-2} - 3^{-2}$ $\frac{5}{36}$

15. $(30 - 1)(30 + 1)$ 899

16. $(30 - 1) \div (1 - 30)$ -1

Perform the indicated operations.

17. $(x + 3)(x + 5)$
$x^2 + 8x + 15$

18. $x + 3(x + 5)$
$4x + 15$

19. $-5t^3v \cdot 3t^2v^6$
$-15t^5v^7$

20. $(-10t^3v^2) \div (-2t^2v)$
$5tv$

21. $(x^2 + 8x + 15) + (x + 5)$ $x^2 + 9x + 20$

22. $(x^2 + 8x + 15) - (x + 5)$ $x^2 + 7x + 10$

23. $(x^2 + 8x + 15) \div (x + 5)$ $x + 3$

24. $(x^2 + 8x + 15)(x + 5)$ $x^3 + 13x^2 + 55x + 75$

25. $(-6y^3 + 8y^2) \div (-2y^2)$ $3y - 4$

26. $(18y^4 - 12y^3 + 3y^2) \div (3y^2)$ $6y^2 - 4y + 1$

Solve each equation.

27. $2x + 1 = 0$ $\left\{-\frac{1}{2}\right\}$

28. $x - 7 = 0$ $\{7\}$

29. $\frac{3}{4}x - 3 = \frac{1}{2}$ $\left\{\frac{14}{3}\right\}$

30. $\frac{x}{2} - \frac{3}{4} = \frac{1}{8}$ $\left\{\frac{7}{4}\right\}$

31. $2(x - 3) = 3(x - 2)$ $\{0\}$

32. $2(3x - 3) = 3(2x - 2)$ All real numbers

Solve.

33. Find the x-intercept for the line $y = 2x + 1$. $\left(-\frac{1}{2}, 0\right)$

34. Find the y-intercept for the line $y = x - 7$. $(0, -7)$

35. Find the slope of the line $y = 2x + 1$. 2

36. Find the slope of the line that goes through $(0, 0)$ and $\left(\frac{1}{2}, \frac{1}{3}\right)$. $\frac{2}{3}$

37. If $y = \frac{3}{4}x - 3$ and y is $\frac{1}{2}$, then what is x? $\frac{14}{3}$

38. Find y if $y = \frac{x}{2} - \frac{3}{4}$ and x is $\frac{1}{2}$. $-\frac{1}{2}$

Solve the problem.

39. ***Average cost.*** Pineapple Recording plans to spend \$100,000 to record a new CD by the Woozies and \$2.25 per CD to manufacture the disks. The polynomial $2.25n + 100{,}000$ represents the total cost in dollars for recording and manufacturing n disks. Find an expression that represents the average cost per disk by dividing the total cost by n. Find the average cost per disk for $n = 1000$, 100,000, and 1,000,000. What happens to the large initial investment of \$100,000 if the company sells one million CDs?
$\frac{2.25n + 100{,}000}{n}$, \$102.25, \$3.25, \$2.35, It averages out to 10 cents per disk.

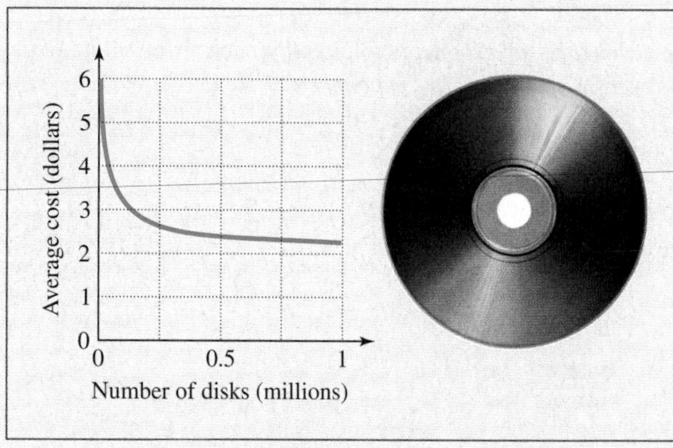

FIGURE FOR EXERCISE 39

CHAPTER

6

Factoring

The sport of skydiving was born in the 1930s soon after the military began using parachutes as a means of deploying troops. Today, skydiving is a popular sport around the world.

With as little as 8 hours of ground instruction, first-time jumpers can be ready to make a solo jump. Without the assistance of oxygen, sky divers can jump from as high as 14,000 feet and reach speeds of more than 100 miles per hour as they fall toward the earth. Jumpers usually open their parachutes between 2000 and 3000 feet and then gradually glide down to their landing area. If the jump and the parachute are handled correctly, the landing can be as gentle as jumping off two steps.

Making a jump and floating to earth are only part of the sport of skydiving. For example, in an activity called "relative work skydiving," a team of as many as 920 free-falling sky divers join together to make geometrically shaped formations. In a related exercise called "canopy relative work," the team members form geometric patterns after their parachutes or canopies have opened. This kind of skydiving takes skill and practice, and teams are not always successful in their attempts.

The amount of time a sky diver has for a free fall depends on the height of the jump and how much the sky diver uses the air to slow the fall. In Exercises 67 and 68 of Section 6.6 we find the amount of time that it takes a sky diver to fall from a given height.

In This Section

- Prime Factorization of Integers
- Greatest Common Factor
- Finding the Greatest Common Factor for Monomials
- Factoring Out the Greatest Common Factor
- Factoring Out the Opposite of the GCF

6.1 FACTORING OUT COMMON FACTORS

In Chapter 5 you learned how to multiply a monomial and a polynomial. In this section you will learn how to reverse that multiplication by finding the greatest common factor for the terms of a polynomial and then factoring the polynomial.

Prime Factorization of Integers

To **factor** an expression means to write the expression as a product. For example, if we start with 12 and write $12 = 4 \cdot 3$, we have factored 12. Both 4 and 3 are **factors** or **divisors** of 12. There are other factorizations of 12:

$$12 = 2 \cdot 6 \qquad 12 = 1 \cdot 12 \qquad 12 = 2 \cdot 2 \cdot 3 = 2^2 \cdot 3$$

The one that is most useful to us is $12 = 2^2 \cdot 3$, because it expresses 12 as a product of *prime numbers.*

Prime Number

A positive integer larger than 1 that has no integral factors other than itself and 1 is called a **prime number.**

The numbers 2, 3, 5, 7, 11, 13, 17, 19, and 23 are the first nine prime numbers. A positive integer larger than 1 that is not a prime is a **composite number.** The numbers 4, 6, 8, 9, 10, and 12 are the first six composite numbers. Every composite number is a product of prime numbers. The **prime factorization** for 12 is $2^2 \cdot 3$.

EXAMPLE 1 **Prime factorization**

Find the prime factorization for 36.

Helpful Hint

The prime factorization of 36 can be found also with a *factoring tree:*

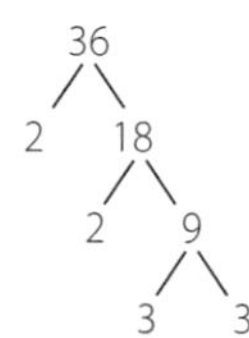

So $36 = 2 \cdot 2 \cdot 3 \cdot 3$.

Solution

We start by writing 36 as a product of two integers:

$36 = 2 \cdot 18$	Write 36 as $2 \cdot 18$.
$= 2 \cdot 2 \cdot 9$	Replace 18 by $2 \cdot 9$.
$= 2 \cdot 2 \cdot 3 \cdot 3$	Replace 9 by $3 \cdot 3$.
$= 2^2 \cdot 3^2$	Use exponential notation.

The prime factorization of 36 is $2^2 \cdot 3^2$. ■

For larger numbers it is helpful to use the method shown in Example 2.

EXAMPLE 2 **Factoring a large number**

Find the prime factorization for 420.

Solution

Start by dividing 420 by the smallest prime number that will divide into it evenly (without remainder). The smallest prime divisor of 420 is 2.

$$2\overline{)420}\ \ \text{quotient } 210$$

Helpful Hint

The fact that every composite number has a unique prime factorization is known as the fundamental theorem of arithmetic.

Now find the smallest prime that will divide evenly into the quotient, 210. The smallest prime divisor of 210 is 2. Continue this procedure, as follows, until the quotient is a prime number:

$$\begin{array}{rrl} & 7 & 35 \div 5 = 7 \\ & 5\overline{)35} & 105 \div 3 = 35 \\ & 3\overline{)105} & 210 \div 2 = 105 \\ & 2\overline{)210} & \\ \text{Start here} \rightarrow & 2\overline{)420} & \end{array}$$

The prime factorization of 420 is $2 \cdot 2 \cdot 3 \cdot 5 \cdot 7$, or $2^2 \cdot 3 \cdot 5 \cdot 7$. Note that it is really not necessary to divide by the smallest prime divisor at each step. We obtain the same factorization if we divide by any prime divisor at each step. ■

Greatest Common Factor

The largest integer that is a factor of two or more integers is called the **greatest common factor (GCF)** of the integers. For example, 1, 2, 3, and 6 are common factors of 18 and 24. Because 6 is the largest, 6 is the GCF of 18 and 24. We can use prime factorizations to find the GCF. For example, to find the GCF of 8 and 12, we first factor 8 and 12:

$$8 = 2 \cdot 2 \cdot 2 = 2^3 \qquad 12 = 2 \cdot 2 \cdot 3 = 2^2 \cdot 3$$

We see that the factor 2 appears twice in both 8 and 12. So 2^2, or 4, is the GCF of 8 and 12. Notice that 2 is a factor in both 2^3 and $2^2 \cdot 3$ and that 2^2 is the smallest power of 2 in these factorizations. In general, we can use the following strategy to find the GCF.

Strategy for Finding the GCF for Positive Integers

1. Find the prime factorization of each integer.
2. Determine which primes appear in all of the factorizations and the smallest exponent that appears on each of the common prime factors.
3. The GCF is the product of the common prime factors using the exponents from part (2).

If two integers have no common prime factors, then their greatest common factor is 1, because 1 is a factor of every integer. For example, 6 and 35 have no common prime factors because $6 = 2 \cdot 3$ and $35 = 5 \cdot 7$. However, because $6 = 1 \cdot 6$ and $35 = 1 \cdot 35$, the GCF for 6 and 35 is 1.

EXAMPLE 3 **Greatest common factor**

Find the GCF for each group of numbers.

a) 150, 225 **b)** 216, 360, 504 **c)** 55, 168

Solution

a) First find the prime factorization for each number:

$$\begin{array}{r} 5 \\ 5\overline{)25} \\ 3\overline{)75} \\ 2\overline{)150} \end{array} \qquad \begin{array}{r} 5 \\ 5\overline{)25} \\ 3\overline{)75} \\ 3\overline{)225} \end{array}$$

$$150 = 2 \cdot 3 \cdot 5^2 \qquad 225 = 3^2 \cdot 5^2$$

Because 2 is not a factor of 225, it is not a common factor of 150 and 225. Only 3 and 5 appear in both factorizations. Looking at both $2 \cdot 3 \cdot 5^2$ and $3^2 \cdot 5^2$, we see that the smallest power of 5 is 2 and the smallest power of 3 is 1. So the GCF of 150 and 225 is $3 \cdot 5^2$, or 75.

b) First find the prime factorization for each number:

$$216 = 2^3 \cdot 3^3 \qquad 360 = 2^3 \cdot 3^2 \cdot 5 \qquad 504 = 2^3 \cdot 3^2 \cdot 7$$

The only common prime factors are 2 and 3. The smallest power of 2 in the factorizations is 3, and the smallest power of 3 is 2. So the GCF is $2^3 \cdot 3^2$, or 72.

c) First find the prime factorization for each number:

$$55 = 5 \cdot 11 \qquad 168 = 2^3 \cdot 3 \cdot 7$$

Because there are no common factors other than 1, the GCF is 1. ■

Finding the Greatest Common Factor for Monomials

To find the GCF for a group of monomials, we use the same procedure as that used for integers.

Strategy for Finding the GCF for Monomials

1. Find the GCF for the coefficients of the monomials.
2. Form the product of the GCF of the coefficients and each variable that is common to all of the monomials, where the exponent on each variable is the smallest power of that variable in any of the monomials.

EXAMPLE 4 **Greatest common factor of monomials**

Find the greatest common factor for each group of monomials.

a) $15x^2, 9x^3$ **b)** $12x^2y^2, 30x^2yz, 42x^3y$

Solution

a) The GCF for 15 and 9 is 3, and the smallest power of x is 2. So the GCF for the monomials is $3x^2$. If we write these monomials as

$$15x^2 = 5 \cdot 3 \cdot x \cdot x \qquad \text{and} \qquad 9x^3 = 3 \cdot 3 \cdot x \cdot x \cdot x,$$

we can see that $3x^2$ is the GCF.

b) The GCF for 12, 30, and 42 is 6. For the common variables x and y, 2 is the smallest power of x and 1 is the smallest power of y. So the GCF for the monomials is $6x^2y$. ■

Study Tip

Success in school depends on effective time management, and effective time management is all about goals. Write down your long-term, short-term, and daily goals. Assess them, develop methods for meeting them, and reward yourself when you do.

Factoring Out the Greatest Common Factor

In Chapter 4 we used the distributive property to multiply monomials and polynomials. For example,

$$6(5x - 3) = 30x - 18.$$

If we start with $30x - 18$ and write

$$30x - 18 = 6(5x - 3),$$

we have factored $30x - 18$. Because multiplication is the last operation to be performed in $6(5x - 3)$, the expression $6(5x - 3)$ is a product. Because 6 is the GCF of 30 and 18, we have **factored out** the GCF.

EXAMPLE 5 **Factoring out the greatest common factor**

Factor the following polynomials by factoring out the GCF.

a) $25a^2 + 40a$ **b)** $6x^4 - 12x^3 + 3x^2$

c) $x^2y^5 + x^6y^3$

Study Tip

The keys to college success are motivation and time management. Anyone who tells you that they are making great grades without studying is probably not telling the truth. Success in college takes effort.

Solution

a) The GCF of the coefficients 25 and 40 is 5. Because the smallest power of the common factor a is 1, we can factor $5a$ out of each term:

$$25a^2 + 40a = 5a \cdot 5a + 5a \cdot 8$$
$$= 5a(5a + 8)$$

b) The GCF of 6, 12, and 3 is 3. We can factor x^2 out of each term, since the smallest power of x in the three terms is 2. So factor $3x^2$ out of each term as follows:

$$6x^4 - 12x^3 + 3x^2 = 3x^2 \cdot 2x^2 - 3x^2 \cdot 4x + 3x^2 \cdot 1$$
$$= 3x^2(2x^2 - 4x + 1)$$

Check by multiplying: $3x^2(2x^2 - 4x + 1) = 6x^4 - 12x^3 + 3x^2$.

c) The GCF of the numerical coefficients is 1. Both x and y are common to each term. Using the lowest powers of x and y, we get

$$x^2y^5 + x^6y^3 = x^2y^3 \cdot y^2 + x^2y^3 \cdot x^4$$
$$= x^2y^3(y^2 + x^4).$$

Check by multiplying. ■

CAUTION If the GCF is one of the terms of the polynomial, then you must remember to leave a 1 in place of that term when the GCF is factored out. For example,

$$ab + b = a \cdot b + 1 \cdot b = b(a + 1).$$

You should always check your answer by multiplying the factors.

In Example 6 the greatest common factor is a binomial. This type of factoring will be used in factoring trinomials in Section 6.2.

EXAMPLE 6 **A binomial factor**

Factor out the greatest common factor.

a) $(a + b)w + (a + b)6$ **b)** $x(x + 2) + 3(x + 2)$

c) $y(y - 3) - (y - 3)$

Solution

a) The greatest common factor is $a + b$:

$$(a + b)w + (a + b)6 = (a + b)(w + 6)$$

b) The greatest common factor is $x + 2$:

$$x(x + 2) + 3(x + 2) = (x + 3)(x + 2)$$

c) The greatest common factor is $y - 3$:

$$y(y - 3) - (y - 3) = y(y - 3) - 1(y - 3)$$
$$= (y - 1)(y - 3)$$ ■

Factoring Out the Opposite of the GCF

The greatest common factor for $-4x + 2xy$ is $2x$. Note that you can factor out the GCF ($2x$) or the opposite of the GCF ($-2x$):

$$-4x + 2xy = 2x(-2 + y) \qquad -4x + 2xy = -2x(2 - y)$$

It is useful to know both of these factorizations. For example, if we wanted to show that $2 - y$ is a factor of $-4x + 2xy$, then the second factorization shows it. Factoring out the opposite of the GCF will be used in factoring by grouping in Section 6.2 and in factoring trinomials with negative leading coefficients in Section 6.4. Remember to check all factoring by multiplying the factors to see if you get the original polynomial.

EXAMPLE 7 **Factoring out the opposite of the GCF**

Factor each polynomial twice. First factor out the greatest common factor, and then factor out the opposite of the GCF.

a) $3x - 3y$ **b)** $a - b$ **c)** $-x^3 + 2x^2 - 8x$

Solution

a) $3x - 3y = 3(x - y)$ Factor out 3.

$= -3(-x + y)$ Factor out -3.

Note that the signs of the terms in parentheses change when -3 is factored out. Check the answers by multiplying.

b) $a - b = 1(a - b)$ Factor out 1, the GCF of a and b.

$= -1(-a + b)$ Factor out -1.

We can also write $a - b = -1(b - a)$.

c) $-x^3 + 2x^2 - 8x = x(-x^2 + 2x - 8)$ Factor out x.

$= -x(x^2 - 2x + 8)$ Factor out $-x$. ■

CAUTION Be sure to change the sign of each term in parentheses when you factor out the opposite of the greatest common factor.

WARM-UPS

True or false? Explain your answer.

1. There are only nine prime numbers. False
2. The prime factorization of 32 is $2^3 \cdot 3$. False
3. The integer 51 is a prime number. False
4. The GCF of the integers 12 and 16 is 4. True
5. The GCF of the integers 10 and 21 is 1. True
6. The GCF of the polynomial $x^5y^3 - x^4y^7$ is x^4y^3. True
7. For the polynomial $2x^2y - 6xy^2$ we can factor out either $2xy$ or $-2xy$. True
8. The greatest common factor of the polynomial $8a^3b - 12a^2b$ is $4ab$. False
9. $x - 7 = 7 - x$ for any real number x. False
10. $-3x^2 + 6x = -3x(x - 2)$ for any real number x. True

6.1 EXERCISES

Reading and Writing *After reading this section, write out the answers to these questions. Use complete sentences.*

1. What does it mean to factor an expression?
   To factor means to write as a product.
2. What is a prime number?
   A prime number is an integer greater than 1 that has no factors besides itself and 1.
3. How do you find the prime factorization of a number?
   You can find the prime factorization by dividing by prime factors until the result is prime.
4. What is the greatest common factor for two numbers?
   The GCF for two numbers is the largest number that is a factor of both.
5. What is the greatest common factor for two monomials?
   The GCF for two monomials consists of the GCF of their coefficients and every variable that they have in common raised to the lowest power that appears on the variable.
6. How can you check if you have factored an expression correctly?
   You can check all factoring by multiplying the factors.

Find the prime factorization of each integer. See Examples 1 and 2.

7. 18 $2 \cdot 3^2$
8. 20 $2^2 \cdot 5$
9. 52 $2^2 \cdot 13$
10. 76 $2^2 \cdot 19$
11. 98 $2 \cdot 7^2$
12. 100 $2^2 \cdot 5^2$
13. 460 $2^2 \cdot 5 \cdot 23$
14. 345 $3 \cdot 5 \cdot 23$
15. 924 $2^2 \cdot 3 \cdot 7 \cdot 11$
16. 585 $3^2 \cdot 5 \cdot 13$

Find the greatest common factor (GCF) for each group of integers. See Example 3.

17. 8, 20 4
18. 18, 42 6
19. 36, 60 12
20. 42, 70 14
21. 40, 48, 88 8
22. 15, 35, 45 5
23. 76, 84, 100 4
24. 66, 72, 120 6
25. 39, 68, 77 1
26. 81, 200, 539 1

Find the greatest common factor (GCF) for each group of monomials. See Example 4.

27. $6x, 8x^3$ $2x$
28. $12x^2, 4x^3$ $4x^2$
29. $12x^3, 4x^2, 6x$ $2x$
30. $3y^5, 9y^4, 15y^3$ $3y^3$
31. $3x^2y, 2xy^2$ xy
32. $7a^2x^3, 5a^3x$ a^2x
33. $24a^2bc, 60ab^2$ $12ab$
34. $30x^2yz^3, 75x^3yz^6$ $15x^2yz^3$
35. $12u^3v^2, 25s^2t^4$ 1
36. $45m^2n^5, 56a^4b^8$ 1
37. $18a^3b, 30a^2b^2, 54ab^3$ $6ab$
38. $16x^2z, 40xz^2, 72z^3$ $8z$

Complete the factoring of each monomial.

39. $27x = 9(3x)$
40. $51y = 3y(17)$
41. $24t^2 = 8t(3t)$
42. $18u^2 = 3u(6u)$
43. $36y^5 = 4y^2(9y^3)$
44. $42z^4 = 3z^2(14z^2)$
45. $u^4v^3 = uv(u^3v^2)$
46. $x^5y^3 = x^2y(x^3y^2)$
47. $-14m^4n^3 = 2m^4(-7n^3)$
48. $-8y^3z^4 = 4z^3(-2y^3z)$
49. $-33x^4y^3z^2 = -3x^3yz(11xy^2z)$
50. $-96a^3b^4c^5 = -12ab^3c^3(8a^2bc^2)$

Factor out the GCF in each expression. See Example 5.

51. $x^3 - 6x$ $x(x^2 - 6)$
52. $10y^4 - 30y^2$ $10y^2(y^2 - 3)$
53. $5ax + 5ay$ $5a(x + y)$
54. $6wz + 15wa$ $3w(2z + 5a)$
55. $h^5 + h^3$ $h^3(h^2 + 1)$
56. $y^6 + y^5$ $y^5(y + 1)$
57. $-2k^7m^4 + 4k^3m^6$ $2k^3m^4(-k^4 + 2m^2)$
58. $-6h^5t^2 + 3h^3t^6$ $3h^3t^2(-2h^2 + t^4)$
59. $2x^3 - 6x^2 + 8x$ $2x(x^2 - 3x + 4)$
60. $6x^3 + 18x^2 + 24x$ $6x(x^2 + 3x + 4)$
61. $12x^4t + 30x^3t - 24x^2t^2$ $6x^2t(2x^2 + 5x - 4t)$
62. $15x^2y^2 - 9xy^2 + 6x^2y$ $3xy(5xy - 3y + 2x)$

Factor out the GCF in each expression. See Example 6.

63. $(x - 3)a + (x - 3)b$ $(x - 3)(a + b)$
64. $(y + 4)3 + (y + 4)z$ $(y + 4)(3 + z)$
65. $x(x - 1) - 5(x - 1)$ $(x - 5)(x - 1)$
66. $a(a + 1) - 3(a + 1)$ $(a - 3)(a + 1)$
67. $m(m + 9) + (m + 9)$ $(m + 1)(m + 9)$
68. $(x - 2)x - (x - 2)$ $(x - 2)(x - 1)$
69. $a(y + 1)^2 + b(y + 1)^2$ $(a + b)(y + 1)^2$
70. $w(w + 2)^2 + 8(w + 2)^2$ $(w + 8)(w + 2)^2$

First factor out the GCF, and then factor out the opposite of the GCF. See Example 7.

71. $8x - 8y$ $8(x - y), -8(-x + y)$
72. $2a - 6b$ $2(a - 3b), -2(-a + 3b)$
73. $-4x + 8x^2$ $4x(-1 + 2x), -4x(1 - 2x)$
74. $-5x^2 + 10x$ $5x(-x + 2), -5x(x - 2)$
75. $x - 5$ $1(x - 5), -1(-x + 5)$
76. $a - 6$ $1(a - 6), -1(-a + 6)$
77. $4 - 7a$ $1(4 - 7a), -1(-4 + 7a)$
78. $7 - 5b$ $1(7 - 5b), -1(-7 + 5b)$
79. $-24a^3 + 16a^2$ $8a^2(-3a + 2), -8a^2(3a - 2)$
80. $-30b^4 + 75b^3$ $15b^3(-2b + 5), -15b^3(2b - 5)$
81. $-12x^2 - 18x$ $6x(-2x - 3), -6x(2x + 3)$
82. $-20b^2 - 8b$ $4b(-5b - 2), -4b(5b + 2)$
83. $-2x^3 - 6x^2 + 14x$
$2x(-x^2 - 3x + 7), -2x(x^2 + 3x - 7)$
84. $-8x^4 + 6x^3 - 2x^2$
$2x^2(-4x^2 + 3x - 1), -2x^2(4x^2 - 3x + 1)$
85. $4a^3b - 6a^2b^2 - 4ab^3$
$2ab(2a^2 - 3ab - 2b^2), -2ab(-2a^2 + 3ab + 2b^2)$
86. $12u^5v^6 + 18u^2v^3 - 15u^4v^5$
$3u^2v^3(4u^3v^3 + 6 - 5u^2v^2), -3u^2v^3(-4u^3v^3 - 6 + 5u^2v^2)$

Solve each problem by factoring.

87. ***Uniform motion.*** Helen traveled a distance of $20x + 40$ miles at 20 miles per hour on the Yellowhead Highway. Find a binomial that represents the time that she traveled.
$x + 2$ hours

88. ***Area of a painting.*** A rectangular painting with a width of x centimeters has an area of $x^2 + 50x$ square centimeters. Find a binomial that represents the length.
$x + 50$ cm

FIGURE FOR EXERCISE 88

89. ***Tomato soup.*** The amount of metal S (in square inches) that it takes to make a can for tomato soup depends on the radius r and height h:

$$S = 2\pi r^2 + 2\pi rh$$

a) Rewrite this formula by factoring out the greatest common factor on the right-hand side.
$S = 2\pi r(r + h)$

b) Let $h = 5$ in. and write a formula that expresses S in terms of r. $S = 2\pi r^2 + 10\pi r$

c) The accompanying graph shows S for r between 1 in. and 3 in. (with $h = 5$ in.). Which of these r-values gives the maximum surface area? 3 in.

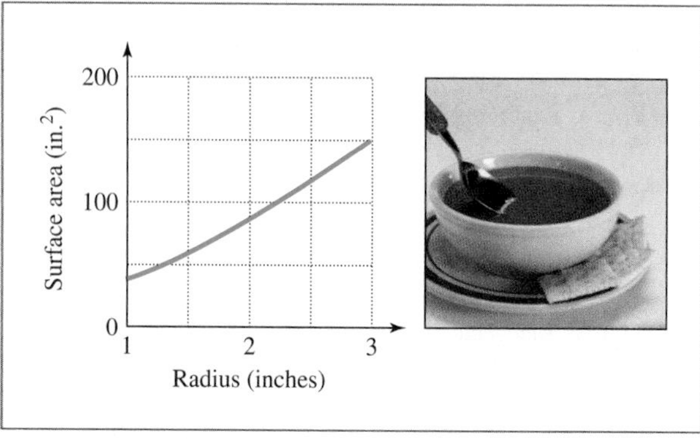

FIGURE FOR EXERCISE 89

90. ***Amount of an investment.*** The amount of an investment of P dollars for t years at simple interest rate r is given by $A = P + Prt$.

a) Rewrite this formula by factoring out the greatest common factor on the right-hand side.
$A = P(1 + rt)$

b) Find A if \$8300 is invested for 3 years at a simple interest rate of 15%. \$12,035

GETTING MORE INVOLVED

91. ***Discussion.*** Is the greatest common factor of $-6x^2 + 3x$ positive or negative? Explain.
The GCF is an algebraic expression.

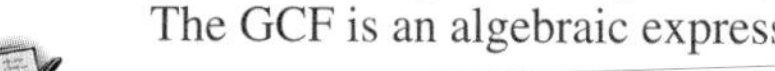

92. ***Writing.*** Explain in your own words why you use the smallest power of each common prime factor when finding the GCF of two or more integers.

6.2 FACTORING THE SPECIAL PRODUCTS AND FACTORING BY GROUPING

In This Section

- Factoring a Difference of Two Squares
- Factoring a Perfect Square Trinomial
- Factoring Completely
- Factoring by Grouping

In Section 5.4 you learned how to find the special products: the square of a sum, the square of a difference, and the product of a sum and a difference. In this section you will learn how to reverse those operations.

Factoring a Difference of Two Squares

In Section 5.4 you learned that the product of a sum and a difference is a difference of two squares:

$$(a + b)(a - b) = a^2 - ab + ab - b^2 = a^2 - b^2$$

So a difference of two squares can be factored as a product of a sum and a difference, using the following rule.

Factoring a Difference of Two Squares

For any real numbers a and b,

$$a^2 - b^2 = (a + b)(a - b).$$

Note that the square of an integer is a perfect square. For example, 64 is a perfect square because $64 = 8^2$. The square of a monomial in which the coefficient is an integer is also called a **perfect square** or simply a **square.** For example, $9m^2$ is a perfect square because $9m^2 = (3m)^2$.

EXAMPLE 1 **Factoring a difference of two squares**

Factor each polynomial.

a) $y^2 - 81$ **b)** $9m^2 - 16$ **c)** $4x^2 - 9y^2$

Solution

a) Because $81 = 9^2$, the binomial $y^2 - 81$ is a difference of two squares:

$$\begin{aligned} y^2 - 81 &= y^2 - 9^2 && \text{Rewrite as a difference of two squares.} \\ &= (y + 9)(y - 9) && \text{Factor.} \end{aligned}$$

Check by multiplying.

b) Because $9m^2 = (3m)^2$ and $16 = 4^2$, the binomial $9m^2 - 16$ is a difference of two squares:

$$\begin{aligned} 9m^2 - 16 &= (3m)^2 - 4^2 && \text{Rewrite as a difference of two squares.} \\ &= (3m + 4)(3m - 4) && \text{Factor.} \end{aligned}$$

Check by multiplying.

c) Because $4x^2 = (2x)^2$ and $9y^2 = (3y)^2$, the binomial $4x^2 - 9y^2$ is a difference of two squares:

$$4x^2 - 9y^2 = (2x + 3y)(2x - 3y)$$

■

Factoring a Perfect Square Trinomial

In Section 5.4 you learned how to square a binomial using the rule

$$(a + b)^2 = a^2 + 2ab + b^2.$$

You can reverse this rule to factor a trinomial such as $x^2 + 6x + 9$. Notice that

$$x^2 + 6x + 9 = \underset{a^2}{\underset{\uparrow}{x^2}} + \underbrace{2 \cdot x \cdot 3}_{2ab} + \underset{b^2}{\underset{\uparrow}{3^2}}.$$

So if $a = x$ and $b = 3$, then $x^2 + 6x + 9$ fits the form $a^2 + 2ab + b^2$, and

$$x^2 + 6x + 9 = (x + 3)^2.$$

A trinomial that is of the form $a^2 + 2ab + b^2$ or $a^2 - 2ab + b^2$ is called a **perfect square trinomial.** A perfect square trinomial is the square of a binomial. Perfect square trinomials can be identified by using the following strategy.

Strategy for Identifying a Perfect Square Trinomial

A trinomial is a perfect square trinomial if:

1. the first and last terms are of the form a^2 and b^2 (perfect squares)

2. the middle term is $2ab$ or $-2ab$.

EXAMPLE 2 **Identifying the special products**

Determine whether each binomial is a difference of two squares and whether each trinomial is a perfect square trinomial.

a) $x^2 - 14x + 49$ **b)** $4x^2 - 81$

c) $4a^2 + 24a + 25$ **d)** $9y^2 - 24y - 16$

Study Tip

A lumber mill turns logs into plywood, adding value to the logs. College is like a lumber mill. If you are not changing, growing, and learning, you may not be increasing in value. Everything that you learn increases your value.

Solution

a) The first term is x^2, and the last term is 7^2. The middle term, $-14x$, is $-2 \cdot x \cdot 7$. So this trinomial is a perfect square trinomial.

b) Both terms of $4x^2 - 81$ are perfect squares, $(2x)^2$ and 9^2. So $4x^2 - 81$ is a difference of two squares.

c) The first term of $4a^2 + 24a + 25$ is $(2a)^2$ and the last term is 5^2. However, $2 \cdot 2a \cdot 5$ is $20a$. Because the middle term is $24a$, this trinomial is not a perfect square trinomial.

d) The first and last terms in a perfect square trinomial are both positive. Because the last term in $9y^2 - 24y - 16$ is negative, the trinomial is not a perfect square trinomial. ■

Note that the middle term in a perfect square trinomial may have a positive or a negative coefficient, while the first and last terms must be positive. Any perfect square trinomial can be factored as the square of a binomial by using the following rule.

Factoring Perfect Square Trinomials

For any real numbers a and b,

$$a^2 + 2ab + b^2 = (a + b)^2$$
$$a^2 - 2ab + b^2 = (a - b)^2.$$

EXAMPLE 3 **Factoring perfect square trinomials**

Factor.

a) $x^2 - 4x + 4$ **b)** $a^2 + 16a + 64$ **c)** $4x^2 - 12x + 9$

Solution

a) The first term is x^2, and the last term is 2^2. Because the middle term is $-2 \cdot 2 \cdot x$, or $-4x$, this polynomial is a perfect square trinomial:

$$x^2 - 4x + 4 = (x - 2)^2$$

Check by expanding $(x - 2)^2$.

b) $a^2 + 16a + 64 = (a + 8)^2$

Check by expanding $(a + 8)^2$.

c) The first term is $(2x)^2$, and the last term is 3^2. Because $-2 \cdot 2x \cdot 3 = -12x$, the polynomial is a perfect square trinomial. So

$$4x^2 - 12x + 9 = (2x - 3)^2.$$

Check by expanding $(2x - 3)^2$. ■

Factoring Completely

To factor a polynomial means to write it as a product of simpler polynomials. A polynomial that cannot be factored is called a **prime** or **irreducible polynomial.**

The polynomials $3x$, $w + 1$, and $4m - 5$ are prime polynomials. A polynomial is **factored completely** when it is written as a product of prime polynomials. So $(y - 8)(y + 1)$ is a complete factorization. When factoring polynomials, we usually do not factor integers that occur as common factors. So $6x(x - 7)$ is considered to be factored completely even though 6 could be factored.

Some polynomials have a factor common to all terms. To factor such polynomials completely, it is simpler to factor out the greatest common factor (GCF) and then factor the remaining polynomial. The following example illustrates factoring completely.

EXAMPLE 4 **Factoring completely**

Factor each polynomial completely.

a) $2x^3 - 50x$ **b)** $8x^2y - 32xy + 32y$

Solution

a) The greatest common factor of $2x^3$ and $50x$ is $2x$:

$$2x^3 - 50x = 2x(x^2 - 25) \quad \text{Check this step by multiplying.}$$
$$= 2x(x + 5)(x - 5) \quad \text{Difference of two squares}$$

b) $$8x^2y - 32xy + 32y = 8y(x^2 - 4x + 4) \quad \text{Check this step by multiplying.}$$
$$= 8y(x - 2)^2 \quad \text{Perfect square trinomial}$$

■

Remember that factoring reverses multiplication and *every step of factoring can be checked by multiplication.*

Factoring by Grouping

The product of two binomials may be a polynomial with four terms. For example,

$$(x + a)(x + 3) = (x + a)x + (x + a)3$$
$$= x^2 + ax + 3x + 3a.$$

We can factor a polynomial of this type by simply reversing the steps we used to find the product. To reverse these steps, we factor out common factors from the first two terms and from the last two terms. This procedure is called **factoring by grouping.**

EXAMPLE 5 **Factoring by grouping**

Use grouping to factor each polynomial completely.

a) $xy + 2y + 3x + 6$ **b)** $2x^3 - 3x^2 - 2x + 3$ **c)** $ax + 3y - 3x - ay$

Study Tip

When you take notes, leave space. Go back later and fill in more details, make corrections, or work another problem of the same type.

Solution

a) Notice that the first two terms have a common factor of y and the last two terms have a common factor of 3:

$$xy + 2y + 3x + 6 = (xy + 2y) + (3x + 6) \quad \text{Use the associative property to group the terms.}$$
$$= y(x + 2) + 3(x + 2) \quad \text{Factor out the common factors in each group.}$$
$$= (y + 3)(x + 2) \quad \text{Factor out } x + 2.$$

b) We can factor x^2 out of the first two terms and 1 out of the last two terms:

$$2x^3 - 3x^2 - 2x + 3 = (2x^3 - 3x^2) + (-2x + 3) \quad \text{Group the terms.}$$
$$= x^2(2x - 3) + 1(-2x + 3)$$

However, we cannot proceed any further because $2x - 3$ and $-2x + 3$ are not the same. To get $2x - 3$ as a common factor, we must factor out -1 from the last two terms:

$$2x^3 - 3x^2 - 2x + 3 = x^2(2x - 3) - 1(2x - 3) \quad \text{Factor out the common factors.}$$
$$= (x^2 - 1)(2x - 3) \quad \text{Factor out } 2x - 3.$$
$$= (x - 1)(x + 1)(2x - 3) \quad \text{Difference of two squares}$$

c) In $ax + 3y - 3x - ay$ there are no common factors in the first two or the last two terms. However, if we use the commutative property to rewrite the polynomial as $ax - 3x - ay + 3y$, then we can factor by grouping:

$$ax + 3y - 3x - ay = ax - 3x - ay + 3y \quad \text{Rearrange the terms.}$$
$$= x(a - 3) - y(a - 3) \quad \text{Factor out } x \text{ and } -y.$$
$$= (x - y)(a - 3) \quad \text{Factor out } a - 3.$$

■

WARM-UPS

True or false? Explain your answer.

1. The polynomial $x^2 + 16$ is a difference of two squares. False
2. The polynomial $x^2 - 8x + 16$ is a perfect square trinomial. True
3. The polynomial $9x^2 + 21x + 49$ is a perfect square trinomial. False
4. $4x^2 + 4 = (2x + 2)^2$ for any real number x. False
5. A difference of two squares is equal to a product of a sum and a difference. True
6. The polynomial $16y + 1$ is a prime polynomial. True
7. The polynomial $x^2 + 9$ can be factored as $(x + 3)(x + 3)$. False
8. The polynomial $4x^2 - 4$ is factored completely as $4(x^2 - 1)$. False
9. $y^2 - 2y + 1 = (y - 1)^2$ for any real number y. True
10. $2x^2 - 18 = 2(x - 3)(x + 3)$ for any real number x. True

6.2 EXERCISES

Reading and Writing *After reading this section, write out the answers to these questions. Use complete sentences.*

1. What is a perfect square?
A perfect square is a square of an integer or an algebraic expression.

2. How do we factor a difference of two squares?
$a^2 - b^2 = (a + b)(a - b)$

3. How can you recognize if a trinomial is a perfect square?
A perfect square trinomial is of the form $a^2 + 2ab + b^2$ or $a^2 - 2ab + b^2$.

4. What is a prime polynomial?
A prime polynomial is a polynomial that cannot be factored.

5. When is a polynomial factored completely?
A polynomial is factored completely when it is a product of prime polynomials.

6. What should you always look for first when attempting to factor a polynomial completely?
Always factor out the GCF first.

Factor each polynomial. See Example 1.

7. $a^2 - 4$ $\quad (a - 2)(a + 2)$

8. $h^2 - 9$ $\quad (h - 3)(h + 3)$

9. $x^2 - 49$ $\quad (x - 7)(x + 7)$

10. $y^2 - 36$ $\quad (y - 6)(y + 6)$

11. $y^2 - 9x^2$ $\quad (y + 3x)(y - 3x)$

12. $16x^2 - y^2$ $(4x - y)(4x + y)$

13. $25a^2 - 49b^2$ $(5a + 7b)(5a - 7b)$

14. $9a^2 - 64b^2$ $(3a - 8b)(3a + 8b)$

15. $121m^2 - 1$ $(11m + 1)(11m - 1)$

16. $144n^2 - 1$ $(12n - 1)(12n + 1)$

17. $9w^2 - 25c^2$ $(3w - 5c)(3w + 5c)$

18. $144w^2 - 121a^2$ $(12w - 11a)(12w + 11a)$

Determine whether each binomial is a difference of two squares and whether each trinomial is a perfect square trinomial. See Example 2.

19. $x^2 - 20x + 100$ Perfect square trinomial

20. $x^2 - 10x - 25$ Neither

21. $y^2 - 40$ Neither

22. $a^2 - 49$ Difference of two squares

23. $4y^2 + 12y + 9$ Perfect square trinomial

24. $9a^2 - 30a - 25$ Neither

25. $x^2 - 8x + 64$ Neither

26. $x^2 + 4x + 4$ Perfect square trinomial

27. $9y^2 - 25c^2$ Difference of two squares

28. $9x^2 + 4$ Neither

29. $9a^2 + 6ab + b^2$ Perfect square trinomial

30. $4x^2 - 4xy + y^2$ Perfect square trinomial

Factor each perfect square trinomial. See Example 3.

31. $x^2 + 12x + 36$ $(x + 6)^2$

32. $y^2 + 14y + 49$ $(y + 7)^2$

33. $a^2 - 4a + 4$ $(a - 2)^2$

34. $b^2 - 6b + 9$ $(b - 3)^2$

35. $4w^2 + 4w + 1$ $(2w + 1)^2$

36. $9m^2 + 6m + 1$ $(3m + 1)^2$

37. $16x^2 - 8x + 1$ $(4x - 1)^2$

38. $25y^2 - 10y + 1$ $(5y - 1)^2$

39. $4t^2 + 20t + 25$ $(2t + 5)^2$

40. $9y^2 - 12y + 4$ $(3y - 2)^2$

41. $9w^2 + 42w + 49$ $(3w + 7)^2$

42. $144x^2 + 24x + 1$ $(12x + 1)^2$

43. $n^2 + 2nt + t^2$ $(n + t)^2$

44. $x^2 - 2xy + y^2$ $(x - y)^2$

Factor each polynomial completely. See Example 4.

45. $5x^2 - 125$ $5(x - 5)(x + 5)$

46. $3y^2 - 27$ $3(y - 3)(y + 3)$

47. $-2x^2 + 18$ $-2(x - 3)(x + 3)$

48. $-5y^2 + 20$ $-5(y - 2)(y + 2)$

49. $a^3 - ab^2$ $a(a - b)(a + b)$

50. $x^2y - y$ $y(x - 1)(x + 1)$

51. $3x^2 + 6x + 3$ $3(x + 1)^2$

52. $12a^2 + 36a + 27$ $3(2a + 3)^2$

53. $-5y^2 + 50y - 125$ $-5(y - 5)^2$

54. $-2a^2 - 16a - 32$ $-2(a + 4)^2$

55. $x^3 - 2x^2y + xy^2$ $x(x - y)^2$

56. $x^3y + 2x^2y^2 + xy^3$ $xy(x + y)^2$

57. $-3x^2 + 3y^2$ $-3(x - y)(x + y)$

58. $-8a^2 + 8b^2$ $-8(a - b)(a + b)$

59. $2ax^2 - 98a$ $2a(x - 7)(x + 7)$

60. $32x^2y - 2y^3$ $2y(4x - y)(4x + y)$

61. $3ab^2 - 18ab + 27a$ $3a(b - 3)^2$

62. $-2a^2b + 8ab - 8b$ $-2b(a - 2)^2$

63. $-4m^3 + 24m^2n - 36mn^2$ $-4m(m - 3n)^2$

64. $10a^3 - 20a^2b + 10ab^2$ $10a(a - b)^2$

Use grouping to factor each polynomial completely. See Example 5.

65. $bx + by + cx + cy$ $(b + c)(x + y)$

66. $3x + 3z + ax + az$ $(3 + a)(x + z)$

67. $x^3 + x^2 - 4x - 4$ $(x - 2)(x + 2)(x + 1)$

68. $x^3 + x^2 - x - 1$ $(x - 1)(x + 1)^2$

69. $3a - 3b - xa + xb$ $(3 - x)(a - b)$

70. $ax - bx - 4a + 4b$ $(x - 4)(a - b)$

71. $a^3 + 3a^2 + a + 3$ $(a^2 + 1)(a + 3)$

72. $y^3 - 5y^2 + 8y - 40$ $(y^2 + 8)(y - 5)$

73. $xa + ay + 3y + 3x$ $(a + 3)(x + y)$

74. $x^3 + ax + 3a + 3x^2$ $(x + 3)(x^2 + a)$

75. $abc - 3 + c - 3ab$ $(c - 3)(ab + 1)$

76. $xa + tb + ba + tx$ $(a + t)(x + b)$

77. $x^2a - b + bx^2 - a$ $(a + b)(x - 1)(x + 1)$

78. $a^2m - b^2n + a^2n - b^2m$ $(m + n)(a - b)(a + b)$

79. $y^2 + y + by + b$ $(y + b)(y + 1)$

80. $ac + mc + aw^2 + mw^2$ $(c + w^2)(a + m)$

Factor each polynomial completely.

81. $6a^3y + 24a^2y^2 + 24ay^3$ $6ay(a + 2y)^2$

82. $8b^5c - 8b^4c^2 + 2b^3c^3$ $2b^3c(2b - c)^2$

83. $24a^3y - 6ay^3$ $6ay(2a - y)(2a + y)$

84. $27b^3c - 12bc^3$ $3bc(3b - 2c)(3b + 2c)$

85. $2a^3y^2 - 6a^2y$ $2a^2y(ay - 3)$

86. $9x^3y - 18x^2y^2$ $9x^2y(x - 2y)$

87. $ab + 2bw - 4aw - 8w^2$ $(b - 4w)(a + 2w)$

88. $3am - 6n - an + 18m$ $(3m - n)(a + 6)$

Use factoring to solve each problem.

89. ***Skydiving.*** The height (in feet) above the earth for a sky diver t seconds after jumping from an airplane at 6400 ft is approximated by the formula $h = -16t^2 + 6400$, provided that $t < 5$. Rewrite the formula with the right-hand side factored completely. Use your revised formula to find h when $t = 2$. $h = -16(t - 20)(t + 20)$, 6336 feet

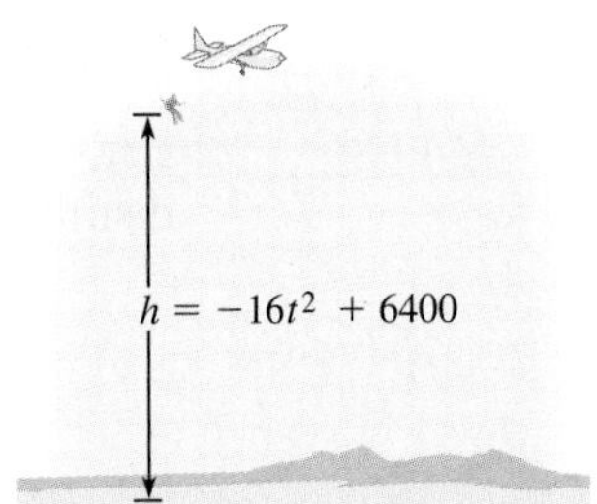

FIGURE FOR EXERCISE 89

90. ***Demand for pools.*** Tropical Pools sells an aboveground model for p dollars each. The monthly revenue from the sale of this model is given by

$$R = -0.08p^2 + 300p.$$

Revenue is the product of the price p and the demand (quantity sold).

a) Factor out the price on the right-hand side of the formula.
$R = p(-0.08p + 300)$

b) What is an expression for the monthly demand?
$-0.08p + 300$

c) What is the monthly demand for this pool when the price is \$3000? 60 pools

d) Use the graph to estimate the price at which the revenue is maximized. Approximately how many pools will be sold monthly at this price?
\$2000, 140 pools

e) What is the approximate maximum revenue?
\$280,000

f) Use the accompanying graph to estimate the price at which the revenue is zero.
\$0 or \$3800

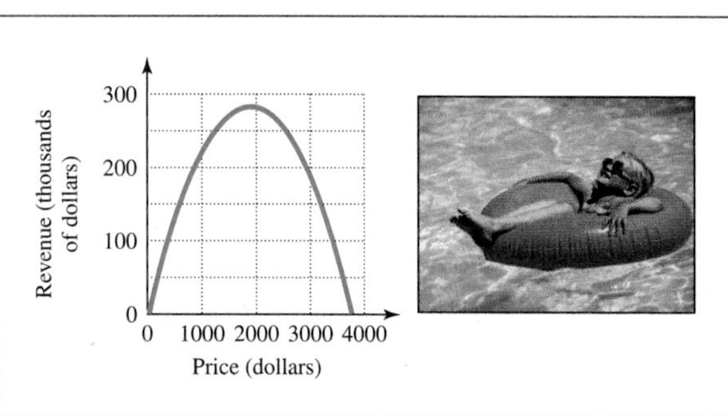

FIGURE FOR EXERCISE 90

91. ***Volume of a tank.*** The volume of a fish tank with a square base and height y is $y^3 - 6y^2 + 9y$ cubic inches. Find the length of a side of the square base.
$y - 3$ inches

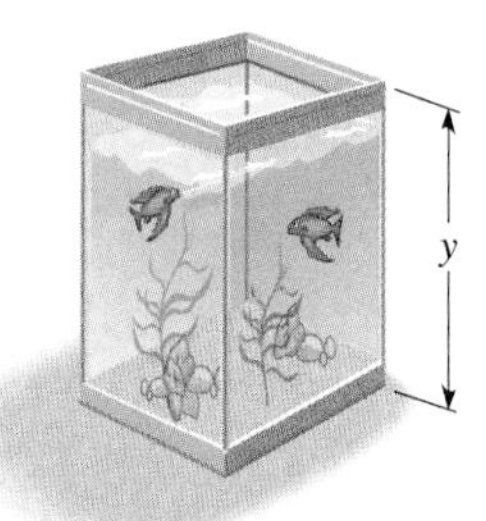

FIGURE FOR EXERCISE 91

GETTING MORE INVOLVED

92. ***Discussion.*** For what real number k, does $3x^2 - k$ factor as $3(x - 2)(x + 2)$? $k = 12$

93. ***Writing.*** Explain in your own words how to factor a four-term polynomial by grouping.

94. ***Writing.*** Explain how you know that $x^2 + 1$ is a prime polynomial.

6.3 FACTORING $ax^2 + bx + c$ WITH $a = 1$

In This Section

- Factoring $ax^2 + bx + c$ with $a = 1$
- Prime Polynomials
- Factoring with Two Variables
- Factoring Completely

In this section we will factor the type of trinomials that result from multiplying two different binomials. We will do this only for trinomials in which the coefficient of x^2, the leading coefficient, is 1. Factoring trinomials with leading coefficient not equal to 1 will be done in Section 6.4.

Factoring $ax^2 + bx + c$ with $a = 1$

Let's look closely at an example of finding the product of two binomials using the distributive property:

$$\begin{aligned}(x + 2)(x + 3) &= (x + 2)x + (x + 2)3 && \text{Distributive property}\\ &= x^2 + 2x + 3x + 6 && \text{Distributive property}\\ &= x^2 + 5x + 6 && \text{Combine like terms.}\end{aligned}$$

To factor $x^2 + 5x + 6$, we reverse these steps as shown in Example 1.

EXAMPLE 1 **Factoring a trinomial**

Factor.

a) $x^2 + 5x + 6$ **b)** $x^2 + 8x + 12$ **c)** $a^2 - 9a + 20$

Solution

a) The coefficient 5 is the sum of two numbers that have a product of 6. The only integers that have a product of 6 and a sum of 5 are 2 and 3. So write $5x$ as $2x + 3x$, then factor by grouping:

$$\begin{aligned} x^2 + 5x + 6 &= x^2 + 2x + 3x + 6 && \text{Replace } 5x \text{ by } 2x + 3x.\\ &= (x^2 + 2x) + (3x + 6) && \text{Group terms together.}\\ &= x(x + 2) + 3(x + 2) && \text{Factor out common factors.}\\ &= (x + 2)(x + 3) && \text{Factor out } x + 2. \end{aligned}$$

b) To factor $x^2 + 8x + 12$, we must find two integers that have a product of 12 and a sum of 8. The pairs of integers with a product of 12 are 1 and 12, 2 and 6, and 3 and 4. Only 2 and 6 have a sum of 8. So write $8x$ as $2x + 6x$ and factor by grouping:

$$\begin{aligned} x^2 + 8x + 12 &= x^2 + 2x + 6x + 12 \\ &= (x + 2)x + (x + 2)6 && \text{Factor out the common factors.}\\ &= (x + 2)(x + 6) && \text{Factor out } x + 2. \end{aligned}$$

Check by using FOIL: $(x + 2)(x + 6) = x^2 + 8x + 12$.

c) To factor $a^2 - 9a + 20$, we need two integers that have a product of 20 and a sum of -9. The integers are -4 and -5. Now replace $-9a$ by $-4a - 5a$ and factor by grouping:

$$\begin{aligned} a^2 - 9a + 20 &= a^2 - 4a - 5a + 20 && \text{Replace } -9a \text{ by } -4a - 5a.\\ &= a(a - 4) - 5(a - 4) && \text{Factor by grouping.}\\ &= (a - 5)(a - 4) && \text{Factor out } a - 4. \end{aligned}$$

■

Study Tip

Effective time management will allow adequate time for school, social life, and free time. However, at times you will have to sacrifice to do well.

After sufficient practice factoring trinomials, you may be able to skip most of the steps shown in these examples. For example, to factor $x^2 + x - 6$, simply find a pair of integers with a product of -6 and a sum of 1. The integers are 3 and -2, so we can write

$$x^2 + x - 6 = (x + 3)(x - 2)$$

and check by using FOIL.

EXAMPLE 2 **Factoring trinomials**

Factor.

a) $x^2 + 5x + 4$ **b)** $y^2 + 6y - 16$ **c)** $w^2 - 5w - 24$

Solution

a) To get a product of 4 and a sum of 5, use 1 and 4:

$$x^2 + 5x + 4 = (x + 1)(x + 4)$$

Check by using FOIL on $(x + 1)(x + 4)$.

b) To get a product of -16 we need a positive number and a negative number. To also get a sum of 6, use 8 and -2:

$$y^2 + 6y - 16 = (y + 8)(y - 2)$$

Check by using FOIL on $(y + 8)(y - 2)$.

c) To get a product of -24 and a sum of -5, use -8 and 3:

$$w^2 - 5w - 24 = (w - 8)(w + 3)$$

Check by using FOIL. ■

Polynomials are easiest to factor when they are in the form $ax^2 + bx + c$. So if a polynomial can be rewritten into that form, rewrite it before attempting to factor it. In Example 3 we factor polynomials that need to be rewritten.

EXAMPLE 3 **Factoring trinomials**

Factor.

a) $2x - 8 + x^2$ **b)** $-36 + t^2 - 9t$

Solution

a) Before factoring, write the trinomial as $x^2 + 2x - 8$. Now, to get a product of -8 and a sum of 2, use -2 and 4:

$$\begin{aligned} 2x - 8 + x^2 &= x^2 + 2x - 8 && \text{Write in } ax^2 + bx + c \text{ form.} \\ &= (x + 4)(x - 2) && \text{Factor and check by multiplying.} \end{aligned}$$

b) Before factoring, write the trinomial as $t^2 - 9t - 36$. Now, to get a product of -36 and a sum of -9, use -12 and 3:

$$\begin{aligned} -36 + t^2 - 9t &= t^2 - 9t - 36 && \text{Write in } ax^2 + bx + c \text{ form.} \\ &= (t - 12)(t + 3) && \text{Factor and check by multiplying.} \end{aligned}$$

■

Prime Polynomials

To factor $x^2 + bx + c$, we try pairs of integers that have a product of c until we find a pair that has a sum of b. If there is no such pair of integers, then the polynomial cannot be factored and it is a prime polynomial. Before you can conclude that a polynomial is prime, you must try *all* possibilities.

EXAMPLE 4 **Prime polynomials**

Factor.

a) $x^2 + 7x - 6$ **b)** $x^2 + 9$

Solution

a) Because the last term is -6, we want a positive integer and a negative integer that have a product of -6 and a sum of 7. Check all possible pairs of integers:

Product	Sum
$-6 = (-1)(6)$	$-1 + 6 = 5$
$-6 = (1)(-6)$	$1 + (-6) = -5$
$-6 = (2)(-3)$	$2 + (-3) = -1$
$-6 = (-2)(3)$	$-2 + 3 = 1$

Study Tip

Stay alert for the entire class period. The first 20 minutes is the easiest and the last 20 minutes the hardest. Some students put down their pencils, fold up their notebooks, and daydream for those last 20 minutes. Don't give in. Recognize when you are losing it and force yourself to stay alert. Think of how much time you will have to spend outside of class figuring out what happened during those last 20 minutes.

None of these possible factors of -6 have a sum of 7, so we can be certain that $x^2 + 7x - 6$ cannot be factored. It is a prime polynomial.

b) Because the x-term is missing in $x^2 + 9$, its coefficient is 0. That is, $x^2 + 9 = x^2 + 0x + 9$. So we seek two positive integers or two negative integers that have a product of 9 and a sum of 0. Check all possibilities:

Product	Sum
$9 = (3)(3)$	$3 + 3 = 6$
$9 = (-3)(-3)$	$-3 + (-3) = -6$
$9 = (9)(1)$	$9 + 1 = 10$
$9 = (-9)(-1)$	$-9 + (-1) = -10$

None of these pairs of integers have a sum of 0, so we can conclude that $x^2 + 9$ is a prime polynomial. Note that $x^2 + 9$ does not factor as $(x + 3)^2$ because $(x + 3)^2$ has a middle term: $(x + 3)^2 = x^2 + 6x + 9$. ■

The prime polynomial $x^2 + 9$ in Example 4(b) is a sum of two squares. It can be shown that any sum of two squares (in which there are no common factors) is a prime polynomial.

Helpful Hint

Don't confuse $a^2 + b^2$ with the difference of two squares $a^2 - b^2$ which is not a prime polynomial:

$$a^2 - b^2 = (a + b)(a - b).$$

Sum of Two Squares

If a sum of two squares, $a^2 + b^2$, has no common factor other than 1, then it is a prime polynomial.

Factoring with Two Variables

In Example 5 we factor polynomials that have two variables using the same technique that we used for one variable.

EXAMPLE 5 **Polynomials with two variables**

Factor.

a) $x^2 + 2xy - 8y^2$ **b)** $a^2 - 7ab + 10b^2$

Solution

a) To get a product of -8 and a sum of 2, use 4 and -2. To get a product of $-8y^2$ use $4y$ and $-2y$:

$$x^2 + 2xy - 8y^2 = (x + 4y)(x - 2y)$$

Check by multiplying $(x + 4y)(x - 2y)$.

b) To get a product of 10 and a sum of -7, use -5 and -2. To get a product of $10b^2$, we use $-5b$ and $-2b$:

$$a^2 - 7ab + 10b^2 = (a - 5b)(a - 2b)$$

Check by multiplying. ■

Factoring Completely

In Section 6.2 you learned that binomials such as $3x - 5$ (with no common factor) are prime polynomials. In Example 4 of this section we saw a trinomial that is a prime polynomial. There are infinitely many prime trinomials. When factoring a polynomial completely, we could have a factor that is a prime trinomial.

EXAMPLE 6 **Factoring completely**

Factor each polynomial completely.

a) $x^3 - 6x^2 - 16x$

b) $4x^3 + 4x^2 + 4x$

Solution

a) $x^3 - 6x^2 - 16x = x(x^2 - 6x - 16)$ Factor out the GCF.

$= x(x - 8)(x + 2)$ Factor $x^2 - 6x - 16$.

b) First factor out $4x$, the greatest common factor:

$$4x^3 + 4x^2 + 4x = 4x(x^2 + x + 1)$$

To factor $x^2 + x + 1$, we would need two integers with a product of 1 and a sum of 1. Because there are no such integers, $x^2 + x + 1$ is prime, and the factorization is complete. ■

WARM-UPS

True or false? Answer true if the correct factorization is given and false if the factorization is incorrect. Explain your answer.

1. $x^2 - 6x + 9 = (x - 3)^2$ True
2. $x^2 + 6x + 9 = (x + 3)^2$ True
3. $x^2 + 10x + 9 = (x - 9)(x - 1)$ False
4. $x^2 - 8x - 9 = (x - 8)(x - 9)$ False
5. $x^2 + 8x - 9 = (x + 9)(x - 1)$ True
6. $x^2 + 8x + 9 = (x + 3)^2$ False
7. $x^2 - 10xy + 9y^2 = (x - y)(x - 9y)$ True
8. $x^2 + x + 1 = (x + 1)(x + 1)$ False
9. $x^2 + xy + 20y^2 = (x + 5y)(x - 4y)$ False
10. $x^2 + 1 = (x + 1)(x + 1)$ False

6.3 EXERCISES

Reading and Writing *After reading this section, write out the answers to these questions. Use complete sentences.*

1. What types of polynomials did we factor in this section?
   We factored $ax^2 + bx + c$ with $a = 1$.
2. How can you check if you have factored a trinomial correctly?
   You can check all factoring by multiplying the factors.
3. How can you determine if $x^2 + bx + c$ is prime?
   If there are no two integers that have a product of c and a sum of b, then $x^2 + bx + c$ is prime.
4. How do you factor a sum of two squares?
   A sum of two squares with no common factor is prime.
5. When is a polynomial factored completely?
   A polynomial is factored completely when all of the factors are prime polynomials.
6. What should you always look for first when attempting to factor a polynomial completely?
   Always look for the GCF first.

Factor each trinomial. Write out all of the steps as shown in Example 1.

7. $x^2 + 4x + 3$
   $(x + 3)(x + 1)$
8. $y^2 + 6y + 5$
   $(y + 5)(y + 1)$
9. $x^2 + 9x + 18$
   $(x + 3)(x + 6)$
10. $w^2 + 6w + 8$
    $(w + 2)(w + 4)$

11. $a^2 - 7a + 12$ $(a - 3)(a - 4)$

12. $m^2 - 9m + 14$ $(m - 2)(m - 7)$

13. $b^2 - 5b - 6$ $(b - 6)(b + 1)$

14. $a^2 + 5a - 6$ $(a + 6)(a - 1)$

Factor each polynomial. See Examples 2–4. If the polynomial is prime, say so.

15. $y^2 + 7y + 10$ $(y + 2)(y + 5)$

16. $x^2 + 8x + 15$ $(x + 3)(x + 5)$

17. $a^2 - 6a + 8$ $(a - 2)(a - 4)$

18. $b^2 - 8b + 15$ $(b - 3)(b - 5)$

19. $m^2 - 10m + 16$ $(m - 8)(m - 2)$

20. $m^2 - 17m + 16$ $(m - 16)(m - 1)$

21. $w^2 + 9w - 10$ $(w + 10)(w - 1)$

22. $m^2 + 6m - 16$ $(m + 8)(m - 2)$

23. $w^2 - 8 - 2w$ $(w - 4)(w + 2)$

24. $-16 + m^2 - 6m$ $(m - 8)(m + 2)$

25. $a^2 - 2a - 12$ Prime

26. $x^2 + 3x + 3$ Prime

27. $15m - 16 + m^2$ $(m + 16)(m - 1)$

28. $3y + y^2 - 10$ $(y + 5)(y - 2)$

29. $a^2 - 4a + 12$ Prime

30. $y^2 - 6y - 8$ Prime

31. $z^2 - 25$ $(z - 5)(z + 5)$

32. $p^2 - 1$ $(p - 1)(p + 1)$

33. $h^2 + 49$ Prime

34. $q^2 + 4$ Prime

35. $m^2 + 12m + 20$ $(m + 2)(m + 10)$

36. $m^2 + 21m + 20$ $(m + 1)(m + 20)$

37. $t^2 - 3t + 10$ Prime

38. $x^2 - 5x - 3$ Prime

39. $m^2 - 18 - 17m$ $(m - 18)(m + 1)$

40. $h^2 - 36 + 5h$ $(h + 9)(h - 4)$

41. $m^2 - 23m + 24$ Prime

42. $m^2 + 23m + 24$ Prime

43. $5t - 24 + t^2$ $(t + 8)(t - 3)$

44. $t^2 - 24 - 10t$ $(t - 12)(t + 2)$

45. $t^2 - 2t - 24$ $(t - 6)(t + 4)$

46. $t^2 + 14t + 24$ $(t + 12)(t + 2)$

47. $t^2 - 10t - 200$ $(t - 20)(t + 10)$

48. $t^2 + 30t + 200$ $(t + 20)(t + 10)$

49. $x^2 - 5x - 150$ $(x - 15)(x + 10)$

50. $x^2 - 25x + 150$ $(x - 15)(x - 10)$

51. $13y + 30 + y^2$ $(y + 3)(y + 10)$

52. $18z + 45 + z^2$ $(z + 3)(z + 15)$

Factor each polynomial. See Example 5.

53. $x^2 + 5ax + 6a^2$ $(x + 3a)(x + 2a)$

54. $a^2 + 7ab + 10b^2$ $(a + 2b)(a + 5b)$

55. $x^2 - 4xy - 12y^2$ $(x - 6y)(x + 2y)$

56. $y^2 + yt - 12t^2$ $(y + 4t)(y - 3t)$

57. $x^2 - 13xy + 12y^2$ $(x - 12y)(x - y)$

58. $h^2 - 9hs + 9s^2$ Prime

59. $x^2 + 4xz - 33z^2$ Prime

60. $x^2 - 5xs - 24s^2$ $(x - 8s)(x + 3s)$

Factor each polynomial completely. Use the methods discussed in Sections 6.1 through 6.3. See Example 6.

61. $w^2 - 8w$ $w(w - 8)$

62. $x^4 - x^3$ $x^3(x - 1)$

63. $2w^2 - 162$ $2(w - 9)(w + 9)$

64. $6w^4 - 54w^2$ $6w^2(w - 3)(w + 3)$

65. $x^2w^2 + 9x^2$ $x^2(w^2 + 9)$

66. $a^4b + a^2b^3$ $a^2b(a^2 + b^2)$

67. $w^2 - 18w + 81$ $(w - 9)^2$

68. $w^2 + 30w + 81$ $(w + 3)(w + 27)$

69. $6w^2 - 12w - 18$ $6(w - 3)(w + 1)$

70. $9w - w^3$ $w(3 - w)(3 + w)$

71. $32x^2 - 2x^4$ $2x^2(4 - x)(4 + x)$

72. $20w^2 + 100w + 40$ $20(w^2 + 5w + 2)$

73. $3w^2 + 27w + 54$ $3(w + 3)(w + 6)$

74. $w^3 - 3w^2 - 18w$ $w(w - 6)(w + 3)$

75. $18w^2 + w^3 + 36w$ $w(w^2 + 18w + 36)$

76. $18a^2 + 3a^3 + 36a$ $3a(a^2 + 6a + 12)$

77. $8vw^2 + 32vw + 32v$ $8v(w + 2)^2$

78. $3h^2t + 6ht + 3t$ $3t(h + 1)^2$

79. $6x^3y + 30x^2y^2 + 36xy^3$ $6xy(x + 3y)(x + 2y)$

80. $3x^3y^2 - 3x^2y^2 + 3xy^2$ $3xy^2(x^2 - x + 1)$

Complete the factoring.

81. $a^2 + 4a - 12 = (a + 6)(a - 2)$

82. $a^2 - a - 12 = (a - 4)(a + 3)$

83. $a^2 - 3a = (a - 3)(a)$

84. $a^2 - 9 = (a - 3)(a + 3)$

85. $a^2 + 12a + 36 = (a + 6)^2$

86. $a^2 - 12a + 36 = (a - 6)^2$

87. $a^4 + 9a^3 - 36a^2 = a^2(a + 12)(a - 3)$

88. $a^5 + a^4 - 12a^3 = (a^3)(a + 4)(a - 3)$

Use factoring to solve each problem.

89. ***Area of a deck.*** A rectangular deck has an area of $x^2 + 6x + 8$ square feet and a width of $x + 2$ feet. Find the length of the deck. $x + 4$ feet

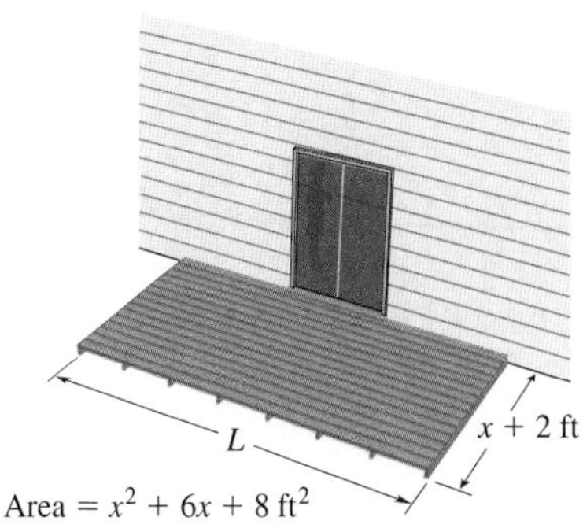

FIGURE FOR EXERCISE 89

90. ***Area of a sail.*** A triangular sail has an area of $x^2 + 5x + 6$ square meters and a height of $x + 3$ meters. Find the length of the sail's base. $2x + 4$ meters

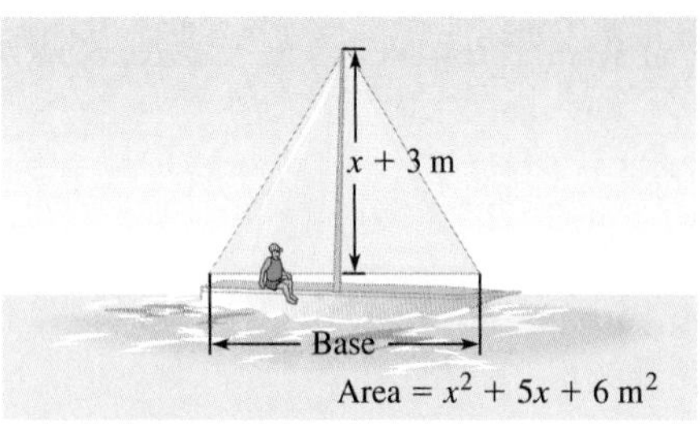

FIGURE FOR EXERCISE 90

91. ***Volume of a cube.*** Hector designed a cubic box with volume x^3 cubic feet. After increasing the dimensions of the bottom, the box has a volume of $x^3 + 8x^2 + 15x$ cubic feet. If each of the dimensions of the bottom was increased by a whole number of feet, then how much was each increase? 3 feet and 5 feet

92. ***Volume of a container.*** A cubic shipping container had a volume of a^3 cubic meters. The height was decreased by a whole number of meters and the width was increased by a whole number of meters so that the volume of the container is now $a^3 + 2a^2 - 3a$ cubic meters. By how many meters were the height and width changed? Height 1 foot smaller, width 3 feet larger

GETTING MORE INVOLVED

93. ***Discussion.*** Which of the following products is not equivalent to the others. Explain your answer.

a) $(2x - 4)(x + 3)$ **b)** $(x - 2)(2x + 6)$

c) $2(x - 2)(x + 3)$ **d)** $(2x - 4)(2x + 6)$ d

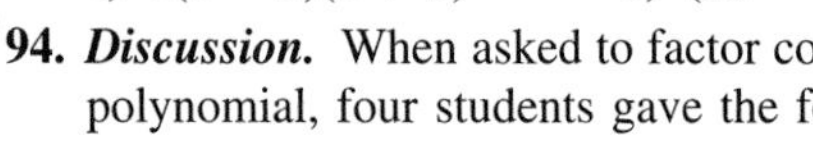

94. ***Discussion.*** When asked to factor completely a certain polynomial, four students gave the following answers. Only one student gave the correct answer. Which one must it be? Explain your answer.

a) $3(x^2 - 2x - 15)$ **b)** $(3x - 5)(5x - 15)$

c) $3(x - 5)(x - 3)$ **d)** $(3x - 15)(x - 3)$ c

In This Section

- The *ac* Method
- Trial and Error
- Factoring Completely

6.4 FACTORING $ax^2 + bx + c$ WITH $a \neq 1$

In Section 6.3 we used grouping to factor trinomials with a leading coefficient of 1. In this section we will also use grouping to factor trinomials with a leading coefficient that is not equal to 1.

The *ac* Method

The first step in factoring $ax^2 + bx + c$ with $a = 1$ is to find two numbers with a product of c and a sum of b. If $a \neq 1$, then the first step is to find two numbers with a product of ac and a sum of b. This method is called the ***ac* method.** The strategy for factoring by the ac method follows. Note that this strategy works whether or not the leading coefficient is 1.

Strategy for Factoring $ax^2 + bx + c$ by the *ac* Method

To factor the trinomial $ax^2 + bx + c$:

1. Find two numbers that have a product equal to ac and a sum equal to b.
2. Replace bx by two terms using the two new numbers as coefficients.
3. Factor the resulting four-term polynomial by grouping.

EXAMPLE 1 **The *ac* method**

Factor each trinomial.

a) $2x^2 + 7x + 6$ **b)** $2x^2 + x - 6$ **c)** $10x^2 + 13x - 3$

Solution

a) In $2x^2 + 7x + 6$ we have $a = 2$, $b = 7$, and $c = 6$. So

$$ac = 2 \cdot 6 = 12.$$

Now we need two integers with a product of 12 and a sum of 7. The pairs of integers with a product of 12 are 1 and 12, 2 and 6, and 3 and 4. Only 3 and 4 have a sum of 7. Replace $7x$ by $3x + 4x$ and factor by grouping:

$$\begin{aligned} 2x^2 + 7x + 6 &= 2x^2 + 3x + 4x + 6 && \text{Replace } 7x \text{ by } 3x + 4x. \\ &= (2x + 3)x + (2x + 3)2 && \text{Factor out the common factors.} \\ &= (2x + 3)(x + 2) && \text{Factor out } 2x + 3. \end{aligned}$$

Check by FOIL.

b) In $2x^2 + x - 6$ we have $a = 2$, $b = 1$, and $c = -6$. So

$$ac = 2(-6) = -12.$$

Now we need two integers with a product of -12 and a sum of 1. We can list the possible pairs of integers with a product of -12 as follows:

1 and -12	2 and -6	3 and -4
-1 and 12	-2 and 6	-3 and 4

Only -3 and 4 have a sum of 1. Replace x by $-3x + 4x$ and factor by grouping:

$$\begin{aligned} 2x^2 + x - 6 &= 2x^2 - 3x + 4x - 6 && \text{Replace } x \text{ by } -3x + 4x. \\ &= (2x - 3)x + (2x - 3)2 && \text{Factor out the common factors.} \\ &= (2x - 3)(x + 2) && \text{Factor out } 2x - 3. \end{aligned}$$

Check by FOIL.

c) Because $ac = 10(-3) = -30$, we need two integers with a product of -30 and a sum of 13. The product is negative, so the integers must have opposite signs. We can list all pairs of factors of -30 as follows:

1 and -30	2 and -15	3 and -10	5 and -6
-1 and 30	-2 and 15	-3 and 10	-5 and 6

The only pair that has a sum of 13 is -2 and 15:

$$\begin{aligned} 10x^2 + 13x - 3 &= 10x^2 - 2x + 15x - 3 && \text{Replace } 13x \text{ by } -2x + 15x. \\ &= (5x - 1)2x + (5x - 1)3 && \text{Factor out the common factors.} \\ &= (5x - 1)(2x + 3) && \text{Factor out } 5x - 1. \end{aligned}$$

Check by FOIL. ■

EXAMPLE 2 **Factoring a trinomial in two variables by the *ac* method**

Factor $8x^2 - 14xy + 3y^2$

Solution

Since $a = 8$, $b = -14$, and $c = 3$, we have $ac = 24$. Two numbers with a product of 24 and a sum of -14 must both be negative. The possible pairs with a product of 24 follow:

-1 and -24	-3 and -8
-2 and -12	-4 and -6

Only -2 and -12 have a sum of -14. Replace $-14xy$ by $-2xy - 12xy$ and factor by grouping:

$$\begin{aligned} 8x^2 - 14xy + 3y^2 &= 8x^2 - 2xy - 12xy + 3y^2 \\ &= (4x - y)2x + (4x - y)(-3y) \\ &= (4x - y)(2x - 3y) \end{aligned}$$

Check by FOIL. ■

Trial and Error

After you have gained some experience at factoring by the *ac* method, you can often find the factors without going through the steps of grouping. For example, consider the polynomial

$$3x^2 + 7x - 6.$$

The factors of $3x^2$ can only be $3x$ and x. The factors of 6 could be 2 and 3 or 1 and 6. We can list all of the possibilities that give the correct first and last terms, without regard to the signs:

$$(3x \quad 3)(x \quad 2) \qquad (3x \quad 2)(x \quad 3) \qquad (3x \quad 6)(x \quad 1) \qquad (3x \quad 1)(x \quad 6)$$

Because the factors of -6 have unlike signs, one binomial factor is a sum and the other binomial is a difference. Now we try some products to see if we get a middle term of $7x$:

$$(3x + 3)(x - 2) = 3x^2 - 3x - 6 \quad \text{Incorrect.}$$
$$(3x - 3)(x + 2) = 3x^2 + 3x - 6 \quad \text{Incorrect.}$$

Actually, there is no need to try $(3x \quad 3)(x \quad 2)$ or $(3x \quad 6)(x \quad 1)$ because each contains a binomial with a common factor. As you can see from the above products, a common factor in the binomial causes a common factor in the product. But $3x^2 + 7x - 6$ has no common factor. So the factors must come from either $(3x \quad 2)(x \quad 3)$ or $(3x \quad 1)(x \quad 6)$. So we try again:

$$(3x + 2)(x - 3) = 3x^2 - 7x - 6 \quad \text{Incorrect.}$$
$$(3x - 2)(x + 3) = 3x^2 + 7x - 6 \quad \text{Correct.}$$

Even though there may be many possibilities in some factoring problems, it is often possible to find the correct factors without writing down every possibility. We can use a bit of guesswork in factoring trinomials. *Try* whichever possibility you think might work. *Check* it by multiplying. If it is not right, then *try again.* That is why this method is called **trial and error.**

Helpful Hint

The *ac* method is more systematic than trial and error. However, trial and error can be faster and easier, especially if your first or second trial is correct.

EXAMPLE 3 **Trial and error**

Factor each trinomial using trial and error.

a) $2x^2 + 5x - 3$ **b)** $3x^2 - 11x + 6$

Solution

a) Because $2x^2$ factors only as $2x \cdot x$ and 3 factors only as $1 \cdot 3$, there are only two possible ways to get the correct first and last terms, without regard to the signs:

$$(2x \quad 1)(x \quad 3) \quad \text{and} \quad (2x \quad 3)(x \quad 1)$$

Study Tip

Have you ever used excuses to avoid studying? ("Before I can study, I have to do my laundry and go to the bank.") Since the average attention span for one task is approximately 20 minutes, it is better to take breaks from studying to run errands and do laundry than to get everything done before you start studying.

Because the last term of the trinomial is negative, one of the missing signs must be $+$, and the other must be $-$. The trinomial is factored correctly as

$$2x^2 + 5x - 3 = (2x - 1)(x + 3).$$

Check by using FOIL.

b) There are four possible ways to factor $3x^2 - 11x + 6$:

$$(3x \quad 1)(x \quad 6) \qquad (3x \quad 2)(x \quad 3)$$
$$(3x \quad 6)(x \quad 1) \qquad (3x \quad 3)(x \quad 2)$$

Because the last term in $3x^2 - 11x + 6$ is positive and the middle term is negative, both signs in the factors must be negative. Because $3x^2 - 11x + 6$ has no common factor, we can rule out $(3x \quad 6)(x \quad 1)$ and $(3x \quad 3)(x \quad 2)$. So the only possibilities left are $(3x - 1)(x - 6)$ and $(3x - 2)(x - 3)$. The trinomial is factored correctly as

$$3x^2 - 11x + 6 = (3x - 2)(x - 3).$$

Check by using FOIL. ■

Factoring by trial and error is not just guessing. In fact, if the trinomial has a positive leading coefficient, we can determine in advance whether its factors are sums or differences.

Using Signs in Trial and Error

1. If the signs of the terms of a trinomial are $+ + +$ then both factors are sums: $x^2 + 5x + 6 = (x + 2)(x + 3)$.
2. If the signs are $+ - +$ then both factors are differences: $x^2 - 5x + 6 = (x - 2)(x - 3)$.
3. If the signs are $+ + -$ or $+ - -$ then one factor is a sum and the other is a difference: $x^2 + x - 6 = (x + 3)(x - 2)$ and $x^2 - x - 6 = (x - 3)(x + 2)$.

In Example 4 we factor a trinomial that has two variables.

EXAMPLE 4 **Factoring a trinomial with two variables by trial and error**

Factor $6x^2 - 7xy + 2y^2$.

Solution

We list the possible ways to factor the trinomial:

$$(3x \quad 2y)(2x \quad y) \qquad (3x \quad y)(2x \quad 2y) \qquad (6x \quad 2y)(x \quad y) \qquad (6x \quad y)(x \quad 2y)$$

Because the last term of the trinomial is positive and the middle term is negative, both factors must contain subtraction symbols. To get the middle term of $-7xy$, we use the first possibility listed:

$$6x^2 - 7xy + 2y^2 = (3x - 2y)(2x - y)$$ ■

Factoring Completely

You can use the latest factoring technique along with the techniques that you learned earlier to factor polynomials completely. Remember always to first factor out the greatest common factor (if it is not 1).

EXAMPLE 5 **Factoring completely**

Factor each polynomial completely.

a) $4x^3 + 14x^2 + 6x$ b) $12x^2y + 6xy + 6y$

Solution

a) $4x^3 + 14x^2 + 6x = 2x(2x^2 + 7x + 3)$ Factor out the GCF, $2x$.

$= 2x(2x + 1)(x + 3)$ Factor $2x^2 + 7x + 3$.

Check by multiplying.

b) $12x^2y + 6xy + 6y = 6y(2x^2 + x + 1)$ Factor out the GCF, $6y$.

To factor $2x^2 + x + 1$ by the *ac* method, we need two numbers with a product of 2 and a sum of 1. Because there are no such numbers, $2x^2 + x + 1$ is prime and the factorization is complete. ■

Our first step in factoring is to factor out the greatest common factor (if it is not 1). If the first term of a polynomial has a negative coefficient, then it is better to factor out the opposite of the GCF so that the resulting polynomial will have a positive leading coefficient.

EXAMPLE 6 **Factoring out the opposite of the GCF**

Factor each polynomial completely.

a) $-18x^3 + 51x^2 - 15x$ b) $-3a^2 + 2a + 21$

Solution

a) The GCF is $3x$. Because the first term has a negative coefficient, we factor out $-3x$:

$-18x^3 + 51x^2 - 15x = -3x(6x^2 - 17x + 5)$ Factor out $-3x$.

$= -3x(3x - 1)(2x - 5)$ Factor $6x^2 - 17x + 5$.

b) The GCF for $-3a^2 + 2a + 21$ is 1. Because the first term has a negative coefficient, factor out -1:

$-3a^2 + 2a + 21 = -1(3a^2 - 2a - 21)$ Factor out -1.

$= -1(3a + 7)(a - 3)$ Factor $3a^2 - 2a - 21$. ■

WARM-UPS

True or false? Answer true if the correct factorization is given and false if the factorization is incorrect. Explain your answer.

1. $2x^2 + 3x + 1 = (2x + 1)(x + 1)$ True

2. $2x^2 + 5x + 3 = (2x + 1)(x + 3)$ False

3. $3x^2 + 10x + 3 = (3x + 1)(x + 3)$ True

4. $15x^2 + 31x + 14 = (3x + 7)(5x + 2)$ False

5. $2x^2 - 7x - 9 = (2x - 9)(x + 1)$ True

6. $2x^2 + 3x - 9 = (2x + 3)(x - 3)$ False

7. $2x^2 - 16x - 9 = (2x - 9)(2x + 1)$ False

8. $8x^2 - 22x - 5 = (4x - 1)(2x + 5)$ False

9. $9x^2 + x - 1 = (5x - 1)(4x + 1)$ False

10. $12x^2 - 13x + 3 = (3x - 1)(4x - 3)$ True

6.4 EXERCISES

Reading and Writing *After reading this section, write out the answers to these questions. Use complete sentences.*

1. What types of polynomials did we factor in this section?
We factored $ax^2 + bx + c$ with $a \neq 1$.

2. What is the *ac* method of factoring?
In the *ac* method we find two integers whose product is equal to *ac* and whose sum is *b*, and then we use factoring by grouping.

3. How can you determine if $ax^2 + bx + c$ is prime?
If there are no two integers whose product is *ac* and whose sum is *b*, then $ax^2 + bx + c$ is prime.

4. What is the trial-and-error method of factoring?
In trial and error, we make an educated guess at the factors and then check by FOIL.

Find the following. See Example 1.

5. Two integers that have a product of 20 and a sum of 12
2 and 10

6. Two integers that have a product of 36 and a sum of -20
-2 and -18

7. Two integers that have a product of -12 and a sum of -4
-6 and 2

8. Two integers that have a product of -8 and a sum of 7
8 and -1

Each of the following trinomials is in the form $ax^2 + bx + c$. For each trinomial, find two integers that have a product of ac and a sum of b. Do not factor the trinomials. See Example 1.

9. $6x^2 + 7x + 2$
3 and 4

10. $5x^2 + 17x + 6$
2 and 15

11. $6y^2 - 11y + 3$
-2 and -9

12. $6z^2 - 19z + 10$
-4 and -15

13. $12w^2 + w - 1$
-3 and 4

14. $15t^2 - 17t - 4$
-20 and 3

Factor each trinomial using the ac method. See Example 1.

15. $2x^2 + 3x + 1$
$(2x + 1)(x + 1)$

16. $2x^2 + 11x + 5$
$(2x + 1)(x + 5)$

17. $2x^2 + 9x + 4$
$(2x + 1)(x + 4)$

18. $2h^2 + 7h + 3$
$(2h + 1)(h + 3)$

19. $3t^2 + 7t + 2$
$(3t + 1)(t + 2)$

20. $3t^2 + 8t + 5$
$(3t + 5)(t + 1)$

21. $2x^2 + 5x - 3$
$(2x - 1)(x + 3)$

22. $3x^2 - x - 2$
$(3x + 2)(x - 1)$

23. $6x^2 + 7x - 3$
$(3x - 1)(2x + 3)$

24. $21x^2 + 2x - 3$
$(3x - 1)(7x + 3)$

25. $3x^2 - 5x + 4$ Prime

26. $6x^2 - 5x + 3$ Prime

27. $2x^2 - 7x + 6$
$(2x - 3)(x - 2)$

28. $3a^2 - 14a + 15$
$(3a - 5)(a - 3)$

29. $5b^2 - 13b + 6$
$(5b - 3)(b - 2)$

30. $7y^2 + 16y - 15$
$(7y - 5)(y + 3)$

31. $4y^2 - 11y - 3$
$(4y + 1)(y - 3)$

32. $35x^2 - 2x - 1$
$(7x + 1)(5x - 1)$

33. $3x^2 + 2x + 1$ Prime

34. $6x^2 - 4x - 5$ Prime

35. $8x^2 - 2x - 1$
$(4x + 1)(2x - 1)$

36. $8x^2 - 10x - 3$
$(4x + 1)(2x - 3)$

37. $9t^2 - 9t + 2$
$(3t - 1)(3t - 2)$

38. $9t^2 + 5t - 4$
$(9t - 4)(t + 1)$

39. $15x^2 + 13x + 2$
$(5x + 1)(3x + 2)$

40. $15x^2 - 7x - 2$
$(5x + 1)(3x - 2)$

Use the ac method to factor each trinomial. See Example 2.

41. $4a^2 + 16ab + 15b^2$
$(2a + 3b)(2a + 5b)$

42. $10x^2 + 17xy + 3y^2$
$(5x + y)(2x + 3y)$

43. $6m^2 - 7mn - 5n^2$
$(3m - 5n)(2m + n)$

44. $3a^2 + 2ab - 21b^2$
$(3a - 7b)(a + 3b)$

45. $3x^2 - 8xy + 5y^2$
$(x - y)(3x - 5y)$

46. $3m^2 - 13mn + 12n^2$
$(m - 3n)(3m - 4n)$

Factor each trinomial using trial and error. See Examples 3 and 4.

47. $5a^2 + 11a + 2$
$(5a + 1)(a + 2)$

48. $3y^2 + 10y + 7$
$(3y + 7)(y + 1)$

49. $4w^2 + 8w + 3$
$(2w + 3)(2w + 1)$

50. $6z^2 + 13z + 5$
$(2z + 1)(3z + 5)$

51. $15x^2 - x - 2$
$(5x - 2)(3x + 1)$

52. $15x^2 + 13x - 2$
$(15x - 2)(x + 1)$

53. $8x^2 - 6x + 1$
$(4x - 1)(2x - 1)$

54. $8x^2 - 22x + 5$
$(4x - 1)(2x - 5)$

55. $15x^2 - 31x + 2$
$(15x - 1)(x - 2)$

56. $15x^2 + 31x + 2$
$(15x + 1)(x + 2)$

57. $4x^2 - 4x + 3$ Prime

58. $4x^2 + 12x - 5$ Prime

59. $2x^2 + 18x - 90$
$2(x^2 + 9x - 45)$

60. $3x^2 + 11x + 10$
$(x + 2)(3x + 5)$

61. $3x^2 + x - 10$
$(3x - 5)(x + 2)$

62. $3x^2 - 17x + 10$
$(3x - 2)(x - 5)$

63. $10x^2 - 3xy - y^2$
$(5x + y)(2x - y)$

64. $8x^2 - 2xy - y^2$
$(4x + y)(2x - y)$

65. $42a^2 - 13ab + b^2$
$(6a - b)(7a - b)$

66. $10a^2 - 27ab + 5b^2$
$(5a - b)(2a - 5b)$

Complete the factoring.

67. $3x^2 + 7x + 2 = (x + 2)(3x + 1)$

68. $2x^2 - x - 15 = (x - 3)(2x + 5)$

69. $5x^2 + 11x + 2 = (5x + 1)(x + 2)$

70. $4x^2 - 19x - 5 = (4x + 1)(x - 5)$

71. $6a^2 - 17a + 5 = (3a - 1)(2a - 5)$

72. $4b^2 - 16b + 15 = (2b - 5)(2b - 3)$

Factor each polynomial completely. See Examples 5 and 6.

73. $81w^3 - w$
$w(9w - 1)(9w + 1)$

74. $81w^3 - w^2$
$w^2(81w - 1)$

75. $4w^2 + 2w - 30$
$2(2w - 5)(w + 3)$

76. $2x^2 - 28x + 98$
$2(x - 7)^2$

77. $27 + 12x^2 + 36x$
$3(2x + 3)^2$

78. $24y + 12y^2 + 12$
$12(y + 1)^2$

79. $6w^2 - 11w - 35$ $(3w + 5)(2w - 7)$

80. $8y^2 - 14y - 15$ $(2y - 5)(4y + 3)$

81. $3x^2z - 3zx - 18z$ $3z(x - 3)(x + 2)$

82. $a^2b + 2ab - 15b$ $b(a + 5)(a - 3)$

83. $9x^3 - 21x^2 + 18x$ $3x(3x^2 - 7x + 6)$

84. $-8x^3 + 4x^2 - 2x$ $-2x(4x^2 - 2x + 1)$

85. $a^2 + 2ab - 15b^2$ $(a + 5b)(a - 3b)$

86. $a^2b^2 - 2a^2b - 15a^2$ $a^2(b - 5)(b + 3)$

87. $2x^2y^2 + xy^2 + 3y^2$ $y^2(2x^2 + x + 3)$

88. $18x^2 - 6x + 6$ $6(3x^2 - x + 1)$

89. $-6t^3 - t^2 + 2t$ $-t(3t + 2)(2t - 1)$

90. $-36t^2 - 6t + 12$ $-6(3t + 2)(2t - 1)$

91. $12t^4 - 2t^3 - 4t^2$ $2t^2(3t - 2)(2t + 1)$

92. $12t^3 + 14t^2 + 4t$ $2t(3t + 2)(2t + 1)$

93. $4x^2y - 8xy^2 + 3y^3$ $y(2x - y)(2x - 3y)$

94. $9x^2 + 24xy - 9y^2$ $3(3x - y)(x + 3y)$

95. $-4w^2 + 7w - 3$ $-1(w - 1)(4w - 3)$

96. $-30w^2 + w + 1$ $-1(5w - 1)(6w + 1)$

97. $-12a^3 + 22a^2b - 6ab^2$ $-2a(2a - 3b)(3a - b)$

98. $-36a^2b + 21ab^2 - 3b^3$ $-3b(3a - b)(4a - b)$

Solve each problem.

99. ***Height of a ball.*** If a ball is thrown upward at 40 feet per second from a rooftop 24 feet above the ground, then its height above the ground t seconds after it is thrown is given by $h = -16t^2 + 40t + 24$. Rewrite this formula with the polynomial on the right-hand side factored completely. Use the factored version of the formula to find h when $t = 3$.
$h = -8(2t + 1)(t - 3)$, 0 feet

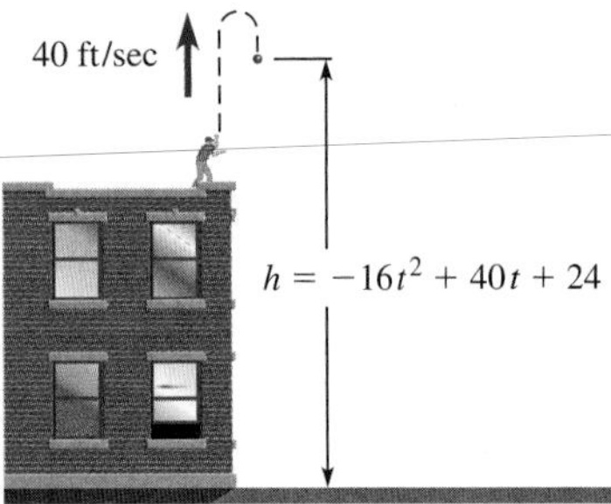

FIGURE FOR EXERCISE 99

100. ***Worker efficiency.*** In a study of worker efficiency at Wong Laboratories it was found that the number of components assembled per hour by the average worker t hours after starting work could be modeled by the formula

$$N(t) = -3t^3 + 23t^2 + 8t.$$

a) Rewrite the formula by factoring the right-hand side completely. $N(t) = -t(3t + 1)(t - 8)$

b) Use the factored version of the formula to find $N(3)$. 150

c) Use the accompanying graph to estimate the time at which the workers are most efficient. 5 hr

d) Use the accompanying graph to estimate the maximum number of components assembled per hour during an 8-hour shift. 250 components

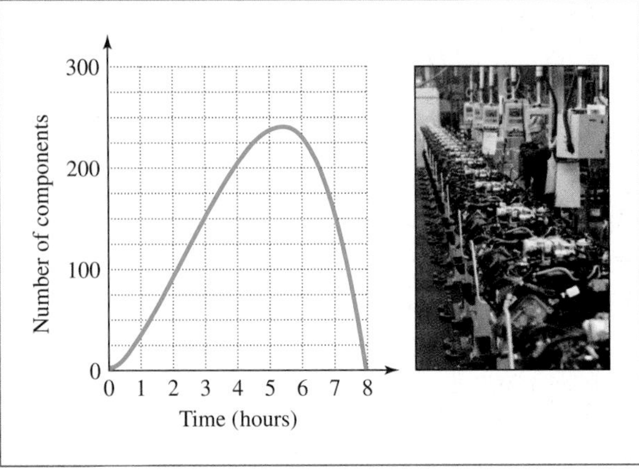

FIGURE FOR EXERCISE 100

GETTING MORE INVOLVED

101. ***Exploration.*** Find all positive and negative integers b for which each polynomial can be factored.

a) $x^2 + bx + 3$ $-4, 4$ **b)** $3x^2 + bx + 5$ $\pm 8, \pm 16$

c) $2x^2 + bx - 15$ $\pm 1, \pm 7, \pm 13, \pm 29$

102. ***Exploration.*** Find two integers c (positive or negative) for which each polynomial can be factored. Many answers are possible.

a) $x^2 + x + c$ $-2, -6$ **b)** $x^2 - 2x + c$ $1, -8$

c) $2x^2 - 3x + c$ $1, -9$

103. ***Cooperative learning.*** Working in groups, cut two large squares, three rectangles, and one small square out of paper that are exactly the same size as shown in the accompanying figure. Then try to place the six figures next to one another so that they form a large rectangle. Do not overlap the pieces or leave any gaps. Explain how factoring $2x^2 + 3x + 1$ can help you solve this puzzle.

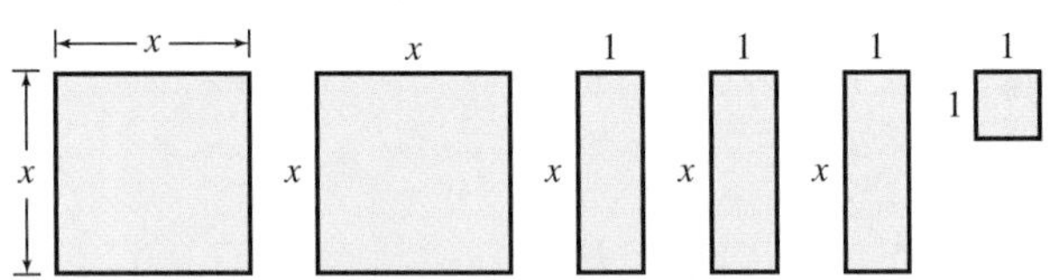

FIGURE FOR EXERCISE 103

104. ***Cooperative learning.*** Working in groups, cut four squares and eight rectangles out of paper as in the previous exercise to illustrate the trinomial $4x^2 + 7x + 3$. Select one group to demonstrate how to arrange the 12 pieces to form a large rectangle. Have another group explain how factoring the trinomial can help you solve this puzzle.

6.5 FACTORING A DIFFERENCE OR SUM OF TWO CUBES

In This Section

- Using Division in Factoring
- Factoring a Difference or Sum of Two Cubes
- The Factoring Strategy

In Sections 6.1 to 6.4 we established the general idea of factoring and some special cases. In this section we will see how division relates to factoring and see two more special cases. We will then summarize all of the factoring that we have done with a factoring strategy.

Using Division in Factoring

To find the prime factorization for a large integer such as 1001, you could divide possible factors (prime numbers) into 1001 until you find one that leaves no remainder. If you are told that 13 is a factor (or make a lucky guess), then you could divide 1001 by 13 to get the quotient 77. With this information you can factor 1001:

$$1001 = 77 \cdot 13$$

Now you can factor 77 to get the prime factorization of 1001:

$$1001 = 7 \cdot 11 \cdot 13$$

We can use this same idea with polynomials that are of higher degree than the ones we have been factoring. If we can guess a factor or if we are given a factor, we can use division to find the other factor and then proceed to factor the polynomial completely. Of course, it is harder to guess a factor of a polynomial than it is to guess a factor of an integer. In the next example we will factor a third-degree polynomial completely, given one factor.

EXAMPLE 1 **Using division in factoring**

Factor the polynomial $x^3 + 2x^2 - 5x - 6$ completely, given that the binomial $x + 1$ is a factor of the polynomial.

Study Tip

Everyone has a different attention span. Start by studying 10–15 minutes at a time and then build up to longer periods over time. In your senior year, you should be able to concentrate on one task for 30–45 minutes without a break. Be realistic. When you can't remember what you have read and can no longer concentrate, take a break.

Solution

Divide the polynomial by the binomial:

$$\begin{array}{r l}
 & x^2 + x - 6 \\
x + 1 \,\big) & x^3 + 2x^2 - 5x - 6 \\
 & \underline{x^3 + x^2} \\
 & \quad x^2 - 5x \\
 & \quad \underline{x^2 + x} \\
 & \qquad -6x - 6 \qquad -5x - x = -6x \\
 & \qquad \underline{-6x - 6} \\
 & \qquad\qquad 0 \qquad -6 - (-6) = 0
\end{array}$$

Because the remainder is 0, the dividend is the divisor times the quotient:

$$x^3 + 2x^2 - 5x - 6 = (x + 1)(x^2 + x - 6)$$

Now we factor the remaining trinomial to get the complete factorization:

$$x^3 + 2x^2 - 5x - 6 = (x + 1)(x + 3)(x - 2)$$ ■

Factoring a Difference or Sum of Two Cubes

We can use division to discover that $a - b$ is a factor of $a^3 - b^3$ (a difference of two cubes) and $a + b$ is a factor of $a^3 + b^3$ (a sum of two cubes):

$$\begin{array}{r} a^2 + ab + b^2 \\ a - b\overline{)a^3 + 0a^2b + 0ab^2 - b^3} \\ \underline{a^3 - a^2b} \qquad\qquad\quad \\ a^2b + 0ab^2 \qquad \\ \underline{a^2b - ab^2} \qquad \\ ab^2 - b^3 \\ \underline{ab^2 - b^3} \\ 0 \end{array} \qquad \begin{array}{r} a^2 - ab + b^2 \\ a + b\overline{)a^3 + 0a^2b + 0ab^2 + b^3} \\ \underline{a^3 + a^2b} \qquad\qquad\quad \\ -a^2b + 0ab^2 \qquad \\ \underline{-a^2b - ab^2} \qquad \\ ab^2 + b^3 \\ \underline{ab^2 + b^3} \\ 0 \end{array}$$

So $a - b$ is a factor of $a^3 - b^3$, and $a + b$ is a factor of $a^3 + b^3$. These results give us two more factoring rules.

Factoring a Difference or Sum of Two Cubes

$$a^3 - b^3 = (a - b)(a^2 + ab + b^2)$$
$$a^3 + b^3 = (a + b)(a^2 - ab + b^2)$$

Note that $a^2 + ab + b^2$ and $a^2 - ab + b^2$ are prime. Do not confuse them with $a^2 + 2ab + b^2$ and $a^2 - 2ab + b^2$, which are not prime because

$$a^2 + 2ab + b^2 = (a + b)^2 \qquad \text{and} \qquad a^2 - 2ab + b^2 = (a - b)^2.$$

These similarities can help you remember the rules for factoring $a^3 - b^3$ and $a^3 + b^3$. Note also how $a^3 - b^3$ compares with $a^2 - b^2$:

$$a^2 - b^2 = (a - b)(a + b)$$
$$a^3 - b^3 = (a - b)(a^2 + ab + b^2)$$

EXAMPLE 2 **Factoring a difference or sum of two cubes**

Factor each polynomial.

a) $w^3 - 8$ **b)** $x^3 + 1$ **c)** $8y^3 - 27$

Solution

a) Because $8 = 2^3$, $w^3 - 8$ is a difference of two cubes. To factor $w^3 - 8$, let $a = w$ and $b = 2$ in the formula $a^3 - b^3 = (a - b)(a^2 + ab + b^2)$:

$$w^3 - 8 = (w - 2)(w^2 + 2w + 4)$$

b) Because $1 = 1^3$, the binomial $x^3 + 1$ is a sum of two cubes. Let $a = x$ and $b = 1$ in the formula $a^3 + b^3 = (a + b)(a^2 - ab + b^2)$:

$$x^3 + 1 = (x + 1)(x^2 - x + 1)$$

c) $8y^3 - 27 = (2y)^3 - 3^3$ This is a difference of two cubes.

$= (2y - 3)(4y^2 + 6y + 9)$ Let $a = 2y$ and $b = 3$ in the formula. ■

Study Tip

Many schools have study skills centers that offer courses, workshops, and individual help on how to study. A search for "study skills" on the world wide web will turn up more information than you could possibly read. If you are not having the success in school that you would like, do something about it. What you do now will affect you the rest of your life.

In Example 2, we used the first three perfect cubes, 1, 8, and 27. You should verify that 1, 8, 27, 64, 125, 216, 343, 512, 729, and 1000 are the first 10 perfect cubes.

CAUTION The polynomial $(a - b)^3$ is not equivalent to $a^3 - b^3$ because if $a = 2$ and $b = 1$, then

$$(a - b)^3 = (2 - 1)^3 = 1^3 = 1$$

and

$$a^3 - b^3 = 2^3 - 1^3 = 8 - 1 = 7.$$

Likewise, $(a + b)^3$ is not equivalent to $a^3 + b^3$.

The Factoring Strategy

The following is a summary of the ideas that we use to factor a polynomial completely.

Strategy for Factoring Polynomials Completely

1. If there are any common factors, factor them out first.
2. When factoring a binomial, check to see whether it is a difference of two squares, a difference of two cubes, or a sum of two cubes. *A sum of two squares does not factor.*
3. When factoring a trinomial, check to see whether it is a perfect square trinomial.
4. When factoring a trinomial that is not a perfect square, use the *ac* method or the trial-and-error method.
5. If the polynomial has four terms, try factoring by grouping.
6. Check to see whether any of the factors can be factored again.

We will use the factoring strategy in Example 3.

EXAMPLE 3 **Factoring polynomials**

Factor each polynomial completely.

a) $2a^2b - 24ab + 72b$ **b)** $3x^3 + 6x^2 - 75x - 150$

Solution

a) $2a^2b - 24ab + 72b = 2b(a^2 - 12a + 36)$ First factor out the GCF, $2b$.

$= 2b(a - 6)^2$ Factor the perfect square trinomial.

b) $3x^3 + 6x^2 - 75x - 150 = 3[x^3 + 2x^2 - 25x - 50]$ Factor out the GCF, 3.

$= 3[x^2(x + 2) - 25(x + 2)]$ Factor out common factors.

$= 3(x^2 - 25)(x + 2)$ Factor by grouping.

$= 3(x + 5)(x - 5)(x + 2)$ Factor the difference of two squares. ■

WARM-UPS

True or false? Explain your answer.

1. $x^2 - 4 = (x - 2)^2$ for any real number x. False
2. The trinomial $4x^2 + 6x + 9$ is a perfect square trinomial. False
3. The polynomial $4y^2 + 25$ is a prime polynomial. True
4. $3y + ay + 3x + ax = (x + y)(3 + a)$ for any values of the variables. True
5. The polynomial $3x^2 + 51$ cannot be factored. False
6. If the GCF is not 1, then you should factor it out first. True
7. $x^2 + 9 = (x + 3)^2$ for any real number x. False
8. The polynomial $x^2 - 3x - 5$ is a prime polynomial. True
9. The polynomial $y^2 - 5y - my + 5m$ can be factored by grouping. True
10. The polynomial $x^2 + ax - 3x + 3a$ can be factored by grouping. False

6.5 EXERCISES

Reading and Writing *After reading this section, write out the answers to these questions. Use complete sentences.*

1. What is the relationship between division and factoring? If there is no remainder, then the dividend factors as the divisor times the quotient.

2. How do we know that $a - b$ is a factor of $a^3 - b^3$? If you divide $a^3 - b^3$ by $a - b$ there will be no remainder.

3. How do we know that $a + b$ is a factor of $a^3 + b^3$? If you divide $a^3 + b^3$ by $a + b$ there will be no remainder.

4. How do you recognize if a polynomial is a sum of two cubes? A sum of two cubes is of the form $a^3 + b^3$.

5. How do you factor a sum of two cubes? $a^3 + b^3 = (a + b)(a^2 - ab + b^2)$

6. How do you factor a difference of two cubes? $a^3 - b^3 = (a - b)(a^2 + ab + b^2)$

Factor each polynomial completely, given that the binomial following it is a factor of the polynomial. See Example 1.

7. $x^3 + 3x^2 - 10x - 24, x + 4$ $(x + 4)(x - 3)(x + 2)$
8. $x^3 - 7x + 6, x - 1$ $(x - 1)(x + 3)(x - 2)$
9. $x^3 + 4x^2 + x - 6, x - 1$ $(x - 1)(x + 3)(x + 2)$
10. $x^3 - 5x^2 - 2x + 24, x + 2$ $(x + 2)(x - 3)(x - 4)$
11. $x^3 - 8, x - 2$ $(x - 2)(x^2 + 2x + 4)$
12. $x^3 + 27, x + 3$ $(x + 3)(x^2 - 3x + 9)$
13. $x^3 + 4x^2 - 3x + 10, x + 5$ $(x + 5)(x^2 - x + 2)$
14. $2x^3 - 5x^2 - x - 6, x - 3$ $(x - 3)(2x^2 + x + 2)$
15. $x^3 + 2x^2 + 2x + 1, x + 1$ $(x + 1)(x^2 + x + 1)$
16. $x^3 + 2x^2 - 5x - 6, x + 3$ $(x + 3)(x - 2)(x + 1)$

Factor each difference or sum of cubes. See Example 2.

17. $m^3 - 1$ $(m - 1)(m^2 + m + 1)$
18. $z^3 - 27$ $(z - 3)(z^2 + 3z + 9)$
19. $x^3 + 8$ $(x + 2)(x^2 - 2x + 4)$
20. $y^3 + 27$ $(y + 3)(y^2 - 3y + 9)$
21. $a^3 + 125$ $(a + 5)(a^2 - 5a + 25)$
22. $b^3 - 216$ $(b - 6)(b^2 + 6b + 36)$
23. $c^3 - 343$ $(c - 7)(c^2 + 7c + 49)$
24. $d^3 + 1000$ $(d + 10)(d^2 - 10d + 100)$
25. $8w^3 + 1$ $(2w + 1)(4w^2 - 2w + 1)$
26. $125m^3 + 1$ $(5m + 1)(25m^2 - 5m + 1)$
27. $8t^3 - 27$ $(2t - 3)(4t^2 + 6t + 9)$
28. $125n^3 - 8$ $(5n - 2)(25n^2 + 10n + 4)$
29. $x^3 - y^3$ $(x - y)(x^2 + xy + y^2)$
30. $m^3 + n^3$ $(m + n)(m^2 - mn + n^2)$
31. $8t^3 + y^3$ $(2t + y)(4t^2 - 2ty + y^2)$
32. $u^3 - 125v^3$ $(u - 5v)(u^2 + 5uv + 25v^2)$

Factor each polynomial completely. If a polynomial is prime, say so. See Example 3.

33. $2x^2 - 18$ $2(x - 3)(x + 3)$
34. $3x^3 - 12x$ $3x(x - 2)(x + 2)$
35. $4x^2 + 8x - 60$ $4(x + 5)(x - 3)$
36. $3x^2 + 18x + 27$ $3(x + 3)^2$
37. $x^3 + 4x^2 + 4x$ $x(x + 2)^2$
38. $a^3 - 5a^2 + 6a$ $a(a - 2)(a - 3)$
39. $5max^2 + 20ma$ $5am(x^2 + 4)$
40. $3bmw^2 - 12bm$ $3bm(w - 2)(w + 2)$
41. $2x^2 - 3x - 1$ Prime
42. $3x^2 - 8x - 5$ Prime
43. $9x^2 + 6x + 1$ $(3x + 1)^2$
44. $9x^2 + 6x + 3$ $3(3x^2 + 2x + 1)$
45. $6x^2y + xy - 2y$ $y(3x + 2)(2x - 1)$
46. $5x^2y^2 - xy^2 - 6y^2$ $y^2(5x - 6)(x + 1)$
47. $y^2 + 10y - 25$ Prime
48. $x^2 - 20x + 25$ Prime
49. $48a^2 - 24a + 3$ $3(4a - 1)^2$
50. $8b^2 + 24b + 18$ $2(2b + 3)^2$
51. $16m^2 - 4m - 2$ $2(4m + 1)(2m - 1)$
52. $32a^2 + 4a - 6$ $2(2a + 1)(8a - 3)$
53. $9a^2 + 24a + 16$ $(3a + 4)^2$
54. $3x^2 - 18x - 48$ $3(x - 8)(x + 2)$
55. $24x^2 - 26x + 6$ $2(3x - 1)(4x - 3)$
56. $4x^2 - 6x - 12$ $2(2x^2 - 3x - 6)$
57. $3a^2 - 27a$ $3a(a - 9)$
58. $a^2 - 25a$ $a(a - 25)$
59. $8 - 2x^2$ $2(2 - x)(2 + x)$
60. $x^3 + 6x^2 + 9x$ $x(x + 3)^2$
61. $w^2 + 4t^2$ Prime
62. $9x^2 + 4y^2$ Prime
63. $6x^3 - 5x^2 + 12x$ $x(6x^2 - 5x + 12)$
64. $x^3 + 2x^2 - x - 2$ $(x - 1)(x + 1)(x + 2)$
65. $a^3b - 4ab$ $ab(a - 2)(a + 2)$
66. $2m^2 - 1800$ $2(m - 30)(m + 30)$
67. $x^3 + 2x^2 - 4x - 8$ $(x - 2)(x + 2)^2$
68. $m^2a + 2ma^2 + a^3$ $a(m + a)^2$
69. $2w^4 - 16w$ $2w(w - 2)(w^2 + 2w + 4)$
70. $m^4n + mn^4$ $mn(m + n)(m^2 - mn + n^2)$
71. $3a^2w - 18aw + 27w$ $3w(a - 3)^2$
72. $8a^3 + 4a$ $4a(2a^2 + 1)$
73. $5x^2 - 500$ $5(x - 10)(x + 10)$

b) Use the accompanying graph to determine whether the sky diver (with no air resistance) falls farther in the first 5 seconds or the last 5 seconds of the fall.
last 5 sec

c) Is the sky diver's velocity increasing or decreasing as she falls? increasing

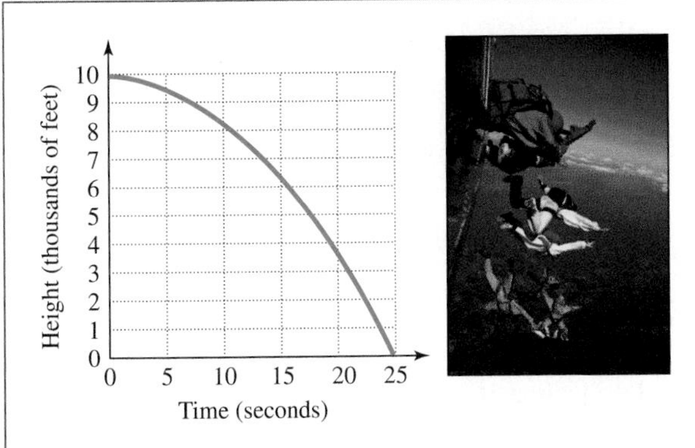

FIGURE FOR EXERCISE 67

68. ***Skydiving.*** If a sky diver jumps from an airplane at a height of 8256 feet, then for the first five seconds, her height above the earth is approximated by the formula $h = -16t^2 + 8256$. How many seconds does it take her to reach 8000 feet.
4 sec

69. ***Throwing a sandbag.*** If a balloonist throws a sandbag downward at 24 feet per second from an altitude of 720 feet, then its height (in feet) above the ground after t seconds is given by $S = -16t^2 - 24t + 720$. How long does it take for the sandbag to reach the earth? (On the ground, $S = 0$.)
6 sec

70. ***Throwing a sandbag.*** If the balloonist of the previous exercise throws his sandbag downward from an altitude of 128 feet with an initial velocity of 32 feet per second, then its altitude after t seconds is given by the formula $S = -16t^2 - 32t + 128$. How long does it take for the sandbag to reach the earth?
2 sec

71. ***Glass prism.*** One end of a glass prism is in the shape of a triangle with a height that is 1 inch longer than twice the base. If the area of the triangle is 39 square inches, then how long are the base and height?
Base 6 in., height 13 in.

72. ***Areas of two circles.*** The radius of a circle is 1 meter longer than the radius of another circle. If their areas differ by 5π square meters, then what is the radius of each?
2 m and 3 m

73. ***Changing area.*** Last year Otto's garden was square. This year he plans to make it smaller by shortening one side 5 feet and the other 8 feet. If the area of the smaller garden will be 180 square feet, then what was the size of Otto's garden last year?
20 ft by 20 ft

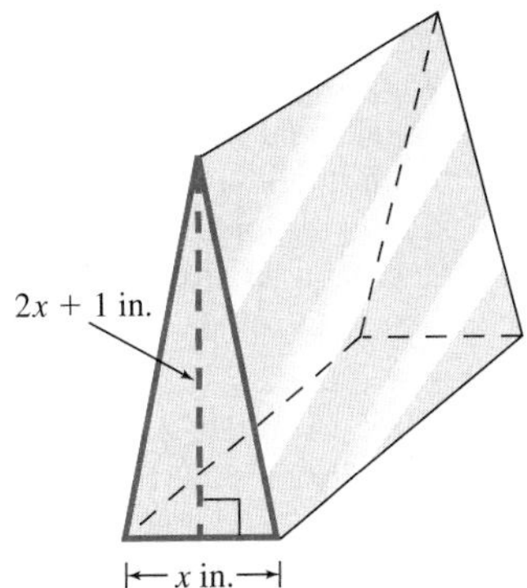

FIGURE FOR EXERCISE 71

74. ***Dimensions of a box.*** Rosita's Christmas present from Carlos is in a box that has a width that is 3 inches shorter than the height. The length of the base is 5 inches longer than the height. If the area of the base is 84 square inches, then what is the height of the package?
9 in.

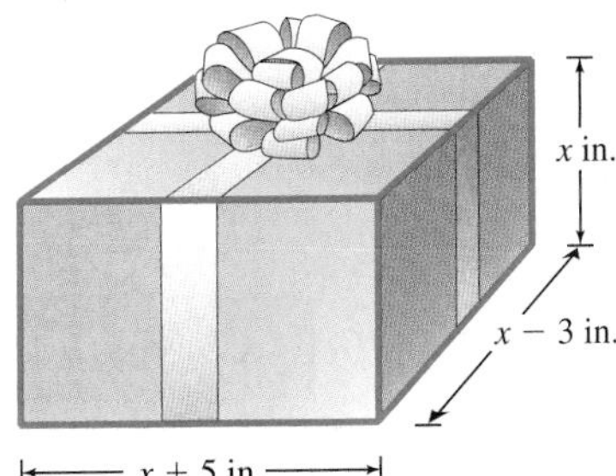

FIGURE FOR EXERCISE 74

75. ***Flying a kite.*** Imelda and Gordon have designed a new kite. While Imelda is flying the kite, Gordon is standing directly below it. The kite is designed so that its altitude is always 20 feet larger than the distance between Imelda and Gordon. What is the altitude of the kite when it is 100 feet from Imelda?
80 ft

76. ***Avoiding a collision.*** A car is traveling on a road that is perpendicular to a railroad track. When the car is 30 meters from the crossing, the car's new collision detector warns the driver that there is a train 50 meters from the car and heading toward the same crossing. How far is the train from the crossing?
40 m

77. ***Carpeting two rooms.*** Virginia is buying carpet for two square rooms. One room is 3 yards wider than the other. If she needs 45 square yards of carpet, then what are the dimensions of each room?
3 yd by 3 yd, 6 yd by 6 yd

78. ***Winter wheat.*** While finding the amount of seed needed to plant his three square wheat fields, Hank observed that the side of one field was 1 kilometer longer than the side of the smallest field and that the side of the largest field was 3 kilometers longer than the side of the smallest field. If the total area of the three fields is 38 square kilometers, then what is the area of each field?
4 km², 9 km², 25 km²

79. ***Sailing to Miami.*** At point A the captain of a ship determined that the distance to Miami was 13 miles. If she sailed north to point B and then west to Miami, the distance would be 17 miles. If the distance from point A to point B is greater than the distance from point B to Miami, then how far is it from point A to point B?
12 mi

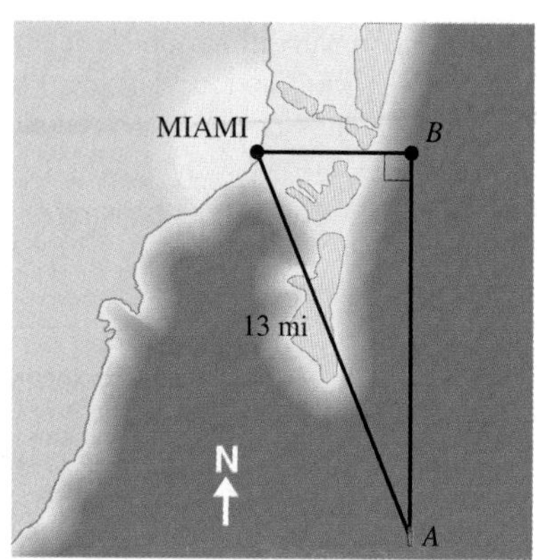

FIGURE FOR EXERCISE 79

80. ***Buried treasure.*** Ahmed has half of a treasure map, which indicates that the treasure is buried in the desert $2x + 6$ paces from Castle Rock. Vanessa has the other half of the map. Her half indicates that to find the treasure, one must get to Castle Rock, walk x paces to the north, and then walk $2x + 4$ paces to the east. If they share their information, then they can find x and save a lot of digging. What is x? 10 paces

81. ***Emerging markets.*** Catarina's investment of \$16,000 in an emerging market fund grew to \$25,000 in two years. Find the average annual rate of return by solving the equation $16{,}000(1 + r)^2 = 25{,}000$.
25%

82. ***Venture capital.*** Henry invested \$12,000 in a new restaurant. When the restaurant was sold two years later, he received \$27,000. Find his average annual return by solving the equation $12{,}000(1 + r)^2 = 27{,}000$. 50%

COLLABORATIVE ACTIVITIES

Hannah's Inheritance

Grouping: Three students per group
Topic: Factoring Polynomials

Part I: Sally, Kelly, and Hannah inherited property from their father. Sally, who married the neighbor to the east, is given a piece of land adjacent to her husband's property. The land is 5 hectometers wide and the length matches that of her husband's property. Kelly who has married the neighbor to the south is given property that is 4 hectometers wide and its length is the common boundary of her spouse's property. Hannah is to have a square piece that is left after her sister's property is taken out. Hannah wants to find out the dimensions of her land. She knows that her father's land totaled 380 square hectometers. Use the diagram below to find the dimensions of Hannah's land.

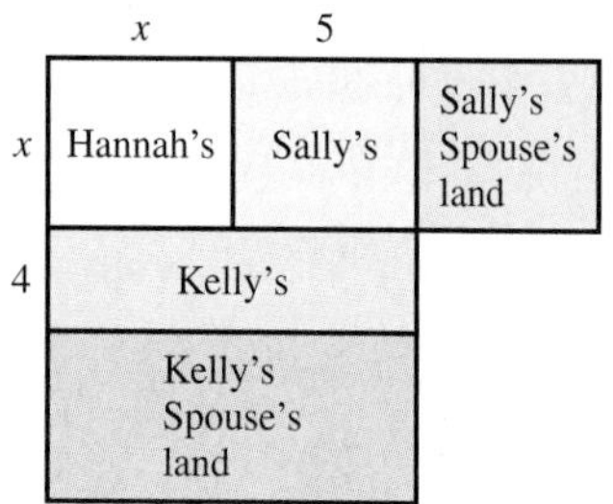

Part II: A few years later Hannah married Harry, a math teacher from town. Now they are getting divorced. As part of the settlement, Harry will get 32% of the land that they now own jointly. However, Harry did not get along with Hannah's sisters and does not want his land to be adjacent to their land. They have agreed that Harry will get a triangular section at the corner as shown in the accompanying figure. Use your answer from Part I to find y and z in the figure.

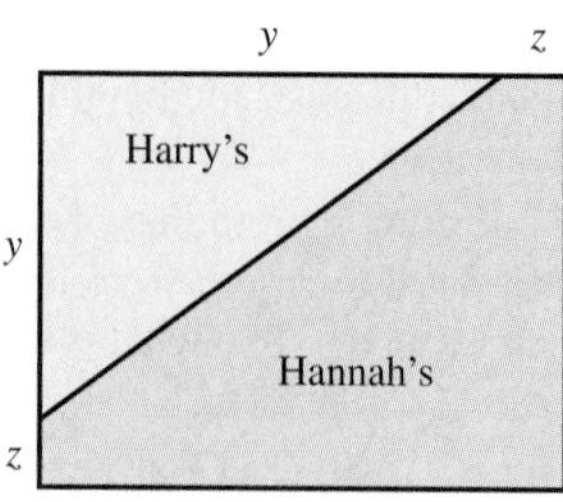

WRAP-UP CHAPTER 6

SUMMARY

Factoring		**Examples**
Prime number	A positive integer larger than 1 that has no integral factors other than 1 and itself	2, 3, 5, 7, 11
Prime polynomial	A polynomial that cannot be factored is prime.	$x^2 + 3$ and $x^2 - x + 5$ are prime.
Strategy for finding the GCF for monomials	1. Find the GCF for the coefficients of the monomials. 2. Form the product of the GCF of the coefficients and each variable that is common to all of the monomials, where the exponent on each variable equals the smallest power of that variable in any of the monomials.	$12x^3yz$, $8x^2y^3$ GCF $= 4x^2y$
Factoring out the GCF	Use the distributive property to factor out the GCF from all terms of a polynomial.	$2x^3 - 4x = 2x(x^2 - 2)$

Special Cases		**Examples**
Difference of two squares	$a^2 - b^2 = (a + b)(a - b)$	$m^2 - 9 = (m - 3)(m + 3)$
Perfect square trinomial	$a^2 + 2ab + b^2 = (a + b)^2$ $a^2 - 2ab + b^2 = (a - b)^2$	$x^2 + 6x + 9 = (x + 3)^2$ $4h^2 - 12h + 9 = (2h - 3)^2$
Difference or sum of two cubes	$a^3 - b^3 = (a - b)(a^2 + ab + b^2)$ $a^3 + b^3 = (a + b)(a^2 - ab + b^2)$	$t^3 - 8 = (t - 2)(t^2 + 2t + 4)$ $p^3 + 1 = (p + 1)(p^2 - p + 1)$

Factoring Polynomials		**Examples**
Factoring by grouping	Factor out common factors from groups of terms.	$6x + 6w + ax + aw$ $= 6(x + w) + a(x + w)$ $= (6 + a)(x + w)$
Strategy for factoring $ax^2 + bx + c$ by the *ac* method	1. Find two numbers that have a product equal to *ac* and a sum equal to *b*. 2. Replace *bx* by two terms using the two new numbers as coefficients. 3. Factor the resulting four-term polynomial by grouping.	$6x^2 + 17x + 12$ $= 6x^2 + 9x + 8x + 12$ $= (2x + 3)3x + (2x + 3)4$ $= (2x + 3)(3x + 4)$
Factoring by trial and error	Try possible factors of the trinomial and check by using FOIL. If incorrect, try again.	$2x^2 + 5x - 12$ $= (2x - 3)(x + 4)$

Strategy for factoring polynomials completely	1. First factor out any common factors. 2. When factoring a binomial, check to see whether it is a difference of two squares, a difference of two cubes, or a sum of two cubes. Remember that a sum of two squares (with no common factor) is prime. 3. When factoring a trinomial, check to see whether it is a perfect square trinomial. 4. When factoring a trinomial that is not a perfect square, use the *ac* method or trial and error. 5. If the polynomial has four terms, try factoring by grouping. 6. Check to see whether any factors can be factored again.	

Solving Equations | | **Examples**

Zero factor property	The equation $a \cdot b = 0$ is equivalent to $a = 0$ or $b = 0$.	$x(x - 1) = 0$ $x = 0$ or $x - 1 = 0$
Strategy for solving an equation by factoring	1. Rewrite the equation with 0 on the right-hand side. 2. Factor the left-hand side completely. 3. Set each factor equal to zero to get linear equations. 4. Solve the linear equations. 5. Check the answers in the original equation. 6. State the solution(s) to the original equation.	$x^2 + 3x = 18$ $x^2 + 3x - 18 = 0$ $(x + 6)(x - 3) = 0$ $x + 6 = 0$ or $x - 3 = 0$ $x = -6$ or $x = 3$

ENRICHING YOUR MATHEMATICAL WORD POWER

For each mathematical term, choose the correct meaning.

1. **factor**
   a. to write an expression as a product
   b. to multiply
   c. what two numbers have in common
   d. to FOIL a

2. **prime number**
   a. a polynomial that cannot be factored
   b. a number with no divisors
   c. an integer between 1 and 10
   d. an integer larger than 1 that has no integral factors other than itself and 1 d

3. **greatest common factor**
   a. the least common multiple
   b. the least common denominator
   c. the largest integer that is a factor of two or more integers
   d. the largest number in a product c

4. **prime polynomial**
   a. a polynomial that has no factors
   b. a product of prime numbers
   c. a first-degree polynomial
   d. a monomial a

5. **factor completely**
   a. to factor by grouping
   b. to factor out a prime number
   c. to write as a product of primes
   d. to factor by trial and error c

6. **sum of two cubes**
   a. $(a + b)^3$
   b. $a^3 + b^3$
   c. $a^3 - b^3$
   d. a^3b^3 b

7. **quadratic equation**
   a. $ax + b = 0$ where $a \neq 0$
   b. $ax + b = cx + d$

c. $ax^2 + bx + c = 0$ where $a \neq 0$
d. any equation with four terms c

8. **zero factor property**
a. If $ab = 0$ then $a = 0$ or $b = 0$
b. $a \cdot 0 = 0$ for any a
c. $a = a + 0$ for any real number a
d. $a + (-a) = 0$ for any real number a a

9. **Pythagorean theorem**
a. $a^2 + b^2 = (a + b)^2$
b. a triangle is a right triangle if and only if it has one right angle
c. the legs of a right triangle meet at a 90° angle
d. a theorem that gives a relationship between the two legs and the hypotenuse of a right triangle d

10. **difference of two squares**
a. $a^3 - b^3$
b. $2a - 2b$
c. $a^2 - b^2$
d. $(a - b)^2$ c

REVIEW EXERCISES

6.1 *Find the prime factorization for each integer.*

1. 144 $2^4 \cdot 3^2$
2. 121 11^2
3. 58 $2 \cdot 29$
4. 76 $2^2 \cdot 19$
5. 150 $2 \cdot 3 \cdot 5^2$
6. 200 $2^3 \cdot 5^2$

Find the greatest common factor for each group.

7. 36, 90 18
8. 30, 42, 78 6
9. $8x$, $12x^2$ $4x$
10. $6a^2b$, $9ab^2$, $15a^2b^2$ $3ab$

Complete the factorization of each binomial.

11. $3x + 6 = 3(x + 2)$
12. $7x^2 + x = x(7x + 1)$
13. $2a - 20 = -2(-a + 10)$
14. $a^2 - a = -a(-a + 1)$

Factor each polynomial by factoring out the GCF.

15. $2a - a^2$ $a(2 - a)$
16. $9 - 3b$ $3(3 - b)$
17. $6x^2y^2 - 9x^5y$ $3x^2y(2y - 3x^3)$
18. $a^3b^5 + a^3b^2$ $a^3b^2(b^3 + 1)$
19. $3x^2y - 12xy - 9y^2$ $3y(x^2 - 4x - 3y)$
20. $2a^2 - 4ab^2 - ab$ $a(2a - 4b^2 - b)$

6.2 *Factor each polynomial completely.*

21. $y^2 - 400$ $(y - 20)(y + 20)$
22. $4m^2 - 9$ $(2m - 3)(2m + 3)$
23. $w^2 - 8w + 16$ $(w - 4)^2$
24. $t^2 + 20t + 100$ $(t + 10)^2$
25. $4y^2 + 20y + 25$ $(2y + 5)^2$
26. $2a^2 - 4a - 2$ $2(a^2 - 2a - 1)$
27. $r^2 - 4r + 4$ $(r - 2)^2$
28. $3m^2 - 75$ $3(m - 5)(m + 5)$
29. $8t^3 - 24t^2 + 18t$ $2t(2t - 3)^2$
30. $t^2 - 9w^2$ $(t - 3w)(t + 3w)$
31. $x^2 + 12xy + 36y^2$ $(x + 6y)^2$
32. $9y^2 - 12xy + 4x^2$ $(3y - 2x)^2$
33. $x^2 + 5x - xy - 5y$ $(x - y)(x + 5)$
34. $x^2 + xy + ax + ay$ $(x + a)(x + y)$

6.3 *Factor each polynomial.*

35. $b^2 + 5b - 24$ $(b + 8)(b - 3)$
36. $a^2 - 2a - 35$ $(a - 7)(a + 5)$
37. $r^2 - 4r - 60$ $(r - 10)(r + 6)$
38. $x^2 + 13x + 40$ $(x + 8)(x + 5)$
39. $y^2 - 6y - 55$ $(y - 11)(y + 5)$
40. $a^2 + 6a - 40$ $(a + 10)(a - 4)$
41. $u^2 + 26u + 120$ $(u + 20)(u + 6)$
42. $v^2 - 22v - 75$ $(v - 25)(v + 3)$

Factor completely.

43. $3t^3 + 12t^2$ $3t^2(t + 4)$
44. $-4m^4 - 36m^2$ $-4m^2(m^2 + 9)$
45. $5w^3 + 25w^2 + 25w$ $5w(w^2 + 5w + 5)$
46. $-3t^3 + 3t^2 - 6t$ $-3t(t^2 - t + 2)$
47. $2a^3b + 3a^2b^2 + ab^3$ $ab(2a + b)(a + b)$
48. $6x^2y^2 - xy^3 - y^4$ $y^2(2x - y)(3x + y)$
49. $9x^3 - xy^2$ $x(3x - y)(3x + y)$
50. $h^4 - 100h^2$ $h^2(h - 10)(h + 10)$

6.4 *Factor each polynomial completely.*

51. $14t^2 + t - 3$ $(7t - 3)(2t + 1)$
52. $15x^2 - 22x - 5$ $(5x + 1)(3x - 5)$
53. $6x^2 - 19x - 7$ $(3x + 1)(2x - 7)$
54. $2x^2 - x - 10$ $(x + 2)(2x - 5)$
55. $6p^2 + 5p - 4$ $(3p + 4)(2p - 1)$
56. $3p^2 + 2p - 5$ $(p - 1)(3p + 5)$
57. $-30p^3 + 8p^2 + 8p$ $-2p(5p + 2)(3p - 2)$
58. $-6q^2 - 40q - 50$ $-2(3q + 5)(q + 5)$
59. $6x^2 - 29xy - 5y^2$ $(6x + y)(x - 5y)$
60. $10a^2 + ab - 2b^2$ $(5a - 2b)(2a + b)$
61. $32x^2 + 16xy + 2y^2$ $2(4x + y)^2$
62. $8a^2 + 40ab + 50b^2$ $2(2a + 5b)^2$

6.5 *Factor completely.*

63. $5x^3 + 40x$ $5x(x^2 + 8)$
64. $w^2 + 6w + 9$ $(w + 3)^2$

65. $9x^2 + 3x - 2$ $(3x - 1)(3x + 2)$

66. $ax^3 + ax$ $ax(x^2 + 1)$

67. $x^3 + 2x^2 - x - 2$ $(x + 2)(x - 1)(x + 1)$

68. $16x^2 - 2x - 3$ $(8x + 3)(2x - 1)$

69. $x^2y - 16xy^2$ $xy(x - 16y)$

70. $-3x^2 + 27$ $-3(x - 3)(x + 3)$

71. $a^2 + 2a + 1$ $(a + 1)^2$

72. $-2w^2 - 12w - 18$ $-2(w + 3)^2$

73. $x^3 - x^2 + x - 1$ $(x^2 + 1)(x - 1)$

74. $9x^2y^2 - 9y^2$ $9y^2(x - 1)(x + 1)$

75. $a^2 + ab + 2a + 2b$ $(a + 2)(a + b)$

76. $4m^2 + 20m + 25$ $(2m + 5)^2$

77. $-2x^2 + 16x - 24$ $-2(x - 6)(x - 2)$

78. $6x^2 + 21x - 45$ $3(2x - 3)(x + 5)$

79. $m^3 - 1000$ $(m - 10)(m^2 + 10m + 100)$

80. $8p^3 + 1$ $(2p + 1)(4p^2 - 2p + 1)$

Factor each polynomial completely, given that the binomial following it is a factor of the polynomial.

81. $x^3 + x + 10, x + 2$ $(x + 2)(x^2 - 2x + 5)$

82. $x^3 - 5x - 12, x - 3$ $(x - 3)(x^2 + 3x + 4)$

83. $x^3 + 6x^2 - 7x - 60, x + 4$ $(x + 4)(x + 5)(x - 3)$

84. $x^3 - 4x^2 - 3x - 10, x - 5$ $(x - 5)(x^2 + x + 2)$

6.6 *Solve each equation.*

85. $x^3 = 5x^2$ $0, 5$

86. $2m^2 + 10m = -12$ $-3, -2$

87. $(a - 2)(a - 3) = 6$ $0, 5$

88. $(w - 2)(w + 3) = 50$ $-8, 7$

89. $2m^2 - 9m - 5 = 0$ $-\frac{1}{2}, 5$

90. $12x^2 + 5x - 3 = 0$ $-\frac{3}{4}, \frac{1}{3}$

91. $m^3 + 4m^2 - 9m = 36$ $-4, -3, 3$

92. $w^3 + 5w^2 - w = 5$ $-5, -1, 1$

93. $(x + 3)^2 + x^2 = 5$ $-2, -1$

94. $(h - 2)^2 + (h + 1)^2 = 9$ $-1, 2$

95. $p^2 + \frac{1}{4}p - \frac{1}{8} = 0$ $-\frac{1}{2}, \frac{1}{4}$

96. $t^2 + 1 = \frac{13}{6}t$ $\frac{2}{3}, \frac{3}{2}$

Solve each problem.

97. ***Positive numbers.*** Two positive numbers differ by 6, and their squares differ by 96. Find the numbers. 5, 11

98. ***Consecutive integers.*** Find three consecutive integers such that the sum of their squares is 77. $-6, -5, -4$ or 4, 5, 6

99. ***Dimensions of a notebook.*** The perimeter of a notebook is 28 inches, and the diagonal measures 10 inches. What are the length and width of the notebook? 6 in. by 8 in.

100. ***Two numbers.*** The sum of two numbers is 8.5, and their product is 18. Find the numbers. 4 and 4.5

101. ***Poiseuille's law.*** According to the nineteenth-century physician Poiseuille, the velocity (in centimeters per second) of blood r centimeters from the center of an artery of radius R centimeters is given by $v = kR^2 - kr^2$, where k is a constant. Rewrite the formula by factoring the right-hand side completely. $v = k(R - r)(R + r)$

102. ***Racquetball.*** The volume of rubber (in cubic centimeters) in a hollow rubber ball used in racquetball is given by

$$V = \frac{4}{3}\pi R^3 - \frac{4}{3}\pi r^3,$$

where the inside radius is r centimeters and the outside radius is R centimeters.

a) Rewrite the formula by factoring the right-hand side completely. $V = \frac{4}{3}\pi(R - r)(R^2 + Rr + r^2)$

b) The accompanying graph shows the relationship between r and V when $R = 3$. Use the graph to estimate the value of r for which $V = 100$ cm^3. 1.5 cm

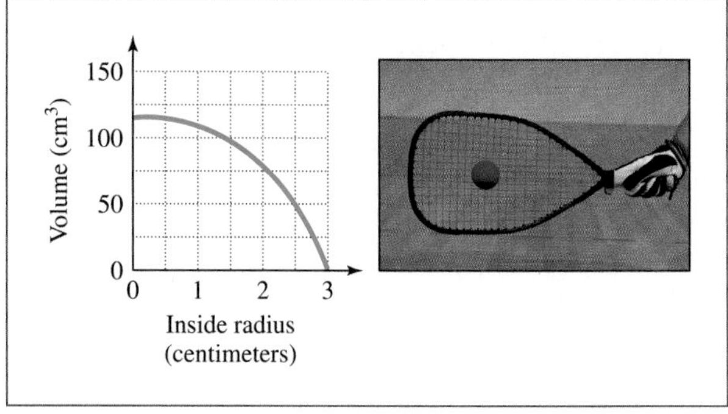

FIGURE FOR EXERCISE 102

103. ***Leaning ladder.*** A 10-foot ladder is placed against a building so that the distance from the bottom of the ladder to the building is 2 feet less than the distance from the top of the ladder to the ground. What is the distance from the bottom of the ladder to the building? 6 ft

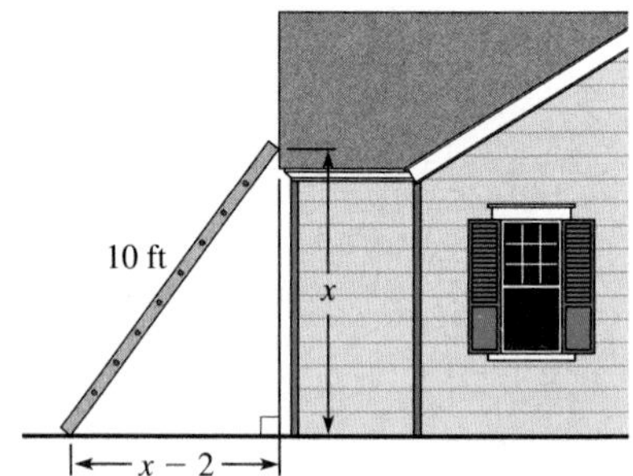

FIGURE FOR EXERCISE 103

104. ***Towering antenna.*** A guy wire of length 50 feet is attached to the ground and to the top of an antenna. The height of the antenna is 10 feet larger than the distance from the base of the antenna to the point where the guy wire is attached to the ground. What is the height of the antenna? 40 ft

CHAPTER 6 TEST

Give the prime factorization for each integer.

1. 66 $2 \cdot 3 \cdot 11$

2. 336 $2^4 \cdot 3 \cdot 7$

Find the greatest common factor (GCF) for each group.

3. 48, 80 16

4. 42, 66, 78 6

5. $6y^2, 15y^3$ $3y^2$

6. $12a^2b, 18ab^2, 24a^3b^3$ $6ab$

Factor each polynomial completely.

7. $5x^2 - 10x$ $5x(x - 2)$

8. $6x^2y^2 + 12xy^2 + 12y^2$ $6y^2(x^2 + 2x + 2)$

9. $3a^3b - 3ab^3$ $3ab(a - b)(a + b)$

10. $a^2 + 2a - 24$ $(a + 6)(a - 4)$

11. $4b^2 - 28b + 49$ $(2b - 7)^2$

12. $3m^3 + 27m$ $3m(m^2 + 9)$

13. $ax - ay + bx - by$ $(a + b)(x - y)$

14. $ax - 2a - 5x + 10$ $(a - 5)(x - 2)$

15. $6b^2 - 7b - 5$ $(3b - 5)(2b + 1)$

16. $m^2 + 4mn + 4n^2$ $(m + 2n)^2$

17. $2a^2 - 13a + 15$ $(2a - 3)(a - 5)$

18. $z^3 + 9z^2 + 18z$ $z(z + 3)(z + 6)$

Factor the polynomial completely, given that $x - 1$ is a factor.

19. $x^3 - 6x^2 + 11x - 6$ $(x - 1)(x - 2)(x - 3)$

Solve each equation.

20. $2x^2 + 5x - 12 = 0$ $\frac{3}{2}, -4$

21. $3x^3 = 12x$ $0, -2, 2$

22. $(2x - 1)(3x + 5) = 5$ $-2, \frac{5}{6}$

Write a complete solution to each problem.

23. If the length of a rectangle is 3 feet longer than the width and the diagonal is 15 feet, then what are the length and width?
Length 12 ft, width 9 ft

24. The sum of two numbers is 4, and their product is -32. Find the numbers. -4 and 8

MAKING CONNECTIONS CHAPTERS 1–6

Simplify each expression.

1. $\dfrac{91-17}{17-91}$ -1
2. $\dfrac{4-18}{-6-1}$ 2
3. $5-2(7-3)$ -3
4. $3^2-4(6)(-2)$ 57
5. 2^5-2^4 16
6. $0.07(37)+0.07(63)$ 7

Perform the indicated operations.

7. $x \cdot 2x$ $2x^2$
8. $x+2x$ $3x$
9. $\dfrac{6+2x}{2}$ $3+x$
10. $\dfrac{6 \cdot 2x}{2}$ $6x$
11. $2 \cdot 3y \cdot 4z$ $24yz$
12. $2(3y+4z)$ $6y+8z$
13. $2-(3-4z)$ $4z-1$
14. $t^8 \div t^2$ t^6
15. $t^8 \cdot t^2$ t^{10}
16. $\dfrac{8t^8}{2t^2}$ $4t^6$

Solve and graph each inequality.

17. $2x-5>3x+4$ $x<-9$

    −13 −12 −11 −10 −9 −8 −7

18. $4-5x \le -11$ $x \ge 3$

    1 2 3 4 5 6 7

19. $-\dfrac{2}{3}x+3<-5$ $x>12$

    10 11 12 13 14 15 16

20. $0.05(x-120)-24<0$ $x<600$

    0 200 400 600 800

Solve each equation.

21. $2x-3=0$ $\dfrac{3}{2}$
22. $2x+1=0$ $-\dfrac{1}{2}$
23. $(x-3)(x+5)=0$ $3, -5$
24. $(2x-3)(2x+1)=0$ $\dfrac{3}{2}, -\dfrac{1}{2}$
25. $3x(x-3)=0$ $0, 3$
26. $x^2=x$ $0, 1$
27. $3x-3x=0$ All real numbers
28. $3x-3x=1$ No solution
29. $0.01x-x+14.9=0.5x$ 10
30. $0.05x+0.04(x-40)=2$ 40
31. $2x^2=18$ $-3, 3$
32. $2x^2+7x-15=0$ $-5, \dfrac{3}{2}$

Solve the problem.

33. ***Another ace.*** Professional tennis players can serve a tennis ball at speeds over 120 mph into a rectangular region that has a perimeter of 69 feet and an area of 283.5 square feet. Find the length and width of the service region.
    Length 21 ft, width 13.5 ft

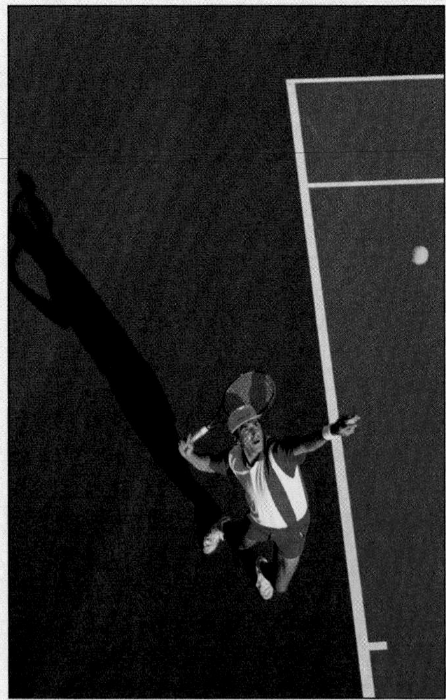

Quotient Rule

Suppose $a \neq 0$, and m and n are positive integers.

$$\text{If } m \geq n, \text{ then } \frac{a^m}{a^n} = a^{m-n}.$$

$$\text{If } m < n, \text{ then } \frac{a^m}{a^n} = \frac{1}{a^{n-m}}.$$

EXAMPLE 4 **Using the quotient rule in reducing**

Reduce to lowest terms.

a) $\dfrac{3a^{15}}{6a^7}$ **b)** $\dfrac{6x^4y^2}{4xy^5}$

Solution

a)
$$\frac{3a^{15}}{6a^7} = \frac{\not{3}a^{15}}{\not{3} \cdot 2a^7} \quad \text{Factor.}$$
$$= \frac{a^{15-7}}{2} \quad \text{Quotient rule}$$
$$= \frac{a^8}{2}$$

b)
$$\frac{6x^4y^2}{4xy^5} = \frac{\not{2} \cdot 3x^4y^2}{\not{2} \cdot 2xy^5} \quad \text{Factor.}$$
$$= \frac{3x^{4-1}}{2y^{5-2}} \quad \text{Quotient rule}$$
$$= \frac{3x^3}{2y^3}$$

■

The essential part of reducing is getting a complete factorization for the numerator and denominator. To get a complete factorization, you must use the techniques for factoring from Chapter 6. If there are large integers in the numerator and denominator, you can use the technique shown in Section 6.1 to get a prime factorization of each integer.

EXAMPLE 5 **Reducing expressions involving large integers**

Reduce $\frac{420}{616}$ to lowest terms.

Solution

Use the method of Section 6.1 to get a prime factorization of 420 and 616:

$$\begin{array}{rr} 7 & 11 \\ 5\overline{)35} & 7\overline{)77} \\ 3\overline{)105} & 2\overline{)154} \\ 2\overline{)210} & 2\overline{)308} \\ \text{Start here} \rightarrow 2\overline{)420} & 2\overline{)616} \end{array}$$

The complete factorization for 420 is $2^2 \cdot 3 \cdot 5 \cdot 7$, and the complete factorization for 616 is $2^3 \cdot 7 \cdot 11$. To reduce the fraction, we divide out the common factors:

$$\frac{420}{616} = \frac{2^2 \cdot 3 \cdot 5 \cdot 7}{2^3 \cdot 7 \cdot 11} = \frac{3 \cdot 5}{2 \cdot 11} = \frac{15}{22}$$

■

Dividing $a - b$ by $b - a$

In Section 5.2 you learned that $a - b = -(b - a) = -1(b - a)$. So if $a - b$ is divided by $b - a$, the quotient is -1:

$$\frac{a-b}{b-a} = \frac{-1(b-a)}{b-a} = -1$$

We will use this fact in Example 6.

EXAMPLE 6 **Expressions with $a - b$ and $b - a$**

Reduce to lowest terms.

a) $\dfrac{5x - 5y}{4y - 4x}$ **b)** $\dfrac{m^2 - n^2}{n - m}$

Solution

a) $\dfrac{5x - 5y}{4y - 4x} = \dfrac{5(x - y)}{4(y - x)}$ Factor.

$= \dfrac{5}{4} \cdot (-1)$ $\quad \dfrac{x-y}{y-x} = -1$

$= -\dfrac{5}{4}$

b) $\dfrac{m^2 - n^2}{n - m} = \dfrac{\overset{-1}{\cancel{(m - n)}}(m + n)}{\cancel{n - m}}$ Factor.

$= -1(m + n)$ $\quad \dfrac{m-n}{n-m} = -1$

$= -m - n$

■

CAUTION We can reduce $\dfrac{a-b}{b-a}$ to -1, but we cannot reduce $\dfrac{a-b}{a+b}$. There is no factor that is common to the numerator and denominator of $\dfrac{a-b}{a+b}$.

Factoring Out the Opposite of a Common Factor

If we can factor out a common factor, we can also factor out the opposite of that common factor. For example, from $-3x - 6y$ we can factor out the common factor 3 or the common factor -3:

$$-3x - 6y = 3(-x - 2y) \quad \text{or} \quad -3x - 6y = -3(x + 2y)$$

To reduce an expression, it is sometimes necessary to factor out the opposite of a common factor.

EXAMPLE 7 **Factoring out the opposite of a common factor**

Reduce $\frac{-3w - 3w^2}{w^2 - 1}$ to lowest terms.

Solution

We can factor $3w$ or $-3w$ from the numerator. If we factor out $-3w$, we get a common factor in the numerator and denominator:

$$\frac{-3w - 3w^2}{w^2 - 1} = \frac{-3w(1 + w)}{(w - 1)(w + 1)} \quad \text{Factor.}$$

$$= \frac{-3w}{w - 1} \quad \text{Since } 1 + w = w + 1 \text{, we divide out } w + 1.$$

$$= \frac{3w}{1 - w} \quad \text{Multiply numerator and denominator by } -1.$$

The last step in this reduction is not absolutely necessary, but we usually perform it to make the answer look a little simpler. ■

The main points to remember for reducing rational expressions are summarized in the following reducing strategy.

Strategy for Reducing Rational Expressions

1. Reducing is done by dividing out all common factors.
2. Factor the numerator and denominator completely to see the common factors.
3. Use the quotient rule to reduce a ratio of two monomials.
4. You may have to factor out a common factor with a negative sign to get identical factors in the numerator and denominator.
5. The quotient of $a - b$ and $b - a$ is -1.

Writing Rational Expressions

Rational expressions occur naturally in applications involving rates.

EXAMPLE 8 **Writing rational expressions**

Answer each question with a rational expression.

a) If a trucker drives 500 miles in $x + 1$ hours, then what is his average speed?

b) If a wholesaler buys 100 pounds of shrimp for x dollars, then what is the price per pound?

c) If a painter completes an entire house in $2x$ hours, then at what rate is she painting?

Solution

a) Because $R = \frac{D}{T}$, he is averaging $\frac{500}{x + 1}$ mph.

b) At x dollars for 100 pounds, the wholesaler is paying $\frac{x}{100}$ dollars per pound or $\frac{x}{100}$ dollars/pound.

c) By completing 1 house in $2x$ hours, her rate is $\frac{1}{2x}$ house/hour. ■

WARM-UPS

True or false? Explain your answer.

1. A complete factorization of 3003 is $2 \cdot 3 \cdot 7 \cdot 11 \cdot 13$. False

2. A complete factorization of 120 is $2^3 \cdot 3 \cdot 5$. True

3. Any number can be used in place of x in the expression $\frac{x - 2}{5}$. True

4. We cannot replace x by -1 or 3 in the expression $\frac{x + 1}{x - 3}$. False

5. The rational expression $\frac{x + 2}{2}$ reduces to x. False

6. $\frac{2x}{2} = x$ for any real number x. True

7. $\frac{x^{13}}{x^{20}} = \frac{1}{x^7}$ for any nonzero value of x. True

8. $\frac{a^2 + b^2}{a + b}$ reduced to lowest terms is $a + b$. False

9. If $a \neq b$, then $\frac{a - b}{b - a} = 1$. False

10. The expression $\frac{-3x - 6}{x + 2}$ reduces to -3. True

7.1 EXERCISES

Reading and Writing *After reading this section, write out the answers to these questions. Use complete sentences.*

1. What is a rational number?
A rational number is a ratio of two integers with the denominator not 0.

2. What is a rational expression?
A rational expression is a ratio of two polynomials with the denominator not 0.

3. How do you reduce a rational number to lowest terms?
A rational number is reduced to lowest terms by dividing the numerator and denominator by the GCF.

4. How do you reduce a rational expression to lowest terms?
A rational expression is reduced to lowest terms by dividing the numerator and denominator by the GCF.

5. How is the quotient rule used in reducing rational expressions?
The quotient rule is used in reducing ratios of monomials.

6. What is the relationship between $a - b$ and $b - a$?
The expressions $a - b$ and $b - a$ are opposites.

Evaluate each rational expression. See Example 1.

7. Evaluate $\frac{3x - 3}{x + 5}$ for $x = -2$. -3

8. Evaluate $\frac{3x + 1}{4x - 4}$ for $x = 5$. 1

9. If $R(x) = \frac{2x + 9}{x}$, find $R(3)$. 5

10. If $R(x) = \frac{-20x - 2}{x - 8}$, find $R(-1)$. -2

11. If $R(x) = \frac{x - 5}{x + 3}$, find $R(2)$, $R(-4)$, $R(-3.02)$, and $R(-2.96)$. $-0.6, 9, 401, -199$

12. If $R(x) = \frac{x^2 - 2x - 3}{x - 2}$, find $R(3)$, $R(5)$, $R(2.05)$, and $R(1.999)$. $0, 4, -57.95, 3001.999$

Which numbers cannot be used in place of the variable in each rational expression? See Example 2.

13. $\frac{x}{x + 1}$ -1

14. $\frac{3x}{x - 7}$ 7

c) $\dfrac{a+b}{6a} \cdot \dfrac{8a^2}{a^2+2ab+b^2} = \dfrac{a+b}{2 \cdot 3a} \cdot \dfrac{2 \cdot 4a^2}{(a+b)^2}$ Factor.

$= \dfrac{4a}{3(a+b)}$ Reduce.

$= \dfrac{4a}{3a+3b}$ ■

Division of Rational Numbers

Division of rational numbers can be accomplished by multiplying by the reciprocal of the divisor.

Division of Rational Numbers

If $b \neq 0$, $c \neq 0$, and $d \neq 0$, then

$$\frac{a}{b} \div \frac{c}{d} = \frac{a}{b} \cdot \frac{d}{c}.$$

EXAMPLE 4 **Dividing rational numbers**

Find each quotient.

a) $5 \div \dfrac{1}{2}$

b) $\dfrac{6}{7} \div \dfrac{3}{14}$

Solution

a) $5 \div \dfrac{1}{2} = 5 \cdot 2 = 10$

b) $\dfrac{6}{7} \div \dfrac{3}{14} = \dfrac{6}{7} \cdot \dfrac{14}{3} = \dfrac{2 \cdot \cancel{3}}{\cancel{7}} \cdot \dfrac{2 \cdot \cancel{7}}{\cancel{3}} = 4$ ■

Division of Rational Expressions

We divide rational expressions in the same way we divide rational numbers: Invert the divisor and multiply.

EXAMPLE 5 **Dividing rational expressions**

Find each quotient.

a) $\dfrac{5}{3x} \div \dfrac{5}{6x}$

b) $\dfrac{x^7}{2} \div (2x^2)$

c) $\dfrac{4-x^2}{x^2+x} \div \dfrac{x-2}{x^2-1}$

Solution

a) $\dfrac{5}{3x} \div \dfrac{5}{6x} = \dfrac{5}{3x} \cdot \dfrac{6x}{5}$ Invert the divisor and multiply.

$= \dfrac{\cancel{5}}{\cancel{3x}} \cdot \dfrac{2 \cdot \cancel{3x}}{\cancel{5}}$ Factor.

$= 2$ Divide out the common factors.

Helpful Hint

A doctor told a nurse to give a patient half of the usual dose of a certain medicine. The nurse figured, "dividing in half means dividing by 1/2 which means multiply by 2." So the patient got four times the prescribed amount and died (true story). There is a big difference between dividing a quantity in half and dividing by one-half.

b) $\dfrac{x^7}{2} \div (2x^2) = \dfrac{x^7}{2} \cdot \dfrac{1}{2x^2}$ Invert and multiply.

$= \dfrac{x^5}{4}$ Quotient rule

c) $\dfrac{4 - x^2}{x^2 + x} \div \dfrac{x - 2}{x^2 - 1} = \dfrac{4 - x^2}{x^2 + x} \cdot \dfrac{x^2 - 1}{x - 2}$ Invert and multiply.

$= \dfrac{\overset{-1}{\cancel{(2 - x)}}(2 + x)}{x\cancel{(x + 1)}} \cdot \dfrac{\cancel{(x + 1)}(x - 1)}{\cancel{x - 2}}$ Factor.

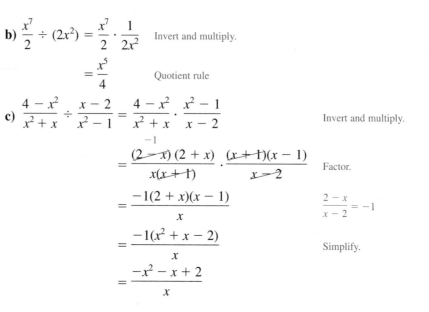

$= \dfrac{-1(2 + x)(x - 1)}{x}$ $\dfrac{2 - x}{x - 2} = -1$

$= \dfrac{-1(x^2 + x - 2)}{x}$ Simplify.

$= \dfrac{-x^2 - x + 2}{x}$ ■

We sometimes write division of rational expressions using the fraction bar. For example, we can write

$$\frac{a + b}{3} \div \frac{1}{6} \quad \text{as} \quad \frac{\frac{a + b}{3}}{\frac{1}{6}}.$$

No matter how division is expressed, we invert the divisor and multiply.

EXAMPLE 6 **Division expressed with a fraction bar**

Find each quotient.

a) $\dfrac{\frac{a + b}{3}}{\frac{1}{6}}$ b) $\dfrac{\frac{x^2 - 1}{2}}{\frac{x - 1}{3}}$

c) $\dfrac{\frac{a^2 + 5}{3}}{2}$

Solution

a) $\dfrac{\frac{a + b}{3}}{\frac{1}{6}} = \dfrac{a + b}{3} \div \dfrac{1}{6}$ Rewrite as division.

$= \dfrac{a + b}{3} \cdot \dfrac{6}{1}$ Invert and multiply.

$= \dfrac{a + b}{\cancel{3}} \cdot \dfrac{2 \cdot \cancel{3}}{1}$ Factor.

$= (a + b)2$ Reduce.

$= 2a + 2b$

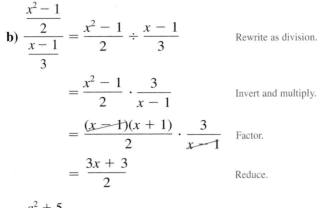

b) $$\frac{\frac{x^2-1}{2}}{\frac{x-1}{3}} = \frac{x^2-1}{2} \div \frac{x-1}{3} \quad \text{Rewrite as division.}$$

$$= \frac{x^2-1}{2} \cdot \frac{3}{x-1} \quad \text{Invert and multiply.}$$

$$= \frac{(x-1)(x+1)}{2} \cdot \frac{3}{x-1} \quad \text{Factor.}$$

$$= \frac{3x+3}{2} \quad \text{Reduce.}$$

c) $$\frac{\frac{a^2+5}{3}}{2} = \frac{a^2+5}{3} \div 2 \quad \text{Rewrite as division.}$$

$$= \frac{a^2+5}{3} \cdot \frac{1}{2} = \frac{a^2+5}{6}$$

■

Applications

We saw in Section 7.1 that rational expressions can be used to represent rates. Note that there are several ways to write rates. For example, miles per hour is written mph, mi/hr, or $\frac{\text{mi}}{\text{hr}}$. The last way is best when doing operations with rates because it helps us reconcile our answers. Notice how hours "cancels" when we multiply miles per hour and hours in the next example, giving an answer in miles, as it should be.

EXAMPLE 7 **Writing rational expressions**

Answer each question with a rational expression.

a) Shasta averaged $\frac{200}{x}$ mph for x hours before she had lunch. How many miles did she drive in the first 3 hours after lunch assuming that she continued to average $\frac{200}{x}$ mph?

b) If a bathtub can be filled in x minutes, then the rate at which it is filling is $\frac{1}{x}$ tub/min. How much of the tub is filled in 10 minutes?

Solution

a) Because $R \cdot T = D$, the distance she traveled after lunch is the product of the rate and time:

$$\frac{200}{x}\frac{\text{mi}}{\text{hr}} \cdot 3\ \text{hr} = \frac{600}{x}\ \text{mi}$$

b) Because $R \cdot T = W$, the work completed is the product of the rate and time:

$$\frac{1}{x}\frac{\text{tub}}{\text{min}} \cdot 10\ \text{min} = \frac{10}{x}\ \text{tub}$$

■

WARM-UPS

True or false? Explain your answer.

1. $\frac{2}{3} \cdot \frac{5}{3} = \frac{10}{9}$. True
2. The product of $\frac{x-7}{3}$ and $\frac{6}{7-x}$ is -2. True
3. Dividing by 2 is equivalent to multiplying by $\frac{1}{2}$. True
4. $3 \div x = \frac{1}{3} \cdot x$ for any nonzero number x. False
5. Factoring polynomials is essential in multiplying rational expressions. True
6. One-half of one-fourth is one-sixth. False
7. One-half divided by three is three-halves. False
8. The quotient of $(839 - 487)$ and $(487 - 839)$ is -1. True
9. $\frac{a}{3} \div 3 = \frac{a}{9}$ for any value of a. True
10. $\frac{a}{b} \cdot \frac{b}{a} = 1$ for any nonzero values of a and b. True

7.2 EXERCISES

Reading and Writing *After reading this section, write out the answers to these questions. Use complete sentences.*

1. How do you multiply rational numbers?
   Rational numbers are multiplied by multiplying their numerators and their denominators.
2. How do you multiply rational expressions?
   Rational expressions are multiplied by multiplying their numerators and their denominators.
3. What can be done to simplify the process of multiplying rational numbers or rational expressions?
   Reducing can be done before multiplying rational numbers or expressions.
4. How do you divide rational numbers or rational expressions?
   To divide rational expressions, invert the divisor and multiply.

Perform the indicated operation. See Example 1.

5. $\frac{8}{15} \cdot \frac{35}{24}$ $\frac{7}{9}$
6. $\frac{3}{4} \cdot \frac{8}{21}$ $\frac{2}{7}$
7. $\frac{12}{17} \cdot \frac{51}{10}$ $\frac{18}{5}$
8. $\frac{25}{48} \cdot \frac{56}{35}$ $\frac{5}{6}$
9. $24 \cdot \frac{7}{20}$ $\frac{42}{5}$
10. $\frac{3}{10} \cdot 35$ $\frac{21}{2}$

Perform the indicated operation. See Example 2.

11. $\frac{5a}{12b} \cdot \frac{3ab}{55a}$ $\frac{a}{44}$
12. $\frac{3m}{7p} \cdot \frac{35p}{6mp}$ $\frac{5}{2p}$
13. $\frac{-2x^6}{7a^5} \cdot \frac{21a^2}{6x}$ $\frac{-x^5}{a^3}$
14. $\frac{5z^3w}{-9y^3} \cdot \frac{-6y^5}{20z^9}$ $\frac{wy^2}{6z^6}$
15. $\frac{15t^3y^5}{20w^7} \cdot 24t^5w^3y^2$ $\frac{18t^8y^7}{w^4}$
16. $22x^2y^3z \cdot \frac{6x^5}{33y^3z^4}$ $\frac{4x^7}{z^3}$

Perform the indicated operation. See Example 3.

17. $\frac{3a+3b}{15} \cdot \frac{10a}{a^2-b^2}$ $\frac{2a}{a-b}$
18. $\frac{b^3+b}{5} \cdot \frac{10}{b^2+b}$ $\frac{2b^2+2}{b+1}$
19. $(x^2-6x+9) \cdot \frac{3}{x-3}$ $3x-9$
20. $\frac{12}{4x+10} \cdot (4x^2+20x+25)$ $12x+30$
21. $\frac{16a+8}{5a^2+5} \cdot \frac{2a^2+a-1}{4a^2-1}$ $\frac{8a+8}{5a^2+5}$
22. $\frac{6x-18}{2x^2-5x-3} \cdot \frac{4x^2+4x+1}{6x+3}$ 2

Perform the indicated operation. See Example 4.

23. $12 \div \frac{2}{5}$ 30
24. $32 \div \frac{1}{4}$ 128
25. $\frac{5}{7} \div \frac{15}{14}$ $\frac{2}{3}$
26. $\frac{3}{4} \div \frac{15}{2}$ $\frac{1}{10}$
27. $\frac{40}{3} \div 12$ $\frac{10}{9}$
28. $\frac{22}{9} \div 9$ $\frac{22}{81}$

Perform the indicated operation. See Example 5.

29. $\frac{5x^2}{3} \div \frac{10x}{21}$ $\frac{7x}{2}$
30. $\frac{4u^2}{3v} \div \frac{14u}{15v^6}$ $\frac{10uv^5}{7}$
31. $\frac{8m^3}{n^4} \div (12mn^2)$ $\frac{2m^2}{3n^6}$
32. $\frac{2p^4}{3q^3} \div (4pq^5)$ $\frac{p^3}{6q^8}$
33. $\frac{y-6}{2} \div \frac{6-y}{6}$ -3
34. $\frac{4-a}{5} \div \frac{a^2-16}{3}$ $\frac{-3}{5a+20}$

35. $\dfrac{x^2+4x+4}{8} \div \dfrac{(x+2)^3}{16}$ $\quad \dfrac{2}{x+2}$

36. $\dfrac{a^2+2a+1}{3} \div \dfrac{a^2-1}{a}$ $\quad \dfrac{a^2+a}{3a-3}$

37. $\dfrac{t^2+3t-10}{t^2-25} \div (4t-8)$ $\quad \dfrac{1}{4t-20}$

38. $\dfrac{w^2-7w+12}{w^2-4w} \div (w^2-9)$ $\quad \dfrac{1}{w^2+3w}$

39. $(2x^2-3x-5) \div \dfrac{2x-5}{x-1}$ $\quad x^2-1$

40. $(6y^2-y-2) \div \dfrac{2y+1}{3y-2}$ $\quad 9y^2-12y+4$

Perform the indicated operation. See Example 6.

41. $\dfrac{\frac{x-2y}{5}}{\frac{1}{10}}$ $\quad 2x-4y$

42. $\dfrac{\frac{3m+6n}{8}}{\frac{3}{4}}$ $\quad \dfrac{m+2n}{2}$

43. $\dfrac{\frac{x^2-4}{12}}{\frac{x-2}{6}}$ $\quad \dfrac{x+2}{2}$

44. $\dfrac{\frac{6a^2+6}{5}}{\frac{6a+6}{5}}$ $\quad \dfrac{a^2+1}{a+1}$

45. $\dfrac{\frac{x^2+9}{3}}{5}$ $\quad \dfrac{x^2+9}{15}$

46. $\dfrac{\frac{1}{a-3}}{4}$ $\quad \dfrac{1}{4a-12}$

47. $\dfrac{\frac{x^2-y^2}{x-y}}{9}$ $\quad 9x+9y$

48. $\dfrac{x^2+6x+8}{\frac{x+2}{x+1}}$ $\quad x^2+5x+4$

Perform the indicated operation.

49. $\dfrac{x-1}{3} \cdot \dfrac{9}{1-x}$ $\quad -3$

50. $\dfrac{2x-2y}{3} \cdot \dfrac{1}{y-x}$ $\quad -\dfrac{2}{3}$

51. $\dfrac{3a+3b}{a} \cdot \dfrac{1}{3}$ $\quad \dfrac{a+b}{a}$

52. $\dfrac{a-b}{2b-2a} \cdot \dfrac{2}{5}$ $\quad -\dfrac{1}{5}$

53. $\dfrac{\frac{b}{a}}{\frac{1}{2}}$ $\quad \dfrac{2b}{a}$

54. $\dfrac{\frac{2g}{3h}}{\frac{1}{h}}$ $\quad \dfrac{2g}{3}$

55. $\dfrac{6y}{3} \div (2x)$ $\quad \dfrac{y}{x}$

56. $\dfrac{8x}{9} \div (18x)$ $\quad \dfrac{4}{81}$

57. $\dfrac{a^3b^4}{-2ab^2} \cdot \dfrac{a^5b^7}{ab}$ $\quad \dfrac{-a^6b^8}{2}$

58. $\dfrac{-2a^2}{3a^2} \cdot \dfrac{20a}{15a^3}$ $\quad \dfrac{-8}{9a^2}$

59. $\dfrac{2mn^4}{6mn^2} \div \dfrac{3m^5n^7}{m^2n^4}$ $\quad \dfrac{1}{9m^3n}$

60. $\dfrac{rt^2}{rt^2} \div \dfrac{rt^2}{r^3t^2}$ $\quad r^2$

61. $\dfrac{3x^2+16x+5}{x} \cdot \dfrac{x^2}{9x^2-1}$ $\quad \dfrac{x^2+5x}{3x-1}$

62. $\dfrac{x^2+6x+5}{x} \cdot \dfrac{x^4}{3x+3}$ $\quad \dfrac{x^4+5x^3}{3}$

63. $\dfrac{a^2-2a+4}{a^2-4} \cdot \dfrac{(a+2)^3}{2a+4}$ $\quad \dfrac{a^3+8}{2a-4}$

64. $\dfrac{w^2-1}{(w-1)^2} \cdot \dfrac{w-1}{w^2+2w+1}$ $\quad \dfrac{1}{w+1}$

65. $\dfrac{2x^2+19x-10}{x^2-100} \div \dfrac{4x^2-1}{2x^2-19x-10}$ $\quad 1$

66. $\dfrac{x^3-1}{x^2+1} \div \dfrac{9x^2+9x+9}{x^2-x}$ $\quad \dfrac{x^3-2x^2+x}{9x^2+9}$

67. $\dfrac{9+6m+m^2}{9-6m+m^2} \cdot \dfrac{m^2-9}{m^2+mk+3m+3k}$ $\quad \dfrac{(m+3)^2}{(m-3)(m+k)}$

68. $\dfrac{3x+3w+bx+bw}{x^2-w^2} \cdot \dfrac{6-2b}{9-b^2}$ $\quad \dfrac{2}{x-w}$

Solve each problem. Answers could be rational expressions. Be sure to give your answer with appropriate units. See Example 7.

69. ***Distance.*** Florence averaged $\frac{26.2}{x}$ mph for the x hours in which she ran the Boston Marathon. If she ran at that same rate for $\frac{1}{2}$ hour in the Manchac Fun Run, then how many miles did she run at Manchac?
$\dfrac{13.1}{x}$ mi

70. ***Work.*** Henry sold 120 magazine subscriptions in $x+2$ days. If he sold at the same rate for another week, then how many magazines did he sell in the extra week?
$\dfrac{840}{x+2}$ magazines

71. ***Area of a rectangle.*** If the length of a rectangular flag is x meters and its width is $\frac{5}{x}$ meters, then what is the area of the rectangle? 5 square meters

FIGURE FOR EXERCISE 71

72. ***Area of a triangle.*** If the base of a triangle is $8x+16$ yards and its height is $\frac{1}{x+2}$ yards, then what is the area of the triangle?
4 yd^2

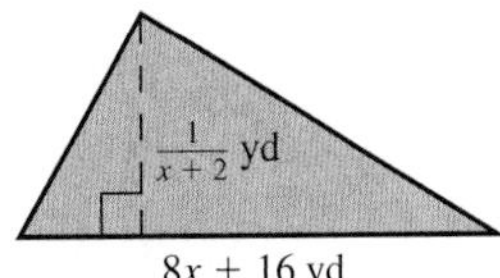

FIGURE FOR EXERCISE 72

GETTING MORE INVOLVED

73. *Discussion.* Evaluate each expression.

a) One-half of $\frac{1}{4}$ **b)** One-third of 4

c) One-half of $\frac{4x}{3}$ **d)** One-half of $\frac{3x}{2}$

a) $\frac{1}{8}$ b) $\frac{4}{3}$ c) $\frac{2x}{3}$ d) $\frac{3x}{4}$

74. *Exploration.* Let $R = \frac{6x^2 + 23x + 20}{24x^2 + 29x - 4}$ and $H = \frac{2x + 5}{8x - 1}$.

a) Find R when $x = 2$ and $x = 3$. Find H when $x = 2$ and $x = 3$.

b) How are these values of R and H related and why?

a) $R = \frac{3}{5}, R = \frac{11}{23}, H = \frac{3}{5}, H = \frac{11}{23}$

b) R reduces to H.

7.3 FINDING THE LEAST COMMON DENOMINATOR

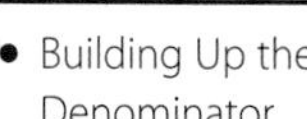

In This Section

- Building Up the Denominator
- Finding the Least Common Denominator
- Converting to the LCD

Every rational expression can be written in infinitely many equivalent forms. Because we can add or subtract only fractions with identical denominators, we must be able to change the denominator of a fraction. You have already learned how to change the denominator of a fraction by reducing. In this section you will learn the opposite of reducing, which is called **building up the denominator.**

Building Up the Denominator

To convert the fraction $\frac{2}{3}$ into an equivalent fraction with a denominator of 21, we factor 21 as $21 = 3 \cdot 7$. Because $\frac{2}{3}$ already has a 3 in the denominator, multiply the numerator and denominator of $\frac{2}{3}$ by the missing factor 7 to get a denominator of 21:

$$\frac{2}{3} = \frac{2}{3} \cdot \frac{7}{7} = \frac{14}{21}$$

For rational expressions the process is the same. To convert the rational expression

$$\frac{5}{x + 3}$$

into an equivalent rational expression with a denominator of $x^2 - x - 12$, first factor $x^2 - x - 12$:

$$x^2 - x - 12 = (x + 3)(x - 4)$$

From the factorization we can see that the denominator $x + 3$ needs only a factor of $x - 4$ to have the required denominator. So multiply the numerator and denominator by the missing factor $x - 4$:

$$\frac{5}{x + 3} = \frac{5(x - 4)}{(x + 3)(x - 4)} = \frac{5x - 20}{x^2 - x - 12}$$

EXAMPLE 1 **Building up the denominator**

Build each rational expression into an equivalent rational expression with the indicated denominator.

a) $3 = \frac{?}{12}$ **b)** $\frac{3}{w} = \frac{?}{wx}$ **c)** $\frac{2}{3y^3} = \frac{?}{12y^8}$

Solution

a) Because $3 = \frac{3}{1}$, we get a denominator of 12 by multiplying the numerator and denominator by 12:

$$3 = \frac{3}{1} = \frac{3 \cdot 12}{1 \cdot 12} = \frac{36}{12}$$

b) Multiply the numerator and denominator by x:

$$\frac{3}{w} = \frac{3 \cdot x}{w \cdot x} = \frac{3x}{wx}$$

c) To build the denominator $3y^3$ up to $12y^8$, multiply by $4y^5$:

$$\frac{2}{3y^3} = \frac{2 \cdot 4y^5}{3y^3 \cdot 4y^5} = \frac{8y^5}{12y^8}$$

■

In Example 2 we must factor the original denominator before building up the denominator.

EXAMPLE 2 **Building up the denominator**

Build each rational expression into an equivalent rational expression with the indicated denominator.

a) $\frac{7}{3x - 3y} = \frac{?}{6y - 6x}$ **b)** $\frac{x - 2}{x + 2} = \frac{?}{x^2 + 8x + 12}$

Helpful Hint

Notice that reducing and building up are exactly the opposite of each other. In reducing you remove a factor that is common to the numerator and denominator, and in building up you put a common factor into the numerator and denominator.

Solution

a) Because $3x - 3y = 3(x - y)$, we factor -6 out of $6y - 6x$. This will give a factor of $x - y$ in each denominator:

$$3x - 3y = 3(x - y)$$
$$6y - 6x = -6(x - y) = -2 \cdot 3(x - y)$$

To get the required denominator, we multiply the numerator and denominator by -2 only:

$$\frac{7}{3x - 3y} = \frac{7(-2)}{(3x - 3y)(-2)}$$
$$= \frac{-14}{6y - 6x}$$

b) Because $x^2 + 8x + 12 = (x + 2)(x + 6)$, we multiply the numerator and denominator by $x + 6$, the missing factor:

$$\frac{x - 2}{x + 2} = \frac{(x - 2)(x + 6)}{(x + 2)(x + 6)}$$
$$= \frac{x^2 + 4x - 12}{x^2 + 8x + 12}$$

■

CAUTION When building up a denominator, *both* the numerator and the denominator must be multiplied by the appropriate expression, because that is how we build up fractions.

Finding the Least Common Denominator

We can use the idea of building up the denominator to convert two fractions with different denominators into fractions with identical denominators. For example,

$$\frac{5}{6} \text{ and } \frac{1}{4}$$

can both be converted into fractions with a denominator of 12, since $12 = 2 \cdot 6$ and $12 = 3 \cdot 4$:

$$\frac{5}{6} = \frac{5 \cdot 2}{6 \cdot 2} = \frac{10}{12} \qquad \frac{1}{4} = \frac{1 \cdot 3}{4 \cdot 3} = \frac{3}{12}$$

The smallest number that is a multiple of all of the denominators is called the **least common denominator (LCD).** The LCD for the denominators 6 and 4 is 12.

To find the LCD in a systematic way, we look at a complete factorization of each denominator. Consider the denominators 24 and 30:

$$24 = 2 \cdot 2 \cdot 2 \cdot 3 = 2^3 \cdot 3$$
$$30 = 2 \cdot 3 \cdot 5$$

Study Tip

Studying in a quiet place is better than studying in a noisy place. There are very few people who can listen to music or a conversation and study at the same time.

Any multiple of 24 must have three 2's in its factorization, and any multiple of 30 must have one 2 as a factor. So a number with three 2's in its factorization will have enough to be a multiple of both 24 and 30. The LCD must also have one 3 and one 5 in its factorization. *We use each factor the maximum number of times it appears in either factorization.* So the LCD is $2^3 \cdot 3 \cdot 5$:

$$2^3 \cdot 3 \cdot 5 = \overbrace{2 \cdot 2 \cdot 2 \cdot 3}^{24} \cdot 5 = 120, \quad \text{where } \underbrace{2 \cdot 3 \cdot 5}_{30}$$

If we omitted any one of the factors in $2 \cdot 2 \cdot 2 \cdot 3 \cdot 5$, we would not have a multiple of both 24 and 30. That is what makes 120 the *least* common denominator. To find the LCD for two polynomials, we use the same strategy.

Strategy for Finding the LCD for Polynomials

1. Factor each denominator completely. Use exponent notation for repeated factors.
2. Write the product of all of the different factors that appear in the denominators.
3. On each factor, use the highest power that appears on that factor in any of the denominators.

EXAMPLE 3 **Finding the LCD**

If the given expressions were used as denominators of rational expressions, then what would be the LCD for each group of denominators?

a) 20, 50 **b)** x^3yz^2, x^5y^2z, xyz^5

c) $a^2 + 5a + 6$, $a^2 + 4a + 4$

Solution

a) First factor each number completely:

$$20 = 2^2 \cdot 5 \qquad 50 = 2 \cdot 5^2$$

The highest power of 2 is 2, and the highest power of 5 is 2. So the LCD of 20 and 50 is $2^2 \cdot 5^2$, or 100.

b) The expressions x^3yz^2, x^5y^2z, and xyz^5 are already factored. For the LCD, use the highest power of each variable. So the LCD is $x^5y^2z^5$.

c) First factor each polynomial.

$$a^2 + 5a + 6 = (a + 2)(a + 3) \qquad a^2 + 4a + 4 = (a + 2)^2$$

The highest power of $(a + 3)$ is 1, and the highest power of $(a + 2)$ is 2. So the LCD is $(a + 3)(a + 2)^2$. ■

Converting to the LCD

When adding or subtracting rational expressions, we must convert the expressions into expressions with identical denominators. To keep the computations as simple as possible, we use the least common denominator.

EXAMPLE 4 **Converting to the LCD**

Find the LCD for the rational expressions, and convert each expression into an equivalent rational expression with the LCD as the denominator.

a) $\dfrac{4}{9xy}, \dfrac{2}{15xz}$ **b)** $\dfrac{5}{6x^2}, \dfrac{1}{8x^3y}, \dfrac{3}{4y^2}$

Helpful Hint

What is the difference between LCD, GCF, CBS, and NBC? The LCD for the denominators 4 and 6 is 12. The *least* common denominator is *greater than* or equal to both numbers. The GCF for 4 and 6 is 2. The *greatest* common factor is *less than* or equal to both numbers. CBS and NBC are TV networks.

Solution

a) Factor each denominator completely:

$$9xy = 3^2xy \qquad 15xz = 3 \cdot 5xz$$

The LCD is $3^2 \cdot 5xyz$. Now convert each expression into an expression with this denominator. We must multiply the numerator and denominator of the first rational expression by $5z$ and the second by $3y$:

$$\left.\begin{aligned} \frac{4}{9xy} &= \frac{4 \cdot 5z}{9xy \cdot 5z} = \frac{20z}{45xyz} \\ \frac{2}{15xz} &= \frac{2 \cdot 3y}{15xz \cdot 3y} = \frac{6y}{45xyz} \end{aligned}\right\} \text{Same denominator}$$

b) Factor each denominator completely:

$$6x^2 = 2 \cdot 3x^2 \qquad 8x^3y = 2^3x^3y \qquad 4y^2 = 2^2y^2$$

The LCD is $2^3 \cdot 3 \cdot x^3y^2$ or $24x^3y^2$. Now convert each expression into an expression with this denominator:

$$\frac{5}{6x^2} = \frac{5 \cdot 4xy^2}{6x^2 \cdot 4xy^2} = \frac{20xy^2}{24x^3y^2}$$

$$\frac{1}{8x^3y} = \frac{1 \cdot 3y}{8x^3y \cdot 3y} = \frac{3y}{24x^3y^2}$$

$$\frac{3}{4y^2} = \frac{3 \cdot 6x^3}{4y^2 \cdot 6x^3} = \frac{18x^3}{24x^3y^2}$$

■

EXAMPLE 5 **Converting to the LCD**

Find the LCD for the rational expressions

$$\frac{5x}{x^2 - 4} \quad \text{and} \quad \frac{3}{x^2 + x - 6}$$

and convert each into an equivalent rational expression with that denominator.

Solution

First factor the denominators:

$$x^2 - 4 = (x - 2)(x + 2)$$
$$x^2 + x - 6 = (x - 2)(x + 3)$$

The LCD is $(x - 2)(x + 2)(x + 3)$. Now we multiply the numerator and denominator of the first rational expression by $(x + 3)$ and those of the second rational expression by $(x + 2)$. Because each denominator already has one factor of $(x - 2)$, there is no reason to multiply by $(x - 2)$. We multiply each denominator by the factors in the LCD that are missing from that denominator:

$$\frac{5x}{x^2 - 4} = \frac{5x(x + 3)}{(x - 2)(x + 2)(x + 3)} = \frac{5x^2 + 15x}{(x - 2)(x + 2)(x + 3)}$$
$$\frac{3}{x^2 + x - 6} = \frac{3(x + 2)}{(x - 2)(x + 3)(x + 2)} = \frac{3x + 6}{(x - 2)(x + 2)(x + 3)}$$

Same denominator

■

Note that in Example 5 we multiplied the expressions in the numerators but left the denominators in factored form. The numerators are simplified because it is the numerators that must be added when we add rational expressions in Section 7.4. Because we can add rational expressions with identical denominators, there is no need to multiply the denominators.

WARM-UPS

True or false? Explain your answer.

1. To convert $\frac{2}{3}$ into an equivalent fraction with a denominator of 18, we would multiply only the denominator of $\frac{2}{3}$ by 6. False
2. Factoring has nothing to do with finding the least common denominator. False
3. $\frac{3}{2ab^2} = \frac{15a^2b^2}{10a^3b^4}$ for any nonzero values of a and b. True
4. The LCD for the denominators $2^5 \cdot 3$ and $2^4 \cdot 3^2$ is $2^5 \cdot 3^2$. True
5. The LCD for the fractions $\frac{1}{6}$ and $\frac{1}{10}$ is 60. False
6. The LCD for the denominators $6a^2b$ and $4ab^3$ is $2ab$. False
7. The LCD for the denominators $a^2 + 1$ and $a + 1$ is $a^2 + 1$. False
8. $\frac{x}{2} = \frac{x + 7}{2 + 7}$ for any real number x. False
9. The LCD for the rational expressions $\frac{1}{x - 2}$ and $\frac{3}{x + 2}$ is $x^2 - 4$. True
10. $x = \frac{3x}{3}$ for any real number x. True

7.3 EXERCISES

Reading and Writing *After reading this section, write out the answers to these questions. Use complete sentences.*

1. What is building up the denominator?
We can build up a denominator by multiplying the numerator and denominator of a fraction by the same nonzero number.
2. How do we build up the denominator of a rational expression?
To build up the denominator of a rational expression, we can multiply the numerator and denominator by the same polynomial.
3. What is the least common denominator for fractions?
For fractions, the LCD is the smallest number that is a multiple of all of the denominators.
4. How do you find the LCD for two polynomial denominators?
For polynomial denominators, the LCD consists of every factor that appears, raised to the highest power that appears on the factor.

Build each rational expression into an equivalent rational expression with the indicated denominator. See Example 1.

5. $\frac{1}{3} = \frac{?}{27}$ $\quad \frac{9}{27}$

6. $\frac{2}{5} = \frac{?}{35}$ $\quad \frac{14}{35}$

7. $7 = \frac{?}{2x}$ $\quad \frac{14x}{2x}$

8. $6 = \frac{?}{4y}$ $\quad \frac{24y}{4y}$

9. $\frac{5}{b} = \frac{?}{3bt}$ $\quad \frac{15t}{3bt}$

10. $\frac{7}{2ay} = \frac{?}{2ayz}$ $\quad \frac{7z}{2ayz}$

11. $\frac{-9z}{2aw} = \frac{?}{8awz}$ $\quad \frac{-36z^2}{8awz}$

12. $\frac{-7yt}{3x} = \frac{?}{18xyt}$ $\quad \frac{-42y^2t^2}{18xyt}$

13. $\frac{2}{3a} = \frac{?}{15a^3}$ $\quad \frac{10a^2}{15a^3}$

14. $\frac{7b}{12c^5} = \frac{?}{36c^8}$ $\quad \frac{21bc^3}{36c^8}$

15. $\frac{4}{5xy^2} = \frac{?}{10x^2y^5}$ $\quad \frac{8xy^3}{10x^2y^5}$

16. $\frac{5y^2}{8x^3z} = \frac{?}{24x^5z^3}$ $\quad \frac{15x^2y^2z^2}{24x^5z^3}$

Build each rational expression into an equivalent rational expression with the indicated denominator. See Example 2.

17. $\frac{5}{2x+2} = \frac{?}{-8x-8}$ $\quad \frac{-20}{-8x-8}$

18. $\frac{3}{m-n} = \frac{?}{2n-2m}$ $\quad \frac{-6}{2n-2m}$

19. $\frac{8a}{5b^2-5b} = \frac{?}{20b^2-20b^3}$ $\quad \frac{-32ab}{20b^2-20b^3}$

20. $\frac{5x}{-6x-9} = \frac{?}{18x^2+27x}$ $\quad \frac{-15x^2}{18x^2+27x}$

21. $\frac{3}{x+2} = \frac{?}{x^2-4}$ $\quad \frac{3x-6}{x^2-4}$

22. $\frac{a}{a+3} = \frac{?}{a^2-9}$ $\quad \frac{a^2-3a}{a^2-9}$

23. $\frac{3x}{x+1} = \frac{?}{x^2+2x+1}$ $\quad \frac{3x^2+3x}{x^2+2x+1}$

24. $\frac{-7x}{2x-3} = \frac{?}{4x^2-12x+9}$ $\quad \frac{-14x^2+21x}{4x^2-12x+9}$

25. $\frac{y-6}{y-4} = \frac{?}{y^2+y-20}$ $\quad \frac{y^2-y-30}{y^2+y-20}$

26. $\frac{z-6}{z+3} = \frac{?}{z^2-2z-15}$ $\quad \frac{z^2-11z+30}{z^2-2z-15}$

If the given expressions were used as denominators of rational expressions, then what would be the LCD for each group of denominators? See Example 3.

27. 12, 16 $\quad$ 48

28. 28, 42 $\quad$ 84

29. 12, 18, 20 $\quad$ 180

30. 24, 40, 48 $\quad$ 240

31. $6a^2$, $15a$ $\quad 30a^2$

32. $18x^2$, $20xy$ $\quad 180x^2y$

33. $2a^4b$, $3ab^6$, $4a^3b^2$ $\quad 12a^4b^6$

34. $4m^3nw$, $6mn^5w^8$, $9m^6nw$ $\quad 36m^6n^5w^8$

35. x^2-16, $x^2+8x+16$ $\quad (x-4)(x+4)^2$

36. x^2-9, x^2+6x+9 $\quad (x-3)(x+3)^2$

37. x, $x+2$, $x-2$ $\quad x(x+2)(x-2)$

38. y, $y-5$, $y+2$ $\quad y(y-5)(y+2)$

39. x^2-4x, x^2-16, $2x$ $\quad 2x(x-4)(x+4)$

40. y, y^2-3y, $3y$ $\quad 3y(y-3)$

Find the LCD for the given rational expressions, and convert each rational expression into an equivalent rational expression with the LCD as the denominator. See Example 4.

41. $\frac{1}{6}, \frac{3}{8}$ $\quad \frac{4}{24}, \frac{9}{24}$

42. $\frac{5}{12}, \frac{3}{20}$ $\quad \frac{25}{60}, \frac{9}{60}$

43. $\frac{3}{84a}, \frac{5}{63b}$ $\quad \frac{9b}{252ab}, \frac{20a}{252ab}$

44. $\frac{4b}{75a}, \frac{6}{105ab}$ $\quad \frac{28b^2}{525ab}, \frac{30}{525ab}$

45. $\frac{1}{3x^2}, \frac{3}{2x^5}$ $\quad \frac{2x^3}{6x^5}, \frac{9}{6x^5}$

46. $\frac{3}{8a^3b^9}, \frac{5}{6a^2c}$ $\quad \frac{9c}{24a^3b^9c}, \frac{20ab^9}{24a^3b^9c}$

47. $\frac{x}{9y^5z}, \frac{y}{12x^3}, \frac{1}{6x^2y}$ $\quad \frac{4x^4}{36x^3y^5z}, \frac{3y^6z}{36x^3y^5z}, \frac{6xy^4z}{36x^3y^5z}$

48. $\frac{5}{12a^6b}, \frac{3b}{14a^3}, \frac{1}{2ab^3}$ $\quad \frac{35b^2}{84a^6b^3}, \frac{18a^3b^4}{84a^6b^3}, \frac{42a^5}{84a^6b^3}$

In Exercises 49–60, find the LCD for the given rational expressions, and convert each rational expression into an equivalent rational expression with the LCD as the denominator. See Example 5.

49. $\frac{2x}{x-3}, \frac{5x}{x+2}$ $\quad \frac{2x^2+4x}{(x-3)(x+2)}, \frac{5x^2-15x}{(x-3)(x+2)}$

50. $\frac{2a}{a-5}, \frac{3a}{a+2}$ $\quad \frac{2a^2+4a}{(a-5)(a+2)}, \frac{3a^2-15a}{(a-5)(a+2)}$

51. $\dfrac{4}{a-6}, \dfrac{5}{6-a}$ $\quad \dfrac{4}{a-6}, \dfrac{-5}{a-6}$

52. $\dfrac{4}{x-y}, \dfrac{5x}{2y-2x}$ $\quad \dfrac{8}{2x-2y}, \dfrac{-5x}{2x-2y}$

53. $\dfrac{x}{x^2-9}, \dfrac{5x}{x^2-6x+9}$

$\dfrac{x^2-3x}{(x-3)^2(x+3)}, \dfrac{5x^2+15x}{(x-3)^2(x+3)}$

54. $\dfrac{5x}{x^2-1}, \dfrac{-4}{x^2-2x+1}$

$\dfrac{5x^2-5x}{(x+1)(x-1)^2}, \dfrac{-4x-4}{(x+1)(x-1)^2}$

55. $\dfrac{w+2}{w^2-2w-15}, \dfrac{-2w}{w^2-4w-5}$

$\dfrac{w^2+3w+2}{(w-5)(w+3)(w+1)}, \dfrac{-2w^2-6w}{(w-5)(w+3)(w+1)}$

56. $\dfrac{z-1}{z^2+6z+8}, \dfrac{z+1}{z^2+5z+6}$

$\dfrac{z^2+2z-3}{(z+2)(z+4)(z+3)}, \dfrac{z^2+5z+4}{(z+2)(z+4)(z+3)}$

57. $\dfrac{-5}{6x-12}, \dfrac{x}{x^2-4}, \dfrac{3}{2x+4}$

$\dfrac{-5x-10}{6(x-2)(x+2)}, \dfrac{6x}{6(x-2)(x+2)}, \dfrac{9x-18}{6(x-2)(x+2)}$

58. $\dfrac{3}{4b^2-9}, \dfrac{2b}{2b+3}, \dfrac{-5}{2b^2-3b}$

$\dfrac{3b}{b(2b-3)(2b+3)}, \dfrac{4b^3-6b^2}{b(2b-3)(2b+3)},$

$\dfrac{-10b-15}{b(2b-3)(2b+3)}$

59. $\dfrac{2}{2q^2-5q-3}, \dfrac{3}{2q^2+9q+4}, \dfrac{4}{q^2+q-12}$

$\dfrac{2q+8}{(2q+1)(q-3)(q+4)}, \dfrac{3q-9}{(2q+1)(q-3)(q+4)},$

$\dfrac{8q+4}{(2q+1)(q-3)(q+4)}$

60. $\dfrac{-3}{2p^2+7p-15}, \dfrac{p}{2p^2-11p+12}, \dfrac{2}{p^2+p-20}$

$\dfrac{-3p+12}{(2p-3)(p+5)(p-4)}, \dfrac{p^2+5p}{(2p-3)(p+5)(p-4)},$

$\dfrac{4p-6}{(2p-3)(p+5)(p-4)}$

GETTING MORE INVOLVED

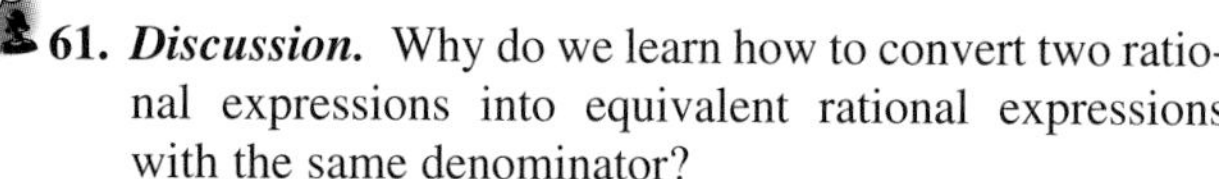

61. ***Discussion.*** Why do we learn how to convert two rational expressions into equivalent rational expressions with the same denominator?
Identical denominators are needed for addition and subtraction.

62. ***Discussion.*** Which expression is the LCD for

$$\frac{3x-1}{2^2 \cdot 3 \cdot x^2(x+2)} \quad \text{and} \quad \frac{2x+7}{2 \cdot 3^2 \cdot x(x+2)^2}?$$

a) $2 \cdot 3 \cdot x(x+2)$
b) $36x(x+2)$
c) $36x^2(x+2)^2$
d) $2^3 \cdot 3^3x^3(x+2)^2$ c

In This Section

- Addition and Subtraction of Rational Numbers
- Addition and Subtraction of Rational Expressions
- Applications

7.4 ADDITION AND SUBTRACTION

In Section 7.3 you learned how to find the LCD and build up the denominators of rational expressions. In this section we will use that knowledge to add and subtract rational expressions with different denominators.

Addition and Subtraction of Rational Numbers

We can add or subtract rational numbers (or fractions) only with identical denominators according to the following definition.

Addition and Subtraction of Rational Numbers

If $b \neq 0$, then

$$\frac{a}{b} + \frac{c}{b} = \frac{a+c}{b} \quad \text{and} \quad \frac{a}{b} - \frac{c}{b} = \frac{a-c}{b}.$$

EXAMPLE 1 **Adding or subtracting fractions with the same denominator**

Perform the indicated operations. Reduce answers to lowest terms.

a) $\frac{1}{12} + \frac{7}{12}$ **b)** $\frac{1}{4} - \frac{3}{4}$

Solution

a) $\frac{1}{12} + \frac{7}{12} = \frac{8}{12} = \frac{\cancel{4} \cdot 2}{\cancel{4} \cdot 3} = \frac{2}{3}$ **b)** $\frac{1}{4} - \frac{3}{4} = \frac{-2}{4} = -\frac{1}{2}$ ■

If the rational numbers have different denominators, we must convert them to equivalent rational numbers that have identical denominators and then add or subtract. Of course, it is most efficient to use the least common denominator (LCD), as in the following example.

EXAMPLE 2 **Adding or subtracting fractions with different denominators**

Find each sum or difference.

a) $\frac{3}{20} + \frac{7}{12}$ **b)** $\frac{1}{6} - \frac{4}{15}$

Helpful Hint

Note how all of the operations with rational expressions are performed according to the rules for fractions. So keep thinking of how you perform operations with fractions and you will improve your skills with fractions and with rational expressions.

Solution

a) Because $20 = 2^2 \cdot 5$ and $12 = 2^2 \cdot 3$, the LCD is $2^2 \cdot 3 \cdot 5$, or 60. Convert each fraction to an equivalent fraction with a denominator of 60:

$$\frac{3}{20} + \frac{7}{12} = \frac{3 \cdot 3}{20 \cdot 3} + \frac{7 \cdot 5}{12 \cdot 5} \quad \text{Build up the denominators.}$$

$$= \frac{9}{60} + \frac{35}{60} \quad \text{Simplify numerators and denominators.}$$

$$= \frac{44}{60} \quad \text{Add the fractions.}$$

$$= \frac{4 \cdot 11}{4 \cdot 15} \quad \text{Factor.}$$

$$= \frac{11}{15} \quad \text{Reduce.}$$

b) Because $6 = 2 \cdot 3$ and $15 = 3 \cdot 5$, the LCD is $2 \cdot 3 \cdot 5$ or 30:

$$\frac{1}{6} - \frac{4}{15} = \frac{1}{2 \cdot 3} - \frac{4}{3 \cdot 5} \quad \text{Factor the denominators.}$$

$$= \frac{1 \cdot 5}{2 \cdot 3 \cdot 5} - \frac{4 \cdot 2}{3 \cdot 5 \cdot 2} \quad \text{Build up the denominators.}$$

$$= \frac{5}{30} - \frac{8}{30} \quad \text{Simplify the numerators and denominators.}$$

$$= \frac{-3}{30} \quad \text{Subtract.}$$

$$= \frac{-1 \cdot 3}{10 \cdot 3} \quad \text{Factor.}$$

$$= -\frac{1}{10} \quad \text{Reduce.}$$

■

Addition and Subtraction of Rational Expressions

Rational expressions are added or subtracted just like rational numbers. We can add or subtract only rational expressions that have identical denominators.

EXAMPLE 3 **Rational expressions with the same denominator**

Perform the indicated operations and reduce answers to lowest terms.

a) $\frac{2}{3y} + \frac{4}{3y}$ b) $\frac{2x}{x + 2} + \frac{4}{x + 2}$

c) $\frac{x^2 + 2x}{(x - 1)(x + 3)} - \frac{2x + 1}{(x - 1)(x + 3)}$

Study Tip

Eliminate the obvious distractions when you study. Disconnect the telephone and put away newspapers, magazines, and unfinished projects. Even the sight of a textbook from another class might keep reminding you of how far behind you are in that class.

Solution

a) $\frac{2}{3y} + \frac{4}{3y} = \frac{6}{3y}$ Add the fractions.

$= \frac{2}{y}$ Reduce.

b) $\frac{2x}{x + 2} + \frac{4}{x + 2} = \frac{2x + 4}{x + 2}$ Add the fractions.

$= \frac{2(x + 2)}{x + 2}$ Factor the numerator.

$= 2$ Reduce.

c) $\frac{x^2 + 2x}{(x - 1)(x + 3)} - \frac{2x + 1}{(x - 1)(x + 3)} = \frac{x^2 + 2x - (2x + 1)}{(x - 1)(x + 3)}$ Subtract the fractions.

$= \frac{x^2 + 2x - 2x - 1}{(x - 1)(x + 3)}$ Remove parentheses.

$= \frac{x^2 - 1}{(x - 1)(x + 3)}$ Combine like terms.

$= \frac{(x - 1)(x + 1)}{(x - 1)(x + 3)}$ Factor.

$= \frac{x + 1}{x + 3}$ Reduce. ■

CAUTION When subtracting a numerator containing more than one term, be sure to enclose it in parentheses, as in Example 3(c). Because that numerator is a binomial, the sign of each of its terms must be changed for the subtraction.

In the next example the rational expressions have different denominators.

EXAMPLE 4 **Rational expressions with different denominators**

Perform the indicated operations.

a) $\frac{5}{2x} + \frac{2}{3}$ b) $\frac{4}{x^3y} + \frac{2}{xy^3}$

c) $\frac{a + 1}{6} - \frac{a - 2}{8}$

Helpful Hint

You can remind yourself of the difference between addition and multiplication of fractions with a simple example: If you and your spouse each own 1/7 of Microsoft, then together you own 2/7 of Microsoft. If you own 1/7 of Microsoft, and give 1/7 of your stock to your child, then your child owns 1/49 of Microsoft.

Solution

a) The LCD for $2x$ and 3 is $6x$:

$$\frac{5}{2x} + \frac{2}{3} = \frac{5 \cdot 3}{2x \cdot 3} + \frac{2 \cdot 2x}{3 \cdot 2x} \quad \text{Build up both denominators to } 6x.$$

$$= \frac{15}{6x} + \frac{4x}{6x} \quad \text{Simplify numerators and denominators.}$$

$$= \frac{15 + 4x}{6x} \quad \text{Add the rational expressions.}$$

b) The LCD is x^3y^3.

$$\frac{4}{x^3y} + \frac{2}{xy^3} = \frac{4 \cdot y^2}{x^3y \cdot y^2} + \frac{2 \cdot x^2}{xy^3 \cdot x^2} \quad \text{Build up both denominators to the LCD.}$$

$$= \frac{4y^2}{x^3y^3} + \frac{2x^2}{x^3y^3} \quad \text{Simplify numerators and denominators.}$$

$$= \frac{4y^2 + 2x^2}{x^3y^3} \quad \text{Add the rational expressions.}$$

c) Because $6 = 2 \cdot 3$ and $8 = 2^3$, the LCD is $2^3 \cdot 3$, or 24:

$$\frac{a+1}{6} - \frac{a-2}{8} = \frac{(a+1)4}{6 \cdot 4} - \frac{(a-2)3}{8 \cdot 3} \quad \text{Build up both denominators to the LCD 24.}$$

$$= \frac{4a+4}{24} - \frac{3a-6}{24} \quad \text{Simplify numerators and denominators.}$$

$$= \frac{4a + 4 - (3a - 6)}{24} \quad \text{Subtract the rational expressions.}$$

$$= \frac{4a + 4 - 3a + 6}{24} \quad \text{Remove the parentheses.}$$

$$= \frac{a + 10}{24} \quad \text{Combine like terms.}$$

■

EXAMPLE 5 **Rational expressions with different denominators**

Perform the indicated operations:

a) $\dfrac{1}{x^2 - 9} + \dfrac{2}{x^2 + 3x}$ **b)** $\dfrac{4}{5 - a} - \dfrac{2}{a - 5}$

Helpful Hint

Once the denominators are factored as in Example 5(a), you can simply look at each denominator and ask, "What factor does the other denominator(s) have that is missing from this one." Then use the missing factor to build up the denominator. Repeat until all denominators are identical and you will have the LCD.

Solution

a)

$$\frac{1}{x^2 - 9} + \frac{2}{x^2 + 3x} = \underbrace{\frac{1}{(x-3)(x+3)}}_{\text{Needs } x} + \underbrace{\frac{2}{x(x+3)}}_{\text{Needs } x - 3} \quad \text{The LCD is } x(x-3)(x+3).$$

$$= \frac{1 \cdot x}{(x-3)(x+3)x} + \frac{2(x-3)}{x(x+3)(x-3)}$$

$$= \frac{x}{x(x-3)(x+3)} + \frac{2x-6}{x(x-3)(x+3)}$$

$$= \frac{3x-6}{x(x-3)(x+3)} \quad \text{We usually leave the denominator in factored form.}$$

b) Because $-1(5 - a) = a - 5$, we can get identical denominators by multiplying only the first expression by -1 in the numerator and denominator:

$$\frac{4}{5-a} - \frac{2}{a-5} = \frac{4(-1)}{(5-a)(-1)} - \frac{2}{a-5}$$

$$= \frac{-4}{a-5} - \frac{2}{a-5}$$

$$= \frac{-6}{a-5} \quad -4 - 2 = -6$$

$$= -\frac{6}{a-5}$$

In Example 6 we combine three rational expressions by addition and subtraction.

EXAMPLE 6 **Rational expressions with different denominators**

Perform the indicated operations.

$$\frac{x+1}{x^2+2x} + \frac{2x+1}{6x+12} - \frac{1}{6}$$

Solution

The LCD for $x(x + 2)$, $6(x + 2)$, and 6 is $6x(x + 2)$.

$$\frac{x+1}{x^2+2x} + \frac{2x+1}{6x+12} - \frac{1}{6} = \frac{x+1}{x(x+2)} + \frac{2x+1}{6(x+2)} - \frac{1}{6} \quad \text{Factor denominators.}$$

$$= \frac{6(x+1)}{6x(x+2)} + \frac{x(2x+1)}{6x(x+2)} - \frac{1x(x+2)}{6x(x+2)} \quad \text{Build up to the LCD.}$$

$$= \frac{6x+6}{6x(x+2)} + \frac{2x^2+x}{6x(x+2)} - \frac{x^2+2x}{6x(x+2)} \quad \text{Simplify numerators.}$$

$$= \frac{6x+6+2x^2+x-x^2-2x}{6x(x+2)} \quad \text{Combine the numerators.}$$

$$= \frac{x^2+5x+6}{6x(x+2)} \quad \text{Combine like terms.}$$

$$= \frac{(x+3)(x+2)}{6x(x+2)} \quad \text{Factor.}$$

$$= \frac{x+3}{6x} \quad \text{Reduce.}$$

Applications

We have seen how rational expressions can occur in problems involving rates. In Example 7 we see an applied situation in which we add rational expressions.

EXAMPLE 7 **Adding work**

Harry takes twice as long as Lucy to proofread a manuscript. Write a rational expression for the amount of work they do in 3 hours working together on a manuscript.

Solution

Let x = the number of hours it would take Lucy to complete the manuscript alone and $2x$ = the number of hours it would take Harry to complete the manuscript alone. Make a table showing rate, time, and work completed:

	Rate	Time	Work
Lucy	$\frac{1}{x}\frac{\text{msp}}{\text{hr}}$	3 hr	$\frac{3}{x}$ msp
Harry	$\frac{1}{2x}\frac{\text{msp}}{\text{hr}}$	3 hr	$\frac{3}{2x}$ msp

Now find the sum of each person's work.

$$\frac{3}{x} + \frac{3}{2x} = \frac{2 \cdot 3}{2 \cdot x} + \frac{3}{2x}$$
$$= \frac{6}{2x} + \frac{3}{2x}$$
$$= \frac{9}{2x}$$

So in 3 hours working together they will complete $\frac{9}{2x}$ of the manuscript. ■

WARM-UPS

True or false? Explain your answer.

1. $\frac{1}{2} + \frac{1}{3} = \frac{2}{5}$ False
2. $\frac{7}{12} - \frac{1}{12} = \frac{1}{2}$ True
3. $\frac{3}{5} + \frac{4}{3} = \frac{29}{15}$ True
4. $\frac{4}{5} - \frac{5}{7} = \frac{3}{35}$ True
5. $\frac{5}{20} + \frac{3}{4} = 1$ True
6. $\frac{2}{x} + 1 = \frac{3}{x}$ for any nonzero value of x. False
7. $1 + \frac{1}{a} = \frac{a + 1}{a}$ for any nonzero value of a. True
8. $a - \frac{1}{4} = \frac{3}{4}a$ for any value of a. False
9. $\frac{a}{2} + \frac{b}{3} = \frac{3a + 2b}{6}$ for any values of a and b. True
10. The LCD for the rational expressions $\frac{1}{x}$ and $\frac{3x}{x - 1}$ is $x^2 - 1$. False

7.4 EXERCISES

Reading and Writing *After reading this section, write out the answers to these questions. Use complete sentences.*

1. How do you add or subtract rational numbers?
   We can add rational numbers with identical denominators as follows: $\frac{a}{c} + \frac{b}{c} = \frac{a+b}{c}$.
2. How do you add or subtract rational expressions?
   Rational expressions with identical denominators are added in the same manner as rational numbers.
3. What is the least common denominator?
   The LCD is the smallest number that is a multiple of all denominators.
4. Why do we use the *least* common denominator when adding rational expressions?
   We use the least common denominator to keep the addition process as simple as possible.

Perform the indicated operation. Reduce each answer to lowest terms. See Example 1.

5. $\frac{1}{10} + \frac{1}{10}$ $\quad \frac{1}{5}$
6. $\frac{1}{8} + \frac{3}{8}$ $\quad \frac{1}{2}$
7. $\frac{7}{8} - \frac{1}{8}$ $\quad \frac{3}{4}$
8. $\frac{4}{9} - \frac{1}{9}$ $\quad \frac{1}{3}$
9. $\frac{1}{6} - \frac{5}{6}$ $\quad -\frac{2}{3}$
10. $-\frac{3}{8} - \frac{7}{8}$ $\quad -\frac{5}{4}$
11. $-\frac{7}{8} + \frac{1}{8}$ $\quad -\frac{3}{4}$
12. $-\frac{9}{20} + \left(-\frac{3}{20}\right)$ $\quad -\frac{3}{5}$

Perform the indicated operation. Reduce each answer to lowest terms. See Example 2.

13. $\frac{1}{3} + \frac{2}{9}$ $\quad \frac{5}{9}$
14. $\frac{1}{4} + \frac{5}{6}$ $\quad \frac{13}{12}$
15. $\frac{7}{16} + \frac{5}{18}$ $\quad \frac{103}{144}$
16. $\frac{7}{6} + \frac{4}{15}$ $\quad \frac{43}{30}$
17. $\frac{1}{8} - \frac{9}{10}$ $\quad -\frac{31}{40}$
18. $\frac{2}{15} - \frac{5}{12}$ $\quad -\frac{17}{60}$
19. $-\frac{1}{6} - \left(-\frac{3}{8}\right)$ $\quad \frac{5}{24}$
20. $-\frac{1}{5} - \left(-\frac{1}{7}\right)$ $\quad -\frac{2}{35}$

Perform the indicated operation. Reduce each answer to lowest terms. See Example 3.

21. $\frac{1}{2x} + \frac{1}{2x}$ $\quad \frac{1}{x}$
22. $\frac{1}{3y} + \frac{2}{3y}$ $\quad \frac{1}{y}$
23. $\frac{3}{2w} + \frac{7}{2w}$ $\quad \frac{5}{w}$
24. $\frac{5x}{3y} + \frac{7x}{3y}$ $\quad \frac{4x}{y}$
25. $\frac{3a}{a+5} + \frac{15}{a+5}$ $\quad 3$
26. $\frac{a+7}{a-4} + \frac{9-5a}{a-4}$ $\quad -4$
27. $\frac{q-1}{q-4} - \frac{3q-9}{q-4}$ $\quad -2$
28. $\frac{3-a}{3} - \frac{a-5}{3}$ $\quad \frac{8-2a}{3}$
29. $\frac{4h-3}{h(h+1)} - \frac{h-6}{h(h+1)}$ $\quad \frac{3}{h}$
30. $\frac{2t-9}{t(t-3)} - \frac{t-9}{t(t-3)}$ $\quad \frac{1}{t-3}$
31. $\frac{x^2-x-5}{(x+1)(x+2)} + \frac{1-2x}{(x+1)(x+2)}$ $\quad \frac{x-4}{x+2}$
32. $\frac{2x-5}{(x-2)(x+6)} + \frac{x^2-2x+1}{(x-2)(x+6)}$ $\quad \frac{x+2}{x+6}$

Perform the indicated operation. Reduce each answer to lowest terms. See Example 4.

33. $\frac{1}{a} + \frac{1}{2a}$ $\quad \frac{3}{2a}$
34. $\frac{1}{3w} + \frac{2}{w}$ $\quad \frac{7}{3w}$
35. $\frac{x}{3} + \frac{x}{2}$ $\quad \frac{5x}{6}$
36. $\frac{y}{4} + \frac{y}{2}$ $\quad \frac{3y}{4}$
37. $\frac{m}{5} + m$ $\quad \frac{6m}{5}$
38. $\frac{y}{4} + 2y$ $\quad \frac{9y}{4}$
39. $\frac{3}{2a} + \frac{1}{5a}$ $\quad \frac{17}{10a}$
40. $\frac{5}{6y} - \frac{3}{8y}$ $\quad \frac{11}{24y}$
41. $\frac{w-3}{9} - \frac{w-4}{12}$ $\quad \frac{w}{36}$
42. $\frac{y+4}{10} - \frac{y-2}{14}$ $\quad \frac{y+19}{35}$
43. $\frac{b^2}{4a} - c$ $\quad \frac{b^2-4ac}{4a}$
44. $y + \frac{3}{7b}$ $\quad \frac{7by+3}{7b}$
45. $\frac{2}{wz^2} + \frac{3}{w^2z}$ $\quad \frac{2w+3z}{w^2z^2}$
46. $\frac{1}{a^5b} - \frac{5}{ab^3}$ $\quad \frac{b^2-5a^4}{a^5b^3}$

Perform the indicated operation. Reduce each answer to lowest terms. See Examples 5 and 6.

47. $\frac{2}{x+1} - \frac{3}{x}$ $\quad \frac{-x-3}{x(x+1)}$
48. $\frac{1}{a-1} - \frac{2}{a}$ $\quad \frac{2-a}{a(a-1)}$
49. $\frac{2}{a-b} + \frac{1}{a+b}$ $\quad \frac{3a+b}{(a-b)(a+b)}$
50. $\frac{3}{x+1} + \frac{2}{x-1}$ $\quad \frac{5x-1}{(x+1)(x-1)}$
51. $\frac{3}{x^2+x} - \frac{4}{5x+5}$ $\quad \frac{15-4x}{5x(x+1)}$
52. $\frac{3}{a^2+3a} - \frac{2}{5a+15}$ $\quad \frac{15-2a}{5a(a+3)}$
53. $\frac{2a}{a^2-9} + \frac{a}{a-3}$ $\quad \frac{a^2+5a}{(a-3)(a+3)}$
54. $\frac{x}{x^2-1} + \frac{3}{x-1}$ $\quad \frac{4x+3}{(x-1)(x+1)}$
55. $\frac{4}{a-b} + \frac{4}{b-a}$ $\quad 0$

56. $\dfrac{2}{x-3} + \dfrac{3}{3-x}$ $\dfrac{1}{3-x}$

57. $\dfrac{3}{2a-2} - \dfrac{2}{1-a}$ $\dfrac{7}{2a-2}$

58. $\dfrac{5}{2x-4} - \dfrac{3}{2-x}$ $\dfrac{11}{2x-4}$

59. $\dfrac{1}{x^2-4} - \dfrac{3}{x^2-3x-10}$ $\dfrac{-2x+1}{(x-5)(x+2)(x-2)}$

60. $\dfrac{2x}{x^2-9} + \dfrac{3x}{x^2+4x+3}$ $\dfrac{5x^2-7x}{(x-3)(x+3)(x+1)}$

61. $\dfrac{3}{x^2+x-2} + \dfrac{4}{x^2+2x-3}$ $\dfrac{7x+17}{(x+2)(x-1)(x+3)}$

62. $\dfrac{x-1}{x^2-x-12} + \dfrac{x+4}{x^2+5x+6}$

$\dfrac{2x^2+x-18}{(x+2)(x+3)(x-4)}$

63. $\dfrac{2}{x} - \dfrac{1}{x-1} + \dfrac{1}{x+2}$ $\dfrac{2x^2-x-4}{x(x-1)(x+2)}$

64. $\dfrac{1}{a} - \dfrac{2}{a+1} + \dfrac{3}{a-1}$ $\dfrac{2a^2+5a-1}{a(a-1)(a+1)}$

65. $\dfrac{5}{3a-9} - \dfrac{3}{2a} + \dfrac{4}{a^2-3a}$ $\dfrac{a+51}{6a(a-3)}$

66. $\dfrac{3}{4c+2} - \dfrac{c-4}{2c^2+c} - \dfrac{5}{6c}$ $\dfrac{-7c+19}{6c(2c+1)}$

Match each sum in Exercises 67–74 with the equivalent rational expression in (a)–(h).

67. $\dfrac{1}{y} + 2$ g

68. $\dfrac{1}{y} + \dfrac{2}{y}$ b

69. $\dfrac{1}{y} + \dfrac{1}{2}$ f

70. $\dfrac{1}{y} + \dfrac{1}{2y}$ c

71. $\dfrac{2}{y} + 1$ e

72. $\dfrac{y}{2} + 1$ d

73. $\dfrac{1}{2} + y$ h

74. $\dfrac{1}{2y} + \dfrac{1}{2y}$ a

(a) $\dfrac{1}{y}$ **(b)** $\dfrac{3}{y}$ **(c)** $\dfrac{3}{2y}$

(d) $\dfrac{y+2}{2}$ **(e)** $\dfrac{y+2}{y}$ **(f)** $\dfrac{y+2}{2y}$

(g) $\dfrac{2y+1}{y}$ **(h)** $\dfrac{2y+1}{2}$

Solve each problem. See Example 7.

75. ***Perimeter of a rectangle.*** Suppose that the length of a rectangle is $\frac{3}{x}$ feet and its width is $\frac{5}{2x}$ feet. Find a rational expression for the perimeter of the rectangle.
$\dfrac{11}{x}$ feet

76. ***Perimeter of a triangle.*** The lengths of the sides of a triangle are $\frac{1}{x}$, $\frac{1}{2x}$, and $\frac{2}{3x}$ meters. Find a rational expression for the perimeter of the triangle.
$\dfrac{13}{6x}$ meters

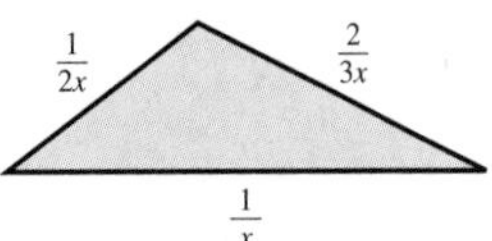

FIGURE FOR EXERCISE 76

77. ***Traveling time.*** Janet drove 120 miles at x mph before 6:00 A.M. After 6:00 A.M., she increased her speed by 5 mph and drove 195 additional miles. Use the fact that $T = \frac{D}{R}$ to complete the following table.

	Rate	Time	Distance
Before	$x\ \dfrac{\text{mi}}{\text{hr}}$		120 mi
After	$x+5\ \dfrac{\text{mi}}{\text{hr}}$		195 mi

Write a rational expression for her total traveling time. Evaluate the expression for $x = 60$.
$\dfrac{315x+600}{x(x+5)}$ hours, 5 hours

78. ***Traveling time.*** After leaving Moose Jaw, Hanson drove 200 kilometers at x km/hr and then decreased his speed by 20 km/hr and drove 240 additional kilometers. Make a table like the one in Exercise 77. Write a rational expression for his total traveling time. Evaluate the expression for $x = 100$.
$\dfrac{440x-4000}{x(x-20)}$ hours, 5 hours

79. ***House painting.*** Kent can paint a certain house by himself in x days. His helper Keith can paint the same house by himself in $x + 3$ days. Suppose that they work together on the job for 2 days. To complete the table, use the fact that the work completed is the product of the rate and the time.

	Rate	Time	Work
Kent	$\dfrac{1}{x}\ \dfrac{\text{job}}{\text{day}}$	2 days	
Keith	$\dfrac{1}{x+3}\ \dfrac{\text{job}}{\text{day}}$	2 days	

Write a rational expression for the fraction of the house that they complete by working together for 2 days. Evaluate the expression for $x = 6$.
$\dfrac{4x+6}{x(x+3)}$ job, $\dfrac{5}{9}$ job

80. ***Barn painting.*** Melanie can paint a certain barn by herself in x days. Her helper Melissa can paint the same barn by herself in $2x$ days. Write a rational expression for the fraction of the barn that they complete in one day by working together. Evaluate the expression for $x = 5$.
$\frac{3}{2x}$ barn, $\frac{3}{10}$ barn

FIGURE FOR EXERCISE 80

GETTING MORE INVOLVED

81. ***Writing.*** Write a step-by-step procedure for adding rational expressions.

82. ***Writing.*** Explain why fractions must have the same denominator to be added. Use real-life examples.

In This Section

- Complex Fractions
- Using the LCD to Simplify Complex Fractions
- Applications

7.5 COMPLEX FRACTIONS

In this section we will use the idea of least common denominator to simplify complex fractions. Also we will see how complex fractions can arise in applications.

Complex Fractions

A **complex fraction** is a fraction having rational expressions in the numerator, denominator, or both. Consider the following complex fraction:

$$\frac{\frac{1}{2} + \frac{2}{3}}{\frac{1}{4} - \frac{5}{8}} \quad \begin{array}{l} \leftarrow \text{Numerator of complex fraction} \\ \leftarrow \text{Denominator of complex fraction} \end{array}$$

To simplify it, we can combine the fractions in the numerator as follows:

$$\frac{1}{2} + \frac{2}{3} = \frac{1 \cdot 3}{2 \cdot 3} + \frac{2 \cdot 2}{3 \cdot 2} = \frac{3}{6} + \frac{4}{6} = \frac{7}{6}$$

We can combine the fractions in the denominator as follows:

$$\frac{1}{4} - \frac{5}{8} = \frac{1 \cdot 2}{4 \cdot 2} - \frac{5}{8} = \frac{2}{8} - \frac{5}{8} = -\frac{3}{8}$$

Now divide the numerator by the denominator:

$$\frac{\frac{1}{2} + \frac{2}{3}}{\frac{1}{4} - \frac{5}{8}} = \frac{\frac{7}{6}}{-\frac{3}{8}} = \frac{7}{6} \div \left(-\frac{3}{8}\right)$$

$$= \frac{7}{6} \cdot \left(-\frac{8}{3}\right)$$

$$= -\frac{56}{18} = -\frac{28}{9}$$

23. $\dfrac{\dfrac{1}{3-x}-5}{\dfrac{1}{x-3}-2}$ $\quad \dfrac{5x-14}{2x-7}$

24. $\dfrac{\dfrac{2}{x-5}-x}{\dfrac{3x}{5-x}-1}$ $\quad \dfrac{x^2-5x-2}{4x-5}$

25. $\dfrac{1-\dfrac{5}{a-1}}{3-\dfrac{2}{1-a}}$ $\quad \dfrac{a-6}{3a-1}$

26. $\dfrac{\dfrac{1}{3}-\dfrac{2}{9-x}}{\dfrac{1}{6}-\dfrac{1}{x-9}}$ $\quad \dfrac{2x-6}{x-15}$

27. $\dfrac{\dfrac{1}{m-3}-\dfrac{4}{m}}{\dfrac{3}{m-3}+\dfrac{1}{m}}$ $\quad \dfrac{-3m+12}{4m-3}$

28. $\dfrac{\dfrac{1}{y+3}-\dfrac{4}{y}}{\dfrac{1}{y}-\dfrac{2}{y+3}}$ $\quad \dfrac{3y+12}{y-3}$

29. $\dfrac{\dfrac{2}{w-1}-\dfrac{3}{w+1}}{\dfrac{4}{w+1}+\dfrac{5}{w-1}}$ $\quad \dfrac{-w+5}{9w+1}$

30. $\dfrac{\dfrac{1}{x+2}-\dfrac{3}{x+3}}{\dfrac{2}{x+3}+\dfrac{3}{x+2}}$ $\quad \dfrac{-2x-3}{5x+13}$

31. $\dfrac{\dfrac{1}{a-b}-\dfrac{1}{a+b}}{\dfrac{1}{b-a}+\dfrac{1}{b+a}}$ $\quad -1$

32. $\dfrac{\dfrac{1}{2+x}-\dfrac{1}{2-x}}{\dfrac{1}{x+2}-\dfrac{1}{x-2}}$ $\quad -\dfrac{x}{2}$

Simplify each complex fraction.

33. $\dfrac{\dfrac{2x-9}{6}}{\dfrac{2x-3}{9}}$ $\quad \dfrac{6x-27}{4x-6}$

34. $\dfrac{\dfrac{a-5}{12}}{\dfrac{a+2}{15}}$ $\quad \dfrac{5a-25}{4a+8}$

35. $\dfrac{\dfrac{2x-4y}{xy^2}}{\dfrac{3x-6y}{x^3y}}$ $\quad \dfrac{2x^2}{3y}$

36. $\dfrac{\dfrac{ab+b^2}{4ab^5}}{\dfrac{a+b}{6a^2b^4}}$ $\quad \dfrac{3a}{2}$

37. $\dfrac{\dfrac{a^2+2a-24}{a+1}}{\dfrac{a^2-a-12}{(a+1)^2}}$ $\quad \dfrac{a^2+7a+6}{a+3}$

38. $\dfrac{\dfrac{y^2-3y-18}{y^2-4}}{\dfrac{y^2+5y+6}{y-2}}$ $\quad \dfrac{y-6}{(y+2)^2}$

39. $\dfrac{\dfrac{x}{x+1}}{\dfrac{1}{x^2-1}-\dfrac{1}{x-1}}$ $\quad 1-x$

40. $\dfrac{\dfrac{a}{a^2-b^2}}{\dfrac{1}{a+b}+\dfrac{1}{a-b}}$ $\quad \dfrac{1}{2}$

Solve each problem. See Example 4.

41. ***Sophomore math.*** A survey of college sophomores showed that $\frac{5}{6}$ of the males were taking a mathematics class and $\frac{3}{4}$ of the females were taking a mathematics class. One-third of the males were enrolled in calculus, and $\frac{1}{5}$ of the females were enrolled in calculus. If just as many males as females were surveyed, then what fraction of the surveyed students taking mathematics were enrolled in calculus? Rework this problem assuming that the number of females in the survey was twice the number of males. $\frac{32}{95}, \frac{11}{35}$

42. ***Commuting students.*** At a well-known university, $\frac{1}{4}$ of the undergraduate students commute, and $\frac{1}{3}$ of the graduate students commute. One-tenth of the undergraduate students drive more than 40 miles daily, and $\frac{1}{6}$ of the graduate students drive more than 40 miles daily. If there are twice as many undergraduate students as there are graduate students, then what fraction of the commuters drive more than 40 miles daily? $\frac{11}{25}$

FIGURE FOR EXERCISE 42

GETTING MORE INVOLVED

43. ***Exploration.*** Simplify

$$\frac{1}{1+\frac{1}{2}},\quad \frac{1}{1+\frac{1}{1+\frac{1}{2}}},\quad \text{and}\quad \frac{1}{1+\frac{1}{1+\frac{1}{1+\frac{1}{2}}}}.$$

a) Are these fractions getting larger or smaller as the fractions become more complex?

b) Continuing the pattern, find the next two complex fractions and simplify them.

c) Now what can you say about the values of all five complex fractions?

44. ***Discussion.*** A complex fraction can be simplified by writing the numerator and denominator as single fractions and then dividing them or by multiplying the numerator and denominator by the LCD. Simplify the complex fraction

$$\frac{\frac{4}{xy^2}-\frac{6}{xy}}{\frac{2}{x^2}+\frac{4}{x^2y}}$$

by using each of these methods. Compare the number of steps used in each method, and determine which method requires fewer steps.

7.6 SOLVING EQUATIONS WITH RATIONAL EXPRESSIONS

In This Section

- Equations with Rational Expressions
- Extraneous Solutions

Many problems in algebra can be solved by using equations involving rational expressions. In this section you will learn how to solve equations that involve rational expressions, and in Sections 7.7 and 7.8 you will solve problems using these equations.

Equations with Rational Expressions

We solved some equations involving fractions in Section 2.3. In that section the equations had only integers in the denominators. Our first step in solving those equations was to multiply by the LCD to eliminate all of the denominators.

EXAMPLE 1 **Integers in the denominators**

Solve $\frac{1}{2} - \frac{x-2}{3} = \frac{1}{6}$.

Solution

The LCD for 2, 3, and 6 is 6. Multiply each side of the equation by 6:

$$\frac{1}{2} - \frac{x-2}{3} = \frac{1}{6} \qquad \text{Original equation}$$

$$6\left(\frac{1}{2} - \frac{x-2}{3}\right) = 6 \cdot \frac{1}{6} \qquad \text{Multiply each side by 6.}$$

$$6 \cdot \frac{1}{2} - \overset{2}{\cancel{6}} \cdot \frac{x-2}{\cancel{3}} = \cancel{6} \cdot \frac{1}{\cancel{6}} \qquad \text{Distributive property}$$

$$3 - 2(x-2) = 1 \qquad \text{Simplify.}$$

$$3 - 2x + 4 = 1 \qquad \text{Distributive property}$$

$$-2x = -6 \qquad \text{Subtract 7 from each side.}$$

$$x = 3 \qquad \text{Divide each side by } -2.$$

Check $x = 3$ in the original equation:

$$\frac{1}{2} - \frac{3-2}{3} = \frac{1}{2} - \frac{1}{3} = \frac{3}{6} - \frac{2}{6} = \frac{1}{6}$$

The solution to the equation is 3. ■

Helpful Hint

Note that it is not necessary to convert each fraction into an equivalent fraction with a common denominator here. Since we can multiply both sides of an equation by any expression we choose, we choose to multiply by the LCD. This tactic eliminates the fractions in one step.

CAUTION When a numerator contains a binomial, as in Example 1, the numerator must be enclosed in parentheses when the denominator is eliminated.

To solve an equation involving rational expressions, we usually multiply each side of the equation by the LCD for all the denominators involved, just as we do for an equation with fractions.

EXAMPLE 2 **Variables in the denominators**

Solve $\frac{1}{x} + \frac{1}{6} = \frac{1}{4}$.

Solution

We multiply each side of the equation by $12x$, the LCD for 4, 6, and x:

$$\frac{1}{x} + \frac{1}{6} = \frac{1}{4} \quad \text{Original equation}$$

$$12x\left(\frac{1}{x} + \frac{1}{6}\right) = 12x\left(\frac{1}{4}\right) \quad \text{Multiply each side by } 12x.$$

$$12x \cdot \frac{1}{x} + \overset{2}{12}x \cdot \frac{1}{6} = \overset{3}{12}x \cdot \frac{1}{4} \quad \text{Distributive property}$$

$$12 + 2x = 3x \quad \text{Simplify.}$$

$$12 = x \quad \text{Subtract } 2x \text{ from each side.}$$

Check that 12 satisfies the original equation:

$$\frac{1}{12} + \frac{1}{6} = \frac{1}{12} + \frac{2}{12} = \frac{3}{12} = \frac{1}{4}$$

■

EXAMPLE 3 **An equation with two solutions**

Solve the equation $\frac{100}{x} + \frac{100}{x+5} = 9$.

Study Tip

Your mood for studying should match the mood in which you are tested. Being too relaxed during studying will not match the increased level of activation you attain during a test. Likewise, if you get too tensed-up during a test, you will not do well because your test-taking mood will not match your studying mood.

Solution

The LCD for the denominators x and $x + 5$ is $x(x + 5)$:

$$\frac{100}{x} + \frac{100}{x+5} = 9 \quad \text{Original equation}$$

$$x(x+5)\frac{100}{x} + x(x+5)\frac{100}{x+5} = x(x+5)9 \quad \text{Multiply each side by } x(x+5).$$

$$(x+5)100 + x(100) = (x^2 + 5x)9 \quad \text{All denominators are eliminated.}$$

$$100x + 500 + 100x = 9x^2 + 45x \quad \text{Simplify.}$$

$$500 + 200x = 9x^2 + 45x$$

$$0 = 9x^2 - 155x - 500 \quad \text{Get 0 on one side.}$$

$$0 = (9x + 25)(x - 20) \quad \text{Factor.}$$

$$9x + 25 = 0 \quad \text{or} \quad x - 20 = 0 \quad \text{Zero factor property}$$

$$x = -\frac{25}{9} \quad \text{or} \quad x = 20$$

A check will show that both $-\frac{25}{9}$ and 20 satisfy the original equation. ■

Extraneous Solutions

In a rational expression we can replace the variable only by real numbers that do not cause the denominator to be 0. When solving equations involving rational expressions, we must check every solution to see whether it causes 0 to appear in a denominator. If a number causes the denominator to be 0, then it cannot be a solution to the equation. A number that appears to be a solution but causes 0 in a denominator is called an **extraneous solution.**

EXAMPLE 4 **An equation with an extraneous solution**

Solve the equation $\frac{1}{x-2} = \frac{x}{2x-4} + 1$.

Solution

Because the denominator $2x - 4$ factors as $2(x - 2)$, the LCD is $2(x - 2)$.

$$2(x-2)\frac{1}{x-2} = 2(x-2)\frac{x}{2(x-2)} + 2(x-2) \cdot 1 \quad \text{Multiply each side of the original equation by } 2(x-2).$$

$$2 = x + 2x - 4 \quad \text{Simplify.}$$

$$2 = 3x - 4$$

$$6 = 3x$$

$$2 = x$$

Check 2 in the original equation:

$$\frac{1}{2-2} = \frac{2}{2 \cdot 2 - 4} + 1$$

The denominator $2 - 2$ is 0. So 2 does not satisfy the equation, and it is an extraneous solution. The equation has no solutions. ■

EXAMPLE 5 **Another extraneous solution**

Solve the equation $\frac{1}{x} + \frac{1}{x-3} = \frac{x-2}{x-3}$.

Solution

The LCD for the denominators x and $x - 3$ is $x(x - 3)$:

$$\frac{1}{x} + \frac{1}{x-3} = \frac{x-2}{x-3} \quad \text{Original equation}$$

$$x(x-3) \cdot \frac{1}{x} + x(x-3) \cdot \frac{1}{x-3} = x(x-3) \cdot \frac{x-2}{x-3} \quad \text{Multiply each side by } x(x-3).$$

$$x - 3 + x = x(x - 2)$$

$$2x - 3 = x^2 - 2x$$

$$0 = x^2 - 4x + 3$$

$$0 = (x - 3)(x - 1)$$

$$x - 3 = 0 \quad \text{or} \quad x - 1 = 0$$

$$x = 3 \quad \text{or} \quad x = 1$$

If $x = 3$, then the denominator $x - 3$ has a value of 0. If $x = 1$, the original equation is satisfied. The only solution to the equation is 1. ■

CAUTION Always be sure to check your answers in the original equation to determine whether they are extraneous solutions.

WARM-UPS

True or false? Explain your answers.

1. The LCD is not used in solving equations with rational expressions. False
2. To solve the equation $x^2 = 8x$, we divide each side by x. False
3. An extraneous solution is an irrational number. False

Use the following equations for Questions 4–10.

a) $\frac{3}{x} + \frac{5}{x-2} = \frac{2}{3}$ b) $\frac{1}{x} + \frac{1}{2} = \frac{3}{4}$ c) $\frac{1}{x-1} + 2 = \frac{1}{x+1}$

4. To solve Eq. (a), we must add the expressions on the left-hand side. False
5. Both 0 and 2 satisfy Eq. (a). False
6. To solve Eq. (a), we multiply each side by $3x^2 - 6x$. True
7. The only solution to Eq. (b) is 4. True
8. Equation (b) is equivalent to $4 + 2x = 3x$. True
9. To solve Eq. (c), we multiply each side by $x^2 - 1$. True
10. The numbers 1 and -1 do not satisfy Eq. (c). True

7.6 EXERCISES

Reading and Writing *After reading this section, write out the answers to these questions. Use complete sentences.*

1. What is the typical first step for solving an equation involving rational expressions?
The first step is usually to multiply each side by the LCD.
2. What is the difference in procedure for solving an equation involving rational expressions and adding rational expressions?
In adding rational expressions we build up each expression to get the LCD as the common denominator.
3. What is an extraneous solution?
An extraneous solution is a number that appears when we solve an equation, but it does not check in the original equation.
4. Why do extraneous solutions sometimes occur for equations with rational expressions?
Extraneous solutions occur because we multiply by an expression involving a variable and that variable expression might have a value of zero.

Solve each equation. See Example 1.

5. $\frac{x}{3} - 5 = \frac{x}{2} - 7$ 12
6. $\frac{x}{3} - \frac{x}{2} = \frac{x}{5} - 11$ 30
7. $\frac{y}{5} - \frac{2}{3} = \frac{y}{6} + \frac{1}{3}$ 30
8. $\frac{z}{6} + \frac{5}{4} = \frac{z}{2} - \frac{3}{4}$ 6
9. $\frac{3}{4} - \frac{t-4}{3} = \frac{t}{12}$ 5
10. $\frac{4}{5} - \frac{v-1}{10} = \frac{v-5}{30}$ 8
11. $\frac{1}{5} - \frac{w+10}{15} = \frac{1}{10} - \frac{w+1}{6}$ 4
12. $\frac{q}{5} - \frac{q-1}{2} = \frac{13}{20} - \frac{q+1}{4}$ 2

Solve each equation. See Example 2.

13. $\frac{1}{x} + \frac{1}{2} = \frac{3}{4}$ 4
14. $\frac{3}{x} + \frac{1}{4} = \frac{5}{8}$ 8
15. $\frac{2}{3x} + \frac{1}{2x} = \frac{7}{24}$ 4
16. $\frac{1}{6x} - \frac{1}{8x} = \frac{1}{72}$ 3
17. $\frac{1}{2} + \frac{a-2}{a} = \frac{a+2}{2a}$ 3
18. $\frac{1}{b} + \frac{1}{5} = \frac{b-1}{5b} + \frac{3}{10}$ 4
19. $\frac{1}{3} - \frac{k+3}{6k} = \frac{1}{3k} - \frac{k-1}{2k}$ 2
20. $\frac{3}{p} - \frac{p+3}{3p} = \frac{2p-1}{2p} - \frac{5}{6}$ 5

Solve each equation. See Example 3.

21. $\frac{x}{2} = \frac{5}{x+3}$ $-5, 2$
22. $\frac{x}{3} = \frac{4}{x+1}$ $-4, 3$
23. $\frac{x}{x+1} = \frac{6}{x+7}$ $-3, 2$
24. $\frac{x}{x+3} = \frac{2}{x-3}$ $-1, 6$

25. $\frac{2}{x+1} = \frac{1}{x} + \frac{1}{6}$ 2, 3

26. $\frac{1}{w+1} - \frac{1}{2w} = \frac{3}{40}$ $\frac{5}{3}$, 4

27. $\frac{a-1}{a^2-4} + \frac{1}{a-2} = \frac{a+4}{a+2}$ −3, 3

28. $\frac{b+17}{b^2-1} - \frac{1}{b+1} = \frac{b-2}{b-1}$ −4, 5

Solve each equation. Watch for extraneous solutions. See Examples 4 and 5.

29. $\frac{1}{x-1} + \frac{2}{x} = \frac{x}{x-1}$ 2

30. $\frac{4}{x} + \frac{3}{x-3} = \frac{x}{x-3} - \frac{1}{3}$ 6

31. $\frac{5}{x+2} + \frac{2}{x-3} = \frac{x-1}{x-3}$ No solution

32. $\frac{6}{y-2} + \frac{7}{y-8} = \frac{y-1}{y-8}$ No solution

33. $1 + \frac{3y}{y-2} = \frac{6}{y-2}$ No solution

34. $\frac{5}{y-3} = \frac{y+7}{2y-6} + 1$ No solution

35. $\frac{z}{z+1} - \frac{1}{z+2} = \frac{2z+5}{z^2+3z+2}$ 3

36. $\frac{z}{z-2} - \frac{1}{z+5} = \frac{7}{z^2+3z-10}$ 1

In Exercises 37–58, solve each equation.

37. $\frac{a}{4} = \frac{5}{2}$ 10

38. $\frac{y}{3} = \frac{6}{5}$ $\frac{18}{5}$

39. $\frac{w}{6} = \frac{3w}{11}$ 0

40. $\frac{2m}{3} = \frac{3m}{2}$ 0

41. $\frac{5}{x} = \frac{x}{5}$ −5, 5

42. $\frac{-3}{x} = \frac{x}{-3}$ −3, 3

43. $\frac{x-3}{5} = \frac{x-3}{x}$ 3, 5

44. $\frac{a+4}{2} = \frac{a+4}{a}$ −4, 2

45. $\frac{1}{x+2} = \frac{x}{x+2}$ 1

46. $\frac{-3}{w+2} = \frac{w}{w+2}$ −3

47. $\frac{1}{2x-4} + \frac{1}{x-2} = \frac{3}{2}$ 3

48. $\frac{7}{3x-9} - \frac{1}{x-3} = \frac{4}{3}$ 4

49. $\frac{3}{a^2-a-6} = \frac{2}{a^2-4}$ 0

50. $\frac{8}{a^2+a-6} = \frac{6}{a^2-9}$ 6

51. $\frac{4}{c-2} - \frac{1}{2-c} = \frac{25}{c+6}$ 4

52. $\frac{3}{x+1} - \frac{1}{1-x} = \frac{10}{x^2-1}$ 3

53. $\frac{1}{x^2-9} + \frac{3}{x+3} = \frac{4}{x-3}$ −20

54. $\frac{3}{x-2} - \frac{5}{x+3} = \frac{1}{x^2+x-6}$ 9

55. $\frac{3}{2x+4} - \frac{1}{x+2} = \frac{1}{3x+1}$ 3

56. $\frac{5}{2m+6} - \frac{1}{m+1} = \frac{1}{m+3}$ 3

57. $\frac{2t-1}{3t+3} + \frac{3t-1}{6t+6} = \frac{t}{t+1}$ 3

58. $\frac{4w-1}{3w+6} - \frac{w-1}{3} = \frac{w-1}{w+2}$ 2

Solve each problem.

59. ***Lens equation.*** The focal length f for a camera lens is related to the object distance o and the image distance i by the formula

$$\frac{1}{f} = \frac{1}{o} + \frac{1}{i}.$$

See the accompanying figure. The image is in focus at distance i from the lens. For an object that is 600 mm from a 50-mm lens, use $f = 50$ mm and $o = 600$ mm to find i.

$54\frac{6}{11}$ mm

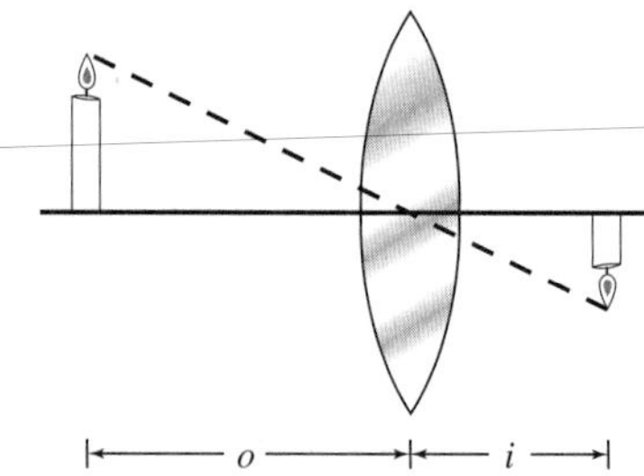

FIGURE FOR EXERCISE 59

60. ***Telephoto lens.*** Use the formula from Exercise 59 to find the image distance i for an object that is 2,000,000 mm from a 250-mm telephoto lens. 250.03 mm

FIGURE FOR EXERCISE 60

7.7 APPLICATIONS OF RATIOS AND PROPORTIONS

In This Section

- Ratios
- Proportions

In this section we will use the ideas of rational expressions in ratio and proportion problems. We will solve proportions in the same way we solved equations in Section 7.6.

Ratios

In Chapter 1 we defined a rational number as the *ratio of two integers*. We will now give a more general definition of ratio. If a and b are any real numbers (not just integers), with $b \neq 0$, then the expression $\frac{a}{b}$ is called the **ratio of a and b** or the **ratio of a to b.** The ratio of a to b is also written as $a:b$. A ratio is a comparison of two numbers. Some examples of ratios are

$$\frac{3}{4}, \quad \frac{4.2}{2.1}, \quad \frac{\frac{1}{4}}{\frac{1}{2}}, \quad \frac{3.6}{5}, \quad \text{and} \quad \frac{100}{1}.$$

Ratios are treated just like fractions. We can reduce ratios, and we can build them up. We generally express ratios as ratios of integers. When possible, we will convert a ratio into an equivalent ratio of integers in lowest terms.

EXAMPLE 1 **Finding equivalent ratios**

Find an equivalent ratio of integers in lowest terms for each ratio.

a) $\frac{4.2}{2.1}$ **b)** $\frac{\frac{1}{4}}{\frac{1}{2}}$ **c)** $\frac{3.6}{5}$

Solution

a) Because both the numerator and the denominator have one decimal place, we will multiply the numerator and denominator by 10 to eliminate the decimals:

$$\frac{4.2}{2.1} = \frac{4.2(10)}{2.1(10)} = \frac{42}{21} = \frac{21 \cdot 2}{21 \cdot 1} = \frac{2}{1} \quad \text{Do not omit the 1 in a ratio.}$$

So the ratio of 4.2 to 2.1 is equivalent to the ratio 2 to 1.

b) This ratio is a complex fraction. We can simplify this expression using the LCD method as shown in Section 7.5. Multiply the numerator and denominator of this ratio by 4:

$$\frac{\frac{1}{4}}{\frac{1}{2}} = \frac{\frac{1}{4} \cdot 4}{\frac{1}{2} \cdot 4} = \frac{1}{2}$$

c) We can get a ratio of integers if we multiply the numerator and denominator by 10.

$$\frac{3.6}{5} = \frac{3.6(10)}{5(10)} = \frac{36}{50}$$

$$= \frac{18}{25} \quad \text{Reduce to lowest terms.}$$

■

In Example 2 a ratio is used to compare quantities.

EXAMPLE 2 **Nitrogen to potash**

In a 50-pound bag of lawn fertilizer there are 8 pounds of nitrogen and 12 pounds of potash. What is the ratio of nitrogen to potash?

Solution

The nitrogen and potash occur in this fertilizer in the ratio of 8 pounds to 12 pounds:

$$\frac{8}{12} = \frac{2 \cdot \not{4}}{3 \cdot \not{4}} = \frac{2}{3}$$

So the ratio of nitrogen to potash is 2 to 3. ■

EXAMPLE 3 **Males to females**

In a class of 50 students, there were exactly 20 male students. What was the ratio of males to females in this class?

Solution

Because there were 20 males in the class of 50, there were 30 females. The ratio of males to females was 20 to 30, or 2 to 3. ■

Ratios give us a means of comparing the size of two quantities. For this reason *the numbers compared in a ratio should be expressed in the same units.* For example, if one dog is 24 inches high and another is 1 foot high, then the ratio of their heights is 2 to 1, not 24 to 1.

EXAMPLE 4 **Quantities with different units**

What is the ratio of length to width for a poster with a length of 30 inches and a width of 2 feet?

Solution

Because the width is 2 feet, or 24 inches, the ratio of length to width is 30 to 24. Reduce as follows:

$$\frac{30}{24} = \frac{5 \cdot 6}{4 \cdot 6} = \frac{5}{4}$$

So the ratio of length to width is 5 to 4. ■

Proportions

A **proportion** is any statement expressing the equality of two ratios. The statement

$$\frac{a}{b} = \frac{c}{d} \quad \text{or} \quad a:b = c:d$$

is a proportion. In any proportion the numbers in the positions of a and d above are called the **extremes.** The numbers in the positions of b and c above are called the **means.** In the proportion

$$\frac{30}{24} = \frac{5}{4},$$

the means are 24 and 5, and the extremes are 30 and 4.

Study Tip

To get the "big picture," survey the chapter that you are studying. Read the headings to get the general idea of the chapter content. Read the chapter summary to see what is important in the chapter. Repeat this survey procedure several times while you are working in a chapter.

If we multiply each side of the proportion

$$\frac{a}{b} = \frac{c}{d}$$

by the LCD, *bd*, we get

$$\frac{a}{b} \cdot bd = \frac{c}{d} \cdot bd$$

or

$$a \cdot d = b \cdot c.$$

We can express this result by saying *that the product of the extremes is equal to the product of the means*. We call this fact the **extremes-means property** or **cross-multiplying.**

Extremes-Means Property (Cross-Multiplying)

Suppose a, b, c, and d are real numbers with $b \neq 0$ and $d \neq 0$. If

$$\frac{a}{b} = \frac{c}{d}, \quad \text{then} \quad ad = bc.$$

We use the extremes-means property to solve proportions.

EXAMPLE 5

Using the extremes-means property

Solve the proportion $\frac{3}{x} = \frac{5}{x + 5}$ for x.

Helpful Hint

The extremes-means property or cross-multiplying is nothing new. You can accomplish the same thing by multiplying each side of the equation by the LCD.

Solution

Instead of multiplying each side by the LCD, we use the extremes-means property:

$$\begin{aligned} \frac{3}{x} &= \frac{5}{x + 5} && \text{Original proportion} \\ 3(x + 5) &= 5x && \text{Extremes-means property} \\ 3x + 15 &= 5x && \text{Distributive property} \\ 15 &= 2x \\ \frac{15}{2} &= x \end{aligned}$$

Check:

$$\frac{3}{\frac{15}{2}} = 3 \cdot \frac{2}{15} = \frac{2}{5}$$

$$\frac{5}{\frac{15}{2} + 5} = \frac{5}{\frac{25}{2}} = 5 \cdot \frac{2}{25} = \frac{2}{5}$$

So $\frac{15}{2}$ is the solution to the equation or the solution to the proportion. ■

EXAMPLE 6

Solving a proportion

The ratio of men to women at Brighton City College is 2 to 3. If there are 894 men, then how many women are there?

Solution

Because the ratio of men to women is 2 to 3, we have

$$\frac{\text{Number of men}}{\text{Number of women}} = \frac{2}{3}.$$

If x represents the number of women, then we have the following proportion:

$$\frac{894}{x} = \frac{2}{3}$$

$$2x = 2682 \quad \text{Extremes-means property}$$

$$x = 1341$$

The number of women is 1341. ■

Note that any proportion can be solved by multiplying each side by the LCD as we did when we solved other equations involving rational expressions. The extremes-means property gives us a shortcut for solving proportions.

EXAMPLE 7 **Solving a proportion**

In a conservative portfolio the ratio of the amount invested in bonds to the amount invested in stocks should be 3 to 1. A conservative investor invested \$2850 more in bonds than she did in stocks. How much did she invest in each category?

Solution

Because the ratio of the amount invested in bonds to the amount invested in stocks is 3 to 1, we have

$$\frac{\text{Amount invested in bonds}}{\text{Amount invested in stocks}} = \frac{3}{1}.$$

If x represents the amount invested in stocks and $x + 2850$ represents the amount invested in bonds, then we can write and solve the following proportion:

$$\frac{x + 2850}{x} = \frac{3}{1}$$

$$3x = x + 2850 \quad \text{Extremes-means property}$$

$$2x = 2850$$

$$x = 1425$$

$$x + 2850 = 4275$$

So she invested \$4275 in bonds and \$1425 in stocks. Note that these amounts are in the ratio of 3 to 1. ■

Example 8 shows how conversions from one unit of measurement to another can be done by using proportions.

EXAMPLE 8 **Converting measurements**

There are 3 feet in 1 yard. How many feet are there in 12 yards?

Solution

Let x represent the number of feet in 12 yards. There are two proportions that we can write to solve the problem:

$$\frac{3 \text{ feet}}{x \text{ feet}} = \frac{1 \text{ yard}}{12 \text{ yards}} \qquad \frac{3 \text{ feet}}{1 \text{ yard}} = \frac{x \text{ feet}}{12 \text{ yards}}$$

MATH AT WORK

$x^2 + (x+1)^2 = 5^2$

SALES ANALYST

Did you ever wonder how your local store calculates how much of your favorite cosmetic to stock on the shelf? Mike Pittman, National Account Manager for a major cosmetic company, is responsible for providing more than 2000 stores across the United States with personal care products such as skin lotions, fragrances, and cosmetics.

Data on what has been sold is transmitted from the point of sale across a number of satellite dishes and computers to Mr. Pittman. The data usually includes size, color, and other pertinent facts. The information is then combined with demographics for certain geographic areas and movement data, as well as advertising and promotional information to answer questions such as: What color is selling best? Is it time to stock sunscreen? Is this a trend-setting area of the country? On the basis of his analysis of these and many other questions, Mr. Pittman recommends changes in packaging, promotional programs, and the quantities of products to be shipped.

Mr. Pittman's job requires a unique blend of sales, marketing, and quantitative skills. Of course, knowledge of computers and an understanding of people help. So the next time you see a whole aisle of personal care and cosmetic products, think of all the information that has been analyzed to put it there.

In Exercise 59 of this section you will see how Mr. Pittman uses a proportion to determine the quantity of mascara needed in a warehouse.

The ratios in the second proportion violate the rule of comparing only measurements that are expressed in the same units. Note that each side of the second proportion is actually the ratio 1 to 1, since 3 feet = 1 yard and x feet = 12 yards. For doing conversions we can use ratios like this to compare measurements in different units. Applying the extremes-means property to either proportion gives

$$3 \cdot 12 = x \cdot 1,$$

or

$$x = 36.$$

So there are 36 feet in 12 yards. ■

WARM-UPS

True or false? Explain your answer.

1. The ratio of 40 men to 30 women can be expressed as the ratio 4 to 3. True
2. The ratio of 3 feet to 2 yards can be expressed as the ratio 3 to 2. False
3. If the ratio of men to women in the Chamber of Commerce is 3 to 2 and there are 20 men, then there must be 30 women. False
4. The ratio of 1.5 to 2 is equivalent to the ratio of 3 to 4. True
5. A statement that two ratios are equal is called a proportion. True
6. The product of the extremes is equal to the product of the means. True

WARM-UPS

(continued)

7. If $\frac{2}{x} = \frac{3}{5}$, then $5x = 6$. False

8. The ratio of the height of a 12-inch cactus to the height of a 3-foot cactus is 4 to 1. False

9. If 30 out of 100 lawyers preferred aspirin and the rest did not, then the ratio of lawyers that preferred aspirin to those who did not is 30 to 100. False

10. If $\frac{x+5}{x} = \frac{2}{3}$, then $3x + 15 = 2x$. True

7.7 EXERCISES

Reading and Writing *After reading this section, write out the answers to these questions. Use complete sentences.*

1. What is a ratio?
A ratio is a comparison of two numbers.

2. What are the different ways of expressing a ratio?
The ratio of a to b is also written as $\frac{a}{b}$ or $a : b$.

3. What are equivalent ratios?
Equivalent ratios are ratios that are equivalent as fractions.

4. What is a proportion?
A proportion is an equation that expresses equality of two ratios.

5. What are the means and what are the extremes?
In the proportion $\frac{a}{b} = \frac{c}{d}$ the means are b and c and the extremes are a and d.

6. What is the extremes-means property?
The extremes-means property says that if $\frac{a}{b} = \frac{c}{d}$ then $ad = bc$.

For each ratio, find an equivalent ratio of integers in lowest terms. See Example 1.

7. $\frac{2.5}{3.5}$ $\frac{5}{7}$

8. $\frac{4.8}{1.2}$ $\frac{4}{1}$

9. $\frac{0.32}{0.6}$ $\frac{8}{15}$

10. $\frac{0.05}{0.8}$ $\frac{1}{16}$

11. $\frac{35}{10}$ $\frac{7}{2}$

12. $\frac{88}{33}$ $\frac{8}{3}$

13. $\frac{4.5}{7}$ $\frac{9}{14}$

14. $\frac{3}{2.5}$ $\frac{6}{5}$

15. $\frac{\frac{1}{2}}{\frac{1}{5}}$ $\frac{5}{2}$

16. $\frac{\frac{2}{3}}{\frac{3}{4}}$ $\frac{8}{9}$

17. $\frac{5}{\frac{1}{3}}$ $\frac{15}{1}$

18. $\frac{4}{\frac{1}{4}}$ $\frac{16}{1}$

Find a ratio for each of the following, and write it as a ratio of integers in lowest terms. See Examples 2–4.

19. ***Men and women.*** Find the ratio of men to women in a bowling league containing 12 men and 8 women.
3 to 2

20. ***Coffee drinkers.*** Among 100 coffee drinkers, 36 said that they preferred their coffee black and the rest did not prefer their coffee black. Find the ratio of those who prefer black coffee to those who prefer nonblack coffee.
9 to 16

FIGURE FOR EXERCISE 20

21. ***Smokers.*** A life insurance company found that among its last 200 claims, there were six dozen smokers. What is the ratio of smokers to nonsmokers in this group of claimants? 9 to 16

22. ***Hits and misses.*** A woman threw 60 darts and hit the target a dozen times. What is her ratio of hits to misses?
1 to 4

23. ***Violence and kindness.*** While watching television for one week, a consumer group counted 1240 acts of violence and 40 acts of kindness. What is the violence to kindness ratio for television, according to this group?
31 to 1

24. ***Length to width.*** What is the ratio of length to width for the rectangle shown?
8 to 5

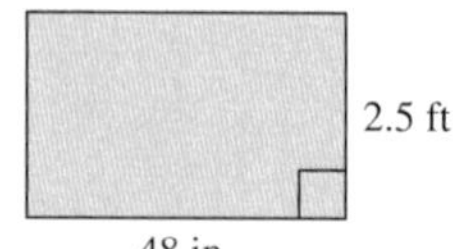

FIGURE FOR EXERCISE 24

25. ***Rise to run.*** What is the ratio of rise to run for the stairway shown in the accompanying figure? 2 to 3

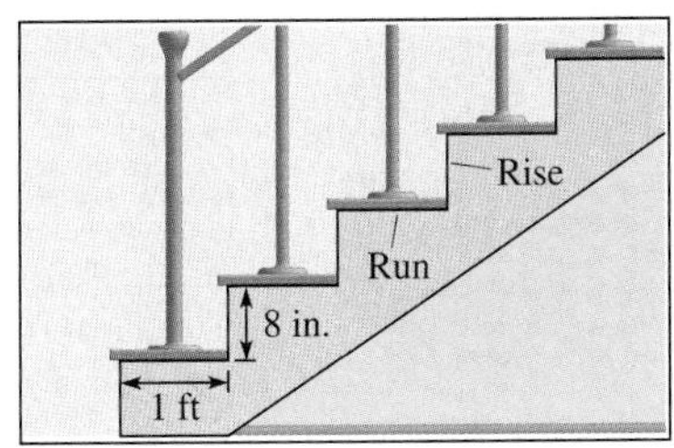

FIGURE FOR EXERCISE 25

26. ***Rise and run.*** If the rise is $\frac{3}{2}$ and the run is 5, then what is the ratio of the rise to the run? 3 to 10

Solve each proportion. See Example 5.

27. $\frac{4}{x} = \frac{2}{3}$ 6

28. $\frac{9}{x} = \frac{3}{2}$ 6

29. $\frac{a}{2} = \frac{-1}{5}$ $-\frac{2}{5}$

30. $\frac{b}{3} = \frac{-3}{4}$ $-\frac{9}{4}$

31. $-\frac{5}{9} = \frac{3}{x}$ $-\frac{27}{5}$

32. $-\frac{3}{4} = \frac{5}{x}$ $-\frac{20}{3}$

33. $\frac{10}{x} = \frac{34}{x + 12}$ 5

34. $\frac{x}{3} = \frac{x + 1}{2}$ -3

35. $\frac{a}{a + 1} = \frac{a + 3}{a}$ $-\frac{3}{4}$

36. $\frac{c + 3}{c - 1} = \frac{c + 2}{c - 3}$ -7

37. $\frac{m - 1}{m - 2} = \frac{m - 3}{m + 4}$ $\frac{5}{4}$

38. $\frac{h}{h - 3} = \frac{h}{h - 9}$ 0

Use a proportion to solve each problem. See Examples 6–8.

39. ***New shows and reruns.*** The ratio of new shows to reruns on cable TV is 2 to 27. If Frank counted only eight new shows one evening, then how many reruns were there? 108

40. ***Fast food.*** If four out of five doctors prefer fast food, then at a convention of 445 doctors, how many prefer fast food? 356

41. ***Voting.*** If 220 out of 500 voters surveyed said that they would vote for the incumbent, then how many votes could the incumbent expect out of the 400,000 voters in the state? 176,000

FIGURE FOR EXERCISE 41

42. ***New product.*** A taste test with 200 randomly selected people found that only three of them said that they would buy a box of new Sweet Wheats cereal. How many boxes could the manufacturer expect to sell in a country of 280 million people?
4.2 million

43. ***Basketball blowout.*** As the final buzzer signaled the end of the basketball game, the Lions were 34 points ahead of the Tigers. If the Lions scored 5 points for every 3 scored by the Tigers, then what was the final score?
Lions 85, Tigers 51

44. ***The golden ratio.*** The ancient Greeks thought that the most pleasing shape for a rectangle was one for which the ratio of the length to the width was 8 to 5, the golden ratio. If the length of a rectangular painting is 2 ft longer than its width, then for what dimensions would the length and width have the golden ratio?
Length $\frac{16}{3}$ ft, width $\frac{10}{3}$ ft

45. ***Automobile sales.*** The ratio of sports cars to luxury cars sold in Wentworth one month was 3 to 2. If 20 more sports cars were sold than luxury cars, then how many of each were sold that month?
40 luxury cars, 60 sports cars

46. ***Foxes and rabbits.*** The ratio of foxes to rabbits in the Deerfield Forest Preserve is 2 to 9. If there are 35 fewer foxes than rabbits, then how many of each are there?
45 rabbits, 10 foxes

47. ***Inches and feet.*** If there are 12 inches in 1 foot, then how many inches are there in 7 feet? 84 in.

48. ***Feet and yards.*** If there are 3 feet in 1 yard, then how many yards are there in 28 feet? $\frac{28}{3}$ yd

49. ***Minutes and hours.*** If there are 60 minutes in 1 hour, then how many minutes are there in 0.25 hour?
15 min

50. ***Meters and kilometers.*** If there are 1000 meters in 1 kilometer, then how many meters are there in 2.33 kilometers?
2330 m

51. ***Miles and hours.*** If Alonzo travels 230 miles in 3 hours, then how many miles does he travel in 7 hours?
536.7 mi

52. ***Hiking time.*** If Evangelica can hike 19 miles in 2 days on the Appalachian Trail, then how many days will it take her to hike 63 miles?
$\frac{126}{19}$ days

53. ***Force on basketball shoes.*** The designers of Converse shoes know that the force exerted on shoe soles in a jump shot is proportional to the weight of the person jumping. If a 70-pound boy exerts a force of 980 pounds on his shoe soles when he returns to the court after a jump, then what force does a 6 ft 8 in. professional ball player weighing 280 pounds exert on the soles of his

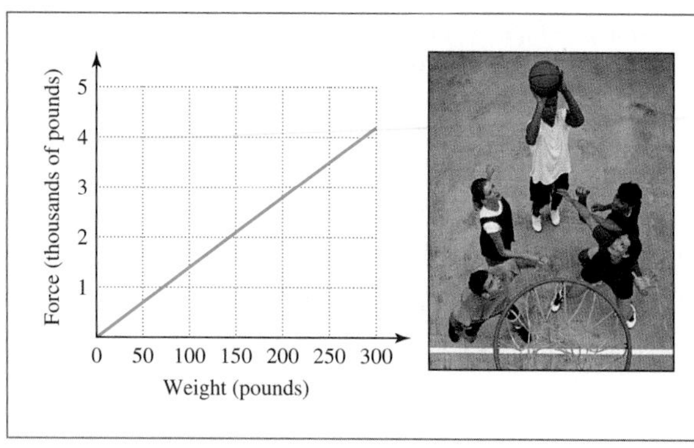

FIGURE FOR EXERCISE 53

shoes when he returns to the court after a jump? Use the accompanying graph to estimate the force for a 150-pound player.
3920 lbs, 2000 lbs

54. ***Force on running shoes.*** The designers of Converse shoes know that the ratio of the force on the shoe soles to the weight of a runner is 3 to 1. What force does a 130-pound jogger exert on the soles of her shoes?
390 lbs

55. ***Capture-recapture.*** To estimate the number of trout in Trout Lake, rangers used the capture-recapture method. They caught, tagged, and released 200 trout. One week later, they caught a sample of 150 trout and found that 5 of them were tagged. Assuming that the ratio of tagged trout to the total number of trout in the lake is the same as the ratio of tagged trout in the sample to the number of trout in the sample, find the number of trout in the lake. 6000

56. ***Bear population.*** To estimate the size of the bear population on the Keweenaw Peninsula, conservationists captured, tagged, and released 50 bears. One year later, a random sample of 100 bears included only 2 tagged bears. What is the conservationist's estimate of the size of the bear population?
2500

57. ***Fast-food waste.*** The accompanying figure shows the typical distribution of waste at a fast-food restaurant (U.S. Environmental Protection Agency, www.epa.gov).

a) What is the ratio of customer waste to food waste?
3 to 17

b) If a typical McDonald's generates 67 more pounds of food waste than customer waste per day, then how many pounds of customer waste does it generate?
14.4 lbs

58. ***Corrugated waste.*** Use the accompanying figure to find the ratio of waste from corrugated shipping boxes to waste not from corrugated shipping boxes. If a typical McDonald's generates 81 pounds of waste per day from corrugated shipping boxes, then how many pounds of waste per day does it generate that is not from corrugated shipping boxes?
17 to 33, 157.2 lbs

WASTE GENERATION AT A FAST-FOOD RESTAURANT

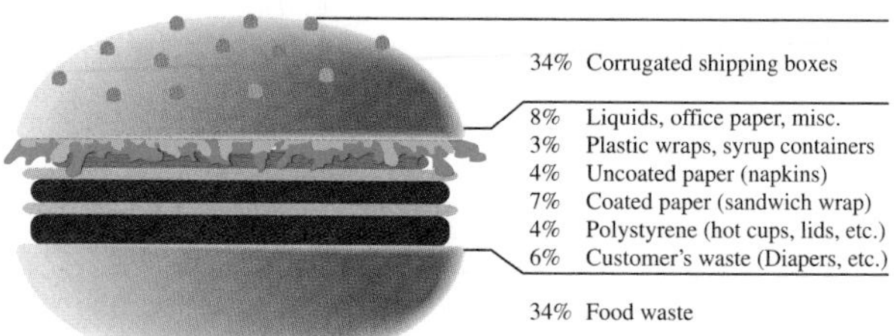

FIGURE FOR EXERCISES 57 AND 58

59. ***Mascara needs.*** In determining warehouse needs for a particular mascara for a chain of 2000 stores, Mike Pittman first determines a need B based on sales figures for the past 52 weeks. He then determines the actual need A from the equation $\frac{A}{B} = k$, where

$$k = 1 + V + C + X - D.$$

He uses $V = 0.22$ if there is a national TV ad and $V = 0$ if not, $C = 0.26$ if there is a national coupon and $C = 0$ if not, $X = 0.36$ if there is a chain-specific ad and $X = 0$ if not, and $D = 0.29$ if there is a special display in the chain and $D = 0$ if not. (D is subtracted because less product is needed in the warehouse when more is on display in the store.) If $B = 4200$ units and there is a special display and a national coupon but no national TV ad and no chain-specific ad, then what is the value of A?
4074

GETTING MORE INVOLVED

60. ***Discussion.*** Which of the following equations is not a proportion? Explain.

a) $\frac{1}{2} = \frac{1}{2}$ b) $\frac{x}{x+2} = \frac{4}{5}$

c) $\frac{x}{4} = \frac{9}{x}$ d) $\frac{8}{x+2} - 1 = \frac{5}{x+2}$

d

61. ***Discussion.*** Find all of the errors in the following solution to an equation.

$$\frac{7}{x} = \frac{8}{x+3} + 1$$
$$7(x+3) = 8x + 1$$
$$7x + 3 = 8x$$
$$-x = -3$$
$$x = 3$$

7.8 APPLICATIONS OF RATIONAL EXPRESSIONS

In This Section

- Formulas
- Uniform Motion Problems
- Work Problems
- Purchasing Problems

In this section we will study additional applications of rational expressions.

Formulas

Many formulas involve rational expressions. When solving a formula of this type for a certain variable, we usually multiply each side by the LCD to eliminate the denominators.

EXAMPLE 1

An equation of a line

The equation for the line through $(-2, 4)$ with slope $3/2$ can be written as

$$\frac{y-4}{x+2} = \frac{3}{2}.$$

We studied equations of this type in Chapter 3. Solve this equation for y.

Helpful Hint

When this equation was written in the form

$$y - y_1 = m(x - x_1)$$

in Chapter 3, we called it the point-slope formula for the equation of a line.

Solution

To isolate y on the left-hand side of the equation, we multiply each side by $x + 2$:

$$\frac{y-4}{x+2} = \frac{3}{2} \quad \text{Original equation}$$

$$(x+2)\cdot\frac{y-4}{x+2} = (x+2)\cdot\frac{3}{2} \quad \text{Multiply by } x + 2.$$

$$y - 4 = \frac{3}{2}x + 3 \quad \text{Simplify.}$$

$$y = \frac{3}{2}x + 7 \quad \text{Add 4 to each side.}$$

Because the original equation is a proportion, we could have used the extremes-means property to solve it for y. ■

EXAMPLE 2

Distance, rate, and time

Solve the formula $\frac{D}{T} = R$ for T.

Study Tip

As you study from the text, think about the material. Ask yourself questions. If you were the professor, what questions would you ask on the test?

Solution

Because the only denominator is T, we multiply each side by T:

$$\frac{D}{T} = R \quad \text{Original formula}$$

$$T\cdot\frac{D}{T} = T\cdot R \quad \text{Multiply each side by } T.$$

$$D = TR$$

$$\frac{D}{R} = \frac{TR}{R} \quad \text{Divide each side by } R.$$

$$\frac{D}{R} = T \quad \text{Simplify.}$$

The formula solved for T is $T = \frac{D}{R}$. ■

In Example 3, different subscripts are used on a variable to indicate that they are different variables. Think of R_1 as the first resistance, R_2 as the second resistance, and R as a combined resistance.

EXAMPLE 3 **Total resistance**

The formula

$$\frac{1}{R} = \frac{1}{R_1} + \frac{1}{R_2}$$

(from physics) expresses the relationship between different amounts of resistance in a parallel circuit. Solve it for R_2.

Solution

The LCD for R, R_1, and R_2 is RR_1R_2:

$$\frac{1}{R} = \frac{1}{R_1} + \frac{1}{R_2} \quad \text{Original formula}$$

$$RR_1R_2 \cdot \frac{1}{R} = RR_1R_2 \cdot \frac{1}{R_1} + RR_1R_2 \cdot \frac{1}{R_2} \quad \text{Multiply each side by the LCD, } RR_1R_2.$$

$$R_1R_2 = RR_2 + RR_1 \quad \text{All denominators are eliminated.}$$

$$R_1R_2 - RR_2 = RR_1 \quad \text{Get all terms involving } R_2 \text{ onto the left side.}$$

$$R_2(R_1 - R) = RR_1 \quad \text{Factor out } R_2.$$

$$R_2 = \frac{RR_1}{R_1 - R} \quad \text{Divide each side by } R_1 - R.$$

■

EXAMPLE 4 **Finding the value of a variable**

In the formula of Example 1, find x if $y = -3$.

Solution

Substitute $y = -3$ into the formula, then solve for x:

$$\frac{y - 4}{x + 2} = \frac{3}{2} \quad \text{Original formula}$$

$$\frac{-3 - 4}{x + 2} = \frac{3}{2} \quad \text{Replace } y \text{ by } -3.$$

$$\frac{-7}{x + 2} = \frac{3}{2} \quad \text{Simplify.}$$

$$3x + 6 = -14 \quad \text{Extremes-means property}$$

$$3x = -20$$

$$x = -\frac{20}{3}$$

■

Uniform Motion Problems

In uniform motion problems we use the formula $D = RT$. In some problems in which the time is unknown, we can use the formula $T = \frac{D}{R}$ to get an equation involving rational expressions.

EXAMPLE 5 **Driving to Florida**

Susan drove 1500 miles to Daytona Beach for spring break. On the way back she averaged 10 miles per hour less, and the drive back took her 5 hours longer. Find Susan's average speed on the way to Daytona Beach.

Solution

If x represents her average speed going there, then $x - 10$ is her average speed for the return trip. See Fig. 7.1. We use the formula $T = \frac{D}{R}$ to make the following table.

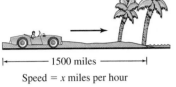

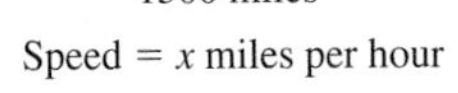

Speed = x miles per hour

Speed = $x - 10$ miles per hour

FIGURE 7.1

	D	R	T	
Going	1500	x	$\frac{1500}{x}$	← Shorter time
Returning	1500	$x - 10$	$\frac{1500}{x - 10}$	← Longer time

Because the difference between the two times is 5 hours, we have

$$\text{longer time} - \text{shorter time} = 5.$$

Using the time expressions from the table, we get the following equation:

$$\frac{1500}{x - 10} - \frac{1500}{x} = 5$$

$$x(x - 10)\frac{1500}{x - 10} - x(x - 10)\frac{1500}{x} = x(x - 10)5 \quad \text{Multiply by } x(x - 10).$$

$$1500x - 1500(x - 10) = 5x^2 - 50x$$

$$15{,}000 = 5x^2 - 50x \quad \text{Simplify.}$$

$$3000 = x^2 - 10x \quad \text{Divide each side by 5.}$$

$$0 = x^2 - 10x - 3000$$

$$(x + 50)(x - 60) = 0 \quad \text{Factor.}$$

$$x + 50 = 0 \quad \text{or} \quad x - 60 = 0$$

$$x = -50 \quad \text{or} \quad x = 60$$

The answer $x = -50$ is a solution to the equation, but it cannot indicate the average speed of the car. Her average speed going to Daytona Beach was 60 mph. ■

Helpful Hint

Notice that a work rate is the same as a slope from Chapter 3. The only difference is that the work rates here can contain a variable.

Work Problems

If you can complete a job in 3 hours, then you are working at the rate of $\frac{1}{3}$ of the job per hour. If you work for 2 hours at the rate of $\frac{1}{3}$ of the job per hour, then you will complete $\frac{2}{3}$ of the job. The product of the rate and time is the amount of work completed. For problems involving work, we will always assume that the work is done at a constant rate. So if a job takes x hours to complete, then the rate is $\frac{1}{x}$ of the job per hour.

EXAMPLE 6 **Shoveling snow**

After a heavy snowfall, Brian can shovel all of the driveway in 30 minutes. If his younger brother Allen helps, the job takes only 20 minutes. How long would it take Allen to do the job by himself?

Helpful Hint

The secret to work problems is remembering that the individual rates or the amounts of work can be added when people work together. If your painting rate is 1/10 of the house per day and your helper's rate is 1/5 of the house per day, then your rate together will be 3/10 of the house per day. In 2 days you will paint 2/10 of the house and your helper will paint 2/5 of the house for a total of 3/5 of the house completed.

Solution

Let x represent the number of minutes it would take Allen to do the job by himself. Brian's rate for shoveling is $\frac{1}{30}$ of the driveway per minute, and Allen's rate for shoveling is $\frac{1}{x}$ of the driveway per minute. We organize all of the information in a table like the table in Example 5.

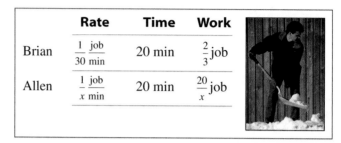

	Rate	Time	Work
Brian	$\frac{1 \text{ job}}{30 \text{ min}}$	20 min	$\frac{2}{3}$ job
Allen	$\frac{1 \text{ job}}{x \text{ min}}$	20 min	$\frac{20}{x}$ job

If Brian works for 20 min at the rate $\frac{1}{30}$ of the job per minute, then he does $\frac{20}{30}$ or $\frac{2}{3}$ of the job, as shown in Fig. 7.2. The amount of work that each boy does is a fraction of the whole job. So the expressions for work in the last column of the table have a sum of 1:

$$\frac{2}{3} + \frac{20}{x} = 1$$

$$3x \cdot \frac{2}{3} + 3x \cdot \frac{20}{x} = 3x \cdot 1 \quad \text{Multiply each side by } 3x.$$

$$2x + 60 = 3x$$

$$60 = x$$

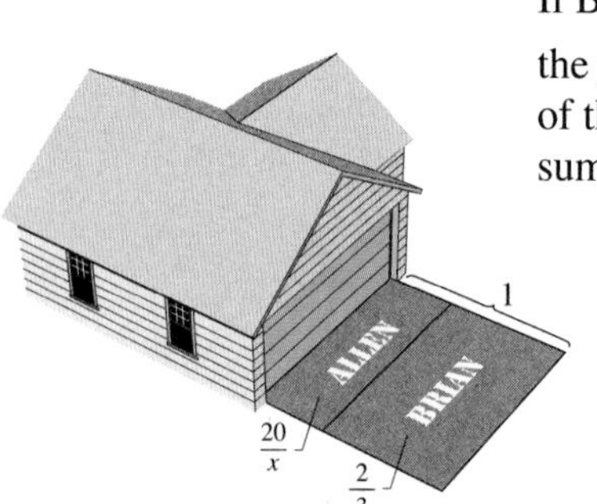

FIGURE 7.2

If it takes Allen 60 min to do the job by himself, then he works at the rate of $\frac{1}{60}$ of the job per minute. In 20 minutes he does $\frac{1}{3}$ of the job while Brian does $\frac{2}{3}$. So it would take Allen 60 minutes to shovel the driveway by himself. ■

Notice the similarities between the uniform motion problem in Example 5 and the work problem in Example 6. In both cases it is beneficial to make a table. We use $D = R \cdot T$ in uniform motion problems and $W = R \cdot T$ in work problems. The main points to remember when solving work problems are summarized in the following strategy.

Strategy for Solving Work Problems

1. If a job is completed in x hours, then the rate is $\frac{1}{x}$ job/hr.
2. Make a table showing rate, time, and work completed ($W = R \cdot T$) for each person or machine.
3. The total work completed is the sum of the individual amounts of work completed.
4. If the job is completed, then the total work done is 1 job.

Purchasing Problems

Rates are used in uniform motion and work problems. But rates also occur in purchasing problems. If gasoline is 99.9 cents/gallon, then that is the rate at which your bill is increasing as you pump the gallons into your tank. In purchasing problems the product of the rate and the quantity purchased is the total cost.

EXAMPLE 7 **Oranges and grapefruit**

Tamara bought 50 pounds of fruit consisting of Florida oranges and Texas grapefruit. She paid twice as much per pound for the grapefruit as she did for the oranges. If Tamara bought \$12 worth of oranges and \$16 worth of grapefruit, then how many pounds of each did she buy?

Solution

Let x represent the number of pounds of oranges and $50 - x$ represent the number of pounds of grapefruit. See Fig. 7.3. Make a table.

x lb

Oranges

$50 - x$ lb

Grapefruit

FIGURE 7.3

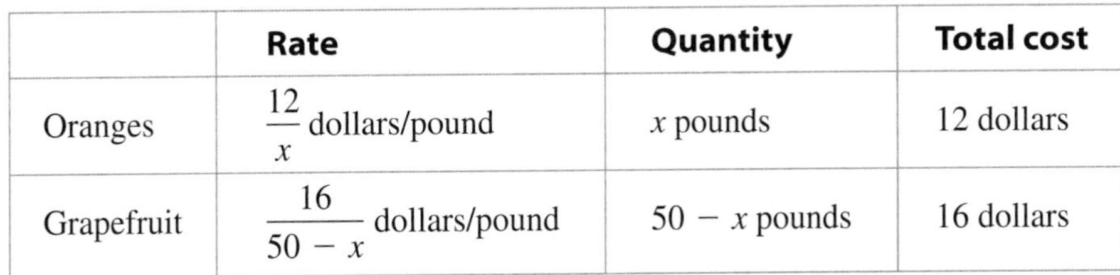

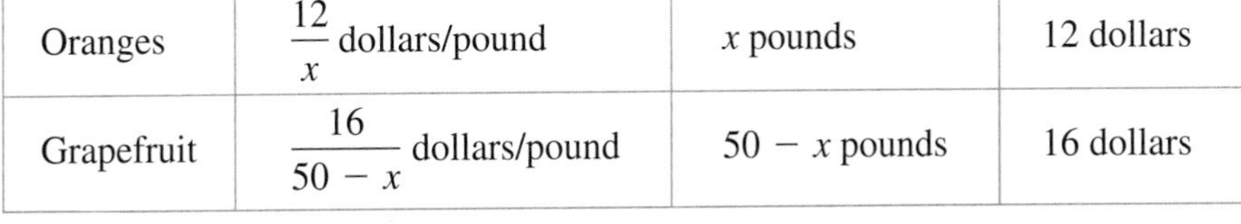

	Rate	**Quantity**	**Total cost**
Oranges	$\frac{12}{x}$ dollars/pound	x pounds	12 dollars
Grapefruit	$\frac{16}{50 - x}$ dollars/pound	$50 - x$ pounds	16 dollars

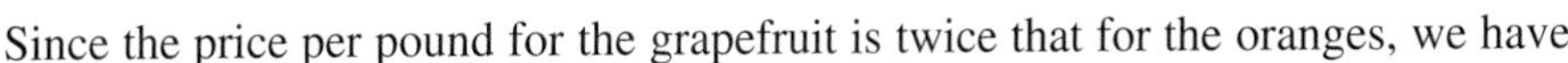

Since the price per pound for the grapefruit is twice that for the oranges, we have:

$$2(\text{price per pound for oranges}) = \text{price per pound for grapefruit}$$

$$2\left(\frac{12}{x}\right) = \frac{16}{50 - x}$$

$$\frac{24}{x} = \frac{16}{50 - x}$$

$$16x = 1200 - 24x \quad \text{Extremes-means property}$$

$$40x = 1200$$

$$x = 30$$

$$50 - x = 20$$

If Tamara purchased 20 pounds of grapefruit for \$16, then she paid \$0.80 per pound. If she purchased 30 pounds of oranges for \$12, then she paid \$0.40 per pound. Because \$0.80 is twice \$0.40, we can be sure that she purchased 20 pounds of grapefruit and 30 pounds of oranges. ■

WARM-UPS

True or false? Explain your answer.

1. The formula $t = \frac{1 - t}{m}$, solved for m, is $m = \frac{1 - t}{t}$. True
2. To solve $\frac{1}{m} + \frac{1}{n} = \frac{1}{2}$ for m, we multiply each side by $2mn$. True
3. If Fiona drives 300 miles in x hours, then her average speed is $\frac{x}{300}$ mph. False
4. If Miguel drives 20 hard bargains in x hours, then he is driving $\frac{20}{x}$ hard bargains per hour. True
5. If Fred can paint a house in y days, then he paints $\frac{1}{y}$ of the house per day. True

WARM-UPS

(continued)

6. If $\frac{1}{x}$ is 1 less than $\frac{2}{x+3}$, then $\frac{1}{x} - 1 = \frac{2}{x+3}$. False

7. If a and b are nonzero and $a = \frac{m}{b}$, then $b = am$. False

8. If $D = RT$, then $T = \frac{D}{R}$. True

9. Solving $P + Prt = I$ for P gives $P = I - Prt$. False

10. To solve $3R + yR = m$ for R, we must first factor the left-hand side. True

7.8 EXERCISES

Solve each equation for y. See Example 1.

1. $\frac{y-1}{x-3} = 2$
$y = 2x - 5$

2. $\frac{y-2}{x-4} = -2$
$y = -2x + 10$

3. $\frac{y-1}{x+6} = -\frac{1}{2}$
$y = -\frac{1}{2}x - 2$

4. $\frac{y+5}{x-2} = -\frac{1}{2}$
$y = -\frac{1}{2}x - 4$

5. $\frac{y+a}{x-b} = m$
$y = mx - mb - a$

6. $\frac{y-h}{x+k} = a$
$y = ax + ak + h$

7. $\frac{y-1}{x+4} = -\frac{1}{3}$
$y = -\frac{1}{3}x - \frac{1}{3}$

8. $\frac{y-1}{x+3} = -\frac{3}{4}$
$y = -\frac{3}{4}x - \frac{5}{4}$

Solve each formula for the indicated variable. See Examples 2 and 3.

9. $A = \frac{B}{C}$ for C
$C = \frac{B}{A}$

10. $P = \frac{A}{C+D}$ for A
$A = PC + PD$

11. $\frac{1}{a} + m = \frac{1}{p}$ for p
$p = \frac{a}{1+am}$

12. $\frac{2}{f} + t = \frac{3}{m}$ for m
$m = \frac{3f}{2+ft}$

13. $F = k\frac{m_1m_2}{r^2}$ for m_1
$m_1 = \frac{r^2F}{km_2}$

14. $F = \frac{mv^2}{r}$ for r
$r = \frac{mv^2}{F}$

15. $\frac{1}{a} + \frac{1}{b} = \frac{1}{f}$ for a
$a = \frac{bf}{b-f}$

16. $\frac{1}{R} = \frac{1}{R_1} + \frac{1}{R_2}$ for R
$R = \frac{R_1R_2}{R_1+R_2}$

17. $S = \frac{a}{1-r}$ for r
$r = \frac{S-a}{S}$

18. $I = \frac{E}{R+r}$ for R
$R = \frac{E-Ir}{I}$

19. $\frac{P_1V_1}{T_1} = \frac{P_2V_2}{T_2}$ for P_2
$P_2 = \frac{P_1V_1T_2}{T_1V_2}$

20. $\frac{P_1V_1}{T_1} = \frac{P_2V_2}{T_2}$ for T_1
$T_1 = \frac{P_1V_1T_2}{P_2V_2}$

21. $V = \frac{4}{3}\pi r^2h$ for h
$h = \frac{3V}{4\pi r^2}$

22. $h = \frac{S - 2\pi r^2}{2\pi r}$ for S
$S = 2\pi rh + 2\pi r^2$

Find the value of the indicated variable. See Example 4.

23. In the formula of Exercise 9, if $A = 12$ and $B = 5$, find C.
$\frac{5}{12}$

24. In the formula of Exercise 10, if $A = 500$, $P = 100$, and $C = 2$, find D.
3

25. In the formula of Exercise 11, if $p = 6$ and $m = 4$, find a.
$-\frac{6}{23}$

26. In the formula of Exercise 12, if $m = 4$ and $t = 3$, find f.
$-\frac{8}{9}$

27. In the formula of Exercise 13, if $F = 32$, $r = 4$, $m_1 = 2$, and $m_2 = 6$, find k.
$\frac{128}{3}$

28. In the formula of Exercise 14, if $F = 10$, $v = 8$, and $r = 6$, find m.
$\frac{15}{16}$

29. In the formula of Exercise 15, if $f = 3$ and $a = 2$, find b.
-6

3. **building up the denominator**
   a. the opposite of reducing a fraction
   b. finding the least common denominator
   c. adding the same number to the numerator and denominator
   d. writing a fraction larger a

4. **least common denominator**
   a. the largest number that is a multiple of all denominators
   b. the sum of the denominators
   c. the product of the denominators
   d. the smallest number that is a multiple of all denominators d

5. **extraneous solution**
   a. a number that appears to be a solution to an equation but does not satisfy the equation
   b. an extra solution to an equation
   c. the second solution
   d. a nonreal solution a

6. **ratio of a to b**
   a. b/a
   b. a/b
   c. $a/(a+b)$
   d. ab b

7. **proportion**
   a. a ratio
   b. two ratios
   c. the product of the means equals the product of the extremes
   d. a statement expressing the equality of two ratios d

8. **extremes**
   a. a and d in $a/b = c/d$
   b. b and c in $a/b = c/d$
   c. the extremes-means property
   d. if $a/b = c/d$ then $ad = bc$ a

9. **means**
   a. the average of a, b, c, and d
   b. a and d in $a/b = c/d$
   c. b and c in $a/b = c/d$
   d. if $a/b = c/d$, then $(a+b)/2 = (c+d)/2$ c

10. **cross-multiplying**
   a. $ab = ba$ for any real numbers a and b
   b. $(a-b)^2 = (b-a)^2$ for any real numbers a and b
   c. if $a/b = c/d$, then $ab = cd$
   d. if $a/b = c/d$, then $ad = bc$ d

REVIEW EXERCISES

7.1 *Reduce each rational expression to lowest terms.*

1. $\frac{24}{28}$ $\frac{6}{7}$
2. $\frac{42}{18}$ $\frac{7}{3}$
3. $\frac{2a^3c^3}{8a^5c}$ $\frac{c^2}{4a^2}$
4. $\frac{39x^6}{15x}$ $\frac{13x^5}{5}$
5. $\frac{6w-9}{9w-12}$ $\frac{2w-3}{3w-4}$
6. $\frac{3t-6}{8-4t}$ $-\frac{3}{4}$
7. $\frac{x^2-1}{3-3x}$ $-\frac{x+1}{3}$
8. $\frac{3x^2-9x+6}{10-5x}$ $\frac{3-3x}{5}$

7.2 *Perform the indicated operation.*

9. $\frac{1}{6k} \cdot 3k^2$ $\frac{1}{2}k$
10. $\frac{1}{15abc} \cdot 5a^3b^5c^2$ $\frac{1}{3}a^2b^4c$
11. $\frac{2xy}{3} \div y^2$ $\frac{2x}{3y}$
12. $4ab \div \frac{1}{2a^4}$ $8a^5b$
13. $\frac{a^2-9}{a-2} \cdot \frac{a^2-4}{a+3}$ a^2-a-6
14. $\frac{x^2-1}{3x} \cdot \frac{6x}{2x-2}$ $x+1$
15. $\frac{w-2}{3w} \div \frac{4w-8}{6w}$ $\frac{1}{2}$
16. $\frac{2y+2x}{x-xy} \div \frac{x^2+2xy+y^2}{y^2-y}$ $\frac{-2y}{x^2+xy}$

7.3 *Find the least common denominator for each group of denominators.*

17. 36, 54 108
18. 10, 15, 35 210
19. $6ab^3$, $8a^7b^2$ $24a^7b^3$
20. $20u^4v$, $18uv^5$, $12u^2v^3$ $180u^4v^5$
21. $4x$, $6x-6$ $12x(x-1)$
22. $8a$, $6a$, $2a^2+2a$ $24a(a+1)$
23. x^2-4, x^2-x-2 $(x+1)(x-2)(x+2)$
24. x^2-9, x^2+6x+9 $(x-3)(x+3)^2$

Convert each rational expression into an equivalent rational expression with the indicated denominator.

25. $\frac{5}{12} = \frac{?}{36}$ $\frac{15}{36}$
26. $\frac{2a}{15} = \frac{?}{45}$ $\frac{6a}{45}$
27. $\frac{2}{3xy} = \frac{?}{15x^2y}$ $\frac{10x}{15x^2y}$
28. $\frac{3z}{7x^2y} = \frac{?}{42x^3y^8}$ $\frac{18xy^7z}{42x^3y^8}$
29. $\frac{5}{y-6} = \frac{?}{12-2y}$ $\frac{-10}{12-2y}$
30. $\frac{-3}{2-t} = \frac{?}{2t-4}$ $\frac{6}{2t-4}$
31. $\frac{x}{x-1} = \frac{?}{x^2-1}$ $\frac{x^2+x}{x^2-1}$
32. $\frac{t}{t-3} = \frac{?}{t^2+2t-15}$ $\frac{t^2+5t}{t^2+2t-15}$

7.4 *Perform the indicated operation.*

33. $\frac{5}{36}+\frac{9}{28}$ $\frac{29}{63}$

34. $\frac{7}{30}-\frac{11}{42}$ $-\frac{1}{35}$

35. $3-\frac{4}{x}$ $\frac{3x-4}{x}$

36. $1+\frac{3a}{2b}$ $\frac{2b+3a}{2b}$

37. $\frac{2}{ab^2}-\frac{1}{a^2b}$ $\frac{2a-b}{a^2b^2}$

38. $\frac{3}{4x^3}+\frac{5}{6x^2}$ $\frac{10x+9}{12x^3}$

39. $\frac{9a}{2a-3}+\frac{5}{3a-2}$ $\frac{27a^2-8a-15}{(2a-3)(3a-2)}$

40. $\frac{3}{x-2}-\frac{5}{x+3}$ $\frac{-2x+19}{(x-2)(x+3)}$

41. $\frac{1}{a-8}-\frac{2}{8-a}$ $\frac{3}{a-8}$

42. $\frac{5}{x-14}+\frac{4}{14-x}$ $\frac{1}{x-14}$

43. $\frac{3}{2x-4}+\frac{1}{x^2-4}$ $\frac{3x+8}{2(x+2)(x-2)}$

44. $\frac{x}{x^2-2x-3}-\frac{3x}{x^2-9}$ $\frac{-2x^2}{(x-3)(x+3)(x+1)}$

7.5 *Simplify each complex fraction.*

45. $\frac{\frac{1}{2}-\frac{3}{4}}{\frac{2}{3}+\frac{1}{2}}$ $-\frac{3}{14}$

46. $\frac{\frac{2}{3}+\frac{5}{8}}{\frac{1}{2}-\frac{3}{8}}$ $\frac{31}{3}$

47. $\frac{\frac{1}{a}+\frac{2}{3b}}{\frac{1}{2b}-\frac{3}{a}}$ $\frac{6b+4a}{3a-18b}$

48. $\frac{\frac{3}{xy}-\frac{1}{3y}}{\frac{1}{6x}-\frac{3}{5y}}$ $\frac{90-10x}{5y-18x}$

49. $\frac{\frac{1}{x-2}-\frac{3}{x+3}}{\frac{2}{x+3}+\frac{1}{x-2}}$ $\frac{-2x+9}{3x-1}$

50. $\frac{\frac{4}{a+1}+\frac{5}{a^2-1}}{\frac{1}{a^2-1}-\frac{3}{a-1}}$ $\frac{4a+1}{-3a-2}$

51. $\frac{\frac{x-1}{x-3}}{\frac{1}{x^2-x-6}-\frac{4}{x+2}}$ $\frac{x^2+x-2}{-4x+13}$

52. $\frac{\frac{6}{a^2+5a+6}-\frac{8}{a+2}}{\frac{2}{a+3}-\frac{4}{a+2}}$ $\frac{4a+9}{a+4}$

7.6 *Solve each equation.*

53. $\frac{-2}{5}=\frac{3}{x}$ $-\frac{15}{2}$

54. $\frac{3}{x}+\frac{5}{3x}=1$ $\frac{14}{3}$

55. $\frac{14}{a^2-1}+\frac{1}{a-1}=\frac{3}{a+1}$ 9

56. $2+\frac{3}{y-5}=\frac{2y}{y-5}$ No solution

57. $z-\frac{3z}{2-z}=\frac{6}{z-2}$ -3

58. $\frac{1}{x}+\frac{1}{3}=\frac{1}{2}$ 6

7.7 *Solve each proportion.*

59. $\frac{3}{x}=\frac{2}{7}$ $\frac{21}{2}$

60. $\frac{4}{x}=\frac{x}{4}$ $-4, 4$

61. $\frac{2}{w-3}=\frac{5}{w}$ 5

62. $\frac{3}{t-3}=\frac{5}{t+4}$ $\frac{27}{2}$

Solve each problem by using a proportion.

63. ***Taxis in Times Square.*** The ratio of taxis to private automobiles in Times Square at 6:00 P.M. on New Year's Eve was estimated to be 15 to 2. If there were 60 taxis, then how many private automobiles were there? 8

FIGURE FOR EXERCISE 63

64. ***Student-teacher ratio.*** The student-teacher ratio for Washington High was reported to be 27.5 to 1. If there are 42 teachers, then how many students are there?
1155

65. ***Water and rice.*** At Wong's Chinese Restaurant the secret recipe for white rice calls for a 2 to 1 ratio of water to rice. In one batch the chef used 28 more cups of water than rice. How many cups of each did he use?
56 cups water, 28 cups rice

FIGURE FOR EXERCISE 65

66. ***Oil and gas.*** An outboard motor calls for a fuel mixture that has a gasoline-to-oil ratio of 50 to 1. How many pints of oil should be added to 6 gallons of gasoline?
$\frac{24}{25}$ pints

7.8 *Solve each formula for the indicated variable.*

67. $\frac{y - b}{m} = x$ for y $\quad y = mx + b$

68. $\frac{A}{h} = \frac{a + b}{2}$ for a $\quad a = \frac{2A - hb}{h}$

69. $F = \frac{mv + 1}{m}$ for m $\quad m = \frac{1}{F - v}$

70. $m = \frac{r}{1 + rt}$ for r $\quad r = \frac{m}{1 - mt}$

71. $\frac{y + 1}{x - 3} = 4$ for y $\quad y = 4x - 13$

72. $\frac{y - 3}{x + 2} = \frac{-1}{3}$ for y $\quad y = -\frac{1}{3}x + \frac{7}{3}$

Solve each problem.

73. ***Making a puzzle.*** Tracy, Stacy, and Fred assembled a very large puzzle together in 40 hours. If Stacy worked twice as fast as Fred and Tracy worked just as fast as Stacy, then how long would it have taken Fred to assemble the puzzle alone?
200 hours

74. ***Going skiing.*** Leon drove 270 miles to the lodge in the same time as Pat drove 330 miles to the lodge. If Pat drove 10 miles per hour faster than Leon, then how fast did each of them drive? Leon 45 mph, Pat 55 mph

FIGURE FOR EXERCISE 74

75. ***Merging automobiles.*** When Bert and Ernie merged their automobile dealerships, Bert had 10 more cars than Ernie. While 36% of Ernie's stock consisted of new cars, only 25% of Bert's stock consisted of new cars. If they had 33 new cars on the lot after the merger, then how many cars did each one have before the merger?
Bert 60 cars, Ernie 50 cars

76. ***Magazine sales.*** A company specializing in magazine sales over the telephone found that in 2500 phone calls, 360 resulted in sales and were made by male callers, and 480 resulted in sales and were made by female callers. If the company gets twice as many sales per call with a woman's voice than with a man's voice, then how many of the 2500 calls were made by females? 1000

77. ***Distribution of waste.*** The accompanying figure shows the distribution of the total municipal solid waste into various categories in 2000 (U.S. Environmental Protection Agency, www.epa.gov). If the paper waste was 59.8 million tons greater than the yard waste, then what was the amount of yard waste generated? 27.83 million tons

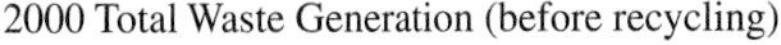

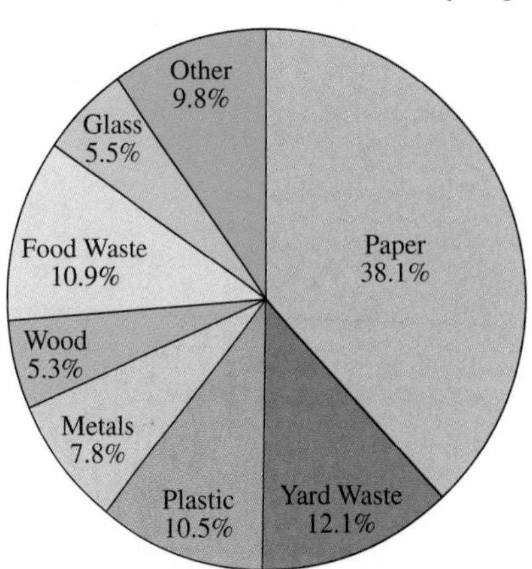

FIGURE FOR EXERCISES 77 AND 78

78. ***Total waste.*** Use the information given in Exercise 77 to find the total waste generated in 2000 and the amount of food waste.
230 million tons, 25.07 million tons

MISCELLANEOUS

In place of each question mark, put an expression that makes each equation an identity.

79. $\frac{5}{x} = \frac{?}{2x}$ $\frac{10}{2x}$

80. $\frac{?}{a} = \frac{6}{3a}$ $\frac{2}{a}$

81. $\frac{2}{a-5} = \frac{?}{5-a}$ $\frac{-2}{5-a}$

82. $\frac{-1}{a-7} = \frac{1}{?}$ $\frac{1}{7-a}$

83. $3 = \frac{?}{x}$ $\frac{3x}{x}$

84. $2a = \frac{?}{b}$ $\frac{2ab}{b}$

85. $m \div \frac{1}{2} = ?$ $2m$

86. $5x \div \frac{1}{x} = ?$ $5x^2$

87. $2a \div ? = 12a$ $\frac{1}{6}$

88. $10x \div ? = 20x^2$ $\frac{1}{2x}$

89. $\frac{a-1}{a^2-1} = \frac{1}{?}$ $\frac{1}{a+1}$

90. $\frac{?}{x^2-9} = \frac{1}{x-3}$ $\frac{x+3}{x^2-9}$

91. $\frac{1}{a} - \frac{1}{5} = ?$ $\frac{5-a}{5a}$

92. $\frac{3}{7} - \frac{2}{b} = ?$ $\frac{3b-14}{7b}$

93. $\frac{a}{2} - 1 = \frac{?}{2}$ $\frac{a-2}{2}$

94. $\frac{1}{a} - 1 = \frac{?}{a}$ $\frac{1-a}{a}$

95. $(a - b) \div (-1) = ?$ $b - a$

96. $(a - 7) \div (7 - a) = ?$ -1

97. $\frac{\frac{1}{5a}}{2} = ?$ $\frac{1}{10a}$

98. $\frac{3a}{\frac{1}{2}} = ?$ $6a$

For each expression in Exercises 99–118, either perform the indicated operation or solve the equation, whichever is appropriate.

99. $\frac{1}{x} + \frac{1}{2x}$ $\frac{3}{2x}$

100. $\frac{1}{y} + \frac{1}{3y} = 2$ $\frac{2}{3}$

101. $\frac{2}{3xy} + \frac{1}{6x}$ $\frac{4+y}{6xy}$

102. $\frac{3}{x-1} - \frac{3}{x}$ $\frac{3}{x(x-1)}$

103. $\frac{5}{a-5} - \frac{3}{5-a}$ $\frac{8}{a-5}$

104. $\frac{2}{x-2} - \frac{3}{x} = \frac{-1}{x}$ No solution

105. $\frac{2}{x-1} - \frac{2}{x} = 1$ $-1, 2$

106. $\frac{2}{x-2} \cdot \frac{6x-12}{14}$ $\frac{6}{7}$

107. $\frac{-3}{x+2} \cdot \frac{5x+10}{9}$ $-\frac{5}{3}$

108. $\frac{3}{10} = \frac{5}{x}$ $\frac{50}{3}$

109. $\frac{1}{-3} = \frac{-2}{x}$ 6

110. $\frac{x^2-4}{x} \div \frac{4x-8}{x}$ $\frac{x+2}{4}$

111. $\frac{ax+am+3x+3m}{a^2-9} \div \frac{2x+2m}{a-3}$ $\frac{1}{2}$

112. $\frac{-2}{x} = \frac{3}{x+2}$ $-\frac{4}{5}$

113. $\frac{2}{x^2-25} + \frac{1}{x^2-4x-5}$ $\frac{3x+7}{(x-5)(x+5)(x+1)}$

114. $\frac{4}{a^2-1} + \frac{1}{2a+2}$ $\frac{a+7}{2(a+1)(a-1)}$

115. $\frac{-3}{a^2-9} - \frac{2}{a^2+5a+6}$ $\frac{-5a}{(a-3)(a+3)(a+2)}$

116. $\frac{-5}{a^2-4} - \frac{2}{a^2-3a+2}$ $\frac{-7a+1}{(a+2)(a-2)(a-1)}$

117. $\frac{1}{a^2-1} + \frac{2}{1-a} = \frac{3}{a+1}$ $\frac{2}{5}$

118. $3 + \frac{1}{x-2} = \frac{2x-3}{x-2}$ No solution

CHAPTER 7 TEST

What numbers cannot be used for x in each rational expression?

1. $\frac{2x-1}{x^2-1}$ $-1, 1$

2. $\frac{5}{2-3x}$ $\frac{2}{3}$

3. $\frac{1}{x}$ 0

Perform the indicated operation. Write each answer in lowest terms.

4. $\frac{2}{15} - \frac{4}{9}$ $-\frac{14}{45}$

5. $\frac{1}{y} + 3$ $\frac{1+3y}{y}$

6. $\frac{3}{a-2} - \frac{1}{2-a}$ $\frac{4}{a-2}$

7. $\frac{2}{x^2-4} - \frac{3}{x^2+x-2}$ $\frac{-x+4}{(x+2)(x-2)(x-1)}$

8. $\frac{m^2-1}{(m-1)^2} \cdot \frac{2m-2}{3m+3}$ $\frac{2}{3}$

9. $\frac{a-b}{3} \div \frac{b^2-a^2}{6}$ $\frac{-2}{a+b}$

10. $\frac{5a^2b}{12a} \cdot \frac{2a^3b}{15ab^6}$ $\frac{a^3}{18b^4}$

Simplify each complex fraction.

11. $\dfrac{\frac{2}{3}+\frac{4}{5}}{\frac{2}{5}-\frac{3}{2}}$ $-\frac{4}{3}$

12. $\dfrac{\frac{2}{x}+\frac{1}{x-2}}{\frac{1}{x-2}-\frac{3}{x}}$ $\frac{3x-4}{-2x+6}$

Solve each equation.

13. $\frac{3}{x} = \frac{7}{5}$ $\frac{15}{7}$

14. $\frac{x}{x-1} - \frac{3}{x} = \frac{1}{2}$ $2, 3$

15. $\frac{1}{x} + \frac{1}{6} = \frac{1}{4}$ 12

Solve each formula for the indicated variable.

16. $\frac{y-3}{x+2} = \frac{-1}{5}$ for y $y = -\frac{1}{5}x + \frac{13}{5}$

17. $M = \frac{1}{3}b(c+d)$ for c $c = \frac{3M-bd}{b}$

Solve each problem.

18. If $R(x) = \frac{x+2}{1-x}$, then what is $R(0.9)$? 29

19. When all of the grocery carts escape from the supermarket, it takes Reginald 12 minutes to round them up and bring them back. Because Norman doesn't make as much per hour as Reginald, it takes Norman 18 minutes to do the same job. How long would it take them working together to complete the roundup?
7.2 minutes

20. Brenda and her husband Randy bicycled cross-country together. One morning, Brenda rode 30 miles. By traveling only 5 miles per hour faster and putting in one more hour, Randy covered twice the distance Brenda covered. What was the speed of each cyclist?
Brenda 15 mph and Randy 20 mph, or Brenda 10 mph and Randy 15 mph

21. For a certain time period the ratio of the dollar value of exports to the dollar value of imports for the United States was 2 to 3. If the value of exports during that time period was 48 billion dollars, then what was the value of imports?
$72 billion

MAKING CONNECTIONS CHAPTERS 1–7

Solve each equation.

1. $3x - 2 = 5$
$\frac{7}{3}$

2. $\frac{3}{5}x = -2$
$-\frac{10}{3}$

3. $2(x - 2) = 4x$
-2

4. $2(x - 2) = 2x$
No solution

5. $2(x + 3) = 6x + 6$
0

6. $2(3x + 4) + x^2 = 0$
$-4, -2$

7. $4x - 4x^3 = 0$
$-1, 0, 1$

8. $\frac{3}{x} = \frac{-2}{5}$
$-\frac{15}{2}$

9. $\frac{3}{x} = \frac{x}{12}$
$-6, 6$

10. $\frac{x}{2} = \frac{4}{x - 2}$
$-2, 4$

11. $\frac{w}{18} - \frac{w - 1}{9} = \frac{4 - w}{6}$
5

12. $\frac{x}{x + 1} + \frac{1}{2x + 2} = \frac{7}{8}$
3

Solve each equation for y.

13. $2x + 3y = c$
$y = \frac{c - 2x}{3}$

14. $\frac{y - 3}{x - 5} = \frac{1}{2}$
$y = \frac{1}{2}x + \frac{1}{2}$

15. $2y = ay + c$
$y = \frac{c}{2 - a}$

16. $\frac{A}{y} = \frac{C}{B}$
$y = \frac{AB}{C}$

17. $\frac{A}{y} + \frac{1}{3} = \frac{B}{y}$
$y = 3B - 3A$

18. $\frac{A}{y} - \frac{1}{2} = \frac{1}{3}$
$y = \frac{6A}{5}$

19. $3y - 5ay = 8$
$y = \frac{8}{3 - 5a}$

20. $y^2 - By = 0$
$y = 0$ or $y = B$

21. $A = \frac{1}{2}h(b + y)$
$y = \frac{2A - hb}{h}$

22. $2(b + y) = b$
$y = -\frac{b}{2}$

Calculate the value of $b^2 - 4ac$ for each choice of a, b, and c.

23. $a = 1, b = 2, c = -15$
64

24. $a = 1, b = 8, c = 12$
16

25. $a = 2, b = 5, c = -3$
49

26. $a = 6, b = 7, c = -3$
121

Perform each indicated operation.

27. $(3x - 5) - (5x - 3)$
$-2x - 2$

28. $(2a - 5)(a - 3)$
$2a^2 - 11a + 15$

29. $x^7 \div x^3$
x^4

30. $\frac{x - 3}{5} + \frac{x + 4}{5}$
$\frac{2x + 1}{5}$

31. $\frac{1}{2} \cdot \frac{1}{x}$
$\frac{1}{2x}$

32. $\frac{1}{2} + \frac{1}{x}$
$\frac{x + 2}{2x}$

33. $\frac{1}{2} \div \frac{1}{x}$
$\frac{x}{2}$

34. $\frac{1}{2} - \frac{1}{x}$
$\frac{x - 2}{2x}$

35. $\frac{x - 3}{5} - \frac{x + 4}{5}$
$-\frac{7}{5}$

36. $\frac{3a}{2} \div 2$
$\frac{3a}{4}$

37. $(x - 8)(x + 8)$
$x^2 - 64$

38. $3x(x^2 - 7)$
$3x^3 - 21x$

39. $2a^5 \cdot 5a^9$
$10a^{14}$

40. $x^2 \cdot x^8$
x^{10}

41. $(k - 6)^2$
$k^2 - 12k + 36$

42. $(j + 5)^2$
$j^2 + 10j + 25$

43. $(g - 3) \div (3 - g)$
-1

44. $(6x^3 - 8x^2) \div (2x)$
$3x^2 - 4x$

Solve.

45. ***Present value.*** An investor is interested in the amount or present value that she would have to invest today to receive periodic payments in the future. The present value of \$1 in one year and \$1 in two years with interest rate r compounded annually is given by the formula

$$P = \frac{1}{1 + r} + \frac{1}{(1 + r)^2}.$$

a) Rewrite the formula so that the right-hand side is a single rational expression.

b) Find P if $r = 7\%$.

c) The present value of \$1 per year for the next 10 years is given by the formula

$$P = \frac{1}{1 + r} + \frac{1}{(1 + r)^2} + \frac{1}{(1 + r)^3} + \cdots + \frac{1}{(1 + r)^{10}}.$$

Use this formula to find P if $r = 5\%$.

$P = \frac{r + 2}{(1 + r)^2}$, \$1.81, \$7.72

CHAPTER 8

Powers and Roots

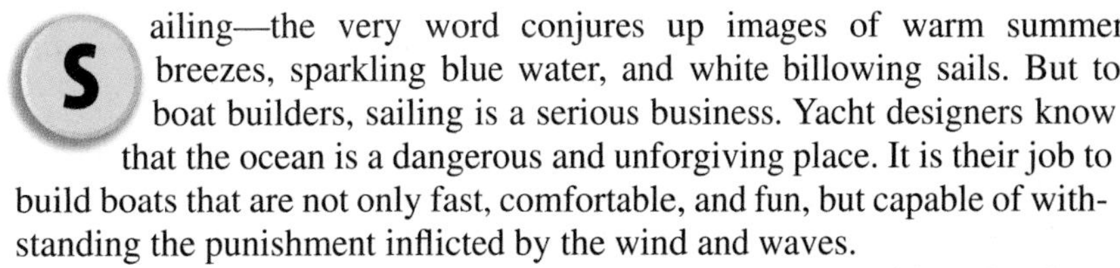

Sailing—the very word conjures up images of warm summer breezes, sparkling blue water, and white billowing sails. But to boat builders, sailing is a serious business. Yacht designers know that the ocean is a dangerous and unforgiving place. It is their job to build boats that are not only fast, comfortable, and fun, but capable of withstanding the punishment inflicted by the wind and waves.

Designing sailboats is a technical balancing act. A good boat has just the right combination of length, width (or beam), sail area, and displacement. A boat can always be made faster by increasing the sail area, but too much sail area increases the chance of capsize. Making the boat wider decreases the chance of capsize, but causes more resistance from the water and slows down the boat.

There are four ratios commonly used to measure performance and safety for a yacht: ballast-displacement ratio, displacement-length ratio, sail area-displacement ratio, and capsize screening value. The formulas for these ratios involve powers and roots, which is the subject of this chapter. In Exercise 71 of Section 8.5 you will find the sail area-displacement ratio for a sailboat.

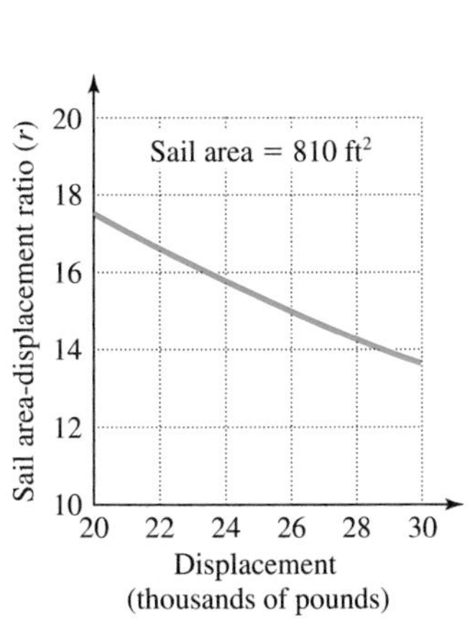

In This Section

- Fundamentals
- Roots and Variables
- Product Rule for Radicals
- Quotient Rule for Radicals

8.1 ROOTS, RADICALS, AND RULES

In Section 5.6 you learned the basic facts about powers. In this section you will study roots and see how powers and roots are related.

Fundamentals

We use the idea of roots to reverse powers. Because $3^2 = 9$ and $(-3)^2 = 9$, both 3 and -3 are square roots of 9. Because $2^4 = 16$ and $(-2)^4 = 16$, both 2 and -2 are fourth roots of 16. Because $2^3 = 8$ and $(-2)^3 = -8$, there is only one real cube root of 8 and only one real cube root of -8. The cube root of 8 is 2 and the cube root of -8 is -2.

*n*th Roots

If $a = b^n$ for a positive integer n, then b is an ***n*th root of *a*.** If $a = b^2$, then b is a **square root** of a. If $a = b^3$, then b is the **cube root** of a.

The following box contains three more definitions concerning roots.

Even Roots, Principal Roots, and Odd Roots

If n is a positive *even* integer and a is positive, then there are two real nth roots of a. We call these roots **even roots.**

The positive even root of a positive number is called the **principal root.**

If n is a positive *odd* integer and a is any real number, then there is only one real nth root of a. We call that root an **odd root.**

So the principal square root of 9 is 3 and the principal fourth root of 16 is 2 and these roots are even roots. Because $2^5 = 32$, the fifth root of 32 is 2 and 2 is an odd root.

We use the **radical symbol** $\sqrt{\ }$ to signify roots.

$\sqrt[n]{a}$

If n is a positive *even* integer and a is positive, then $\sqrt[n]{a}$ denotes the *principal nth root of a.*

If n is a positive *odd* integer, then $\sqrt[n]{a}$ denotes the nth root of a.

If n is any positive integer, then $\sqrt[n]{0} = 0$.

We read $\sqrt[n]{a}$ as "the nth root of a." In the notation $\sqrt[n]{a}$, n is the **index of the radical** and a is the **radicand.** For square roots the index is omitted, and we simply write $\sqrt{a}$.

EXAMPLE 1 **Evaluating radical expressions**

Find the following roots:

a) $\sqrt{25}$ **b)** $\sqrt[3]{-27}$ **c)** $\sqrt[6]{64}$ **d)** $-\sqrt{4}$

Solution

a) Because $5^2 = 25$, $\sqrt{25} = 5$.

b) Because $(-3)^3 = -27$, $\sqrt[3]{-27} = -3$.

c) Because $2^6 = 64$, $\sqrt[6]{64} = 2$.

d) Because $\sqrt{4} = 2$, $-\sqrt{4} = -(\sqrt{4}) = -2$. ■

Calculator Close-Up

We can use the radical symbol to find a square root on a graphing calculator, but for other roots we use the xth root symbol as shown. The xth root symbol is in the MATH menu.

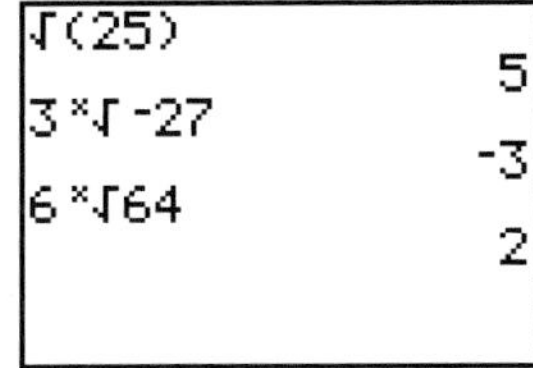

CAUTION In radical notation, $\sqrt{4}$ represents the *principal square root of* 4, so $\sqrt{4} = 2$. Note that -2 is also a square root of 4, but $\sqrt{4} \neq -2$.

Note that even roots of negative numbers are omitted from the definition of nth roots because even powers of real numbers are never negative. So no real number can be an even root of a negative number. Expressions such as

$$\sqrt{-9}, \quad \sqrt[4]{-81}, \quad \text{and} \quad \sqrt[6]{-64}$$

are not real numbers. Square roots of negative numbers will be discussed in Section 9.5 when we discuss the imaginary numbers.

Roots and Variables

Consider the result of squaring a power of x:

$$(x^1)^2 = x^2, \quad (x^2)^2 = x^4, \quad (x^3)^2 = x^6, \quad \text{and} \quad (x^4)^2 = x^8.$$

When a power of x is squared, the exponent is multiplied by 2. So any even power of x is a perfect square.

Perfect Squares

The following expressions are perfect squares:

$$x^2, \quad x^4, \quad x^6, \quad x^8, \quad x^{10}, \quad x^{12}, \quad \ldots$$

Calculator Close-Up

A calculator can provide numerical support for this discussion of roots. Note that $\sqrt{(-3)^2} = 3$ not -3 because $\sqrt{x^2} \neq x$ when x is negative. Note also that the calculator will not evaluate $\sqrt{-3^2}$ because $\sqrt{-3^2} = \sqrt{-9}$.

```
√((-3)²)
                3
√(-3²)
```

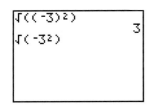

Since taking a square root reverses the operation of squaring, the square root of an even power of x is found by dividing the exponent by 2. Provided x is nonnegative (see Caution below), we have:

$$\sqrt{x^2} = x^1 = x, \quad \sqrt{x^4} = x^2, \quad \sqrt{x^6} = x^3, \quad \text{and} \quad \sqrt{x^8} = x^4.$$

CAUTION If x is negative, equations like $\sqrt{x^2} = x$ and $\sqrt{x^6} = x^3$ are not correct because the radical represents the nonnegative square root but x and x^3 are negative. That is why we assume x is nonnegative.

If a power of x is cubed, the exponent is multiplied by 3:

$$(x^1)^3 = x^3, \quad (x^2)^3 = x^6, \quad (x^3)^3 = x^9, \quad \text{and} \quad (x^4)^3 = x^{12}.$$

So if the exponent is a multiple of 3, we have a perfect cube.

Perfect Cubes

The following expressions are perfect cubes:

$$x^3, \quad x^6, \quad x^9, \quad x^{12}, \quad x^{15}, \quad \ldots$$

Since the cube root reverses the operation of cubing, the cube root of any of these perfect cubes is found by dividing the exponent by 3:

$$\sqrt[3]{x^3} = x^1 = x, \quad \sqrt[3]{x^6} = x^2, \quad \sqrt[3]{x^9} = x^3, \quad \text{and} \quad \sqrt[3]{x^{12}} = x^4.$$

If the exponent is divisible by 4, we have a perfect fourth power, and so on.

EXAMPLE 2 **Roots of exponential expressions**

Find each root. Assume that all variables represent nonnegative real numbers.

a) $\sqrt{x^{22}}$ **b)** $\sqrt[3]{t^{18}}$ **c)** $\sqrt[5]{s^{30}}$

Solution

a) $\sqrt{x^{22}} = x^{11}$ because $(x^{11})^2 = x^{22}$.

b) $\sqrt[3]{t^{18}} = t^6$ because $(t^6)^3 = t^{18}$.

c) $\sqrt[5]{s^{30}} = s^6$ because one-fifth of 30 is 6. ■

Calculator Close-Up

You can illustrate the product rule for radicals with a calculator.

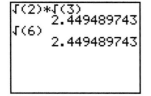

```
√(2)*√(3)
      2.449489743
√(6)
      2.449489743
```

Product Rule for Radicals

Consider the expression $\sqrt{2} \cdot \sqrt{3}$. If we square this product, we get

$$(\sqrt{2} \cdot \sqrt{3})^2 = (\sqrt{2})^2(\sqrt{3})^2 \quad \text{Power of a product rule}$$
$$= 2 \cdot 3 \quad (\sqrt{2})^2 = 2 \text{ and } (\sqrt{3})^2 = 3$$
$$= 6.$$

The number $\sqrt{6}$ is the unique positive number whose square is $\sqrt{6}$. Because we squared $\sqrt{2} \cdot \sqrt{3}$ and obtained 6, we must have $\sqrt{6} = \sqrt{2} \cdot \sqrt{3}$. This example illustrates the product rule for radicals.

Product Rule for Radicals

The nth root of a product is equal to the product of the nth roots. In symbols,

$$\sqrt[n]{ab} = \sqrt[n]{a} \cdot \sqrt[n]{b},$$

provided all of these roots are real numbers.

EXAMPLE 3 **Using the product rule for radicals**

Simplify each radical. Assume that all variables represent positive real numbers.

a) $\sqrt{4y}$ **b)** $\sqrt{3y^8}$

Solution

a) $\sqrt{4y} = \sqrt{4} \cdot \sqrt{y}$ Product rule for radicals

$= 2\sqrt{y}$ Simplify.

b) $\sqrt{3y^8} = \sqrt{3} \cdot \sqrt{y^8}$ Product rule for radicals

$= \sqrt{3} \cdot y^4$ $\sqrt{y^8} = y^4$

$= y^4\sqrt{3}$ A radical is usually written last in a product. ■

Calculator Close-Up

You can illustrate the quotient rule for radicals with a calculator.

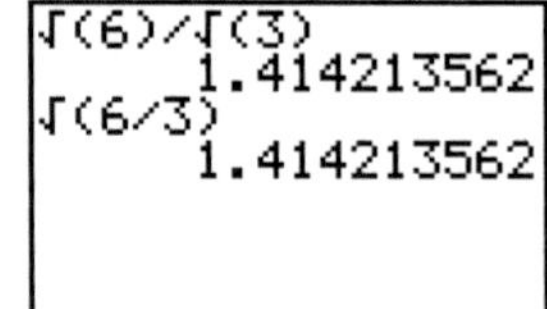

Quotient Rule for Radicals

Because $\sqrt{2} \cdot \sqrt{3} = \sqrt{6}$, we have $\sqrt{6} \div \sqrt{3} = \sqrt{2}$, or

$$\sqrt{2} = \sqrt{\frac{6}{3}} = \frac{\sqrt{6}}{\sqrt{3}}.$$

This example illustrates the quotient rule for radicals.

Quotient Rule for Radicals

The nth root of a quotient is equal to the quotient of the nth roots. In symbols,

$$\sqrt[n]{\frac{a}{b}} = \frac{\sqrt[n]{a}}{\sqrt[n]{b}},$$

provided that all of these roots are real numbers and $b \neq 0$.

In Example 4 we use the quotient rule to simplify radical expressions.

EXAMPLE 4 **Using the quotient rule for radicals**

Simplify each radical. Assume that all variables represent positive real numbers.

a) $\sqrt{\frac{t}{9}}$ **b)** $\sqrt[3]{\frac{x^{21}}{y^6}}$

Solution

a) $\sqrt{\frac{t}{9}} = \frac{\sqrt{t}}{\sqrt{9}}$ Quotient rule for radicals

$= \frac{\sqrt{t}}{3}$

b) $\sqrt[3]{\frac{x^{21}}{y^6}} = \frac{\sqrt[3]{x^{21}}}{\sqrt[3]{y^6}}$ Quotient rule for radicals

$= \frac{x^7}{y^2}$ ■

MATH AT WORK

COMPUTER SYSTEMS DESIGNER

3–2–1–Blast off! Joseph Bursavich watches each Space Shuttle mission with particular attention. He is a Senior Computer Systems Designer for Martin Marietta, writing programs that support the building of the external tank that is the structural backbone of the Space Shuttle for NASA. Each tank, made up of three major parts, is 157 feet long and takes almost two years to build. To date, 74 tanks have been completed, and each has done its job in carrying cargo into space.

Currently, Mr. Bursavich supports a team whose objective is to develop a superlightweight tank made of a mixture of aluminum and other metals. Reducing the weight of the tank is vital to the space station program because a lighter tank means that each mission can carry a greater payload into space. Because of economic considerations, as many as ten external tanks might be produced at one time. Mr. Bursavich writes programs to determine the economic order quantity (EOQ) for components of the external tank. The EOQ depends on setup costs, labor costs, the quantity of the component to be used in one year, the cost of holding stock for one year, and maintenance costs.

In Exercise 83 of this section you will use the formula that Mr. Bursavich uses to determine the EOQ for component parts of the tanks.

WARM-UPS

True or false? Explain your answer.

1. $\sqrt{2} \cdot \sqrt{2} = 2$ True
2. $\sqrt[3]{2} \cdot \sqrt[3]{2} = 2$ False
3. $\sqrt[3]{-27} = -3$ True
4. $\sqrt{-25} = -5$ False
5. $\sqrt[4]{16} = 2$ True
6. $\sqrt{9} = 3$ True
7. $\sqrt{2^9} = 2^3$ False
8. $\sqrt{17} \cdot \sqrt{17} = 289$ False
9. If $w \geq 0$, then $\sqrt{w^2} = w$. True
10. If $t \geq 0$, then $\sqrt[4]{t^{12}} = t^3$. True

8.1 EXERCISES

Reading and Writing *After reading this section, write out the answers to these questions. Use complete sentences.*

1. How do you know if b is an nth root of a?
   If $b^n = a$, then b is an nth root of a.
2. What is a principal root?
   The principal root is the positive even root of a positive number.
3. What is the difference between an even root and an odd root?
   If $b^n = a$, then b is an even root provided n is even or an odd root provided n is odd.
4. What symbol is used to indicate an nth root?
   The nth root of a is written as $\sqrt[n]{a}$.
5. What is the product rule for radicals?
   The product rule for radicals says that $\sqrt[n]{a} \cdot \sqrt[n]{b} = \sqrt[n]{ab}$ provided all of these roots are real.
6. What is the quotient rule for radicals?
   The quotient rule for radicals says that $\sqrt[n]{a}/\sqrt[n]{b} = \sqrt[n]{a/b}$ provided all of these roots are real.

For all of the exercises in this section assume that all variables represent positive real numbers.

Find each root. See Example 1.

7. $\sqrt{36}$ 6
8. $\sqrt{49}$ 7
9. $\sqrt[5]{32}$ 2
10. $\sqrt[4]{81}$ 3
11. $\sqrt[3]{1000}$ 10
12. $\sqrt[4]{16}$ 2
13. $\sqrt[4]{-16}$ Not a real number
14. $\sqrt{-1}$ Not a real number
15. $\sqrt{0}$ 0
16. $\sqrt{1}$ 1
17. $\sqrt[3]{-1}$ -1
18. $\sqrt[3]{0}$ 0
19. $\sqrt[3]{1}$ 1
20. $\sqrt[4]{1}$ 1
21. $\sqrt[4]{-81}$ Not a real number
22. $\sqrt[6]{-64}$ Not a real number
23. $\sqrt[6]{64}$ 2
24. $\sqrt[7]{128}$ 2
25. $\sqrt[5]{-32}$ -2
26. $\sqrt[3]{-125}$ -5
27. $-\sqrt{100}$ -10
28. $-\sqrt{36}$ -6
29. $\sqrt[4]{-50}$ Not a real number
30. $-\sqrt{-144}$ Not a real number

Find each root. See Example 2.

31. $\sqrt{m^2}$ m
32. $\sqrt{m^6}$ m^3
33. $\sqrt[5]{y^{15}}$ y^3
34. $\sqrt[4]{m^8}$ m^2
35. $\sqrt[3]{y^{15}}$ y^5
36. $\sqrt{m^8}$ m^4
37. $\sqrt[3]{m^3}$ m
38. $\sqrt[4]{x^4}$ x
39. $\sqrt{3^6}$ 27
40. $\sqrt{4^2}$ 4
41. $\sqrt{2^{10}}$ 32
42. $\sqrt[3]{2^{99}}$ 2^{33}
43. $\sqrt{5^9}$ 125
44. $\sqrt[3]{10^{18}}$ 1,000,000
45. $\sqrt{10^{20}}$ 10^{10}
46. $\sqrt{10^{18}}$ 10^9

Use the product rule for radicals to simplify each expression. See Example 3.

47. $\sqrt{9y}$ $3\sqrt{y}$
48. $\sqrt{16n}$ $4\sqrt{n}$
49. $\sqrt{4a^2}$ $2a$
50. $\sqrt{36n^2}$ $6n$
51. $\sqrt{x^4y^2}$ x^2y
52. $\sqrt{w^6t^2}$ w^3t
53. $\sqrt{5m^{12}}$ $m^6\sqrt{5}$
54. $\sqrt{7z^{16}}$ $z^8\sqrt{7}$
55. $\sqrt[3]{8y}$ $2\sqrt[3]{y}$
56. $\sqrt[3]{27z^2}$ $3\sqrt[3]{z^2}$
57. $\sqrt[3]{-27w^3}$ $-3w$
58. $\sqrt[3]{-125m^6}$ $-5m^2$
59. $\sqrt[4]{16s}$ $2\sqrt[4]{s}$
60. $\sqrt[4]{81w}$ $3\sqrt[4]{w}$
61. $\sqrt[3]{-125a^9y^6}$ $-5a^3y^2$
62. $\sqrt[3]{-27z^3w^{15}}$ $-3zw^5$

Simplify each radical. See Example 4.

63. $\sqrt{\frac{t}{4}}$ $\frac{\sqrt{t}}{2}$
64. $\sqrt{\frac{w}{36}}$ $\frac{\sqrt{w}}{6}$
65. $\sqrt{\frac{625}{16}}$ $\frac{25}{4}$
66. $\sqrt{\frac{9}{144}}$ $\frac{1}{4}$
67. $\sqrt[3]{\frac{t}{8}}$ $\frac{\sqrt[3]{t}}{2}$
68. $\sqrt[3]{\frac{a}{27}}$ $\frac{\sqrt[3]{a}}{3}$
69. $\sqrt[3]{\frac{-8x^6}{y^3}}$ $\frac{-2x^2}{y}$
70. $\sqrt[3]{\frac{-27y^{36}}{1000}}$ $\frac{-3y^{12}}{10}$

EXAMPLE 2 **Rationalizing denominators**

Simplify each expression by rationalizing its denominator.

a) $\frac{3}{\sqrt{5}}$ **b)** $\frac{\sqrt{3}}{\sqrt{7}}$

Solution

a) Because $\sqrt{5} \cdot \sqrt{5} = 5$, we multiply numerator and denominator by $\sqrt{5}$:

$$\frac{3}{\sqrt{5}} = \frac{3 \cdot \sqrt{5}}{\sqrt{5} \cdot \sqrt{5}} \quad \text{Multiply numerator and denominator by } \sqrt{5}.$$

$$= \frac{3\sqrt{5}}{5} \quad \sqrt{5} \cdot \sqrt{5} = 5$$

b) Because $\sqrt{7} \cdot \sqrt{7} = 7$, multiply the numerator and denominator by $\sqrt{7}$:

$$\frac{\sqrt{3}}{\sqrt{7}} = \frac{\sqrt{3} \cdot \sqrt{7}}{\sqrt{7} \cdot \sqrt{7}} \quad \text{Multiply numerator and denominator by } \sqrt{7}.$$

$$= \frac{\sqrt{21}}{7} \quad \text{Product rule for radicals}$$

■

Simplified Form of a Square Root

When we simplify any expression, we try to write a "simpler" expression that is equivalent to the original. However, one person's idea of simpler is sometimes different from another person's. For a square root the expression must satisfy three conditions to be in simplified form. These three conditions provide specific rules to follow for simplifying square roots.

Simplified Form for Square Roots

An expression involving a square root is in **simplified form** if it has

1. *no* perfect-square factors inside the radical,
2. *no* fractions inside the radical, and
3. *no* radicals in the denominator.

Because a decimal is a form of a fraction, a simplified square root should not contain any decimal numbers. Also, a simplified expression should use the fewest number of radicals possible. So we write $\sqrt{6}$ rather than $\sqrt{2} \cdot \sqrt{3}$ even though both $\sqrt{2}$ and $\sqrt{3}$ are both in simplified form.

EXAMPLE 3 **Simplified form for square roots**

Write each radical expression in simplified form.

a) $\sqrt{300}$ **b)** $\sqrt{\frac{2}{5}}$ **c)** $\frac{\sqrt{10}}{\sqrt{6}}$

Solution

a) We must remove the perfect square factor of 100 from inside the radical:

$$\sqrt{300} = \sqrt{100 \cdot 3} = \sqrt{100} \cdot \sqrt{3} = 10\sqrt{3}$$

Calculator Close-Up

Using a calculator to check simplification problems will help you to understand the concepts.

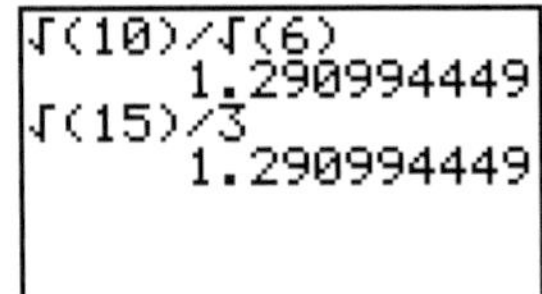

b) We first use the quotient rule to remove the fraction $\frac{2}{5}$ from inside the radical:

$$\sqrt{\frac{2}{5}} = \frac{\sqrt{2}}{\sqrt{5}} \quad \text{Quotient rule for radicals}$$

$$= \frac{\sqrt{2} \cdot \sqrt{5}}{\sqrt{5} \cdot \sqrt{5}} \quad \text{Rationalize the denominator.}$$

$$= \frac{\sqrt{10}}{5} \quad \text{Product rule for radicals}$$

c) The numerator and denominator have a common factor of $\sqrt{2}$:

$$\frac{\sqrt{10}}{\sqrt{6}} = \frac{\sqrt{2} \cdot \sqrt{5}}{\sqrt{2} \cdot \sqrt{3}} \quad \text{Product rule for radicals}$$

$$= \frac{\sqrt{5}}{\sqrt{3}} \quad \text{Reduce.}$$

$$= \frac{\sqrt{5} \cdot \sqrt{3}}{\sqrt{3} \cdot \sqrt{3}} \quad \text{Rationalize the denominator.}$$

$$= \frac{\sqrt{15}}{3} \quad \text{Product rule for radicals}$$

Note that we could have simplified $\frac{\sqrt{10}}{\sqrt{6}}$ by first using the quotient rule to get $\frac{\sqrt{10}}{\sqrt{6}} = \sqrt{\frac{10}{6}}$ and then reducing $\frac{10}{6}$. Another way to simplify $\frac{\sqrt{10}}{\sqrt{6}}$ is to first multiply the numerator and denominator by $\sqrt{6}$. You should try these alternatives. Of course, the simplified form is $\frac{\sqrt{15}}{3}$ by any method. ■

In Example 4 we simplify some expressions involving variables. Remember that *any exponential expression with an even exponent is a perfect square.*

EXAMPLE 4 **Radicals containing variables**

Simplify each expression. All variables represent nonnegative real numbers.

a) $\sqrt{x^3}$

b) $\sqrt{8a^9}$

c) $\sqrt{18a^4b^7}$

Solution

a) $\sqrt{x^3} = \sqrt{x^2 \cdot x}$ — The largest perfect square factor of x^3 is x^2.

$= \sqrt{x^2} \cdot \sqrt{x}$ — Product rule for radicals

$= x\sqrt{x}$ — For any nonnegative x, $\sqrt{x^2} = x$.

b) $\sqrt{8a^9} = \sqrt{4a^8} \cdot \sqrt{2a}$ — The largest perfect square factor of $8a^9$ is $4a^8$.

$= 2a^4\sqrt{2a}$ — $\sqrt{4a^8} = 2a^4$

c) $\sqrt{18a^4b^7} = \sqrt{9a^4b^6} \cdot \sqrt{2b}$ — Factor out the perfect squares.

$= 3a^2b^3\sqrt{2b}$ — $\sqrt{9a^4b^6} = 3a^2b^3$ ■

If square roots of variables appear in the denominator, then we rationalize the denominator.

EXAMPLE 5 **Radicals containing variables**

Simplify each expression. All variables represent positive real numbers.

a) $\dfrac{5}{\sqrt{a}}$

b) $\sqrt{\dfrac{a}{b}}$

c) $\dfrac{\sqrt{2}}{\sqrt{6a}}$

Helpful Hint

If you are going to compute the value of a radical expression with a calculator, it doesn't matter if the denominator is rational. However, rationalizing the denominator provides another opportunity to practice building up the denominator of a fraction and multiplying radicals.

Solution

a) $\dfrac{5}{\sqrt{a}} = \dfrac{5 \cdot \sqrt{a}}{\sqrt{a} \cdot \sqrt{a}}$ Multiply numerator and denominator by $\sqrt{a}$.

$= \dfrac{5\sqrt{a}}{a}$ $\sqrt{a} \cdot \sqrt{a} = a$

b) $\sqrt{\dfrac{a}{b}} = \dfrac{\sqrt{a}}{\sqrt{b}}$ Quotient rule for radicals

$= \dfrac{\sqrt{a} \cdot \sqrt{b}}{\sqrt{b} \cdot \sqrt{b}}$ Rationalize the denominator.

$= \dfrac{\sqrt{ab}}{b}$ Product rule for radicals

c) $\dfrac{\sqrt{2}}{\sqrt{6a}} = \dfrac{\sqrt{2} \cdot \sqrt{6a}}{\sqrt{6a} \cdot \sqrt{6a}}$ Rationalize the denominator.

$= \dfrac{\sqrt{12a}}{6a}$ Product rule for radicals

$= \dfrac{\sqrt{4} \cdot \sqrt{3a}}{6a}$ Factor out the perfect square.

$= \dfrac{2\sqrt{3a}}{6a}$ $\sqrt{4} = 2$

$= \dfrac{2\sqrt{3a}}{2 \cdot 3a}$ Factor the denominator.

$= \dfrac{\sqrt{3a}}{3a}$ Divide out the common factor 2. ■

CAUTION Do not attempt to reduce an expression like the one in Example 5(c):

$$\frac{\sqrt{3a}}{3a}$$

You cannot divide out common factors when one is inside a radical.

WARM-UPS

True or false? Explain your answer.

1. $\sqrt{20} = 2\sqrt{5}$ True **2.** $\sqrt{18} = 9\sqrt{2}$ False

3. $\frac{1}{\sqrt{3}} = \frac{\sqrt{3}}{3}$ True **4.** $\frac{9}{4} = \frac{3}{2}$ False

5. $\sqrt{a^3} = a\sqrt{a}$ for any positive value of a. True

6. $\sqrt{a^9} = a^3$ for any positive value of a. False

7. $\sqrt{y^{17}} = y^8\sqrt{y}$ for any positive value of y. True

8. $\frac{\sqrt{6}}{2} = \sqrt{3}$ False **9.** $\sqrt{4} = \sqrt{2}$ False **10.** $\sqrt{283} = 17$ False

8.2 EXERCISES

Reading and Writing *After reading this section, write out the answers to these questions. Use complete sentences.*

1. How do we simplify a radical with the product rule?
We use the product rule to factor out a perfect square from inside a square root.

2. Which integers are perfect squares?
The perfect squares are 1, 4, 9, 16, 25, and so on.

3. What does it mean to rationalize a denominator?
To rationalize a denominator means to rewrite the expression so that the denominator is a rational number.

4. What is simplified form for a square root?
A square root in simplified form has no perfect squares or fractions inside the radical and no radicals in the denominator.

5. How do you simplify a square root that contains a variable?
To simplify square roots containing variables use the same techniques as we use on square roots of numbers.

6. How can you tell if an exponential expression is a perfect square?
Any even power of a variable is a perfect square.

Assume that all variables in the exercises represent positive real numbers.

Simplify each radical. See Example 1.

7. $\sqrt{8}$ $2\sqrt{2}$ **8.** $\sqrt{20}$ $2\sqrt{5}$ **9.** $\sqrt{24}$ $2\sqrt{6}$

10. $\sqrt{75}$ $5\sqrt{3}$ **11.** $\sqrt{28}$ $2\sqrt{7}$ **12.** $\sqrt{40}$ $2\sqrt{10}$

13. $\sqrt{90}$ $3\sqrt{10}$ **14.** $\sqrt{200}$ $10\sqrt{2}$ **15.** $\sqrt{500}$ $10\sqrt{5}$

16. $\sqrt{98}$ $7\sqrt{2}$ **17.** $\sqrt{150}$ $5\sqrt{6}$ **18.** $\sqrt{120}$ $2\sqrt{30}$

Simplify each expression by rationalizing the denominator. See Example 2.

19. $\frac{1}{\sqrt{5}}$ $\frac{\sqrt{5}}{5}$ **20.** $\frac{1}{\sqrt{6}}$ $\frac{\sqrt{6}}{6}$ **21.** $\frac{3}{\sqrt{2}}$ $\frac{3\sqrt{2}}{2}$

22. $\frac{4}{\sqrt{3}}$ $\frac{4\sqrt{3}}{3}$ **23.** $\frac{\sqrt{3}}{\sqrt{2}}$ $\frac{\sqrt{6}}{2}$ **24.** $\frac{\sqrt{7}}{\sqrt{6}}$ $\frac{\sqrt{42}}{6}$

25. $\frac{-3}{\sqrt{10}}$ $\frac{-3\sqrt{10}}{10}$ **26.** $\frac{-4}{\sqrt{5}}$ $-\frac{4\sqrt{5}}{5}$ **27.** $\frac{-10}{\sqrt{17}}$ $\frac{-10\sqrt{17}}{17}$

28. $\frac{-3}{\sqrt{19}}$ $-\frac{3\sqrt{19}}{19}$ **29.** $\frac{\sqrt{11}}{\sqrt{7}}$ $\frac{\sqrt{77}}{7}$ **30.** $\frac{\sqrt{10}}{\sqrt{3}}$ $\frac{\sqrt{30}}{3}$

Write each radical expression in simplified form. See Example 3.

31. $\sqrt{63}$ $3\sqrt{7}$ **32.** $\sqrt{48}$ $4\sqrt{3}$ **33.** $\sqrt{\frac{3}{2}}$ $\frac{\sqrt{6}}{2}$

34. $\sqrt{\frac{3}{5}}$ $\frac{\sqrt{15}}{5}$ **35.** $\sqrt{\frac{5}{8}}$ $\frac{\sqrt{10}}{4}$ **36.** $\sqrt{\frac{5}{18}}$ $\frac{\sqrt{10}}{6}$

37. $\dfrac{\sqrt{6}}{\sqrt{10}}$ $\dfrac{\sqrt{15}}{5}$

38. $\dfrac{\sqrt{12}}{\sqrt{20}}$ $\dfrac{\sqrt{15}}{5}$

39. $\dfrac{\sqrt{75}}{\sqrt{3}}$ 5

40. $\dfrac{\sqrt{45}}{\sqrt{5}}$ 3

41. $\dfrac{\sqrt{15}}{\sqrt{10}}$ $\dfrac{\sqrt{6}}{2}$

42. $\dfrac{\sqrt{30}}{\sqrt{21}}$ $\dfrac{\sqrt{70}}{7}$

Simplify each expression. See Example 4.

43. $\sqrt{a^8}$ a^4

44. $\sqrt{y^{10}}$ y^5

45. $\sqrt{a^9}$ $a^4\sqrt{a}$

46. $\sqrt{t^{11}}$ $t^5\sqrt{t}$

47. $\sqrt{8a^6}$ $2a^3\sqrt{2}$

48. $\sqrt{18w^9}$ $3w^4\sqrt{2w}$

49. $\sqrt{20a^4b^9}$ $2a^2b^4\sqrt{5b}$

50. $\sqrt{12x^2y^3}$ $2xy\sqrt{3y}$

51. $\sqrt{27x^3y^3}$ $3xy\sqrt{3xy}$

52. $\sqrt{45x^5y^3}$ $3x^2y\sqrt{5xy}$

53. $\sqrt{27a^3b^8c^2}$ $3ab^4c\sqrt{3a}$

54. $\sqrt{125x^3y^9z^4}$ $5xy^4z^2\sqrt{5xy}$

Simplify each expression. See Example 5.

55. $\dfrac{1}{\sqrt{x}}$ $\dfrac{\sqrt{x}}{x}$

56. $\dfrac{1}{\sqrt{2x}}$ $\dfrac{\sqrt{2x}}{2x}$

57. $\dfrac{\sqrt{2}}{\sqrt{3a}}$ $\dfrac{\sqrt{6a}}{3a}$

58. $\dfrac{\sqrt{5}}{\sqrt{2b}}$ $\dfrac{\sqrt{10b}}{2b}$

59. $\dfrac{\sqrt{3}}{\sqrt{15y}}$ $\dfrac{\sqrt{5y}}{5y}$

60. $\dfrac{\sqrt{5}}{\sqrt{10x}}$ $\dfrac{\sqrt{2x}}{2x}$

61. $\sqrt{\dfrac{3x}{2y}}$ $\dfrac{\sqrt{6xy}}{2y}$

62. $\sqrt{\dfrac{6}{5w}}$ $\dfrac{\sqrt{30w}}{5w}$

63. $\sqrt{\dfrac{10y}{15x}}$ $\dfrac{\sqrt{6xy}}{3x}$

64. $\sqrt{\dfrac{6x}{4y}}$ $\dfrac{\sqrt{6xy}}{2y}$

65. $\sqrt{\dfrac{8x^3}{y}}$ $\dfrac{2x\sqrt{2xy}}{y}$

66. $\sqrt{\dfrac{8s^5}{t}}$ $\dfrac{2s^2\sqrt{2st}}{t}$

Simplify each expression.

67. $\sqrt{80x^3}$ $4x\sqrt{5x}$

68. $\sqrt{90y^{80}}$ $3y^{40}\sqrt{10}$

69. $\sqrt{9y^9x^{15}}$ $3y^4x^7\sqrt{yx}$

70. $\sqrt{48x^2y^7}$ $4xy^3\sqrt{3y}$

71. $\dfrac{20x^6}{\sqrt{5x^5}}$ $4x^3\sqrt{5x}$

72. $\dfrac{7x^7y}{\sqrt{7x^9}}$ $x^2y\sqrt{7x}$

73. $\dfrac{-22p^2}{p\sqrt{6pq}}$ $\dfrac{-11\sqrt{6pq}}{3q}$

74. $\dfrac{-30t^5}{t^2\sqrt{3t}}$ $-10t^2\sqrt{3t}$

75. $\dfrac{a^3b^7\sqrt{a^2b^3c^4}}{\sqrt{abc}}$ $a^3b^8c\sqrt{ac}$

76. $\dfrac{3n^4b^5\sqrt{n^2b^2c^7}}{\sqrt{nbc}}$ $3n^4b^5c^3\sqrt{nb}$

77. $\dfrac{\sqrt{4xy^2}}{x^9y^3\sqrt{6xy^3}}$ $\dfrac{\sqrt{6y}}{3x^9y^4}$

78. $\dfrac{\sqrt{8m^3n^2}}{m^3n^2\sqrt{6mn^3}}$ $\dfrac{2\sqrt{3n}}{3m^2n^3}$

Use a calculator to evaluate each expression.

79. $\dfrac{1}{\sqrt{2}} - \dfrac{\sqrt{2}}{2}$ 0

80. $\dfrac{\sqrt{2}}{\sqrt{3}} - \dfrac{\sqrt{6}}{3}$ 0

81. $\dfrac{\sqrt{6}}{\sqrt{2}} - \sqrt{3}$ 0

82. $2 - \dfrac{\sqrt{20}}{\sqrt{5}}$ 0

Solve each problem.

83. ***Economic order quantity.*** The formula for economic order quantity

$$E = \sqrt{\frac{2AS}{I}}$$

was used in Exercise 83 of Section 8.1.

a) Express the right-hand side in simplified form.
$E = \dfrac{\sqrt{2AIS}}{I}$

b) Find E when $A = 23$, $S = \$4566$, and $I = \$80$.
51.2

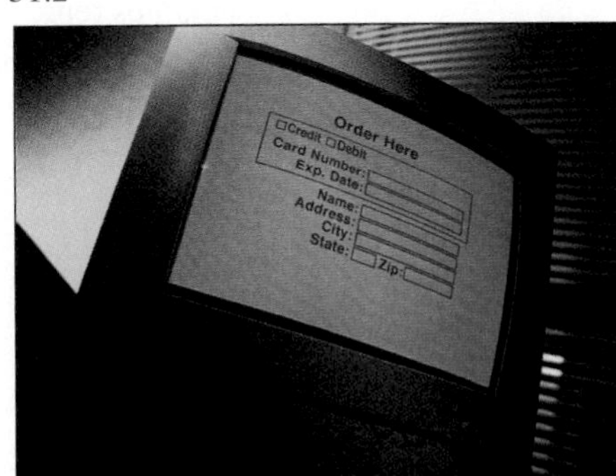

FIGURE FOR EXERCISE 83

84. ***Landing speed.*** Aircraft design engineers determine the proper landing speed V (in ft/sec) by using the formula

$$V = \sqrt{\frac{841L}{CS}},$$

where L is the gross weight of the aircraft in pounds, C is the coefficient of lift, and S is the wing surface area in square feet.

a) Express the right-hand side in simplified form.
$V = \frac{29\sqrt{LCS}}{CS}$

b) Find V when $L = 8600$ pounds, $C = 2.81$, and $S = 200$ square feet.
113.4

FIGURE FOR EXERCISE 84

8.3 OPERATIONS WITH RADICALS

In This Section

- Adding and Subtracting Radicals
- Multiplying Radicals
- Dividing Radicals

In this section you will learn how to perform the basic operations of arithmetic with radical expressions.

Adding and Subtracting Radicals

Consider the sum $2\sqrt{3} + 5\sqrt{3}$. When you studied like terms, you learned that

$$2x + 5x = 7x$$

is true for any value of x. If $x = \sqrt{3}$, then we get

$$2\sqrt{3} + 5\sqrt{3} = 7\sqrt{3}.$$

The expressions $2\sqrt{3}$ and $5\sqrt{3}$ are called **like radicals** because they can be combined just as we combine like terms. We can also add or subtract other radicals as long as they have *the same index and the same radicand.* For example,

$$8\sqrt[3]{w} - 6\sqrt[3]{w} = 2\sqrt[3]{w}.$$

EXAMPLE 1 **Combining like radicals**

Simplify the following expressions by combining like radicals. Assume that the variables represent nonnegative numbers.

a) $2\sqrt{5} + 7\sqrt{5}$ **b)** $3\sqrt[3]{2} - 9\sqrt[3]{2}$ **c)** $\sqrt{2} - 5\sqrt{a} + 4\sqrt{2} - 3\sqrt{a}$

Solution

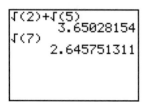

a) $2\sqrt{5} + 7\sqrt{5} = 9\sqrt{5}$

b) $3\sqrt[3]{2} - 9\sqrt[3]{2} = -6\sqrt[3]{2}$

c) $\sqrt{2} - 5\sqrt{a} + 4\sqrt{2} - 3\sqrt{a} = \sqrt{2} + 4\sqrt{2} - 5\sqrt{a} - 3\sqrt{a}$
$= 5\sqrt{2} - 8\sqrt{a}$ Combine like radicals only. ■

Calculator Close-Up

A calculator can show you when a rule is applied incorrectly. For example, we cannot say that $\sqrt{2} + \sqrt{5}$ is equal to $\sqrt{7}$.

```
√(2)+√(5)
        3.65028154
√(7)
       2.645751311
```

CAUTION We cannot combine the terms in the expressions

$$\sqrt{2} + \sqrt{5}, \quad \sqrt[3]{x} - \sqrt{x}, \quad \text{or} \quad 3\sqrt{2} + \sqrt[4]{6}$$

because the radicands are unequal in the first expression, the indices are unequal in the second, and both the radicands and the indices are unequal in the third.

In Example 2 we simplify the radicals before adding or subtracting.

EXAMPLE 2 **Simplifying radicals before combining like terms**

Simplify. Assume that all variables represent nonnegative real numbers.

a) $\sqrt{12} + \sqrt{75}$ **b)** $\sqrt{8x^3} + x\sqrt{18x}$ **c)** $\dfrac{4}{\sqrt{2}} - \dfrac{\sqrt{3}}{\sqrt{6}}$

Solution

a) $\sqrt{12} + \sqrt{75} = \sqrt{4} \cdot \sqrt{3} + \sqrt{25} \cdot \sqrt{3}$ Product rule for radicals

$= 2\sqrt{3} + 5\sqrt{3}$ Simplify.

$= 7\sqrt{3}$ Combine like radicals.

b) $\sqrt{8x^3} + x\sqrt{18x} = \sqrt{4x^2} \cdot \sqrt{2x} + x\sqrt{9} \cdot \sqrt{2x}$ Product rule for radicals

$= 2x\sqrt{2x} + 3x\sqrt{2x}$ Simplify.

$= 5x\sqrt{2x}$ Combine like radicals.

c) $\dfrac{4}{\sqrt{2}} - \dfrac{\sqrt{3}}{\sqrt{6}} = \dfrac{4 \cdot \sqrt{2}}{\sqrt{2} \cdot \sqrt{2}} - \dfrac{\sqrt{3} \cdot \sqrt{6}}{\sqrt{6} \cdot \sqrt{6}}$ Rationalize the denominators.

$= \dfrac{4\sqrt{2}}{2} - \dfrac{\sqrt{18}}{6}$ Simplify.

$= \dfrac{4\sqrt{2}}{2} - \dfrac{3\sqrt{2}}{6}$ $\sqrt{18} = \sqrt{9} \cdot \sqrt{2} = 3\sqrt{2}$

$= \dfrac{4\sqrt{2}}{2} - \dfrac{\sqrt{2}}{2}$ Reduce $\dfrac{3\sqrt{2}}{6}$ to $\dfrac{\sqrt{2}}{2}$.

$= \dfrac{3\sqrt{2}}{2}$ $4\sqrt{2} - \sqrt{2} = 3\sqrt{2}$

Multiplying Radicals

We have been using the product rule for radicals

$$\sqrt[n]{ab} = \sqrt[n]{a} \cdot \sqrt[n]{b}$$

to express a root of a product as a product of the roots of the factors. When we rationalized denominators in Section 8.2, we used the product rule to multiply radicals. We will now study multiplication of radicals in more detail.

EXAMPLE 3 **Multiplying radical expressions**

Multiply and simplify. Assume that variables represent positive numbers.

a) $\sqrt{2} \cdot \sqrt{5}$ **b)** $2\sqrt{5} \cdot 3\sqrt{6}$ **c)** $\sqrt{2a^2} \cdot \sqrt{6a}$ **d)** $\sqrt[3]{4} \cdot \sqrt[3]{2}$

Solution

a) $\sqrt{2} \cdot \sqrt{5} = \sqrt{2 \cdot 5}$ Product rule for radicals

$= \sqrt{10}$

b) $2\sqrt{5} \cdot 3\sqrt{6} = 2 \cdot 3 \cdot \sqrt{5} \cdot \sqrt{6}$

$= 6\sqrt{30}$ Product rule for radicals

c) $\sqrt{2a^2} \cdot \sqrt{6a} = \sqrt{12a^3}$ Product rule for radicals

$= \sqrt{4a^2} \cdot \sqrt{3a}$ Factor out the perfect square.

$= 2a\sqrt{3a}$ Simplify.

d) $\sqrt[3]{4} \cdot \sqrt[3]{2} = \sqrt[3]{8}$ Product rule for radicals

$= 2$

Helpful Hint

Students often write

$$\sqrt{15} \cdot \sqrt{15} = \sqrt{225} = 15.$$

Although this is correct, you should get used to the idea that

$$\sqrt{15} \cdot \sqrt{15} = 15.$$

Because the definition of square root, $\sqrt{a} \cdot \sqrt{a} = a$ for any positive number a.

A sum such as $\sqrt{6} + \sqrt{2}$ is in its simplest form, and so it is treated like a binomial when it occurs in a product.

EXAMPLE 4 **Using the distributive property with radicals**

Find the product: $3\sqrt{3}(\sqrt{6} + \sqrt{2})$

Solution

$$
\begin{aligned}
3\sqrt{3}(\sqrt{6} + \sqrt{2}) &= 3\sqrt{3} \cdot \sqrt{6} + 3\sqrt{3} \cdot \sqrt{2} && \text{Distributive property} \\
&= 3\sqrt{18} + 3\sqrt{6} && \text{Product rule for radicals} \\
&= 3 \cdot 3\sqrt{2} + 3\sqrt{6} && \sqrt{18} = \sqrt{9} \cdot \sqrt{2} = 3\sqrt{2} \\
&= 9\sqrt{2} + 3\sqrt{6}
\end{aligned}
$$

In Example 5 we use the FOIL method to find products of expressions involving radicals.

EXAMPLE 5 **Using FOIL to multiply radicals**

Multiply and simplify.

a) $(\sqrt{3} + 5)(\sqrt{3} - 2)$ **b)** $(\sqrt{5} - 2)(\sqrt{5} + 2)$

c) $(2\sqrt{3} + \sqrt{5})(\sqrt{3} - \sqrt{5})$

Solution

a)
$$
\begin{aligned}
(\sqrt{3} + 5)(\sqrt{3} - 2) &= \overset{F}{3} - \overset{O}{2\sqrt{3}} + \overset{I}{5\sqrt{3}} - \overset{L}{10} \\
&= 3\sqrt{3} - 7 && \text{Add the like terms.}
\end{aligned}
$$

b) The product $(\sqrt{5} - 2)(\sqrt{5} + 2)$ is the product of a sum and a difference. Recall that $(a - b)(a + b) = a^2 - b^2$.

$$
\begin{aligned}
(\sqrt{5} - 2)(\sqrt{5} + 2) &= (\sqrt{5})^2 - 2^2 \\
&= 5 - 4 \\
&= 1
\end{aligned}
$$

c)
$$
\begin{aligned}
(2\sqrt{3} + \sqrt{5})(\sqrt{3} - \sqrt{5}) &= \overset{F}{2\sqrt{3}\sqrt{3}} - \overset{O}{2\sqrt{3}\sqrt{5}} + \overset{I}{\sqrt{5}\sqrt{3}} - \overset{L}{\sqrt{5}\sqrt{5}} \\
&= 6 - 2\sqrt{15} + \sqrt{15} - 5 \\
&= 1 - \sqrt{15}
\end{aligned}
$$

Study Tip

Review, review, review! Don't wait until the end of a chapter to review. Do a little review every time you study for this course.

Dividing Radicals

In Section 8.1 we used the quotient rule for radicals to write a square root of a quotient as a quotient of square roots. We can also use the quotient rule for radicals to divide radicals of the same index. For example,

$$\frac{\sqrt{10}}{\sqrt{2}} = \sqrt{\frac{10}{2}} = \sqrt{5}.$$

Division of radicals is simplest when the quotient of the radicands is a whole number, as it was in the example $\sqrt{10} \div \sqrt{2} = \sqrt{5}$. If the quotient of the radicands is not a whole number, then we divide by rationalizing the denominator, as shown in Example 6.

EXAMPLE 6 **Dividing radicals**

Divide and simplify.

a) $\sqrt{30} \div \sqrt{3}$ **b)** $(5\sqrt{2}) \div (2\sqrt{5})$ **c)** $(15\sqrt{6}) \div (3\sqrt{2})$

Solution

a) $\sqrt{30} \div \sqrt{3} = \dfrac{\sqrt{30}}{\sqrt{3}} = \sqrt{10}$

b) $(5\sqrt{2}) \div (2\sqrt{5}) = \dfrac{5\sqrt{2}}{2\sqrt{5}} = \dfrac{5\sqrt{2} \cdot \sqrt{5}}{2\sqrt{5} \cdot \sqrt{5}}$ Rationalize the denominator.

$= \dfrac{5\sqrt{10}}{2 \cdot 5}$ Product rule for radicals

$= \dfrac{\sqrt{10}}{2}$ Reduce.

Note that $\sqrt{10} \div 2 \neq \sqrt{5}$.

c) $(15\sqrt{6}) \div (3\sqrt{2}) = \dfrac{15\sqrt{6}}{3\sqrt{2}} = 5\sqrt{3}$ $\sqrt{6} \div \sqrt{2} = \sqrt{3}$ ■

CAUTION You can use the quotient rule to divide roots of the same index only. For example,

$$\frac{\sqrt{14}}{\sqrt{2}} = \sqrt{7} \quad \text{but} \quad \frac{\sqrt{14}}{2} \neq \sqrt{7}.$$

In Example 7 we simplify expressions with radicals in the numerator and whole numbers in the denominator.

EXAMPLE 7 **Simplifying radical expressions**

Simplify.

a) $\dfrac{4 - \sqrt{20}}{4}$ **b)** $\dfrac{-6 + \sqrt{27}}{3}$

Helpful Hint

The expressions in Example 7 are the types of expression that you must simplify when you learn the quadratic formula in Chapter 9.

Solution

a) $\dfrac{4 - \sqrt{20}}{4} = \dfrac{4 - 2\sqrt{5}}{4}$ $\sqrt{20} = \sqrt{4} \cdot \sqrt{5} = 2\sqrt{5}$

$= \dfrac{\not{2}(2 - \sqrt{5})}{\not{2} \cdot 2}$ Factor out the GCF, 2.

$= \dfrac{2 - \sqrt{5}}{2}$ Reduce.

b) $\dfrac{-6 + \sqrt{27}}{3} = \dfrac{-6 + 3\sqrt{3}}{3} = \dfrac{\not{3}(-2 + \sqrt{3})}{\not{3}} = -2 + \sqrt{3}$ ■

CAUTION In the expression $\frac{2 - \sqrt{5}}{2}$ you cannot divide out the remaining 2's because 2 is not a *factor* of the numerator.

In Example 5(b) we used the rule for the product of a sum and a difference to get $(\sqrt{5} - 2)(\sqrt{5} + 2) = 1$. If we apply the same rule to other products of this type, we also get a rational number as the result. For example,

$$(\sqrt{7} + \sqrt{2})(\sqrt{7} - \sqrt{2}) = 7 - 2 = 5.$$

Expressions such as $\sqrt{5} + 2$ and $\sqrt{5} - 2$ are called **conjugates** of each other. The conjugate of $\sqrt{7} + \sqrt{2}$ is $\sqrt{7} - \sqrt{2}$. We can use conjugates to simplify a radical expression that has a sum or a difference in its denominator.

EXAMPLE 8 **Rationalizing the denominator using conjugates**

Simplify each expression.

a) $\dfrac{\sqrt{3}}{\sqrt{7} - \sqrt{2}}$ **b)** $\dfrac{4}{6 + \sqrt{2}}$

Helpful Hint

The word conjugate is used in many contexts in mathematics. According to the dictionary, conjugate means joined together, especially as in a pair.

Solution

a) $\dfrac{\sqrt{3}}{\sqrt{7} - \sqrt{2}} = \dfrac{\sqrt{3}(\sqrt{7} + \sqrt{2})}{(\sqrt{7} - \sqrt{2})(\sqrt{7} + \sqrt{2})}$ Multiply by $\sqrt{7} + \sqrt{2}$, the conjugate of $\sqrt{7} - \sqrt{2}$.

$= \dfrac{\sqrt{21} + \sqrt{6}}{7 - 2}$

$= \dfrac{\sqrt{21} + \sqrt{6}}{5}$

b) $\dfrac{4}{6 + \sqrt{2}} = \dfrac{4(6 - \sqrt{2})}{(6 + \sqrt{2})(6 - \sqrt{2})}$ Multiply by $6 - \sqrt{2}$, the conjugate of $6 + \sqrt{2}$.

$= \dfrac{24 - 4\sqrt{2}}{36 - 2}$

$= \dfrac{24 - 4\sqrt{2}}{34}$

$= \dfrac{\not{2}(12 - 2\sqrt{2})}{\not{2} \cdot 17} = \dfrac{12 - 2\sqrt{2}}{17}$ ■

WARM-UPS

True or false? Explain your answer.

1. $\sqrt{9} + \sqrt{16} = \sqrt{25}$ False

2. $\dfrac{5}{\sqrt{5}} = \sqrt{5}$ True

3. $\sqrt{10} \div 2 = \sqrt{5}$ False

4. $3\sqrt{2} \cdot 3\sqrt{2} = 9\sqrt{2}$ False

5. $3\sqrt{5} \cdot 3\sqrt{2} = 9\sqrt{10}$ True

6. $\sqrt{5} + 3\sqrt{5} = 4\sqrt{10}$ False

7. $\dfrac{\sqrt{15}}{3} = \sqrt{5}$ False

8. $\sqrt{2} \div \sqrt{6} = \sqrt{3}$ False

9. $\dfrac{\sqrt{27}}{\sqrt{3}} = 3$ True

10. $(\sqrt{3} - 1)(\sqrt{3} + 1) = 2$ True

8.3 EXERCISES

Reading and Writing *After reading this section, write out the answers to these questions. Use complete sentences.*

1. What are like radicals?
Like radicals have the same index and same radicand.

2. How do we combine like radicals?
We combine like radicals just like we combine like terms.

3. What operations can be performed with radicals?
Radicals can be added, subtracted, multiplied, and divided.

4. What method can we use to multiply a sum of two square roots by a sum of two square roots?
We use the FOIL method for multiplying sums of radicals.

5. What radical expressions are conjugates of each other?
Radical expressions such as $\sqrt{a} + \sqrt{b}$ and $\sqrt{a} - \sqrt{b}$ are conjugates.

6. How do you rationalize a denominator that contains a sum of two radicals?
Rationalizing a denominator that contains a sum can be done by multiplying the numerator and denominator by the conjugate of the denominator.

Assume that all variables in these exercises represent only positive real numbers.

Simplify each expression by combining like radicals. See Example 1.

7. $4\sqrt{5} + 3\sqrt{5}$ $7\sqrt{5}$
8. $\sqrt{2} + \sqrt{2}$ $2\sqrt{2}$
9. $\sqrt[3]{2} + \sqrt[3]{2}$ $2\sqrt[3]{2}$
10. $4\sqrt[3]{6} - 7\sqrt[3]{6}$ $-3\sqrt[3]{6}$
11. $3u\sqrt{11} + 5u\sqrt{11}$ $8u\sqrt{11}$
12. $9m\sqrt{5} - 12m\sqrt{5}$ $-3m\sqrt{5}$
13. $\sqrt{2} + \sqrt{3} - 5\sqrt{2} + 3\sqrt{3}$ $4\sqrt{3} - 4\sqrt{2}$
14. $8\sqrt{6} - \sqrt{2} - 3\sqrt{6} + 5\sqrt{2}$ $5\sqrt{6} + 4\sqrt{2}$
15. $3\sqrt{y} - \sqrt{x} - 4\sqrt{y} - 3\sqrt{x}$ $-4\sqrt{x} - \sqrt{y}$
16. $5\sqrt{7} - \sqrt{a} + 3\sqrt{7} - 5\sqrt{a}$ $8\sqrt{7} - 6\sqrt{a}$
17. $3x\sqrt{y} - \sqrt{a} + 2x\sqrt{y} + 3\sqrt{a}$ $5x\sqrt{y} + 2\sqrt{a}$
18. $a\sqrt{b} + 5a\sqrt{b} - 2\sqrt{a} + 3\sqrt{a}$ $6a\sqrt{b} + \sqrt{a}$

Simplify each expression. See Example 2.

19. $\sqrt{24} + \sqrt{54}$
$5\sqrt{6}$
20. $\sqrt{12} + \sqrt{27}$
$5\sqrt{3}$
21. $2\sqrt{27} - 4\sqrt{75}$
$-14\sqrt{3}$
22. $\sqrt{2} - \sqrt{18}$
$-2\sqrt{2}$
23. $\sqrt{3a} - \sqrt{12a}$
$-\sqrt{3a}$
24. $\sqrt{5w} - \sqrt{45w}$
$-2\sqrt{5w}$
25. $\sqrt{x^3} + x\sqrt{4x}$
$3x\sqrt{x}$
26. $\sqrt{27x^3} + 5x\sqrt{12x}$
$13x\sqrt{3x}$
27. $\dfrac{1}{\sqrt{3}} + \dfrac{\sqrt{2}}{\sqrt{6}}$
$\dfrac{2\sqrt{3}}{3}$
28. $\dfrac{3}{\sqrt{5}} + \dfrac{\sqrt{2}}{\sqrt{10}}$
$\dfrac{4\sqrt{5}}{5}$
29. $\dfrac{1}{\sqrt{3}} + \sqrt{12}$
$\dfrac{7\sqrt{3}}{3}$
30. $\dfrac{1}{\sqrt{2}} + 3\sqrt{8}$
$\dfrac{13\sqrt{2}}{2}$

Multiply and simplify. See Example 3.

31. $\sqrt{7} \cdot \sqrt{11}$
$\sqrt{77}$
32. $\sqrt{3} \cdot \sqrt{13}$
$\sqrt{39}$
33. $2\sqrt{6} \cdot 3\sqrt{6}$
36
34. $4\sqrt{2} \cdot 3\sqrt{2}$
24
35. $-3\sqrt{5} \cdot 4\sqrt{2}$
$-12\sqrt{10}$
36. $-8\sqrt{3} \cdot 3\sqrt{2}$
$-24\sqrt{6}$
37. $\sqrt{2a^3} \cdot \sqrt{6a^5}$
$2a^4\sqrt{3}$
38. $\sqrt{3w^7} \cdot \sqrt{w^9}$
$w^8\sqrt{3}$
39. $\sqrt[3]{9} \cdot \sqrt[3]{3}$
3
40. $\sqrt[3]{-25} \cdot \sqrt[3]{5}$
-5
41. $\sqrt[3]{-4m^2} \cdot \sqrt[3]{2m}$
$-2m$
42. $\sqrt[3]{100m^4} \cdot \sqrt[3]{10m^2}$
$10m^2$

Multiply and simplify. See Example 4.

43. $\sqrt{2}(\sqrt{2} + \sqrt{3})$
$2 + \sqrt{6}$
44. $\sqrt{3}(\sqrt{3} - \sqrt{2})$
$3 - \sqrt{6}$
45. $3\sqrt{2}(2\sqrt{6} + \sqrt{10})$
$12\sqrt{3} + 6\sqrt{5}$
46. $2\sqrt{3}(\sqrt{6} + 2\sqrt{15})$
$6\sqrt{2} + 12\sqrt{5}$
47. $2\sqrt{5}(\sqrt{5} - 3\sqrt{10})$
$10 - 30\sqrt{2}$
48. $\sqrt{6}(\sqrt{24} - 6)$
$12 - 6\sqrt{6}$

Multiply and simplify. See Example 5.

49. $(\sqrt{5} - 4)(\sqrt{5} + 3)$ $-7 - \sqrt{5}$
50. $(\sqrt{6} - 2)(\sqrt{6} - 3)$ $12 - 5\sqrt{6}$
51. $(\sqrt{3} - 1)(\sqrt{3} + 1)$ 2
52. $(\sqrt{6} + 2)(\sqrt{6} - 2)$ 2
53. $(\sqrt{5} - \sqrt{2})(\sqrt{5} + \sqrt{2})$ 3
54. $(\sqrt{3} - \sqrt{6})(\sqrt{3} + \sqrt{6})$ -3
55. $(2\sqrt{5} + 1)(3\sqrt{5} - 2)$ $28 - \sqrt{5}$
56. $(2\sqrt{2} + 3)(4\sqrt{2} + 4)$ $28 + 20\sqrt{2}$
57. $(2\sqrt{3} - 3\sqrt{5})(3\sqrt{3} + 4\sqrt{5})$ $-42 - \sqrt{15}$
58. $(4\sqrt{3} + 3\sqrt{7})(2\sqrt{3} + 4\sqrt{7})$ $108 + 22\sqrt{21}$
59. $(2\sqrt{3} + 5)^2$ $37 + 20\sqrt{3}$
60. $(3\sqrt{2} + \sqrt{6})^2$ $24 + 12\sqrt{3}$

Divide and simplify. See Example 6.

61. $\sqrt{10} \div \sqrt{5}$
$\sqrt{2}$
62. $\sqrt{14} \div \sqrt{2}$
$\sqrt{7}$
63. $\sqrt{5} \div \sqrt{3}$
$\dfrac{\sqrt{15}}{3}$
64. $\sqrt{3} \div \sqrt{2}$
$\dfrac{\sqrt{6}}{2}$
65. $(4\sqrt{5}) \div (3\sqrt{6})$
$\dfrac{2\sqrt{30}}{9}$
66. $(3\sqrt{7}) \div (4\sqrt{3})$
$\dfrac{\sqrt{21}}{4}$
67. $(5\sqrt{14}) \div (3\sqrt{2})$
$\dfrac{5\sqrt{7}}{3}$
68. $(4\sqrt{15}) \div (5\sqrt{2})$
$\dfrac{2\sqrt{30}}{5}$

Simplify each expression. See Example 7.

69. $\dfrac{2 + \sqrt{8}}{2}$ $1 + \sqrt{2}$
70. $\dfrac{3 + \sqrt{18}}{3}$ $1 + \sqrt{2}$
71. $\dfrac{-4 + \sqrt{20}}{2}$ $-2 + \sqrt{5}$
72. $\dfrac{-6 + \sqrt{45}}{3}$ $-2 + \sqrt{5}$

73. $\dfrac{4 - \sqrt{20}}{6}$ $\dfrac{2 - \sqrt{5}}{3}$

74. $\dfrac{-6 - \sqrt{27}}{6}$ $\dfrac{-2 - \sqrt{3}}{2}$

75. $\dfrac{-4 - \sqrt{24}}{-6}$ $\dfrac{2 + \sqrt{6}}{3}$

76. $\dfrac{-3 - \sqrt{27}}{-3}$ $1 + \sqrt{3}$

77. $\dfrac{3 + \sqrt{12}}{6}$ $\dfrac{3 + 2\sqrt{3}}{6}$

78. $\dfrac{3 + \sqrt{8}}{3}$ $\dfrac{3 + 2\sqrt{2}}{3}$

Simplify each expression. See Example 8.

79. $\dfrac{5}{\sqrt{3} - \sqrt{2}}$ $5\sqrt{3} + 5\sqrt{2}$

80. $\dfrac{3}{\sqrt{6} + \sqrt{2}}$ $\dfrac{3\sqrt{6} - 3\sqrt{2}}{4}$

81. $\dfrac{\sqrt{3}}{\sqrt{5} - \sqrt{3}}$ $\dfrac{\sqrt{15} + 3}{2}$

82. $\dfrac{2}{\sqrt{2} + \sqrt{5}}$ $\dfrac{-2 + \sqrt{10}}{3}$

83. $\dfrac{2 + \sqrt{3}}{5 - \sqrt{3}}$ $\dfrac{13 + 7\sqrt{3}}{22}$

84. $\dfrac{\sqrt{2} - \sqrt{3}}{\sqrt{3} - 1}$ $\dfrac{\sqrt{6} - 3 + \sqrt{2} - \sqrt{3}}{2}$

85. $\dfrac{\sqrt{7} - 5}{2\sqrt{7} + 1}$ $\dfrac{19 - 11\sqrt{7}}{27}$

86. $\dfrac{\sqrt{5} + 4}{3\sqrt{2} - \sqrt{5}}$ $\dfrac{3\sqrt{10} + 12\sqrt{2} + 5 + 4\sqrt{5}}{13}$

Simplify.

87. $\sqrt{5a} + \sqrt{20a}$ $3\sqrt{5a}$

88. $a\sqrt{6} \cdot a\sqrt{12}$ $6a^2\sqrt{2}$

89. $\sqrt{75} \div \sqrt{6}$ $\dfrac{5\sqrt{2}}{2}$

90. $\sqrt{24} - \sqrt{150}$ $-3\sqrt{6}$

91. $(5 + 3\sqrt{5})^2$ $70 + 30\sqrt{5}$

92. $(\sqrt{6} - \sqrt{5})(\sqrt{6} + \sqrt{5})$ 1

93. $\sqrt{5} + \dfrac{\sqrt{20}}{3}$ $\dfrac{5\sqrt{5}}{3}$

94. $\dfrac{5}{\sqrt{8} - \sqrt{3}}$ $2\sqrt{2} + \sqrt{3}$

Use a calculator to find the approximate value of each expression to three decimal places.

95. $\dfrac{2 + \sqrt{3}}{2}$ 1.866

96. $\dfrac{-2 + \sqrt{3}}{-6}$ 0.045

97. $\dfrac{-4 - \sqrt{6}}{5 - \sqrt{3}}$ −1.974

98. $\dfrac{-5 - \sqrt{2}}{\sqrt{3} + \sqrt{7}}$ −1.465

Solve each problem.

99. Find the exact area and perimeter of the given rectangle. 12 ft^2, $10\sqrt{2}$ ft

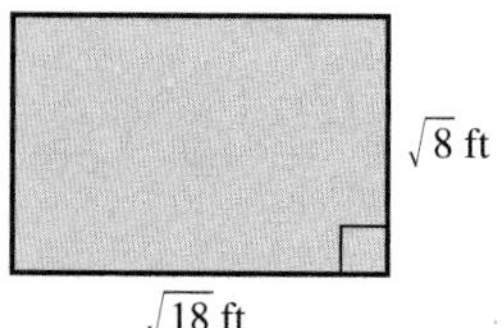

FIGURE FOR EXERCISE 99

100. Find the exact area and perimeter of the given triangle. 9, $5\sqrt{3} + \sqrt{39}$

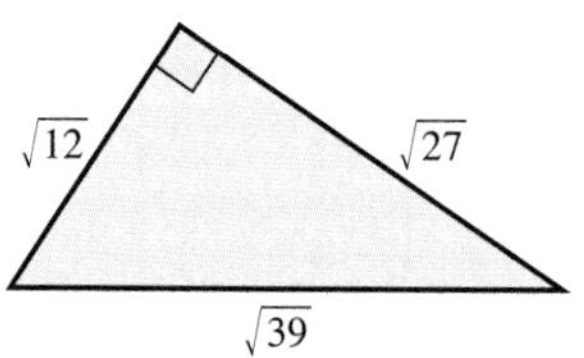

FIGURE FOR EXERCISE 100

101. Find the exact volume in cubic meters of the given rectangular box. 6 m^3

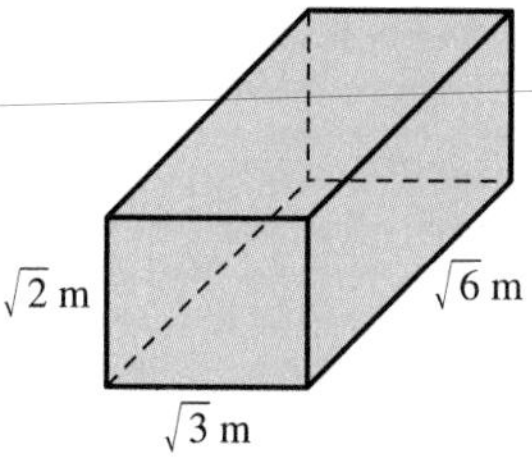

FIGURE FOR EXERCISE 101

102. Find the exact surface area of the box in Exercise 101. $2\sqrt{6} + 4\sqrt{3} + 6\sqrt{2} \text{ m}^2$

In This Section

- The Square Root Property
- Obtaining Equivalent Equations
- Squaring Each Side of an Equation
- Solving for the Indicated Variable
- Applications

8.4 SOLVING EQUATIONS WITH RADICALS AND EXPONENTS

Equations involving radicals and exponents occur in many applications. In this section you will learn to solve equations of this type, and you will see how these equations occur in applications.

The Square Root Property

An equation of the form $x^2 = k$ can have two solutions, one solution, or no solutions, depending on the value of k. For example,

$$x^2 = 4$$

has two solutions because $(2)^2 = 4$ and $(-2)^2 = 4$. So $x^2 = 4$ is equivalent to the compound equation

$$x = 2 \quad \text{or} \quad x = -2,$$

which is also written as $x = \pm 2$ and is read as "x equals positive or negative 2." The only solution to $x^2 = 0$ is 0. The equation

$$x^2 = -4$$

has no real solution because the square of every real number is nonnegative.

These examples illustrate the **square root property.**

Square Root Property (Solving $x^2 = k$)

For $k > 0$ the equation $x^2 = k$ is equivalent to the compound equation

$$x = \sqrt{k} \quad \text{or} \quad x = -\sqrt{k}. \qquad (\text{Also written } x = \pm\sqrt{k}.)$$

For $k = 0$ the equation $x^2 = k$ is equivalent to $x = 0$.
For $k < 0$ the equation $x^2 = k$ has no real solution.

CAUTION The expression $\sqrt{9}$ has a value of 3 only, but the equation $x^2 = 9$ has two solutions: 3 and -3.

EXAMPLE 1 **Using the square root property**

Solve each equation.

a) $x^2 = 12$ **b)** $2(x + 1)^2 - 18 = 0$
c) $x^2 = -9$ **d)** $(x - 16)^2 = 0$

Helpful Hint

We do not say "take the square root of each side." We are not doing the same thing to each side of $x^2 = 12$ when we write $x = \pm\sqrt{12}$. This is the second time that we have seen a rule for obtaining an equivalent equation without "doing the same thing to each side." (What was the first?)

Solution

a) $x^2 = 12$

$x = \pm\sqrt{12}$ Square root property

$x = \pm 2\sqrt{3}$

Check: $(2\sqrt{3})^2 = 4 \cdot 3 = 12$, and $(-2\sqrt{3})^2 = 4 \cdot 3 = 12$.

b) $2(x + 1)^2 - 18 = 0$

$2(x + 1)^2 = 18$ Add 18 to each side.

$(x + 1)^2 = 9$ Divide each side by 2.

$x + 1 = \pm\sqrt{9}$ Square root property

$x + 1 = 3 \quad \text{or} \quad x + 1 = -3$

$x = 2 \quad \text{or} \quad x = -4$

Check 2 and -4 in the original equation. Both -4 and 2 are solutions to the equation.

c) The equation $x^2 = -9$ has no real solution because no real number has a square that is negative.

d) $(x - 16)^2 = 0$

$x - 16 = 0$ Square root property

$x = 16$

Check: $(16 - 16)^2 = 0$. The equation has only one solution, 16. ■

Obtaining Equivalent Equations

When solving equations, we use a sequence of equivalent equations with each one simpler than the last. To get an equivalent equation, we can

1. add the same number to each side,
2. subtract the same number from each side,
3. multiply each side by the same nonzero number, or
4. divide each side by the same nonzero number.

However, "doing the same thing to each side" is not the only way to obtain an equivalent equation. In Chapter 6 we used the zero factor property to obtain equivalent equations. For example, by the zero factor property the equation

$$(x - 3)(x + 2) = 0$$

is equivalent to the compound equation

$$x - 3 = 0 \quad \text{or} \quad x + 2 = 0.$$

In this section you just learned how to obtain equivalent equations by the square root property. This property tells us how to write an equation that is equivalent to the equation $x^2 = k$. Note that the square root property does not tell us to "take the square root of each side." Because a real number might have two, one, or no square roots, taking *the* square root of each side can lead to errors. To become proficient at solving equations, we must understand these methods. One of our main goals in algebra is to keep expanding our skills for solving equations.

Study Tip

Make sure that you know what your instructor expects from you. You can determine what your instructor feels is important by looking at the examples that your instructor works in class and the homework assigned. When in doubt, ask your instructor what you will be responsible for and write down the answer.

Squaring Each Side of an Equation

Some equations involving radicals can be solved by squaring each side:

$$\sqrt{x} = 5$$

$$(\sqrt{x})^2 = 5^2 \quad \text{Square each side.}$$

$$x = 25$$

All three of these equations are equivalent. Because $\sqrt{25} = 5$ is correct, 25 satisfies the original equation.

However, squaring each side does not necessarily produce an equivalent equation. For example, consider the equation

$$x = 3.$$

Squaring each side, we get

$$x^2 = 9.$$

Both 3 and -3 satisfy $x^2 = 9$, but only 3 satisfies the original equation $x = 3$. So $x^2 = 9$ is not equivalent to $x = 3$. The extra solution to $x^2 = 9$ is called an **extraneous solution.**

These two examples illustrate the **squaring property of equality.**

Squaring Property of Equality

When we square each side of an equation, the solutions to the new equation include all of the solutions to the original equation. However, the new equation might have extraneous solutions.

This property means that *we may square each side of an equation, but we must check all of our answers to eliminate extraneous solutions.*

EXAMPLE 2 **Using the squaring property of equality**

Solve each equation.

a) $\sqrt{x^2 - 16} = 3$ **b)** $x = \sqrt{2x + 3}$

Study Tip

Most instructors believe that what they do in class is important. If you miss class, you miss what is important to your instructor. You miss what is most likely to appear on the test.

Solution

a)

$$\sqrt{x^2 - 16} = 3$$

$$(\sqrt{x^2 - 16})^2 = 3^2 \quad \text{Square each side.}$$

$$x^2 - 16 = 9$$

$$x^2 = 25$$

$$x = \pm 5 \quad \text{Square root property}$$

Check each solution:

$$\sqrt{5^2 - 16} = \sqrt{25 - 16} = \sqrt{9} = 3$$

$$\sqrt{(-5)^2 - 16} = \sqrt{25 - 16} = \sqrt{9} = 3$$

So both 5 and -5 are solutions to the equation.

b)

$$x = \sqrt{2x + 3}$$

$$x^2 = (\sqrt{2x + 3})^2 \quad \text{Square each side.}$$

$$x^2 = 2x + 3$$

$$x^2 - 2x - 3 = 0 \quad \text{Solve by factoring.}$$

$$(x - 3)(x + 1) = 0 \quad \text{Factor.}$$

$$x - 3 = 0 \quad \text{or} \quad x + 1 = 0 \quad \text{Zero factor property}$$

$$x = 3 \quad \text{or} \quad x = -1$$

Check in the original equation:

Check $x = 3$:	Check $x = -1$:
$3 = \sqrt{2 \cdot 3 + 3}$	$-1 = \sqrt{2(-1) + 3}$
$3 = \sqrt{9}$ Correct	$-1 = \sqrt{1}$ Incorrect

Because -1 does not satisfy the original equation, -1 is an extraneous solution. The only solution is 3. ■

The equations in Example 3 have radicals on both sides of the equation.

EXAMPLE 3 **Radicals on both sides**

Solve each equation.

a) $\sqrt{x - 3} = \sqrt{2x + 5}$ **b)** $\sqrt{x^2 - 4x} = \sqrt{2 - 3x}$

Solution

a)

$$\begin{aligned} \sqrt{x-3} &= \sqrt{2x+5} \\ (\sqrt{x-3})^2 &= (\sqrt{2x+5})^2 && \text{Square each side.} \\ x-3 &= 2x+5 && \text{Simplify.} \\ x-8 &= 2x \\ -8 &= x \end{aligned}$$

Check $x = -8$ in the original equation:

$$\begin{aligned} \sqrt{-8-3} &= \sqrt{2(-8)+5} \\ \sqrt{-11} &= \sqrt{-11} \end{aligned}$$

Because $\sqrt{-11}$ is not a real number -8 does not satisfy the equation and the equation has no solution.

b)

$$\begin{aligned} \sqrt{x^2-4x} &= \sqrt{2-3x} \\ x^2-4x &= 2-3x && \text{Square each side.} \\ x^2-x-2 &= 0 \\ (x-2)(x+1) &= 0 \end{aligned}$$

$$\begin{aligned} x-2 = 0 \quad &\text{or} \quad x+1 = 0 && \text{Zero factor property} \\ x = 2 \quad &\text{or} \quad x = -1 \end{aligned}$$

Check each solution in the original equation:

Check $x = 2$:

$$\begin{aligned} \sqrt{2^2-4\cdot 2} &= \sqrt{2-3\cdot 2} \\ \sqrt{-4} &= \sqrt{-4} \end{aligned}$$

Check $x = -1$:

$$\begin{aligned} \sqrt{(-1)^2-4(-1)} &= \sqrt{2-3(-1)} \\ \sqrt{5} &= \sqrt{5} \end{aligned}$$

Because $\sqrt{-4}$ is not a real number, 2 is an extraneous solution. The only solution to the original equation is -1. ■

In Example 4, one of the sides of the equation is a binomial. When we square each side, we must be sure to square the binomial properly.

EXAMPLE 4 **Squaring each side of an equation**

Solve the equation $x + 2 = \sqrt{-2-3x}$.

Solution

$$\begin{aligned} x+2 &= \sqrt{-2-3x} \\ (x+2)^2 &= (\sqrt{-2-3x})^2 && \text{Square each side.} \\ x^2+4x+4 &= -2-3x && \text{Square the binomial on the left side.} \\ x^2+7x+6 &= 0 \\ (x+6)(x+1) &= 0 && \text{Factor.} \end{aligned}$$

$$\begin{aligned} x+6 = 0 \quad &\text{or} \quad x+1 = 0 \\ x = -6 \quad &\text{or} \quad x = -1 \end{aligned}$$

Check these solutions in the original equation:

Check $x = -6$:	Check $x = -1$:
$-6 + 2 = \sqrt{-2 - 3(-6)}$	$-1 + 2 = \sqrt{-2 - 3(-1)}$
$-4 = \sqrt{16}$ Incorrect	$1 = \sqrt{1}$ Correct

The solution -6 does not check. The only solution to the original equation is -1. ■

Solving for the Indicated Variable

In Example 5 we use the square root property to solve a formula for an indicated variable.

EXAMPLE 5 **Solving for a variable**

Solve the formula $A = \pi r^2$ for r.

Solution

$$A = \pi r^2$$

$$\frac{A}{\pi} = r^2 \quad \text{Divide each side by } \pi.$$

$$\pm\sqrt{\frac{A}{\pi}} = r \quad \text{Square root property}$$

The formula solved for r is

$$r = \pm\sqrt{\frac{A}{\pi}}.$$

If r is the radius of a circle with area A, then r is positive and

$$r = \sqrt{\frac{A}{\pi}}.$$

■

Applications

Equations involving exponents occur in many applications. If the exact answer to a problem is an irrational number in radical notation, it is usually helpful to find a decimal approximation for the answer.

EXAMPLE 6 **Finding the side of a square with a given diagonal**

If the diagonal of a square window is 10 feet long, then what are the exact and approximate lengths of a side? Round the approximate answer to two decimal places.

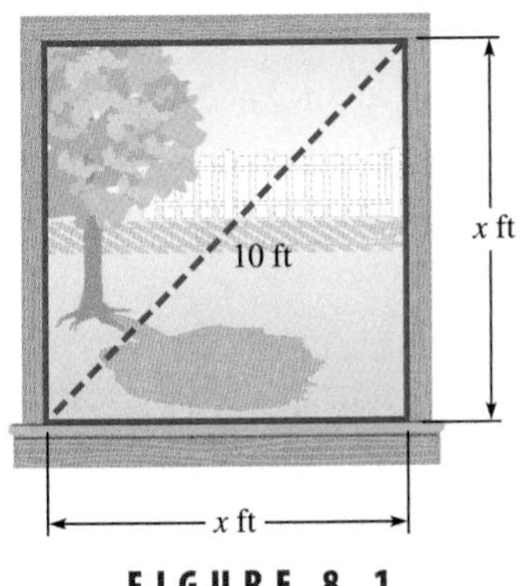

FIGURE 8.1

Solution

First make a sketch as in Fig. 8.1. Let x be the length of a side. The Pythagorean theorem tells us that the sum of the squares of the sides is equal to the diagonal squared:

$$x^2 + x^2 = 10^2$$
$$2x^2 = 100$$
$$x^2 = 50$$
$$x = \pm\sqrt{50}$$
$$= \pm 5\sqrt{2}$$

Because the length of a side must be positive, we disregard the negative solution. The exact length of a side is $5\sqrt{2}$ feet. Use a calculator to get $5\sqrt{2} \approx 7.07$. The symbol $\approx$ means "is approximately equal to." The approximate length of a side is 7.07 feet. ■

WARM-UPS

True or false? Explain your answer.

1. The equation $x^2 = 9$ is equivalent to the equation $x = 3$. False
2. The equation $x^2 = -16$ has no real solution. True
3. The equation $a^2 = 0$ has no solution. False
4. Both $-\sqrt{5}$ and $\sqrt{5}$ are solutions to $x^2 + 5 = 0$. False
5. The equation $-x^2 = 9$ has no real solution. True
6. To solve $\sqrt{x + 4} = \sqrt{2x - 9}$, first take the square root of each side. False
7. All extraneous solutions give us a denominator of zero. False
8. Squaring both sides of $\sqrt{x} = -1$ will produce an extraneous solution. True
9. The equation $x^2 - 3 = 0$ is equivalent to $x = \pm\sqrt{3}$. True
10. The equation $-2 = \sqrt{6x^2 - x - 8}$ has no solution. True

8.4 EXERCISES

Reading and Writing *After reading this section, write out the answers to these questions. Use complete sentences.*

1. What is the square root property?
   The square root property says that $x^2 = k$ for $k > 0$ is equivalent to $x = \pm\sqrt{k}$.
2. When do we take the square root of each side of an equation?
   We do not take the square root of each side of an equation.
3. What new techniques were introduced in this section for solving equations?
   The square root property and squaring each side are two new techniques used for solving equations.
4. Is there any way to obtain an equivalent equation other than doing the same thing to each side?
   The square root property and the zero factor property produce equivalent equations without doing the same thing to each side.
5. Which property for solving equations can give extraneous roots?
   Squaring each side can produce extraneous roots.
6. Which property of equality does not always give you an equivalent equation?
   Squaring each side does not always give equivalent equations.

Solve each equation. See Example 1.

7. $x^2 = 16$
   $-4, 4$
8. $x^2 = 49$
   $-7, 7$
9. $x^2 - 40 = 0$
   $-2\sqrt{10}, 2\sqrt{10}$
10. $x^2 - 24 = 0$
    $-2\sqrt{6}, 2\sqrt{6}$
11. $3x^2 = 2$
    $-\frac{\sqrt{6}}{3}, \frac{\sqrt{6}}{3}$
12. $2x^2 = 3$
    $-\frac{\sqrt{6}}{2}, \frac{\sqrt{6}}{2}$
13. $9x^2 = -4$
    No solution
14. $25x^2 + 1 = 0$
    No solution
15. $(x - 1)^2 = 4$
    $-1, 3$
16. $(x + 3)^2 = 9$
    $-6, 0$
17. $2(x - 5)^2 + 1 = 7$
    $5 - \sqrt{3}, 5 + \sqrt{3}$
18. $3(x - 6)^2 - 4 = 11$
    $6 - \sqrt{5}, 6 + \sqrt{5}$
19. $(x + 19)^2 = 0$
    -19
20. $5x^2 + 5 = 5$
    0

Solve each equation. See Examples 2 and 3.

21. $\sqrt{x - 9} = 9$
    90
22. $\sqrt{x + 3} = 4$
    13
23. $\sqrt{2x - 3} = -4$
    No solution
24. $\sqrt{3x - 5} = -9$
    No solution

For example, to evaluate $8^{-2/3}$ mentally, we find the cube root of 8 (which is 2), square 2 to get 4, then find the reciprocal of 4 to get $\frac{1}{4}$. In print $8^{-2/3}$ could be written for evaluation as $((8^{1/3})^2)^{-1}$ or $\frac{1}{(8^{1/3})^2}$. We prefer the latter because it looks a little simpler.

EXAMPLE 4 **Rational exponents**

Evaluate each expression.

a) $27^{2/3}$ **b)** $4^{-3/2}$ **c)** $81^{-3/4}$ **d)** $(-8)^{-5/3}$

Calculator Close-Up

A negative fractional exponent indicates a reciprocal, a root, and a power. To find $4^{-3/2}$ you can find the reciprocal first, the square root first, or the third power first as shown here.

```
(1/4)^(3/2)
            .125
(√(4))^-3
            .125
(4³)^(-1/2)
            .125
```

Solution

a) Because the exponent is 2/3, we find the cube root of 27, which is 3, and then square it to get 9. In symbols,

$$27^{2/3} = (27^{1/3})^2 = 3^2 = 9.$$

b) Because the exponent is $-3/2$, we find the principal square root of 4, which is 2, cube it to get 8, and then find the reciprocal to get $\frac{1}{8}$. In symbols,

$$4^{-3/2} = \frac{1}{(4^{1/2})^3} = \frac{1}{2^3} = \frac{1}{8}.$$

c) Because the exponent is $-3/4$, we find the principal fourth root of 81, which is 3, cube it to get 27, and then find the reciprocal to get $\frac{1}{27}$. In symbols,

$$81^{-3/4} = \frac{1}{(81^{1/4})^3} = \frac{1}{3^3} = \frac{1}{27}.$$

d) $(-8)^{-5/3} = \dfrac{1}{((-8)^{1/3})^5} = \dfrac{1}{(-2)^5} = \dfrac{1}{-32} = -\dfrac{1}{32}$ ■

CAUTION An expression with a negative base and a negative exponent can have a positive or a negative value. For example,

$$(-8)^{-5/3} = -\frac{1}{32} \quad \text{and} \quad (-8)^{-2/3} = \frac{1}{4}.$$

Using the Rules

As we mentioned earlier, the advantage of using exponents to express roots is that the rules for integral exponents can also be used for rational exponents. We will use those rules in Example 5.

EXAMPLE 5 **Using the rules of exponents**

Simplify each expression. Write answers with positive exponents. Assume that all variables represent positive real numbers.

a) $2^{1/2} \cdot 2^{3/2}$ **b)** $\dfrac{x}{x^{2/3}}$ **c)** $(b^{1/2})^{1/3}$

d) $(x^4y^{-6})^{1/2}$ **e)** $\left(\dfrac{x^6}{y^3}\right)^{-2/3}$

Helpful Hint

Look what happens when we apply legitimate rules to an illegitimate expression:

$$(-1)^{2/2} = (-1)^1 = -1$$

and

$$(-1)^{2/2} = ((-1)^2)^{1/2} = 1^{1/2} = 1$$

So we conclude that $1 = -1$! However, $(-1)^{2/2}$ is not a legal expression because there is no real square root of -1.

Solution

a) $2^{1/2} \cdot 2^{3/2} = 2^{1/2+3/2}$ Product rule

$= 2^2$ $\quad \frac{1}{2} + \frac{3}{2} = \frac{4}{2} = 2$

$= 4$

b) $\dfrac{x}{x^{2/3}} = x^{1-2/3}$ Quotient rule

$= x^{1/3}$ $\quad 1 - \frac{2}{3} = \frac{3}{3} - \frac{2}{3} = \frac{1}{3}$

c) $(b^{1/2})^{1/3} = b^{(1/2)\cdot(1/3)}$ Power rule

$= b^{1/6}$ $\quad \frac{1}{2}\cdot\frac{1}{3} = \frac{1}{6}$

d) $(x^4y^{-6})^{1/2} = (x^4)^{1/2}(y^{-6})^{1/2}$ Power of a product rule

$= x^2y^{-3}$ Power rule

$= \dfrac{x^2}{y^3}$ Definition of negative exponent

e) $\left(\dfrac{x^6}{y^3}\right)^{-2/3} = \left(\dfrac{y^3}{x^6}\right)^{2/3}$ Negative exponent rule

$= \dfrac{(y^3)^{2/3}}{(x^6)^{2/3}}$ Power of a quotient rule

$= \dfrac{y^2}{x^4}$ Power rule ■

WARM-UPS

True or false? Explain your answer.

1. $9^{1/3} = \sqrt[3]{9}$ True

2. $8^{5/3} = \sqrt[5]{8^3}$ False

3. $(-16)^{1/2} = -16^{1/2}$ False

4. $9^{-3/2} = \dfrac{1}{27}$ True

5. $6^{-1/2} = \dfrac{\sqrt{6}}{6}$ True

6. $\dfrac{2}{2^{1/2}} = 2^{1/2}$ True

7. $2^{1/2} \cdot 2^{1/2} = 4^{1/2}$ True

8. $16^{-1/4} = -2$ False

9. $6^{1/6} \cdot 6^{1/6} = 6^{1/3}$ True

10. $(2^8)^{3/4} = 2^6$ True

8.5 EXERCISES

Reading and Writing *After reading this section, write out the answers to these questions. Use complete sentences.*

1. How do we indicate an nth root using exponents?
The nth root of a is $a^{1/n}$.

2. How do we indicate the mth power of the nth root using exponents?
The mth power of the nth root of a is $a^{m/n}$.

3. What is the meaning of a negative rational exponent?
The expression $a^{-m/n}$ means $\frac{1}{a^{m/n}}$.

4. Which rules of exponents hold for rational exponents?
All of the rules of exponents hold for rational exponents.

5. In what order must you perform the operations indicated by a negative rational exponent?
The operations can be performed in any order, but the easiest is usually root, power, and then reciprocal.

6. When is $a^{-m/n}$ a real number?
The expression $a^{-m/n}$ is a real number except when n is even and a is negative, or when $a = 0$.

Write each radical expression using exponent notation and each exponential expression using radical notation. See Example 1.

7. $\sqrt[4]{7}$ $7^{1/4}$

8. $\sqrt[3]{cbs}$ $(cbs)^{1/3}$

9. $9^{1/5}$ $\sqrt[5]{9}$

10. $3^{1/2}$ $\sqrt{3}$

11. $\sqrt{5x}$ $(5x)^{1/2}$

12. $\sqrt{3y}$ $(3y)^{1/2}$

13. $a^{1/2}$ $\sqrt{a}$

14. $(-b)^{1/5}$ $\sqrt[5]{-b}$

Evaluate each expression. See Example 2.

15. $25^{1/2}$ 5

16. $16^{1/2}$ 4

17. $(-125)^{1/3}$ -5

18. $(-32)^{1/5}$ -2

19. $16^{1/4}$ 2

20. $8^{1/3}$ 2

21. $(-4)^{1/2}$ Not a real number

22. $(-16)^{1/4}$ Not a real number

Write each radical expression using exponent notation and each exponential expression using radical notation. See Example 3.

23. $\sqrt[3]{w^7}$ $w^{7/3}$

24. $\sqrt{a^5}$ $a^{5/2}$

25. $\frac{1}{\sqrt[3]{2^{10}}}$ $2^{-10/3}$

26. $\sqrt[3]{\frac{1}{a^2}}$ $a^{-2/3}$

27. $w^{-3/4}$ $\sqrt[4]{\frac{1}{w^3}}$

28. $6^{-5/3}$ $\sqrt[3]{\frac{1}{6^5}}$

29. $(ab)^{3/2}$ $\sqrt{(ab)^3}$

30. $(3m)^{-1/5}$ $\sqrt[5]{\frac{1}{3m}}$

Evaluate each expression. See Example 4.

31. $125^{2/3}$ 25

32. $1000^{2/3}$ 100

33. $25^{3/2}$ 125

34. $16^{3/2}$ 64

35. $27^{-4/3}$ $\frac{1}{81}$

36. $16^{-3/4}$ $\frac{1}{8}$

37. $4^{-3/2}$ $\frac{1}{8}$

38. $25^{-3/2}$ $\frac{1}{125}$

39. $(-27)^{-1/3}$ $-\frac{1}{3}$

40. $(-8)^{-4/3}$ $\frac{1}{16}$

41. $(-16)^{-1/4}$ Not a real number

42. $(-100)^{-3/2}$ Not a real number

Simplify each expression. Write answers with positive exponents only. See Example 5.

43. $x^{1/4}x^{1/4}$ $x^{1/2}$

44. $y^{1/3}y^{2/3}$ y

45. $n^{1/2}n^{-1/3}$ $n^{1/6}$

46. $w^{-1/4}w^{3/5}$ $w^{7/20}$

47. $\frac{x^2}{x^{1/2}}$ $x^{3/2}$

48. $\frac{a^{1/2}}{a^{1/3}}$ $a^{1/6}$

49. $\frac{8t^{1/2}}{4t^{1/4}}$ $2t^{1/4}$

50. $\frac{6w^{1/4}}{3w^{1/3}}$ $\frac{2}{w^{1/12}}$

51. $(x^6)^{1/3}$ x^2

52. $(y^{-4})^{1/2}$ $\frac{1}{y^2}$

53. $(5^{-1/4})^{-1/2}$ $5^{1/8}$

54. $(7^{-3/4})^6$ $\frac{1}{7^{9/2}}$

55. $(x^2y^6)^{1/2}$ xy^3

56. $(t^3w^6)^{1/3}$ tw^2

57. $(9x^{-2}y^8)^{-1/2}$ $\frac{x}{3y^4}$

58. $(4w^{-2}t^{-4})^{-1/2}$ $\frac{wt^2}{2}$

Evaluate each expression.

59. $16^{-1/2} + 2^{-1}$ $\frac{3}{4}$

60. $4^{-1/2} - 8^{-2/3}$ $\frac{1}{4}$

61. $27^{-1/6} \cdot 27^{-1/2}$ $\frac{1}{9}$

62. $32^{-1/10} \cdot 32^{-1/10}$ $\frac{1}{2}$

63. $\frac{81^{5/6}}{81^{1/12}}$ 27

64. $\frac{25^{-3/4}}{25^{3/4}}$ $\frac{1}{125}$

65. $(3^{-4} \cdot 6^8)^{-1/4}$ $\frac{1}{12}$

66. $(-2^{-9} \cdot 3^6)^{-1/3}$ $-\frac{8}{9}$

Solve each problem.

67. ***Yacht dimensions.*** Since 1988, a yacht competing for the America's Cup must satisfy the inequality

$$L + 1.25S^{1/2} - 9.8D^{1/3} \le 16.296,$$

where L is the boat's length in meters, S is the sail area in square meters, and D is the displacement in cubic meters (www.americascupnews.com). Does a boat with a displacement of 21.8 m^3, a sail area of 305.4 m^2, and a length of 21.5 m satisfy the inequality? If the length and displacement are not changed, then what is the maximum number of square meters of sail that could be added and still have the boat satisfy the inequality?
Yes, 9.2 m^2

FIGURE FOR EXERCISE 67

68. ***Surface area.*** If A is the surface area of a cube and V is its volume, then

$$A = 6V^{2/3}.$$

Find the surface area of a cube that has a volume of 27 cubic centimeters. What is the surface area of a cube whose sides measure 5 centimeters each?
54 cm^2, 150 cm^2

69. ***Average annual return.*** The average annual return on an investment, r, is given by the formula

$$r = \left(\frac{S}{P}\right)^{1/n} - 1$$

where P is the original investment and S is the value of the investment after n years. An investment of \$10,000 in 1992 in T. Rowe Price's Equity Income Fund amounted to \$36,427 in 2002 (www.troweprice.com). What was the average annual return to the nearest tenth of a percent?
13.8%

70. ***Population growth.*** The U.S. population grew from 248.7 million in 1990 to 286.6 million in 2002 (U.S. Census Bureau, www.census.gov). Use the formula from Exercise 69 to find the average annual rate of growth to the nearest tenth of a percent.
1.2%

71. ***Sail area-displacement ratio.*** The sail area-displacement ratio r for a boat with sail area A (in square feet) and displacement d (in pounds) is given by

$$r = A(d/64)^{-2/3}.$$

The Tartan 4100 has a sail area of 810 ft² and a displacement of 23,245 pounds.

a) Find r for this boat.
15.9

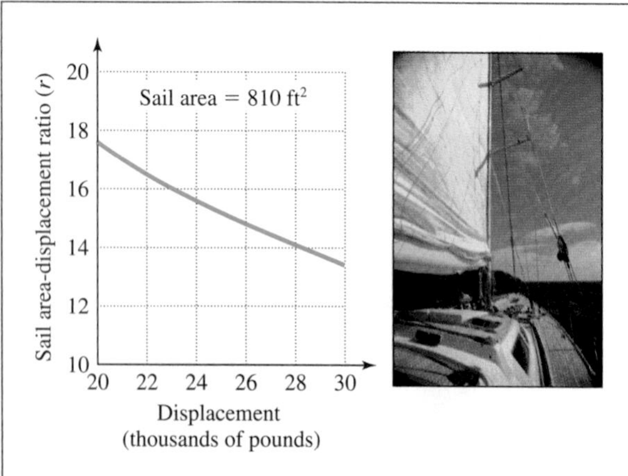

FIGURE FOR EXERCISE 71

b) Use the accompanying graph to determine whether the ratio is increasing or decreasing as the displacement increases, with sail area fixed at 810 ft².
Decreasing

c) Estimate the displacement for $r = 14$ using the accompanying graph.
28,000 pounds

72. ***Piano tuning.*** The note middle C on a piano is tuned so that the string vibrates at 262 cycles per second, or 262 Hz (Hertz). The C note that is one octave higher is tuned to 524 Hz. Tuning for the 11 notes in between using the method of *equal temperament* is $262 \cdot 2^{n/12}$, where n takes the values 1 through 11. Find the tuning rounded to the nearest whole Hertz for those 11 notes.
278, 294, 312, 330, 350, 371, 393, 416, 441, 467, 495 Hz

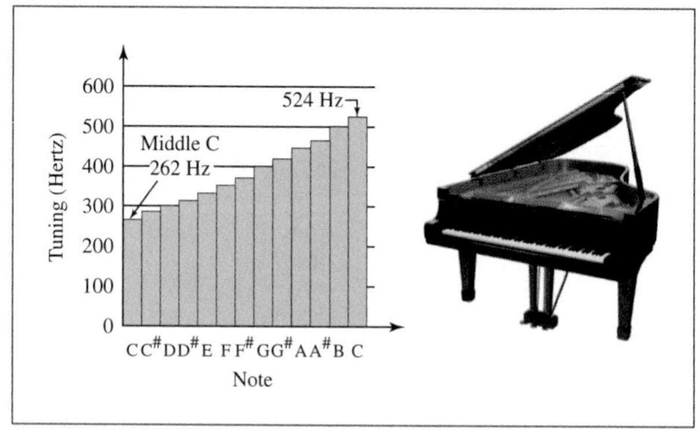

FIGURE FOR EXERCISE 72

GETTING MORE INVOLVED

73. ***Discussion.*** If $a^{-m/n} < 0$, then what can you conclude about the values of a, m, and n?
$a < 0$ and m and n are odd

74. ***Discussion.*** If $a^{-m/n}$ is not a real number, then what can you conclude about the values of a, m, and n?
$a < 0$ and n even, or $a = 0$

75. ***Exploration.*** Arrange the following expressions in order from smallest to largest. Use a calculator if you need it.

$$3^0, 3^{-8/9}, 3^{8/9}, 3^{-2/3}, 3^{-1/2}, 3^{2/3}, 3^{1/2}$$

Now arrange the following expressions in order from smallest to largest. Do not use a calculator.

$$7^{-3/4}, 7^0, 7^{1/3}, 7^{-1/3}, 7^{3/4}, 7^{9/10}, 7^{-9/10}$$

Use a calculator to check both arrangements. If $5^m < 5^n$, what can you say about m and n?
$3^{-8/9} < 3^{-2/3} < 3^{-1/2} < 3^0 < 3^{1/2} < 3^{2/3} < 3^{8/9}$, $7^{-9/10} < 7^{-3/4} < 7^{-1/3} < 7^0 < 7^{1/3} < 7^{3/4} < 7^{9/10}$, $m < n$

COLLABORATIVE ACTIVITIES

Ray's Rafters

Grouping: Four students per group

Topic: Square roots, Pythagorean theorem

Robyn and Ray have started a small business making roof trusses. Each truss has a width w and a height h as shown in the accompanying diagram. The rafter length r includes the part above the building as well as the *lookout,* which is two feet long and hangs over the edge of the building to create the eaves.

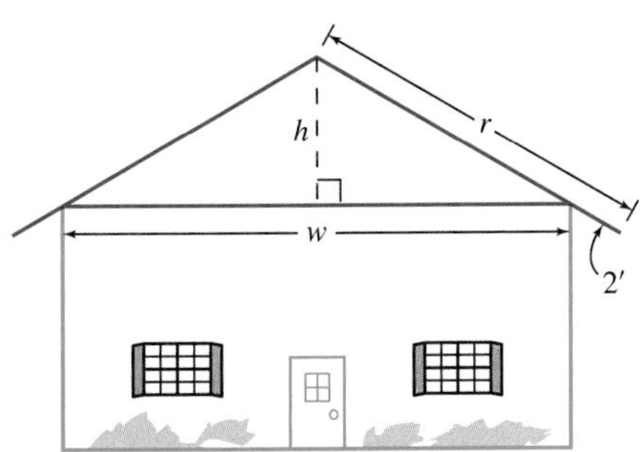

The customer specifies the width and pitch of the roof. Robyn and Ray must determine h and r. For example, a customer orders trusses with a width of 36 feet and a 3/12 pitch. To find the height, Robyn sets up the proportion

$$\frac{3}{12} = \frac{h}{18}$$

and solves it to get $h = \frac{9}{2}$. Why did Robyn use 18 in the proportion? Ray then uses the Pythagorean theorem to find the hypotenuse of the triangle:

$$\sqrt{\left(\frac{9}{2}\right)^2 + 18^2} = \sqrt{\frac{1377}{4}} = \frac{9\sqrt{17}}{2}$$

He adds 2 feet for the lookout:

$$\frac{9\sqrt{17}}{2} + \frac{4}{2} = \frac{9\sqrt{17} + 4}{2}$$

Ray then calculates that $r \approx 20.6$ feet.

To save time Robyn and Ray want to make some tables that they could use to find h and r for various widths and pitches. Your job is to fill in the tables for Robyn and Ray. In each case find r as a simplified radical and find a decimal approximation to the nearest tenth of a foot.

Table for 3/12 pitch

w	*h*	*r* (radical)	*r* (decimal)
36			
48			
60			

Table for 4/12 pitch

w	*h*	*r* (radical)	*r* (decimal)
24			
36			
60			

WRAP-UP CHAPTER 8

SUMMARY

Powers and Roots		**Examples**
nth roots	If $a = b^n$ for a positive integer n, then b is an nth root of a.	2 and -2 are fourth roots of 16.
Principal root	The positive even root of a positive number	The principal fourth root of 16 is 2.

Radical notation	If n is a positive even integer and a is positive, then the symbol $\sqrt[n]{a}$ denotes the principal nth root of a.	$\sqrt[4]{16} = 2$ $\sqrt[4]{16} \neq -2$
	If n is a positive odd integer, then the symbol $\sqrt[n]{a}$ denotes the nth root of a.	$\sqrt[3]{-8} = -2$, $\sqrt[3]{8} = 2$
	If n is any positive integer, then $\sqrt[n]{0} = 0$.	$\sqrt[5]{0} = 0$, $\sqrt[6]{0} = 0$
Definition of $a^{1/n}$	If n is any positive integer, then $a^{1/n} = \sqrt[n]{a}$ provided that $\sqrt[n]{a}$ is a real number.	$8^{1/3} = \sqrt[3]{8} = 2$ $(-4)^{1/2}$ is not real.
Definition of $a^{m/n}$	If m and n are positive integers, then $a^{m/n} = (a^{1/n})^m$, provided that $a^{1/n}$ is a real number.	$8^{2/3} = (8^{1/3})^2 = 2^2 = 4$ $(-16)^{3/4}$ is not real.
Definition of $a^{-m/n}$	If m and n are positive integers and $a \neq 0$, then $a^{-m/n} = \frac{1}{a^{m/n}}$, provided that $a^{1/n}$ is a real number.	$8^{-2/3} = \frac{1}{8^{2/3}} = \frac{1}{4}$

Rules of Exponents **Examples**

The following rules hold for any rational numbers m and n and nonzero real numbers a and b, provided that all expressions represent real numbers.

Zero exponent	$a^0 = 1$	$8^0 = 1$, $(3x - y)^0 = 1$
Product rule	$a^m a^n = a^{m+n}$	$3^3 \cdot 3^4 = 3^7$, $x^5 x^{-2} = x^3$
Quotient rule	$\frac{a^m}{a^n} = a^{m-n}$	$\frac{3^5}{3^7} = 3^{-2}$, $\frac{x}{x^{1/4}} = x^{3/4}$
Power rule	$(a^m)^n = a^{mn}$	$(2^2)^3 = 2^6$ $(w^{3/4})^4 = w^3$
Power of a product rule	$(ab)^n = a^n b^n$	$(2t)^{1/2} = 2^{1/2} t^{1/2}$
Power of a quotient rule	$\left(\frac{a}{b}\right)^n = \frac{a^n}{b^n}$	$\left(\frac{x}{3}\right)^{-3} = \frac{x^{-3}}{3^{-3}}$

Rules for Radicals **Examples**

The following rules hold, provided that all roots are real numbers and n is a positive integer.

Product rule for radicals	$\sqrt[n]{ab} = \sqrt[n]{a} \cdot \sqrt[n]{b}$	$\sqrt{2} \cdot \sqrt{3} = \sqrt{6}$ $\sqrt{9y} = 3\sqrt{y}$
Quotient rule for radicals	$\sqrt[n]{\frac{a}{b}} = \frac{\sqrt[n]{a}}{\sqrt[n]{b}}$	$\sqrt{\frac{5}{4}} = \frac{\sqrt{5}}{2}$, $\frac{\sqrt{6}}{\sqrt{2}} = \sqrt{\frac{6}{2}} = \sqrt{3}$

Simplified form for square roots	A square root expression is in simplified form if it has 1. *no* perfect square factors inside the radical,	$\sqrt{12} = \sqrt{4 \cdot 3} = 2\sqrt{3}$
	2. *no* fractions inside the radical, and	$\sqrt{\frac{5}{2}} = \frac{\sqrt{5}}{\sqrt{2}}$
	3. *no* radicals in the denominator.	$\frac{\sqrt{5}}{\sqrt{2}} = \frac{\sqrt{5} \cdot \sqrt{2}}{\sqrt{2} \cdot \sqrt{2}} = \frac{\sqrt{10}}{2}$

Solving Equations Involving Squares and Square Roots — **Examples**

Square root property (solving $x^2 = k$)	If $k > 0$, the equation $x^2 = k$ is equivalent to $x = \sqrt{k}$ or $x = -\sqrt{k}$ (also written $x = \pm\sqrt{k}$).	$x^2 = 6$ $x = \pm\sqrt{6}$
	If $k = 0$, the equation $x^2 = k$ is equivalent to $x = 0$.	$t^2 = 0$ $t = 0$
	If $k < 0$, the equation $x^2 = k$ has no real solution.	$x^2 = -8$, no solution
Squaring property of equality	Squaring each side of an equation may introduce extraneous solutions. We must check all of our answers.	$\sqrt{x} = -3$ $(\sqrt{x})^2 = (-3)^2$ $x = 9$ Extraneous solution

ENRICHING YOUR MATHEMATICAL WORD POWER

For each mathematical term, choose the correct meaning.

1. **nth root of a**
   a. a square root
   b. the root of a^n
   c. a number b such that $a^n = b$
   d. a number b such that $b^n = a$ d

2. **square of a**
   a. a number b such that $b^2 = a$
   b. a^2
   c. $|a|$
   d. $\sqrt{a}$ b

3. **cube root of a**
   a. a^3
   b. a number b such that $b^3 = a$
   c. $a/3$
   d. a number b such that $b = a^3$ b

4. **principal root**
   a. the main root
   b. the positive even root of a positive number
   c. the positive odd root of a negative number
   d. the negative odd root of a negative number b

5. **odd root of a**
   a. the number b such that $b^n = a$ where a is an odd number
   b. the opposite of the even root of a
   c. the nth root of a
   d. the number b such that $b^n = a$ where n is an odd number d

6. **index of a radical**
   a. the number n in $n\sqrt{a}$
   b. the number n in $\sqrt[n]{a}$
   c. the number n in a^n
   d. the number n in $\sqrt{a^n}$ b

7. **like radicals**
   a. radicals with the same index
   b. radicals with the same radicand
   c. radicals with the same radicand and the same index
   d. radicals with even indices c

8. **rational exponent**
   a. an exponent that produces a rational number
   b. an integral exponent
   c. an exponent that is a real number
   d. an exponent that is a rational number d

REVIEW EXERCISES

8.1 *Find each root.*

1. $\sqrt[5]{32}$ 2
2. $\sqrt[3]{-27}$ −3
3. $\sqrt[3]{1000}$ 10
4. $\sqrt{100}$ 10
5. $\sqrt{x^{12}}$ x^6
6. $\sqrt{a^{10}}$ a^5
7. $\sqrt[3]{x^6}$ x^2
8. $\sqrt[3]{a^9}$ a^3
9. $\sqrt{4x^2}$ $2x$
10. $\sqrt{9y^4}$ $3y^2$
11. $\sqrt[3]{125x^6}$ $5x^2$
12. $\sqrt[3]{8y^{12}}$ $2y^4$
13. $\sqrt{\frac{4x^{16}}{y^{14}}}$ $\frac{2x^8}{y^7}$
14. $\sqrt{\frac{9y^8}{t^{10}}}$ $\frac{3y^4}{t^5}$
15. $\sqrt{\frac{w^2}{16}}$ $\frac{w}{4}$
16. $\sqrt{\frac{a^4}{25}}$ $\frac{a^2}{5}$

8.2 *Write each expression in simplified form.*

17. $\sqrt{72}$ $6\sqrt{2}$
18. $\sqrt{48}$ $4\sqrt{3}$
19. $\frac{1}{\sqrt{3}}$ $\frac{\sqrt{3}}{3}$
20. $\frac{2}{\sqrt{5}}$ $\frac{2\sqrt{5}}{5}$
21. $\sqrt{\frac{3}{5}}$ $\frac{\sqrt{15}}{5}$
22. $\sqrt{\frac{5}{6}}$ $\frac{\sqrt{30}}{6}$
23. $\frac{\sqrt{33}}{\sqrt{3}}$ $\sqrt{11}$
24. $\frac{\sqrt{50}}{\sqrt{5}}$ $\sqrt{10}$
25. $\frac{\sqrt{3}}{\sqrt{8}}$ $\frac{\sqrt{6}}{4}$
26. $\frac{\sqrt{2}}{\sqrt{18}}$ $\frac{1}{3}$
27. $\sqrt{y^6}$ y^3
28. $\sqrt{z^{10}}$ z^5
29. $\sqrt{24t^9}$ $2t^4\sqrt{6t}$
30. $\sqrt{8p^7}$ $2p^3\sqrt{2p}$
31. $\sqrt{12m^5t^3}$ $2m^2t\sqrt{3mt}$
32. $\sqrt{18p^3q^7}$ $3pq^3\sqrt{2pq}$
33. $\frac{\sqrt{2}}{\sqrt{x}}$ $\frac{\sqrt{2x}}{x}$
34. $\frac{\sqrt{5}}{\sqrt{y}}$ $\frac{\sqrt{5y}}{y}$
35. $\sqrt{\frac{3a^5}{2s}}$ $\frac{a^2\sqrt{6as}}{2s}$
36. $\sqrt{\frac{5x^7}{3w}}$ $\frac{x^3\sqrt{15xw}}{3w}$

8.3 *Perform each computation and simplify.*

37. $2\sqrt{7} + 8\sqrt{7}$ $10\sqrt{7}$
38. $3\sqrt{6} - 5\sqrt{6}$ $-2\sqrt{6}$
39. $\sqrt{12} - \sqrt{27}$ $-\sqrt{3}$
40. $\sqrt{18} + \sqrt{50}$ $8\sqrt{2}$
41. $2\sqrt{3} \cdot 5\sqrt{3}$ 30
42. $-3\sqrt{6} \cdot 2\sqrt{6}$ −36
43. $-3\sqrt{6} \cdot 5\sqrt{3}$ $-45\sqrt{2}$
44. $4\sqrt{12} \cdot 6\sqrt{8}$ $96\sqrt{6}$
45. $-3\sqrt{3}(5 + \sqrt{3})$ $-15\sqrt{3} - 9$
46. $4\sqrt{2}(6 + \sqrt{8})$ $24\sqrt{2} + 16$
47. $-\sqrt{3}(\sqrt{6} - \sqrt{15})$ $-3\sqrt{2} + 3\sqrt{5}$
48. $-\sqrt{2}(\sqrt{6} - \sqrt{2})$ $2 - 2\sqrt{3}$
49. $(\sqrt{3} - 5)(\sqrt{3} + 5)$ −22
50. $(\sqrt{2} + \sqrt{7})(\sqrt{2} - \sqrt{7})$ −5
51. $(2\sqrt{5} - \sqrt{6})^2$ $26 - 4\sqrt{30}$
52. $(3\sqrt{2} + \sqrt{6})^2$ $24 + 12\sqrt{3}$
53. $(4 - 3\sqrt{6})(5 - \sqrt{6})$ $38 - 19\sqrt{6}$
54. $(\sqrt{3} - 2\sqrt{5})(\sqrt{3} + 4\sqrt{5})$ $-37 + 2\sqrt{15}$
55. $3\sqrt{5} \div (6\sqrt{2})$ $\frac{\sqrt{10}}{4}$
56. $6\sqrt{5} \div (4\sqrt{3})$ $\frac{\sqrt{15}}{2}$
57. $\frac{2 - \sqrt{20}}{10}$ $\frac{1 - \sqrt{5}}{5}$
58. $\frac{6 - \sqrt{12}}{-2}$ $-3 + \sqrt{3}$
59. $\frac{3}{1 - \sqrt{5}}$ $-\frac{3 + 3\sqrt{5}}{4}$
60. $\frac{2}{\sqrt{6} + \sqrt{3}}$ $\frac{2\sqrt{3} - \sqrt{6}}{3}$

8.4 *Solve each equation.*

61. $x^2 = 400$ −20, 20
62. $x^2 = 121$ −11, 11
63. $7x^2 = 3$ $-\frac{\sqrt{21}}{7}, \frac{\sqrt{21}}{7}$
64. $3x^2 - 7 = 0$ $-\frac{\sqrt{21}}{3}, \frac{\sqrt{21}}{3}$
65. $(x - 4)^2 - 18 = 0$ $4 - 3\sqrt{2}, 4 + 3\sqrt{2}$
66. $2(x + 1)^2 - 40 = 0$ $-1 - 2\sqrt{5}, -1 + 2\sqrt{5}$
67. $\sqrt{x} = 9$ 81
68. $\sqrt{x} - 20 = 0$ 400
69. $x = \sqrt{36 - 5x}$ 4
70. $x = \sqrt{2 - x}$ 1
71. $x + 2 = \sqrt{52 + 2x}$ 6
72. $x - 4 = \sqrt{x - 4}$ 4, 5

Solve each formula for t.

73. $t^2 - 8sw = 0$ $t = \pm 2\sqrt{2sw}$
74. $(t + b)^2 = b^2 - 4ac$ $t = -b \pm \sqrt{b^2 - 4ac}$
75. $3a = \sqrt{bt}$ $t = \frac{9a^2}{b}$
76. $a - \sqrt{t} = w$ $t = (a - w)^2$

8.5 *Simplify each expression. Answers with exponents should have positive exponents only.*

77. $25^{-3/2}$ $\frac{1}{125}$

78. $9^{-5/2}$ $\frac{1}{243}$

79. $25^{1/2}$ 5

80. $9^{3/2}$ 27

81. $64^{-1/2}$ $\frac{1}{8}$

82. $125^{-2/3}$ $\frac{1}{25}$

83. $x^{-3/5}x^{-2/5}$ $\frac{1}{x}$

84. $t^{-1/3}t^{1/2}$ $t^{1/6}$

85. $(-8x^{-6})^{-1/3}$ $-\frac{1}{2}x^2$

86. $(-27x^{-9})^{-2/3}$ $\frac{1}{9}x^6$

87. $w^{-3/2} \div w^{-7/2}$ w^2

88. $m^{1/3} \div m^{-1/4}$ $m^{7/12}$

89. $\left(\frac{9t^{-6}}{s^{-4}}\right)^{-1/2}$ $\frac{t^3}{3s^2}$

90. $\left(\frac{8y^{-3}}{x^6}\right)^{-2/3}$ $\frac{x^4y^2}{4}$

91. $\left(\frac{8x^{-12}}{y^{30}}\right)^{2/3}$ $\frac{4}{x^8y^{20}}$

92. $\left(\frac{16y^{-3/4}}{t^{1/2}}\right)^{-2}$ $\frac{y^{3/2}t}{256}$

Solve each problem.

93. ***Depreciation of a Lumina.*** If the cost of a piece of equipment was C dollars and it is sold for S dollars after n years, then the annual depreciation rate is given by

$$r = 1 - \left(\frac{S}{C}\right)^{1/n}.$$

A 1994 Chevrolet Lumina that sold new for \$15,446 sells for \$2965 in 2002 (www.edmunds.com).

a) Find the depreciation rate for this car to the nearest tenth of a percent.

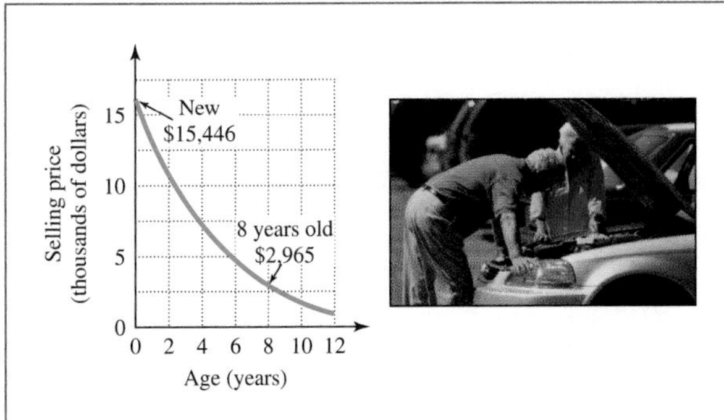

FIGURE FOR EXERCISE 93

b) Use the accompanying graph to determine whether this car depreciated more during the first two years or the last two years shown.
a) 18.6%
b) First two years

94. ***Depreciation of a Thunderbird.*** A 1996 Ford Thunderbird that sold new for \$16,892 sells for \$7212 in 2002 (www.edmunds.com). Use the formula from the previous exercise to find the depreciation rate to the nearest tenth of a percent. 13.2%

95. ***Radius of a drop.*** The amount of water in a large raindrop is 0.25 cm³. Use the formula

$$r = \left(\frac{3V}{4\pi}\right)^{1/3}$$

to find the radius of the spherical drop to the nearest tenth of a centimeter. 0.4 cm

96. ***Radius of a circle.*** Solve the formula $A = \pi r^2$ for r.
$r = \sqrt{\frac{A}{\pi}}$

97. ***Waffle cones.*** A large waffle cone has a height of 6 in. and a radius of 2 in. as shown in the accompanying figure. Find the exact amount of waffle in a cone this size. The formula $A = \pi r\sqrt{r^2 + h^2}$ gives the lateral surface area of a right circular cone with radius r and height h. Be sure to simplify the radical. Use a calculator to find the answer to the nearest square inch. $4\pi\sqrt{10}$ or 40 in².

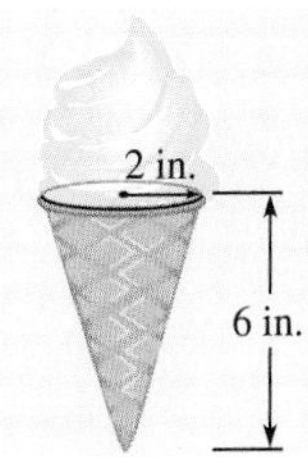

FIGURE FOR EXERCISE 97

98. ***Salting the roads.*** A city manager wants to find the amount of canvas required to cover a conical salt pile that is stored for the winter. The height of the pile is 10 yards, and the diameter of the base is 24 yards. Use the formula in Exercise 97 to find the exact number of square yards of canvas needed. Simplify the radical. Use a calculator to find the answer to the nearest square yard.
$24\pi\sqrt{61}$ or 589 yd²

99. ***Wide screen TV.*** The screen on a new Panasonic widescreen flat panel television measures 10.8 in. by 19.2 in. Find the diagonal measure of the screen to the nearest inch. 22 in.

100. ***Wider screen TV.*** The diagonal measure of the screen on a flat panel Samsung television is 40 in. The aspect ratio (ratio of the length to the width) of the screen is 16 to 9. Find the length and width to the nearest tenth of an inch. 34.9 in. by 19.6 in.

CHAPTER 8 TEST

Simplify each expression.

1. $\sqrt{36}$
6

2. $\sqrt{144}$
12

3. $\sqrt[3]{-27}$
-3

4. $\sqrt[5]{32}$
2

5. $16^{1/4}$
2

6. $\sqrt{24}$
$2\sqrt{6}$

7. $\sqrt{\frac{3}{8}}$
$\frac{\sqrt{6}}{4}$

8. $(-4)^{3/2}$
Not a real number

9. $\sqrt{8} + \sqrt{2}$
$3\sqrt{2}$

10. $(2 + \sqrt{3})^2$
$7 + 4\sqrt{3}$

11. $(3\sqrt{2} - \sqrt{7})(3\sqrt{2} + \sqrt{7})$ 11

12. $\sqrt{21} \div \sqrt{3}$ $\sqrt{7}$

13. $\sqrt{20} \div \sqrt{3}$
$\frac{2\sqrt{15}}{3}$

14. $\frac{2 + \sqrt{8}}{2}$
$1 + \sqrt{2}$

15. $27^{4/3}$
81

16. $\sqrt{3}(\sqrt{6} - \sqrt{3})$
$3\sqrt{2} - 3$

Simplify. Assume that all variables represent positive real numbers, and write answers with positive exponents only.

17. $y^{1/2} \cdot y^{1/4}$
$y^{3/4}$

18. $\frac{6x}{2x^{1/3}}$
$3x^{2/3}$

19. $(x^3y^9)^{1/3}$
xy^3

20. $\left(\frac{125w^3}{u^{-12}}\right)^{-1/3}$
$\frac{1}{5u^4w}$

21. $\sqrt{\frac{3}{t}}$
$\frac{\sqrt{3t}}{t}$

22. $\sqrt{4y^6}$
$2y^3$

23. $\sqrt[3]{8y^{12}}$
$2y^4$

24. $\sqrt{18t^7}$
$3t^3\sqrt{2t}$

Solve each equation.

25. $(x + 3)^2 = 36$
$-9, 3$

26. $\sqrt{x + 7} = 5$
18

27. $5x^2 = 2$
$-\frac{\sqrt{10}}{5}, \frac{\sqrt{10}}{5}$

28. $(3x - 4)^2 = 0$
$\frac{4}{3}$

29. $x - 3 = \sqrt{5x + 9}$
11

Solve the equation for the specified variable.

30. $S = \pi r^2 h$ for r
$r = \pm\sqrt{\frac{S}{\pi h}}$

31. $a^2 + b^2 = c^2$ for b
$b = \sqrt{c^2 - a^2}$

Show a complete solution to each problem.

32. Find the exact length of the side of a square whose diagonal is 5 meters.
$\frac{5\sqrt{2}}{2}$ meters

33. To utilize a center-pivot irrigation system, a farmer planted his crop in a circular field of 100,000 square meters. Find the radius of the circular field to the nearest tenth of a meter.
178.4 meters

MAKING CONNECTIONS CHAPTERS 1–8

Solve each equation or inequality. For the inequalities, also sketch the graph of the inequality.

1. $2x + 3 = 0$ $-\frac{3}{2}$

2. $2x = 3$ $\frac{3}{2}$

3. $2x + 3 > 0$ $x > -\frac{3}{2}$

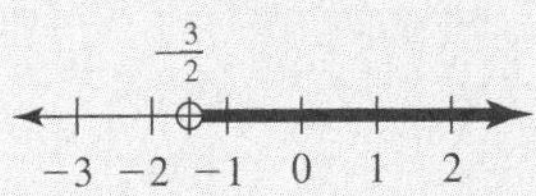

4. $-2x + 3 > 0$ $x < \frac{3}{2}$

$\frac{3}{2}$

−2 −1 0 1 2 3

5. $2(x + 3) = 0$ -3

6. $2x^2 = 3$ $-\frac{\sqrt{6}}{2}, \frac{\sqrt{6}}{2}$

7. $\frac{x}{3} = \frac{2}{x}$ $-\sqrt{6}, \sqrt{6}$

8. $\frac{x-1}{x} = \frac{x}{x-2}$ $\frac{2}{3}$

9. $(2x + 3)^2 = 0$ $-\frac{3}{2}$

10. $(2x + 3)(x - 3) = 0$ $-\frac{3}{2}, 3$

11. $2x^2 + 3 = 0$ No solution

12. $(2x + 3)^2 = 1$ $-2, -1$

13. $(2x + 3)^2 = -1$ No solution

14. $\sqrt{2x^2 - 14} = x - 1$ 3

Let $a = 2$, $b = -3$, and $c = -9$. Find the value of each algebraic expression.

15. b^2 9

16. $-4ac$ 72

17. $b^2 - 4ac$ 81

18. $\sqrt{b^2 - 4ac}$ 9

19. $-b + \sqrt{b^2 - 4ac}$ 12

20. $-b - \sqrt{b^2 - 4ac}$ -6

21. $\frac{-b + \sqrt{b^2 - 4ac}}{2a}$ 3

22. $\frac{-b - \sqrt{b^2 - 4ac}}{2a}$ $-\frac{3}{2}$

Factor each trinomial completely.

23. $x^2 - 6x + 9$ $(x - 3)^2$

24. $x^2 + 10x + 25$ $(x + 5)^2$

25. $x^2 + 12x + 36$ $(x + 6)^2$

26. $x^2 - 20x + 100$ $(x - 10)^2$

27. $2x^2 - 8x + 8$ $2(x - 2)^2$

28. $3x^2 + 6x + 3$ $3(x + 1)^2$

Perform the indicated operation.

29. $(3 + 2x) - (6 - 5x)$ $7x - 3$

30. $(5 + 3t)(4 - 5t)$ $-15t^2 - 13t + 20$

31. $(8 - 6j)(3 + 4j)$ $-24j^2 + 14j + 24$

32. $(1 - u) + (5 + 7u)$ $6u + 6$

33. $(3 - 4v) - (2 - 5v)$ $v + 1$

34. $(2 + t)^2$ $t^2 + 4t + 4$

35. $(t - 7)(t + 7)$ $t^2 - 49$

36. $(3 - 2n)(3 + 2n)$ $-4n^2 + 9$

37. $(1 - m)^2$ $m^2 - 2m + 1$

38. $(-4 - 6t) - (-3 - 8t)$ $2t - 1$

39. $(1 + r)(3 - 4r)$ $-4r^2 - r + 3$

40. $(2 - 6y)(1 + 3y)$ $-18y^2 + 2$

41. $(1 - 2j) + (-6 + 5j)$ $3j - 5$

42. $(-2 - j) + (4 - 5j)$ $-6j + 2$

43. $\frac{4 - 6x}{2}$ $2 - 3x$

44. $\frac{-3 - 9p}{3}$ $-1 - 3p$

45. $\frac{8 - 12q}{-4}$ $-2 + 3q$

46. $\frac{20 - 5z}{-5}$ $-4 + z$

Solve the problem.

47. *Oxygen uptake.* In studying the oxygen uptake rate for marathon runners, Costill and Fox calculate the power expended P in kilocalories per minute using the formula $P = M(av - b)$, where M is the mass of the runner in kilograms and v is the speed in meters per minute (*Medicine and Science in Sports,* Vol. 1). The constants a and b have values $a = 1.02 \times 10^{-3}$ and $b = 2.62 \times 10^{-2}$.

a) Find P for a 60-kg runner who is running at 300 m/min. 16.788 kilocalories per minute

b) Find the velocity of a 55-kg runner who is expending 14 kcal/min. 275.24 m/min

c) Judging from the accompanying graph of velocity and power expenditure for a 55-kg runner, is power expenditure increasing or decreasing as the velocity increases? Increasing

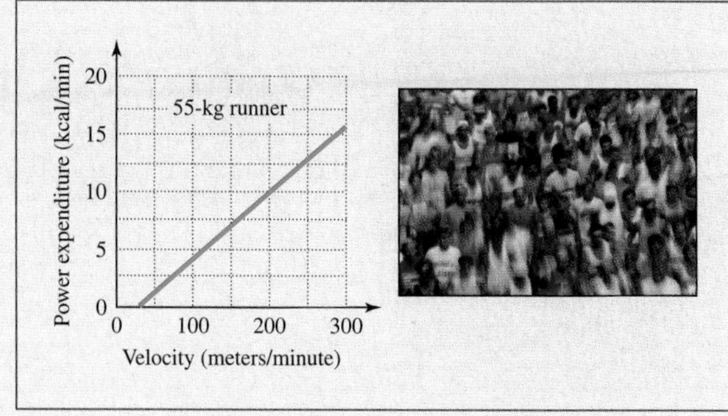

FIGURE FOR EXERCISE 47

CHAPTER 9

Quadratic Equations, Parabolas, and Functions

Throughout time, humans have been building bridges over waterways. Primitive people threw logs across streams or attached ropes to branches to cross the waters. Later, the Romans built stone structures to span rivers and chasms. Throughout the centuries, bridges have been made of wood and stone and later from cast iron, concrete, and steel. Today's bridges are among the most beautiful and complex creations of modern engineering. Whether the bridge spans a small creek or a four-mile-wide stretch of water, mathematics is a part of its very foundation.

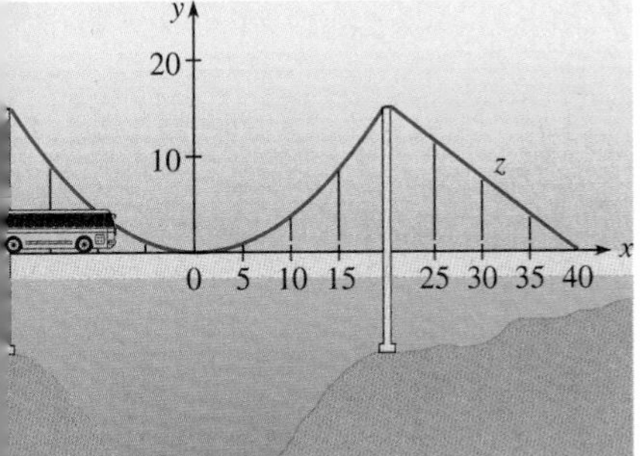

The function of a bridge, the length it must span, and the load it must carry often determine the type of bridge that is built. Some common types designed by civil engineers are cantilevered, arch, cable-stayed, and suspension bridges. The military is known for building trestle bridges and floating or pontoon bridges.

New technology has enabled engineers to build bridges that are stronger, lighter, and less expensive than in the past, as well as being esthetically pleasing. Currently, some engineers are working on making bridges earthquake resistant. Another idea that is being explored is incorporating carbon fibers in cement to warn of small cracks through electronic signals.

In Exercise 45 of Section 9.6 you will see how quadratic equations are used in designing suspension bridges.

9.1 THE SQUARE ROOT PROPERTY AND FACTORING

In This Section

- Definition
- Using the Square Root Property
- Solving Equations by Factoring

We solved some quadratic equations in Chapters 6 and 7 by factoring. In Chapter 8 we solved some quadratic equations using the square root property. In this section we will review the types that you have already learned to solve. In Section 9.2 you will learn a method by which you can solve any quadratic equation.

Definition

We saw the definition of a quadratic equation in Chapter 6, but we will repeat it here.

Quadratic Equation

A **quadratic equation** is an equation of the form

$$ax^2 + bx + c = 0,$$

where a, b, and c are real numbers with $a \neq 0$.

Equations that can be written in the form of the definition may also be called quadratic equations. In Chapters 6, 7, and 8 we solved quadratic equations such as

$$x^2 = 10, \qquad 5(x - 2)^2 = 20, \qquad \text{and} \qquad x^2 - 5x = -6.$$

Using the Square Root Property

If $b = 0$ in $ax^2 + bx + c = 0$, then the quadratic equation can be solved by using the square root property.

EXAMPLE 1 **Using the square root property**

Solve the equations.

a) $x^2 - 9 = 0$ **b)** $2x^2 - 3 = 0$ **c)** $-3(x + 1)^2 = -6$

Solution

a) Solve the equation for x^2, and then use the square root property:

$$x^2 - 9 = 0$$

$$x^2 = 9 \quad \text{Add 9 to each side.}$$

$$x = \pm 3 \quad \text{Square root property}$$

Check 3 and -3 in the original equation. Both 3 and -3 are solutions to $x^2 - 9 = 0$.

b) $2x^2 - 3 = 0$

$$2x^2 = 3$$

$$x^2 = \frac{3}{2}$$

$$x = \pm\sqrt{\frac{3}{2}} \quad \text{Square root property}$$

$$x = \pm\frac{\sqrt{3} \cdot \sqrt{2}}{\sqrt{2} \cdot \sqrt{2}} \quad \text{Rationalize the denominator.}$$

$$x = \pm\frac{\sqrt{6}}{2}$$

Check. The solutions to $2x^2 - 3 = 0$ are $\frac{\sqrt{6}}{2}$ and $-\frac{\sqrt{6}}{2}$.

Study Tip

Effective studying involves actively digging into the subject. Make sure that you are making steady progress. At the end of each week, take note of the progress that you have made. What do you know on Friday that you didn't know on Monday?

c) This equation is a bit different from the previous two equations. If we actually squared the quantity $(x + 1)$, then we would get a term involving x. Then b would not be equal to zero as it is in the other equations. However, we can solve this equation like the others if we do not square $x + 1$.

$$-3(x + 1)^2 = -6$$
$$(x + 1)^2 = 2 \quad \text{Divide each side by } -3.$$
$$x + 1 = \pm\sqrt{2} \quad \text{Square root property}$$
$$x = -1 \pm \sqrt{2} \quad \text{Subtract 1 from each side.}$$

Check $x = -1 \pm \sqrt{2}$ in the original equation:

$$-3(-1 \pm \sqrt{2} + 1)^2 = -3(\pm\sqrt{2})^2 = -3(2) = -6$$

The solutions are $-1 + \sqrt{2}$ and $-1 - \sqrt{2}$. ■

EXAMPLE 2 **A quadratic equation with no real solution**

Solve $x^2 + 12 = 0$.

Solution

The equation $x^2 + 12 = 0$ is equivalent to $x^2 = -12$. Because the square of any real number is nonnegative, this equation has no real solution. ■

Solving Equations by Factoring

In Chapter 6 you learned to factor trinomials and to use factoring to solve some quadratic equations. Recall that quadratic equations are solved by factoring as follows.

Strategy for Solving Quadratic Equations by Factoring

1. Write the equation with 0 on one side of the equal sign.
2. Factor the other side.
3. Use the zero factor property. (Set each factor equal to 0.)
4. Solve the two linear equations.
5. Check the answers in the original quadratic equation.

EXAMPLE 3 **Solving a quadratic equation by factoring**

Solve by factoring.

a) $x^2 + 2x = 8$ **b)** $3x^2 + 13x - 10 = 0$ **c)** $\frac{1}{6}x^2 - \frac{1}{2}x = 3$

Helpful Hint

After you have factored the quadratic polynomial, use FOIL to check that you have factored correctly before proceeding to the next step.

Solution

a)

$$x^2 + 2x = 8$$
$$x^2 + 2x - 8 = 0 \quad \text{Get 0 on the right-hand side.}$$
$$(x + 4)(x - 2) = 0 \quad \text{Factor.}$$
$$x + 4 = 0 \quad \text{or} \quad x - 2 = 0 \quad \text{Zero factor property}$$
$$x = -4 \quad \text{or} \quad x = 2 \quad \text{Solve the linear equations.}$$

Check in the original equation:

$$(-4)^2 + 2(-4) = 16 - 8 = 8$$
$$2^2 + 2 \cdot 2 = 4 + 4 = 8$$

Both -4 and 2 are solutions to the equation.

b) $3x^2 + 13x - 10 = 0$

$$(3x - 2)(x + 5) = 0 \qquad \text{Factor.}$$

$$3x - 2 = 0 \quad \text{or} \quad x + 5 = 0 \qquad \text{Zero factor property}$$

$$3x = 2 \quad \text{or} \quad x = -5$$

$$x = \frac{2}{3} \quad \text{or} \quad x = -5$$

Check in the original equation. Both -5 and $\frac{2}{3}$ are solutions to the equation.

c) $\frac{1}{6}x^2 - \frac{1}{2}x = 3$

$$x^2 - 3x = 18 \qquad \text{Multiply each side by 6.}$$

$$x^2 - 3x - 18 = 0 \qquad \text{Get 0 on the right-hand side.}$$

$$(x - 6)(x + 3) = 0 \qquad \text{Factor.}$$

$$x - 6 = 0 \quad \text{or} \quad x + 3 = 0 \qquad \text{Zero factor property}$$

$$x = 6 \quad \text{or} \quad x = -3$$

Check in the original equation. The solutions are -3 and 6. ■

CAUTION You can set each factor equal to zero only when the product of the factors is zero. Note that $x^2 - 3x = 18$ is equivalent to $x(x - 3) = 18$, but you can make no conclusion about two factors that have a product of 18.

MATH AT WORK

$x^2 + (x+7)^2 = 5^2$

STRUCTURAL ENGINEER

Even as a child, Gregory Brown was building tunnels and bridges in the sand. Now as a structural engineer for Stone and Webster, he is designing and analyzing real-life bridges and buildings. Many factors must be considered in analyzing a new or existing structure. For example, in the northern locations the effects of the weight of snowfall on a roof must be considered, while in the southern areas the strength of hurricane winds must be taken into account. Of course, earthquakes pose yet another consideration.

At the present time, Mr. Brown is working on bridge ratings. To rate a bridge, he first studies the design and construction of the bridge. Then he examines the structure for any kind of deterioration such as rust, cracks, or holes. In addition to the traffic load a bridge carries, the weight and size of trucks using the bridge are evaluated. When all this information is collected, collated, and analyzed, a bridge rating report is submitted. This report notes the extent of the deterioration, provides recommendations regarding vehicle weight limitations, and provides suggested repairs to strengthen the structure.

In Exercises 57 and 58 of this section you will see how an engineer can use a quadratic equation to find the length of a diagonal brace on a bridge.

WARM-UPS

True or false? Explain your answer.

1. Both -4 and 4 satisfy the equation $x^2 - 16 = 0$. True
2. The equation $(x - 3)^2 = 8$ is equivalent to $x - 3 = 2\sqrt{2}$. False
3. Every quadratic equation can be solved by factoring. False
4. Both -5 and 4 are solutions to $(x - 4)(x + 5) = 0$. True
5. The quadratic equation $x^2 = -3$ has no real solutions. True
6. The equation $x^2 = 0$ has no real solutions. False
7. The equation $(2x + 3)(4x - 5) = 0$ is equivalent to $x = \frac{3}{2}$ or $x = \frac{5}{4}$. False
8. The only solution to the equation $(x + 2)^2 = 0$ is -2. True
9. $(x - 3)(x - 5) = 4$ is equivalent to $x - 3 = 2$ or $x - 5 = 2$. False
10. All quadratic equations have two distinct solutions. False

9.1 EXERCISES

Reading and Writing *After reading this section, write out the answers to these questions. Use complete sentences.*

1. What is a quadratic equation?
A quadratic equation is an equation of the form $ax^2 + bx + c = 0$, where $a \neq 0$.
2. What property do we use to solve quadratic equations in which $b = 0$?
If $b = 0$, a quadratic equation can be solved by the square root property.
3. How can a quadratic equation in which $b = 0$ fail to have a real solution?
If $b = 0$, we can get the square root of a negative number and no real solution.
4. What method is discussed for solving quadratic equations in which $b \neq 0$?
If $b \neq 0$, some quadratics can be solved by factoring.
5. When do you need to solve linear equations to find the solutions to a quadratic equation?
After applying the zero factor property we will have linear equations to solve.
6. What new material is presented in this section?
This section reviews material about quadratic equations that was given earlier in this text.

Solve each equation. See Examples 1 and 2.

7. $x^2 - 36 = 0$
$-6, 6$
8. $x^2 - 81 = 0$
$-9, 9$
9. $x^2 + 10 = 0$
No real solution
10. $x^2 + 4 = 0$
No real solution
11. $5x^2 = 50$
$-\sqrt{10}, \sqrt{10}$
12. $7x^2 = 14$
$-\sqrt{2}, \sqrt{2}$
13. $3t^2 - 5 = 0$
$-\frac{\sqrt{15}}{3}, \frac{\sqrt{15}}{3}$
14. $5y^2 - 7 = 0$
$-\frac{\sqrt{35}}{5}, \frac{\sqrt{35}}{5}$
15. $-3y^2 + 8 = 0$
$-\frac{2\sqrt{6}}{3}, \frac{2\sqrt{6}}{3}$
16. $-5w^2 + 12 = 0$
$-\frac{2\sqrt{15}}{5}, \frac{2\sqrt{15}}{5}$
17. $(x - 3)^2 = 4$
$1, 5$
18. $(x + 5)^2 = 9$
$-8, -2$
19. $(y - 2)^2 = 18$
$2 - 3\sqrt{2}, 2 + 3\sqrt{2}$
20. $(m - 5)^2 = 20$
$5 - 2\sqrt{5}, 5 + 2\sqrt{5}$
21. $2(x + 1)^2 = \frac{1}{2}$
$-\frac{3}{2}, -\frac{1}{2}$
22. $-3(x - 1)^2 = -\frac{3}{4}$
$\frac{3}{2}, \frac{1}{2}$
23. $(x - 1)^2 = \frac{1}{2}$
$\frac{2 - \sqrt{2}}{2}, \frac{2 + \sqrt{2}}{2}$
24. $(y + 2)^2 = \frac{1}{2}$
$\frac{-4 - \sqrt{2}}{2}, \frac{-4 + \sqrt{2}}{2}$
25. $\left(x + \frac{1}{2}\right)^2 = \frac{1}{2}$
$\frac{-1 - \sqrt{2}}{2}, \frac{-1 + \sqrt{2}}{2}$
26. $\left(x - \frac{1}{2}\right)^2 = \frac{3}{2}$
$\frac{1 - \sqrt{6}}{2}, \frac{1 + \sqrt{6}}{2}$
27. $(x - 11)^2 = 0$
11
28. $(x + 45)^2 = 0$
-45

Solve each equation by factoring. See Example 3.

29. $x^2 - 2x - 15 = 0$
$-3, 5$
30. $x^2 - x - 12 = 0$
$-3, 4$
31. $x^2 + 6x + 9 = 0$
-3
32. $x^2 + 10x + 25 = 0$
-5

33. $4x^2 - 4x = 8$
$-1, 2$

34. $3x^2 + 3x = 90$
$-6, 5$

35. $3x^2 - 6x = 0$
$0, 2$

36. $-5x^2 + 10x = 0$
$0, 2$

37. $-4t^2 + 6t = 0$
$0, \frac{3}{2}$

38. $-6w^2 + 15w = 0$
$0, \frac{5}{2}$

39. $2x^2 + 11x - 21 = 0$
$-7, \frac{3}{2}$

40. $2x^2 - 5x + 2 = 0$
$2, \frac{1}{2}$

41. $x^2 - 10x + 25 = 0$
5

42. $x^2 - 4x + 4 = 0$
2

43. $x^2 - \frac{7}{2}x = 15$
$-\frac{5}{2}, 6$

44. $3x^2 - \frac{2}{5}x = \frac{1}{5}$
$-\frac{1}{5}, \frac{1}{3}$

45. $\frac{1}{10}a^2 - a + \frac{12}{5} = 0$
$4, 6$

46. $\frac{2}{9}w^2 + \frac{5}{3}w - 3 = 0$
$-9, \frac{3}{2}$

Solve each equation.

47. $x^2 - 2x = 2(3 - x)$
$-\sqrt{6}, \sqrt{6}$

48. $x^2 + 2x = \frac{1 + 4x}{2}$
$-\frac{\sqrt{2}}{2}, \frac{\sqrt{2}}{2}$

49. $x = \frac{27}{12 - x}$
$3, 9$

50. $x = \frac{6}{x + 1}$
$-3, 2$

51. $\sqrt{3x - 8} = x - 2$
$3, 4$

52. $\sqrt{3x - 14} = x - 4$
$5, 6$

Solve each problem.

53. ***Side of a square.*** If the diagonal of a square is 5 meters, then what is the length of a side?
$\frac{5\sqrt{2}}{2}$ meters

54. ***Diagonal of a square.*** If the side of a square is 5 meters, then what is the length of the diagonal? $5\sqrt{2}$ meters

55. ***Howard's journey.*** Howard walked eight blocks east and then four blocks north to reach the public library. How far was he then from where he started? $4\sqrt{5}$ blocks

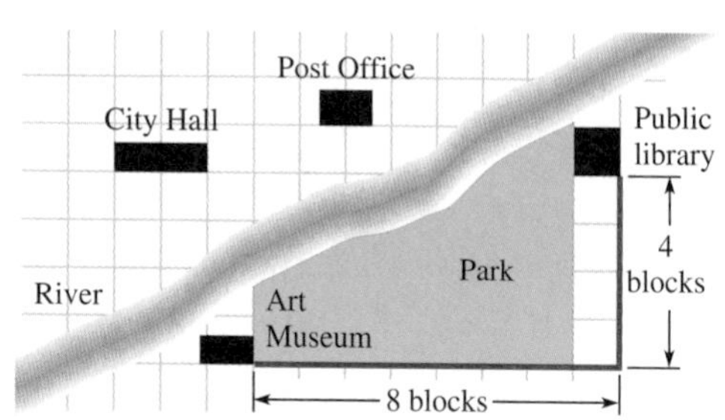

FIGURE FOR EXERCISE 55

56. ***Side and diagonal.*** Each side of a square has length s, and its diagonal has length d. Write a formula for s in terms of d.
$s = \frac{d\sqrt{2}}{2}$

57. ***Designing a bridge.*** Find the length d of the diagonal brace shown in the accompanying diagram.
$2\sqrt{61}$ feet

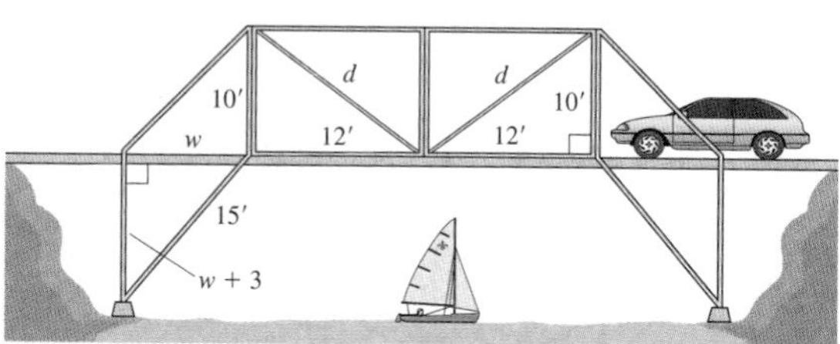

FIGURE FOR EXERCISES 57 AND 58

58. ***Designing a bridge.*** Find the length labeled w in the accompanying diagram. 9 feet

59. ***Two years of interest.*** Tasha deposited \$500 into an account that paid interest compounded annually. At the end of two years she had \$565. Solve the equation $565 = 500(1 + r)^2$ to find the annual rate r. 6.3%

60. ***Rate of increase.*** The price of a new 2000 Dodge Viper convertible was \$68,275 and a new 2002 Viper convertible was \$71,725 (www.edmunds.com). Find the average annual rate of increase (to the nearest tenth of a percent) for that time period by solving the equation

$$71{,}725 = 68{,}275(1 + r)^2. \quad 2.5\%$$

61. ***Projectile motion.*** If an object is projected upward with initial velocity v_0 ft/sec from an initial height of s_0 feet, then its height s (in feet) t seconds after it is projected is given by the formula $s = -16t^2 + v_0t + s_0$.

a) If a baseball is hit upward at 80 ft/sec from a height of 6 feet, then for what values of t is the baseball 102 feet above the ground?

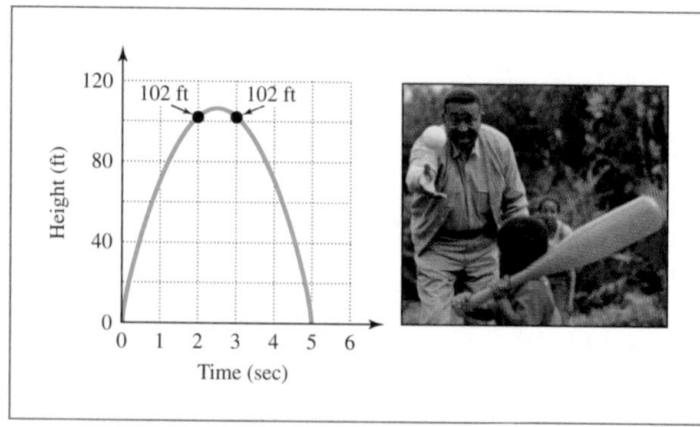

FIGURE FOR EXERCISE 61

b) For what value of t is the baseball back at a height of 6 feet?

a) 2 sec and 3 sec b) 5 sec

62. *Diving time.* A springboard diver can perform complicated maneuvers in a short period of time. If a diver springs into the air at 24 ft/sec from a board that is 16 feet above the water, then in how many seconds will she hit the water? Use the formula from Exercise 61.

2 sec

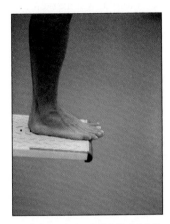

FIGURE FOR EXERCISE 62

63. *Sum of integers*. The formula $S = \frac{n^2 + n}{2}$ gives the sum of the first n positive integers. For what value of n is this sum equal to 45? 9

64. *Serious reading.* Kristy's New Year's resolution is to read one page of *Training Your Boa to Squeeze* on January 1, two pages on January 2, three pages on January 3, and so on. On what date will she finish the 136-page book? See Exercise 63. January 16

GETTING MORE INVOLVED

65. *Writing.* One of the following equations has no real solutions. Find it by inspecting all of the equations (without solving). Explain your answer.

a) $x^2 - 99 = 0$
b) $2(v + 77)^2 = 0$
c) $3(y - 22)^2 + 11 = 0$
d) $5(w - 8)^2 - 9 = 0$

c

66. *Cooperative learning.* For each of three soccer teams A, B, and C to play the other two teams once, it takes three games (AB, AC, and BC). Work in groups to answer the following questions.

a) How many games are required for each team of a four-team league to play every other team once?

6

b) How many games are required in a five-team soccer league?

10

c) Find an expression of the form $an^2 + bn + c$ that gives the number of games required in a soccer league of n teams.

$\frac{1}{2}n^2 - \frac{1}{2}n$

d) The Urban Soccer League has fields available for a 120-game season. If the organizers want each team to play every other team once, then how many teams should be in the league? 16

In This Section

- Perfect Square Trinomials
- Solving a Quadratic Equation by Completing the Square
- Applications

9.2 COMPLETING THE SQUARE

The quadratic equations in Section 9.1 were solved by factoring or the square root property, but some quadratic equations cannot be solved by either of those methods. In this section you will learn a method that works on *any* quadratic equation.

Perfect Square Trinomials

The new method for solving any quadratic equation depends on perfect square trinomials. Recall that a perfect square trinomial is the square of a binomial. Just as we recognize the numbers

$$1, \quad 4, \quad 9, \quad 16, \quad 25, \quad 36, \quad \ldots$$

as being the squares of the positive integers, we can recognize a perfect square trinomial. The following is a list of some perfect square trinomials with a leading coefficient of 1:

$$x^2 + 2x + 1 = (x + 1)^2 \qquad x^2 - 2x + 1 = (x - 1)^2$$
$$x^2 + 4x + 4 = (x + 2)^2 \qquad x^2 - 4x + 4 = (x - 2)^2$$
$$x^2 + 6x + 9 = (x + 3)^2 \qquad x^2 - 6x + 9 = (x - 3)^2$$
$$x^2 + 8x + 16 = (x + 4)^2 \qquad x^2 - 8x + 16 = (x - 4)^2$$

To solve quadratic equations using perfect square trinomials, we must be able to determine the last term of a perfect square trinomial when given the first two terms. This process is called **completing the square.** For example, the perfect square trinomial whose first two terms are $x^2 + 6x$ is $x^2 + 6x + 9$.

If the coefficient of x^2 is 1, there is a simple rule for finding the last term in a perfect square trinomial.

Finding the Last Term

The last term of a perfect square trinomial is the square of one-half of the coefficient of the middle term. In symbols, the perfect square trinomial whose first two terms are $x^2 + bx$ is $x^2 + bx + \left(\frac{b}{2}\right)^2$.

EXAMPLE 1 **Completing the square**

Find the perfect square trinomial whose first two terms are given, and factor the trinomial.

a) $x^2 + 10x$ **b)** $x^2 - 20x$

c) $x^2 + 3x$ **d)** $x^2 - x$

Helpful Hint

Review the rule for squaring a binomial: square the first term, find twice the product of the two terms, then square the last term. If you are still using FOIL to find the square of a binomial, it is time to learn a more efficient rule.

Solution

a) One-half of 10 is 5, and 5 squared is 25. So the perfect square trinomial is $x^2 + 10x + 25$. Factor as follows:

$$x^2 + 10x + 25 = (x + 5)^2$$

b) One-half of -20 is -10, and -10 squared is 100. So the perfect square trinomial is $x^2 - 20x + 100$. Factor as follows:

$$x^2 - 20x + 100 = (x - 10)^2$$

c) One-half of 3 is $\frac{3}{2}$, and $\frac{3}{2}$ squared is $\frac{9}{4}$. So the perfect square trinomial is $x^2 + 3x + \frac{9}{4}$. Factor as follows:

$$x^2 + 3x + \frac{9}{4} = \left(x + \frac{3}{2}\right)^2$$

d) One-half of -1 is $-\frac{1}{2}$, and $\left(-\frac{1}{2}\right)^2 = \frac{1}{4}$. So the perfect square is $x^2 - x + \frac{1}{4}$. Factor as follows:

$$x^2 - x + \frac{1}{4} = \left(x - \frac{1}{2}\right)^2$$

■

Solving a Quadratic Equation by Completing the Square

To complete the squares in Example 1, we simply found the missing last terms. In Examples 2, 3, and 4 we use that process along with the square root property to solve equations of the form $ax^2 + bx + c = 0$. When we use completing the square to solve an equation, we add the appropriate last term to both sides of the equation to obtain an equivalent equation. We first consider an equation in which the coefficient of x^2 is 1.

EXAMPLE 2 **Solving by completing the square ($a = 1$)**

Solve $x^2 + 6x - 7 = 0$ by completing the square.

Helpful Hint

Always check that you have completed the square correctly by squaring the binomial. In this case $(x + 3)^2 = x^2 + 6x + 9$ is correct.

Solution

Add 7 to each side of the equation to isolate $x^2 + 6x$:

$$x^2 + 6x = 7$$

Now complete the square for $x^2 + 6x$. One-half of 6 is 3, and $3^2 = 9$.

$$x^2 + 6x + 9 = 7 + 9 \qquad \text{Add 9 to each side.}$$

$$(x + 3)^2 = 16 \qquad \text{Factor the left side, and simplify the right side.}$$

$$x + 3 = \pm 4 \qquad \text{Square root property}$$

$$x = -3 \pm 4$$

$$x = -3 + 4 \quad \text{or} \quad x = -3 - 4$$

$$x = 1 \quad \text{or} \quad x = -7$$

Check these answers in the original equation. The solutions are -7 and 1. ■

All of the perfect square trinomials in Examples 1 and 2 have 1 as the leading coefficient. If the leading coefficient is not 1, then we must divide each side of the equation by the leading coefficient to get an equation with a leading coefficient of 1. The steps to follow in solving a quadratic equation by completing the square are summarized as follows.

Strategy for Solving a Quadratic Equation by Completing the Square

1. The coefficient of x^2 must be 1.
2. Write the equation with only the x^2-terms and the x-terms on the left-hand side.
3. Complete the square on the left-hand side by adding the square of $\frac{1}{2}$ the coefficient of x to both sides of the equation.
4. Factor the perfect square trinomial as the square of a binomial.
5. Apply the square root property.
6. Solve for x and simplify the answer.
7. Check in the original equation.

In Example 3 we solve a quadratic equation in which the coefficient of x^2 is not 1.

EXAMPLE 3 **Solving by completing the square ($a \neq 1$)**

Solve $2x^2 - 5x - 3 = 0$ by completing the square.

Study Tip

Personal issues can have a tremendous affect on your progress in any course. So do not hesitate to deal with personal problems. If you need help, get it. Most schools have counseling centers that can help you overcome personal issues that are affecting your studies.

Solution

Our perfect square trinomial must begin with x^2 and not $2x^2$:

$$\frac{2x^2 - 5x - 3}{2} = \frac{0}{2} \quad \text{Divide each side by 2 to get 1 for the coefficient of } x^2.$$

$$x^2 - \frac{5}{2}x - \frac{3}{2} = 0 \quad \text{Simplify.}$$

$$x^2 - \frac{5}{2}x = \frac{3}{2} \quad \text{Write only the } x^2\text{- and } x\text{-terms on the left-hand side.}$$

$$x^2 - \frac{5}{2}x + \frac{25}{16} = \frac{3}{2} + \frac{25}{16} \quad \text{Complete the square: } \frac{1}{2}\left(-\frac{5}{2}\right) = -\frac{5}{4}, \left(-\frac{5}{4}\right)^2 = \frac{25}{16}$$

$$\left(x - \frac{5}{4}\right)^2 = \frac{49}{16} \quad \text{Factor the left-hand side.}$$

$$x - \frac{5}{4} = \pm\frac{7}{4} \quad \text{Square root property}$$

$$x = \frac{5}{4} \pm \frac{7}{4}$$

$$x = \frac{5}{4} + \frac{7}{4} \quad \text{or} \quad x = \frac{5}{4} - \frac{7}{4}$$

$$x = \frac{12}{4} \quad \text{or} \quad x = -\frac{2}{4}$$

$$x = 3 \quad \text{or} \quad x = -\frac{1}{2}$$

Check these answers in the original equation. The solutions to the equation are $-\frac{1}{2}$ and 3. ■

The equations in Examples 2 and 3 could have been solved by factoring. The quadratic equation in Example 4 cannot be solved by factoring, but it can be solved by completing the square. In fact, every quadratic equation can be solved by completing the square.

EXAMPLE 4 **A quadratic equation with irrational solutions**

Solve $x^2 + 4x - 3 = 0$ by completing the square.

Solution

$$x^2 + 4x - 3 = 0 \quad \text{Original equation}$$

$$x^2 + 4x = 3 \quad \text{Add 3 to each side to isolate the } x^2\text{- and } x\text{-terms.}$$

$$x^2 + 4x + 4 = 3 + 4 \quad \text{Complete the square by adding 4 to both sides.}$$

$$(x + 2)^2 = 7 \quad \text{Factor the left-hand side.}$$

$$x + 2 = \pm\sqrt{7} \quad \text{Square root property}$$

$$x = -2 \pm \sqrt{7}$$

$$x = -2 + \sqrt{7} \quad \text{or} \quad x = -2 - \sqrt{7}$$

Calculator Close-Up

A good way to check an irrational solution with a calculator is to use the answer key (ANS). The value of ANS is the last value calculated by the calculator.

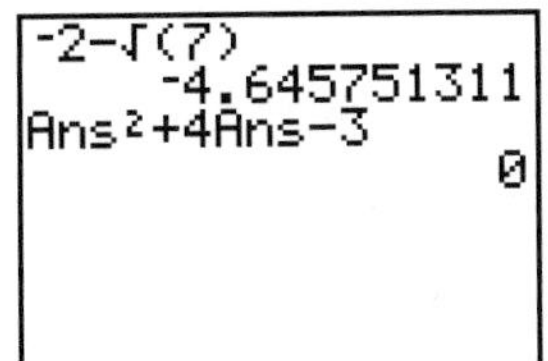

Checking answers involving radicals can be done by using the operations with radicals that you learned in Chapter 8. Replace x with $-2 + \sqrt{7}$ in $x^2 + 4x - 3$:

$$\begin{aligned}(-2 + \sqrt{7})^2 + 4(-2 + \sqrt{7}) - 3 &= 4 - 4\sqrt{7} + 7 - 8 + 4\sqrt{7} - 3 \\ &= 0\end{aligned}$$

You should check $-2 - \sqrt{7}$. Both $-2 + \sqrt{7}$ and $-2 - \sqrt{7}$ satisfy the equation. ■

Applications

In Example 5 we use completing the square to solve a geometric problem.

EXAMPLE 5 **A geometric problem**

The sum of the lengths of the two legs of a right triangle is 8 feet. If the area of the right triangle is 5 square feet, then what are the lengths of the legs?

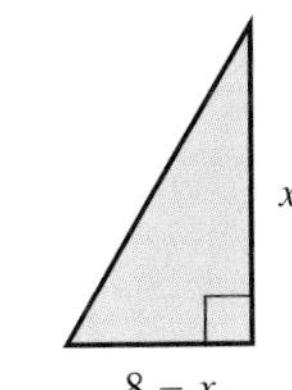

FIGURE 9.1

Solution

If x represents the length of one leg, then $8 - x$ represents the length of the other. See Fig. 9.1. The area of a triangle is given by the formula $A = \frac{1}{2}bh$. Let $A = 5$, $b = x$, and $h = 8 - x$ in this formula:

$$5 = \frac{1}{2}x(8 - x)$$

$$2 \cdot 5 = 2 \cdot \frac{1}{2}x(8 - x) \quad \text{Multiply each side by 2.}$$

$$10 = 8x - x^2$$

$$x^2 - 8x + 10 = 0$$

$$x^2 - 8x = -10 \quad \text{Subtract 10 from each side to isolate the } x^2\text{- and } x\text{-terms.}$$

$$x^2 - 8x + 16 = -10 + 16 \quad \text{Complete the square: } \frac{1}{2}(-8) = -4,\ (-4)^2 = 16$$

$$(x - 4)^2 = 6 \quad \text{Factor.}$$

$$x - 4 = \pm\sqrt{6}$$

$$x - 4 = \sqrt{6} \quad \text{or} \quad x - 4 = -\sqrt{6}$$

$$x = 4 + \sqrt{6} \quad \text{or} \quad x = 4 - \sqrt{6}$$

If $x = 4 + \sqrt{6}$, then

$$8 - x = 8 - (4 + \sqrt{6}) = 4 - \sqrt{6}.$$

If $x = 4 - \sqrt{6}$, then

$$8 - x = 8 - (4 - \sqrt{6}) = 4 + \sqrt{6}.$$

So there is only one pair of possible lengths for the legs: $4 + \sqrt{6}$ ft and $4 - \sqrt{6}$ ft. Check that the area is 5 square feet:

$$A = \frac{1}{2}(4 + \sqrt{6})(4 - \sqrt{6}) = \frac{1}{2}(16 - 6) = \frac{1}{2}(10) = 5$$

■

WARM-UPS

True or false? Explain your answer.

1. Completing the square is used for finding the area of a square. False
2. The polynomial $x^2 + \frac{2}{3}x + \frac{4}{9}$ is a perfect square trinomial. False
3. Every quadratic equation can be solved by factoring. False
4. The polynomial $x^2 - x + 1$ is a perfect square trinomial. False
5. Every quadratic equation can be solved by completing the square. True
6. The solutions to the equation $x - 2 = \pm\sqrt{3}$ are $2 + \sqrt{3}$ and $2 - \sqrt{3}$. True
7. There are no real numbers that satisfy $(x + 7)^2 = -5$. True
8. To solve $x^2 - 5x = 4$ by completing the square, we can add $\frac{25}{4}$ to each side. True
9. One-half of four-fifths is two-fifths. True
10. One-half of three-fourths is three-eighths. True

9.2 EXERCISES

Reading and Writing *After reading this section, write out the answers to these questions. Use complete sentences.*

1. Can every quadratic equation be solved by factoring or the square root property?
   Not every quadratic can be solved by factoring or the square root property.
2. What method can be used to solve any quadratic equation?
   Any quadratic can be solved by completing the square.
3. How do we find the last term in a perfect square trinomial when we know the first two terms?
   The last term is the square of one-half of the coefficient of the middle term.
4. How do you solve a quadratic equation by completing the square when $a \neq 1$?
   If $a \neq 1$, divide each side by a.

Find the perfect square trinomial whose first two terms are given, then factor the trinomial. See Example 1.

5. $x^2 + 6x$ $x^2 + 6x + 9 = (x + 3)^2$
6. $x^2 - 4x$ $x^2 - 4x + 4 = (x - 2)^2$
7. $x^2 + 14x$ $x^2 + 14x + 49 = (x + 7)^2$
8. $x^2 + 16x$ $x^2 + 16x + 64 = (x + 8)^2$
9. $x^2 - 16x$ $x^2 - 16x + 64 = (x - 8)^2$
10. $x^2 - 14x$ $x^2 - 14x + 49 = (x - 7)^2$
11. $t^2 - 18t$ $t^2 - 18t + 81 = (t - 9)^2$
12. $w^2 + 18w$ $w^2 + 18w + 81 = (w + 9)^2$
13. $m^2 + 3m$ $m^2 + 3m + \frac{9}{4} = \left(m + \frac{3}{2}\right)^2$
14. $n^2 - 5n$ $n^2 - 5n + \frac{25}{4} = \left(n - \frac{5}{2}\right)^2$
15. $z^2 + z$ $z^2 + z + \frac{1}{4} = \left(z + \frac{1}{2}\right)^2$
16. $v^2 - v$ $v^2 - v + \frac{1}{4} = \left(v - \frac{1}{2}\right)^2$
17. $x^2 - \frac{1}{2}x$ $x^2 - \frac{1}{2}x + \frac{1}{16} = \left(x - \frac{1}{4}\right)^2$
18. $y^2 + \frac{1}{3}y$ $y^2 + \frac{1}{3}y + \frac{1}{36} = \left(y + \frac{1}{6}\right)^2$
19. $y^2 + \frac{1}{4}y$ $y^2 + \frac{1}{4}y + \frac{1}{64} = \left(y + \frac{1}{8}\right)^2$
20. $z^2 - \frac{4}{3}z$ $z^2 - \frac{4}{3}z + \frac{4}{9} = \left(z - \frac{2}{3}\right)^2$

Factor each perfect square trinomial as the square of a binomial.

21. $x^2 + 10x + 25$
    $(x + 5)^2$
22. $x^2 - 6x + 9$
    $(x - 3)^2$
23. $m^2 - 2m + 1$
    $(m - 1)^2$
24. $n^2 + 4n + 4$
    $(n + 2)^2$
25. $x^2 + x + \frac{1}{4}$
    $\left(x + \frac{1}{2}\right)^2$
26. $y^2 - y + \frac{1}{4}$
    $\left(y - \frac{1}{2}\right)^2$

27. $t^2 + \frac{1}{3}t + \frac{1}{36}$
$\left(t + \frac{1}{6}\right)^2$

28. $v^2 - \frac{2}{3}v + \frac{1}{9}$
$\left(v - \frac{1}{3}\right)^2$

29. $x^2 + \frac{2}{5}x + \frac{1}{25}$
$\left(x + \frac{1}{5}\right)^2$

30. $y^2 - \frac{1}{4}y + \frac{1}{64}$
$\left(y - \frac{1}{8}\right)^2$

Solve each quadratic equation by completing the square. See Examples 2 and 3.

31. $x^2 + 2x - 15 = 0$
$-5, 3$

32. $x^2 + 2x - 24 = 0$
$-6, 4$

33. $x^2 - 4x - 21 = 0$
$-3, 7$

34. $x^2 - 4x - 12 = 0$
$-2, 6$

35. $x^2 + 6x + 9 = 0$
-3

36. $x^2 - 10x + 25 = 0$
5

37. $2t^2 - 3t + 1 = 0$
$\frac{1}{2}, 1$

38. $2t^2 - 3t - 2 = 0$
$-\frac{1}{2}, 2$

39. $2w^2 - 7w + 6 = 0$
$\frac{3}{2}, 2$

40. $4t^2 + 5t - 6 = 0$
$\frac{3}{4}, -2$

41. $3x^2 + 2x - 1 = 0$
$-1, \frac{1}{3}$

42. $3x^2 - 8x - 3 = 0$
$-\frac{1}{3}, 3$

Solve each quadratic equation by completing the square. See Example 4.

43. $x^2 + 2x - 6 = 0$
$-1 - \sqrt{7}, -1 + \sqrt{7}$

44. $x^2 + 4x - 4 = 0$
$-2 - 2\sqrt{2}, -2 + 2\sqrt{2}$

45. $x^2 + 6x + 1 = 0$
$-3 - 2\sqrt{2}, -3 + 2\sqrt{2}$

46. $x^2 - 6x - 3 = 0$
$3 - 2\sqrt{3}, 3 + 2\sqrt{3}$

47. $y^2 - y - 3 = 0$
$\frac{1 - \sqrt{13}}{2}, \frac{1 + \sqrt{13}}{2}$

48. $t^2 + t - 1 = 0$
$\frac{-1 - \sqrt{5}}{2}, \frac{-1 + \sqrt{5}}{2}$

49. $v^2 + 3v - 3 = 0$
$\frac{-3 - \sqrt{21}}{2}, \frac{-3 + \sqrt{21}}{2}$

50. $u^2 - 3u + 1 = 0$
$\frac{3 - \sqrt{5}}{2}, \frac{3 + \sqrt{5}}{2}$

51. $2m^2 - m - 4 = 0$
$\frac{1 - \sqrt{33}}{4}, \frac{1 + \sqrt{33}}{4}$

52. $4q^2 + 2q - 1 = 0$
$\frac{-1 - \sqrt{5}}{4}, \frac{-1 + \sqrt{5}}{4}$

Solve each equation by whichever method is appropriate.

53. $(x - 5)^2 = 7$
$5 - \sqrt{7}, 5 + \sqrt{7}$

54. $x^2 + x = 12$
$-4, 3$

55. $3n^2 - 5 = 0$
$-\frac{\sqrt{15}}{3}, \frac{\sqrt{15}}{3}$

56. $2m^2 + 16 = 0$
No real solution

57. $3x^2 + 1 = 0$
No real solution

58. $x^2 + 6x + 7 = 0$
$-3 - \sqrt{2}, -3 + \sqrt{2}$

59. $x^2 + 5 = 8x - 3$
$4 - 2\sqrt{2}, 4 + 2\sqrt{2}$

60. $2x^2 + 3x = 42 - 2x$
$-6, \frac{7}{2}$

61. $(2x - 7)^2 = 0$
$\frac{7}{2}$

62. $x^2 - 7 = 0$
$-\sqrt{7}, \sqrt{7}$

63. $y^2 + 6y = 11$
$-3 - 2\sqrt{5}, -3 + 2\sqrt{5}$

64. $y^2 + 6y = 0$
$-6, 0$

65. $\frac{1}{4}w^2 + \frac{1}{2} = w$
$2 \pm \sqrt{2}$

66. $\frac{1}{2}z^2 + \frac{1}{2} = 2z$
$2 \pm \sqrt{3}$

67. $t^2 + 0.2t = 0.24$
$-0.6, 0.4$

68. $p^2 - 0.9p + 0.18 = 0$
$0.3, 0.6$

Use a quadratic equation and completing the square to solve each problem. See Example 5.

69. ***Area of a triangle.*** The sum of the measures of the base and height of a triangle is 10 inches. If the area of the triangle is 11 square inches, then what are the measures of the base and height? $5 - \sqrt{3}$ in. and $5 + \sqrt{3}$ in.

70. ***Dimensions of a rectangle.*** A rectangle has a perimeter of 12 inches and an area of 6 square inches. What are the length and width of the rectangle?
Length $3 + \sqrt{3}$ in., width $3 - \sqrt{3}$ in.

71. ***Missing numbers.*** The sum of two numbers is 12, and their product is 34. What are the numbers?
$6 - \sqrt{2}$ and $6 + \sqrt{2}$

72. ***More missing numbers.*** The sum of two numbers is 8, and their product is 11. What are the numbers?
$4 - \sqrt{5}$ and $4 + \sqrt{5}$

73. ***Saving candles.*** Joan has saved the candles from her birthday cake for every year of her life. If Joan has 78 candles, then how old is Joan? (See Exercise 63 of Section 9.1.)
12 years old

74. ***Raffle tickets.*** A charitable organization is selling chances to win a used Corvette. If the tickets are x dollars each, then the members will sell $5000 - 200x$

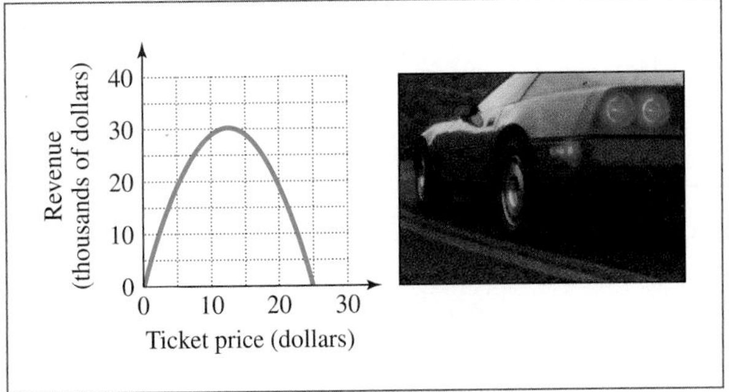

FIGURE FOR EXERCISE 74

tickets. So the total revenue for the tickets is given by $R = x(5000 - 200x)$.

a) What is the revenue if the tickets are sold at \$8 each?

b) For what ticket price is the revenue \$30,000?

c) Use the accompanying graph to estimate the ticket price that will produce the maximum revenue.

a) \$27,200 **b)** \$10 and \$15 **c)** \$12.50

GETTING MORE INVOLVED

75. ***Exploration.***

a) Find the product $[x - (5 + \sqrt{3})][x - (5 - \sqrt{3})]$.

b) Use completing the square to solve the quadratic equation formed by setting the answer to part (a) equal to zero.

c) Write a quadratic equation (in the form $ax^2 + bx + c = 0$) that has solutions $\frac{3 + \sqrt{2}}{2}$ and $\frac{3 - \sqrt{2}}{2}$.

d) Explain how to find a quadratic equation in the form $ax^2 + bx + c = 0$ for any two given solutions.

a) $x^2 - 10x + 22$
b) $5 \pm \sqrt{3}$
c) $x^2 - 3x + \frac{7}{4} = 0$

9.3 THE QUADRATIC FORMULA

In This Section

- Developing the Quadratic Formula
- The Discriminant
- Which Method to Use

In Section 9.2 you learned that every quadratic equation can be solved by completing the square. In this section we use completing the square to get a formula, the quadratic formula, for solving any quadratic equation.

Developing the Quadratic Formula

To develop a formula for solving any quadratic equation, we start with the general quadratic equation

$$ax^2 + bx + c = 0$$

and solve it by completing the square. Assume that a is positive for now, and divide each side by a:

$$\frac{ax^2 + bx + c}{a} = \frac{0}{a} \quad \text{Divide by } a \text{ to get 1 for the coefficient of } x^2.$$

$$x^2 + \frac{b}{a}x + \frac{c}{a} = 0 \quad \text{Simplify.}$$

$$x^2 + \frac{b}{a}x = -\frac{c}{a} \quad \text{Isolate the } x^2\text{- and } x\text{-terms.}$$

Now complete the square on the left. One-half of $\frac{b}{a}$ is $\frac{b}{2a}$, and $\left(\frac{b}{2a}\right)^2 = \frac{b^2}{4a^2}$.

$$x^2 + \frac{b}{a}x + \frac{b^2}{4a^2} = \frac{b^2}{4a^2} - \frac{c}{a} \quad \text{Add } \frac{b^2}{4a^2} \text{ to each side.}$$

$$\left(x + \frac{b}{2a}\right)^2 = \frac{b^2}{4a^2} - \frac{4ac}{4a^2} \quad \text{Factor on the left-hand side, and get a common denominator on the right-hand side.}$$

$$\left(x + \frac{b}{2a}\right)^2 = \frac{b^2 - 4ac}{4a^2}$$

$$x + \frac{b}{2a} = \pm\sqrt{\frac{b^2 - 4ac}{4a^2}} \quad \text{Square root property}$$

$$x = -\frac{b}{2a} \pm \frac{\sqrt{b^2 - 4ac}}{2a} \quad \text{Because } a > 0,\ \sqrt{4a^2} = 2a.$$

$$x = \frac{-b \pm \sqrt{b^2 - 4ac}}{2a} \quad \text{Combine the two expressions.}$$

Study Tip

Try changing subjects or tasks every hour when you study. The brain does not easily assimilate the same material hour after hour. You will learn and remember more from working on it one hour per day than seven hours on Saturday.

We assumed that a was positive so that $\sqrt{4a^2} = 2a$ would be correct. If a is negative, then $\sqrt{4a^2} = -2a$. Either way, the result is the same. It is called the **quadratic formula.** The formula gives x in terms of the coefficients a, b, and c. The quadratic formula is generally used instead of completing the square to solve a quadratic equation that cannot be factored.

The Quadratic Formula

The solutions to $ax^2 + bx + c = 0$, where $a \neq 0$, are given by

$$x = \frac{-b \pm \sqrt{b^2 - 4ac}}{2a}.$$

EXAMPLE 1 **Equations with rational solutions**

Use the quadratic formula to solve each equation.

a) $x^2 + 2x - 3 = 0$ **b)** $4x^2 = -9 + 12x$

Solution

a) To use the formula, we first identify a, b, and c. For the equation

$$\underset{a}{\underset{\uparrow}{1}}x^2 + \underset{b}{\underset{\uparrow}{2}}x - \underset{c}{\underset{\uparrow}{3}} = 0,$$

$a = 1$, $b = 2$, and $c = -3$. Now use these values in the quadratic formula:

$$x = \frac{-b \pm \sqrt{b^2 - 4ac}}{2a}$$

$$x = \frac{-2 \pm \sqrt{2^2 - 4(1)(-3)}}{2(1)} \qquad 2^2 - 4(1)(-3) = 4 + 12 = 16$$

$$x = \frac{-2 \pm \sqrt{16}}{2}$$

$$x = \frac{-2 \pm 4}{2}$$

$$x = \frac{-2 + 4}{2} \quad \text{or} \quad x = \frac{-2 - 4}{2}$$

$$x = 1 \quad \text{or} \quad x = -3$$

Check these answers in the original equation. The solutions are -3 and 1.

b) Write the equation in the form $ax^2 + bx + c = 0$ to identify a, b, and c:

$$4x^2 = -9 + 12x$$

$$4x^2 - 12x + 9 = 0$$

Now $a = 4$, $b = -12$, and $c = 9$. Use these values in the formula:

$$x = \frac{-(-12) \pm \sqrt{(-12)^2 - 4(4)(9)}}{2(4)}$$

$$x = \frac{12 \pm \sqrt{0}}{8} = \frac{12}{8} = \frac{3}{2}$$

Check. The only solution to the equation is $\frac{3}{2}$. ■

The equations in Example 1 could have been solved by factoring. (Try it.) The quadratic equation in Example 2 has an irrational solution and cannot be solved by factoring.

EXAMPLE 2 **An equation with an irrational solution**

Solve $3x^2 - 6x + 1 = 0$.

Calculator Close-Up

Check irrational solutions using the answer key as shown here.

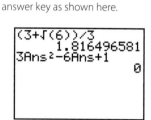

```
(3+√(6))/3
      1.816496581
3Ans²-6Ans+1
                0
```

Solution

For this equation, $a = 3$, $b = -6$, and $c = 1$:

$$x = \frac{-(-6) \pm \sqrt{(-6)^2 - 4(3)(1)}}{2(3)}$$

$$= \frac{6 \pm \sqrt{24}}{6}$$

$$= \frac{6 \pm 2\sqrt{6}}{6} \qquad \sqrt{24} = \sqrt{4}\sqrt{6} = 2\sqrt{6}$$

$$= \frac{2(3 \pm \sqrt{6})}{2(3)} \qquad \text{Numerator and denominator have 2 as a common factor.}$$

$$= \frac{3 \pm \sqrt{6}}{3}$$

The two solutions are the irrational numbers $\frac{3 + \sqrt{6}}{3}$ and $\frac{3 - \sqrt{6}}{3}$. ■

We have seen quadratic equations such as $x^2 = -9$ that do not have any real number solutions. In general, you can conclude that a quadratic equation has no real number solutions if you get a square root of a negative number in the quadratic formula.

EXAMPLE 3 **A quadratic equation with no real number solutions**

Solve $5x^2 - x + 1 = 0$.

Solution

For this equation we have $a = 5$, $b = -1$, and $c = 1$:

$$x = \frac{1 \pm \sqrt{(-1)^2 - 4(5)(1)}}{2(5)} \qquad b = -1,\ -b = 1$$

$$x = \frac{1 \pm \sqrt{-19}}{10}$$

The equation has no real solutions because $\sqrt{-19}$ is not real. ■

The Discriminant

A quadratic equation can have two real solutions, one real solution, or no real number solutions, depending on the value of $b^2 - 4ac$. If $b^2 - 4ac$ is positive, as in Example 1(a) and Example 2, we get two solutions. If $b^2 - 4ac$ is 0, we get only one solution, as in Example 1(b). If $b^2 - 4ac$ is negative, there are no real number solutions, as in Example 3. Table 9.1 summarizes these facts.

Value of $b^2 - 4ac$	Number of real solutions to $ax^2 + bx + c = 0$
Positive	2
Zero	1
Negative	0

TABLE 9.1

The quantity $b^2 - 4ac$ is called the **discriminant** because its value determines the number of real solutions to the quadratic equation.

EXAMPLE 4 **The number of real solutions**

Find the value of the discriminant, and determine the number of real solutions to each equation.

a) $3x^2 - 5x + 1 = 0$

b) $x^2 + 6x + 9 = 0$

c) $2x^2 + 1 = x$

Solution

a) For the equation $3x^2 - 5x + 1 = 0$ we have $a = 3$, $b = -5$, and $c = 1$. Now find the value of the discriminant:

$$b^2 - 4ac = (-5)^2 - 4(3)(1) = 25 - 12 = 13$$

Because the discriminant is positive, there are two real solutions to this quadratic equation.

b) For the equation $x^2 + 6x + 9 = 0$, we have $a = 1$, $b = 6$, and $c = 9$:

$$b^2 - 4ac = (6)^2 - 4(1)(9) = 36 - 36 = 0$$

Since the discriminant is zero, there is only one real solution to the equation.

c) We must first rewrite the equation:

$$2x^2 + 1 = x$$

$$2x^2 - x + 1 = 0 \quad \text{Subtract } x \text{ from each side.}$$

Now $a = 2$, $b = -1$, and $c = 1$.

$$b^2 - 4ac = (-1)^2 - 4(2)(1) = 1 - 8 = -7$$

Because the discriminant is negative, the equation has no real number solutions. ■

Which Method to Use

If the quadratic equation is simple enough, we can solve it by factoring or by the square root property. These methods should be considered first. *All quadratic equations can be solved by the quadratic formula.* Remember that the quadratic formula is just a shortcut to completing the square and is usually easier to use. However, you should learn completing the square because it is used elsewhere in algebra. The available methods are summarized as follows.

Helpful Hint

If our only intent is to get the answer, then we would probably use a calculator that is programmed with the quadratic formula. However, by learning different methods we gain insight into the problem and get valuable practice with algebra. So be sure you learn all of the methods.

Solving the Quadratic Equation $ax^2 + bx + c = 0$

Method	Comments	Examples
Square root property	Use when $b = 0$.	If $x^2 = 7$, then $x = \pm\sqrt{7}$. If $(x - 2)^2 = 9$, then $x - 2 = \pm 3$.
Factoring	Use when the polynomial can be factored.	$x^2 + 5x + 6 = 0$ $(x + 2)(x + 3) = 0$
Quadratic formula	Use when the first two methods do not apply.	$x^2 + 2x - 6 = 0$ $x = \dfrac{-2 \pm \sqrt{2^2 - 4 \cdot 1 \cdot (-6)}}{2 \cdot 1}$
Completing the square	Use only when this method is specified.	$x^2 + 4x - 9 = 0$ $x^2 + 4x + 4 = 9 + 4$ $(x + 2)^2 = 13$

WARM-UPS

True or false? Explain your answer.

1. Completing the square is used to develop the quadratic formula. True
2. For the equation $x^2 - x + 1 = 0$, we have $a = 1$, $b = -x$, and $c = 1$. False
3. For the equation $x^2 - 3 = 5x$, we have $a = 1$, $b = -3$, and $c = 5$. False
4. The quadratic formula can be expressed as $x = -b \pm \dfrac{\sqrt{b^2 - 4ac}}{2a}$. False
5. The quadratic equation $2x^2 - 6x = 0$ has two real solutions. True
6. All quadratic equations have two distinct real solutions. False
7. Some quadratic equations cannot be solved by the quadratic formula. False
8. We could solve $2x^2 - 6x = 0$ by factoring, completing the square, or the quadratic formula. True
9. The equation $x^2 = x$ is equivalent to $\left(x - \frac{1}{2}\right)^2 = \frac{1}{4}$. True
10. The only solution to $x^2 + 6x + 9 = 0$ is -3. True

9.3 EXERCISES

Reading and Writing *After reading this section, write out the answers to these questions. Use complete sentences.*

1. What method presented here can be used to solve any quadratic equation?
The quadratic formula solves any quadratic equation.
2. What is the quadratic formula?
The quadratic formula is $x = \dfrac{-b \pm \sqrt{b^2 - 4ac}}{2a}$.
3. What is the quadratic formula used for?
The quadratic formula is used to solve $ax^2 + bx + c = 0$.
4. What is the discriminant?
The discriminant is $b^2 - 4ac$.
5. How can you determine whether there are no real solutions to a quadratic equation?
If $b^2 - 4ac < 0$, then there are no real solutions.
6. What methods have we studied for solving quadratic equations?
We have solved quadratic equations by factoring, the square root property, completing the square, and the quadratic formula.

Solve by the quadratic formula. See Examples 1–3.

7. $x^2 + 2x - 15 = 0$
$-5, 3$

8. $x^2 - 3x - 18 = 0$
$6, -3$

9. $x^2 + 10x + 25 = 0$
-5

10. $x^2 - 12x + 36 = 0$
6

11. $2x^2 + x - 6 = 0$
$-2, \frac{3}{2}$

12. $2x^2 + x - 15 = 0$
$-3, \frac{5}{2}$

13. $4x^2 + 4x - 3 = 0$
$-\frac{3}{2}, \frac{1}{2}$

14. $4x^2 + 8x + 3 = 0$
$-\frac{3}{2}, -\frac{1}{2}$

15. $2y^2 - 6y + 3 = 0$
$\frac{3 - \sqrt{3}}{2}, \frac{3 + \sqrt{3}}{2}$

16. $3y^2 + 6y + 2 = 0$
$\frac{-3 - \sqrt{3}}{3}, \frac{-3 + \sqrt{3}}{3}$

17. $2t^2 + 4t = -1$
$\frac{-2 - \sqrt{2}}{2}, \frac{-2 + \sqrt{2}}{2}$

18. $w^2 + 2 = 4w$
$2 - \sqrt{2}, 2 + \sqrt{2}$

19. $2x^2 - 2x + 3 = 0$
No real solution

20. $-2x^2 + 3x - 9 = 0$
No real solution

21. $8x^2 = 4x$
$0, \frac{1}{2}$

22. $9y^2 + 3y = -6y$
$-1, 0$

23. $5w^2 - 3 = 0$
$-\frac{\sqrt{15}}{5}, \frac{\sqrt{15}}{5}$

24. $4 - 7z^2 = 0$
$\frac{-2\sqrt{7}}{7}, \frac{2\sqrt{7}}{7}$

25. $\frac{1}{2}h^2 + 7h + \frac{1}{2} = 0$
$-7 + 4\sqrt{3}, -7 - 4\sqrt{3}$

26. $\frac{1}{4}z^2 - 6z + 3 = 0$
$12 - 2\sqrt{33}, 12 + 2\sqrt{33}$

Find the value of the discriminant, and state how many real solutions there are to each quadratic equation. See Example 4.

27. $4x^2 - 4x + 1 = 0$
0, one

28. $9x^2 + 6x + 1 = 0$
0, one

29. $6x^2 - 7x + 4 = 0$
-47, none

30. $-3x^2 + 5x - 7 = 0$
-59, none

31. $-5t^2 - t + 9 = 0$
181, two

32. $-2w^2 - 6w + 5 = 0$
76, two

33. $4x^2 - 12x + 9 = 0$
0, one

34. $9x^2 + 12x + 4 = 0$
0, one

35. $x^2 + x + 4 = 0$
-15, none

36. $y^2 - y + 2 = 0$
-7, none

37. $x - 5 = 3x^2$
-59, none

38. $4 - 3x = x^2$
25, two

Use the method of your choice to solve each equation.

39. $x^2 + \frac{3}{2}x = 1$
$-2, \frac{1}{2}$

40. $x^2 - \frac{7}{2}x = 2$
$-\frac{1}{2}, 4$

41. $(x - 1)^2 + (x - 2)^2 = 5$
$0, 3$

42. $x^2 + (x - 3)^2 = 29$
$-2, 5$

43. $\frac{1}{x} + \frac{1}{x + 2} = \frac{5}{12}$
$-\frac{6}{5}, 4$

44. $\frac{1}{x} + \frac{1}{x + 1} = \frac{5}{6}$
$-\frac{3}{5}, 2$

45. $x^2 + 6x + 8 = 0$
$-4, -2$

46. $2x^2 - 5x - 3 = 0$
$-\frac{1}{2}, 3$

47. $x^2 - 9x = 0$
$0, 9$

48. $x^2 - 9 = 0$
$-3, 3$

49. $(x + 5)^2 = 9$ $\quad -8, -2$

50. $(3x - 1)^2 = 0$ $\quad \frac{1}{3}$

51. $x(x - 3) = 2 - 3(x + 4)$ No real solution

52. $(x - 1)(x + 4) = (2x - 4)^2$ $\quad \frac{4}{3}, 5$

53. $\frac{x}{3} = \frac{x + 2}{x}$ $\quad \frac{3 - \sqrt{33}}{2}, \frac{3 + \sqrt{33}}{2}$

54. $\frac{x - 2}{x} = \frac{5}{x + 2}$ $\quad \frac{5 - \sqrt{41}}{2}, \frac{5 + \sqrt{41}}{2}$

55. $2x^2 - 3x = 0$ $\quad 0, \frac{3}{2}$

56. $x^2 = 5$ $\quad -\sqrt{5}, \sqrt{5}$

Use a calculator to find the approximate solutions to each quadratic equation. Round answers to two decimal places.

57. $x^2 - 3x - 3 = 0$
$-0.79, 3.79$

58. $x^2 - 2x - 2 = 0$
$-0.73, 2.73$

59. $x^2 - x - 3.2 = 0$
$-1.36, 2.36$

60. $x^2 - 4.3x + 3 = 0$
$0.88, 3.42$

61. $5.29x^2 - 3.22x + 0.49 = 0$ $\quad 0.30$

62. $2.6x^2 + 3.1x - 5 = 0$ $\quad -2.11, 0.91$

Use a calculator to solve each problem.

63. ***Concert revenues.*** A promoter uses the formula $R = -500x^2 + 20{,}000x$ to predict the concert revenue in dollars when the price of a ticket is x dollars.

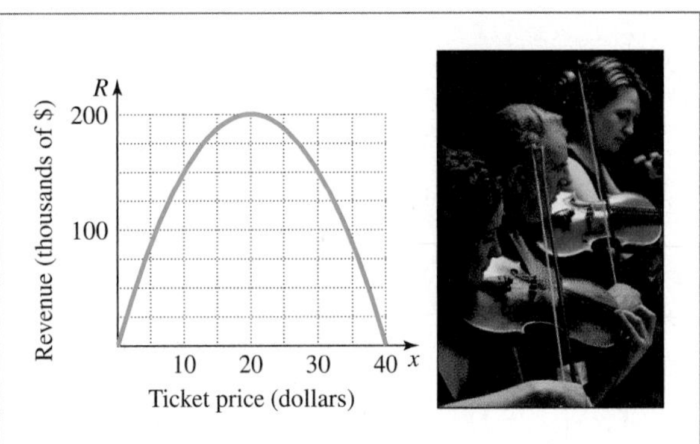

FIGURE FOR EXERCISE 63

a) What is the revenue when the ticket price is \$10?
\$150,000

b) What ticket prices would produce zero revenue?
\$0 and \$40

c) What is the ticket price if the revenue is \$196,875?
\$17.50 or \$22.50

d) Examine the accompanying graph to determine the ticket price that produces the maximum revenue.
\$20

64. ***Raffle tickets.*** The formula $R = -200x^2 + 5000x$ was used in Exercise 74 of Section 9.2 to predict the revenue when raffle tickets are sold for x dollars each. For what ticket price is the revenue \$25,000? \$6.91 and \$18.09

9.4 APPLICATIONS OF QUADRATIC EQUATIONS

In This Section

- Geometric Applications
- Work Problems
- Vertical Motion

In this section we will solve problems that involve quadratic equations.

Geometric Applications

Quadratic equations can be used to solve problems involving area.

EXAMPLE 1 **Dimensions of a rectangle**

The length of a rectangular flower bed is 2 feet longer than the width. If the area is 6 square feet, then what are the exact length and width? Also find the approximate dimensions of the rectangle to the nearest tenth of a foot.

Solution

Let x represent the width, and $x + 2$ represent the length as shown in Fig. 9.2. Write an equation using the formula for the area of a rectangle, $A = LW$:

$$x(x + 2) = 6 \quad \text{The area is 6 square feet.}$$
$$x^2 + 2x - 6 = 0$$

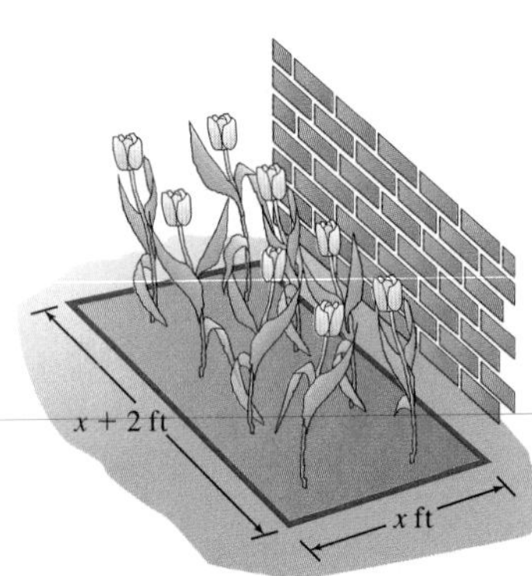

FIGURE 9.2

We use the quadratic formula to solve the equation:

$$x = \frac{-2 \pm \sqrt{2^2 - 4(1)(-6)}}{2(1)} = \frac{-2 \pm \sqrt{28}}{2}$$
$$= \frac{-2 \pm 2\sqrt{7}}{2} = \frac{2(-1 \pm \sqrt{7})}{2} = -1 \pm \sqrt{7}$$

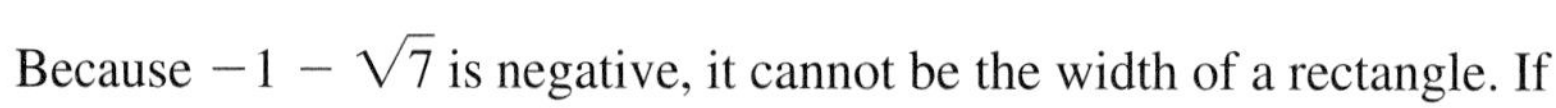

Because $-1 - \sqrt{7}$ is negative, it cannot be the width of a rectangle. If

$$x = -1 + \sqrt{7},$$

then

$$x + 2 = -1 + \sqrt{7} + 2 = 1 + \sqrt{7}.$$

So the exact width is $-1 + \sqrt{7}$ feet, and the exact length is $1 + \sqrt{7}$ feet. We can check that these dimensions give an area of 6 square feet as follows:

$$LW = (1 + \sqrt{7})(-1 + \sqrt{7}) = -1 - \sqrt{7} + \sqrt{7} + 7 = 6$$

Use a calculator to find the approximate dimensions of 1.6 and 3.6 feet. ■

Work Problems

The work problems in this section are similar to the work problems that you solved in Chapter 6. However, you will need the quadratic formula to solve the work problems presented in this section.

9.5 COMPLEX NUMBERS

In This Section

- Definition
- Operations with Complex Numbers
- Square Roots of Negative Numbers
- Complex Solutions to Quadratic Equations

In this chapter we have seen quadratic equations that have no solution in the set of real numbers. In this section you will learn that the set of real numbers is contained in the set of complex numbers. *Quadratic equations that have no real number solutions have solutions that are complex numbers.*

Definition

If we try to solve the equation $x^2 + 1 = 0$ by the square root property, we get

$$x^2 = -1 \quad \text{or} \quad x = \pm\sqrt{-1}.$$

Because $\sqrt{-1}$ has no meaning in the real number system, the equation has no real solution. However, there is an extension of the real number system, called the *complex numbers,* in which $x^2 = -1$ has two solutions.

The complex numbers are formed by adding the **imaginary unit** i to the real number system. We make the definitions that

$$i = \sqrt{-1} \qquad \text{and} \qquad i^2 = -1.$$

In the complex number system, $x^2 = -1$ has two solutions: i and $-i$.

The set of complex numbers is defined as follows.

Helpful Hint

All numbers are ideas. They exist only in our minds. So in a sense we "imagine" all numbers. However, the imaginary numbers are a bit harder to imagine than the real numbers and so only they are called imaginary.

Complex Numbers

The set of complex numbers is the set of all numbers of the form

$$a + bi,$$

where a and b are real numbers, $i = \sqrt{-1}$, and $i^2 = -1$.

In the complex number $a + bi$, a is called the **real part** and b is called the **imaginary part.** If $b \neq 0$, the number $a + bi$ is called an **imaginary number.**

In dealing with complex numbers, we treat $a + bi$ as if it were a binomial with variable i. Thus we would write $2 + (-3)i$ as $2 - 3i$. We agree that $2 + i3$, $3i + 2$, and $i3 + 2$ are just different ways of writing $2 + 3i$. Some examples of complex numbers are

$$2 + 3i, \quad -2 - 5i, \quad 0 + 4i, \quad 9 + 0i, \quad \text{and} \quad 0 + 0i.$$

For simplicity we will write only $4i$ for $0 + 4i$. The complex number $9 + 0i$ is the real number 9, and $0 + 0i$ is the real number 0. Any complex number with $b = 0$, is a real number. The diagram in Fig. 9.3 shows the relationships between the complex numbers, the real numbers, and the imaginary numbers.

Complex numbers

Real numbers	Imaginary numbers
$-3,\ \pi,\ \frac{5}{2}, 0, -9, \sqrt{2}$	$i, 2 + 3i, \sqrt{-5}, -3 - 8i$

FIGURE 9.3

Operations with Complex Numbers

We define addition and subtraction of complex numbers as follows.

Addition and Subtraction of Complex Numbers

$$(a + bi) + (c + di) = (a + c) + (b + d)i$$
$$(a + bi) - (c + di) = (a - c) + (b - d)i$$

According to the definition, we perform these operations as if the complex numbers were binomials with i a variable.

EXAMPLE 1 **Adding and subtracting complex numbers**

Perform the indicated operations.

a) $(2 + 3i) + (4 + 5i)$ **b)** $(2 - 3i) + (-1 - i)$

c) $(3 + 4i) - (1 + 7i)$ **d)** $(2 - 3i) - (-2 - 5i)$

Study Tip

Stay calm and confident. Take breaks when you study. Get six to eight hours of sleep every night and keep reminding yourself that working hard all of the semester will really pay off.

Solution

a) $(2 + 3i) + (4 + 5i) = 6 + 8i$

b) $(2 - 3i) + (-1 - i) = 1 - 4i$

c) $(3 + 4i) - (1 + 7i) = 3 + 4i - 1 - 7i$
$= 2 - 3i$

d) $(2 - 3i) - (-2 - 5i) = 2 - 3i + 2 + 5i$
$= 4 + 2i$ ■

We define multiplication of complex numbers as follows.

Multiplication of Complex Numbers

$$(a + bi)(c + di) = (ac - bd) + (ad + bc)i$$

There is no need to memorize the definition of multiplication of complex numbers. We multiply complex numbers in the same way we multiply polynomials: We use the distributive property. We can also use the FOIL method and the fact that $i^2 = -1$.

EXAMPLE 2 **Multiplying complex numbers**

Perform the indicated operations.

a) $2i(1 - 3i)$ **b)** $(-2 - 5i)(6 - 7i)$

c) $(5i)^2$ **d)** $(-5i)^2$

e) $(3 - 2i)(3 + 2i)$

Solution

a) $2i(1 - 3i) = 2i - 6i^2$ Distributive property
$= 2i - 6(-1)$ $i^2 = -1$
$= 6 + 2i$

Calculator Close-Up

Many calculators can perform operations with complex numbers.

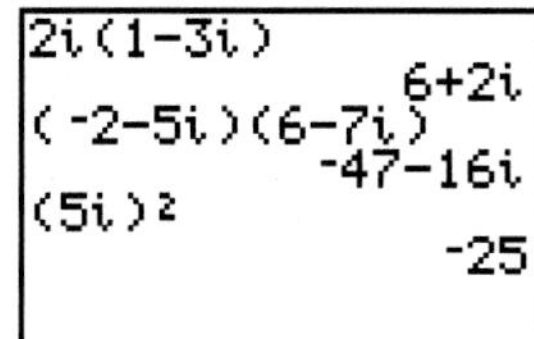

b) Use FOIL to multiply these complex numbers:

$$\begin{aligned}(-2 - 5i)(6 - 7i) &= -12 + 14i - 30i + 35i^2 \\ &= -12 - 16i + 35(-1) \quad i^2 = -1 \\ &= -12 - 16i - 35 \\ &= -47 - 16i\end{aligned}$$

c) $(5i)^2 = 25i^2 = 25(-1) = -25$

d) $(-5i)^2 = (-5)^2i^2 = 25(-1) = -25$

e) $(3 - 2i)(3 + 2i) = 9 - 4i^2$

$$\begin{aligned}&= 9 - 4(-1) \\ &= 9 + 4 \\ &= 13\end{aligned}$$

■

Notice that the product of the imaginary numbers $3 - 2i$ and $3 + 2i$ in Example 2(e) is a real number. We call $3 - 2i$ and $3 + 2i$ *complex conjugates* of each other.

Complex Conjugates

The complex numbers $a + bi$ and $a - bi$ are called **complex conjugates** of each other. Their product is the real number $a^2 + b^2$.

EXAMPLE 3 **Complex conjugates**

Find the product of the given complex number and its conjugate.

a) $4 - 3i$ **b)** $-2 + 5i$

c) $-i$

Helpful Hint

The last time we used "conjugate" was to refer to expressions such as $1 + \sqrt{3}$ and $1 - \sqrt{3}$ as conjugates. Note that we use the rule for the product of a sum and a difference to multiply the radical conjugates or the complex conjugates.

Solution

a) The complex conjugate of $4 - 3i$ is $4 + 3i$:

$$(4 - 3i)(4 + 3i) = 16 - 9i^2 = 16 + 9 = 25$$

b) The conjugate of $-2 + 5i$ is $-2 - 5i$:

$$(-2 + 5i)(-2 - 5i) = 4 - 25i^2 = 4 + 25 = 29$$

c) The conjugate of $-i$ is i:

$$(-i)(i) = -i^2 = -(-1) = 1$$

■

To divide a complex number by a real number, we divide each part by the real number. For example,

$$\frac{4 - 6i}{2} = 2 - 3i.$$

We use the idea of complex conjugates to divide by a complex number. The process is similar to rationalizing the denominator.

Dividing by a Complex Number

To divide by a complex number, multiply the numerator and denominator of the quotient by the complex conjugate of the denominator.

EXAMPLE 4 **Dividing complex numbers**

Perform the indicated operations.

a) $\dfrac{2}{3 - 4i}$ **b)** $\dfrac{6}{2 + i}$ **c)** $\dfrac{3 - 2i}{i}$

Solution

a) Multiply the numerator and denominator by $3 + 4i$, the conjugate of $3 - 4i$:

$$\frac{2}{3 - 4i} = \frac{2(3 + 4i)}{(3 - 4i)(3 + 4i)}$$

$$= \frac{6 + 8i}{9 - 16i^2}$$

$$= \frac{6 + 8i}{25} \quad 9 - 16i^2 = 9 - 16(-1) = 25$$

$$= \frac{6}{25} + \frac{8}{25}i$$

b) Multiply the numerator and denominator by $2 - i$, the conjugate of $2 + i$:

$$\frac{6}{2 + i} = \frac{6(2 - i)}{(2 + i)(2 - i)}$$

$$= \frac{12 - 6i}{4 - i^2}$$

$$= \frac{12 - 6i}{5} \quad 4 - i^2 = 4 - (-1) = 5$$

$$= \frac{12}{5} - \frac{6}{5}i$$

c) Multiply the numerator and denominator by $-i$, the conjugate of i:

$$\frac{3 - 2i}{i} = \frac{(3 - 2i)(-i)}{i(-i)} = \frac{-3i + 2i^2}{-i^2} = \frac{-3i - 2}{1} = -2 - 3i$$

■

Square Roots of Negative Numbers

In Example 2 we saw that both

$$(5i)^2 = -25 \quad \text{and} \quad (-5i)^2 = -25.$$

Because the square of each of these complex numbers is -25, both $5i$ and $-5i$ are square roots of -25. When we use the radical notation, we write

$$\sqrt{-25} = 5i.$$

The square root of a negative number is not a real number, it is a complex number.

Square Root of a Negative Number

For any positive number b,

$$\sqrt{-b} = i\sqrt{b}.$$

For example, $\sqrt{-9} = i\sqrt{9} = 3i$ and $\sqrt{-7} = i\sqrt{7}$. Note that the expression $\sqrt{7}i$ could easily be mistaken for the expression $\sqrt{7i}$, where i is under the radical. For this reason, when the coefficient of i contains a radical, we write i preceding the radical.

EXAMPLE 5 **Square roots of negative numbers**

Write each expression in the form $a + bi$, where a and b are real numbers.

a) $2 + \sqrt{-4}$ **b)** $\dfrac{2 + \sqrt{-12}}{2}$ **c)** $\dfrac{-2 - \sqrt{-18}}{3}$

Solution

a)
$$\begin{aligned} 2 + \sqrt{-4} &= 2 + i\sqrt{4} \\ &= 2 + 2i \end{aligned}$$

b)
$$\begin{aligned} \frac{2 + \sqrt{-12}}{2} &= \frac{2 + i\sqrt{12}}{2} \\ &= \frac{2 + 2i\sqrt{3}}{2} \qquad \sqrt{12} = \sqrt{4} \cdot \sqrt{3} = 2\sqrt{3} \\ &= 1 + i\sqrt{3} \end{aligned}$$

c)
$$\begin{aligned} \frac{-2 - \sqrt{-18}}{3} &= \frac{-2 - i\sqrt{18}}{3} \\ &= \frac{-2 - 3i\sqrt{2}}{3} \qquad \sqrt{18} = \sqrt{9}\sqrt{2} = 3\sqrt{2} \\ &= -\frac{2}{3} - i\sqrt{2} \end{aligned}$$

■

Complex Solutions to Quadratic Equations

Helpful Hint

The fundamental theorem of algebra says that an nth degree polynomial equation has exactly n solutions in the system of complex numbers. They don't all have to be different. For example, a quadratic equation such as $x^2 + 6x + 9 = 0$ has two solutions, both of which are -3.

The equation $x^2 = -4$ has no real number solutions, but it has two complex solutions, which can be found as follows:

$$x^2 = -4$$
$$x = \pm\sqrt{-4} = \pm i\sqrt{4} = \pm 2i$$

Check:

$$(2i)^2 = 4i^2 = 4(-1) = -4$$
$$(-2i)^2 = 4i^2 = -4$$

Both $2i$ and $-2i$ are solutions to the equation.

Consider the general quadratic equation

$$ax^2 + bx + c = 0,$$

where a, b, and c are real numbers. If the discriminant $b^2 - 4ac$ is positive, then the quadratic equation has two real solutions. If the discriminant is 0, then the equation has one real solution. If the discriminant is negative, then the equation has two complex solutions. *In the complex number system, all quadratic equations have solutions.* We can use the quadratic formula to find them.

EXAMPLE 6 **Quadratics with imaginary solutions**

Find the complex solutions to the quadratic equations.

a) $x^2 - 2x + 5 = 0$ **b)** $2x^2 + 3x + 5 = 0$

Calculator Close-Up

The answers in Example 6(b) can be checked with a calculator as shown here.

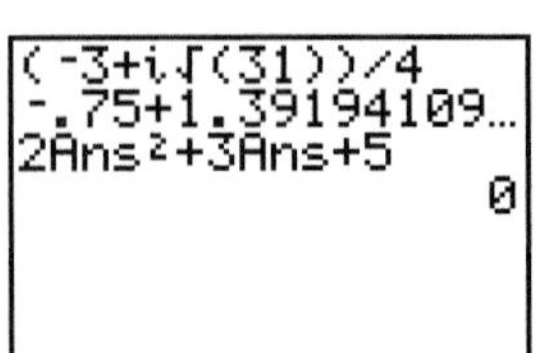

Solution

a) To solve $x^2 - 2x + 5 = 0$, use $a = 1$, $b = -2$, and $c = 5$ in the quadratic formula:

$$x = \frac{2 \pm \sqrt{(-2)^2 - 4(1)(5)}}{2(1)}$$

$$= \frac{2 \pm \sqrt{-16}}{2} = \frac{2 \pm 4i}{2} = 1 \pm 2i$$

We can use the operations with complex numbers to check these solutions:

$$(1 + 2i)^2 - 2(1 + 2i) + 5 = 1 + 4i + 4i^2 - 2 - 4i + 5$$
$$= 1 + 4i - 4 - 2 - 4i + 5 = 0$$

You should verify that $1 - 2i$ also satisfies the equation. The solutions are $1 - 2i$ and $1 + 2i$.

b) To solve $2x^2 + 3x + 5 = 0$, use $a = 2$, $b = 3$, and $c = 5$ in the quadratic formula:

$$x = \frac{-3 \pm \sqrt{3^2 - 4(2)(5)}}{2(2)}$$

$$= \frac{-3 \pm \sqrt{-31}}{4} = \frac{-3 \pm i\sqrt{31}}{4}$$

Check these answers. The solutions are $\frac{-3 + i\sqrt{31}}{4}$ and $\frac{-3 - i\sqrt{31}}{4}$. ■

The following box summarizes the basic facts about complex numbers.

Complex Numbers

1. $i = \sqrt{-1}$ and $i^2 = -1$.
2. A complex number has the form $a + bi$, where a and b are real numbers.
3. The complex number $a + 0i$ is the real number a.
4. If b is a positive real number, then $\sqrt{-b} = i\sqrt{b}$.
5. The complex conjugate of $a + bi$ is $a - bi$.
6. Add, subtract, and multiply complex numbers as if they were binomials with variable i.
7. Divide complex numbers by multiplying the numerator and denominator by the conjugate of the denominator.
8. In the complex number system, all quadratic equations have solutions.

WARM-UPS

True or false? Explain your answer.

1. $(3 + i) + (2 - 4i) = 5 - 3i$ True
2. $(4 - 2i)(3 - 5i) = 2 - 26i$ True
3. $(4 - i)(4 + i) = 17$ True
4. $i^4 = 1$ True
5. $\sqrt{-5} = 5i$ False
6. $\sqrt{-36} = \pm 6i$ False
7. The complex conjugate of $-2 + 3i$ is $2 - 3i$. False
8. Zero is the only real number that is also a complex number. False
9. Both $2i$ and $-2i$ are solutions to the equation $x^2 = 4$. False
10. Every quadratic equation has at least one complex solution. True

9.5 EXERCISES

Reading and Writing *After reading this section, write out the answers to these questions. Use complete sentences.*

1. What is a complex number?
   A complex number is a number of the form $a + bi$, where a and b are real and $i = \sqrt{-1}$.
2. What is the relationship between the complex numbers, the real numbers, and the imaginary numbers.
   The real numbers together with the imaginary numbers make up the set of complex numbers.
3. How do you add or subtract complex numbers?
   Complex numbers are added or subtracted like binomials in which i is the variable.
4. What is a complex conjugate?
   The complex conjugate of $a + bi$ is $a - bi$.
5. What is the square root of a negative number in the complex number system?
   If $b > 0$, then $\sqrt{-b} = i\sqrt{b}$.
6. How many solutions do quadratic equations have in the complex number system?
   In the complex number system all quadratic equations have at least one solution.

Perform the indicated operations. See Example 1.

7. $(3 + 5i) + (2 + 4i)$
   $5 + 9i$
8. $(8 + 3i) + (1 + 2i)$
   $9 + 5i$
9. $(-1 + i) + (2 - i)$
   1
10. $(-2 - i) + (-3 + 5i)$
    $-5 + 4i$
11. $(4 - 5i) - (2 + 3i)$
    $2 - 8i$
12. $(3 - 2i) - (7 + 6i)$
    $-4 - 8i$
13. $(-3 - 5i) - (-2 - i)$
    $-1 - 4i$
14. $(-4 - 8i) - (-2 - 3i)$
    $-2 - 5i$
15. $(8 - 3i) - (9 - 3i)$
    -1
16. $(5 + 6i) - (-3 + 6i)$
    8
17. $\left(\frac{1}{2} + i\right) + \left(\frac{1}{4} - \frac{1}{2}i\right)$
    $\frac{3}{4} + \frac{1}{2}i$
18. $\left(\frac{2}{3} - i\right) - \left(\frac{1}{4} - \frac{1}{2}i\right)$
    $\frac{5}{12} - \frac{1}{2}i$

Perform the indicated operations. See Example 2.

19. $3(2 - 3i)$
    $6 - 9i$
20. $-4(3 - 2i)$
    $-12 + 8i$
21. $(6i)^2$
    -36
22. $(3i)^2$
    -9
23. $(-6i)^2$
    -36
24. $(-3i)^2$
    -9
25. $(2 + 3i)(3 - 5i)$
    $21 - i$
26. $(4 - i)(3 - 6i)$
    $6 - 27i$
27. $(5 - 2i)^2$
    $21 - 20i$
28. $(3 + 4i)^2$
    $-7 + 24i$
29. $(4 - 3i)(4 + 3i)$
    25
30. $(-3 + 5i)(-3 - 5i)$
    34
31. $(1 - i)(1 + i)$
    2
32. $(3 - i)(3 + i)$
    10

Find the product of the given complex number and its conjugate. See Example 3.

33. $2 + 5i$ 29
34. $3 + 4i$ 25
35. $4 - 6i$ 52
36. $2 - 7i$ 53
37. $-3 + 2i$ 13
38. $-4 - i$ 17
39. i 1
40. $-2i$ 4

Perform the indicated operations. See Example 4.

41. $(2 - 6i) \div 2$ $1 - 3i$
42. $(-3 + 6i) \div (-3)$ $1 - 2i$
43. $\dfrac{-2 + 8i}{2}$ $-1 + 4i$

44. $\dfrac{6 - 9i}{-3}$ $-2 + 3i$

45. $\dfrac{4 + 6i}{-2i}$ $-3 + 2i$

46. $\dfrac{3 - 8i}{i}$ $-8 - 3i$

47. $\dfrac{4i}{3 + 2i}$ $\dfrac{8}{13} + \dfrac{12}{13}i$

48. $\dfrac{5}{4 - 5i}$ $\dfrac{20}{41} + \dfrac{25}{41}i$

49. $\dfrac{2 + i}{2 - i}$ $\dfrac{3}{5} + \dfrac{4}{5}i$

50. $\dfrac{i - 5}{5 - i}$ -1

51. $\dfrac{4 - 12i}{3 + i}$ $-4i$

52. $\dfrac{-4 + 10i}{5 - i}$ $-\dfrac{15}{13} + \dfrac{23}{13}i$

Write each expression in the form a + bi, where a and b are real numbers. See Example 5.

53. $5 + \sqrt{-9}$
$5 + 3i$

54. $6 + \sqrt{-16}$
$6 + 4i$

55. $-3 - \sqrt{-7}$
$-3 - i\sqrt{7}$

56. $2 - \sqrt{-3}$
$2 - i\sqrt{3}$

57. $\dfrac{-2 + \sqrt{-12}}{2}$
$-1 + i\sqrt{3}$

58. $\dfrac{-6 + \sqrt{-18}}{3}$
$-2 + i\sqrt{2}$

59. $\dfrac{-8 - \sqrt{-20}}{-4}$
$2 + \dfrac{1}{2}i\sqrt{5}$

60. $\dfrac{6 + \sqrt{-24}}{-2}$
$-3 - i\sqrt{6}$

61. $\dfrac{-4 + \sqrt{-28}}{6}$
$-\dfrac{2}{3} + \dfrac{1}{3}i\sqrt{7}$

62. $\dfrac{6 - \sqrt{-45}}{6}$
$1 - \dfrac{1}{2}i\sqrt{5}$

63. $\dfrac{-2 + \sqrt{-100}}{-10}$
$\dfrac{1}{5} - i$

64. $\dfrac{-3 + \sqrt{-81}}{-9}$
$\dfrac{1}{3} - i$

Find the complex solutions to each quadratic equation. See Example 6.

65. $x^2 + 81 = 0$
$-9i, 9i$

66. $x^2 + 100 = 0$
$-10i, 10i$

67. $x^2 + 5 = 0$
$-i\sqrt{5}, i\sqrt{5}$

68. $x^2 + 6 = 0$
$-i\sqrt{6}, i\sqrt{6}$

69. $3y^2 + 2 = 0$
$-i\dfrac{\sqrt{6}}{3}, i\dfrac{\sqrt{6}}{3}$

70. $5y^2 + 3 = 0$
$-i\dfrac{\sqrt{15}}{5}, i\dfrac{\sqrt{15}}{5}$

71. $x^2 - 4x + 5 = 0$
$2 - i, 2 + i$

72. $x^2 - 6x + 10 = 0$
$3 - i, 3 + i$

73. $y^2 + 13 = 6y$
$3 - 2i, 3 + 2i$

74. $y^2 + 29 = 4y$
$2 - 5i, 2 + 5i$

75. $x^2 - 4x + 7 = 0$
$2 - i\sqrt{3}, 2 + i\sqrt{3}$

76. $x^2 - 10x + 27 = 0$
$5 - i\sqrt{2}, 5 + i\sqrt{2}$

77. $9y^2 - 12y + 5 = 0$
$\dfrac{2 - i}{3}, \dfrac{2 + i}{3}$

78. $2y^2 - 2y + 1 = 0$
$\dfrac{1}{2} - \dfrac{1}{2}i, \dfrac{1}{2} + \dfrac{1}{2}i$

79. $x^2 - x + 1 = 0$
$\dfrac{1 - i\sqrt{3}}{2}, \dfrac{1 + i\sqrt{3}}{2}$

80. $4x^2 - 20x + 27 = 0$
$\dfrac{5}{2} - \dfrac{1}{2}i\sqrt{2}, \dfrac{5}{2} + \dfrac{1}{2}i\sqrt{2}$

81. $-4x^2 + 8x - 9 = 0$
$\dfrac{2 - i\sqrt{5}}{2}, \dfrac{2 + i\sqrt{5}}{2}$

82. $-9x^2 + 12x - 10 = 0$
$\dfrac{2}{3} - \dfrac{1}{3}i\sqrt{6}, \dfrac{2}{3} + \dfrac{1}{3}i\sqrt{6}$

Solve each problem.

83. Evaluate $(2 - 3i)^2 + 4(2 - 3i) - 9$.
$-6 - 24i$

84. Evaluate $(3 + 5i)^2 - 2(3 + 5i) + 5$.
$-17 + 20i$

85. What is the value of $x^2 - 8x + 17$ if $x = 4 - i$?
0

86. What is the value of $x^2 - 6x + 34$ if $x = 3 + 5i$?
0

87. Find the product $[x - (6 - i)][x - (6 + i)]$.
$x^2 - 12x + 37$

88. Find the product $[x - (3 + 7i)][x - (3 - 7i)]$.
$x^2 - 6x + 58$

89. Write a quadratic equation that has $3i$ and $-3i$ as its solutions.
$x^2 + 9 = 0$

90. Write a quadratic equation that has $2 + 5i$ and $2 - 5i$ as its solutions.
$x^2 - 4x + 29 = 0$

GETTING MORE INVOLVED

91. ***Discussion.*** Determine whether each given number is in each of the following sets: the natural numbers, the integers, the rational numbers, the irrational numbers, the real numbers, the imaginary numbers, and the complex numbers.

a) 54 **b)** $-\dfrac{3}{8}$ **c)** $3\sqrt{5}$ **d)** $6i$ **e)** $\pi + i\sqrt{5}$

Natural: 54, integers: 54, rational: 54, $-3/8$, irrational: $3\sqrt{5}$, real: 54, $-3/8$, $3\sqrt{5}$, imaginary: $6i$, $\pi + i\sqrt{5}$, complex: all

92. ***Discussion.*** Which of the following equations have real solutions? Imaginary solutions? Complex solutions?

a) $3x^2 - 2x + 9 = 0$ **b)** $5x^2 - 2x - 10 = 0$

c) $\dfrac{1}{2}x^2 - x + 3 = 0$ **d)** $7w^2 + 12 = 0$

Real: b, imaginary: a, c, d, complex: all

9.6 GRAPHING PARABOLAS

In This Section

- Finding Ordered Pairs
- Graphing Parabolas
- The Vertex and Intercepts
- Applications

The graph of any equation of the form $y = mx + b$ is a straight line. In this section we will see that all equations of the form $y = ax^2 + bx + c$ have graphs that are in the shape of a *parabola*.

Finding Ordered Pairs

It is straightforward to calculate y when given x for an equation of the form $y = ax^2 + bx + c$. However, if we are given y and want to find x, then we must use methods for solving quadratic equations.

EXAMPLE 1 **Finding ordered pairs**

Complete each ordered pair so that it satisfies the given equation. For part (a) the pairs are of the form (x, y) and for part (b) they are of the form (t, s).

a) $y = x^2 - x - 6$; (2,), (, 0)

b) $s = -16t^2 + 48t + 84$; (0,), (, 20)

Solution

a) If $x = 2$, then $y = 2^2 - 2 - 6 = -4$. So the ordered pair is $(2, -4)$. To find x when $y = 0$, replace y by 0 and solve the resulting quadratic equation:

$$x^2 - x - 6 = 0$$
$$(x - 3)(x + 2) = 0$$
$$x - 3 = 0 \quad \text{or} \quad x + 2 = 0$$
$$x = 3 \quad \text{or} \quad x = -2$$

The ordered pairs are $(-2, 0)$ and $(3, 0)$.

b) If $t = 0$, then $s = -16 \cdot 0^2 + 48 \cdot 0 + 84 = 84$. The ordered pair is $(0, 84)$. To find t when $s = 20$, replace s by 20 and solve the equation for t:

$$-16t^2 + 48t + 84 = 20$$
$$-16t^2 + 48t + 64 = 0 \quad \text{Subtract 20 from each side.}$$
$$t^2 - 3t - 4 = 0 \quad \text{Divide each side by } -16.$$
$$(t - 4)(t + 1) = 0 \quad \text{Factor.}$$
$$t - 4 = 0 \quad \text{or} \quad t + 1 = 0 \quad \text{Zero factor property}$$
$$t = 4 \quad \text{or} \quad t = -1$$

The ordered pairs are $(-1, 20)$ and $(4, 20)$. ■

Study Tip

Although you should avoid cramming, there are times when you have no other choice. In this case concentrate on what is in your class notes and the homework assignments. Try to work one or two problems of each type. Instructors often ask some relatively easy questions on a test to see if you have understood the major ideas.

CAUTION When variables other than x and y are used, the independent variable is the first coordinate of an ordered pair, and the dependent variable is the second coordinate. In Example 1(b), t is the independent variable and first coordinate because s depends on t by the formula $s = -16t^2 + 48t + 84$.

Graphing Parabolas

The graph of any equation of the form $y = ax^2 + bx + c$ (with $a \neq 0$) is called a **parabola.** To graph a parabola we simply plot points as we did when we first graphed lines in Section 3.1. Note that any real number may be used in place of x in the equation $y = ax^2 + bx + c$.

EXAMPLE 2 **The simplest parabola**

Graph $y = x^2$.

Calculator Close-Up

This close-up view of $y = x^2$ shows how rounded the curve is at the bottom. When drawing a parabola by hand, be sure to draw it smoothly.

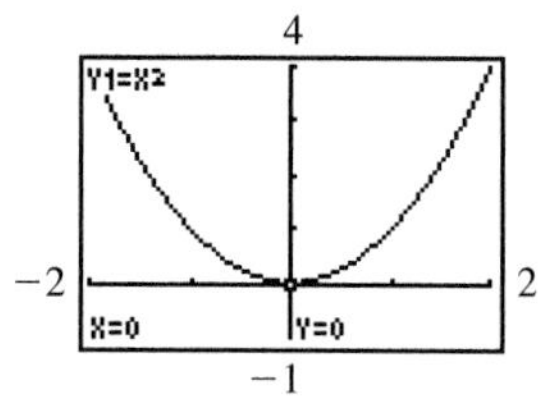

Solution

Arbitrarily select some values for the independent variable x and calculate the corresponding values for the dependent variable y:

If $x = -2$, then $y = (-2)^2 = 4$.
If $x = -1$, then $y = (-1)^2 = 1$.
If $x = 0$, then $y = 0^2 = 0$.
If $x = 1$, then $y = 1^2 = 1$.
If $x = 2$, then $y = 2^2 = 4$.

These values are displayed in the following table:

x	-2	-1	0	1	2
$y = x^2$	4	1	0	1	4

Plot $(-2, 4)$, $(-1, 1)$, $(0, 0)$, $(1, 1)$, and $(2, 4)$, as shown in Fig. 9.4, and draw a parabola through the points. Note that the x-coordinates can be any real numbers, but there are no negative y-coordinates. So $y \geq 0$ in every ordered pair on the graph.

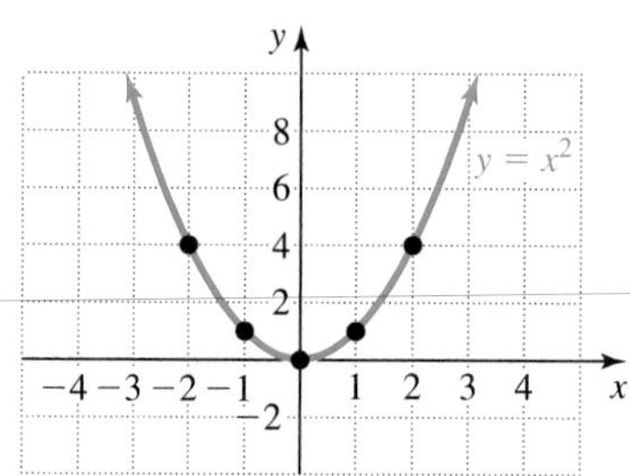

FIGURE 9.4

The parabola in Fig. 9.4 is said to **open upward.** In Example 3 we see a parabola that **opens downward.** If $a > 0$ in the equation $y = ax^2 + bx + c$, then the parabola opens upward. If $a < 0$, then the parabola opens downward.

EXAMPLE 3 **A parabola that opens downward**

Graph $y = 4 - x^2$.

Solution

Find some ordered pairs as follows:

x	-2	-1	0	1	2
$y = 4 - x^2$	0	3	4	3	0

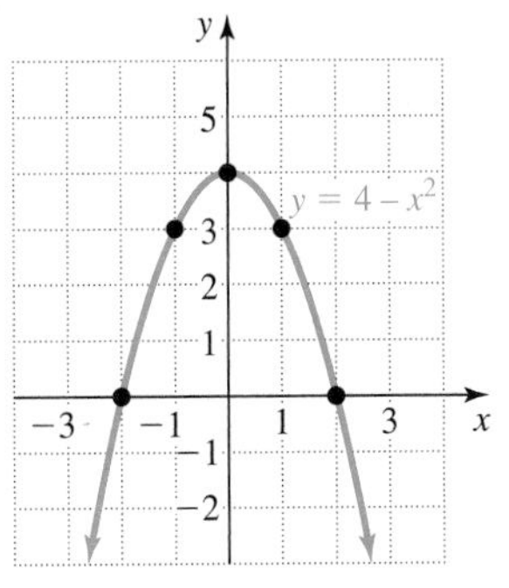

FIGURE 9.5

Plot $(-2, 0)$, $(-1, 3)$, $(0, 4)$, $(1, 3)$, and $(2, 0)$, as shown in Fig. 9.5, and sketch a parabola through the points. Note that the largest y-coordinate on this graph is 4. So $y \le 4$ in every ordered pair on the graph. ■

The Vertex and Intercepts

The lowest point on a parabola that opens upward or the highest point on a parabola that opens downward is called the **vertex.** The y-coordinate of the vertex is the **minimum** y-value on the graph if the parabola opens upward, and it is the **maximum** y-value if the parabola opens downward. For $y = x^2$ the vertex is $(0, 0)$ and 0 is the minimum y-value on the graph. For $y = 4 - x^2$ the vertex is $(0, 4)$ and 4 is the maximum y-value on the graph.

Because the vertex is either the highest or lowest point on a parabola, it is an important feature of the parabola. The vertex can be found using the following fact.

Helpful Hint

To draw a parabola or any curve by hand, use your hand like a compass. The two halves of a parabola should be drawn in two steps. Position your paper so that your hand is approximately at the "center" of the arc you are trying to draw.

Vertex of a Parabola

The x-coordinate of the vertex of $y = ax^2 + bx + c$ is $\frac{-b}{2a}$, provided that $a \ne 0$.

You can remember $\frac{-b}{2a}$ by observing that it is part of the quadratic formula:

$$x = \frac{-b \pm \sqrt{b^2 - 4ac}}{2a}$$

In Examples 2 and 3 we drew the graph by selecting five x-values and calculating y. But how did we know what to select for x? The best way to graph a parabola is to find the vertex first, then select two x-coordinates to the left of the vertex and two to the right of the vertex. This way you will always get a graph that shows the typical parabolic shape.

EXAMPLE 4 **Using the vertex in graphing a parabola**

Graph $y = -x^2 - x + 2$.

Solution

First find the x-coordinate of the vertex:

$$x = \frac{-b}{2a} = \frac{-(-1)}{2(-1)} = \frac{1}{-2} = -\frac{1}{2}$$

Now find y for $x = -\frac{1}{2}$:

$$y = -\left(-\frac{1}{2}\right)^2 - \left(-\frac{1}{2}\right) + 2 = -\frac{1}{4} + \frac{1}{2} + 2 = \frac{9}{4}$$

The vertex is $\left(-\frac{1}{2}, \frac{9}{4}\right)$. Now find a few points on either side of the vertex:

x	-2	-1	$-\frac{1}{2}$	0	1
$y = -x^2 - x + 2$	0	2	$\frac{9}{4}$	2	0

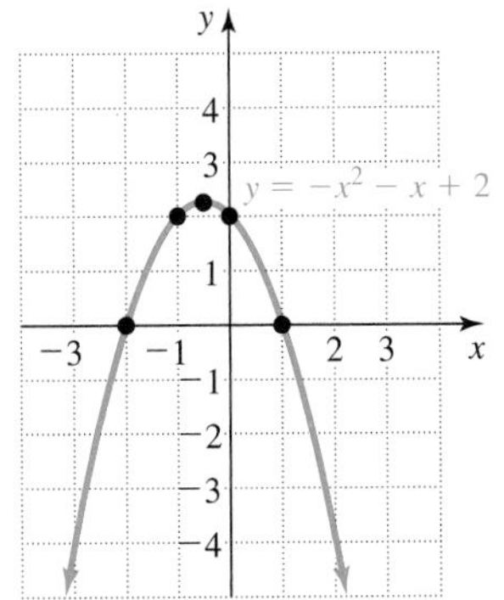

FIGURE 9.6

Sketch a parabola through these points as in Fig. 9.6. ■

The y-intercept of a parabola is the point that has 0 as the first coordinate. The x-intercepts are the points that have 0 as their second coordinates.

EXAMPLE 5 **Using the intercepts in graphing a parabola**

Find the vertex and intercepts, and sketch the graph of each equation.

a) $y = x^2 - 2x - 8$

b) $s = -16t^2 + 64t$

Solution

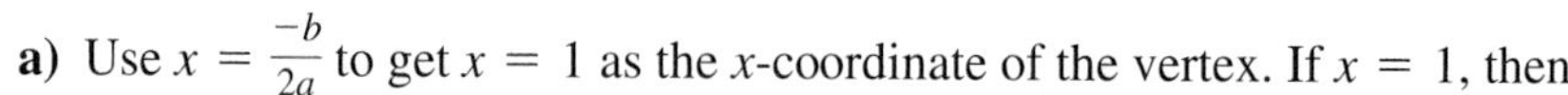

a) Use $x = \frac{-b}{2a}$ to get $x = 1$ as the x-coordinate of the vertex. If $x = 1$, then

$$\begin{aligned} y &= 1^2 - 2 \cdot 1 - 8 \\ &= -9. \end{aligned}$$

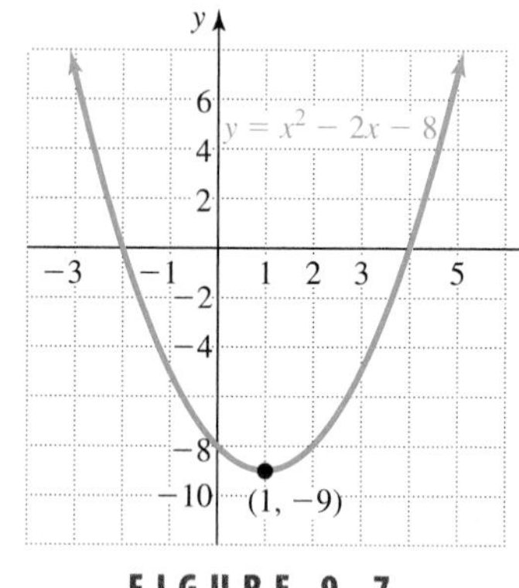

FIGURE 9.7

So the vertex is $(1, -9)$. If $x = 0$, then

$$\begin{aligned} y &= 0^2 - 2 \cdot 0 - 8 \\ &= -8. \end{aligned}$$

The y-intercept is $(0, -8)$. To find the x-intercepts, replace y by 0:

$$x^2 - 2x - 8 = 0$$

$$(x - 4)(x + 2) = 0$$

$$x - 4 = 0 \quad \text{or} \quad x + 2 = 0$$

$$x = 4 \quad \text{or} \quad x = -2$$

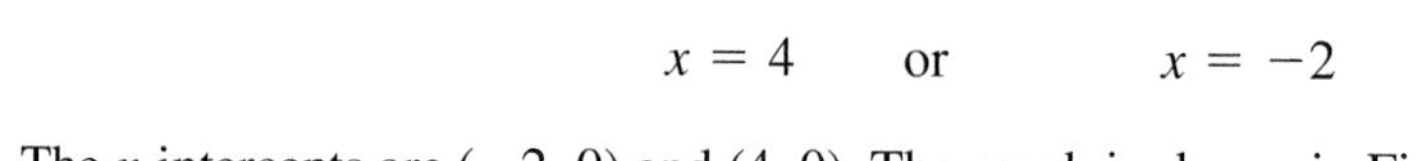

The x-intercepts are $(-2, 0)$ and $(4, 0)$. The graph is shown in Fig. 9.7.

b) Because s is expressed in terms of t, the first coordinate is t. Use $t = \frac{-b}{2a}$ to get

$$t = \frac{-64}{2(-16)} = 2.$$

If $t = 2$, then

$$\begin{aligned} s &= -16 \cdot 2^2 + 64 \cdot 2 \\ &= 64. \end{aligned}$$

So the vertex is $(2, 64)$. If $t = 0$, then

$$\begin{aligned} s &= -16 \cdot 0^2 + 64 \cdot 0 \\ &= 0. \end{aligned}$$

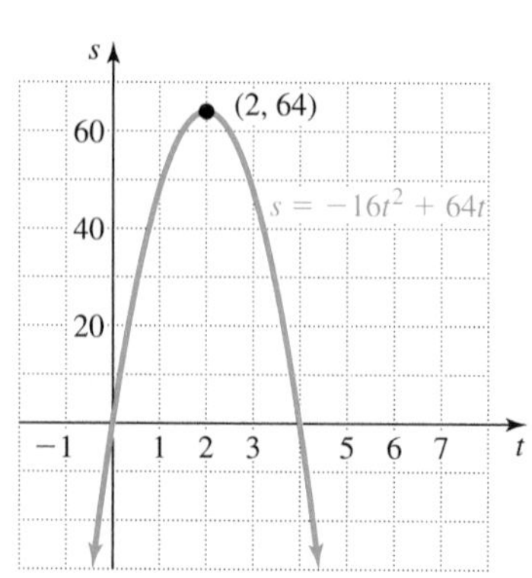

FIGURE 9.8

So the s-intercept is $(0, 0)$. To find the t-intercepts, replace s by 0:

$$-16t^2 + 64t = 0$$

$$-16t(t - 4) = 0$$

$$-16t = 0 \quad \text{or} \quad t - 4 = 0$$

$$t = 0 \quad \text{or} \quad t = 4$$

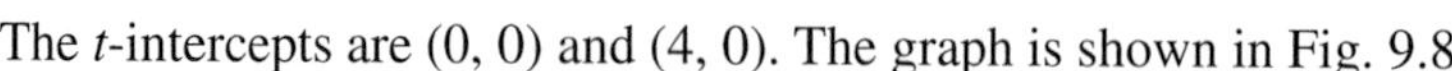

The t-intercepts are $(0, 0)$ and $(4, 0)$. The graph is shown in Fig. 9.8. ■

Calculator Close-Up

You can find the vertex of a parabola with a calculator by using either the maximum or minimum feature. First graph the parabola as shown.

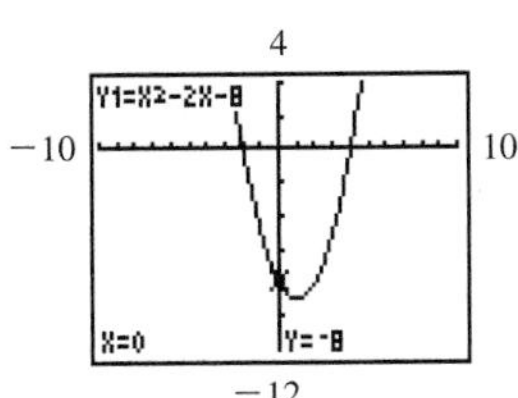

Because this parabola opens upward, the y-coordinate of the vertex is the minimum y-coordinate on the graph. Press CALC and choose minimum.

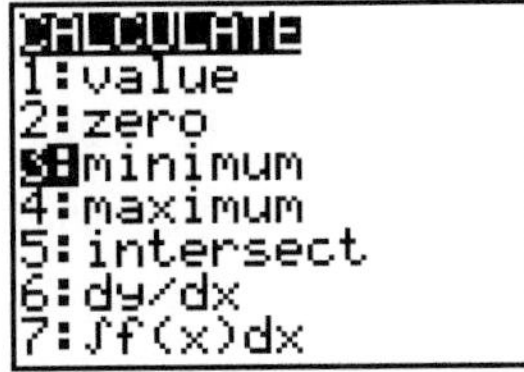

The calculator will ask for a left bound, a right bound, and a guess. For the left bound choose a point to the left of the vertex by moving the cursor to the point and pressing ENTER. For the right bound choose a point to the right of the vertex. For the guess choose a point close to the vertex.

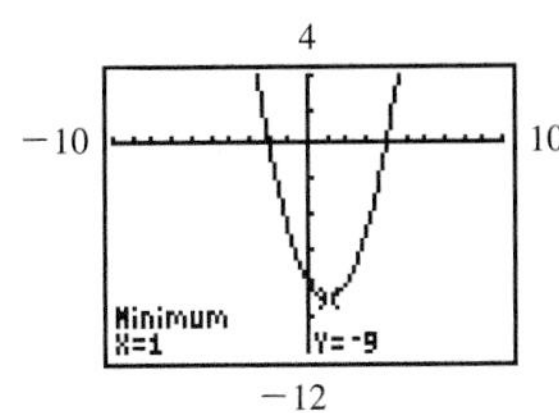

Applications

In applications we are often interested in finding the maximum or minimum value of a variable. If a parabola opens downward, then the maximum value of the dependent variable is the second coordinate of the vertex. If a parabola opens upward, then the minimum value of the dependent variable is the second coordinate of the vertex.

EXAMPLE 6 **Finding the maximum height**

A ball is tossed upward with a velocity of 64 feet per second from a height of 5 feet. What is the maximum height reached by the ball?

Solution

The height s of the ball for any time t is given by $s = -16t^2 + 64t + 5$. Because the maximum height occurs at the vertex of the parabola, we use $t = \frac{-b}{2a}$ to find the vertex:

$$t = \frac{-64}{2(-16)} = 2$$

Now use $t = 2$ to find the second coordinate of the vertex:

$$s = -16(2)^2 + 64(2) + 5 = 69$$

The maximum height reached by the ball is 69 feet. See Fig. 9.9.

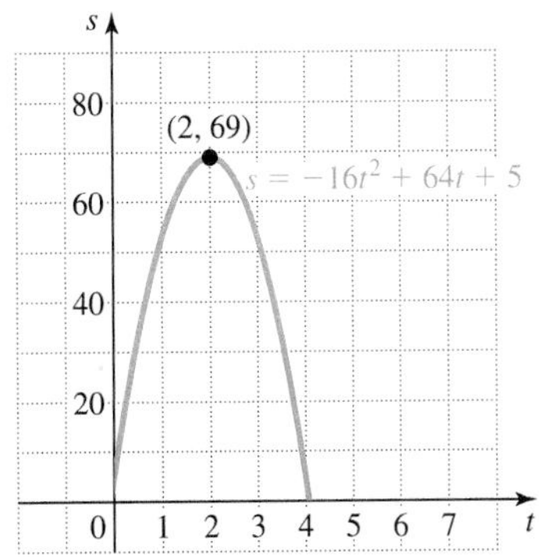

FIGURE 9.9

WARM-UPS

True or false? Explain your answer.

1. The ordered pair $(-2, -1)$ satisfies $y = x^2 - 5$. True
2. The y-intercept for $y = x^2 - 3x + 9$ is $(9, 0)$. False
3. The x-intercepts for $y = x^2 - 5$ are $(\sqrt{5}, 0)$ and $(-\sqrt{5}, 0)$. True
4. The graph of $y = x^2 - 12$ opens upward. True
5. The graph of $y = 4 + x^2$ opens downward. False
6. The vertex of $y = x^2 + 2x$ is $(-1, -1)$. True
7. The parabola $y = x^2 + 1$ has no x-intercepts. True
8. The y-intercept for $y = ax^2 + bx + c$ is $(0, c)$. True
9. If $w = -2v^2 + 9$, then the maximum value of w is 9. True
10. If $y = 3x^2 - 7x + 9$, then the maximum value of y occurs when $x = \frac{7}{6}$. False

9.6 EXERCISES

Reading and Writing *After reading this section, write out the answers to these questions. Use complete sentences.*

1. What type of equation has a straight line as its graph?
The graph of $y = mx + b$ is a straight line.

2. What type of equation has a parabola as its graph?
The graph of $y = ax^2 + bx + c$ with $a \neq 0$ is a parabola.

3. When does a parabola open upward?
The graph of $y = ax^2 + bx + c$ opens upward if $a > 0$.

4. What is the vertex of a parabola?
The vertex is the highest point on a parabola that opens downward or the lowest point on a parabola that opens upward.

5. What is the first coordinate of the vertex for the parabola $y = ax^2 + bx + c$?
The x-coordinate of the vertex is $-b/(2a)$.

6. How do you find the second coordinate of the vertex?
Use $x = -b/(2a)$ to find x and then use the original equation to find y.

Complete each ordered pair so that it satisfies the given equation. See Example 1.

7. $y = x^2 - x - 12$ $(3, \quad), (\quad, 0)$
$(3, -6), (4, 0), (-3, 0)$

8. $y = -\frac{1}{2}x^2 - x + 1$ $(0, \quad), (\quad, -3)$
$(0, 1), (-4, -3), (2, -3)$

9. $s = -16t^2 + 32t$ $(4, \quad), (\quad, 0)$
$(4, -128), (0, 0), (2, 0)$

10. $a = b^2 + 4b + 5$ $(-2, \quad), (\quad, 2)$
$(-2, 1), (-1, 2), (-3, 2)$

Graph each equation. See Examples 2 and 3.

11. $y = x^2 + 2$

12. 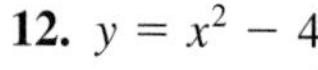 $y = x^2 - 4$

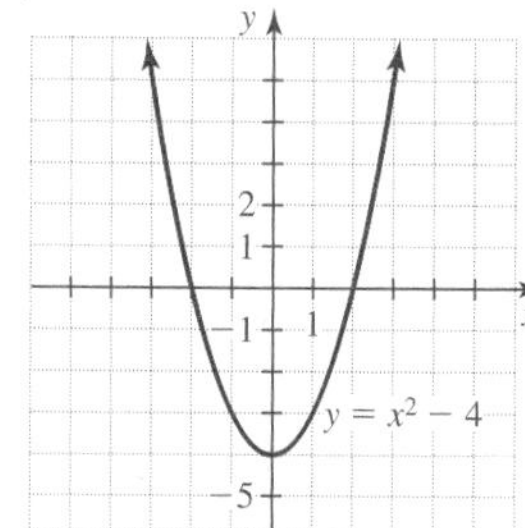

13. 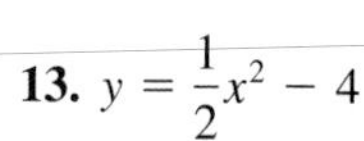 $y = \frac{1}{2}x^2 - 4$

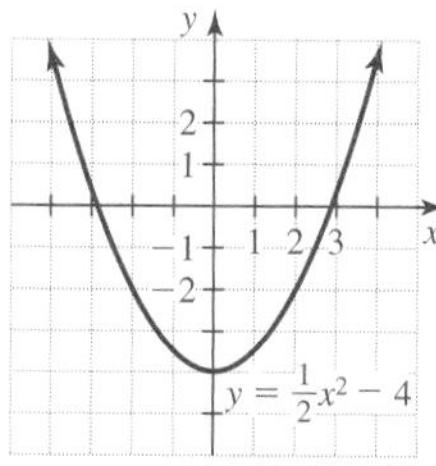

14. $y = \frac{1}{3}x^2 - 6$

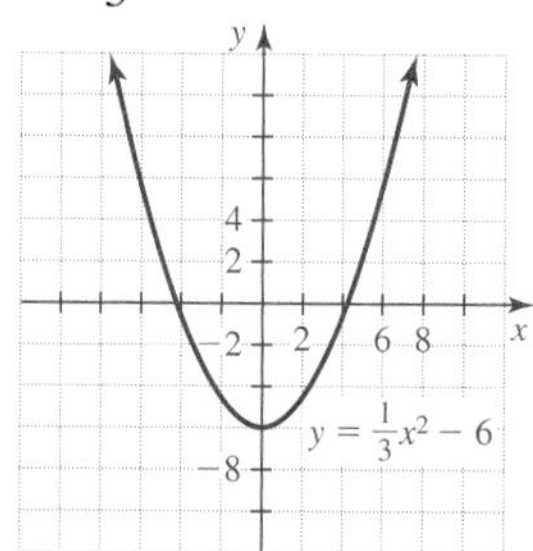

15. $y = -2x^2 + 5$

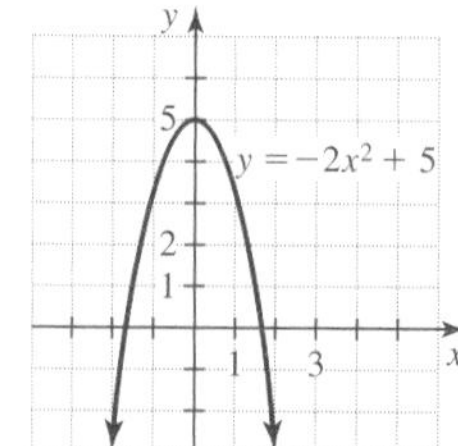

16. $y = -x^2 - 1$

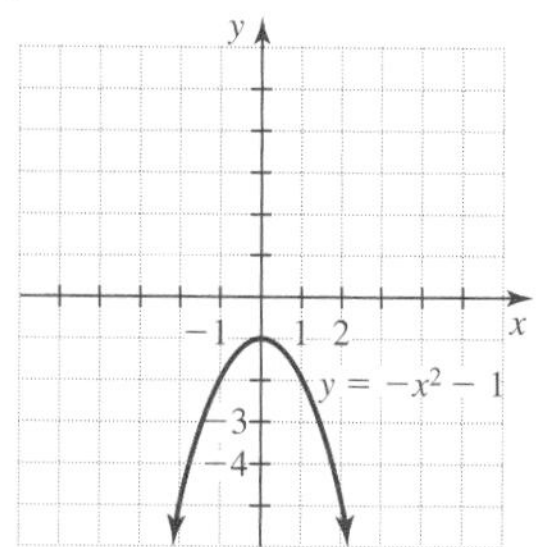

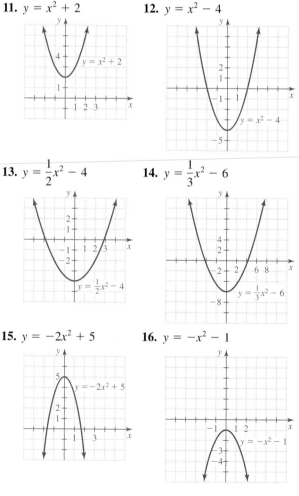

17. $y = -\frac{1}{3}x^2 + 5$

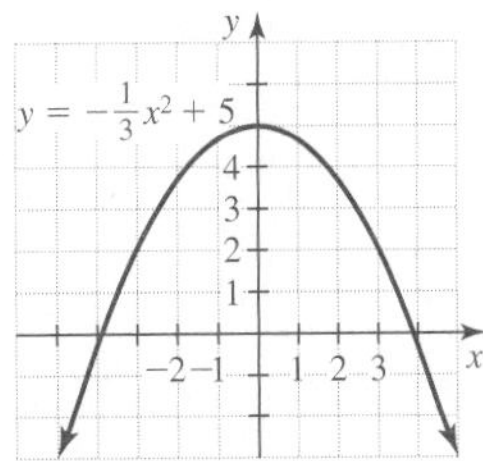

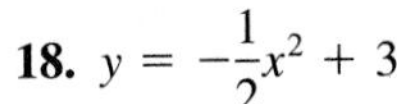

18. $y = -\frac{1}{2}x^2 + 3$

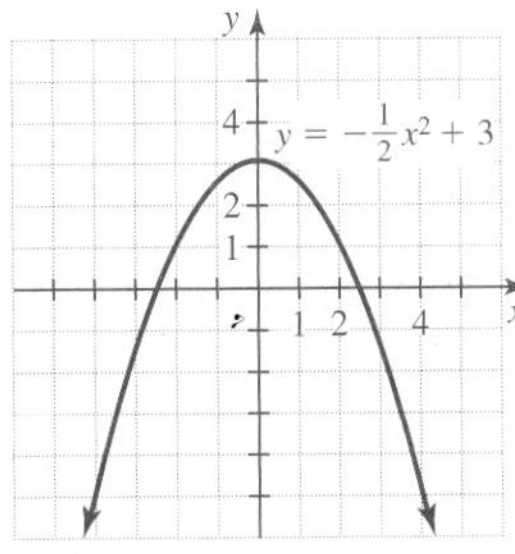

19. $y = (x - 2)^2$

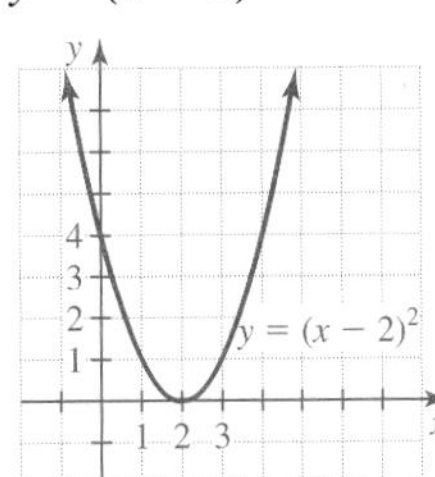

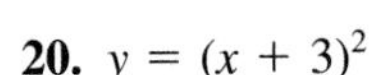

20. $y = (x + 3)^2$

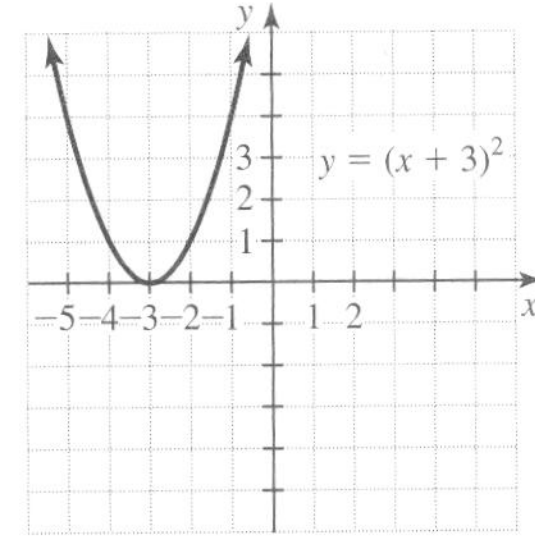

Find the vertex and intercepts for each parabola. Then sketch the graph. See Examples 4 and 5.

21. $y = x^2 - x - 2$
Vertex $\left(\frac{1}{2}, -\frac{9}{4}\right)$, intercepts $(0, -2)$, $(-1, 0)$, $(2, 0)$

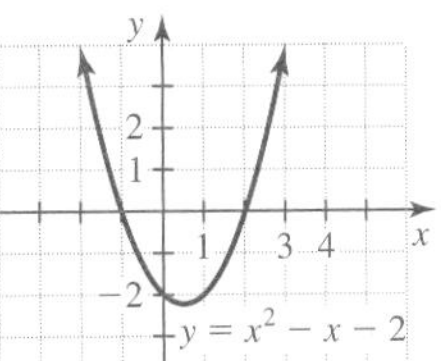

22. $y = x^2 + 2x - 3$
Vertex $(-1, -4)$, intercepts $(0, -3)$, $(-3, 0)$, $(1, 0)$

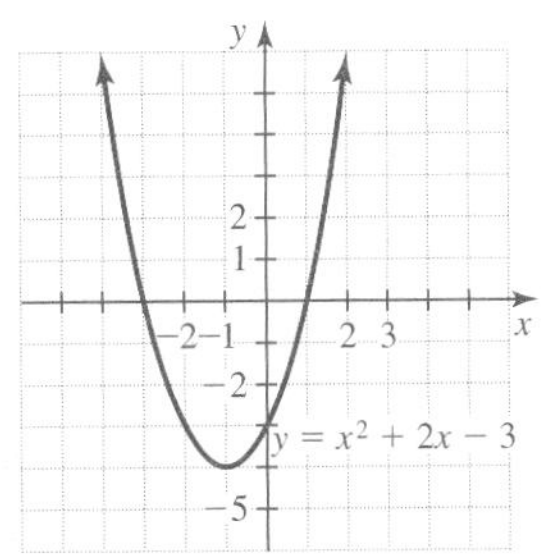

23. $y = x^2 + 2x - 8$
Vertex $(-1, -9)$, intercepts $(0, -8)$, $(-4, 0)$, $(2, 0)$

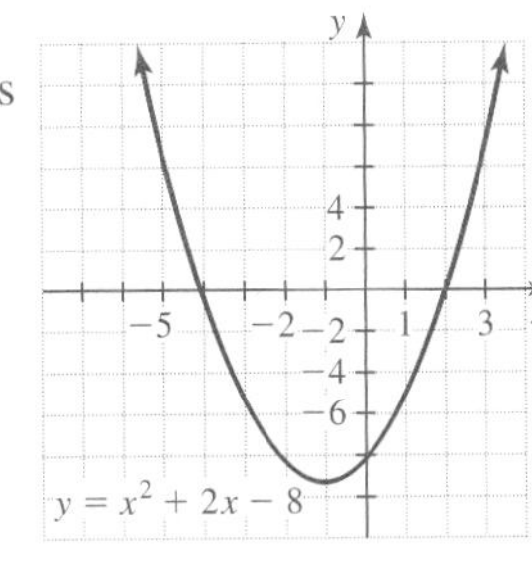

24. $y = x^2 + x - 6$
Vertex $\left(-\frac{1}{2}, -\frac{25}{4}\right)$, intercepts $(0, -6)$, $(-3, 0)$, $(2, 0)$

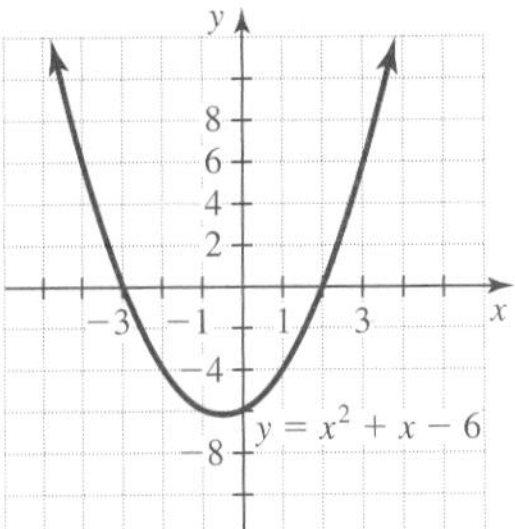

25. $y = -x^2 - 4x - 3$
Vertex $(-2, 1)$, intercepts $(0, -3)$, $(-1, 0)$, $(-3, 0)$

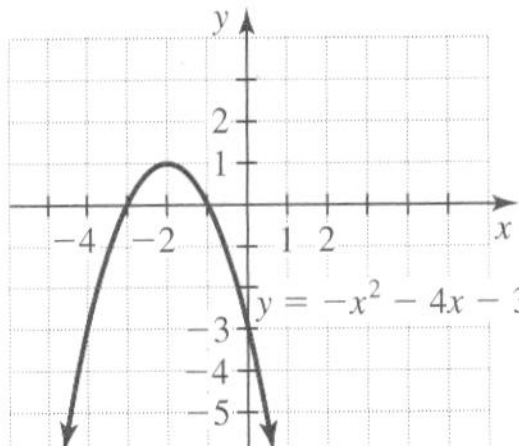

26. $y = -x^2 - 5x - 4$
Vertex $\left(-\frac{5}{2}, \frac{9}{4}\right)$, intercepts $(0, -4)$, $(-1, 0)$, $(-4, 0)$

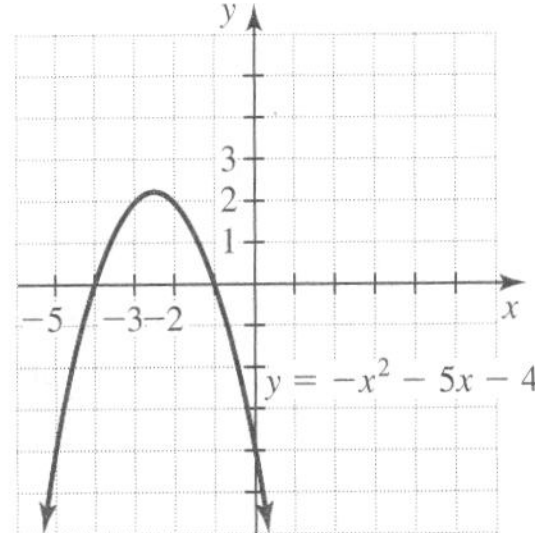

27. $y = -x^2 + 3x + 4$
Vertex $\left(\frac{3}{2}, \frac{25}{4}\right)$, intercepts $(0, 4)$, $(4, 0)$, $(-1, 0)$

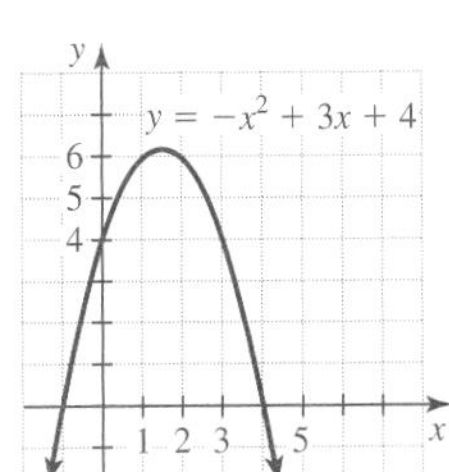

28. $y = -x^2 - 2x + 8$

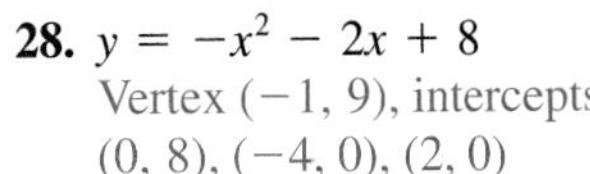
Vertex $(-1, 9)$, intercepts $(0, 8)$, $(-4, 0)$, $(2, 0)$

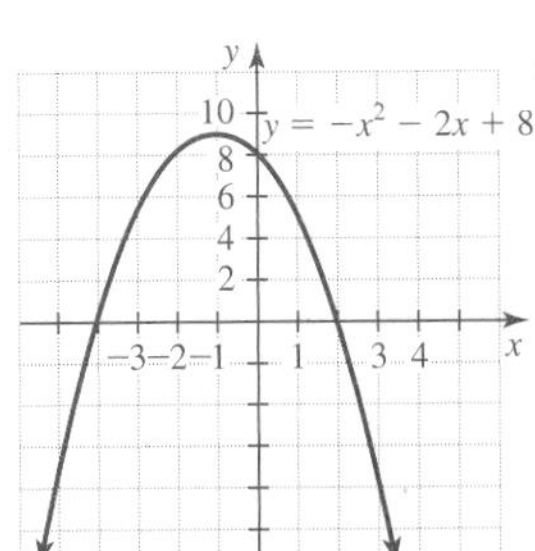

29. $a = b^2 - 6b - 16$
Vertex $(3, -25)$, intercepts $(0, -16)$, $(8, 0)$, $(-2, 0)$

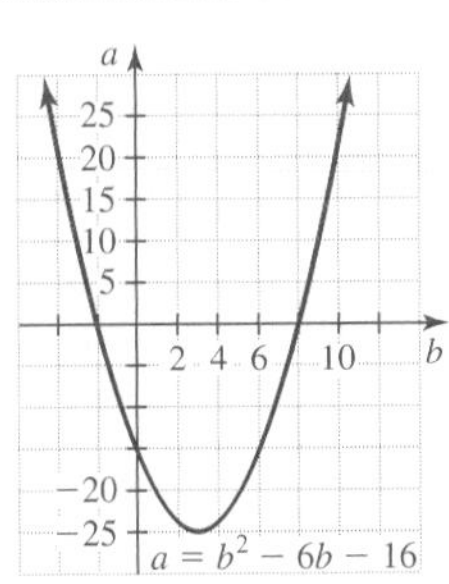

30. $v = -u^2 - 8u + 9$
Vertex $(-4, 25)$, intercepts $(0, 9)$, $(-9, 0)$, $(1, 0)$

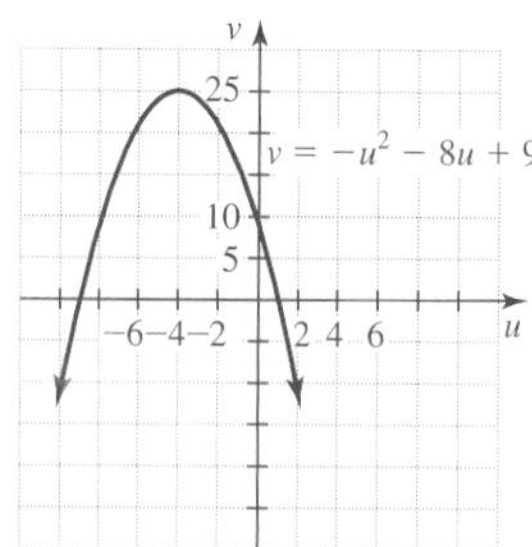

Find the maximum or minimum value for y for each equation.

31. $y = x^2 - 8$
Minimum -8

32. $y = 33 - x^2$
Maximum 33

33. $y = -3x^2 + 14$
Maximum 14

34. $y = 6 + 5x^2$
Minimum 6

35. $y = x^2 + 2x + 3$
Minimum 2

36. $y = x^2 - 2x + 5$
Minimum 4

37. $y = -2x^2 - 4x$
Maximum 2

38. $y = -3x^2 + 24x$
Maximum 48

Solve each problem. See Example 6.

39. ***Maximum height.*** If a baseball is projected upward from ground level with an initial velocity of 64 feet per second, then $s = -16t^2 + 64t$ gives its height in feet s in terms of time in seconds t. Graph this equation for $0 \le t \le 4$. What is the maximum height reached by the ball?
Maximum 64 feet

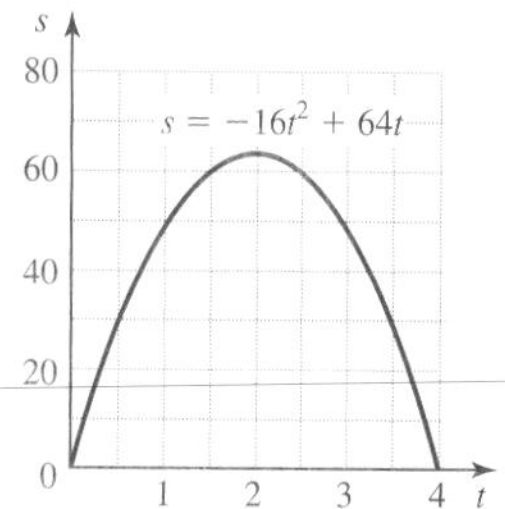

40. ***Maximum height.*** If a soccer ball is kicked straight up with an initial velocity of 32 feet per second, then $s = -16t^2 + 32t$ gives its height in feet s in terms of time in seconds t. Graph this equation for $0 \le t \le 2$. What is the maximum height reached by this ball?
Maximum 16 feet

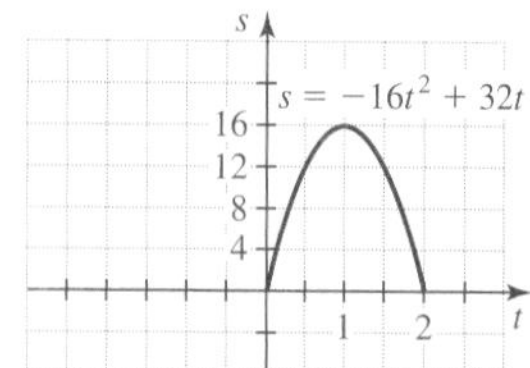

41. ***Maximum area.*** Jason plans to fence a rectangular area with 100 meters of fencing. He has written the formula $A = w(50 - w)$ to express the area in square meters A in terms of the width in meters w. What is the maximum possible area that he can enclose with his fencing?
625 square meters

FIGURE FOR EXERCISE 41

42. ***Minimizing cost.*** A company uses the formula $C = 0.02x^2 - 3.4x + 150$ to model the unit cost in dollars C for producing x stabilizer bars. For what number of bars is the unit cost at its minimum? What is the unit cost at that level of production? 85, \$5.50

43. ***Air pollution.*** The formula $A = -2t^2 + 32t + 12$ gives the amount of nitrogen dioxide A in parts per million (ppm) that was present in the air in the city of Homer on June 14, where t is the number of hours after 6:00 A.M. Use this equation to find the time at which the nitrogen dioxide level was at its maximum. 2 P.M.

44. ***Stabilization ratio.*** The stabilization ratio (births/deaths) for South and Central America can be modeled by the equation

$$y = -0.0012x^2 + 0.074x + 2.69,$$

where y is the number of births divided by the number of deaths in the year $1950 + x$ (World Resources Institute, www.wri.org).

a) Use the accompanying graph to estimate the year in which the stabilization ratio was at its maximum.

b) Use the equation to find the year in which the stabilization ratio was at its maximum.

c) What is the maximum stabilization ratio from part (b)?

d) What is the significance of a stabilization ratio of 1?

a) 1980 **b)** 1981 **c)** 3.83 **d)** stable population

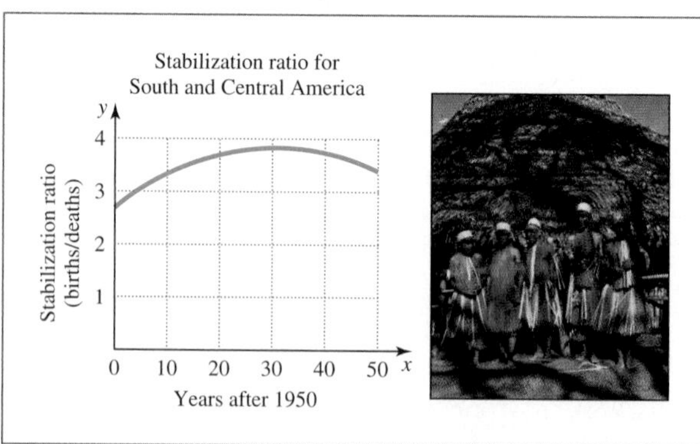

FIGURE FOR EXERCISE 44

45. ***Suspension bridge.*** The cable of the suspension bridge shown in the accompanying figure hangs in the shape of a parabola with equation $y = 0.0375x^2$, where x and y are in meters. What is the height of each tower above the roadway? What is the length z for the cable bracing the tower?
15 meters, 25 meters

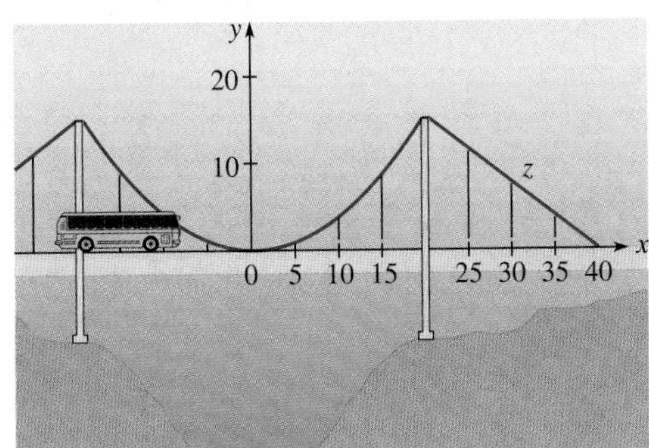

FIGURE FOR EXERCISE 45

GETTING MORE INVOLVED

46. ***Exploration.***

a) Write $y = 3(x - 2)^2 + 6$ in the form $y = ax^2 + bx + c$, and find the vertex of the parabola using the formula $x = \frac{-b}{2a}$.

b) Repeat part (a) with $y = -4(x - 5)^2 - 9$ and $y = 3(x + 2)^2 - 6$.

c) What is the vertex for a parabola that is written in the form $y = a(x - h)^2 + k$? Explain your answer.

a) $y = 3x^2 - 12x + 18$, $(2, 6)$
b) $y = -4x^2 + 40x - 109$, $(5, -9)$, $y = 3x^2 + 12x + 6$, $(-2, -6)$
c) (h, k)

GRAPHING CALCULATOR EXERCISES

47. Graph $y = x^2$, $y = \frac{1}{2}x^2$, and $y = 2x^2$ on the same coordinate system. What can you say about the graph of $y = kx^2$?
The graph of $y = kx^2$ gets narrower as k gets larger.

48. Graph $y = x^2$, $y = (x - 3)^2$, and $y = (x + 3)^2$ on the same coordinate system. How does the graph of $y = (x - k)^2$ compare to the graph of $y = x^2$?
The graph of $y = (x - k)^2$ lies k units to the right of $y = x^2$ if $k > 0$ and $-k$ units to the left of $y = x^2$ if $k < 0$.

49. The equation $x = y^2$ is equivalent to $y = \pm\sqrt{x}$. Graph both $y = \sqrt{x}$ and $y = -\sqrt{x}$ on a graphing calculator. How does the graph of $x = y^2$ compare to the graph of $y = x^2$?
The graph of $y = x^2$ has the same shape as $x = y^2$.

50. Graph each of the following equations by solving for y.

a) $x = y^2 - 1$
$y = \pm\sqrt{x + 1}$

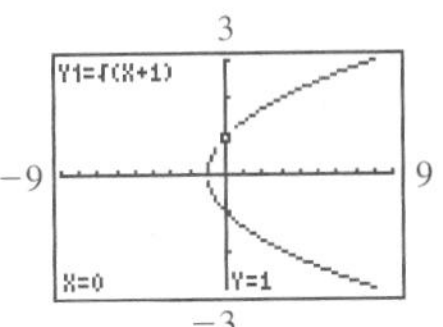

b) $x = -y^2$
$y = \pm\sqrt{-x}$

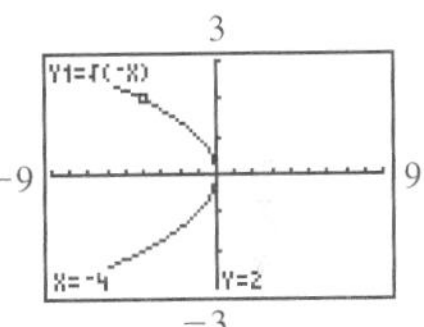

c) $x^2 + y^2 = 4$
$y = \pm\sqrt{4 - x^2}$

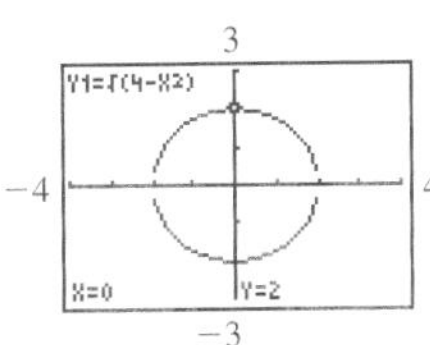

51. Determine the approximate vertex and x-intercepts for each parabola.

a) $y = 3.2x^2 - 5.4x + 1.6$
b) $y = -1.09x^2 + 13x + 7.5$

a) Vertex $(0.84, -0.68)$, x-intercepts $(1.30, 0)$, $(0.38, 0)$
b) Vertex $(5.96, 46.26)$, x-intercepts $(12.48, 0)$, $(-0.55, 0)$

9.7 INTRODUCTION TO FUNCTIONS

In This Section

Even though we have not yet used the term *function,* we have used functions throughout this text. In this section we will study the concept of functions and see that they have been with us from the beginning.

Functions Expressed by Formulas

If you get a speeding ticket, then your speed determines the cost of the ticket. You may not know exactly how the judge determines the cost, but the judge is using some rule to determine a cost from knowing your speed. The cost of the ticket is a function of your speed.

Function (as a Rule)

A function is a rule by which any allowable value of one variable (the **independent variable**) determines a *unique* value of a second variable (the **dependent variable**).

Helpful Hint

According to the dictionary, "determine" means to settle conclusively. If the value of the dependent variable is inconclusive or there is more than one, then the rule is not a function.

One way to express a function is to use a formula. For example, the formula

$$A = \pi r^2$$

gives the area of a circle as a function of its radius. The formula gives us a rule for finding a *unique* area for any given radius. A is the dependent variable, and r is the independent variable. The formula

$$S = -16t^2 + v_0 t + s_0$$

expresses altitude S of a projectile as a function of time t, where v_0 is the initial velocity and s_0 is the initial altitude. S is the dependent variable, and t is the independent variable.

Since a formula can be used as a rule for obtaining the value of the dependent variable from the value of the independent variable, we say that the formula is a function. Formulas describe or **model** relationships between variables. In Example 1, we find a function in a real situation.

EXAMPLE 1 **Writing a formula for a function**

A carpet layer charges \$25 plus \$4 per square yard for installing carpet. Write the total charge C as a function of the number n of square yards of carpet installed.

Solution

At \$4 per square yard, n square yards installed cost $4n$ dollars. If we include the \$25 charge, then the total cost is $4n + 25$ dollars. Thus the equation

$$C = 4n + 25$$

expresses C as a function of n. ■

Any formula that has the form $y = mx + b$ is a **linear function.** So in Example 1, the charge is a linear function of the number of square yards installed and we say that $C = 4n + 25$ is a linear function.

EXAMPLE 2 **A function in geometry**

Express the area of a circle as a function of its diameter.

Solution

The area of a circle is given by $A = \pi r^2$. Because the radius of a circle is one-half of the diameter, we have $r = \frac{d}{2}$. Now replace r by $\frac{d}{2}$ in the formula $A = \pi r^2$:

$$A = \pi\left(\frac{d}{2}\right)^2$$

$$= \frac{\pi d^2}{4}$$

So $A = \frac{\pi}{4}d^2$ expresses the area of a circle as a function of its diameter. ■

Any formula that has the form $y = ax^2 + bx + c$ (with $a \neq 0$) is a **quadratic function.** So in Example 2, the area is a quadratic function of the diameter and we say that $A = \frac{\pi}{4}d^2$ is a quadratic function.

Study Tip

When one student asks another for help, it is often the one who does the explaining that learns the most. When you work together, don't let one person do all of the talking. If you must work alone, then try explaining things to yourself. You might even try talking to yourself.

Functions Expressed by Tables

Another way to express a function is with a table. For example, Table 9.2 can be used to determine the cost at United Freight Service for shipping a package that weighs under 100 pounds. For any *allowable* weight, the table gives us a rule for finding the unique shipping cost. The weight is the independent variable, and the cost is the dependent variable.

Weight in Pounds	Cost
0 to 10	\$4.60
11 to 30	\$12.75
31 to 79	\$32.90
80 to 99	\$55.82

TABLE 9.2

Now consider Table 9.3. It does not look much different from Table 9.2, but there is an important difference. The cost for shipping a 12-pound package according to Table 9.3 is either \$4.60 or \$12.75. Either the table has an error or perhaps \$4.60 and \$12.75 are costs for shipping to different destinations. In any case the weight does not determine a unique cost. So Table 9.3 does not express the cost as a function of the weight.

Weight in Pounds	Cost
0 to 15	\$4.60
10 to 30	\$12.75
31 to 79	\$32.90
80 to 99	\$55.82

TABLE 9.3

EXAMPLE 3 **Functions defined by tables**

Which of the following tables expresses y as a function of x?

a)

x	y
1	3
2	6
3	9
4	12
5	15

b)

x	y
1	1
-1	1
2	2
-2	2
3	3
-3	3

c)

x	y
1988	27,000
1989	27,000
1990	28,500
1991	29,000
1992	30,000
1993	30,750

d)

x	y
23	48
35	27
19	28
23	37
41	56
22	34

Solution

In Tables a), b), and c), every value of x corresponds to only one value of y. Tables a), b), and c) each express y as a function of x. Notice that different values of

x may correspond to the same value of y. In Table d) we have the value of 23 for x corresponding to two different values of y, 48 and 37. So Table d) does not express y as a function of x. ■

Functions Expressed by Ordered Pairs

Helpful Hint

In a function, every value for the independent variable determines conclusively a corresponding value for the dependent variable. If there is more than one possible value for the dependent variable, then the set of ordered pairs is not a function.

A computer at your grocery store determines the price of each item by searching a long list of ordered pairs in which the first coordinate is the universal product code and the second coordinate is the price of the item with that code. For each product code there is a unique price. This process certainly satisfies the rule definition of a function. Since the set of ordered pairs is the essential part of this rule we say that the set of ordered pairs is a function.

Function (as a Set of Ordered Pairs)

A function is a set of ordered pairs of real numbers such that no two ordered pairs have the same first coordinates and different second coordinates.

Note the importance of the phrase "no two ordered pairs have the same first coordinates and different second coordinates." Imagine the problems at the grocery store if the computer gave two different prices for the same universal product code. Note also that the product code is an identification number and it cannot be used in calculations. So the computer can use a function defined by a formula to determine the amount of tax, but it cannot use a formula to determine the price from the product code.

EXAMPLE 4 **Functions expressed by a set of ordered pairs**

Determine whether each set of ordered pairs is a function.

a) $\{(1, 2), (1, 5), (-4, 6)\}$ **b)** $\{(-1, 3), (0, 3), (6, 3), (-3, 2)\}$

Solution

a) This set of ordered pairs is not a function because (1, 2) and (1, 5) have the same first coordinates but different second coordinates.

b) This set of ordered pairs is a function. Note that the same second coordinate with different first coordinates is permitted in a function. ■

If there are infinitely many ordered pairs in a function, then we can use set-builder notation from Chapter 1 along with an equation to express the function. For example,

$$\{(x, y) \mid y = x^2\}$$

is the set of ordered pairs in which the y-coordinate is the square of the x-coordinate. Ordered pairs such as (0, 0), (2, 4), and (−2, 4) belong to this set. This set is a function because every value of x determines only one value of y.

EXAMPLE 5 **Functions expressed by set-builder notation**

Determine whether each set of ordered pairs is a function.

a) $\{(x, y) \mid y = 3x^2 - 2x + 1\}$ **b)** $\{(x, y) \mid y^2 = x\}$ **c)** $\{(x, y) \mid x + y = 6\}$

Solution

a) This set is a function because each value we select for x determines only one value for y.

Helpful Hint

Real-life variables are generally not as simple as the ones we consider. A student's college GPA is not a function of age, because many students with the same age have different GPAs. However, GPA is probably a function of a large number of variables: age, IQ, high school GPA, number of working hours, mother's IQ, and so on.

b) If $x = 9$, then we have $y^2 = 9$. Because both 3 and -3 satisfy $y^2 = 9$, both (9, 3) and (9, -3) belong to this set. So the set is not a function.

c) If we solve $x + y = 6$ for y, we get $y = -x + 6$. Because each value of x determines only one value for y, this set is a function. In fact, this set is a linear function. ■

We often omit the set notation when discussing functions. For example, the equation

$$y = 3x^2 - 2x + 1$$

expresses y as a function of x because the set of ordered pairs determined by the equation is a function. However, the equation

$$y^2 = x$$

does not express y as a function of x because ordered pairs such as (9, 3) and (9, -3) satisfy the equation.

EXAMPLE 6 **Functions expressed by equations**

Determine whether each equation expresses y as a function of x.

a) $y = |x|$ **b)** $y = x^3$ **c)** $x = |y|$

Helpful Hint

To determine whether an equation expresses y as a function of x, always select a number for x (the independent variable) and then see if there is more than one corresponding value for y (the dependent variable). If there is more than one corresponding y-value, then y is not a function of x.

Solution

a) Because every number has a unique absolute value, $y = |x|$ is a function.

b) Because every number has a unique cube, $y = x^3$ is a function.

c) The equation $x = |y|$ does not express y as a function of x because both (4, -4) and (4, 4) satisfy this equation. These ordered pairs have the same first coordinate but different second coordinates. ■

Graphs of Functions

Every function determines a set of ordered pairs, and any set of ordered pairs has a graph in the rectangular coordinate system. For example, the set of ordered pairs determined by the linear function $y = 2x - 1$ is shown in Fig. 9.10.

Every graph illustrates a set of ordered pairs, but not every graph is a graph of a function. For example, the circle in Fig. 9.11 is not a graph of a function because the ordered pairs (0, 4) and (0, -4) are both on the graph, and these two ordered pairs have the same first coordinate and different second coordinates. Whether a graph has such ordered pairs can be determined by a simple visual test called the **vertical-line test.**

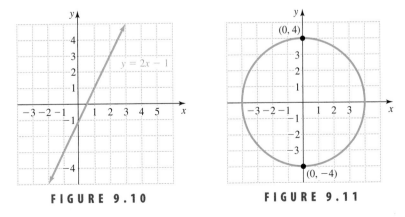

Calculator Close-Up

Most calculators graph only functions. If you enter an equation using the Y= key and the calculator accepts it and draws a graph, the equation defines y as a function of x.

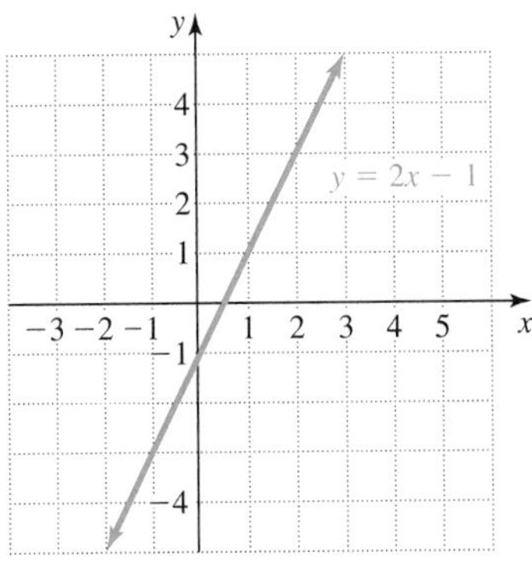

FIGURE 9.10

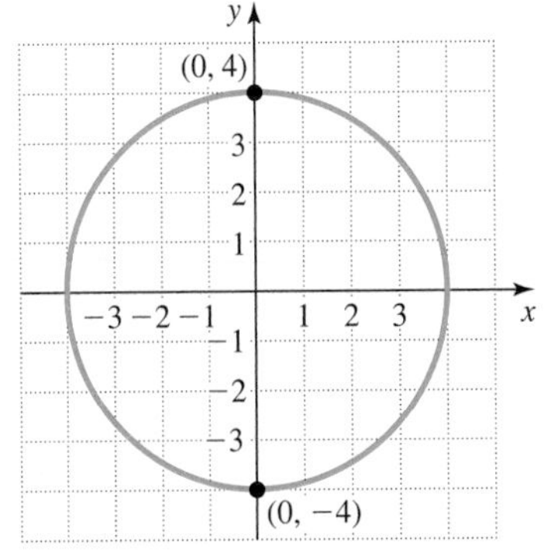

FIGURE 9.11

Vertical-Line Test

If it is possible to draw a vertical line that crosses a graph two or more times, then the graph is not the graph of a function.

If there is a vertical line that crosses a graph twice (or more), then we have two points (or more) with the same x-coordinate and different y-coordinates, and so the graph is not the graph of a function. If you mentally consider every possible vertical line and none of them cross the graph more than once, then you can conclude that the graph is the graph of a function.

EXAMPLE 7 **Using the vertical-line test**

Which of the following graphs are graphs of functions?

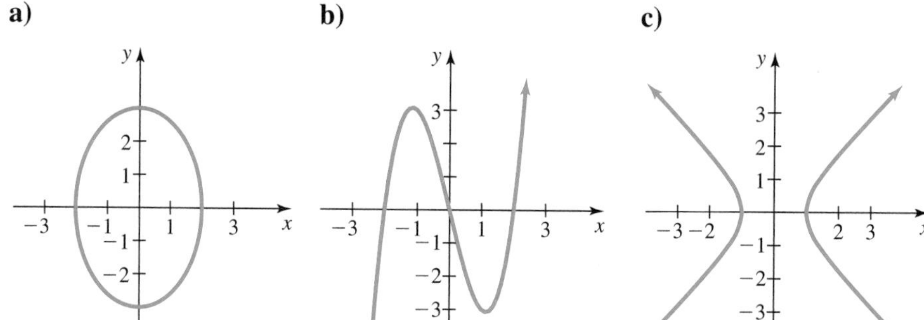

Solution

Neither a) nor c) is the graph of a function, since we can draw vertical lines that cross these graphs twice. Graph b) is the graph of a function, since no vertical line crosses it twice. ■

Helpful Hint

Note that the vertical-line test works in theory, but it is certainly limited by the accuracy of the graph. Pictures are often deceptive. However, in spite of its limitations, the vertical-line test gives us a visual idea of what the graph of a function looks like.

Domain and Range

A function is a set of ordered pairs. The set of all first coordinates of the ordered pairs is the **domain** of the function and the set of all second coordinates of the ordered pairs is the **range** of the function. In Example 8 we identify the domain and range for functions that are given as sets of ordered pairs.

EXAMPLE 8 **Domain and range**

Determine the domain and range of each function.

a) $\{(3, -1), (2, 5), (1, -4)\}$

b) $\{(2, 3), (4, 3), (6, 3), (8, 3)\}$

Solution

a) The domain is the set of numbers that occur as first coordinates, $\{1, 2, 3\}$. The range is the set of second coordinates, $\{-4, -1, 5\}$.

b) The domain is $\{2, 4, 6, 8\}$. Since 3 is the only number used as the second coordinate, the range is $\{3\}$. ■

If a function is defined by an equation, then the domain consists of those real numbers that can be used for the independent variable in the equation. For example, in $y = 3x$ we can use any real number for x. In $y = \frac{1}{x}$ we can use any nonzero real number for x. The domain is often the set of all real numbers, which is abbreviated as R.

The graph of a function is a picture of all ordered pairs of the function. So if a function is defined by an equation, it is usually helpful to graph the function and then use the graph as an aid in determining the domain and range.

EXAMPLE 9 **Finding domain and range from a graph**

Graph each function and then determine its domain and range.

a) $y = 2x - 1$

b) $y = 2x^2 - 8x + 5$

c) $y = -x^2 - 6x$

Solution

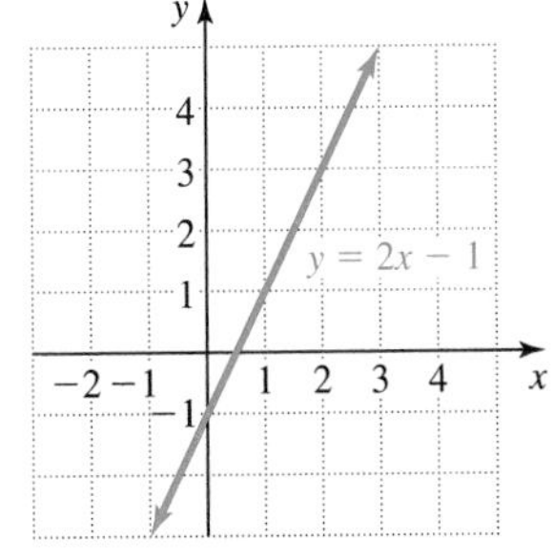

FIGURE 9.12

a) The graph of $y = 2x - 1$ is a line with y-intercept $(0, -1)$ and slope 2 as shown in Fig. 9.12. Any real number can be used for x in $y = 2x - 1$. So the domain is the set of real numbers, R. From the graph we can see that the line extends infinitely upward and downward. So the range is also R.

b) To graph the parabola $y = 2x^2 - 8x + 5$, first find the vertex:

$$x = \frac{-b}{2a} = \frac{8}{4} = 2 \qquad y = 2(2)^2 - 8(2) + 5 = -3$$

The vertex is $(2, -3)$. The parabola also goes through $(1, -1)$ and $(3, -1)$ as shown in Fig. 9.13. Since any number can be used for x in $y = 2x^2 - 8x + 5$, the domain is R. Since the parabola opens upward, the smallest y-coordinate on the graph is -3. So the range is the set of real numbers that are greater than or equal to -3, which is written in set-builder notation as $\{y \mid y \geq -3\}$.

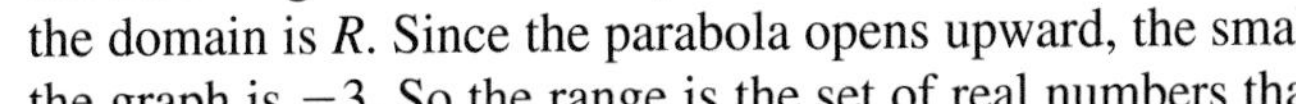

c) To graph the parabola $y = -x^2 - 6x$, first find the vertex:

$$x = \frac{-b}{2a} = \frac{6}{-2} = -3 \qquad y = -(-3)^2 - 6(-3) = 9$$

The vertex is $(-3, 9)$. The parabola also goes through $(-2, 8)$ and $(-4, 8)$ as shown in Fig. 9.14. Since any real number can be used in place of x in $y = -x^2 - 6x$, the domain is R. Since the parabola opens downward, the largest y-coordinate on the graph is 9. So the range is the set of real numbers that are less than or equal to 9, which is written in set notation as $\{y \mid y \leq 9\}$.

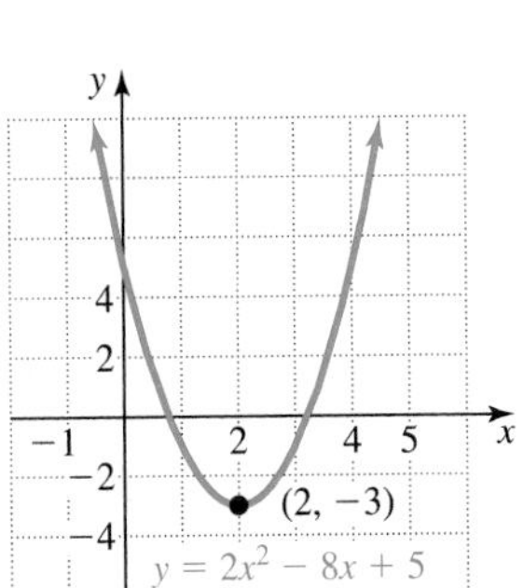

FIGURE 9.13

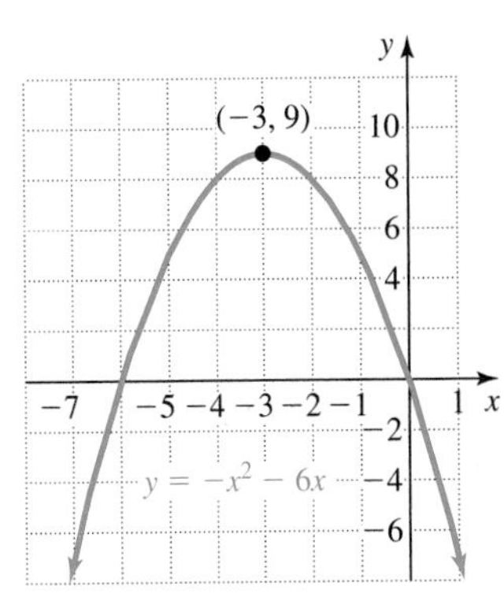

FIGURE 9.14

Function Notation

When the variable y is a function of x, we may use the notation $f(x)$ to represent y. This **function notation** was first used in Section 5.1 with polynomial functions.

The symbol $f(x)$ is read as "f of x." So if x is the independent variable, we may use y or $f(x)$ to represent the dependent variable. For example, the function

$$y = 2x + 3$$

can also be written as

$$f(x) = 2x + 3.$$

We use y and $f(x)$ interchangeably. We think of f as the name of the function. We may use letters other than f. For example, the function $g(x) = 2x + 3$ is the same function as $f(x) = 2x + 3$.

The expression $f(x)$ represents the second coordinate when the first coordinate is x; it does not mean f times x. For example, if we replace x by 4 in $f(x) = 2x + 3$, we get

$$f(4) = 2 \cdot 4 + 3 = 11.$$

So if the first coordinate is 4, then the second coordinate is $f(4)$, or 11. The ordered pair (4, 11) belongs to the function f. This statement means that the function f pairs 4 with 11. We can use the diagram in Fig. 9.15 to picture this situation.

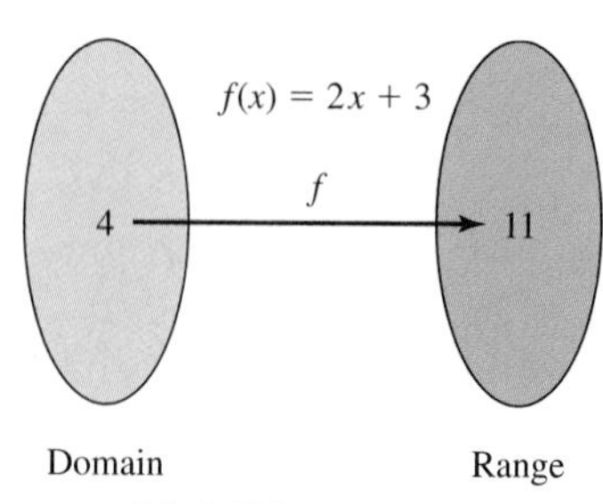

FIGURE 9.15

EXAMPLE 10 **Using function notation**

Suppose $f(x) = x^2 - 1$ and $g(x) = -3x + 2$. Find the following:

a) $f(-2)$ **b)** $f(-1)$

c) $g(0)$ **d)** $g(6)$

Study Tip

Everyone knows that you must practice to be successful with musical instruments, foreign languages, and sports. Success in algebra also requires regular practice. So budget your time so that you have a regular practice period for algebra.

Solution

a) Replace x by -2 in the formula $f(x) = x^2 - 1$:

$$\begin{aligned} f(-2) &= (-2)^2 - 1 \\ &= 4 - 1 \\ &= 3 \end{aligned}$$

So $f(-2) = 3$.

b) Replace x by -1 in the formula $f(x) = x^2 - 1$:

$$\begin{aligned} f(-1) &= (-1)^2 - 1 \\ &= 1 - 1 \\ &= 0 \end{aligned}$$

So $f(-1) = 0$.

c) Replace x by 0 in the formula $g(x) = -3x + 2$:

$$g(0) = -3 \cdot 0 + 2 = 2$$

So $g(0) = 2$.

d) Replace x by 6 in $g(x) = -3x + 2$ to get $g(6) = -16$. ■

EXAMPLE 11 **Using function notation in an application**

The formula $C(n) = 0.10n + 4.95$ gives the monthly cost in dollars for n minutes of long-distance calls. Find $C(40)$ and $C(100)$.

Solution

Replace n with 40 in the formula:

$$\begin{aligned} C(n) &= 0.10n + 4.95 \\ C(40) &= 0.10(40) + 4.95 \\ &= 8.95 \end{aligned}$$

So $C(40) = 8.95$. The cost for 40 minutes of calls is \$8.95. Now

$$C(100) = 0.10(100) + 4.95 = 14.95.$$

So $C(100) = 14.95$. The cost of 100 minutes of calls is \$14.95. ■

CAUTION $C(h)$ is not C times h. In the context of formulas, $C(h)$ represents the value of C corresponding to a value of h.

Calculator Close-Up

A graphing calculator can be used to evaluate a formula in the same manner as in Example 11. To evaluate

$$C = 0.10n + 4.95$$

enter the formula into your calculator as $y_1 = 0.10x + 4.95$ using the Y= key:

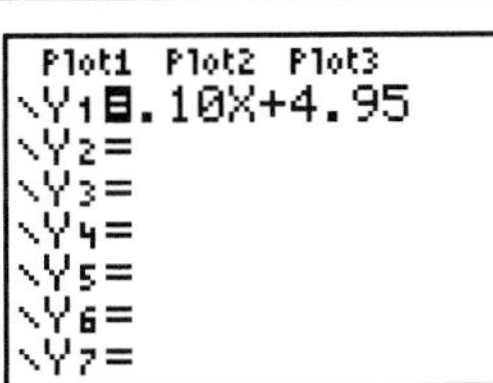

To find the cost of 40 minutes of calls, enter $y_1(40)$ on the home screen and press ENTER:

```
Y1(40)
                8.95
Y1(100)
               14.95
```

WARM-UPS

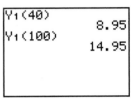

True or false? Explain your answer.

1. Any set of ordered pairs is a function. False
2. The area of a square is a function of the length of a side. True
3. The set $\{(-1, 3), (-3, 1), (-1, -3)\}$ is a function. False
4. The set $\{(1, 5), (3, 5), (7, 5)\}$ is a function. True
5. The domain of $\{(1, 2), (3, 4)\}$, is $\{1, 3\}$. True
6. The domain of $y = x^2$ is R. True
7. The range of $y = x^2$ is $\{y \mid y \geq 0\}$. True
8. The set $\{(x, y) \mid x = 2y\}$ is a function. True
9. The set $\{(x, y) \mid x = y^2\}$ is a function. False
10. If $f(x) = x^2 - 5$, then $f(-2) = -1$. True

9.7 EXERCISES

Reading and Writing *After reading this section, write out the answers to these questions. Use complete sentences.*

1. What is a function?
A function is a set of ordered pairs in which no two have the same first coordinate and different second coordinates.

2. What are the different ways to express functions?
Functions can be expressed by means of a verbal rule, a formula, a table, a list, or a graph.

3. What do all descriptions of functions have in common?
All descriptions of functions involve ordered pairs that satisfy the definition.

4. How can you tell at a glance if a graph is a graph of a function?
A graph is the graph of a function if no vertical line crosses the graph more than once.

5. What is the domain of a function?
The domain is the set of all first coordinates of the ordered pairs.

6. What is function notation?
In function notation x represents the first coordinate and $f(x)$ represents the second coordinate.

Write a formula that describes the function for each of the following. See Examples 1 and 2.

7. A small pizza costs \$5.00 plus 50 cents for each topping. Express the total cost C as a function of the number of toppings t. $C = 0.50t + 5$

8. A developer prices condominiums in Florida at \$20,000 plus \$40 per square foot of living area. Express the cost C as a function of the number of square feet of living area s. $C = 40s + 20{,}000$

9. The sales tax rate on groceries in Mayberry is 9%. Express the total cost T (including tax) as a function of the total price of the groceries S. $T = 1.09S$

10. With a GM MasterCard, 5% of the amount charged is credited toward a rebate on the purchase of a new car. Express the rebate R as a function of the amount charged A. $R = 0.05A$

11. Express the circumference of a circle as a function of its radius. $C = 2\pi r$

12. Express the circumference of a circle as a function of its diameter. $C = \pi d$

13. Express the perimeter P of a square as a function of the length s of a side. $P = 4s$

14. Express the perimeter P of a rectangle with width 10 ft as a function of its length L. $P = 2L + 20$

15. Express the area A of a triangle with a base of 10 m as a function of its height h. $A = 5h$

16. Express the area A of a trapezoid with bases 12 cm and 10 cm as a function of its height h. $A = 11h$

Determine whether each table expresses the second variable as a function of the first variable. See Example 3.

17. Yes

x	y
1	1
4	2
9	3
16	4
25	5
36	6
49	8

18. Yes

x	y
2	4
3	9
4	16
5	25
8	36
9	49
10	100

19. Yes

t	v
2	2
−2	2
3	3
−3	3
4	4
−4	4
5	5

20. Yes

s	W
5	17
6	17
−1	17
−2	17
−3	17
7	17
8	17

21. No

a	P
2	2
2	−2
3	3
3	−3
4	4
4	−4
5	5

22. No

n	r
17	5
17	6
17	−1
17	−2
17	−3
17	−4
17	−5

23. Yes

b	q
1970	0.14
1972	0.18
1974	0.18
1976	0.22
1978	0.25
1980	0.28

24. Yes

c	h
345	0.3
350	0.4
355	0.5
360	0.6
365	0.7
370	0.8
380	0.9

Determine whether each set of ordered pairs is a function. See Example 4.

25. $\{(1, 2), (2, 3), (3, 4)\}$ Yes

26. $\{(1, -3), (1, 3), (2, 12)\}$ No

27. $\{(-1, 4), (2, 4), (3, 4)\}$ Yes

28. $\{(1, 7), (7, 1)\}$ Yes

29. $\{(0, -1), (0, 1)\}$ No

30. $\{(1, 7), (-2, 7), (3, 7), (4, 7)\}$ Yes

31. $\{(50, 50)\}$ Yes

32. $\{(0, 0)\}$ Yes

Determine whether each set is a function. See Example 5.

33. $\{(x, y) \mid y = x - 3\}$ Yes

34. $\{(x, y) \mid y = x^2 - 2x - 1\}$ Yes

35. $\{(x, y) \mid x = |y|\}$ No

36. $\{(x, y) \mid x = y^2 + 1\}$ No

37. $\{(x, y) \mid x = y + 1\}$ Yes

38. $\left\{(x, y) \middle| y = \frac{1}{x}\right\}$ Yes

39. $\{(x, y) \mid x = y^2 - 1\}$ No

40. $\{(x, y) \mid x = 3y\}$ Yes

Determine whether each equation expresses y as a function of x. See Example 6.

41. $x = 4y$ Yes

42. $x = -3y$ Yes

43. $y = \frac{2}{x}$ Yes

44. $y = \frac{x}{2}$ Yes

45. $y = x^3 - 1$ Yes

46. $y = |x - 1|$ Yes

47. $x^2 + y^2 = 25$ No

48. $x^2 - y^2 = 9$ No

Which of the following graphs are graphs of functions? See Example 7.

49. 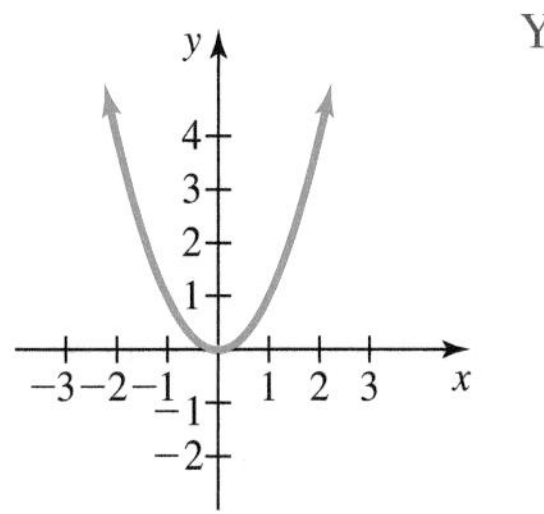 Yes

50. 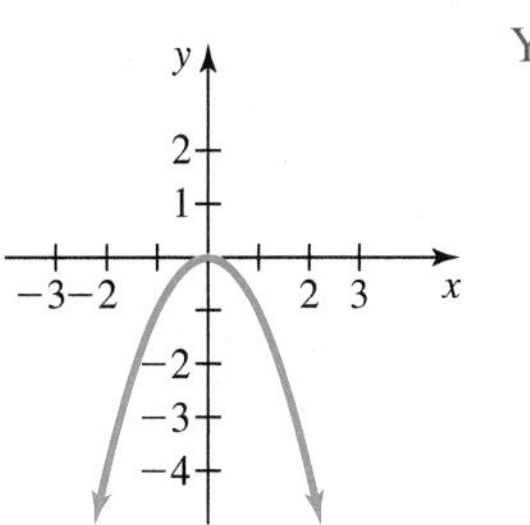 Yes

51. 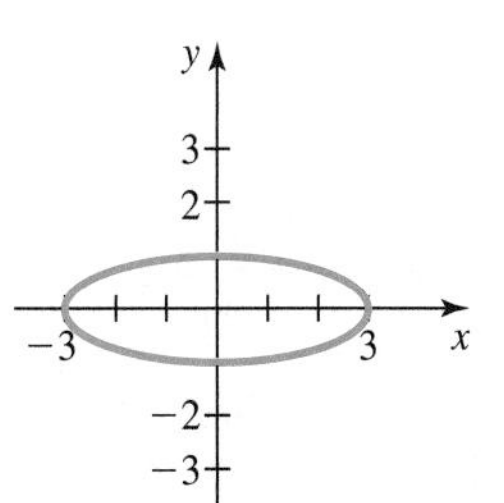 No

52. 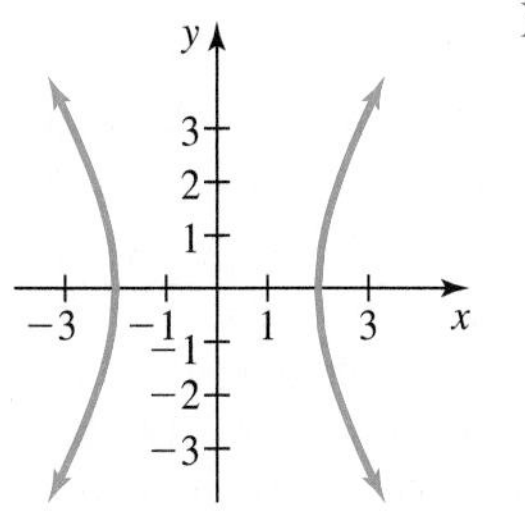 No

53. 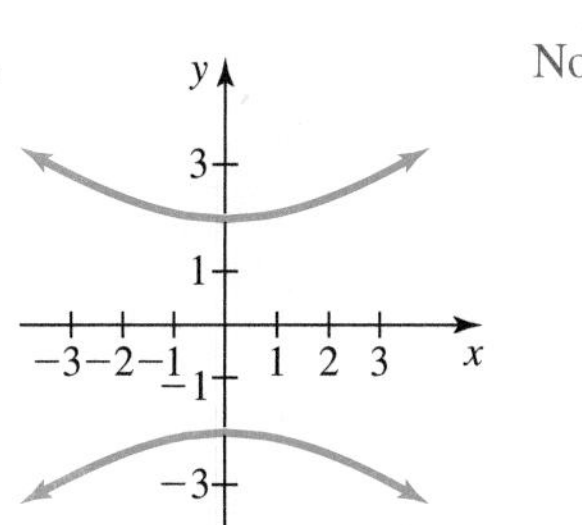 No

54. 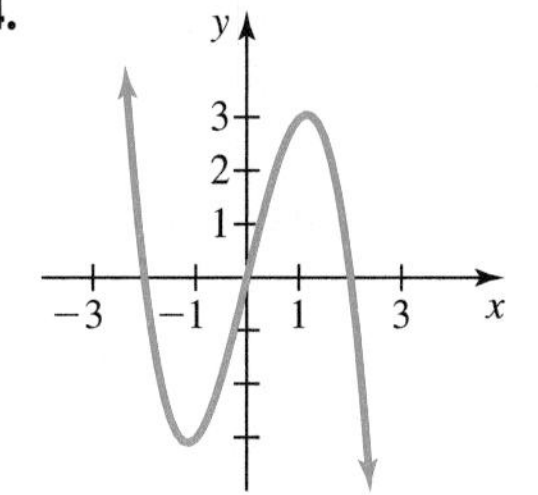 Yes

Determine the domain and range of each function. See Example 8.

55. $\{(3, 3), (2, 5), (1, 7)\}$ $\{1, 2, 3\}, \{3, 5, 7\}$

56. $\{(-1, 4), (3, 5)\}$ $\{-1, 3\}, \{4, 5\}$

57. $\{(0, 1), (2, 1), (4, 1)\}$ $\{0, 2, 4\}, \{1\}$

58. $\{(4, -2), (5, -2), (7, -2), (8, -2)\}$ $\{4, 5, 7, 8\}, \{-2\}$

Graph each function and identify the domain and range. See Example 9.

59. $y = 2x - 6$

R (real numbers), R

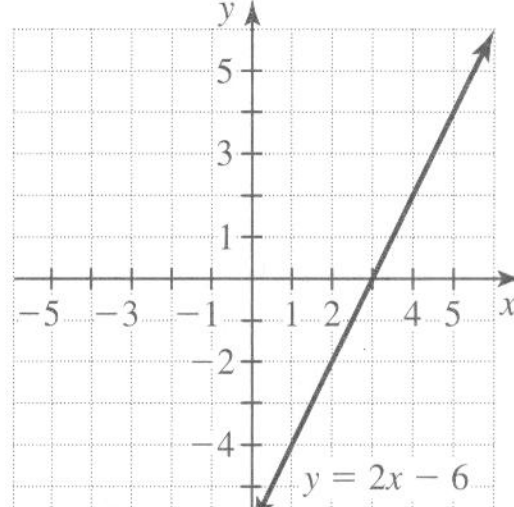

60. $y = 3x + 2$

R, R

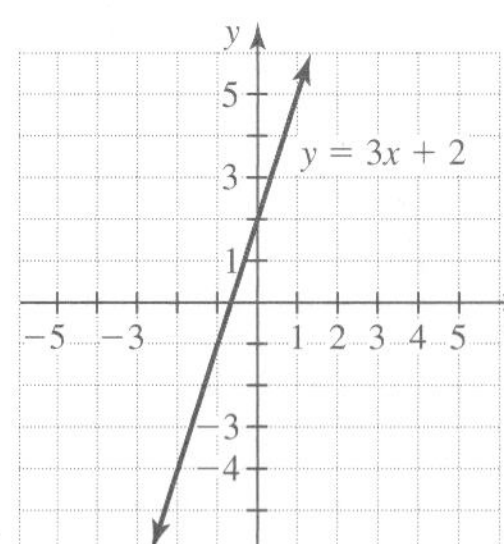

61. $y = -x$

R, R

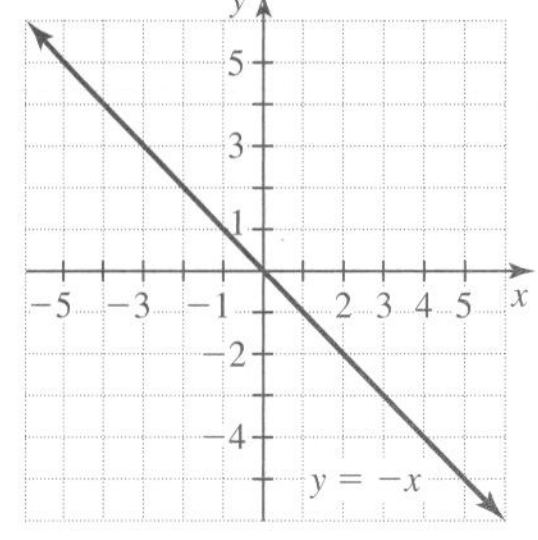

62. $y = 5 - x$
R, R

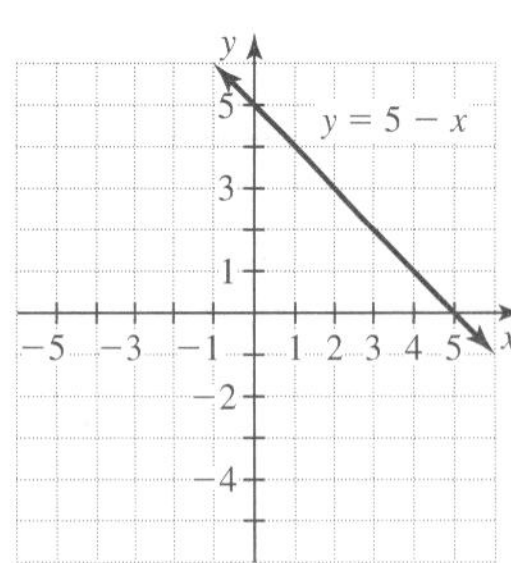

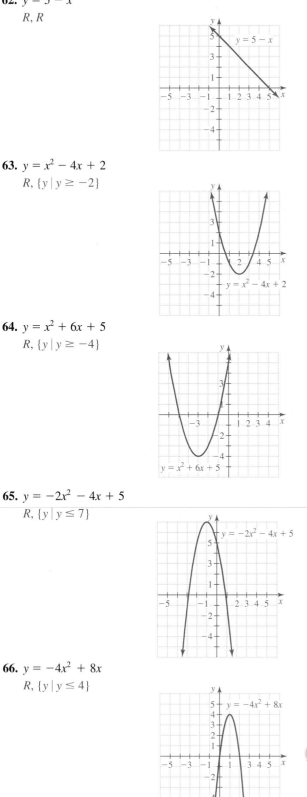

63. $y = x^2 - 4x + 2$
$R, \{y \mid y \geq -2\}$

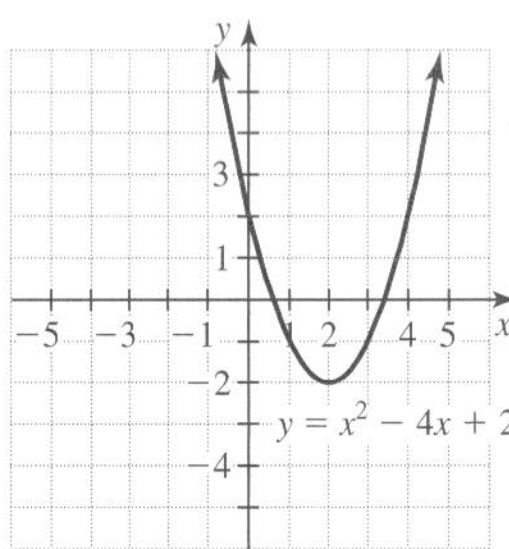

64. $y = x^2 + 6x + 5$
$R, \{y \mid y \geq -4\}$

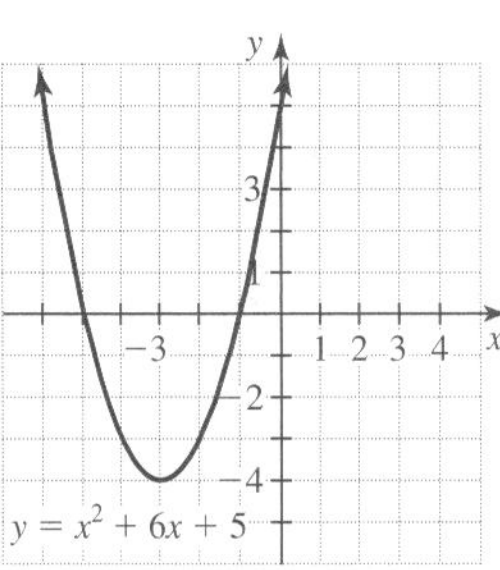

65. $y = -2x^2 - 4x + 5$
$R, \{y \mid y \leq 7\}$

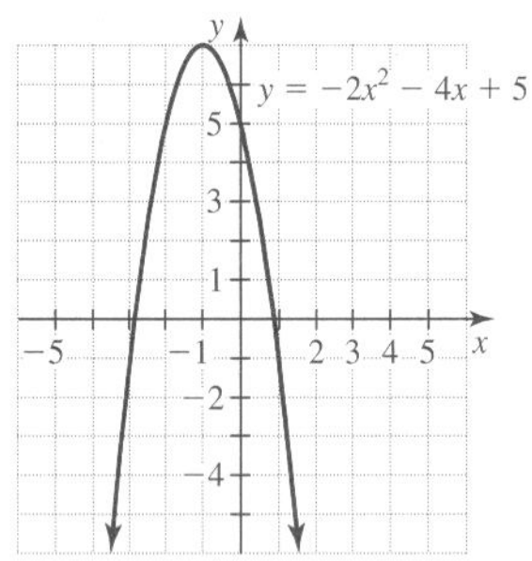

66. $y = -4x^2 + 8x$
$R, \{y \mid y \leq 4\}$

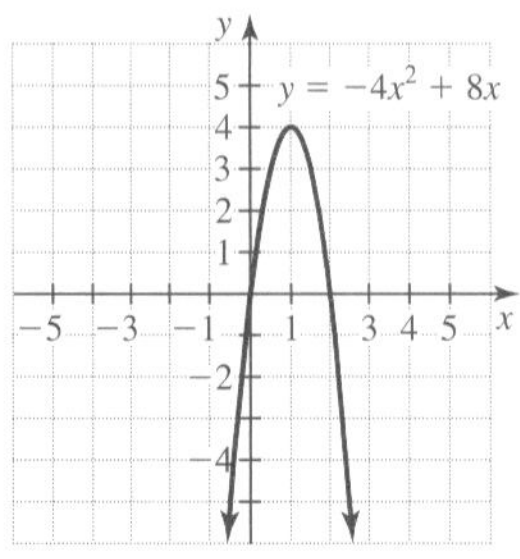

Let $f(x) = 2x - 1$, $g(x) = x^2 - 3$, and $h(x) = |x - 1|$. Find the following. See Example 10.

67. $f(0)$ -1

68. $f(-1)$ -3

69. $f\left(\frac{1}{2}\right)$ 0

70. $f\left(\frac{3}{4}\right)$ $\frac{1}{2}$

71. $g(4)$ 13

72. $g(-4)$ 13

73. $g(0.5)$ -2.75

74. $g(-1.5)$ -0.75

75. $h(3)$ 2

76. $h(-1)$ 2

77. $h(0)$ 1

78. $h(1)$ 0

Let $f(x) = x^3 - x^2$ and $g(x) = x^2 - 4.2x + 2.76$. Find the following. Round each answer to three decimal places.

79. $f(5.68)$ 150.988

80. $g(-2.7)$ 21.39

81. $g(3.5)$ 0.31

82. $f(67.2)$ $298{,}948.608$

Solve each problem. See Example 11.

83. ***Velocity and time.*** If a ball is thrown straight upward into the air with a velocity of 100 ft/sec, then its velocity t seconds later is given by

$$v(t) = -32t + 100.$$

a) Find $v(0)$, $v(1)$, and $v(2)$.
100 ft/sec, 68 ft/sec, 36 ft/sec

b) Is the velocity increasing or decreasing as the time increases? Decreasing

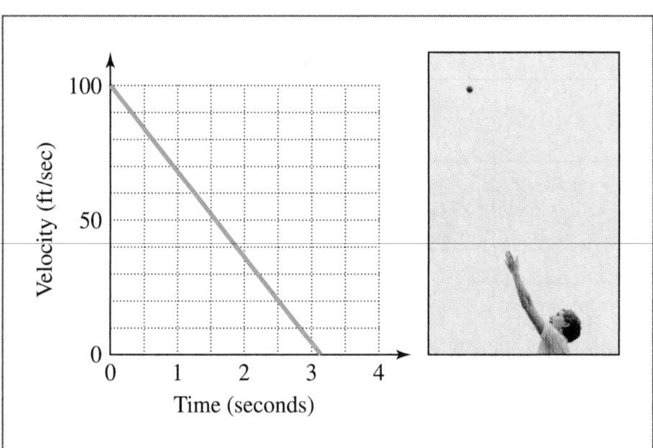

FIGURE FOR EXERCISE 83

84. ***Cost and toppings.*** The cost c in dollars for a pizza with n toppings is given by

$$c(n) = 0.75n + 6.99.$$

a) Find $c(2)$, $c(4)$, and $c(5)$. \$8.49, \$9.99, \$10.74

b) Is the cost increasing or decreasing as the number of toppings increases? Increasing

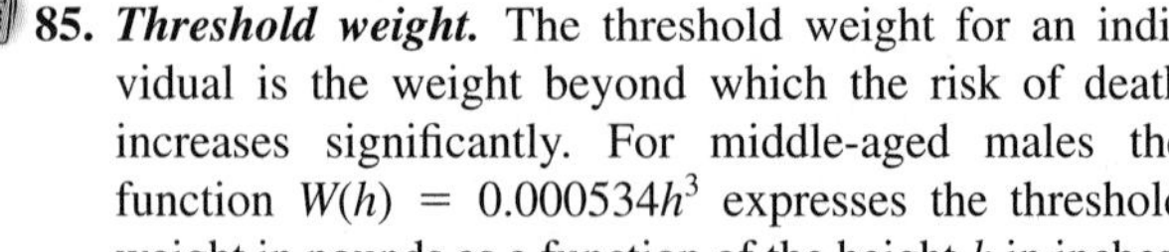

85. ***Threshold weight.*** The threshold weight for an individual is the weight beyond which the risk of death increases significantly. For middle-aged males the function $W(h) = 0.000534h^3$ expresses the threshold weight in pounds as a function of the height h in inches.

Find $W(70)$. Find the threshold weight for a 6′2″ middle-aged male. 183 pounds, 216 pounds

86. *Pole vaulting.* The height a pole vaulter attains is a function of the vaulter's velocity on the runway. The function

$$h(v) = \frac{1}{64}v^2$$

gives the height in feet as a function of the velocity v in feet per second.

a) Find $h(35)$ to the nearest tenth of an inch.
19 ft 1.7 in.

b) Who gains more height from an increase of 1 ft/sec in velocity: a fast runner or a slow runner?
Fast runner

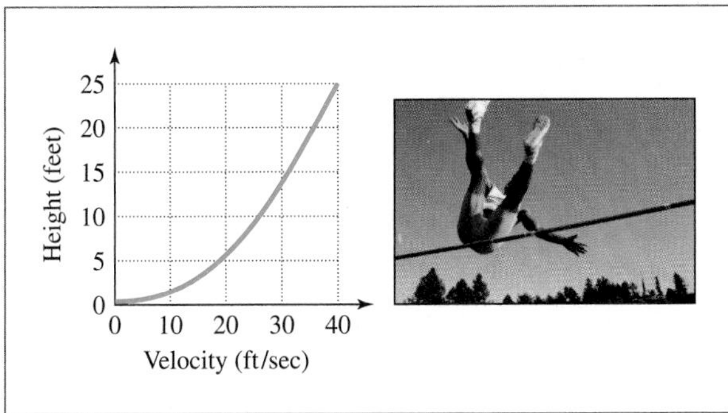

FIGURE FOR EXERCISE 86

87. *Credit card fees.* A certain credit card company gets 4% of each charge, and the retailer receives the rest. At the end of a billing period the retailer receives a statement showing only the retailer's portion of each transaction. Express the original amount charged C as a function of the retailer's portion r. $C = \frac{r}{0.96}$

88. *More credit card fees.* Suppose that the amount charged on the credit card in the previous exercise includes 8% sales tax. The credit card company does not get any of the sales tax. In this case the retailer's portion of each transaction includes sales tax on the original cost of the goods. Express the original amount charged C as a function of the retailer's portion. $C = \frac{27}{26}R$

GETTING MORE INVOLVED

Discussion *In each situation determine whether a is a function of b, b is a function of a, or neither. Answers may vary depending on interpretations.*

89. a = the price per gallon of regular unleaded.
b = the number of gallons that you get for \$10. Both

90. a = the universal product code of an item at Sears.
b = the price of that item. b is a function of a

91. a = a student's score on the last test in this class.
b = the number of hours he/she spent studying.
Neither

92. a = a student's score on the last test in this class.
b = the IQ of the student's mother. Neither

93. a = the weight of a package shipped by UPS.
b = the cost of shipping that package.
Neither

94. a = the Celsius temperature at any time.
b = the Fahrenheit temperature at the same time.
Both

95. a = the weight of a letter.
b = the cost of mailing the letter.
b is a function of a

96. a = the cost of a gallon of milk.
b = the amount of sales tax on that gallon.
b is a function of a

In This Section

- Basic Operations with Functions
- Composition

9.8 COMBINING FUNCTIONS

In this section you will learn how to combine functions to obtain new functions.

Basic Operations with Functions

An entrepreneur plans to rent a stand at a farmers market for \$25 per day to sell strawberries. If she buys x flats of berries for \$5 per flat and sells them for \$9 per flat, then her daily cost in dollars can be written as a function of x:

$$C(x) = 5x + 25$$

Assuming she sells as many flats as she buys, her revenue in dollars is also a function of x:

$$R(x) = 9x$$

Because profit is revenue minus cost, we can find a function for the profit by subtracting the functions for cost and revenue:

$$\begin{aligned} P(x) &= R(x) - C(x) \\ &= 9x - (5x + 25) \\ &= 4x - 25 \end{aligned}$$

The function $P(x) = 4x - 25$ expresses the daily profit as a function of x. Since $P(6) = -1$ and $P(7) = 3$, the profit is negative if 6 or fewer flats are sold and positive if 7 or more flats are sold.

In the example of the entrepreneur we subtracted two functions to find a new function. In other cases we may use addition, multiplication, or division to combine two functions. For any two given functions we can define the sum, difference, product, and quotient functions as follows.

Sum, Difference, Product, and Quotient Functions

Given two functions f and g, the functions $f + g$, $f - g$, $f \cdot g$, and $\frac{f}{g}$ are defined as follows:

Sum function: $(f + g)(x) = f(x) + g(x)$

Difference function: $(f - g)(x) = f(x) - g(x)$

Product function: $(f \cdot g)(x) = f(x) \cdot g(x)$

Quotient function: $\left(\frac{f}{g}\right)(x) = \frac{f(x)}{g(x)}$ provided that $g(x) \neq 0$

The domain of the function $f + g$, $f - g$, $f \cdot g$, or $\frac{f}{g}$ is the intersection of the domain of f and the domain of g. For the function $\frac{f}{g}$ we also rule out any values of x for which $g(x) = 0$.

EXAMPLE 1 **Operations with functions**

Let $f(x) = 4x - 12$ and $g(x) = x - 3$. Find the following.

a) $(f + g)(x)$ **b)** $(f - g)(x)$ **c)** $(f \cdot g)(x)$ **d)** $\left(\frac{f}{g}\right)(x)$

Solution

a) $$\begin{aligned}(f + g)(x) &= f(x) + g(x) \\ &= 4x - 12 + x - 3 \\ &= 5x - 15\end{aligned}$$

b) $$\begin{aligned}(f - g)(x) &= f(x) - g(x) \\ &= 4x - 12 - (x - 3) \\ &= 3x - 9\end{aligned}$$

c) $$\begin{aligned}(f \cdot g)(x) &= f(x) \cdot g(x) \\ &= (4x - 12)(x - 3) \\ &= 4x^2 - 24x + 36\end{aligned}$$

d) $\left(\frac{f}{g}\right)(x) = \frac{f(x)}{g(x)} = \frac{4x - 12}{x - 3} = \frac{4(x - 3)}{x - 3} = 4$ for $x \neq 3$. ■

Helpful Hint

Note that we use $f + g$, $f - g$, $f \cdot g$, and f/g to name these functions only because there is no application in mind here. We generally use a single letter to name functions after they are combined as we did when using P for the profit function rather than $R - C$.

EXAMPLE 2 **Evaluating a sum function**

Let $f(x) = 4x - 12$ and $g(x) = x - 3$. Find $(f + g)(2)$.

Solution

In Example 1(a) we found a general formula for the function $f + g$, namely, $(f + g)(x) = 5x - 15$. If we replace x by 2, we get

$$\begin{aligned}(f + g)(2) &= 5(2) - 15\\ &= -5.\end{aligned}$$

We can also find $(f + g)(2)$ by evaluating each function separately and then adding the results. Because $f(2) = -4$ and $g(2) = -1$, we get

$$\begin{aligned}(f + g)(2) &= f(2) + g(2)\\ &= -4 + (-1)\\ &= -5.\end{aligned}$$

■

Composition

Helpful Hint

The difference between the first four operations with functions and composition is like the difference between parallel and series in electrical connections. Components connected in parallel operate simultaneously and separately. If components are connected in series, then electricity must pass through the first component to get to the second component.

A salesperson's monthly salary is a function of the number of cars he sells: \$1000 plus \$50 for each car sold. If we let S be his salary and n be the number of cars sold, then S in dollars is a function of n:

$$S = 1000 + 50n$$

Each month the dealer contributes \$100 plus 5% of his salary to a profit-sharing plan. If P represents the amount put into profit sharing, then P (in dollars) is a function of S:

$$P = 100 + 0.05S$$

Now P is a function of S, and S is a function of n. Is P a function of n? The value of n certainly determines the value of P. In fact, we can write a formula for P in terms of n by substituting one formula into the other:

$$\begin{aligned}P &= 100 + 0.05S\\ &= 100 + 0.05(1000 + 50n) && \text{Substitute } S = 1000 + 50n.\\ &= 100 + 50 + 2.5n && \text{Distributive property}\\ &= 150 + 2.5n\end{aligned}$$

Now P is written as a function of n, bypassing S. We call this idea **composition of functions.**

EXAMPLE 3 **The composition of two functions**

Given that $y = x^2 - 2x + 3$ and $z = 2y - 5$, write z as a function of x.

Solution

Replace y in $z = 2y - 5$ by $x^2 - 2x + 3$:

$$\begin{aligned}z &= 2y - 5\\ &= 2(x^2 - 2x + 3) - 5 && \text{Replace } y \text{ by } x^2 - 2x + 3.\\ &= 2x^2 - 4x + 1\end{aligned}$$

The equation $z = 2x^2 - 4x + 1$ expresses z as a function of x. ■

The composition of two functions using function notation is defined as follows.

Composition of Functions

The **composition** of f and g is denoted $f \circ g$ and is defined by the equation

$$(f \circ g)(x) = f(g(x)),$$

provided that $g(x)$ is in the domain of f.

The notation $f \circ g$ is read as "the composition of f and g" or "f compose g." The diagram in Fig. 9.16 shows a function g pairing numbers in its domain with numbers in its range. If the range of g is contained in or equal to the domain of f, then f pairs the second coordinates of g with numbers in the range of f. The composition function $f \circ g$ is a rule for pairing numbers in the domain of g directly with numbers in the range of f, bypassing the middle set. The domain of the function $f \circ g$ is the domain of g (or a subset of it) and the range of $f \circ g$ is the range of f (or a subset of it).

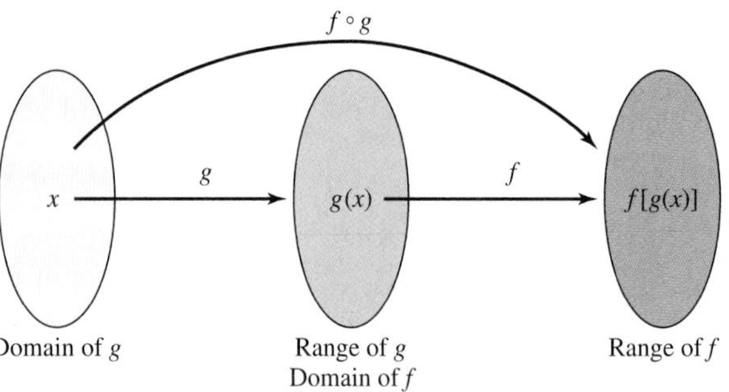

FIGURE 9.16

CAUTION The order in which functions are written is important in composition. For the function $f \circ g$ the function f is applied to $g(x)$. For the function $g \circ f$ the function g is applied to $f(x)$. The function closest to the variable x is applied first.

EXAMPLE 4 **Composition of functions**

Let $f(x) = 3x - 2$ and $g(x) = x^2 + 2x$. Find the following.

a) $(g \circ f)(2)$

b) $(f \circ g)(2)$

c) $(g \circ f)(x)$

d) $(f \circ g)(x)$

Calculator Close-Up

Set $y_1 = 3x - 2$ and $y_2 = x^2 + 2x$. You can find the composition for Examples 4(a) and 4(b) by evaluating $y_2(y_1(2))$ and $y_1(y_2(2))$. Note that the order in which you evaluate the functions is critical.

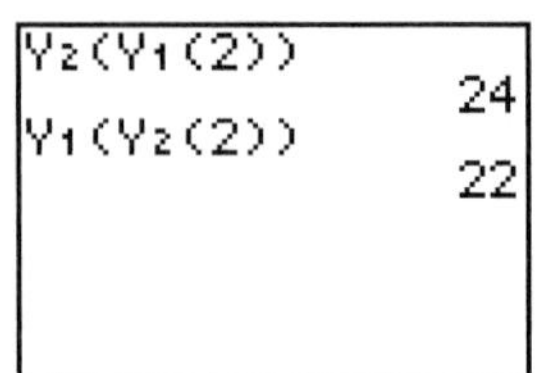

Solution

a) Because $(g \circ f)(2) = g(f(2))$, we first find $f(2)$:

$$f(2) = 3 \cdot 2 - 2 = 4$$

Because $f(2) = 4$, we have

$$(g \circ f)(2) = g(f(2)) = g(4) = 4^2 + 2 \cdot 4 = 24.$$

So $(g \circ f)(2) = 24$.

b) Because $(f \circ g)(2) = f(g(2))$, we first find $g(2)$:

$$g(2) = 2^2 + 2 \cdot 2 = 8$$

Because $g(2) = 8$, we have

$$(f \circ g)(2) = f(g(2)) = f(8) = 3 \cdot 8 - 2 = 22.$$

Thus $(f \circ g)(2) = 22$.

c) $(g \circ f)(x) = g(f(x))$
$= g(3x - 2)$
$= (3x - 2)^2 + 2(3x - 2)$
$= 9x^2 - 12x + 4 + 6x - 4 = 9x^2 - 6x$

So $(g \circ f)(x) = 9x^2 - 6x$.

d) $(f \circ g)(x) = f(g(x))$
$= f(x^2 + 2x)$
$= 3(x^2 + 2x) - 2 = 3x^2 + 6x - 2$

So $(f \circ g)(x) = 3x^2 + 6x - 2$. ■

Helpful Hint

A composition of functions can be viewed as two function machines where the output of the first is the input of the second.

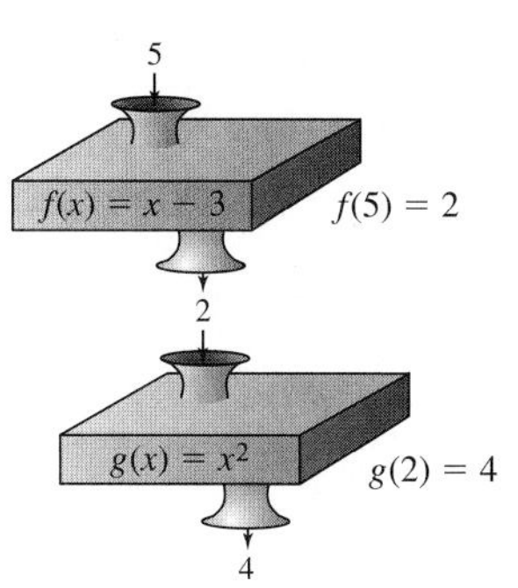

Notice that in Example 4(a) and (b), $(g \circ f)(2) \neq (f \circ g)(2)$. In Example 4(c) and (d) we see that $(g \circ f)(x)$ and $(f \circ g)(x)$ have different formulas defining them. In general, $f \circ g \neq g \circ f$.

It is often useful to view a complicated function as a composition of simpler functions. For example, the function $Q(x) = (x - 3)^2$ consists of two operations, subtracting 3 and squaring. So Q can be described as a composition of the functions $f(x) = x - 3$ and $g(x) = x^2$. To check this, we find $(g \circ f)(x)$:

$$(g \circ f)(x) = g(f(x)) = g(x - 3) = (x - 3)^2$$

We can express the fact that Q is the same as the composition function $g \circ f$ by writing $Q = g \circ f$ or $Q(x) = (g \circ f)(x)$.

EXAMPLE 5 **Expressing a function as a composition of simpler functions**

Let $f(x) = x - 2$, $g(x) = 3x$, and $h(x) = \sqrt{x}$. Write each of the following functions as a composition, using f, g, and h.

a) $F(x) = \sqrt{x - 2}$

b) $H(x) = x - 4$

c) $K(x) = 3x - 6$

Study Tip

Effective studying involves actively digging into the subject. Be sure that you are making steady progress. At the end of each week take note of the progress that you have made. What do you know on Friday that you did not know on Monday?

Solution

a) The function F consists of first subtracting 2 from x and then taking the square root of that result. So $F = h \circ f$. Check this result by finding $(h \circ f)(x)$:

$$(h \circ f)(x) = h(f(x)) = h(x - 2) = \sqrt{x - 2}$$

b) Subtracting 4 from x can be accomplished by subtracting 2 from x and then subtracting 2 from that result. So $H = f \circ f$. Check by finding $(f \circ f)(x)$:

$$(f \circ f)(x) = f(f(x)) = f(x - 2) = x - 2 - 2 = x - 4$$

c) Notice that $K(x) = 3(x - 2)$. The function K consists of subtracting 2 from x and then multiplying the result by 3. So $K = g \circ f$. Check by finding $(g \circ f)(x)$:

$$(g \circ f)(x) = g(f(x)) = g(x - 2) = 3(x - 2) = 3x - 6$$ ■

CAUTION In Example 5(a) we have $F = h \circ f$ because in F we subtract 2 before taking the square root. If we had the function $G(x) = \sqrt{x} - 2$, we would take the square root before subtracting 2. So $G = f \circ h$. Notice how important the order of operations is here.

WARM-UPS

True or false? Explain your answer.

1. If $f(x) = x - 2$ and $g(x) = x + 3$, then $(f - g)(x) = -5$. True
2. If $f(x) = x + 4$ and $g(x) = 3x$, then $\left(\frac{f}{g}\right)(2) = 1$. True
3. The functions $f \circ g$ and $g \circ f$ are always the same. False
4. If $f(x) = x^2$ and $g(x) = x + 2$, then $(f \circ g)(x) = x^2 + 2$. False
5. The functions $f \circ g$ and $f \cdot g$ are always the same. False
6. If $f(x) = \sqrt{x}$ and $g(x) = x - 9$, then $g(f(x)) = f(g(x))$ for every x. False
7. If $f(x) = 3x$ and $g(x) = \frac{x}{3}$, then $(f \circ g)(x) = x$. True
8. If $a = 3b^2 - 7b$, and $c = a^2 + 3a$, then c is a function of b. True
9. The function $F(x) = \sqrt{x - 5}$ is a composition of two functions. True
10. If $F(x) = (x - 1)^2$, $h(x) = x - 1$, and $g(x) = x^2$, then $F = g \circ h$. True

9.8 EXERCISES

Reading and Writing *After reading this section, write out the answers to these questions. Use complete sentences.*

1. What are the basic operations with functions?
   The basic operations of functions are addition, subtraction, multiplication, and division.
2. How do we perform the basic operations with functions?
   We perform the operations with functions by adding, subtracting, multiplying, or dividing the expressions that define the functions.
3. What is the composition of two functions?
   In the composition function the second function is evaluated on the result of the first function.
4. How is the order of operations related to composition of functions?
   Since each operation is a function, the order of operations determines the order in which the functions are composed.

Let $f(x) = 4x - 3$, and $g(x) = x^2 - 2x$. Find the following. See Examples 1 and 2.

5. $(f + g)(x)$ $x^2 + 2x - 3$
6. $(f - g)(x)$ $-x^2 + 6x - 3$
7. $(f \cdot g)(x)$ $4x^3 - 11x^2 + 6x$
8. $\left(\frac{f}{g}\right)(x)$ $\frac{4x - 3}{x^2 - 2x}$
9. $(f + g)(3)$ 12
10. $(f + g)(2)$ 5
11. $(f - g)(-3)$ -30
12. $(f - g)(-2)$ -19
13. $(f \cdot g)(-1)$ -21
14. $(f \cdot g)(-2)$ -88
15. $\left(\frac{f}{g}\right)(4)$ $\frac{13}{8}$
16. $\left(\frac{f}{g}\right)(-2)$ $-\frac{11}{8}$

For Exercises 17–24, use the two functions to write y as a function of x. See Example 3.

17. $y = 3a - 2, a = 2x - 6$ $y = 6x - 20$
18. $y = 2c + 3, c = -3x + 4$ $y = -6x + 11$
19. $y = 2d + 1, d = \frac{x + 1}{2}$ $y = x + 2$
20. $y = -3d + 2, d = \frac{2 - x}{3}$ $y = x$
21. $y = m^2 - 1, m = x + 1$ $y = x^2 + 2x$
22. $y = n^2 - 3n + 1, n = x + 2$ $y = x^2 + x - 1$

23. $y = \dfrac{a-3}{a+2},\ a = \dfrac{2x+3}{1-x}$ $y = x$

24. $y = \dfrac{w+2}{w-5},\ w = \dfrac{5x+2}{x-1}$ $y = x$

Let $f(x) = 2x - 3$, $g(x) = x^2 + 3x$, and $h(x) = \frac{x+3}{2}$. Find the following. See Example 4.

25. $(g \circ f)(1)$ -2

26. $(f \circ g)(-2)$ -7

27. $(f \circ g)(1)$ 5

28. $(g \circ f)(-2)$ 28

29. $(f \circ f)(4)$ 7

30. $(h \circ h)(3)$ 3

31. $(h \circ f)(5)$ 5

32. $(f \circ h)(0)$ 0

33. $(f \circ h)(5)$ 5

34. $(h \circ f)(0)$ 0

35. $(g \circ h)(-1)$ 4

36. $(h \circ g)(-1)$ $\dfrac{1}{2}$

37. $(f \circ g)(2.36)$ 22.2992

38. $(h \circ f)(23.761)$ 23.761

39. $(g \circ f)(x)$
$4x^2 - 6x$

40. $(g \circ h)(x)$
$\dfrac{x^2 + 12x + 27}{4}$

41. $(f \circ g)(x)$
$2x^2 + 6x - 3$

42. $(h \circ g)(x)$
$\dfrac{x^2 + 3x + 3}{2}$

43. $(h \circ f)(x)$
x

44. $(f \circ h)(x)$
x

45. $(f \circ f)(x)$
$4x - 9$

46. $(g \circ g)(x)$
$x^4 + 6x^3 + 12x^2 + 9x$

47. $(h \circ h)(x)$
$\dfrac{x+9}{4}$

48. $(f \circ f \circ f)(x)$
$8x - 21$

Let $f(x) = \sqrt{x}$, $g(x) = x^2$, and $h(x) = x - 3$. Write each of the following functions as a composition using f, g, or h. See Example 5.

49. $F(x) = \sqrt{x-3}$
$F = f \circ h$

50. $N(x) = \sqrt{x} - 3$
$N = h \circ f$

51. $G(x) = x^2 - 6x + 9$
$G = g \circ h$

52. $P(x) = x$ for $x \geq 0$
$P = f \circ g$

53. $H(x) = x^2 - 3$
$H = h \circ g$

54. $M(x) = x^{1/4}$
$M = f \circ f$

55. $J(x) = x - 6$
$J = h \circ h$

56. $R(x) = \sqrt{x^2 - 3}$
$R = f \circ h \circ g$

57. $K(x) = x^4$
$K = g \circ g$

58. $Q(x) = \sqrt{x^2 - 6x + 9}$
$Q = f \circ g \circ h$

Solve each problem.

59. ***Color monitor.*** A color monitor has a square viewing area that has a diagonal measure of 15 inches. Find the area of the viewing area in square inches (in.^{2}). Write a formula for the area of a square as a function of the length of its diagonal.
112.5 in.^{2}, $A = \dfrac{d^2}{2}$

60. ***Perimeter.*** Write a formula for the perimeter of a square as a function of its area.
$P = 4\sqrt{A}$

61. ***Profit function.*** A plastic bag manufacturer has determined that the company can sell as many bags as it can produce each month. If it produces x thousand bags in a month, the revenue is $R(x) = x^2 - 10x + 30$ dollars, and the cost is $C(x) = 2x^2 - 30x + 200$ dollars. Use the fact that profit is revenue minus cost to write the profit as a function of x.
$P(x) = -x^2 + 20x - 170$

62. ***Area of a sign.*** A sign is in the shape of a square with a semicircle of radius x adjoining one side and a semicircle of diameter x removed from the opposite side. If the sides of the square are length $2x$, then write the area of the sign as a function of x.
$A = \dfrac{(32 + 3\pi)x^2}{8}$

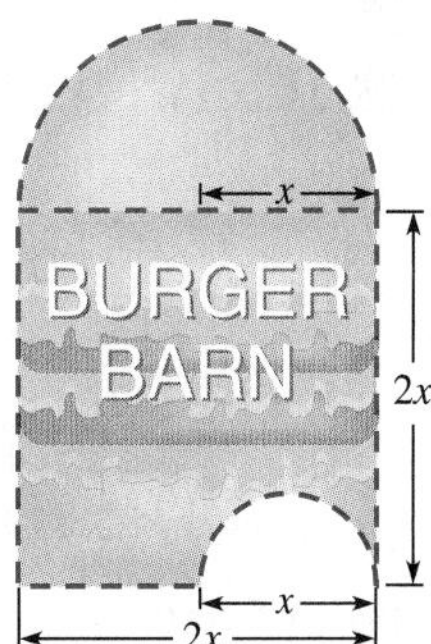

FIGURE FOR EXERCISE 62

63. ***Junk food expenditures.*** Suppose the average family spends 25% of its income on food, $F = 0.25I$, and 10% of each food dollar on junk food, $J = 0.10F$. Write J as a function of I.
$J = 0.025I$

64. ***Area of an inscribed circle.*** A pipe of radius r must pass through a square hole of area M as shown in the figure.

FIGURE FOR EXERCISE 64

Write the cross-sectional area of the pipe A as a function of M.

$A = \pi \frac{M}{4}$

65. ***Displacement-length ratio.*** To find the displacement-length ratio D for a sailboat, first find x, where $x = (L/100)^3$ and L is the length at the water line in feet. Next find D, where $D = (d/2240)/x$ and d is the displacement in pounds.

a) For the Pacific Seacraft 40, $L = 30$ ft 3 in. and $d = 24{,}665$ pounds. Find D.

b) For a boat with a displacement of 25,000 pounds, write D as a function of L.

c) The graph for the function in part (b) is shown in the accompanying figure. For a fixed displacement, does the displacement-length ratio increase or decrease as the length increases?

a) 397.8

b) $D = \frac{1.116 \times 10^7}{L^3}$

c) decreases

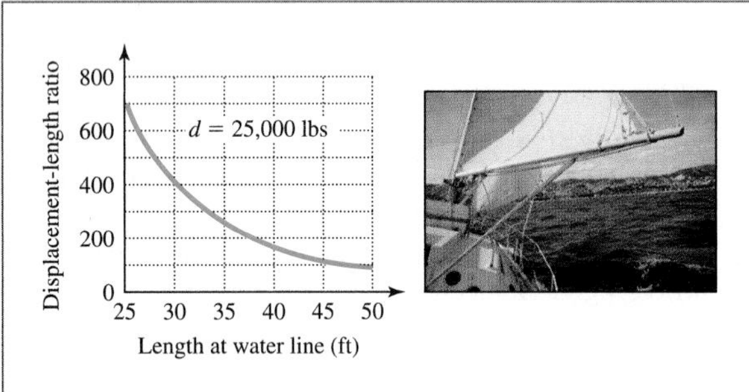

FIGURE FOR EXERCISE 65

66. ***Sail area-displacement ratio.*** To find the sail area-displacement ratio S, first find y, where $y = (d/64)^{2/3}$ and d is the displacement in pounds. Next find S, where $S = A/y$ and A is the sail area in square feet.

a) For the Pacific Seacraft 40, $A = 846$ square feet (ft^2) and $d = 24{,}665$ pounds. Find S.

b) For a boat with a sail area of 900 ft^2, write S as a function of d.

c) For a fixed sail area, does S increase or decrease as the displacement increases?

a) 15.97

b) $S = 14{,}400d^{-2/3}$

c) decreases

COLLABORATIVE ACTIVITIES

Designing a River Park

Grouping: Four students

Topic: Solving quadratic equations

At the last election, voters approved the creation of a park along the river on the edge of the city. You are student representatives on the citizen's committee working with the park designer. The city is allocating 300 acres for the park. The park designer says that the committee needs to decide the dimensions of the park. In terms of the land available, the park dimensions can vary between a length that is twice the width to a length that is 10 times the width. The committee decides to consider the following three cases:

A. The length is twice the width.

B. The length is 10 times the width.

C. The length is 5 times the width.

Because you have all studied algebra, you offer to find the dimensions and draw sketches of the three cases.

The following suggestions will help you in solving this problem.

1. Determine what units of measurement you wish to use (feet, yards, miles, meters or kilometers).

2. A calculator with a conversion feature will make your work easier.

3. Assign roles. You might use the roles Moderator, Recorder, Calculator user, and Sketcher.

In your report to the committee, do the following:

1. For each case, include the equations used, the width and length you found (rounded to the nearest whole number), and your work showing how you found the length and width.

2. Include a small sketch of each case showing the given proportions.

3. Choose one of the three cases that your group feels gives the best dimensions for the park. Considering what features (flowerbeds, playground, soccer field, walking or bike paths, etc.) you would like to have in the park will help you decide which dimensions you would choose. Make a larger-scale drawing of this case and include the features you would want in a park.

4. Write a short paragraph explaining why you chose the case you did.

WRAP-UP CHAPTER 9

SUMMARY

Quadratic Equations		**Examples**
Quadratic equation	An equation of the form $ax^2 + bx + c = 0$, where a, b, and c are real numbers with $a \neq 0$	$x^2 = 10$ $(x + 3)^2 = 8$ $x^2 + 5x - 7 = 0$
Methods for solving quadratic equations	Factoring	$x^2 + 5x + 6 = 0$ $(x + 3)(x + 2) = 0$
	Square root property	$(x - 3)^2 = 6$ $x - 3 = \pm\sqrt{6}$
	Completing the square (works on any quadratic): Take one-half of middle term, square it, then add it to each side.	$x^2 + 6x = 7$ $x^2 + 6x + 9 = 7 + 9$ $(x + 3)^2 = 16$
	Quadratic formula (works on any quadratic): $x = \dfrac{-b \pm \sqrt{b^2 - 4ac}}{2a}$	$2x^2 - 3x - 6 = 0$ $x = \dfrac{3 \pm \sqrt{9 - 4(2)(-6)}}{2(2)}$
Number and types of solutions	Determined by the discriminant $b^2 - 4ac$ $b^2 - 4ac > 0$: two real solutions	$x^2 + 5x - 9 = 0$ has two real solutions because $5^2 - 4(1)(-9) > 0$.
	$b^2 - 4ac = 0$: one real solution	$x^2 + 6x + 9 = 0$ has one real solution because $6^2 - 4(1)(9) = 0$.
	$b^2 - 4ac < 0$: no real solutions (two complex solutions)	$x^2 + 3x + 10 = 0$ has no real solutions because $3^2 - 4(1)(10) < 0$.

Complex Numbers		**Examples**
Complex numbers	Numbers of the form $a + bi$, where a and b are real $i = \sqrt{-1}$ and $i^2 = -1$	$12, -3i, 5 + 4i, \sqrt{2} - i\sqrt{3}$
Imaginary numbers	Numbers of the form $a + bi$, where $b \neq 0$	$5i, 13 + i\sqrt{6}$
Square root of a negative number	If b is a positive real number, then $\sqrt{-b} = i\sqrt{b}$.	$\sqrt{-3} = i\sqrt{3}$ $\sqrt{-4} = i\sqrt{4} = 2i$

Complex conjugates	The complex numbers $a + bi$ and $a - bi$ are called complex conjugates of each other. Their product is a real number.	$(1 + 2i)(1 - 2i) = 1 + 4 = 5$
Complex number operations	Add, subtract, and multiply as if the complex numbers were binomials with variable i. Use the distributive property for multiplication. Remember that $i^2 = -1$.	$(2 + 3i) + (3 - 5i) = 5 - 2i$ $(2 - 5i) - (4 - 2i) = -2 - 3i$ $(3 - 4i)(2 + 5i) = 26 + 7i$
	Divide complex numbers by multiplying the numerator and denominator by the conjugate of the denominator, then simplify.	$\frac{4 - 6i}{5 + 2i} = \frac{(4 - 6i)(5 - 2i)}{(5 + 2i)(5 - 2i)}$

Parabolas		**Examples**
Opening	The parabola $y = ax^2 + bx + c$ opens upward if $a > 0$ or downward if $a < 0$.	$y = x^2 - 2x$ opens upward. $y = -x^2$ opens downward.
Vertex	The first coordinate of the vertex is $x = \frac{-b}{2a}$. The second coordinate of the vertex is the minimum value of y if $a > 0$ or the maximum value of y if $a < 0$.	$y = x^2 - 2x$ opens upward with vertex $(1, -1)$. Minimum y-value is -1.
Intercepts	The x-intercepts are found by solving $ax^2 + bx + c = 0$. The y-intercept is found by replacing x by 0 in $y = ax^2 + bx + c$.	$y = x^2 - 2x - 8$ has x-intercepts $(-2, 0)$ and $(4, 0)$ and y-intercept $(0, -8)$.

Functions		**Examples**
Definition of a function	A function is a rule by which any allowable value of one variable (the independent variable) determines a unique value of a second variable (the dependent variable).	$A = \pi r^2$
Equivalent definition of a function	A function is a set of ordered pairs such that no two ordered pairs have the same first coordinates and different second coordinates. To say that y is a function of x means that y is determined uniquely by x.	$\{(1, 0), (3, 8)\}$ $\{(x, y) \mid y = x^2\}$
Domain	The set of values of the independent variable, x	$y = x^2$ Domain: all real numbers, R
Range	The set of values of the dependent variable, y	$y = x^2$ Range: nonnegative real numbers, $\{y \mid y \geq 0\}$
Linear functions	If $y = mx + b$, we say that y is a linear function of x.	$F = \frac{9}{5}C + 32$
Quadratic function	A function of the form $y = ax^2 + bx + c$, where a, b, and c are real numbers and $a \neq 0$	$y = 3x^2 - 8x + 9$ $p = -3q^2 - 8q + 1$

Function notation	If x is the independent variable, then we use the notation $f(x)$ to represent the dependent variable.	$y = 2x + 3$ $f(x) = 2x + 3$

Combining Functions		**Examples**
Sum	$(f + g)(x) = f(x) + g(x)$	For $f(x) = x^2$ and $g(x) = x + 1$ $(f + g)(x) = x^2 + x + 1$
Difference	$(f - g)(x) = f(x) - g(x)$	$(f - g)(x) = x^2 - x - 1$
Product	$(f \cdot g)(x) = f(x) \cdot g(x)$	$(f \cdot g)(x) = x^3 + x^2$
Quotient	$\left(\frac{f}{g}\right)(x) = \frac{f(x)}{g(x)}$	$\left(\frac{f}{g}\right)(x) = \frac{x^2}{x + 1}$
Composition of functions	$(g \circ f)(x) = g(f(x))$ $(f \circ g)(x) = f(g(x))$	$(g \circ f)(x) = g(x^2) = x^2 + 1$ $(f \circ g)(x) = f(x + 1)$ $= x^2 + 2x + 1$

ENRICHING YOUR MATHEMATICAL WORD POWER

For each mathematical term, choose the correct meaning.

1. **quadratic equation**
   a. $ax + b = c$ with $a \neq 0$
   b. $ax^2 + bx + c = 0$ with $a \neq 0$
   c. $ax + b = 0$ with $a \neq 0$
   d. $a/x^2 + b/x = c$ with $x \neq 0$ b

2. **perfect square trinomial**
   a. a trinomial of the form $a^2 + 2ab + b^2$
   b. a trinomial of the form $a^2 + b^2$
   c. a trinomial of the form $a^2 + ab + b^2$
   d. a trinomial of the form $a^2 - 2ab - b^2$ a

3. **completing the square**
   a. drawing a perfect square
   b. evaluating $(a + b)^2$
   c. drawing the fourth side when given three sides of a square
   d. finding the third term of a perfect square trinomial d

4. **quadratic formula**
   a. $x = \dfrac{-b \pm \sqrt{b^2 - 4ac}}{2}$
   b. $x = -b \pm \dfrac{\sqrt{b^2 - 4ac}}{2a}$
   c. $x = \dfrac{-b \pm \sqrt{b^2 - 4ac}}{2a}$
   d. $x = \dfrac{b \pm \sqrt{b^2 - 4ac}}{2a}$ c

5. **discriminant**
   a. the vertex of a parabola
   b. the radicand in the quadratic formula
   c. the leading coefficient in $ax^2 + bx + c$
   d. to treat unfairly b

6. **complex numbers**
   a. $a + bi$, where a and b are real
   b. irrational numbers
   c. imaginary numbers
   d. $\sqrt{-1}$ a

7. **imaginary unit**
   a. 1
   b. -1
   c. i
   d. $\sqrt{1}$ c

8. **imaginary numbers**
   a. $a + bi$, where a and b are real and $b \neq 0$
   b. i
   c. a complex number
   d. a complex number in which the real part is 0 a

9. **complex conjugates**
   a. i and $\sqrt{-1}$
   b. $a + bi$ and $a - bi$
   c. $(a + b)(a - b)$
   d. i and -1 b

10. **function**
    a. domain and range
    b. a set of ordered pairs
    c. a rule by which any allowable value of one variable determines a unique value of a second variable
    d. a graph c

11. domain
a. the set of first coordinates of a function
b. the set of second coordinates of a function
c. the set of real numbers
d. the integers a

12. range
a. all of the possibilities
b. the coordinates of a function
c. the entire set of numbers
d. the set of second coordinates of a function d

13. quadratic function
a. $y = ax + b$ with $a \neq 0$
b. a parabola
c. $y = ax^2 + bx + c$ with $a \neq 0$
d. the quadratic formula c

14. composition of f and g
a. the function $f \circ g$ where $(f \circ g)(x) = f(g(x))$
b. the function $f \circ g$ where $(f \circ g)(x) = g(f(x))$
c. the function $f \cdot g$ where $(f \cdot g)(x) = f(x) \cdot g(x)$
d. a diagram showing f and g a

15. sum of f and g
a. the function $f \cdot g$ where $(f \cdot g)(x) = f(x) \cdot g(x)$
b. the function $f + g$ where $(f + g)(x) = f(x) + g(x)$
c. the function $f \circ g$ where $(f \circ g)(x) = g(f(x))$
d. the function obtained by adding the domains of f and g b

REVIEW EXERCISES

9.1 *Solve each equation.*

1. $x^2 - 9 = 0$
$-3, 3$

2. $x^2 - 1 = 0$
$-1, 1$

3. $x^2 - 9x = 0$
$0, 9$

4. $x^2 - x = 0$
$0, 1$

5. $x^2 - x = 2$
$-1, 2$

6. $x^2 - 9x = 10$
$-1, 10$

7. $(x - 9)^2 = 10$
$9 - \sqrt{10}, 9 + \sqrt{10}$

8. $(x + 5)^2 = 14$
$-5 - \sqrt{14}, -5 + \sqrt{14}$

9. $4x^2 - 12x + 9 = 0$
$\frac{3}{2}$

10. $9x^2 + 6x + 1 = 0$
$-\frac{1}{3}$

11. $t^2 - 9t + 20 = 0$
$4, 5$

12. $s^2 - 4s + 3 = 0$
$1, 3$

13. $\frac{x}{2} = \frac{7}{x + 5}$
$-7, 2$

14. $\sqrt{x + 4} = \frac{2x - 1}{3}$
5

15. $\frac{1}{2}x^2 + \frac{7}{4}x = 1$
$-4, \frac{1}{2}$

16. $\frac{2}{3}x^2 - 1 = -\frac{1}{3}x$
$-\frac{3}{2}, 1$

9.2 *Solve each equation by completing the square.*

17. $x^2 + 4x - 7 = 0$
$-2 - \sqrt{11}, -2 + \sqrt{11}$

18. $x^2 + 6x - 3 = 0$
$-3 - 2\sqrt{3}, -3 + 2\sqrt{3}$

19. $x^2 + 3x - 28 = 0$
$-7, 4$

20. $x^2 - x - 6 = 0$
$-2, 3$

21. $x^2 + 3x - 5 = 0$
$\frac{-3 - \sqrt{29}}{2}, \frac{-3 + \sqrt{29}}{2}$

22. $x^2 + \frac{4}{3}x - \frac{1}{3} = 0$
$\frac{-2 - \sqrt{7}}{3}, \frac{-2 + \sqrt{7}}{3}$

23. $2x^2 + 9x - 5 = 0$
$-5, \frac{1}{2}$

24. $2x^2 + 6x - 5 = 0$
$\frac{-3 - \sqrt{19}}{2}, \frac{-3 + \sqrt{19}}{2}$

9.3 *Find the value of the discriminant, and tell how many real solutions each equation has.*

25. $25t^2 - 10t + 1 = 0$
0, one

26. $3x^2 + 2 = 0$
-24, none

27. $-3w^2 + 4w - 5 = 0$
-44, none

28. $5x^2 - 7x = 0$
49, two

29. $-3v^2 + 4v = -5$
76, two

30. $49u^2 + 42u + 9 = 0$
0, one

Use the quadratic formula to solve each equation.

31. $6x^2 + x - 2 = 0$
$-\frac{2}{3}, \frac{1}{2}$

32. $-6x^2 + 11x + 10 = 0$
$-\frac{2}{3}, \frac{5}{2}$

33. $x^2 - x = 4$
$\frac{1 - \sqrt{17}}{2}, \frac{1 + \sqrt{17}}{2}$

34. $y^2 - 2y = 4$
$1 - \sqrt{5}, 1 + \sqrt{5}$

35. $5x^2 - 6x - 1 = 0$
$\frac{3 - \sqrt{14}}{5}, \frac{3 + \sqrt{14}}{5}$

36. $t^2 - 6t + 4 = 0$
$3 - \sqrt{5}, 3 + \sqrt{5}$

37. $3x^2 - 5x = 0$
$0, \frac{5}{3}$

38. $2w^2 - w = 15$
$-\frac{5}{2}, 3$

9.4 *For each problem, find the exact and approximate answers. Round the decimal answers to three decimal places.*

39. ***Bird watching.*** Chuck is standing 12 meters from a tree, watching a bird's nest that is 5 meters above eye level. Find the distance from Chuck's eyes to the nest.
13 meters

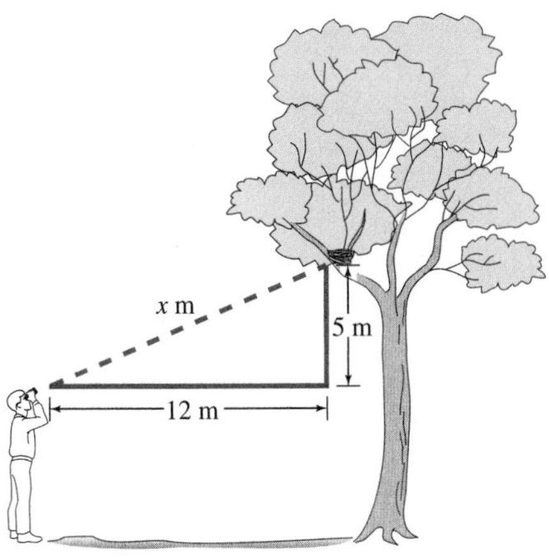

FIGURE FOR EXERCISE 39

40. ***Diagonal of a square.*** Find the diagonal of a square if the length of each side is 20 yards.
$20\sqrt{2}$ or 28.284 yards

41. ***Lengthy legs.*** The hypotenuse of a right triangle measures 5 meters, and one leg is 2 meters longer than the other. Find the lengths of the legs.
$\frac{-2+\sqrt{46}}{2}$ or 2.391 meters and $\frac{2+\sqrt{46}}{2}$ or 4.391 meters

42. ***Width and height.*** The width of a rectangular bookcase is 3 feet shorter than the height. If the diagonal is 7 feet, then what are the dimensions of the bookcase?
Width $\frac{-3+\sqrt{89}}{2}$ or 3.217 feet, height $\frac{3+\sqrt{89}}{2}$ or 6.217 feet

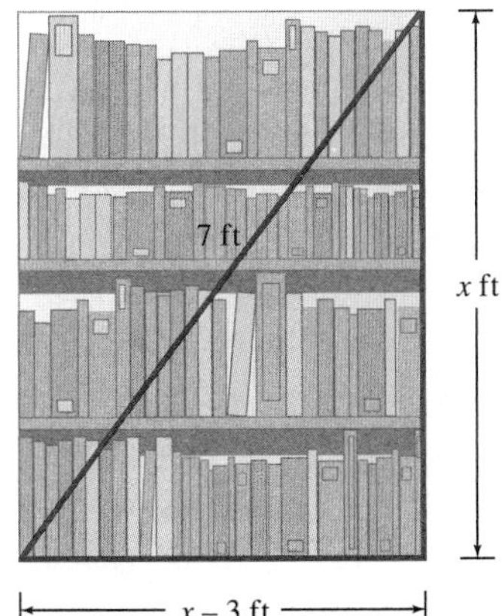

FIGURE FOR EXERCISE 42

43. ***Base and height.*** The base of a triangle is 4 inches longer than the height. If the area of the triangle is 20 square inches, then what are the lengths of the base and height?
Height $-2 + 2\sqrt{11}$ or 4.633 inches, base $2 + 2\sqrt{11}$ or 8.633 inches

44. ***Dimensions of a parallelogram.*** The base of a parallelogram is 1 meter longer than the height. If the area of the parallelogram is 8 square meters, then what are the lengths of the base and height?
Base $\frac{1+\sqrt{33}}{2}$ or 3.372 meters, height $\frac{-1+\sqrt{33}}{2}$ or 2.372 meters

45. ***Unknown numbers.*** Find two positive real numbers whose sum is 6 and whose product is 7.
$3 + \sqrt{2}$ or 4.414 and $3 - \sqrt{2}$ or 1.586

46. ***Dimensions of a rectangle.*** The perimeter of a rectangle is 16 feet, and its area is 13 square feet. What are the dimensions of the rectangle?
Width $4 - \sqrt{3}$ or 2.268 feet, length $4 + \sqrt{3}$ or 5.732 feet

47. ***Printing time.*** The old printer took 2 hours longer than the new printer to print 100,000 mailing labels. With both printers working on the job, the 100,000 labels can be printed in 8 hours. How long would it take each printer working alone to do the job?
New printer $7 + \sqrt{65}$ or 15.062 hours, old printer $9 + \sqrt{65}$ or 17.062 hours

FIGURE FOR EXERCISE 47

48. ***Tilling the garden.*** When Blake uses his old tiller, it takes him 3 hours longer to till the garden than it takes Cassie using her new tiller. If Cassie will not let Blake use her new tiller and they can till the garden together in 6 hours then how long would it take each one working alone?
Blake $\frac{15+3\sqrt{17}}{2}$ or 13.685 hours, Cassie $\frac{9+3\sqrt{17}}{2}$ or 10.685 hours.

FIGURE FOR EXERCISE 48

9.5 *Perform the indicated operations. Write answers in the form $a + bi$.*

49. $(2 + 3i) + (5 - 6i)$
$7 - 3i$

50. $(2 - 5i) + (-9 - 4i)$
$-7 - 9i$

51. $(-5 + 4i) - (-2 - 3i)$
$-3 + 7i$

52. $(1 - i) - (1 + i)$
$-2i$

53. $(2 - 9i)(3 + i)$
$15 - 25i$

54. $2i - 3(6 - 2i)$
$-18 + 8i$

55. $(3 + 8i)^2$
$-55 + 48i$

56. $(-5 - 2i)(-5 + 2i)$
29

57. $\dfrac{-2 - \sqrt{-8}}{2}$ $\quad -1 - i\sqrt{2}$

58. $\dfrac{-6 + \sqrt{-54}}{-3}$ $\quad 2 - i\sqrt{6}$

59. $\dfrac{1 + 3i}{6 - i}$ $\quad \dfrac{3}{37} + \dfrac{19}{37}i$

60. $\dfrac{3i}{8 + 3i}$ $\quad \dfrac{9}{73} + \dfrac{24}{73}i$

61. $\dfrac{5 + i}{4 - i}$ $\quad \dfrac{19}{17} + \dfrac{9}{17}i$

62. $\dfrac{3 + 2i}{i}$ $\quad 2 - 3i$

Find the complex solutions to the quadratic equations.

63. $x^2 + 121 = 0$
$-11i, 11i$

64. $x^2 + 120 = 0$
$-2i\sqrt{30}, 2i\sqrt{30}$

65. $x^2 - 16x + 65 = 0$
$8 - i, 8 + i$

66. $x^2 - 10x + 28 = 0$
$5 - i\sqrt{3}, 5 + i\sqrt{3}$

67. $2x^2 - 3x + 9 = 0$
$\dfrac{3 - 3i\sqrt{7}}{4}, \dfrac{3 + 3i\sqrt{7}}{4}$

68. $3x^2 - 6x + 4 = 0$
$\dfrac{3 - i\sqrt{3}}{3}, \dfrac{3 + i\sqrt{3}}{3}$

9.6 *Find the vertex and intercepts for each parabola and sketch its graph.*

69. $y = x^2 - 6x$
Vertex $(3, -9)$,
intercepts $(0, 0)$, $(6, 0)$

70. $y = x^2 + 4x$
Vertex $(-2, -4)$,
intercepts $(0, 0)$, $(-4, 0)$

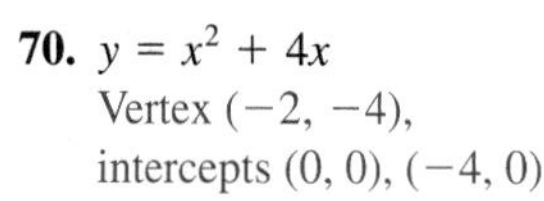

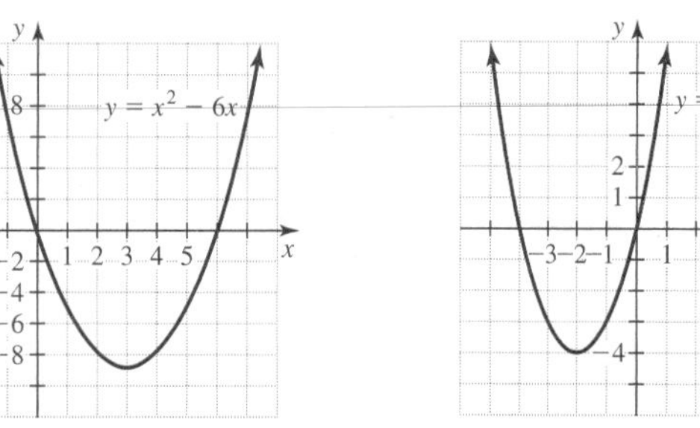

71. $y = x^2 - 4x - 12$
Vertex $(2, -16)$,
intercepts $(0, -12)$,
$(-2, 0)$, and $(6, 0)$

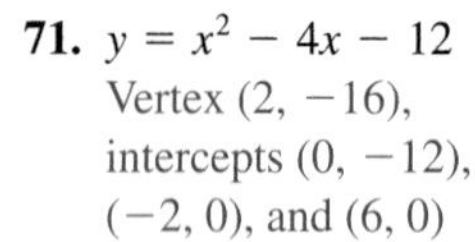

72. $y = x^2 + 2x - 24$
Vertex $(-1, -25)$,
intercepts $(0, -24)$,
$(-6, 0)$, $(4, 0)$

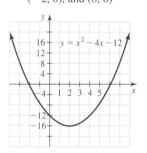

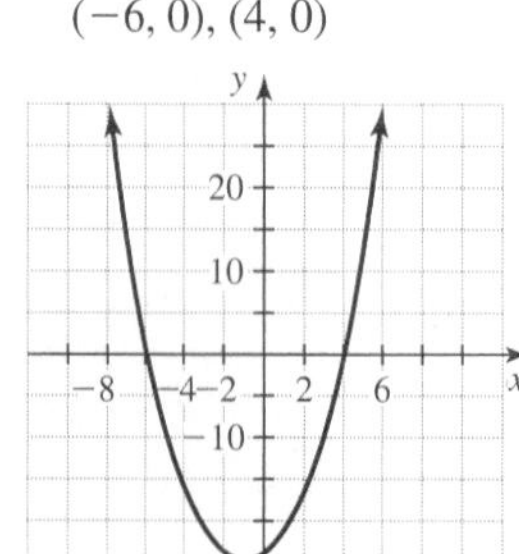

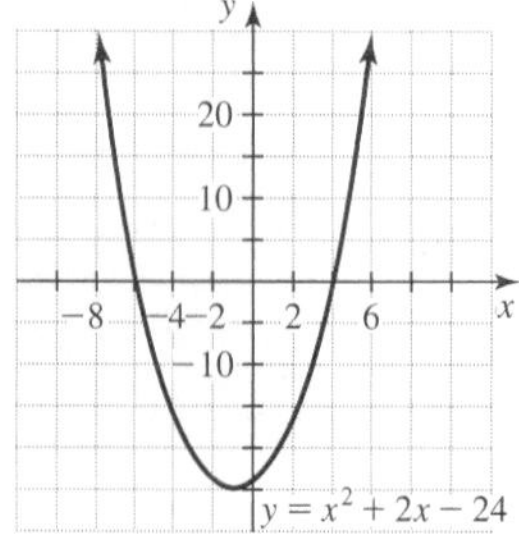

73. $y = -2x^2 + 8x$
Vertex $(2, 8)$,
intercepts $(0, 0)$, $(4, 0)$

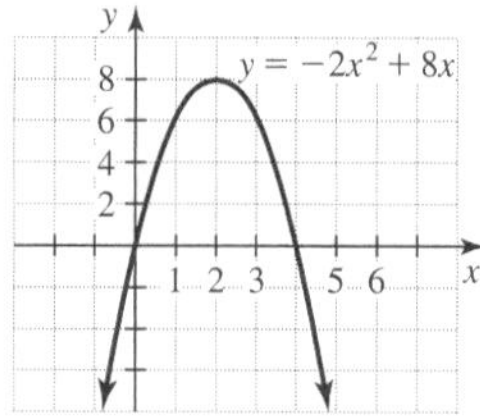

74. $y = -3x^2 + 6x$
Vertex $(1, 3)$,
intercepts $(0, 0)$, $(2, 0)$

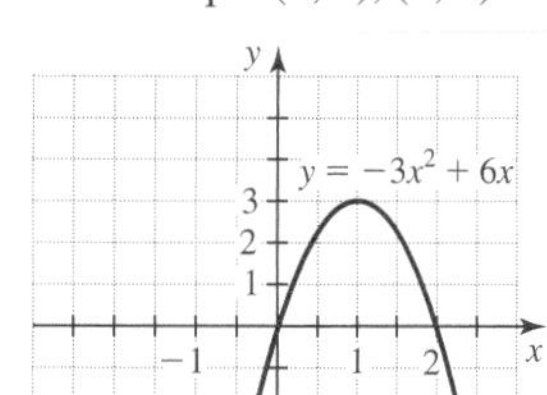

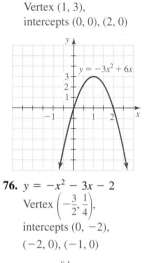

75. $y = -x^2 + 2x + 3$
Vertex $(1, 4)$,
intercepts $(0, 3)$, $(-1, 0)$, $(3, 0)$

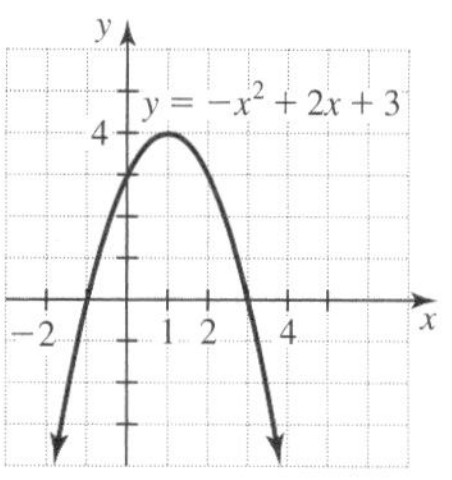

76. $y = -x^2 - 3x - 2$
Vertex $\left(-\dfrac{3}{2}, \dfrac{1}{4}\right)$,
intercepts $(0, -2)$,
$(-2, 0)$, $(-1, 0)$

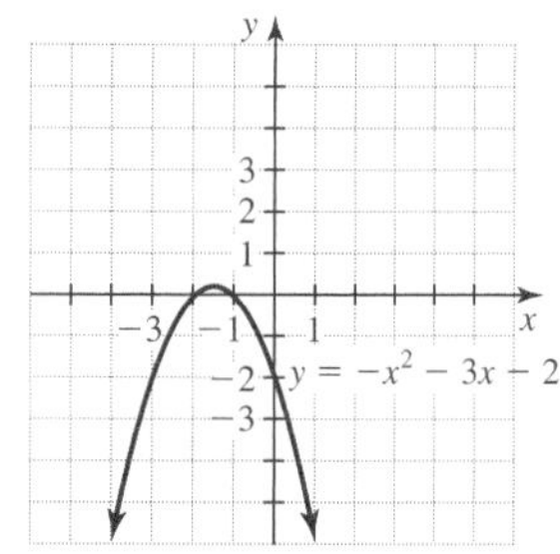

Solve each problem.

77. ***Minimizing cost.*** The unit cost in dollars for manufacturing n starters is given by $C = 0.004n^2 - 3.2n + 660$. What is the unit cost when 390 starters are manufactured? For what number of starters is the unit cost at a minimum?
\$20.40, 400

78. ***Maximizing profit.*** The total profit (in dollars) for sales of x rowing machines is given by $P = -0.2x^2 + 300x - 200$. What is the profit if 500 are sold? For what value of x will the profit be at a maximum? \$99,800, 750

9.7 *Determine whether each set of ordered pairs is a function.*

79. $\{(4, 3), (5, 3)\}$ Yes

80. $\{(0, 0), (0, 1), (0, 2)\}$ No

81. $\{(3, 4), (3, 5)\}$ No

82. $\{(1, 2), (2, 3), (3, 4)\}$ Yes

Determine whether each equation expresses y as a function of x.

83. $y = x^2 + 10$ Yes

84. $y = 2x - 7$ Yes

85. $x^2 + y^2 = 1$ No

86. $x^2 = y^2$ No

Determine the domain and range of each function.

87. $\{(1, 2), (2, 0), (3, 0)\}$
$\{1, 2, 3\}$, $\{0, 2\}$

88. $\{(2, 3), (4, 3), (6, 3)\}$
$\{2, 4, 6\}$, $\{3\}$

Find the domain and range of each quadratic function.

89. $y = x^2 + 4x + 1$ Domain R, range $\{y \mid y \geq -3\}$

90. $y = x^2 - 6x + 2$ Domain R, range $\{y \mid y \geq -7\}$

91. $y = -2x^2 - x + 4$ Domain R, range $\{y \mid y \leq 4.125\}$

92. $y = -3x^2 + 2x + 7$ Domain R, range $\left\{y \mid y \leq \dfrac{22}{3}\right\}$

9.8 *Let* $f(x) = 3x + 5$, $g(x) = x^2 - 2x$, *and* $h(x) = \frac{x-5}{3}$. *Find the following.*

93. $f(-3)$
-4

94. $h(-4)$
-3

95. $(h \circ f)(\sqrt{2})$
$\sqrt{2}$

96. $(f \circ h)(319)$
319

97. $(g \circ f)(2)$
99

98. $(g \circ f)(x)$
$9x^2 + 24x + 15$

99. $(f + g)(3)$
17

100. $(f - g)(x)$
$-x^2 + 5x + 5$

101. $(f \cdot g)(x)$ $3x^3 - x^2 - 10x$

102. $\left(\frac{f}{g}\right)(1)$ -8

103. $(f \circ f)(0)$ 20

104. $(f \circ f)(x)$ $9x + 20$

Let $f(x) = |x|$, $g(x) = x + 2$, *and* $h(x) = x^2$. *Write each of the following functions as a composition of functions, using f, g, or h.*

105. $F(x) = |x + 2|$
$F = f \circ g$

106. $G(x) = |x| + 2$
$G = g \circ f$

107. $H(x) = x^2 + 2$
$H = g \circ h$

108. $K(x) = x^2 + 4x + 4$
$K = h \circ g$

109. $I(x) = x + 4$
$I = g \circ g$

110. $J(x) = x^4 + 2$
$J = g \circ h \circ h$

CHAPTER 9 TEST

Calculate the value of $b^2 - 4ac$ *and state how many real solutions each equation has.*

1. $9x^2 - 12x + 4 = 0$ 0, one

2. $-2x^2 + 3x - 5 = 0$ -31, none

3. $-2x^2 + 5x - 1 = 0$ 17, two

Solve by using the quadratic formula.

4. $5x^2 + 2x - 3 = 0$
$-1, \frac{3}{5}$

5. $2x^2 - 4x - 3 = 0$
$\frac{2 - \sqrt{10}}{2}, \frac{2 + \sqrt{10}}{2}$

Solve by completing the square.

6. $x^2 + 4x - 21 = 0$ $-7, 3$

7. $x^2 + 3x - 5 = 0$ $\frac{-3 - \sqrt{29}}{2}, \frac{-3 + \sqrt{29}}{2}$

Solve by any method.

8. $x(x + 1) = 20$ $-5, 4$

9. $x^2 - 28x + 75 = 0$ $3, 25$

10. $\frac{x-1}{3} = \frac{x+1}{2x}$ $-\frac{1}{2}, 3$

Perform the indicated operations. Write answers in the form $a + bi$.

11. $(2 - 3i) + (8 + 6i)$ $10 + 3i$

12. $(-2 - 5i) - (4 - 12i)$ $-6 + 7i$

13. $(-6i)^2$ -36

14. $(3 - 5i)(4 + 6i)$ $42 - 2i$

15. $(8 - 2i)(8 + 2i)$ 68

16. $(4 - 6i) \div 2$ $2 - 3i$

17. $\frac{-2 + \sqrt{-12}}{2}$ $-1 + i\sqrt{3}$

18. $\frac{6 - \sqrt{-18}}{-3}$ $-2 + i\sqrt{2}$

19. $\frac{5i}{4 + 3i}$ $\frac{3}{5} + \frac{4}{5}i$

Find the complex solutions to the quadratic equations.

20. $x^2 + 6x + 12 = 0$ $-3 - i\sqrt{3}, -3 + i\sqrt{3}$

21. $-5x^2 + 6x - 5 = 0$ $\frac{3}{5} - \frac{4}{5}i, \frac{3}{5} + \frac{4}{5}i$

Graph each quadratic function. State the domain and range.

22. $y = 16 - x^2$
Domain R,
range $\{y \mid y \le 16\}$

23. $y = x^2 - 3x$

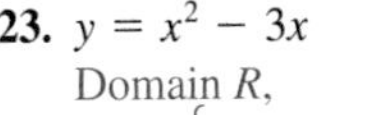

Domain R,

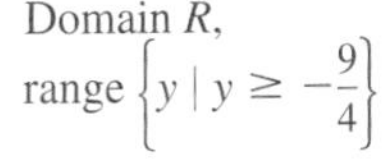

range $\left\{y \mid y \ge -\frac{9}{4}\right\}$

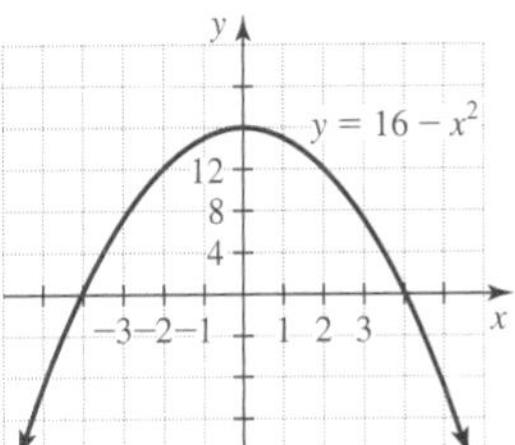

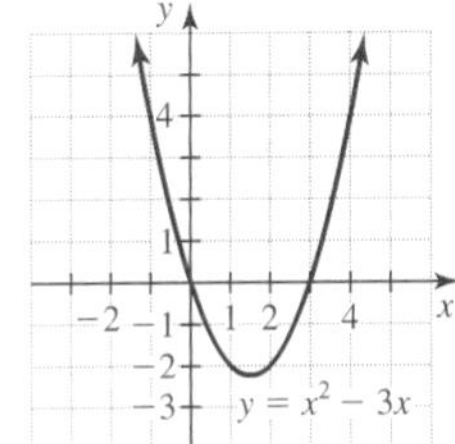

Let $f(x) = -2x + 5$ *and* $g(x) = x^2 + 4$. *Find the following.*

24. $f(-3)$
11

25. $(g \circ f)(-3)$
125

26. $(g + f)(x)$
$x^2 - 2x + 9$

27. $(f \cdot g)(1)$
15

28. $(f/g)(2)$
$\frac{1}{8}$

29. $(f \circ g)(x)$
$-2x^2 - 3$

30. $(g \circ f)(x)$ $4x^2 - 20x + 29$

Let $f(x) = x - 7$ *and* $g(x) = x^2$. *Write each of the following functions as a composition of functions using f and g.*

31. $H(x) = x^2 - 7$
$H = f \circ g$

32. $W(x) = x^2 - 14x + 49$
$W = g \circ f$

Solve each problem.

33. Find the x-intercepts for the parabola $y = x^2 - 6x + 5$.
(1, 0), (5, 0)

34. The height in feet for a ball thrown upward at 48 feet per second is given by $s = -16t^2 + 48t$, where t is the time in seconds after the ball is tossed. What is the maximum height that the ball will reach? 36 feet

35. Find two positive numbers that have a sum of 10 and a product of 23. Give exact answers.
$5 - \sqrt{2}$ and $5 + \sqrt{2}$

MAKING CONNECTIONS CHAPTERS 1–9

Solve each equation.

1. $2x - 1 = 0$

$\frac{1}{2}$

2. $2(x - 1) = 0$

1

3. $2x^2 - 1 = 0$

$-\frac{\sqrt{2}}{2}, \frac{\sqrt{2}}{2}$

4. $(2x - 1)^2 = 8$

$\frac{1 - 2\sqrt{2}}{2}, \frac{1 + 2\sqrt{2}}{2}$

5. $2x^2 - 4x - 1 = 0$

$\frac{2 - \sqrt{6}}{2}, \frac{2 + \sqrt{6}}{2}$

6. $2x^2 - 4x = 0$

$0, 2$

7. $2x^2 + x = 1$

$-1, \frac{1}{2}$

8. $x - 2 = \sqrt{2x - 1}$

5

9. $\frac{1}{x} = \frac{x}{2x - 15}$

$1 - i\sqrt{14}, 1 + i\sqrt{14}$

10. $\frac{1}{x} - \frac{1}{x - 1} = -\frac{1}{2}$

$-1, 2$

Solve each equation for y.

11. $5x - 4y = 8$

$y = \frac{5}{4}x - 2$

12. $3x - y = 9$

$y = 3x - 9$

13. $\frac{y - 4}{x + 2} = \frac{2}{3}$

$y = \frac{2}{3}x + \frac{16}{3}$

14. $ay + b = 0$

$y = -\frac{b}{a}$

15. $ay^2 + by + c = 0$

$y = \frac{-b \pm \sqrt{b^2 - 4ac}}{2a}$

16. $y - 1 = -\frac{2}{3}(x - 9)$

$y = -\frac{2}{3}x + 7$

17. $\frac{2}{3}x + \frac{1}{2}y = \frac{1}{9}$

$y = -\frac{4}{3}x + \frac{2}{9}$

18. $x^2 + y^2 = a^2$

$y = \pm\sqrt{a^2 - x^2}$

Suppose that each side of a square has length s, the diagonal has length d, the area of the square is A, and its perimeter is P.

19. Write P in terms of s.

$P = 4s$

20. Write A in terms of s.

$A = s^2$

21. Write P in terms of d.

$P = 2d\sqrt{2}$

22. Write d in terms of A.

$d = \sqrt{2A}$

Solve each system of equations.

23. $3x - 2y = 12$
$2x + 5y = -11$

$(2, -3)$

24. $y = 3x + 1$
$3x - 0.6y = 3$

$(3, 10)$

Graph each function.

25. $y = x - 3$

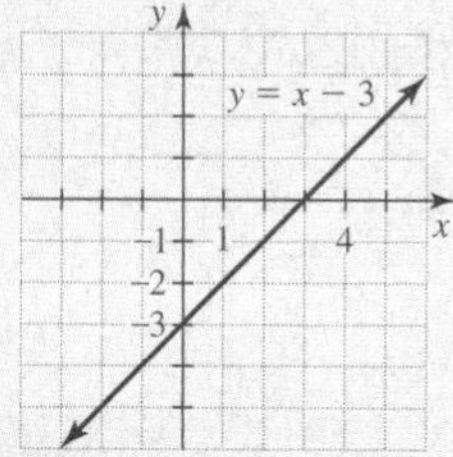

26. $y = 2 - x$

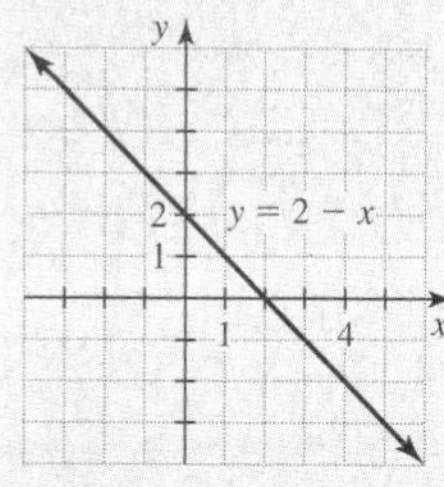

27. $y = x^2 - 3$

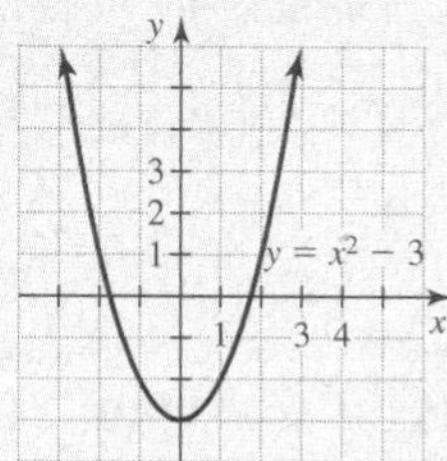

28. $y = 2 - x^2$

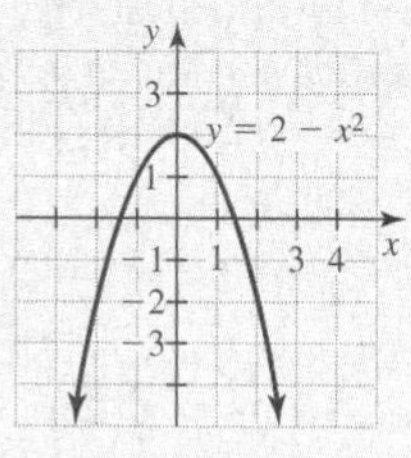

29. $y = \frac{2}{3}x - 4$

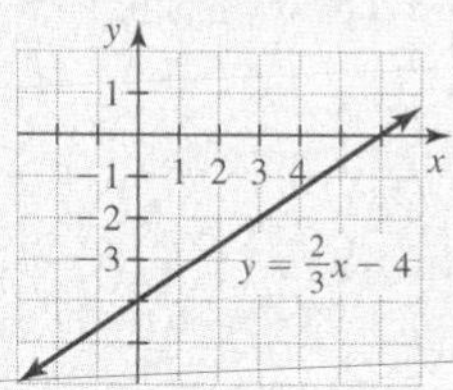

30. $y = -\frac{4}{3}x + 5$

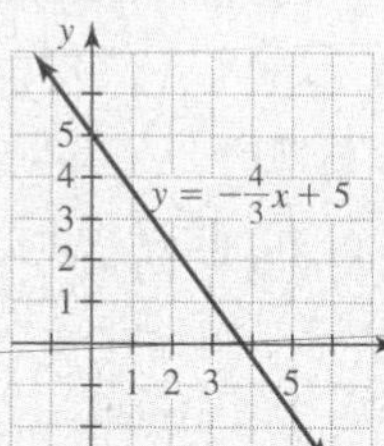

Solve the problem.

31. ***Maximizing revenue.*** For the last three years the Lakeland Air Show has raised the price of its tickets and has sold fewer and fewer tickets, as shown in the table.

Ticket price	\$10	\$12	\$16
Tickets sold	8000	7500	6500

a) Use this information to write the number of tickets sold s as a linear function of the ticket price p.

b) Has the revenue from ticket sales increased or decreased as the ticket price was raised?

c) Write the revenue R as a function of the ticket price p.

d) What ticket price would produce the maximum revenue?

a) $s = -250p + 10{,}500$ **b)** Increased
c) $R = -250p^2 + 10{,}500p$ **d)** \$21

APPENDIX A GEOMETRY REVIEW

Geometric Figures and Formulas

Triangle: A three-sided figure

Area: $A = \frac{1}{2}bh$, Perimeter: $P = a + b + c$

Sum of the measures of the angles is 180°.

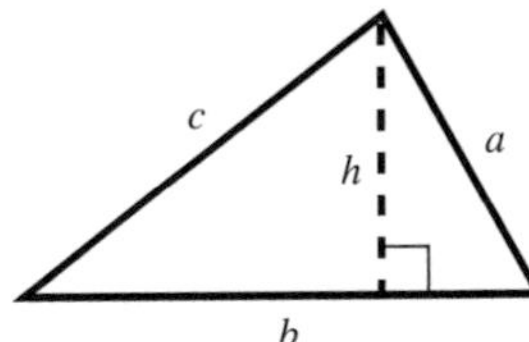

30-60-90 Right Triangle: The side opposite 30° is one-half the length of the hypotenuse.

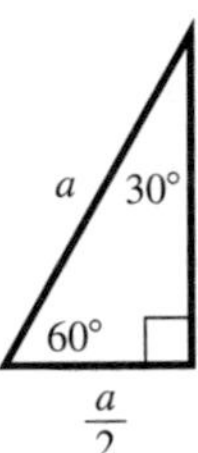

Trapezoid: A four-sided figure with one pair of parallel sides

Area: $A = \frac{1}{2}h(b_1 + b_2)$

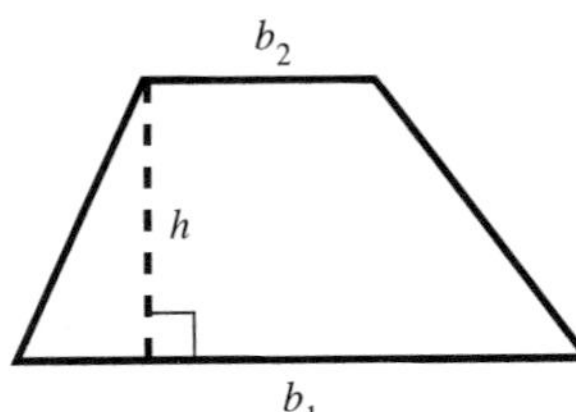

Right Triangle: A triangle with a 90° angle

Area $= \frac{1}{2}ab$, Perimeter: $P = a + b + c$

Pythagorean Theorem: A triangle is a right triangle if and only if $a^2 + b^2 = c^2$.

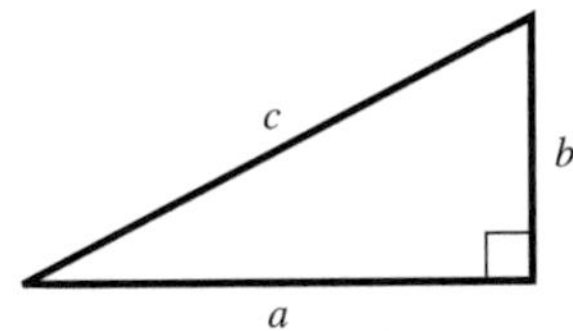

Parallelogram: A four-sided figure with opposite sides parallel

Area: $A = bh$

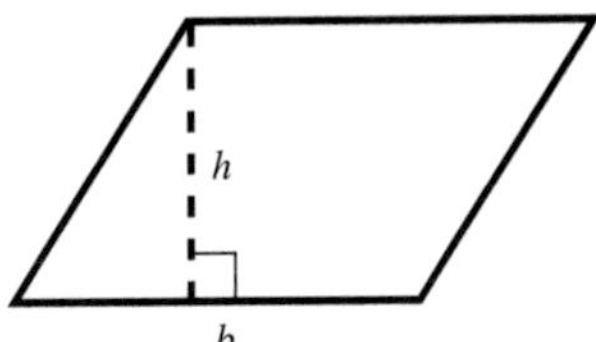

Rhombus: A four-sided figure with four equal sides

Perimeter: $P = 4a$

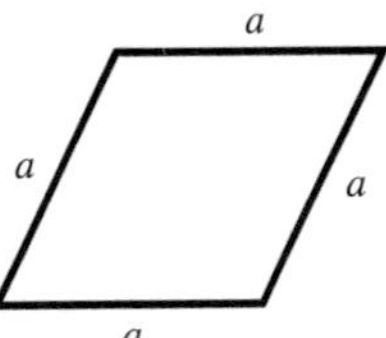

Rectangle: A four-sided figure with four right angles

Area: $A = LW$

Perimeter: $P = 2L + 2W$

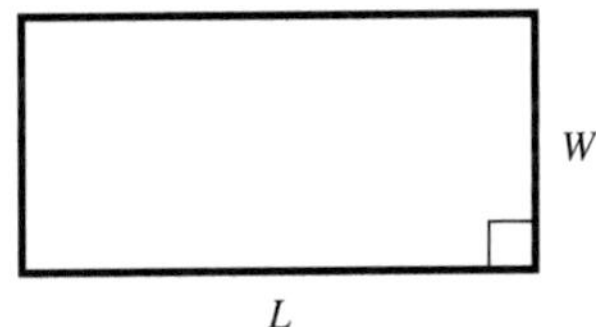

Circle

Area: $A = \pi r^2$

Circumference: $C = 2\pi r$

Diameter: $d = 2r$

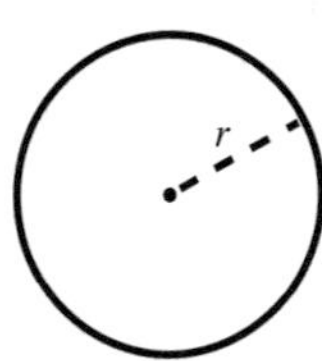

Right Circular Cone

Volume: $V = \frac{1}{3}\pi r^2 h$

Lateral Surface Area: $S = \pi r\sqrt{r^2 + h^2}$

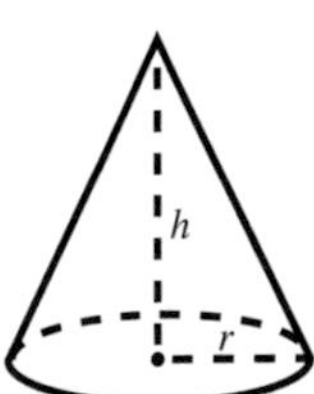

Rectangular Solid
Volume: $V = LWH$
Surface Area: $A = 2LW + 2WH + 2LH$

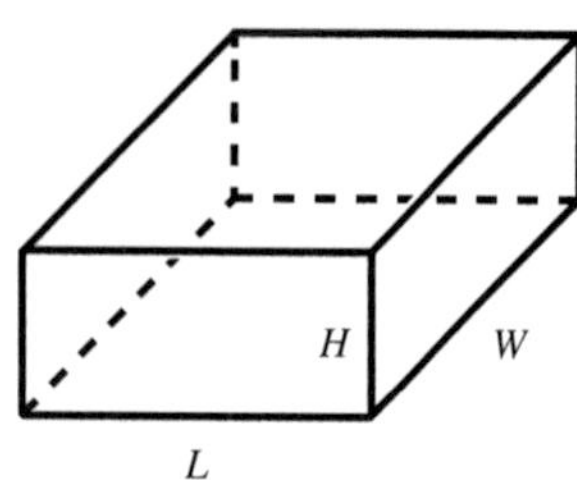

Square: A four-sided figure with four equal sides and four right angles
Area: $A = s^2$
Perimeter: $P = 4s$

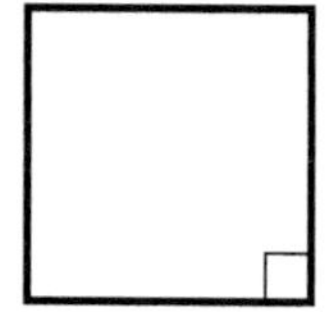

Sphere

Volume: $V = \frac{4}{3}\pi r^3$

Surface Area: $S = 4\pi r^2$

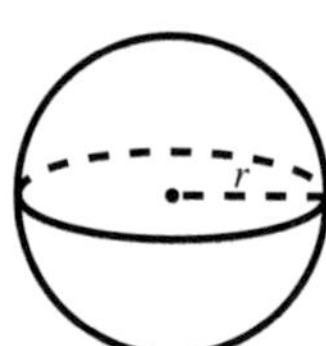

Right Circular Cylinder
Volume: $V = \pi r^2 h$
Lateral Surface Area: $S = 2\pi rh$

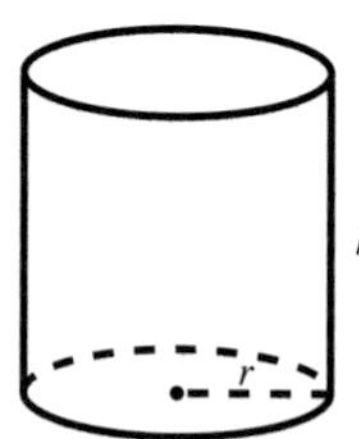

Geometric Terms

An **angle** is a union of two rays with a common endpoint.

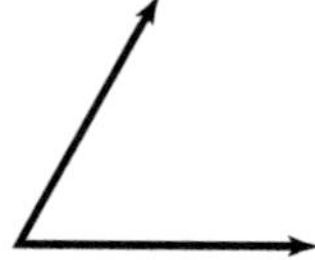

A **right angle** is an angle with a measure of 90°.

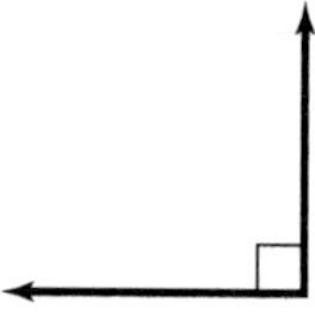

Two angles are **complementary** if the sum of their measures is 90°.

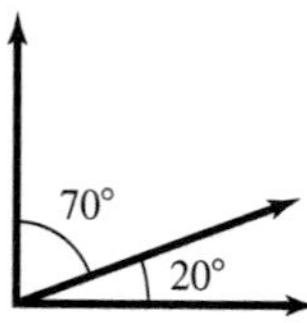

An **isosceles triangle** is a triangle that has two equal sides.

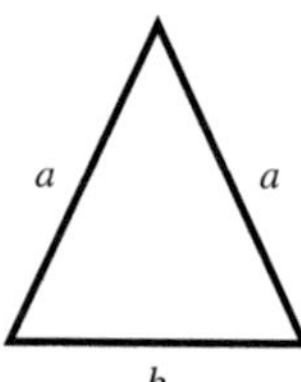

Similar triangles are triangles that have the same shape. Their corresponding angles are equal and corresponding sides are proportional:

$$\frac{a}{d} = \frac{b}{e} = \frac{c}{f}$$

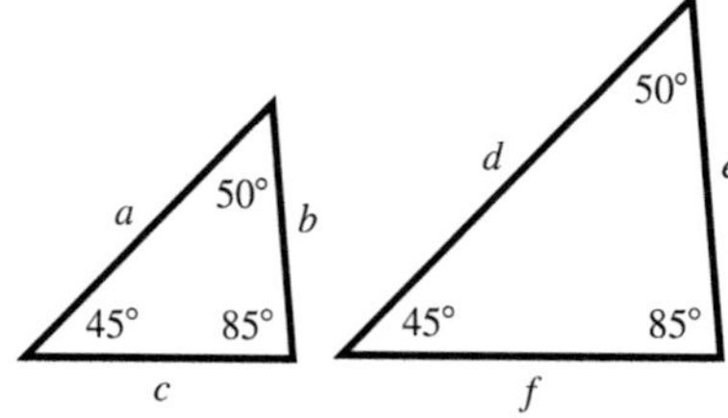

An **acute angle** is an angle with a measure between 0° and 90°.

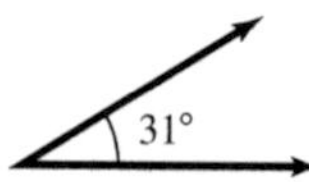

An **obtuse angle** is an angle with a measure between 90° and 180°.

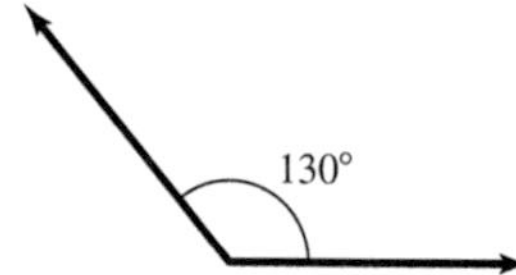

Two angles are **supplementary** if the sum of their measures is 180°.

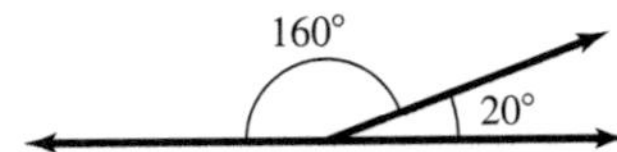

An **equilateral triangle** is a triangle that has three equal sides.

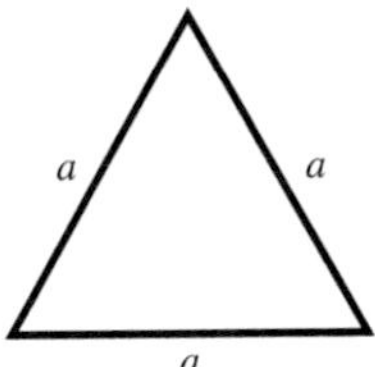

Geometry Review Exercises

(Answers are at the end of the answer section in this text.)

1. Find the perimeter of a triangle whose sides are 3 in., 4 in., and 5 in. 12 in.
2. Find the area of a triangle whose base is 4 ft and height is 12 ft. 24 ft^2
3. If two angles of a triangle are 30° and 90°, then what is the third angle? 60°
4. If the area of a triangle is 36 ft^2 and the base is 12 ft, then what is the height? 6 ft
5. If the side opposite 30° in a 30-60-90 right triangle is 10 cm, then what is the length of the hypotenuse? 20 cm
6. Find the area of a trapezoid whose height is 12 cm and whose parallel sides are 4 cm and 20 cm. 144 cm^2
7. Find the area of the right triangle that has sides of 6 ft, 8 ft, and 10 ft. 24 ft^2
8. If a right triangle has sides of 5 ft, 12 ft, and 13 ft, then what is the length of the hypotenuse? 13 ft
9. If the hypotenuse of a right triangle is 50 cm and the length of one leg is 40 cm, then what is the length of the other leg? 30 cm
10. Is a triangle with sides of 5 ft, 10 ft, and 11 ft a right triangle? No
11. What is the area of a triangle with sides of 7 yd, 24 yd, and 25 yd? 84 yd^2
12. Find the perimeter of a parallelogram in which one side is 9 in. and another side is 6 in. 30 in.
13. Find the area of a parallelogram which has a base of 8 ft and a height of 4 ft. 32 ft^2
14. If one side of a rhombus is 5 km, then what is its perimeter. 20 km
15. Find the perimeter and area of a rectangle whose width is 18 in. and length is 2 ft. 7 ft, 3 ft^2
16. If the width of a rectangle is 8 yd and its perimeter is 60 yd, then what is its length? 22 yd
17. The radius of a circle is 4 ft. Find its area to the nearest tenth of a square foot. 50.3 ft^2
18. The diameter of a circle is 12 ft. Find its circumference to the nearest tenth of a foot. 37.7 ft
19. A right circular cone has radius 4 cm and height 9 cm. Find its volume to the nearest hundredth of a cubic centimeter. 150.80 cm^3
20. A right circular cone has a radius 12 ft and a height of 20 ft. Find its lateral surface area to the nearest hundredth of a square foot. 879.29 ft^2
21. A shoe box has a length of 12 in., a width of 6 in., and a height of 4 in. Find its volume and surface area. 288 $in.^3$, 288 $in.^2$
22. The volume of a rectangular solid is 120 cm^3. If the area of its bottom is 30 cm^2, then what is its height? 4 cm
23. What is the area and perimeter of a square in which one of the sides is 10 mi long? 100 mi^2, 40 mi
24. Find the perimeter of a square whose area is 25 km^2. 20 km
25. Find the area of a square whose perimeter is 26 cm. 42.25 cm^2
26. A sphere has a radius of 2 ft. Find its volume to the nearest thousandth of a cubic foot and its surface area to the nearest thousandth of a square foot. 33.510 ft^3, 50.265 ft^2
27. A can of soup (right circular cylinder) has a radius of 2 in. and a height of 6 in. Find its volume to the nearest tenth of a cubic inch and total surface area to the nearest tenth of a square inch. 75.4 $in.^3$, 100.5 $in.^2$
28. If one of two complementary angles is 34°, then what is the other angle? 56°
29. If the perimeter of an isosceles triangle is 29 cm and one of the equal sides is 12 cm, then what is the length of the shortest side of the triangle? 5 cm
30. A right triangle with sides of 6 in., 8 in., and 10 in., is similar to another right triangle that has a hypotenuse of 25 in. What are the lengths of the other two sides in the second triangle? 15 in. and 20 in.
31. If one of two supplementary angles is 31°, then what is the other angle? 149°
32. Find the perimeter of an equilateral triangle in which one of the sides is 4 km. 12 km
33. Find the length of a side of an equilateral triangle that has a perimeter of 30 yd. 10 yd

APPENDIX B TABLE OF SQUARES AND SQUARE ROOTS

n	n^2	$\sqrt{n}$	n	n^2	$\sqrt{n}$	n	n^2	$\sqrt{n}$
1	1	1.0000	41	1681	6.4031	81	6561	9.0000
2	4	1.4142	42	1764	6.4807	82	6724	9.0554
3	9	1.7321	43	1849	6.5574	83	6889	9.1104
4	16	2.0000	44	1936	6.6332	84	7056	9.1652
5	25	2.2361	45	2025	6.7082	85	7225	9.2195
6	36	2.4495	46	2116	6.7823	86	7396	9.2736
7	49	2.6458	47	2209	6.8557	87	7569	9.3274
8	64	2.8284	48	2304	6.9282	88	7744	9.3808
9	81	3.0000	49	2401	7.0000	89	7921	9.4340
10	100	3.1623	50	2500	7.0711	90	8100	9.4868
11	121	3.3166	51	2601	7.1414	91	8281	9.5394
12	144	3.4641	52	2704	7.2111	92	8464	9.5917
13	169	3.6056	53	2809	7.2801	93	8649	9.6437
14	196	3.7417	54	2916	7.3485	94	8836	9.6954
15	225	3.8730	55	3025	7.4162	95	9025	9.7468
16	256	4.0000	56	3136	7.4833	96	9216	9.7980
17	289	4.1231	57	3249	7.5498	97	9409	9.8489
18	324	4.2426	58	3364	7.6158	98	9604	9.8995
19	361	4.3589	59	3481	7.6811	99	9801	9.9499
20	400	4.4721	60	3600	7.7460	100	10000	10.0000
21	441	4.5826	61	3721	7.8102	101	10201	10.0499
22	484	4.6904	62	3844	7.8740	102	10404	10.0995
23	529	4.7958	63	3969	7.9373	103	10609	10.1489
24	576	4.8990	64	4096	8.0000	104	10816	10.1980
25	625	5.0000	65	4225	8.0623	105	11025	10.2470
26	676	5.0990	66	4356	8.1240	106	11236	10.2956
27	729	5.1962	67	4489	8.1854	107	11449	10.3441
28	784	5.2915	68	4624	8.2462	108	11664	10.3923
29	841	5.3852	69	4761	8.3066	109	11881	10.4403
30	900	5.4772	70	4900	8.3666	110	12100	10.4881
31	961	5.5678	71	5041	8.4261	111	12321	10.5357
32	1024	5.6569	72	5184	8.4853	112	12544	10.5830
33	1089	5.7446	73	5329	8.5440	113	12769	10.6301
34	1156	5.8310	74	5476	8.6023	114	12996	10.6771
35	1225	5.9161	75	5625	8.6603	115	13225	10.7238
36	1296	6.0000	76	5776	8.7178	116	13456	10.7703
37	1369	6.0828	77	5929	8.7750	117	13689	10.8167
38	1444	6.1644	78	6084	8.8318	118	13924	10.8628
39	1521	6.2450	79	6241	8.8882	119	14161	10.9087
40	1600	6.3246	80	6400	8.9443	120	14400	10.9545

ANSWERS

Chapter 1

Section 1.1 Warm-ups T T F F T F T T F F

1. The integers are the numbers in the set $\{\ldots, -3, -2, -1, 0, 1, 2, 3, \ldots\}$.
2. The rational numbers are numbers of the form $\frac{a}{b}$, where a and b are integers and $b \neq 0$.
3. A rational number is a ratio of integers and an irrational number is not.
4. A number line is a line on which there is a point corresponding to every real number.
5. The number a is larger than b if a lies to the right of b on the number line.
6. The ratio of the circumference and diameter of any circle is the number π, which is approximately 3.14.

7. 6 **8.** 7 **9.** 0 **10.** 0 **11.** -2 **12.** -5
13. -12 **14.** -7

15. 1, 2, 3, 4, 5

16. 5, 6, 7, 8, 9, . . .

17. 0, 1, 2, 3, 4
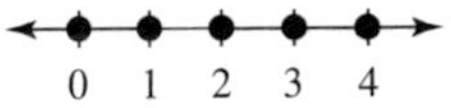

18. $-2, -1, 0, 1, 2$
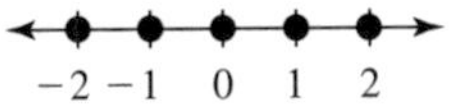

19. 0, 1, 2, 3, 4
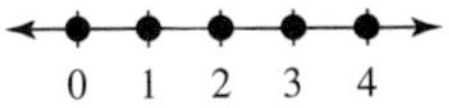

20. $-2, -3, -4, -5, \ldots$
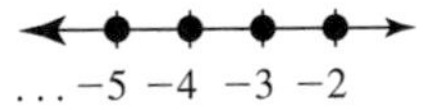

21. 1, 2, 3, 4, 5, . . .
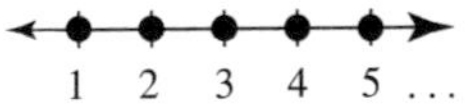

22. 1, 2, 3, 4, 5, 6
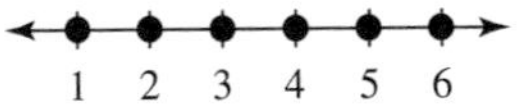

23. 1, 2, 3, 4, 5, . . .
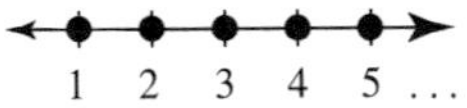

24. 0, 1
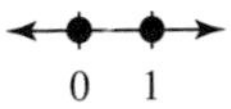

25. True **26.** True **27.** False **28.** False **29.** True
30. True **31.** True **32.** True **33.** True **34.** True
35. False **36.** True **37.** 6 **38.** 4 **39.** 0 **40.** 2
41. 7 **42.** 7 **43.** 9 **44.** 2 **45.** 45 **46.** 30
47. $\frac{3}{4}$ **48.** $\frac{1}{2}$ **49.** 5.09 **50.** 0.00987 **51.** -16
52. -12 **53.** $-\frac{5}{2}$ **54.** $\frac{5}{8}$ **55.** 2 **56.** 0 **57.** 3
58. -4 **59.** -9 **60.** -12 **61.** 16 **62.** -12
63. -4 **64.** 7.9 **65.** -1.99 **66.** 1.99
67. 74 **68.** -74 **69.** -3 and 3 **70.** 0
71. $-4, -3, 3, 4$ **72.** $-4, -3, -2, 2, 3, 4$ **73.** $-1, 0, 1$
74. $-3, -2, -1, 0, 1, 2, 3$ **75.** True **76.** True
77. True **78.** False **79.** True **80.** True

81. What is the probability that a tossed coin turns up heads?

82. a) $\frac{7}{24}$ **b)** -3.115 **c)** 0.66669
d) Add them and divide the result by 2.

83. If a is negative, then $-a$ and $|-a|$ are positive. The rest are negative.

84. Real: all. Irrational: π, $\sqrt{3}$. Rational: $\frac{1}{2}$, -2, $\sqrt{9}$, 6, 0, $-\frac{7}{3}$. Integer: -2, $\sqrt{9}$, 6, 0. Whole: $\sqrt{9}$, 6, 0. Counting: $\sqrt{9}$, 6.

Section 1.2 Warm-ups T T F T T T T T F T

1. If two fractions are identical when reduced to lowest terms, then they are equivalent fractions.
2. Reduce the fraction to lowest terms and then multiply the numerator and denominator by every counting number.
3. To reduce a fraction to lowest terms means to find an equivalent fraction that has no factor common to the numerator and denominator.
4. Common denominators are required for addition and subtraction, because it makes sense to add $\frac{1}{3}$ of a pie and $\frac{1}{3}$ of a pie and get $\frac{2}{3}$ of a pie.
5. Convert a fraction to a decimal by dividing the denominator into the numerator.
6. Convert a percent to a fraction by dividing by 100, as in $4\% = \frac{4}{100}$.

7. $\frac{6}{8}$ **8.** $\frac{15}{21}$ **9.** $\frac{32}{12}$ **10.** $\frac{28}{8}$ **11.** $\frac{10}{2}$ **12.** $\frac{27}{3}$
13. $\frac{75}{100}$ **14.** $\frac{50}{100}$ **15.** $\frac{30}{100}$ **16.** $\frac{40}{100}$ **17.** $\frac{70}{42}$
18. $\frac{70}{98}$ **19.** $\frac{1}{2}$ **20.** $\frac{1}{5}$ **21.** $\frac{2}{3}$ **22.** $\frac{3}{4}$ **23.** 3
24. 3 **25.** $\frac{1}{2}$ **26.** $\frac{1}{200}$ **27.** 2 **28.** $\frac{5}{4}$ **29.** $\frac{3}{8}$
30. $\frac{1}{3}$ **31.** $\frac{13}{21}$ **32.** $\frac{5}{8}$ **33.** $\frac{12}{13}$ **34.** $\frac{11}{12}$ **35.** $\frac{10}{27}$
36. $\frac{1}{64}$ **37.** 5 **38.** 4 **39.** $\frac{7}{10}$ **40.** $\frac{3}{14}$ **41.** $\frac{7}{13}$
42. $\frac{2}{7}$ **43.** $\frac{3}{5}$ **44.** $\frac{3}{10}$ **45.** $\frac{1}{6}$ **46.** $\frac{3}{112}$ **47.** 3
48. $\frac{4}{3}$ **49.** $\frac{1}{15}$ **50.** $\frac{1}{5}$ **51.** 4 **52.** 12 **53.** $\frac{4}{5}$

54. $\frac{7}{5}$ **55.** $\frac{3}{40}$ **56.** $\frac{2}{5}$ **57.** $\frac{1}{2}$ **58.** $\frac{1}{5}$ **59.** $\frac{1}{3}$ **60.** $\frac{6}{7}$
61. $\frac{1}{4}$ **62.** $\frac{1}{2}$ **63.** $\frac{7}{12}$ **64.** $\frac{11}{10}$ **65.** $\frac{1}{12}$ **66.** $\frac{1}{20}$
67. $\frac{19}{24}$ **68.** $\frac{11}{12}$ **69.** $\frac{11}{72}$ **70.** $\frac{11}{80}$ **71.** $\frac{199}{48}$ **72.** $\frac{71}{16}$
73. 60%, 0.6 **74.** 95%, 0.95 **75.** $\frac{9}{100}$, 0.09 **76.** 0.6, $\frac{3}{5}$
77. 8%, $\frac{2}{25}$ **78.** 40%, $\frac{2}{5}$ **79.** 0.75, 75%
80. 0.625, 62.5% **81.** $\frac{1}{50}$, 0.02 **82.** $\frac{6}{5}$, 1.20
83. $\frac{1}{100}$, 1% **84.** $\frac{1}{200}$, 0.5% **85.** 3 **86.** $\frac{49}{12}$ **87.** 1
88. $\frac{3}{32}$ **89.** $\frac{71}{96}$ **90.** $\frac{18}{35}$ **91.** $\frac{17}{120}$ **92.** $\frac{23}{48}$ **93.** $\frac{65}{16}$
94. $\frac{75}{16}$ **95.** $\frac{69}{4}$ **96.** $\frac{13}{7}$ **97.** $\frac{13}{12}$ **98.** $\frac{2}{3}$ **99.** $\frac{1}{8}$
100. $\frac{8}{27}$ **101.** $\frac{19}{96}$ **102.** $\frac{23}{40}$, 57.5%
103. **a)** 1.3 yd^3 **b)** $36\frac{11}{24}$ ft^3 or $1\frac{227}{648}$ yd^3
104. 2625 in.^{2}, 140.7 ft^3
107. Each daughter gets 3 $\text{km}^2 \div 4$ or a $\frac{3}{4}$ km^2 piece of the farm. Divide the farm into 12 equal squares. Give each daughter an L-shaped piece consisting of 3 of those 12 squares.

Section 1.3 Warm-ups T T T F F F T F T F

1. We studied addition and subtraction of signed numbers.
2. The sum of two numbers with the same sign is found by adding their absolute values. The sum is negative if the two numbers are negative.
3. Two numbers are additive inverses of each other if their sum is zero.
4. The sum of two numbers with opposite signs and the same absolute value is zero.
5. To find the sum of two numbers with unlike signs subtract their absolute values. The answer is given the sign of the number with the larger absolute value.
6. Subtraction is defined in terms of addition as $a - b = a + (-b)$.
7. 13 **8.** 100 **9.** -13 **10.** -100 **11.** -1.15
12. -3.15 **13.** $-\frac{1}{2}$ **14.** $\frac{3}{4}$ **15.** 0 **16.** 0 **17.** 0
18. 0 **19.** 2 **20.** -20 **21.** -6 **22.** 12 **23.** 5.6
24. 2.5 **25.** -2.9 **26.** 2.95 **27.** $-\frac{1}{4}$ **28.** $\frac{4}{3}$
29. $8 + (-2)$ **30.** $3.5 + (-1.2)$ **31.** $4 + (-12)$
32. $\frac{1}{2} + \left(-\frac{5}{6}\right)$ **33.** $-3 + 8$ **34.** $-9 + (2.3)$
35. $8.3 + (1.5)$ **36.** $10 + (6)$ **37.** -4 **38.** -16
39. -10 **40.** -15 **41.** 11 **42.** 14 **43.** -11
44. -9 **45.** $-\frac{1}{4}$ **46.** $-\frac{4}{15}$ **47.** $\frac{3}{4}$ **48.** $\frac{5}{6}$
49. 7 **50.** 10 **51.** 0.93 **52.** -0.97 **53.** 9.3
54. -5.25 **55.** -5.03 **56.** 1 **57.** 3 **58.** 4
59. -9 **60.** -25 **61.** -120 **62.** 59 **63.** 78
64. -32 **65.** -27 **66.** 0 **67.** -7 **68.** -24
69. -201 **70.** -17 **71.** -322 **72.** -107
73. -15.97 **74.** -2.81 **75.** -2.92 **76.** -7.2
77. -3.73 **78.** -0.1 **79.** 3.7 **80.** 1.92 **81.** $\frac{3}{20}$
82. $\frac{4}{15}$ **83.** $\frac{7}{24}$ **84.** 0 **85.** -3.49 **86.** -3.822
87. -0.3422 **88.** -5.847 **89.** -48.84 **90.** -0.174
91. -8.85 **92.** 4.537 **93.** $-\$8.85$ **94.** \$67,206
95. -7°C **96.** **a)** -8°F **b)** Below zero
97. When adding signed numbers we add or subtract only positive numbers which are the absolute values of the original numbers. We then determine the appropriate sign for the answer.
98. Subtraction is defined as addition of the opposite, $a - b = a + (-b)$.
99. The distance between x and y is given by either $|x - y|$ or $|y - x|$.

Section 1.4 Warm-ups T F T F T T T F T F

1. We learned to multiply and divide signed numbers.
2. A product is the result of multiplication. The product of a and b is ab. The product of 2 and 4 is 8.
3. To find the product of signed numbers, multiply their absolute values and then affix a negative sign if the two original numbers have opposite signs.
4. Division is defined in terms of multiplication as $a \div b = c$ provided $c \cdot b = a$ and $b \neq 0$.
5. To find the quotient of nonzero numbers divide their absolute values and then affix a negative sign if the two original numbers have opposite signs.
6. Division by zero is undefined because it cannot be made consistent with the definition of division: $a \div b = c$ provided $c \cdot b = a$.
7. -27 **8.** -24 **9.** 132 **10.** 135 **11.** $-\frac{1}{3}$
12. $\frac{4}{7}$ **13.** -0.3 **14.** -0.09 **15.** 144 **16.** 121
17. 0 **18.** 0 **19.** -1 **20.** -3 **21.** 3 **22.** $\frac{1}{2}$
23. $-\frac{2}{3}$ **24.** $\frac{11}{12}$ **25.** $\frac{5}{6}$ **26.** $-\frac{3}{4}$ **27.** Undefined
28. Undefined **29.** 0 **30.** 0 **31.** -80 **32.** -30
33. 0.25 **34.** 1.5 **35.** -100 **36.** -20 **37.** 27
38. 17 **39.** -3 **40.** -43 **41.** -4 **42.** 48
43. -30 **44.** -6 **45.** 19 **46.** -120 **47.** -0.18
48. 0.1 **49.** 0.3 **50.** -0.075 **51.** -6 **52.** -16
53. 1.5 **54.** 70 **55.** 22 **56.** $-\frac{6}{5}$ **57.** $-\frac{1}{3}$ **58.** 3
59. -164.25 **60.** -56.667 **61.** 1529.41 **62.** -26.88
63. 16 **64.** -8 **65.** -8 **66.** 1 **67.** 0 **68.** -16
69. 0 **70.** 0 **71.** -3.9 **72.** -0.4 **73.** -40
74. -4.1 **75.** 0.4 **76.** 3.9 **77.** 0.4 **78.** 0.4
79. -0.2 **80.** -50 **81.** -7.5 **82.** 40 **83.** $-\frac{1}{30}$

84. $-\frac{17}{20}$ **85.** $-\frac{1}{10}$ **86.** 4 **87.** 7.562 **88.** -12.321
89. 19.35 **90.** 4.113 **91.** 0 **92.** 0 **93.** Undefined
94. Undefined

Section 1.5 Warm-ups F F T F F F F T F T

1. An arithmetic expression is the result of writing numbers in a meaningful combination with the ordinary operations of arithmetic.
2. The purpose of grouping symbols is to indicate the order in which to perform operations.
3. An exponential expression is an expression of the form a^n.
4. The value of -3^6 is negative while the value of $(-3)^6$ is positive.
5. The order of operations tells us the order in which to perform operations when grouping symbols are omitted.
6. Grouping symbols used in this section were parentheses, absolute value bars, and the fraction bar.

7. -4 **8.** 10 **9.** 1 **10.** 13 **11.** -8 **12.** 3
13. -7 **14.** 29 **15.** -16 **16.** -4 **17.** -4
18. -1 **19.** 4^4 **20.** 1^5 **21.** $(-5)^4$ **22.** $(-7)^3$
23. $(-y)^3$ **24.** x^5 **25.** $\left(\frac{3}{7}\right)^5$ **26.** $\left(\frac{y}{2}\right)^4$
27. $5 \cdot 5 \cdot 5$ **28.** $(-8)(-8)(-8)(-8)$ **29.** $b \cdot b$
30. $(-a)(-a)(-a)(-a)(-a)$
31. $\left(-\frac{1}{2}\right)\left(-\frac{1}{2}\right)\left(-\frac{1}{2}\right)\left(-\frac{1}{2}\right)\left(-\frac{1}{2}\right)$
32. $\left(-\frac{13}{12}\right)\left(-\frac{13}{12}\right)\left(-\frac{13}{12}\right)$ **33.** $(0.22)(0.22)(0.22)(0.22)$
34. $(1.25)(1.25)(1.25)(1.25)(1.25)(1.25)$
35. 81 **36.** 125 **37.** 0 **38.** 0 **39.** 625
40. -32 **41.** -216 **42.** 144 **43.** 100,000
44. 1,000,000 **45.** -0.001 **46.** 0.04 **47.** $\frac{1}{8}$ **48.** $\frac{8}{27}$
49. $\frac{1}{4}$ **50.** $\frac{4}{9}$ **51.** -64 **52.** -49 **53.** -4096
54. -2401 **55.** 27 **56.** -81 **57.** -13 **58.** -56
59. 36 **60.** 500 **61.** 18 **62.** -44 **63.** -19
64. 14 **65.** -17 **66.** -21 **67.** -44 **68.** 1
69. 18 **70.** -25 **71.** -78 **72.** -180 **73.** 0
74. 7 **75.** 27 **76.** 32 **77.** 1 **78.** -12 **79.** 8
80. 17 **81.** 7 **82.** 29 **83.** 11 **84.** $\frac{1}{2}$ **85.** 111
86. -3 **87.** 21 **88.** 15 **89.** -1 **90.** 29
91. -11 **92.** 25 **93.** 9 **94.** 27 **95.** 16 **96.** -20
97. 28 **98.** 33 **99.** 121 **100.** 144 **101.** -73
102. -81 **103.** 25 **104.** 144 **105.** 0 **106.** -2
107. -2 **108.** -5 **109.** 12 **110.** -13 **111.** 82
112. 0 **113.** -54 **114.** -22 **115.** -79 **116.** 19
117. -24 **118.** 19 **119.** 41.92 **120.** -31.27
121. 184.643547 **122.** 967.032 **123.** 8.0548
124. -0.8021 **125.** **a)** 307.6 million **b)** 2025
126. **a)** 122.3 million **b)** U.S.
127. $(-5)^3 = -(5^3) = -5^3 = -1 \cdot 5^3$ and $-(-5)^3 = 5^3$
128. $-(4^4) = -4^4 = -(-4)^4 = -1 \cdot 4^4$ and $(-4)^4 = 4^4$

Section 1.6 Warm-ups T F T F T F F F T F

1. An algebraic expression is the result of combining numbers and variables with the operations of arithmetic in some meaningful way.
2. An arithmetic expression involves only numbers.
3. An algebraic expression is named according to the last operation to be performed.
4. An algebraic expression is evaluated by replacing the variables with numbers and evaluating the resulting arithmetic expression.
5. An equation is a sentence that expresses equality between two algebraic expressions.
6. If an equation is true when the variable is replaced by a number, then that number is a solution to the equation.

7. Difference **8.** Product **9.** Cube **10.** Sum
11. Sum **12.** Quotient **13.** Difference **14.** Square
15. Product **16.** Difference **17.** Square **18.** Cube
19. The difference of x^2 and a^2 **20.** The sum of a^3 and b^3
21. The square of $x - a$ **22.** The cube of $a + b$
23. The quotient of $x - 4$ and 2
24. The product of 2 and $x - 3$
25. The difference of $\frac{x}{2}$ and 4
26. The difference of $2x$ and 3 **27.** The cube of ab
28. The product of a^3 and b^3 **29.** $2x + 3y$
30. $5xz$ **31.** $8 - 7x$ **32.** $\frac{6}{x + 4}$ **33.** $(a + b)^2$
34. $a^3 - b^3$ **35.** $(x + 9)(x + 12)$ **36.** x^3 **37.** $\frac{x - 7}{7 - x}$
38. $-3(x - 1)$ **39.** 3 **40.** 3 **41.** 3 **42.** -13
43. 16 **44.** 4 **45.** -9 **46.** 35 **47.** -3 **48.** -8
49. -8 **50.** 25 **51.** $-\frac{2}{3}$ **52.** 5 **53.** 4 **54.** 8
55. -1 **56.** -1 **57.** 1 **58.** 5 **59.** -4 **60.** -1
61. 0 **62.** -1 **63.** Yes **64.** Yes **65.** No **66.** Yes
67. Yes **68.** Yes **69.** Yes **70.** No **71.** Yes
72. Yes **73.** Yes **74.** Yes **75.** No **76.** Yes
77. No **78.** No **79.** $5x + 3x = 8x$ **80.** $\frac{y}{2} + 3 = 7$
81. $3(x + 2) = 12$ **82.** $-6(7y) = 13$ **83.** $\frac{x}{3} = 5x$
84. $\frac{x + 3}{5y} = xy$ **85.** $(a + b)^2 = 9$ **86.** $a^2 + b^2 = c^2$
87. $-7, -5, -3, -1, 1$ **88.** 6, 5, 4, 3, 2
89. 4, 8, 16; $\frac{1}{4}, \frac{1}{8}, \frac{1}{16}$; 100, 1000, 10,000; 0.01, 0.001, 0.0001
90. $\frac{1}{3}, \frac{1}{9}, \frac{1}{27}$; 3, 9, 27; 0.1, 0.01, 0.001; 10, 100, 1000
91. 14.65 **92.** 133.21 **93.** 37.12 **94.** 127.93
95. 169.3 cm, 41 cm **96.** 153.6 cm **97.** 13.5, 16, 32.5, 34
98. 450 feet, 109 feet per second **99.** 920 feet
100. 490.9 m^2
102. Using letters, we can make statements that are true for all numbers or statements that are true for only one unknown number.

103. For the square of the sum consider $(2 + 3)^2 = 5^2 = 25$. For the sum of the squares consider $2^2 + 3^2 = 4 + 9 = 13$. So $(2 + 3)^2 \neq 2^2 + 3^2$.

Section 1.7 Warm-ups F F T F T T T T T T

1. The commutative property says that $a + b = b + a$ and the associative property says that $(a + b) + c = a + (b + c)$.
2. The distributive property involves multiplication and addition.
3. Factoring is the process of writing an expression or number as a product.
4. The number 0 is the additive identity and the number 1 is the multiplicative identity.
5. The properties help us to understand the operations and how they are related to each other.
6. If one task is completed in x hours, then the rate is $1/x$ tasks per hour.

7. $r + 9$ **8.** $6 + t$ **9.** $3(x + 2)$ **10.** $P(rt + 1)$
11. $-5x + 4$ **12.** $-2a + b$ **13.** $6x$ **14.** $-9y$
15. $-2(x - 4)$ **16.** $(b + c)a$ **17.** $4 - 8y$ **18.** $9z - 2$
19. $4w^2$ **20.** $2y^2$ **21.** $3a^2b$ **22.** $7x^3$ **23.** $9x^3z$
24. $5y^3w$ **25.** -3 **26.** 0 **27.** -10 **28.** 4
29. -21 **30.** -29 **31.** 0.6 **32.** 13.7 **33.** -22.4
34. -0.03 **35.** $3x - 15$ **36.** $4b - 4$ **37.** $2a + at$
38. $ab + bw$ **39.** $-3w + 18$ **40.** $-3m + 15$
41. $-20 + 4y$ **42.** $-18 + 3p$ **43.** $-a + 7$
44. $-c + 8$ **45.** $-t - 4$ **46.** $-x - 7$ **47.** $2(m + 6)$
48. $3(y + 2)$ **49.** $4(x - 1)$ **50.** $6(y + 1)$
51. $4(y - 4)$ **52.** $5(x + 3)$ **53.** $4(a + 2)$
54. $7(a - 5)$ **55.** 2 **56.** 3 **57.** $-\frac{1}{5}$ **58.** $-\frac{1}{6}$
59. $\frac{1}{7}$ **60.** $\frac{1}{8}$ **61.** 1 **62.** -1 **63.** -4 **64.** $\frac{4}{3}$
65. $\frac{2}{5}$ **66.** $\frac{2}{7}$

67. Commutative property of multiplication
68. Commutative property of addition
69. Distributive property
70. Associative property of multiplication
71. Associative property of multiplication
72. Distributive property
73. Inverse properties
74. Commutative property of addition
75. Commutative property of multiplication
76. Multiplication property of 0
77. Identity property **78.** Inverse property
79. Distributive property **80.** Identity property
81. Inverse property **82.** Identity property
83. Multiplication property of 0
84. Distributive property **85.** Distributive property
86. Distributive property
87. $y + a$ **88.** $6(x + 1)$ **89.** $(5a)w$ **90.** $3 + x$
91. $\frac{1}{2}(x + 1)$ **92.** $-3x + 21$ **93.** $3(2x + 5)$
94. $x + (6 + 1)$ **95.** 1 **96.** $-5 + y$ **97.** 0
98. 1 **99.** $\frac{100}{33}$ **100.** -8
101. a) 45 bricks/hour b) Bricklayer
102. 102.5 balls/hour
103. a) 2.4366 people/second b) 1,473,684 people/week
104. 2740 acres/day
105. The perimeter is twice the sum of the length and width.
106. Due to rounding off, the tax on each item separately does not equal the tax on the total. It looks like the distributive property fails.
107. a) Commutative b) Not commutative

Section 1.8 Warm-ups T F T T F F F F F T

1. Like terms are terms with the same variables and exponents.
2. The coefficient of a term is the number preceding the variable.
3. We can add or subtract like terms.
4. Unlike terms can be multiplied and divided.
5. If a negative sign precedes a set of parentheses, then signs for all terms in the parentheses are changed when the parentheses are removed.
6. Multiplying a number by -1 changes the sign of the number.

7. 7000 **8.** 4500 **9.** 1 **10.** 1 **11.** 356 **12.** 476
13. 350 **14.** 47,800 **15.** 36 **16.** 38 **17.** 36,000
18. 30,000 **19.** 0 **20.** 0 **21.** 98 **22.** -47
23. $11w$ **24.** $14a$ **25.** $3x$ **26.** $-5a$ **27.** $5x$
28. $7b$ **29.** $-a$ **30.** $-4m$ **31.** $-2a$ **32.** 0
33. $10 - 6t$ **34.** $9 - 4w$ **35.** $8x^2$ **36.** $7r^2$
37. $-4x + 2x^2$ **38.** $6w^2 - w$ **39.** $-7mw^2$
40. $-15ab^2$ **41.** $12h$ **42.** $10h$ **43.** $-18b$ **44.** $3m$
45. $-9m^2$ **46.** $-4x^2$ **47.** $12d^2$ **48.** $10t^2$ **49.** y^2
50. $-y^2$ **51.** $-15ab$ **52.** $-21rw$ **53.** $-6a - 3ab$
54. $-6x - 2xy$ **55.** $-k + k^2$ **56.** $-t^2 + t$
57. y **58.** $-t$ **59.** $-3y$ **60.** $-6b$ **61.** y **62.** $2m$
63. $2y^2$ **64.** $4a$ **65.** $2a - 1$ **66.** $-4x + 3$
67. $3x - 2$ **68.** $-2 + x$ **69.** $-2x + 1$ **70.** $2x + 5$
71. $8 - y$ **72.** $-m + 14$ **73.** $m - 6$ **74.** $-10t + 1$
75. $w - 5$ **76.** $-3x - 9$ **77.** $8x + 15$ **78.** $8x + 22$
79. $5x - 1$ **80.** $2x + 3$ **81.** $-2a - 1$ **82.** $3m - 3$
83. $5a - 2$ **84.** $5w + 27$ **85.** $3m - 18$ **86.** $6a - 10$
87. $-3x - 7$ **88.** $k + 7$ **89.** $0.95x - 0.5$
90. $0.98x - 6$ **91.** $-3.2x + 12.71$ **92.** $0.723x - 2.817$
93. $4x - 4$ **94.** $2 - 3x$ **95.** $2y + 4$ **96.** $2a - b + c$
97. $2y + m - 1$ **98.** $2x - 5$ **99.** 3 **100.** $\frac{3}{10}x - 13$
101. $0.15x - 0.4$ **102.** $0.2x + 12$ **103.** $-14k + 23$
104. 15 **105.** 45 **106.** $-2h + 112$
107. a) $0.275x - 5650$ b) \$16,350 c) \$55,000 d) \$275,000
108. a) $0.275x - 3381.25$ b) \$7619 c) Married couple pays \$1112 more.
109. $4x + 80$, 200 feet

110. **a)** $4(2 + x) = 8 + 2x$
b) $4(2x) = (4 \cdot 2)x = 8x$
c) $\frac{4 + x}{2} = \frac{1}{2}(4 + x) = 2 + \frac{1}{2}x$
d) $5 - (x - 3) = 5 - x + 3 = 8 - x$

111. If $x = 5$, then $1/2 \cdot 5 = \frac{1}{2} \cdot 5 = 2.5$ because we do division and multiplication from left to right.

Enriching Your Mathematical Word Power

1. c **2.** b **3.** a **4.** d **5.** b **6.** d **7.** a **8.** d **9.** c **10.** a

Review Exercises

1. 0, 1, 2, 10 **2.** 1, 2, 10 **3.** -2, 0, 1, 2, 10 **4.** -2, 0, 1, 2, 3.14, 10 **5.** $-\sqrt{5}$, π **6.** All of them **7.** True **8.** False **9.** False **10.** True **11.** False **12.** True **13.** True **14.** False **15.** $\frac{17}{24}$ **16.** $\frac{5}{12}$ **17.** 6 **18.** $\frac{3}{50}$ **19.** $\frac{3}{7}$ **20.** 14 **21.** $\frac{14}{3}$ **22.** $\frac{1}{3}$ **23.** $\frac{13}{12}$ **24.** $\frac{1}{12}$ **25.** 2 **26.** -13 **27.** -13 **28.** -12 **29.** -7 **30.** -17 **31.** -7 **32.** 0 **33.** 11.95 **34.** -2.03 **35.** -0.05 **36.** 0 **37.** $-\frac{1}{6}$ **38.** $-\frac{5}{12}$ **39.** $-\frac{11}{15}$ **40.** $\frac{7}{12}$ **41.** -15 **42.** 36 **43.** 4 **44.** -10 **45.** 5 **46.** -6 **47.** $\frac{1}{6}$ **48.** -24 **49.** -0.3 **50.** -14 **51.** -0.24 **52.** 0 **53.** 1 **54.** -6 **55.** 66 **56.** 90 **57.** 49 **58.** 19 **59.** 41 **60.** 4 **61.** 1 **62.** -17 **63.** 50 **64.** 27 **65.** -135 **66.** 12 **67.** -2 **68.** 4 **69.** -16 **70.** 7 **71.** 16 **72.** 9 **73.** 5 **74.** -3 **75.** 9 **76.** 1 **77.** 7 **78.** -9 **79.** $-\frac{1}{3}$ **80.** $\frac{5}{3}$ **81.** 1 **82.** 1 **83.** -9 **84.** -9 **85.** Yes **86.** Yes **87.** No **88.** No **89.** Yes **90.** Yes **91.** No **92.** No
93. Distributive property
94. Associative property of multiplication
95. Inverse property
96. Commutative property
97. Identity property
98. Identity property
99. Associative property of addition
100. Distributive property
101. Commutative property of multiplication
102. Commutative property of addition
103. Inverse property
104. Multiplication property of 0
105. Identity property
106. Commutative property of addition
107. $-a + 12$ **108.** $m + 8$ **109.** $6a^2 - 6a$ **110.** $8a^2 - 5a$ **111.** $-12t + 39$ **112.** $5m - 3$ **113.** $-0.9a - 0.57$ **114.** $-0.9x + 0.93$ **115.** $-0.05x - 4$ **116.** $0.22x - 12$ **117.** $27x^2 + 6x + 5$ **118.** $-7x^2 + 14x + 7$ **119.** $-2a$ **120.** $-2y + 2w$ **121.** $x^2 + 4x - 3$ **122.** $y^2 + y + 3$ **123.** 0 **124.** 88 **125.** 8 **126.** -20 **127.** -21 **128.** -6 **129.** $\frac{1}{2}$ **130.** $\frac{1}{5}$ **131.** -0.5 **132.** 50 **133.** -1 **134.** $-\frac{9}{13}$ **135.** $x + 2$ **136.** $8x$ **137.** $4 + 2x$ **138.** $8 + 4x$ **139.** $2x$ **140.** $-x + 6$ **141.** $-4x + 8$ **142.** $8x^2$ **143.** $6x$ **144.** $x + 6$ **145.** x **146.** $6x$ **147.** $8x$ **148.** $2x$ **149.** 3, 2, 1, 0, -1 **150.** 1, 2, 3, 4, 5 **151.** 25, 125, 625; 16, -64, 256 **152.** $-\frac{1}{3}, \frac{1}{9}, -\frac{1}{27}$; -2, 4, -8 **153.** 18 memberships per hour **154.** **a)** $0.391x - 22{,}889.85$ **b)** \$100,000 **c)** \$8,735,510

Chapter 1 Test

1. 0, 8 **2.** -3, 0, 8 **3.** -3, $-\frac{1}{4}$, 0, 8 **4.** $-\sqrt{3}$, $\sqrt{5}$, π **5.** -21 **6.** -4 **7.** 9 **8.** -7 **9.** -0.95 **10.** -56 **11.** 978 **12.** 13 **13.** -1 **14.** 0 **15.** 9740 **16.** $-\frac{7}{24}$ **17.** -20 **18.** $-\frac{1}{6}$ **19.** -39
20. Distributive property
21. Commutative property of multiplication
22. Associative property of addition
23. Inverse property **24.** Identity property
25. Multiplication property of 0
26. $3(x + 10)$ **27.** $7(w - 1)$ **28.** $6x + 6$ **29.** $4x - 2$ **30.** $7x - 3$ **31.** $0.9x + 7.5$ **32.** $14a^2 + 5a$ **33.** $x + 2$ **34.** $4t$ **35.** $54x^2y^2$ **36.** 41 **37.** 5 **38.** -12 **39.** No **40.** Yes **41.** Yes
42. 9 deliveries per hour
43. $3.66R - 0.06A + 82.205$, 168.905 cm

Chapter 2

Section 2.1 Warm-ups T T F T F T T T T T

1. The addition property of equality says that adding the same number to each side of an equation does not change the solution to the equation.
2. Equivalent equations are equations that have the same solution set.
3. The multiplication property of equality says that multiplying both sides of an equation by the same nonzero number does not change the solution to the equation.
4. A linear equation in one variable is an equation of the form $ax + b = 0$, where $a \neq 0$.
5. Replace the variable in the equation with your solution. If the resulting statement is correct, then the solution is correct.

6. In solving equations, you are not allowed to multiply or divide both sides by 0.

7. {1} 8. {5} 9. {9} 10. {−4} 11. {1} 12. $\left\{\frac{3}{4}\right\}$

13. $\left\{\frac{2}{3}\right\}$ 14. $\left\{\frac{5}{6}\right\}$ 15. {−9} 16. {−7} 17. {−19}

18. {−30} 19. $\left\{\frac{1}{4}\right\}$ 20. $\left\{\frac{2}{3}\right\}$ 21. {0} 22. $\left\{\frac{1}{6}\right\}$

23. {−5} 24. {−2} 25. {−4} 26. {−2} 27. {3}

28. {1.8} 29. $\left\{\frac{1}{4}\right\}$ 30. $\left\{\frac{1}{3}\right\}$ 31. {−8} 32. {−18}

33. {1.8} 34. {4} 35. $\left\{\frac{2}{3}\right\}$ 36. $\left\{\frac{2}{5}\right\}$ 37. $\left\{\frac{1}{2}\right\}$

38. $\left\{\frac{1}{4}\right\}$ 39. {−5} 40. {4} 41. {5} 42. {−6}

43. {1.25} 44. {2.8} 45. $\left\{\frac{1}{4}\right\}$ 46. $\left\{-\frac{1}{9}\right\}$ 47. $\left\{\frac{3}{20}\right\}$

48. $\left\{-\frac{2}{9}\right\}$ 49. {−2} 50. {−12} 51. {120}

52. {16} 53. $\left\{\frac{5}{9}\right\}$ 54. $\left\{\frac{6}{25}\right\}$ 55. $\left\{-\frac{1}{2}\right\}$ 56. $\left\{-\frac{1}{12}\right\}$

57. {−8} 58. {−4} 59. $\left\{\frac{1}{3}\right\}$ 60. $\left\{\frac{7}{8}\right\}$ 61. {−3.4}

62. {−4.9} 63. {99} 64. {17} 65. {−7} 66. {9}
67. {9} 68. {12} 69. {8} 70. {−6} 71. {5}
72. {13} 73. {−5} 74. {−9} 75. {−8}

76. {−18} 77. {2} 78. {0.9} 79. $\left\{\frac{1}{6}\right\}$ 80. $\left\{-\frac{1}{6}\right\}$

81. $\left\{-\frac{1}{3}\right\}$ 82. $\left\{\frac{1}{2}\right\}$ 83. {44} 84. {−55} 85. $\left\{\frac{3}{4}\right\}$

86. $\left\{\frac{4}{9}\right\}$ 87. {7} 88. $\left\{-\frac{1}{2}\right\}$ 89. {−14} 90. {−24}

91. $\left\{\frac{3}{8}\right\}$ 92. $\left\{\frac{7}{9}\right\}$

93. a) 60.6 births per 1000 females
b) 54 births per 1000 females
94. 2.4 trillion metric tons 95. 2877 stocks
96. 4,058,814

Section 2.2 Warm-ups T T T F T F T T T T

1. We can solve $ax + b = 0$ with the addition property and the multiplication property of equality.
2. The multiplication property of equality is usually applied last.
3. Use the multiplication property of equality to solve $-x = 8$.
4. If an equation involves parentheses, then we first remove the parentheses.

5. {2} 6. {−3} 7. {−2} 8. {−6} 9. $\left\{\frac{2}{3}\right\}$

10. $\left\{-\frac{1}{5}\right\}$ 11. $\left\{-\frac{5}{2}\right\}$ 12. $\left\{\frac{8}{9}\right\}$ 13. {6}

14. {−16} 15. {12} 16. {−35} 17. $\left\{\frac{1}{2}\right\}$ 18. $\left\{-\frac{3}{4}\right\}$

19. $\left\{-\frac{1}{6}\right\}$ 20. $\left\{\frac{1}{36}\right\}$ 21. {4} 22. {−2} 23. $\left\{\frac{5}{6}\right\}$

24. $\left\{-\frac{3}{2}\right\}$ 25. {4} 26. {6} 27. {−5} 28. {−1}

29. {34} 30. {−54} 31. {9} 32. {12} 33. {1.2}
34. {0.73} 35. {3} 36. {2} 37. {4} 38. {2}

39. {−3} 40. {3} 41. $\left\{\frac{1}{2}\right\}$ 42. $\left\{\frac{3}{8}\right\}$ 43. {30}

44. {−24} 45. {6} 46. {−3} 47. {−2} 48. {4}
49. {18} 50. {1} 51. {0} 52. {0} 53. {−2}

54. {6} 55. $\left\{\frac{7}{3}\right\}$ 56. $\left\{-\frac{4}{5}\right\}$ 57. {1} 58. {−11}

59. {−6} 60. {−2} 61. {−12} 62. {−1}
63. {−4} 64. {−1} 65. {−13} 66. {−12}
67. {1.7} 68. {1.05} 69. {2} 70. {−14}
71. {4.6} 72. {1} 73. {8} 74. {7} 75. {34}
76. {10} 77. {6} 78. {2} 79. {0} 80. {0}
81. {−10} 82. {−70} 83. {18} 84. {4}

85. {−20} 86. {−35} 87. {−3} 88. $\left\{-\frac{4}{3}\right\}$

89. {−4.3} 90. {7.59} 91. 17 hr 92. 3 hr
93. 20°C 94. a) 100°C b) 158°F 95. 9 ft
96. 3 m, 4 m, 5 m 97. $14,550 98. 3.5 hr

Section 2.3 Warm-ups T T F F F T T F T T

1. If an equation involves fractions we usually multiply each side by the LCD of all of the fractions.
2. If an equation involves decimals we usually multiply each side by a power of 10 to eliminate all decimals.
3. An identity is an equation that is satisfied by all numbers for which both sides are defined.
4. A conditional equation has at least one solution but is not an identity.
5. An inconsistent equation has no solutions.
6. The solution set to an inconsistent equation is the empty set, ∅.

7. {7} 8. {9} 9. {24} 10. {30} 11. {16}
12. {20} 13. {−12} 14. {−4} 15. {60} 16. {20}
17. {24} 18. {150} 19. {90} 20. {70} 21. {6}
22. {−8} 23. {−2} 24. {−20} 25. {80} 26. {30}
27. {60} 28. {90} 29. {200} 30. {350} 31. {800}

32. {200} 33. $\left\{\frac{9}{2}\right\}$ 34. $\left\{-\frac{7}{3}\right\}$ 35. {3} 36. {−4}

37. {25} 38. {6} 39. {−2} 40. {−9} 41. {−3}
42. {4} 43. {5} 44. {−4} 45. {−10} 46. {6}
47. {2} 48. {10} 49. All real numbers, identity
50. All real numbers, identity 51. ∅, inconsistent
52. ∅, inconsistent 53. {0}, conditional
54. {7}, conditional 55. ∅, inconsistent
56. All real numbers, identity 57. ∅, inconsistent
58. ∅, inconsistent 59. {1}, conditional

60. $\left\{\frac{5}{3}\right\}$, conditional **61.** $\varnothing$, inconsistent
62. {4}, conditional **63.** All real numbers, identity
64. All real numbers, identity
65. All nonzero real numbers, identity
66. All real numbers, identity
67. All real numbers, identity
68. All nonzero real numbers, identity
69. {−4} **70.** $\left\{\frac{5}{4}\right\}$ **71.** R **72.** $\varnothing$ **73.** R **74.** {6}
75. {100} **76.** {500} **77.** $\left\{-\frac{3}{2}\right\}$ **78.** $\left\{-\frac{2}{3}\right\}$
79. {30} **80.** {20} **81.** {6} **82.** {40} **83.** {0.5}
84. {450} **85.** {19,608} **86.** {113.642} **87.** $128,000
88. 8 rabbits **89. a)** $140,000 **b)** $139,211
90. $46,153.85

Section 2.4 Warm-ups F F F F F T T T F T
1. A formula is an equation with two or more variables.
2. A literal equation is a formula.
3. To solve for a variable means to find an equivalent equation in which the variable is isolated.
4. If the variable appears on both sides, then get all terms with the variable onto the same side. Then use the distributive property to get one occurrence of the variable.
5. To find the value of a variable in a formula we can solve for the variable then insert values for the other variables, or insert values for the other variables and then solve for the variable.
6. The formula for the perimeter of a rectangle is $P = 2L + 2W$.
7. $R = \frac{D}{T}$ **8.** $W = \frac{A}{L}$ **9.** $D = \frac{C}{\pi}$ **10.** $a = \frac{F}{m}$
11. $P = \frac{I}{rt}$ **12.** $t = \frac{I}{Pr}$ **13.** $C = \frac{5}{9}(F - 32)$
14. $x = \frac{4y + 28}{3}$ **15.** $h = \frac{2A}{b}$ **16.** $b = \frac{2A}{h}$
17. $L = \frac{P - 2W}{2}$ **18.** $W = \frac{P - 2L}{2}$ **19.** $a = 2A - b$
20. $b = 2A - a$ **21.** $r = \frac{S - P}{Pt}$ **22.** $t = \frac{S - P}{Pr}$
23. $a = \frac{2A - hb}{h}$ **24.** $b = \frac{2A - ah}{h}$ **25.** $x = \frac{b - a}{2}$
26. $x = \frac{c + 5b}{5}$ **27.** $x = -7a$ **28.** $x = b - 6a$
29. $x = 12 - a$ **30.** $x = \frac{3w}{5}$ **31.** $x = 7ab$
32. $x = a + 2b$ **33.** $y = -x - 9$ **34.** $y = -3x - 5$
35. $y = -x + 6$ **36.** $y = -4x + 2$ **37.** $y = 2x - 2$
38. $y = x + 3$ **39.** $y = 3x + 4$ **40.** $y = -2x + 5$
41. $y = -\frac{1}{2}x + 2$ **42.** $y = -\frac{3}{2}x + 3$ **43.** $y = x - \frac{1}{2}$
44. $y = \frac{3}{2}x + 3$ **45.** $y = 3x - 14$ **46.** $y = -3x + 6$
47. $y = \frac{1}{2}x$ **48.** $y = -\frac{2}{3}x + 10$ **49.** $y = \frac{3}{2}x + 6$
50. $y = -2x + 2$ **51.** $y = \frac{3}{2}x + \frac{13}{2}$ **52.** $y = \frac{2}{3}x - \frac{16}{3}$
53. $y = -\frac{1}{4}x + \frac{5}{8}$ **54.** $y = -\frac{1}{3}x - \frac{2}{3}$
55. 60, 30, 0, −30, −60 **56.** −60, −40, −20, 0, 20
57. 14, 23, 32, 104, 212 **58.** −40, −10, 0, 15, 30
59. 40, 20, 10, 5, 4 **60.** 100, 20, 5, 2, 1 **61.** 2
62. 1 **63.** 7 **64.** 0 **65.** $-\frac{9}{5}$ **66.** 5 **67.** 1
68. 6 **69.** 1.33 **70.** −67.708 **71.** 4% **72.** 6%
73. 4 years **74.** 3 years **75.** 7 yards **76.** 15 feet
77. 225 feet **78.** 146 feet **79.** $60,500 **80.** $116,000
81. $300 **82.** $5000 **83.** 20% **84.** 8%
85. 160 feet **86.** 72 inches **87.** 24 cubic feet
88. 4 feet **89.** 4 inches **90.** 4 meters **91.** 8 feet
92. 7 meters **93.** 12 inches **94.** 15 centimeters
95. 640 milligrams, age 13 **96.** Age 7
97. 3.75 milliliters **98. a)** $190 billion **b)** 2008 **c)** 2008
99. $L = F\sqrt{S} - 2D + 5.688$

Section 2.5 Warm-ups T T T F T F F F T F
1. To express addition we use words such as plus, sum, increased by, and more than.
2. We can algebraically express two numbers using one variable provided the numbers are related in some known way.
3. Complementary angles have degree measures with a sum of 90°.
4. Supplementary angles have degree measures with a sum of 180°.
5. Distance is the product of rate and time.
6. If x is an even integer, then x and $x + 2$ represent consecutive even integers. If x is an odd integer, then x and $x + 2$ represent consecutive odd integers.
7. $x + 3$ **8.** $x + 2$ **9.** $x - 3$ **10.** $x - 4$ **11.** $5x$
12. $\frac{5}{x}$ **13.** $0.1x$ **14.** $0.08x$ **15.** $\frac{x}{3}$ **16.** $\frac{12}{x}$
17. $\frac{1}{3}x$ **18.** $\frac{3}{4}x$ **19.** x and $x + 15$ **20.** x and $x + 9$
21. x and $6 - x$ **22.** x and $5 - x$ **23.** x and $-4 - x$
24. x and $-8 - x$ **25.** x and $x + 3$ **26.** x and $x + 8$
27. x and $0.05x$ **28.** x and $0.4x$ **29.** x and $1.30x$
30. x and $0.80x$ **31.** x and $90 - x$ **32.** x and $180 - x$
33. x and $120 - x$ **34.** x and $90 - x$
35. n and $n + 2$, where n is an even integer
36. x and $x + 2$, where x is an odd integer
37. x and $x + 1$, where x is an integer
38. x, $x + 2$, and $x + 4$, where x is an even integer
39. x, $x + 2$, and $x + 4$, where x is an odd integer
40. x, $x + 1$, and $x + 2$, where x is an integer
41. x, $x + 2$, $x + 4$, and $x + 6$, where x is an even integer
42. x, $x + 2$, $x + 4$, and $x + 6$, where x is an odd integer
43. $3x$ miles **44.** $5x + 50$ miles **45.** $0.25q$ dollars
46. $0.10t$ yen **47.** $\frac{x}{20}$ hour **48.** $\frac{300}{x + 30}$ hour
49. $\frac{x - 100}{12}$ meters per second **50.** $\frac{200}{x + 3}$ feet per second

51. $5x$ square meters **52.** $b(b - 6)$ square yards
53. $2w + 2(w + 3)$ inches **54.** $2r + 2(r - 1)$ centimeters
55. $150 - x$ feet **56.** $\frac{200}{w}$ feet **57.** $2x + 1$ feet
58. $2w - 3$ feet **59.** $x(x + 5)$ square meters
60. $2(x) + 2(x - 10)$ yards **61.** $0.18(x + 1000)$
62. $0.06(3x)$ **63.** $\frac{16.50}{x}$ dollars per pound
64. $\frac{480}{x}$ dollars per hour **65.** $90 - x$ degrees
66. $180 - x$ degrees
67. x is the smaller number, $x(x + 5) = 8$
68. x is the smaller number, $x(x + 6) = -9$
69. x is the selling price, $x - 0.07x = 84{,}532$
70. x is the selling price, $x - 0.10x = 6570$
71. x is the percent, $500x = 100$
72. x is the percent, $40x = 120$
73. x is the number of nickels, $0.05x + 0.10(x + 2) = 3.80$
74. d is the number of dimes, $0.10d + 0.25(d - 3) = 6.75$
75. x is the number, $x + 5 = 13$
76. x is the number, $x - 12 = -6$
77. x is the smallest integer, $x + (x + 1) + (x + 2) = 42$
78. x is the smallest odd integer, $x + x + 2 + x + 4 = 27$
79. x is the smaller integer, $x(x + 1) = 182$
80. x is the smaller even integer, $x(x + 2) = 168$
81. x is Harriet's income, $0.12x = 3000$
82. x is the number of members, $0.09x = 252$
83. x is the number, $0.05x = 13$
84. x is the number, $0.08x = 300$
85. x is the width, $x(x + 5) = 126$
86. x is the width, $2x + 2(2x - 1) = 298$
87. n is the number of nickels, $5n + 10(n - 1) = 95$
88. q is the number of quarters, $25q + 10(q + 1) + 5(2q) = 90$
89. x is the measure of the larger angle, $x + x - 38 = 180$
90. x is the measure of the smaller angle, $x + x + 16 = 90$
91. **a)** $r + 0.6(220 - (30 + r)) = 144$, where r is the resting heart rate
b) Target heart rate increases as resting heart rate increases.
92. $1.09L = 36$, where L is the inside leg measurement
93. $x(x + 3) = 24$ **94.** $(h + 2)(h + 2) = 24$
95. $w(w - 4) = 24$ **96.** $\frac{1}{2}y(y - 2) = 24$

Section 2.6 Warm-ups F T T F F T T T F T
1. In this section we studied number, geometric, and uniform motion problems.
2. We solve number problems to gain experience at problem solving.
3. Uniform motion is motion at a constant rate of speed.
4. Supplementary angles are angles whose degree measures have a sum of 180°.
5. Complementary angles are angles whose degree measures have a sum of 90°.
6. When solving a geometric problem draw a figure and label the sides.

7. 46, 47, 48 **8.** 36, 38, 40 **9.** 75, 77
10. 27, 29, 31, 33 **11.** 47, 48, 49, 50 **12.** 82, 84, 86, 88
13. Length 50 meters, width 25 meters
14. Length 78 feet, width 36 feet
15. Width 42 inches, length 46 inches
16. 185 miles, 289 miles, 246 miles **17.** 13 inches
18. 18 feet by 26 feet **19.** 35°
20. 24° **21.** 1152 in. **22.** 99 ft **23.** 22.88 km
24. 20.88 mi **25.** 5.31 in. **26.** 106.93 cm
27. 402.57 g **28.** 8.22 oz **29.** 58.67 ft/sec
30. 136.36 mi/hr **31.** 548.53 km/hr **32.** 7.82 mi/min
33. 65 miles per hour **34.** 4 miles per hour
35. 55 miles per hour **36.** 36 miles
37. 4 hours, 2048 miles **38.** 2.5 hours
39. Raiders 32, Vikings 14
40. Exxon \$210 billion, Wal-Mart \$193 billion, GM \$185 billion
41. 3 hours, 106 miles **42.** 4 hours, 138 miles
43. Crawford 1906, Wayne 1907, Stewart 1908
44. Hope 1903, Gable 1901, Fonda 1905
45. 7 ft, 7 ft, 16 ft **46.** 5 ft by 15 ft

Section 2.7 Warm-ups T F T F T T F T F T
1. We studied discount, investment, and mixture problems in this section.
2. The rate of discount is a percentage and the discount is the actual amount that the price is reduced.
3. The product of the rate and the original price gives the amount of discount. The original price minus the discount is the sale price.
4. Both mixture problems and investment problems involve rates.
5. A table helps us to organize the information given in a problem.
6. The product of the interest rate and the amount invested gives the amount of interest.

7. \$320 **8.** \$625 **9.** \$400 **10.** \$31,765
11. \$125,000 **12.** \$900 **13.** \$30.24 **14.** \$14,550
15. 100 Fund \$10,000, 101 Fund \$13,000
16. Brother \$0.50, sister \$1.00
17. Fidelity \$14,000, Price \$11,000
18. Dreyfus \$16,000, Templeton \$14,000
19. 30 gallons **20.** 20 gallons
21. 20 liters of 5% alcohol, 10 liters of 20% alcohol
22. $\frac{10}{3}$ quarts of pure antifreeze, $\frac{50}{3}$ quarts of 40% solution
23. 55,700 **24.** 1320 **25.** \$15,000 **26.** \$40,000
27. 75% **28.** 20% **29.** 600
30. 500 students in the 58% school, 300 students in the 10% school
31. 42 private rooms, 30 semi-private rooms
32. 140 TV ads, 80 radio ads **33.** 12 pounds
34. 32 pounds of premium, 68 pounds of regular
35. 4 nickels, 6 dimes **36.** 30 dimes, 6 quarters
37. 800 gallons **38.** 80 kilograms **39.** $\frac{2}{3}$ gal
40. 21 CD players, 63 VCRs

Section 2.8 Warm-ups T T F T F T F F T F

1. The inequality symbols are $<$, $\leq$, $>$, and $\geq$.
2. The inequalities $a < b$ and $b > a$ have the same meaning.
3. For $\leq$ and $\geq$ use the solid circle and for $<$ and $>$ use the open circle.
4. A compound inequality is a statement involving more than one inequality.
5. The compound inequality $a < b < c$ means $b > a$ and $b < c$, or b is between a and c.
6. "At most" means less than or equal to and "at least" means greater than or equal to.

7. True **8.** True **9.** True **10.** True **11.** False **12.** False **13.** True **14.** True **15.** True **16.** False **17.** True **18.** True **19.** True **20.** True

21. −1 0 1 2 3 4 5

22. −11 −10 −9 −8 −7 −6 −5

23. −4 −3 −2 −1 0 1 2

24. 2 3 4 5 6 7 8

25. −5 −4 −3 −2 −1 0 1

26. −4 −3 −2 −1 0 1 2

27. −4 −3 −2 −1 0 1 2

28. −9 −8 −7 −6 −5 −4 −3

29. $\frac{1}{2}$ −1 0 1 2 3

30. $-\frac{2}{3}$ −3 −2 −1 0 1 2

31. 5.3 1 2 3 4 5 6 7

32. −3.4 −7 −6 −5 −4 −3 −2

33. −4 −3 −2 −1 0 1 2

34. 0 1 2 3 4 5

35. 2 3 4 5 6 7 8

36. −5 −4 −3 −2 −1 0 1

37. −5 −4 −3 −2 −1 0 1

38. −3 −2 −1 0 1 2 3

39. 20 40 60 80 100 120

40. −200 0 200 400 600 800

41. $x > 3$ **42.** $x \leq 4$ **43.** $x \leq 2$ **44.** $0 < x \leq 3$ **45.** $0 < x < 2$ **46.** $-1 \leq x < 3$ **47.** $-5 < x \leq 7$ **48.** $x < 4$ **49.** $x > -4$ **50.** $0 < x \leq 2$ **51.** Yes **52.** No **53.** No **54.** Yes **55.** No **56.** No **57.** Yes **58.** Yes **59.** Yes **60.** Yes **61.** Yes **62.** Yes **63.** No **64.** No **65.** Yes **66.** Yes **67.** No **68.** Yes **69.** 0, 5.1 **70.** −5.1, 0 **71.** 5.1 **72.** −5.1 **73.** 5.1 **74.** −5.1 **75.** −5.1, 0, 5.1 **76.** 5.1, 0 **77.** $0.08p > 1500$

78. $p + 2p + p + 0.25 < 2.00$ **79.** $\frac{44 + 72 + s}{3} \geq 60$

80. $\frac{87 + s}{2} \geq 90$ **81.** $396 < 8R < 453$

82. $399.99 < b + b + 100 < 579.99$

83. $60 < 90 - x < 70$ **84.** $180 - x - (x + 8) \leq 30$

85. $45 + 2(30) + 2h \leq 130$ **86.** $\frac{93 + x}{317 + 20} > 0.300$

87. 79, moderate effort on level ground

Section 2.9 Warm-ups T F F T F T F T T F

1. Equivalent inequalities are inequalities that have the same solutions.
2. The addition property of inequality says that adding any real number to each side of an inequality produces an equivalent inequality.
3. According to the multiplication property of inequality, the inequality symbol is reversed when multiplying (or dividing) by a negative number and not reversed when multiplying (or dividing) by a positive number.
4. For equations or inequalities we try to isolate the variable. The properties of equality and inequality are similar.
5. We solve compound inequalities using the properties of inequality as we do for simple inequalities.
6. The direction of the inequality symbol is reversed when we multiply or divide by a negative number.

7. $x > -7$ **8.** $x < 6$ **9.** $w \geq 3$ **10.** $z \leq 2$ **11.** $k > 1$ **12.** $t < -3$ **13.** $y \leq -8$ **14.** $x \geq -12$

15. $x > -3$
−5 −4 −3 −2 −1 0 1

16. $x \leq -17$
−19 −18 −17 −16 −15

17. $w > -2$
−2 −1 0 1 2 3

18. $w < 21$
19 20 21 22 23

19. $b < 4$
−1 0 1 2 3 4 5

20. $b > 5$
2 3 4 5 6 7

21. $z \geq -\frac{1}{2}$

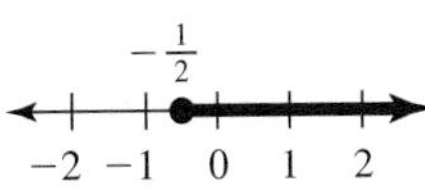

22. $y \leq \frac{5}{2}$
$\frac{5}{2}$
−1 0 1 2 3 4

23. $y < 3$

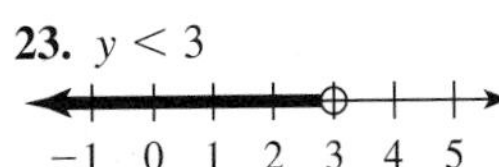

24. $y > -2$

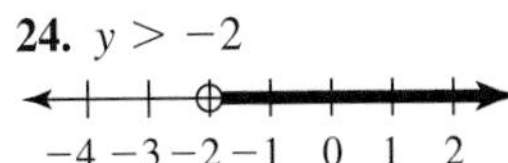

25. $z \geq -\frac{1}{3}$

$-\frac{1}{3}$

−2 −1 0 1 2

26. $z \leq -\frac{4}{3}$

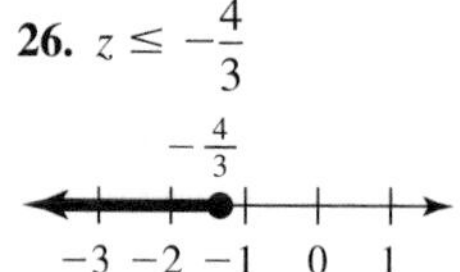

27. $r > -3$

−4 −3 −2 −1 0 1

28. $r \leq 6$

2 3 4 5 6 7 8

29. $p < 13$

9 10 11 12 13 14 15

30. $p < -4$

−8 −7 −6 −5 −4 −3 −2

31. $q \leq 24$

20 22 24 26

32. $q \leq 6$

2 3 4 5 6 7 8

33. $t \leq \frac{7}{2}$

$\frac{7}{2}$

0 1 2 3 4 5

34. $t < \frac{1}{2}$

$\frac{1}{2}$

−2 −1 0 1 2

35. $x < -11$

−15 −14 −13 −12 −11 −10 −9

36. $x < 13$

9 10 11 12 13 14 15

37. $x > -10$

−12 −11 −10 −9 −8 −7 −6

38. $x > 18$

16 17 18 19 20 21 22

39. $x < 614.3$

614.3

40. $x > 3.91$

3.91

2 3 4 5 6 7

41. $8 < x < 10$

6 7 8 9 10 11 12

42. $7 < x < 11$

6 7 8 9 10 11 12

43. $1 < v < \frac{9}{2}$

$\frac{9}{2}$

0 1 2 3 4 5 6

44. $-\frac{7}{3} < v < 1$

$-\frac{7}{3}$

−3 −2 −1 0 1 2 3

45. $-2 \leq k \leq 9$

9

−2 0 2 4 6 8 10

46. $-5 \leq k \leq 1$

−5 −4 −3 −2 −1 0 1

47. $-5 \leq y < 3$

−5 −4 −3 −2 −1 0 1 2 3

48. $-1 < y \leq 1$

−3 −2 −1 0 1 2 3

49. $12 < u < 30$

12 18 24 30

50. $-4 < u < 16$

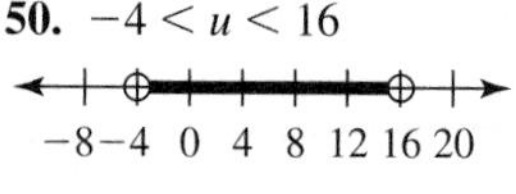

51. $-\frac{1}{2} < m \leq \frac{3}{2}$

$-\frac{1}{2}$ $\frac{3}{2}$

−2 −1 0 1 2 3

52. $-\frac{15}{2} < m \leq \frac{2}{3}$

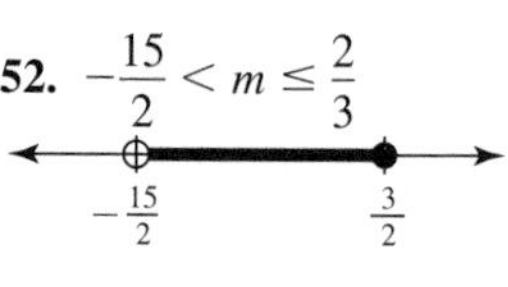

53. $102.1 < x < 108.3$

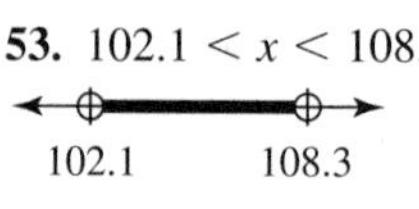

54. $-2.69 < x < -0.91$

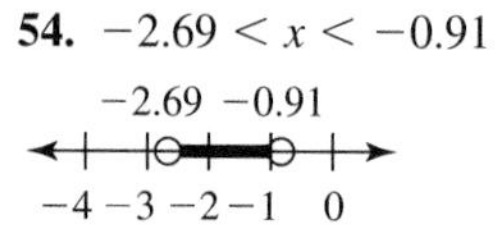

55. $x \leq 6$

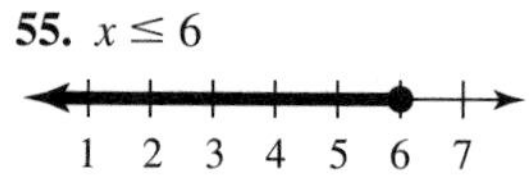

56. $y \leq -8$

−10 −9 −8 −7 −6

57. $x < 0$

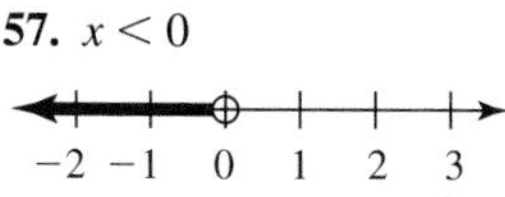

58. $z > 1$

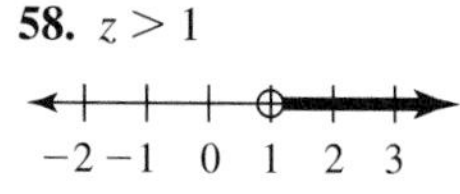

59. $2 < x < 3$

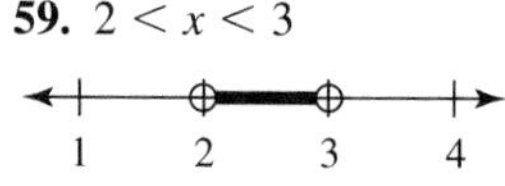

60. $4 < w < 6$

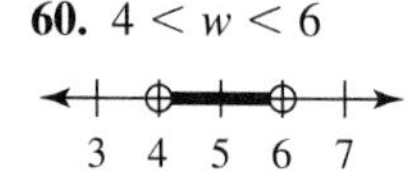

61. At least 28 meters **62.** At most 30 feet
63. Less than \$9358 **64.** At least \$12,250
65. At most \$550 **66.** Less than 90 cents **67.** At least 64
68. At least \$490 **69.** Between 81 and 94.5 inclusive
70. Between 114 and 141 inclusive
71. Between 49.5 and 56.625 miles per hour
72. Between 6 and 9 hours **73.** Between 55° and 85°
74. Between 0° and 65°
75. a) Between 27 and 35 teeth inclusive
b) Between 23.02 in. and 24.79 in.
c) At least 14 teeth
76. a) 7200
b) Less than \$50

Enriching Your Mathematical Word Power

1. b **2.** d **3.** c **4.** c **5.** d **6.** d **7.** a
8. b **9.** c **10.** d

Review Exercises

1. {35} **2.** {−4} **3.** {−6} **4.** {−40} **5.** {−7}
6. {−2} **7.** {13} **8.** {19} **9.** {7} **10.** $\left\{\frac{46}{5}\right\}$
11. {2} **12.** {3} **13.** {7} **14.** {4} **15.** {0}
16. {7} **17.** {−8} **18.** {7} **19.** ∅, inconsistent
20. All real numbers, identity
21. All real numbers, identity **22.** {0}, conditional
23. All nonzero real numbers, identity
24. {1}, conditional **25.** {24}, conditional
26. {2}, conditional **27.** {80}, conditional

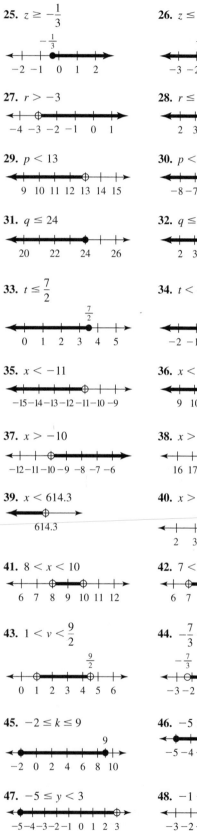

28. {30}, conditional **29.** {1000}, conditional
30. $\left\{\frac{33}{14}\right\}$, conditional **31.** $\left\{\frac{1}{4}\right\}$ **32.** $\left\{\frac{1}{6}\right\}$ **33.** $\left\{\frac{21}{8}\right\}$
34. $\left\{-\frac{5}{3}\right\}$ **35.** $\left\{-\frac{4}{5}\right\}$ **36.** $\left\{-\frac{9}{8}\right\}$ **37.** {4} **38.** {−10}
39. {24} **40.** {−2} **41.** {−100} **42.** {−15}
43. $x = -\frac{b}{a}$ **44.** $x = \frac{t - e}{m}$ **45.** $x = \frac{b + 2}{a}$
46. $x = 5 - b$ **47.** $x = \frac{V}{LM}$ **48.** $x = \frac{2}{y}$ **49.** $x = -\frac{b}{3}$
50. $x = \frac{t}{9}$ **51.** $y = -\frac{5}{2}x + 3$ **52.** $y = \frac{5}{3}x + 3$
53. $y = -\frac{1}{2}x + 4$ **54.** $y = \frac{1}{2}x - 2$ **55.** $y = -2x + 16$
56. $y = \frac{2}{3}x + 2$ **57.** −13 **58.** $\frac{1}{3}$ **59.** $-\frac{2}{5}$ **60.** 2
61. 17 **62.** −17 **63.** 15, 10, 5, 0, −5
64. −4, −2, 0, 2, 4 **65.** −3, −1, 1, 3
66. −100, 0, 100, 200
67. $x + 9$, where x is the number
68. $7x$, where x is the number
69. x and $x + 8$, where x is the smaller number
70. x and $12 - x$, where x is one number
71. $0.65x$, where x is the number
72. $\frac{1}{2}x$, where x is the number
73. $x(x + 5) = 98$, where x is the width
74. $2x + 2(2x + 1) = 56$, where x is the width
75. $2x = 3(x - 10)$, where x is Jim's rate
76. $6(x + 5) + 5x = 840$, where x is Ned's rate
77. $x + x + 2 + x + 4 = 90$, where x is the smallest of three even integers
78. $x + x + 2 = 40$, where x is the smaller of the two odd integers
79. $t + 2t + t - 10 = 180$, where t is the degree measure of an angle
80. $p + 3p - 6 = 90$, where p is the degree measure of an angle
81. 77, 79, 81 **82.** 224, 226
83. Betty 45 mph, Lawanda 60 mph
84. Length 150 feet, width 100 feet
85. Wanda \$36,000, husband \$30,000
86. Aerospace 1800, agriculture 1200
87. No **88.** Yes **89.** No **90.** Yes
91. $x > 1$ **92.** $x < 2$ **93.** $x \ge 2$ **94.** $3 < x < 5$
95. $-3 \le x < 3$ **96.** $x \le 1$ **97.** $x < -1$
98. $-2 \le x < 2$
99. $x > -1$ 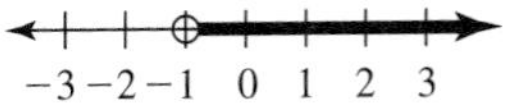
−3 −2 −1 0 1 2 3
100. $x > 10$ 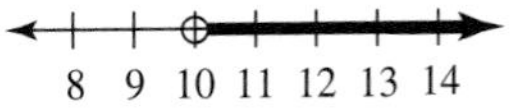
8 9 10 11 12 13 14
101. $x < 3$
−1 0 1 2 3 4
102. $x > 2$ 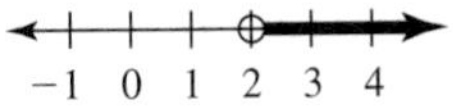
−1 0 1 2 3 4
103. $x \le -4$ 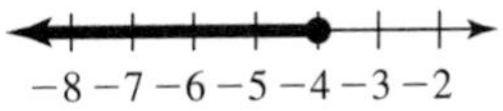
−8 −7 −6 −5 −4 −3 −2
104. $x \ge -15$ 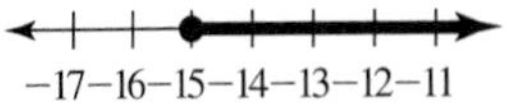
−17 −16 −15 −14 −13 −12 −11
105. $x > -4$ 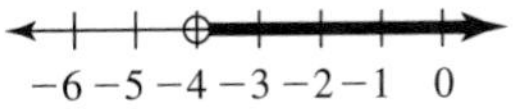
−6 −5 −4 −3 −2 −1 0
106. $x < -10$
−14 −13 −12 −11 −10 −9 −8
107. $-1 < x < 5$
−1 0 1 2 3 4 5 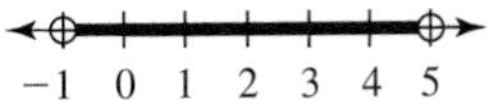
108. $0 \le x < 2$
−1 0 1 2 3
109. $-2 < x \le \frac{1}{2}$
$\frac{1}{2}$
−3 −2 −1 0 1 2 3
110. $-1 \le x < 2$
−2 −1 0 1 2 3
111. $0 \le x \le 3$ 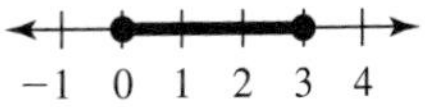
−1 0 1 2 3 4
112. $0 < x < 10$
0 2 4 6 8 10 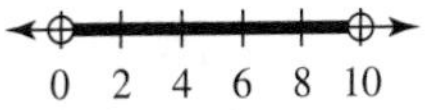
113. $0 < x < 1$
−1 0 1 2
114. $-2 < x \le 2$
−3 −2 −1 0 1 2 3
115. \$537.50 **116.** \$300 **117.** 400 **118.** \$40,000
119. 31° **120.** 70° **121.** Less than 6 feet
122. Between 20 and 25 hours per week inclusive

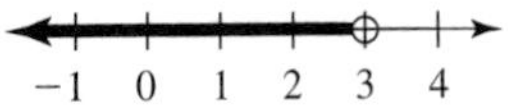

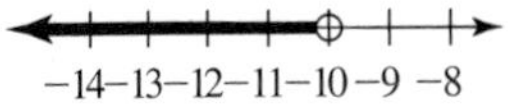

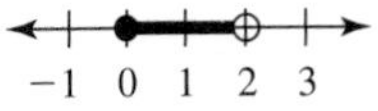

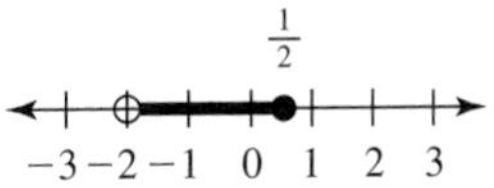

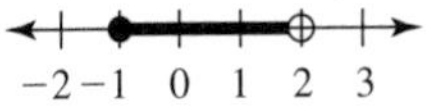

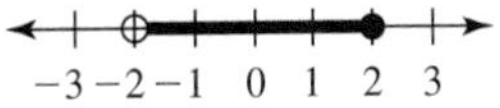

Chapter 2 Test

1. {−7} **2.** {2} **3.** {−9} **4.** {700} **5.** {1} **6.** $\left\{\frac{7}{6}\right\}$
7. {2} **8.** ∅ **9.** All real numbers **10.** $y = \frac{2}{3}x - 3$
11. $a = \frac{m + w}{P}$ **12.** $-3 < x \le 2$ **13.** $x > 1$
14. $w > 19$ 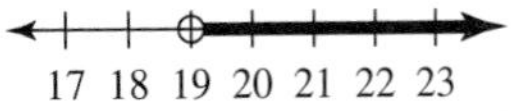
17 18 19 20 21 22 23
15. $-7 < x < -1$
−7 −6 −5 −4 −3 −2 −1 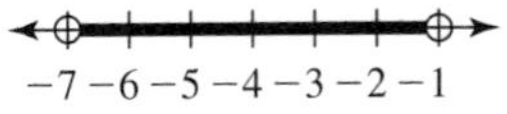
16. $1 < x < 3$
−1 0 1 2 3 4 5
17. $y > -6$ 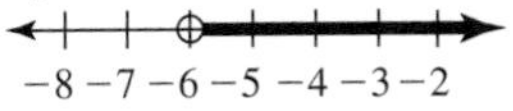
−8 −7 −6 −5 −4 −3 −2
18. 14 meters **19.** 9 in. **20.** 150 liters
21. At most \$2000 **22.** 30°, 60°, 90°

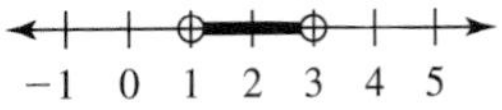

Making Connections Chapters 1–2

1. $8x$ **2.** $15x^2$ **3.** $2x + 1$ **4.** $4x - 7$ **5.** $-2x + 13$
6. 60 **7.** 72 **8.** −10 **9.** $-2x^3$ **10.** −1
11. $\frac{2}{3}$ **12.** $\frac{1}{6}$ **13.** $\frac{1}{9}$ **14.** $\frac{5}{9}$ **15.** 13 **16.** 8
17. $2x + 1$ **18.** $10x - 9$ **19.** $\left\{\frac{2}{3}\right\}$ **20.** $\left\{\frac{1}{6}\right\}$
21. $\left\{\frac{1}{9}\right\}$ **22.** $\left\{\frac{5}{9}\right\}$ **23.** $\left\{\frac{3}{10}\right\}$ **24.** $\left\{\frac{16}{5}\right\}$

25. $\left\{\frac{1}{2}\right\}$ **26.** $\left\{\frac{7}{5}\right\}$ **27.** {1} **28.** All real numbers

29. {0} **30.** {1} **31.** {2} **32.** {2} **33.** $\left\{\frac{13}{2}\right\}$

34. {200} **35. a)** \$13,600 **b)** \$10,000 **c)** \$12,000

Chapter 3

Section 3.1 Warm-ups F F F F T T T F F T

1. An ordered pair is a pair of numbers in which there is a first number and a second number, usually written as (a, b).
2. The rectangular coordinate system is a means of dividing up the plane with two number lines in order to picture all ordered pairs of real numbers.
3. The origin is the point of intersection of the x-axis and y-axis.
4. The graph of an equation is a picture of all ordered pairs that satisfy the equation drawn in the rectangular coordinate system.
5. A linear equation in two variables is an equation of the form $Ax + By = C$, where A and B are not both zero.
6. Intercepts are the points at which a graph crosses the axes.

7. (0, 9), (5, 24), (2, 15) **8.** (8, 21), (−1, 3), (−3, −1)

9. $(0, -7), \left(\frac{1}{3}, -8\right), \left(-\frac{2}{3}, -5\right)$

10. $(-1, 2), \left(-\frac{1}{2}, -\frac{1}{2}\right), \left(-\frac{1}{5}, -2\right)$

11. (0, 5), (10, −115), (−1, 17)

12. (1, 218), (−10, 20), (0, 200) **13.** (3, 0), (0, −2), (12, 6)

14. (−5, 3), (5, −3), (10, −6) **15.** (5, −3), (5, 5), (5, 0)

16. (3, −6), (−1, −6), (4, −6)

17–31 odd

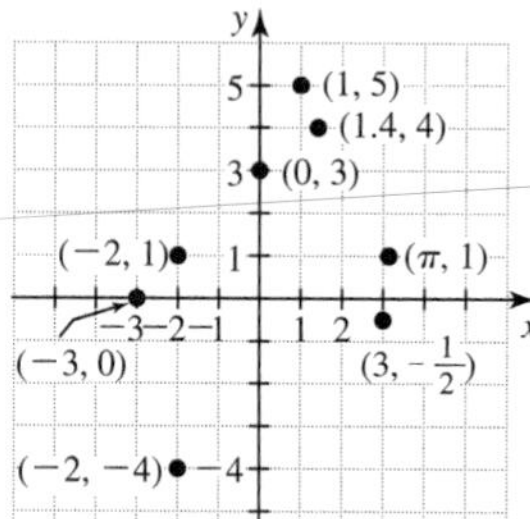

18–32 even

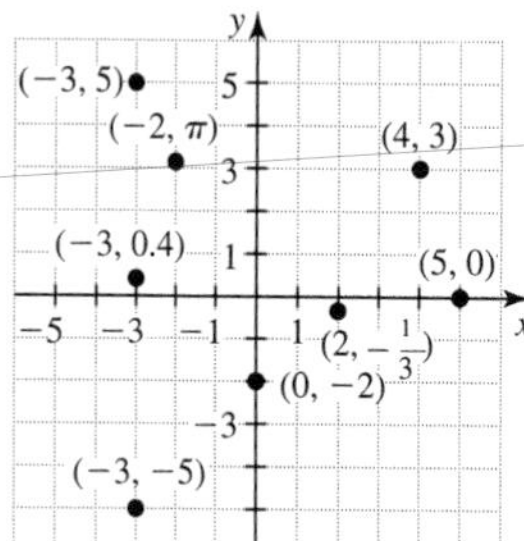

33. (−2, 9), (0, 5), (2, 1), (4, −3), (6, −7)

34. (−2, 6), (0, 4), (2, 2), (4, 0), (6, −2)

35. (−6, 0), (−3, 1), (0, 2), (3, 3)

36. $(-2, 2), \left(-1, \frac{3}{2}\right), (0, 1), \left(1, \frac{1}{2}\right)$

37.

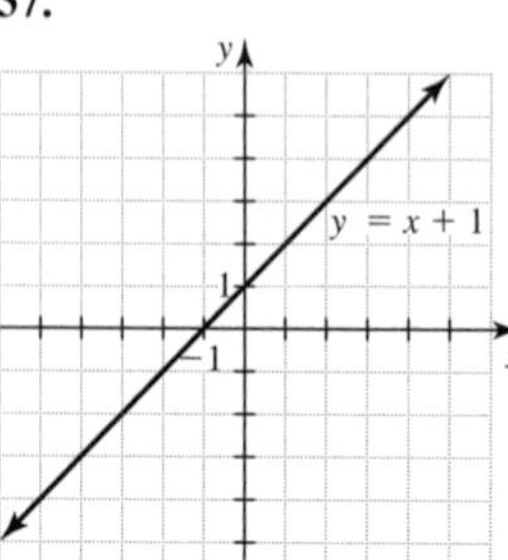

38.

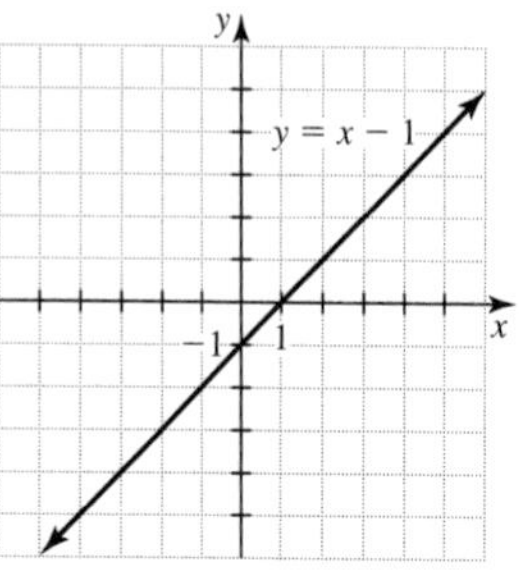

39.

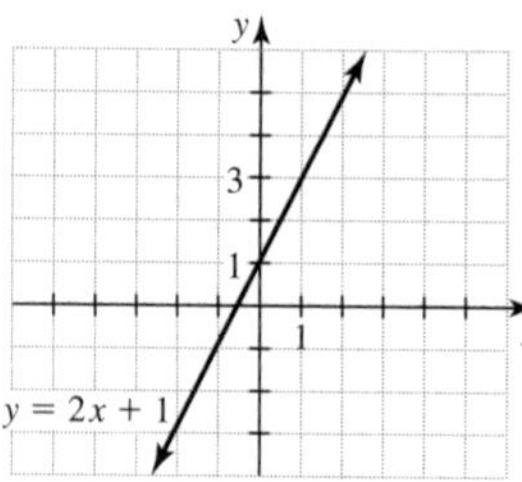

40.

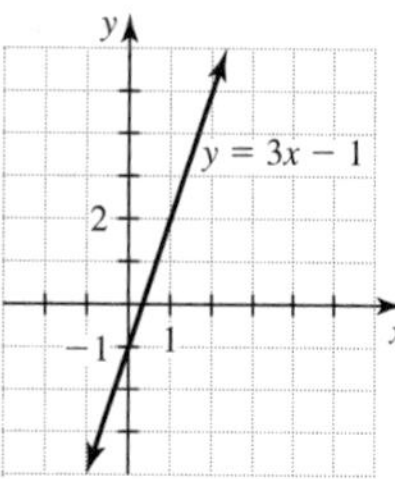

41.

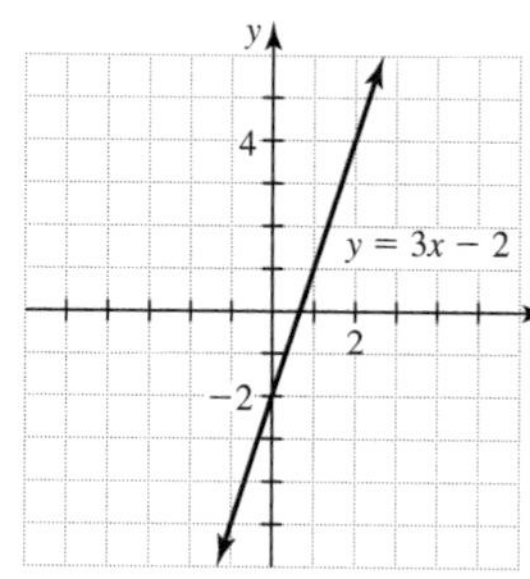

42.

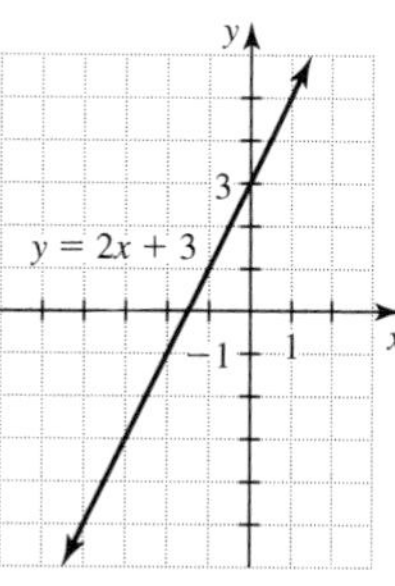

43.

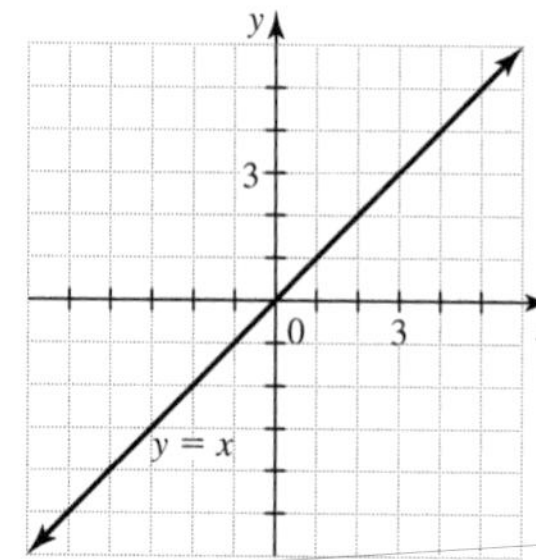

44.

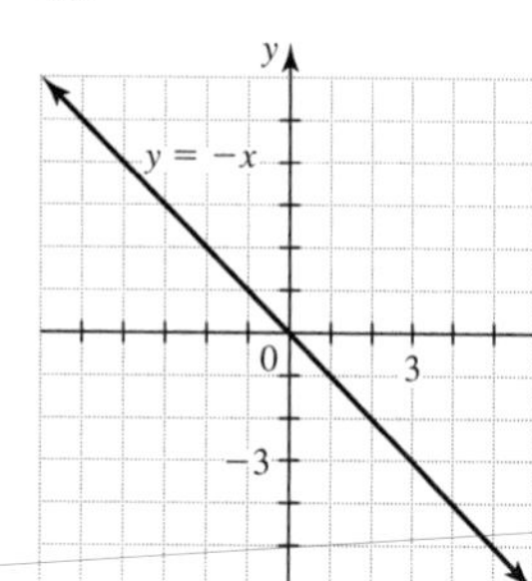

45.

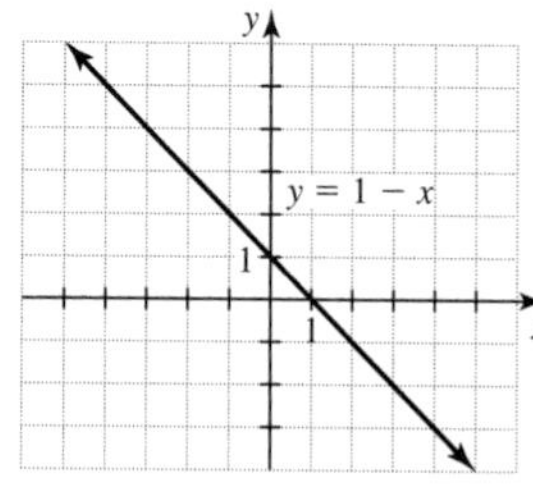

46.

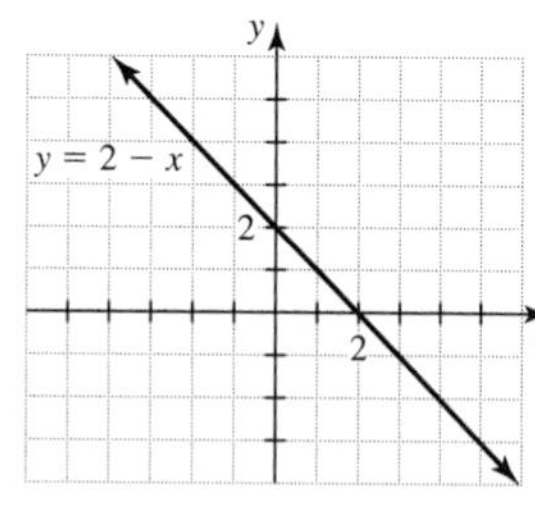

47.

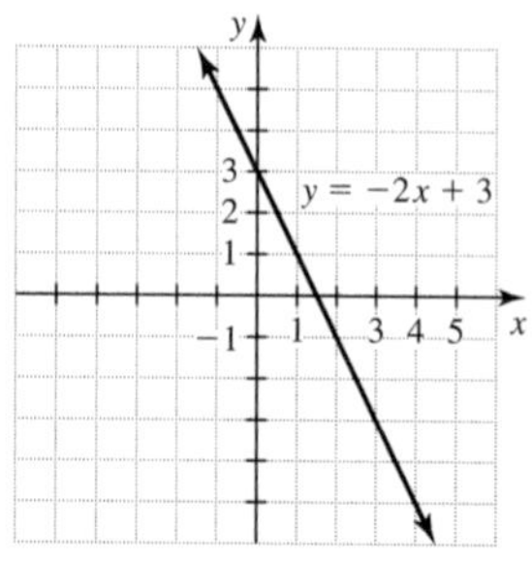

48.

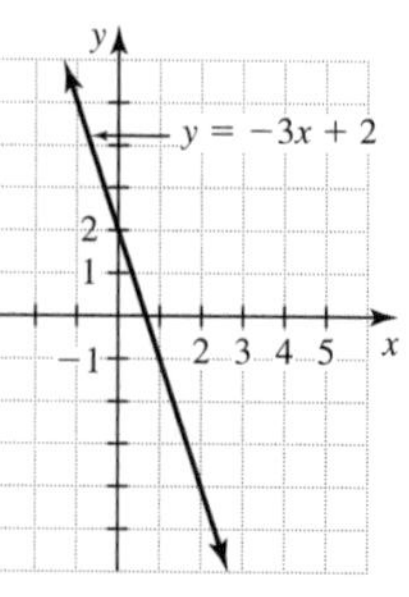

49.

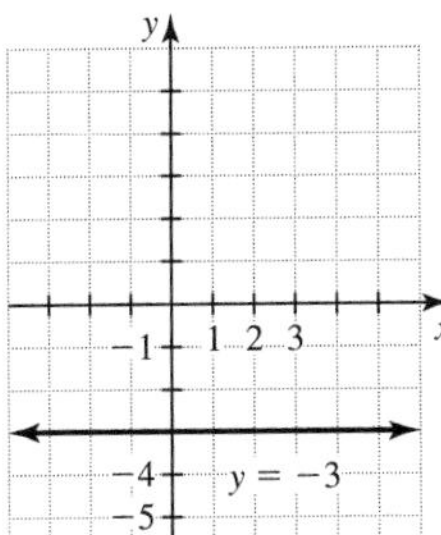

50.

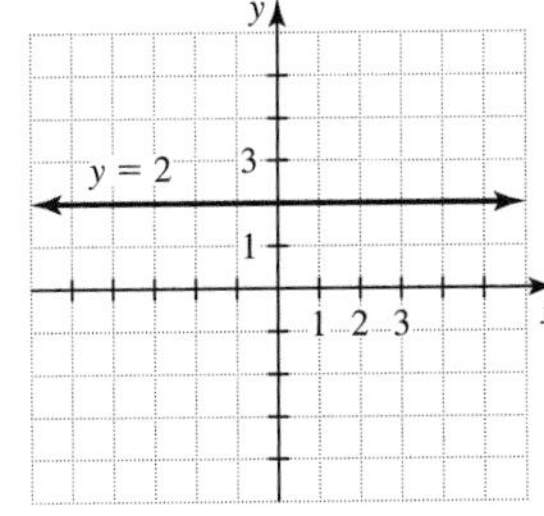

59.

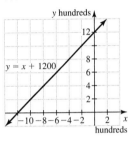

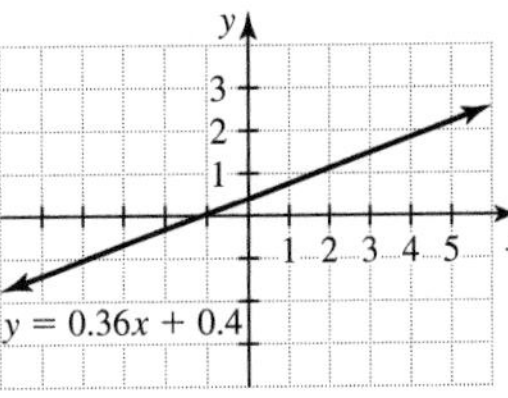

60.

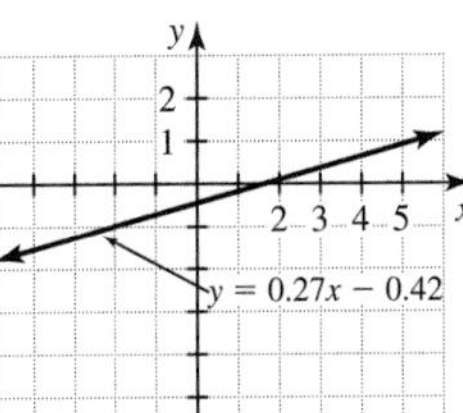

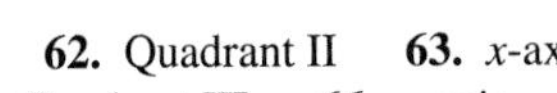

61. Quadrant II **62.** Quadrant II **63.** x-axis
64. y-axis **65.** Quadrant III **66.** x-axis
67. Quadrant I **68.** Quadrant I **69.** Quadrant II
70. Quadrant III **71.** y-axis **72.** x-axis

51.

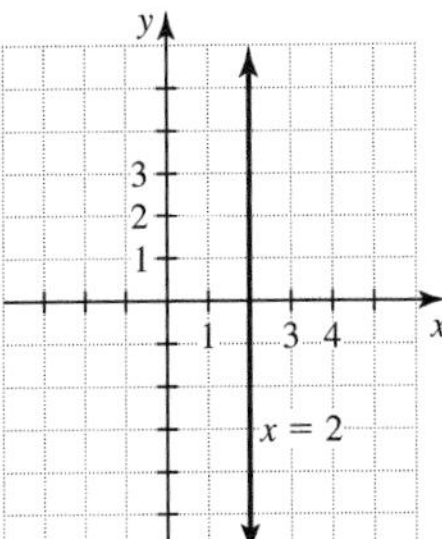

52.

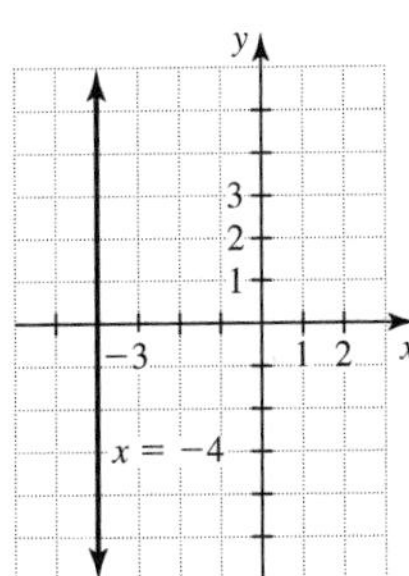

73.

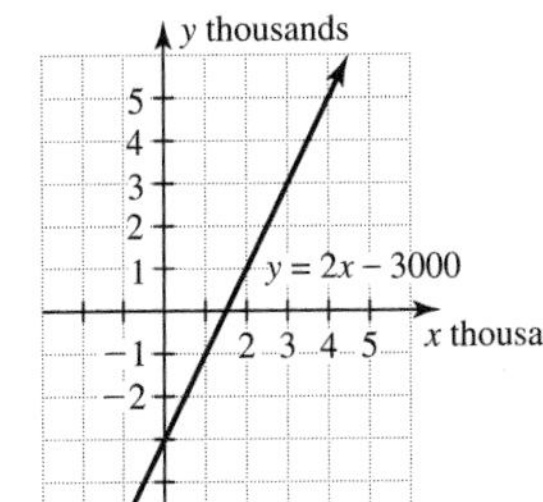

74.

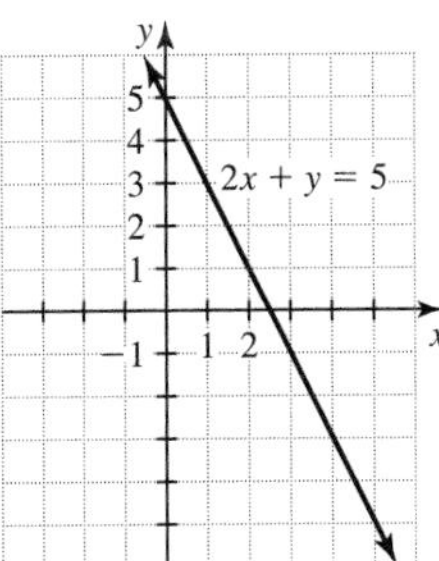

53.

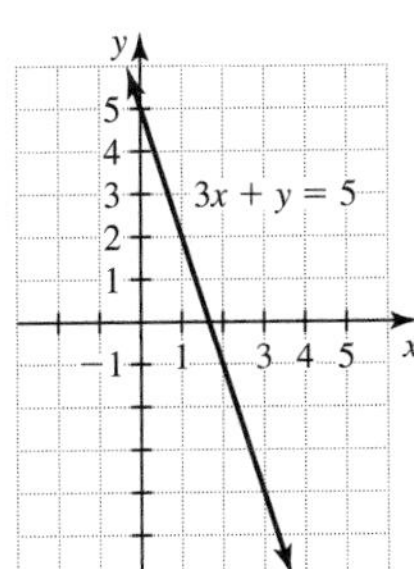

54.

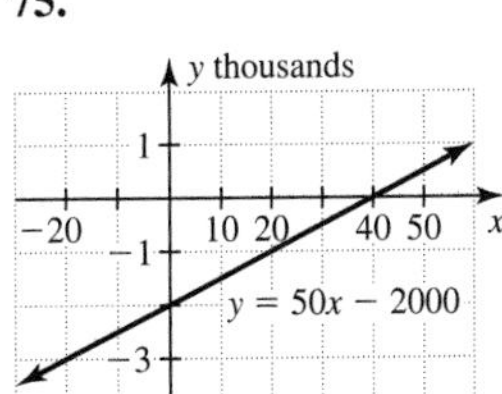

75.

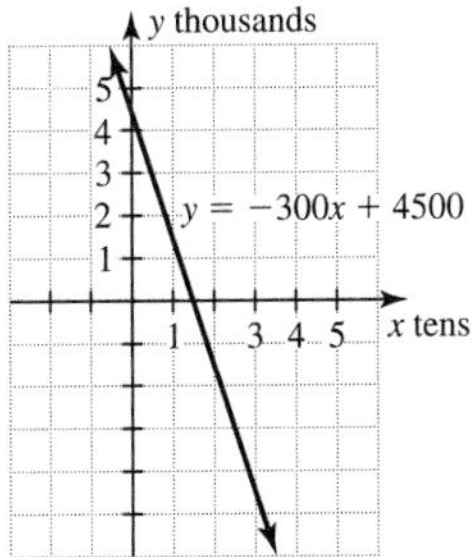

76.

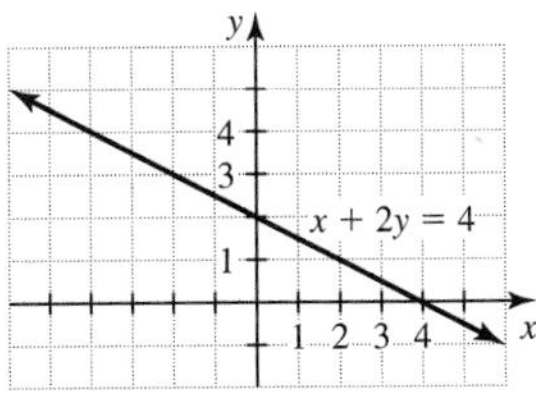

55.

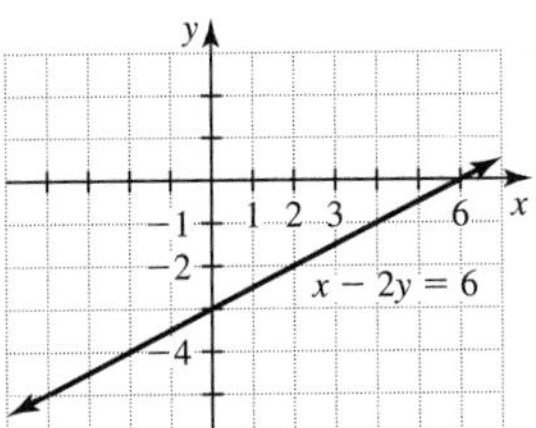

56.

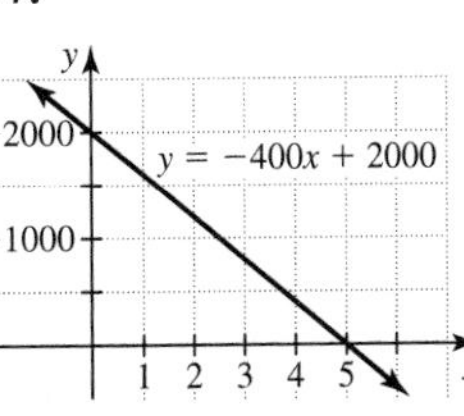

77.

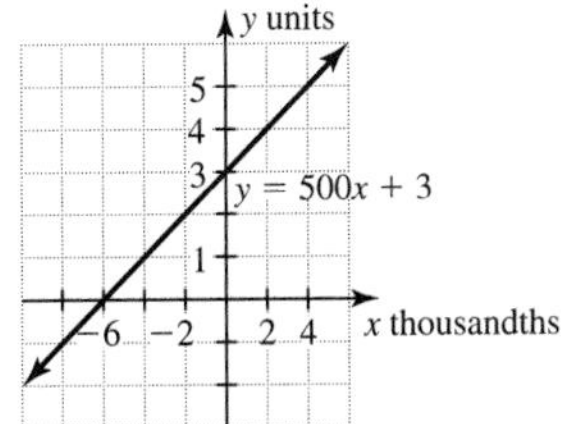

78.

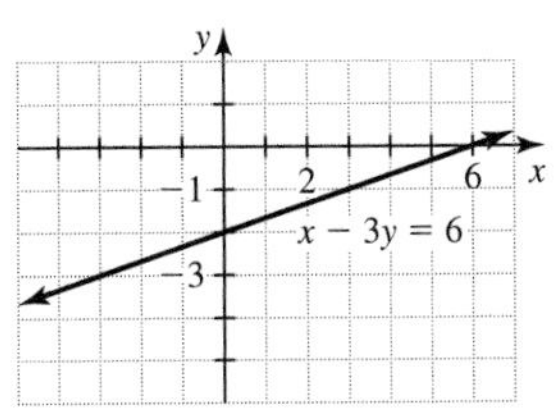

57.

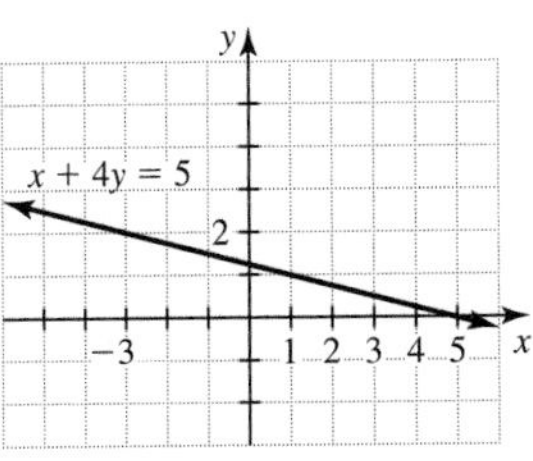

58.

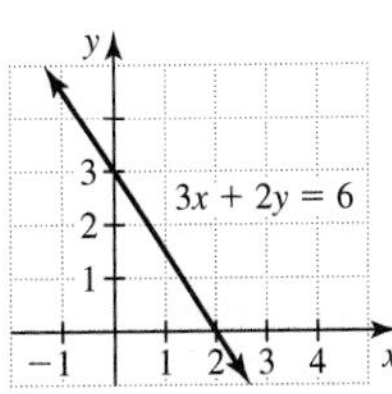

79.

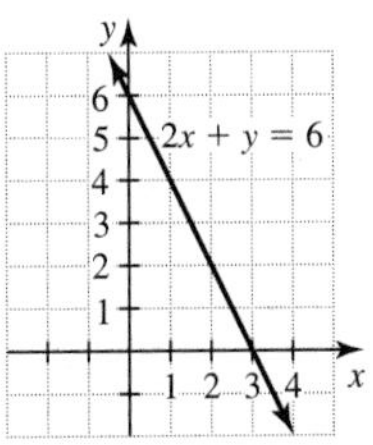

80.

2x + y = 6

81.

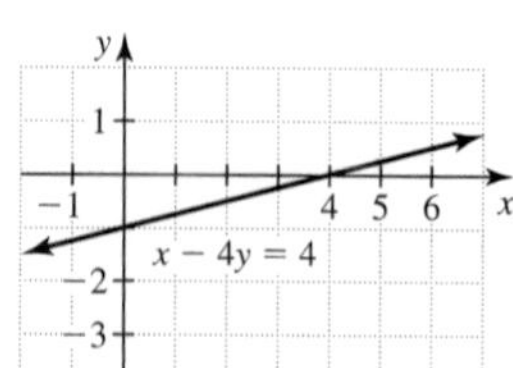

82.

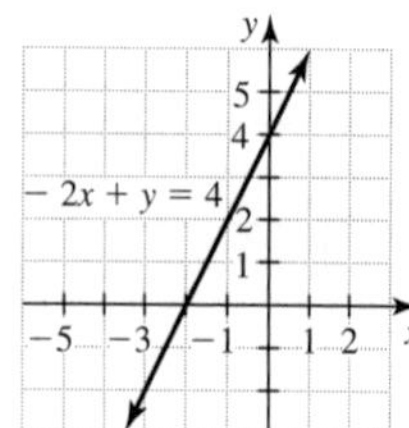

83.

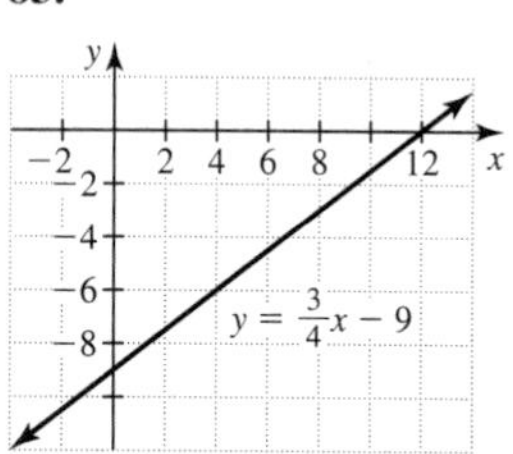

84.

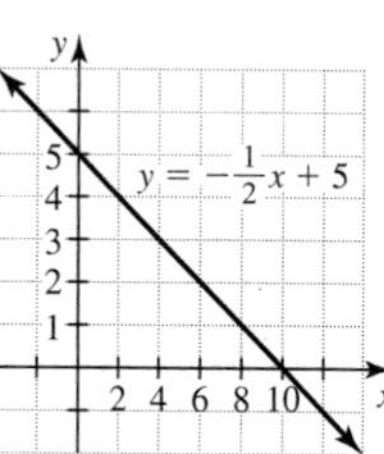

85.

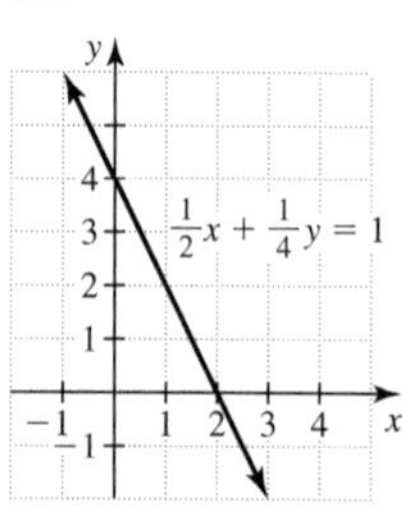

86.

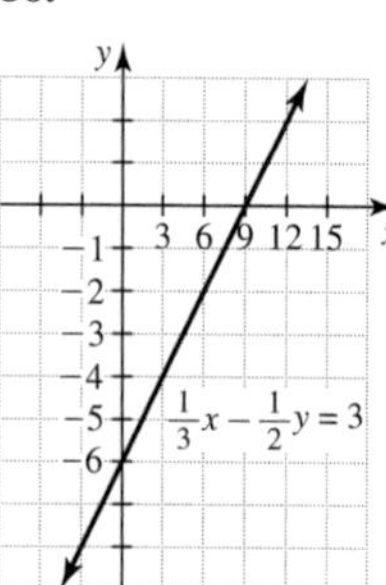

87. 75%, 67, 68 and up

88.

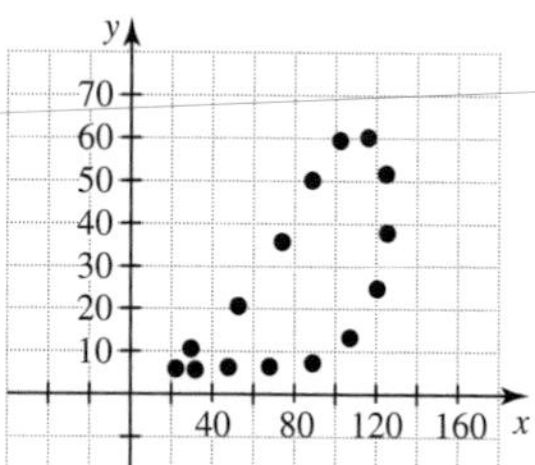

89. a) \$97.3 billion **b)** 2016

c)

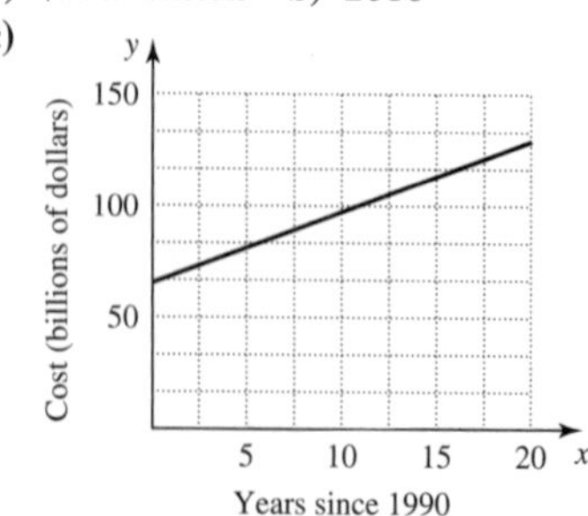

90. a) (0, 30.52); The cost was \$30.52 billion in 1990.
b) (−10.71, 0); The cost was zero dollars in 1979.

c)

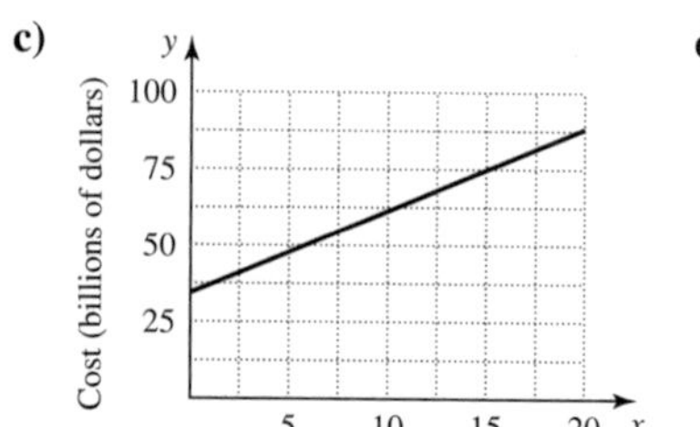

d) 2014

91. a) 4 atm **b)** 130 ft

c)

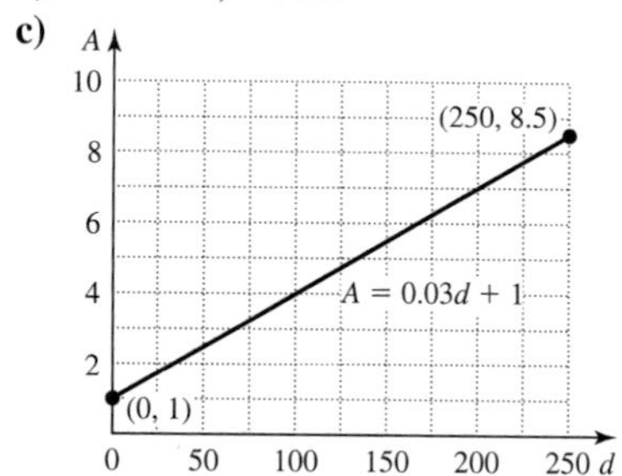

92. a) Less **b)** Goes down by 40 cans

c)

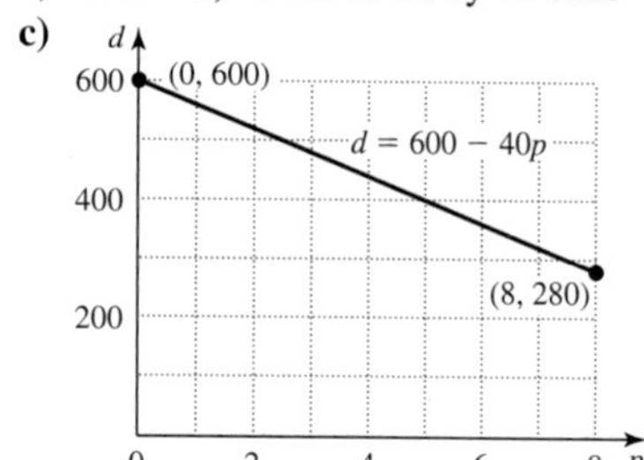

d) \$14.97

93. x = the number of radio ads,
y = the number of TV ads, 21 solutions

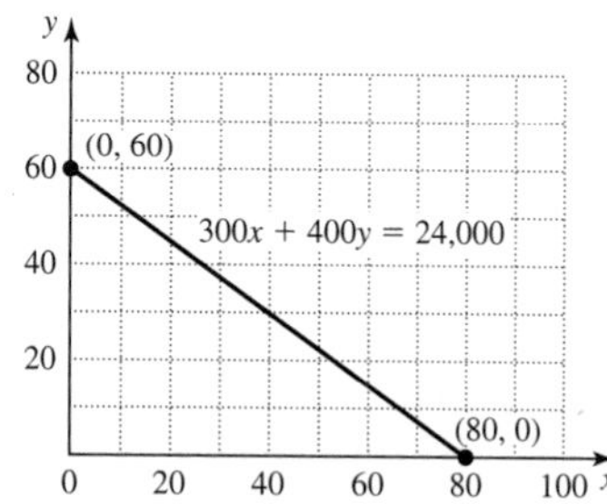

94. x = the number of 8×10 tents,
y = the number of 9×12 tents, 11 solutions

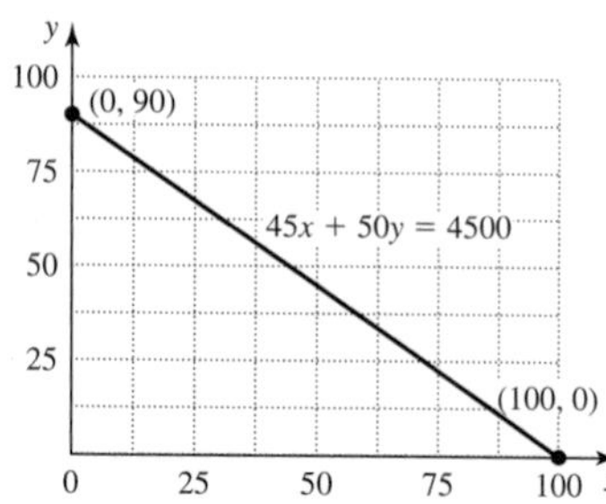

95.

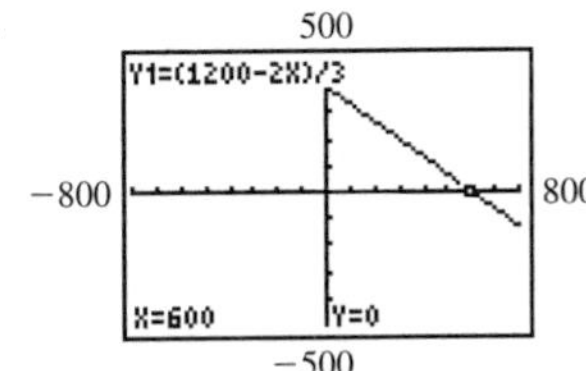

96.

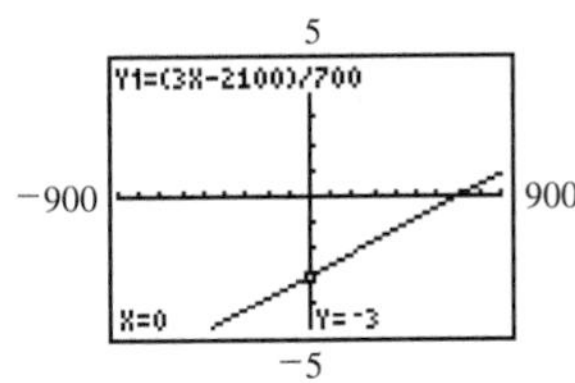

97.

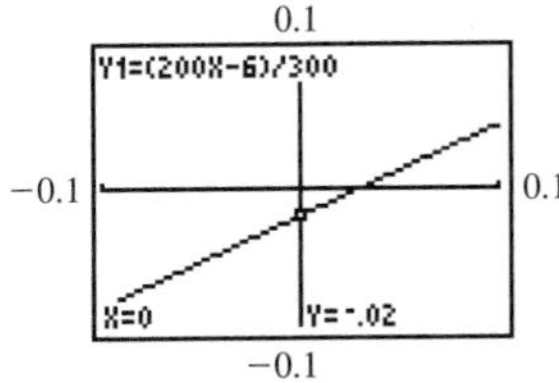

98.

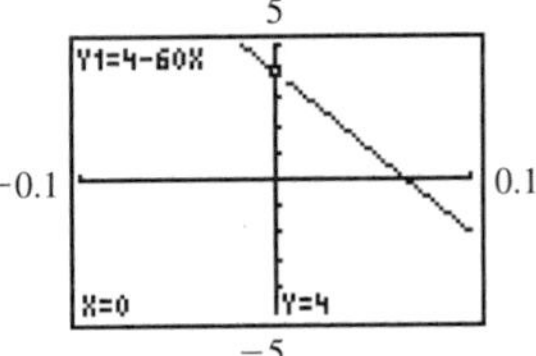

99.

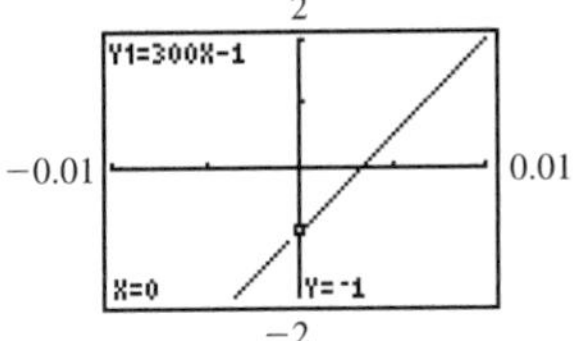

100.

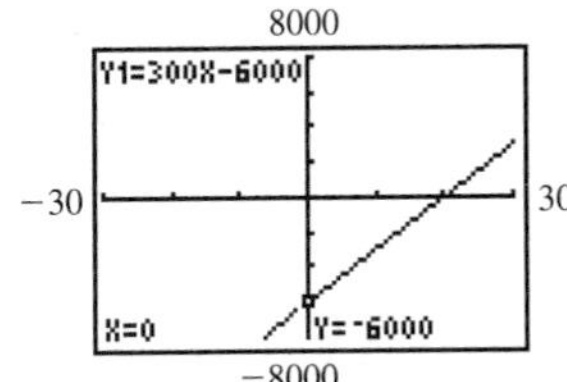

Section 3.2 Warm-ups T T F T F F F F T T

1. The slope of a line is the ratio of its rise and run.
2. Rise is the amount of vertical change and run is the amount of horizontal change.
3. Slope is undefined for vertical lines.
4. Horizontal lines have zero slope.
5. Lines with positive slope are rising as you go from left to right, while lines with negative slope are falling as you go from left to right.
6. If m_1 and m_2 are slopes of perpendicular lines, then $m_1 = \frac{-1}{m_2}$.

7. $-\frac{2}{3}$ **8.** -2 **9.** $\frac{2}{3}$ **10.** 2 **11.** $\frac{3}{2}$ **12.** $-\frac{3}{2}$
13. 0 **14.** 0 **15.** $\frac{2}{5}$ **16.** $-\frac{3}{5}$ **17.** Undefined
18. Undefined **19.** 2 **20.** $\frac{5}{4}$ **21.** $-\frac{5}{3}$ **22.** -1
23. $\frac{5}{7}$ **24.** $\frac{1}{2}$ **25.** $-\frac{4}{3}$ **26.** $\frac{4}{5}$ **27.** -1
28. $-\frac{5}{6}$ **29.** 1 **30.** $\frac{3}{2}$ **31.** Undefined **32.** 0
33. 0 **34.** Undefined **35.** 3 **36.** 1

37.

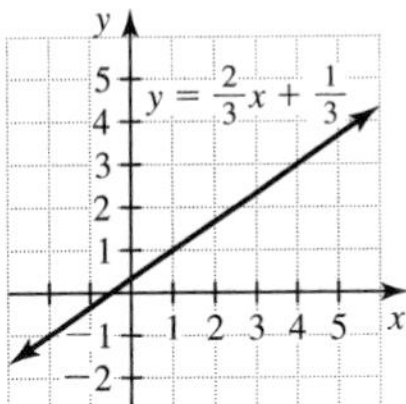

38.

$y = \frac{1}{2}x + 2$

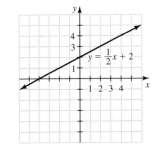

39.

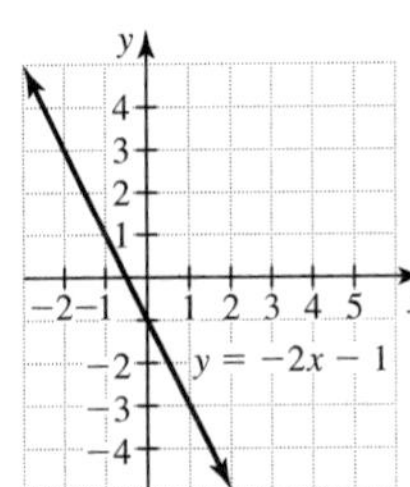

40.

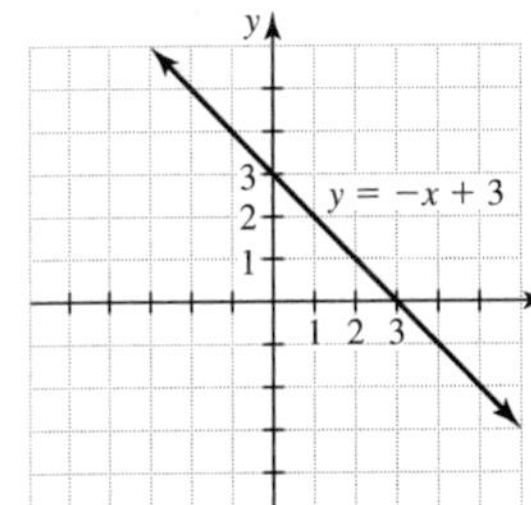

41.

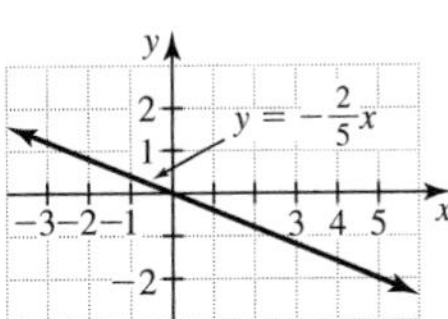

42.

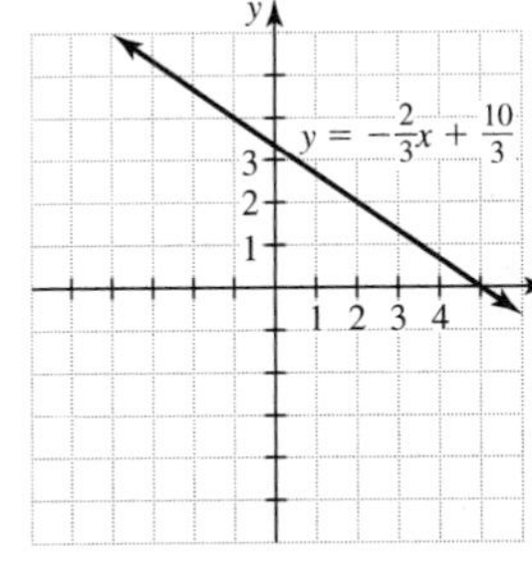

43.

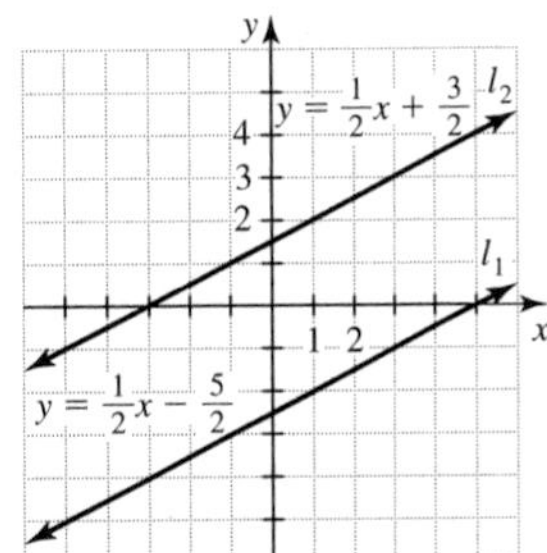

44.

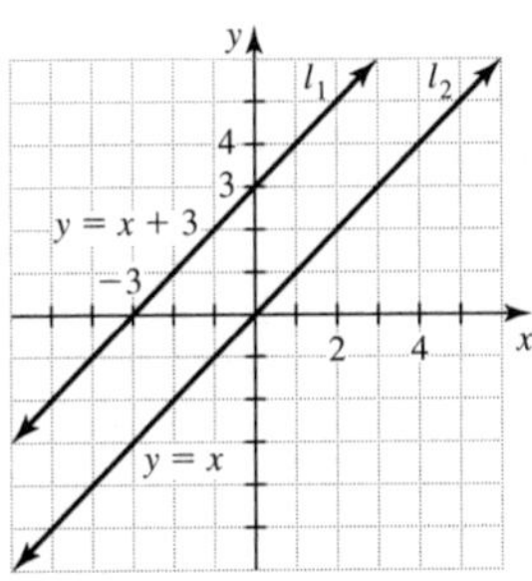

45.

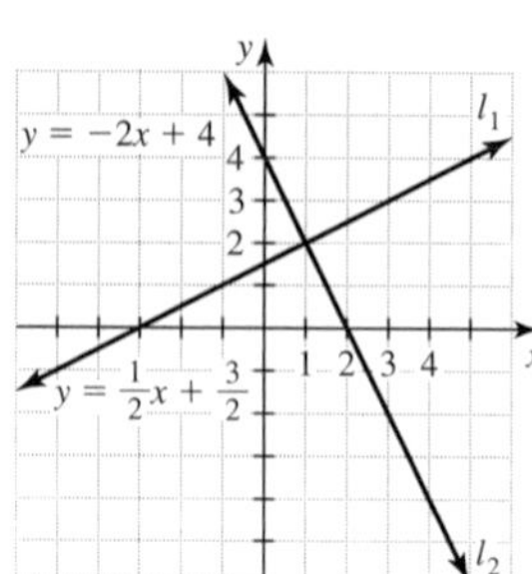

46.

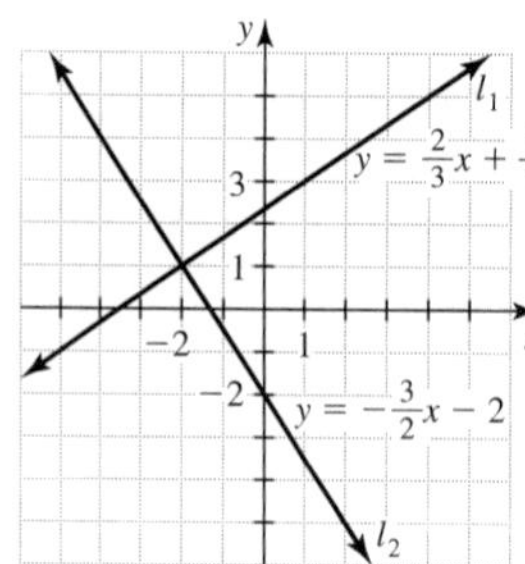

47. $-\frac{4}{3}$

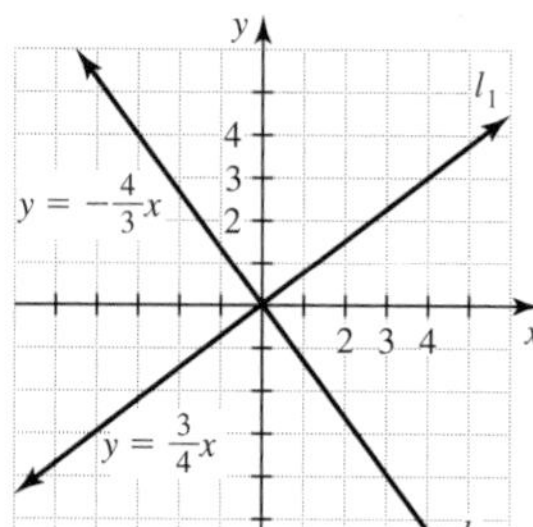

48. 1

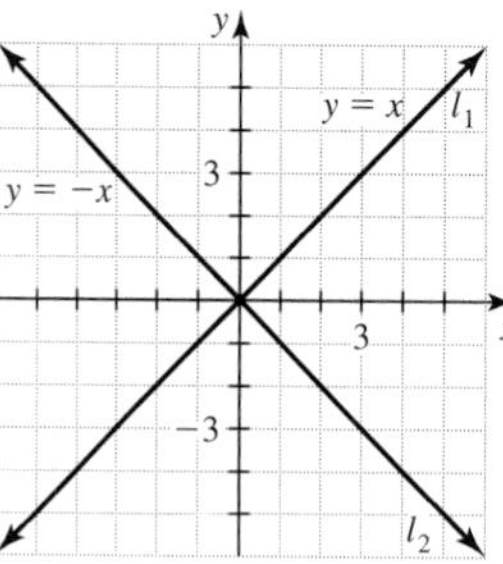

49. $\frac{1}{2}$

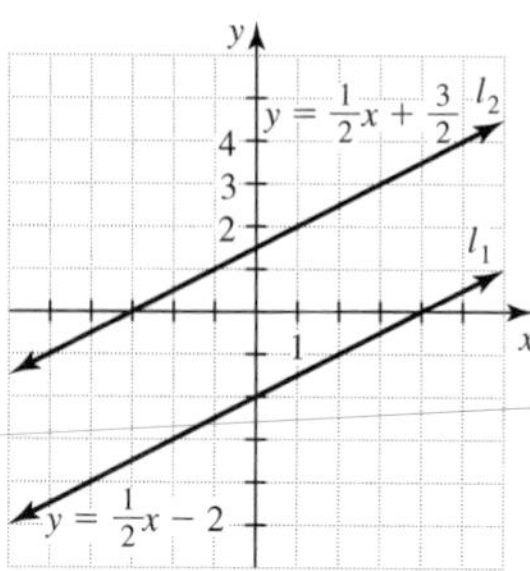

50. $\frac{3}{2}$

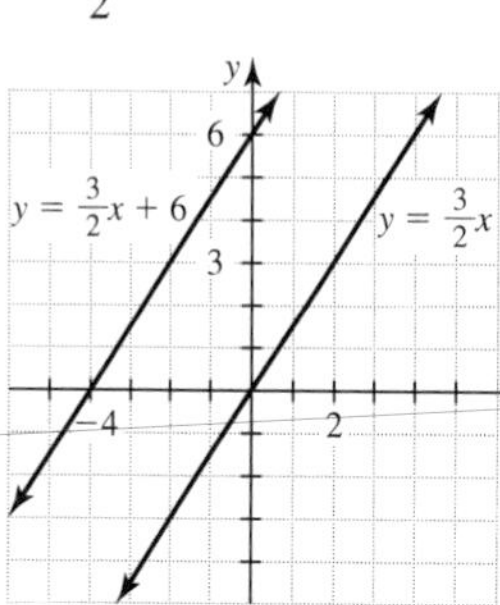

51. 1

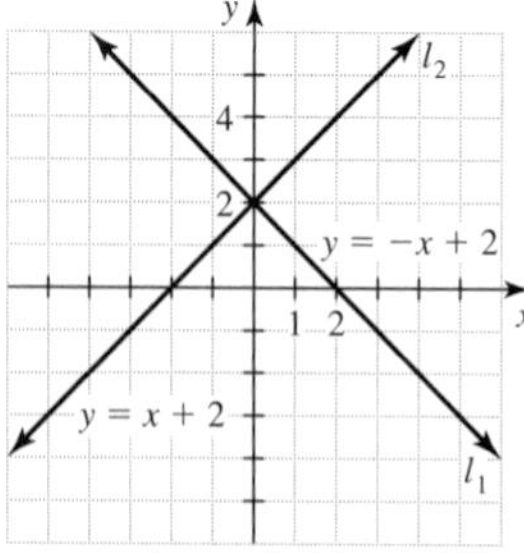

52. -1

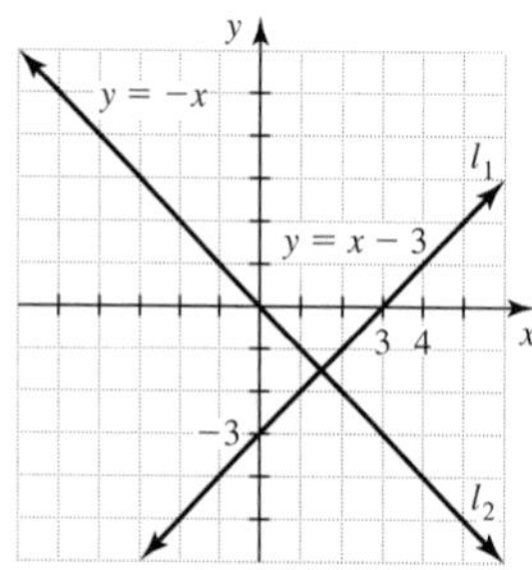

53. **a)** 0.175 slope; Cost is increasing \$175,000 per year. **b)** \$1.65 million, yes **c)** \$3.05 million

54. \$8000, 69, 500, 666.66, 800; The slopes are the yearly increases for each segment.

55. 1 slope; The percentage increases 1% per year.

56. **a)**

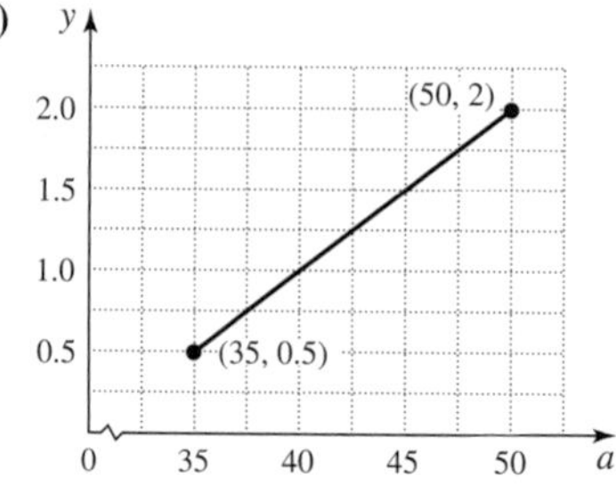

b) 0.1 **c)** 10%

57. (2000, 28,100), (2003, 29,300), (2012, 32,900), (2015, 34,100)

58. (2001, 8400), (2008, 6650), (2011, 5900), (2015, 4900)

59. Yes **60.** Yes **61.** No **62.** No

Section 3.3 Warm-ups T F T T T F F T T F

1. Slope-intercept form is $y = mx + b$.
2. The slope is m and the y-intercept is $(0, b)$.
3. The standard form is $Ax + By = C$.
4. From slope-intercept form, locate the intercept and a second point by counting the rise and run from the y-intercept.
5. The slope-intercept form allows us to write the equation form the y-intercept and the slope.
6. Lines with slopes m and $\frac{-1}{m}$ look perpendicular only if the same unit distance is used on both axes.
7. $y = \frac{3}{2}x + 1$ **8.** $y = -2x + 3$ **9.** $y = -2x + 2$
10. $y = \frac{1}{2}x - 1$ **11.** $y = x - 2$ **12.** $y = -3x - 3$
13. $y = -x$ **14.** $y = 2$ **15.** $y = -1$ **16.** $x = 1$
17. $x = -2$ **18.** $y = x$ **19.** 3, $(0, -9)$
20. -5, $(0, 4)$ **21.** 0, $(0, 4)$ **22.** 0, $(0, -5)$
23. -3, $(0, 0)$ **24.** 2, $(0, 0)$ **25.** -1, $(0, 5)$
26. 1, $(0, -4)$ **27.** $\frac{1}{2}$, $(0, -2)$ **28.** $-\frac{1}{2}$, $\left(0, \frac{3}{2}\right)$
29. $\frac{2}{5}$, $(0, -2)$ **30.** $-\frac{2}{3}$, $(0, 3)$ **31.** 2, $(0, 3)$
32. $\frac{3}{4}$, $(0, -2)$ **33.** Undefined slope, no y-intercept
34. Undefined slope, no y-intercept **35.** $x + y = 2$
36. $3x - y = 5$ **37.** $x - 2y = -6$ **38.** $2x - 3y = 12$
39. $9x - 6y = 2$ **40.** $12x - 15y = -10$
41. $6x + 10y = 7$ **42.** $4x + 6y = -5$ **43.** $x = -10$
44. $x = 18$ **45.** $3y = 10$ **46.** $6y = 1$
47. $5x - 6y = 0$ **48.** $5x + 32y = 0$
49. $x - 50y = -25$ **50.** $20x - 3y = -10$

51.

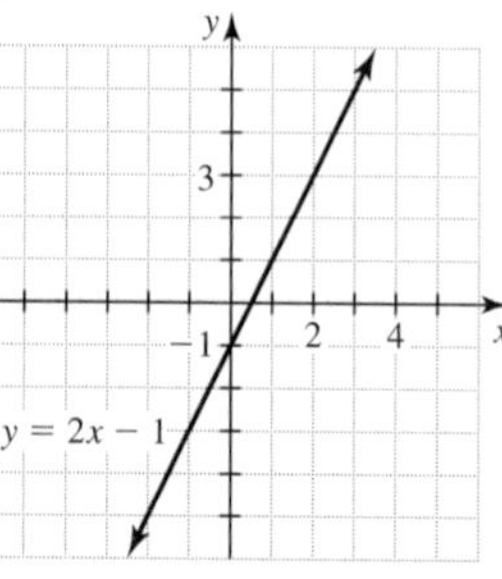

52.

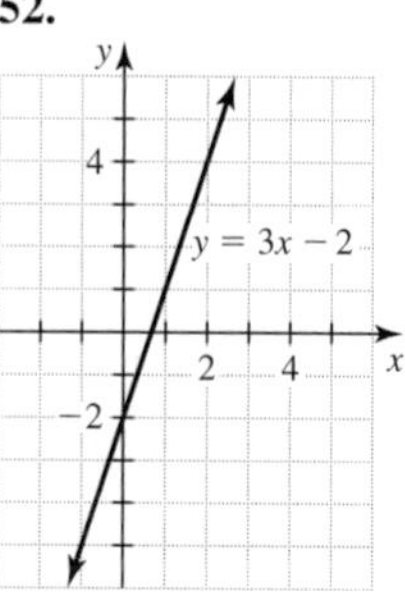

53.

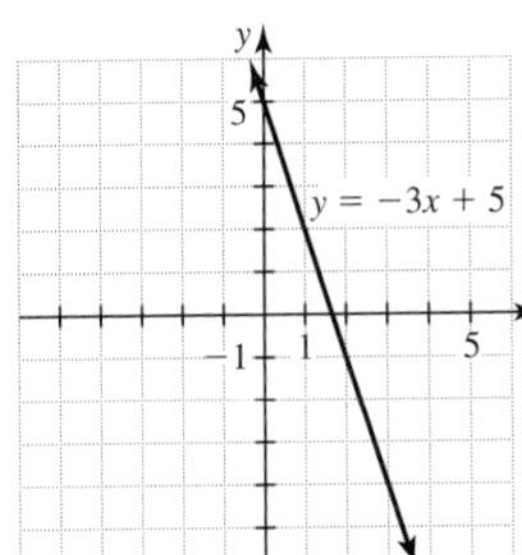

54.

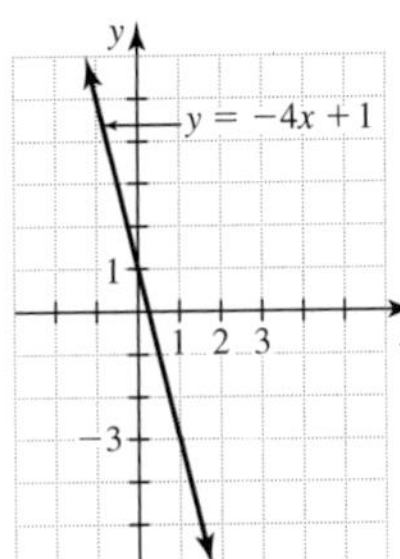

55.

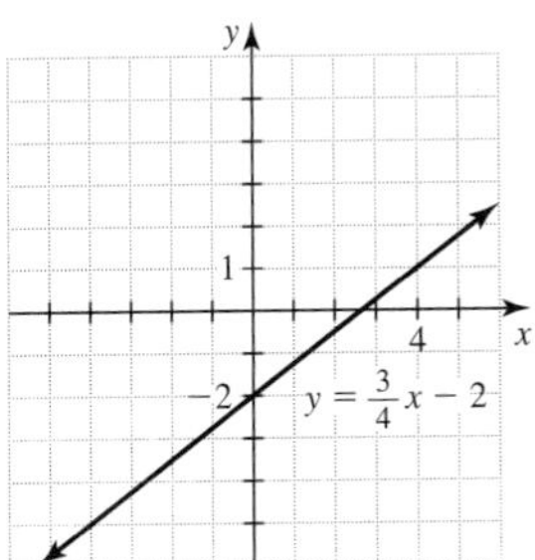

56.

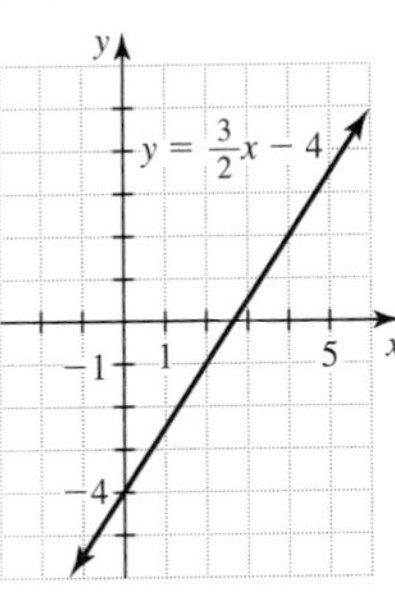

57.

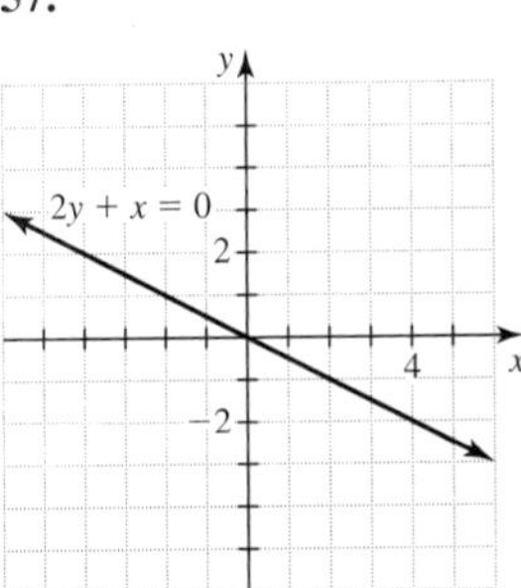

58.

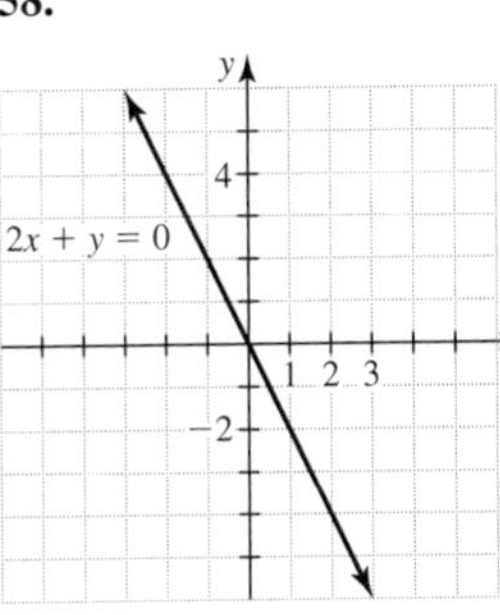

59.

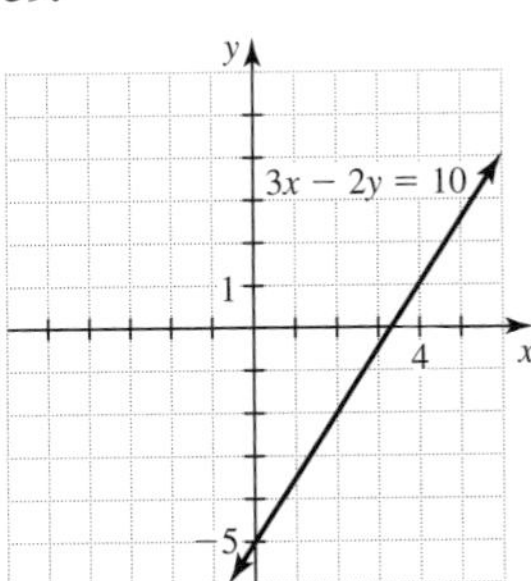

60.

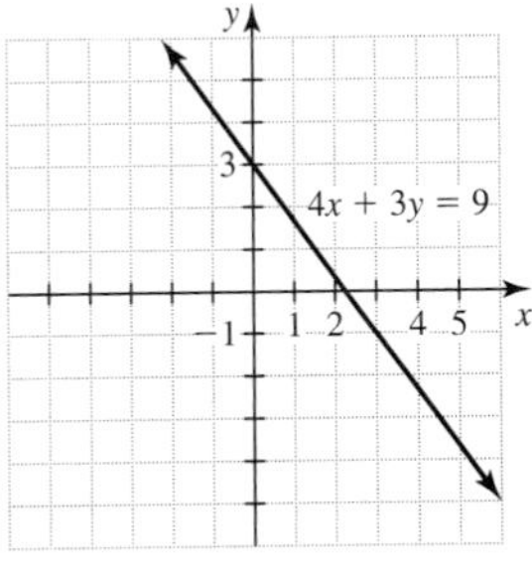

61.

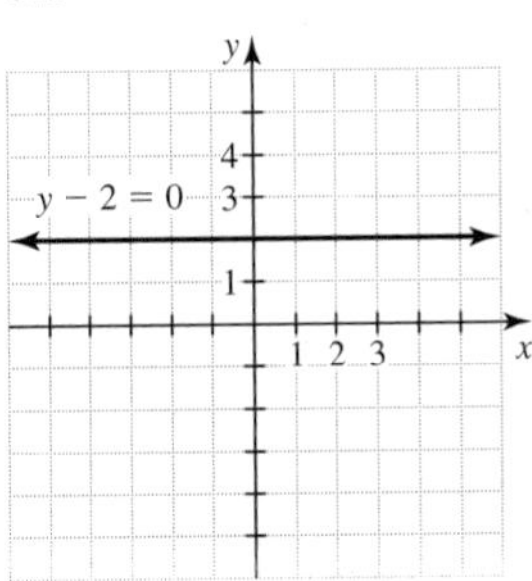

62.

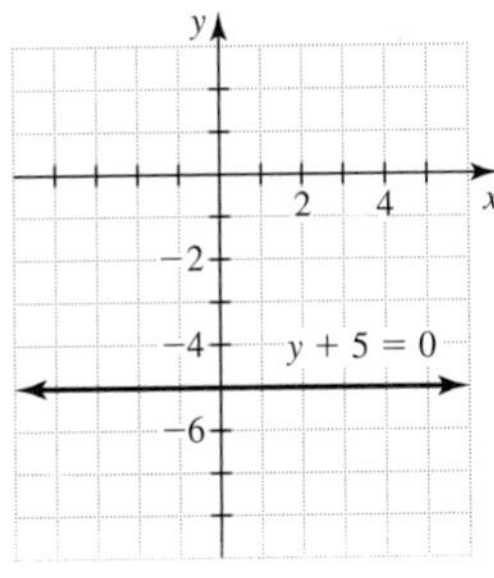

63. $y = \frac{1}{2}x - 4$ **64.** $y = -\frac{1}{2}x + 4$ **65.** $y = 2x + 3$

66. $y = -\frac{1}{3}x - 2$ **67.** $y = -\frac{1}{3}x + 6$ **68.** $y = -x - 1$

69. $y = -2x + 3$ **70.** $y = 3x$ **71.** $y = 3$

72. $x = -3$ **73.** $y = -\frac{3}{2}x + 4$ **74.** $y = -\frac{5}{2}x - 1$

75. $y = -\frac{4}{5}x + 4$ **76.** $y = \frac{3}{4}x - 3$

77. \$1,150,000, \$1,150,200, \$200

78. \$16,000, \$16,120, \$120

79. a) A slope of 1 means that the percentage of workers receiving training is going up 1% per year.
b) $y = x + 5$, where x is the number of years since 1982.
c) The y-intercept (0, 5) means that 5% of the workers received training in 1982. **d)** 33%

80. a) A slope of 0.6 means that the percentage is increasing by 0.6% per year.
b) $y = 0.6x + 55$
c) The y-intercept (0, 55) means that in 1970 55% of the women between 20 and 24 had never been married.
d) 79% **e)** 2045

81. a) $x =$ the number of packs of pansies, $y =$ the number of packs of snapdragons
b)

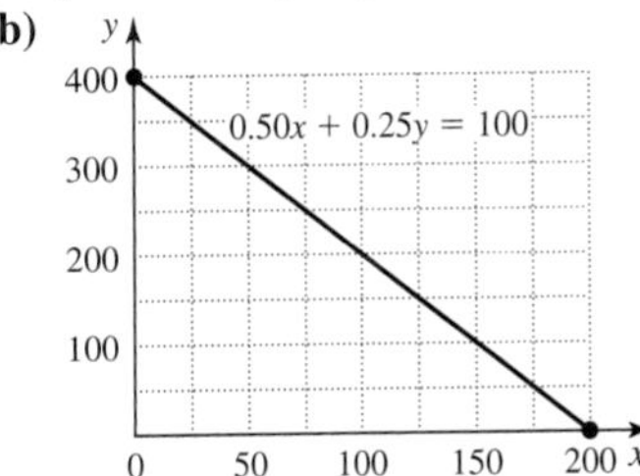

c) $y = -2x + 400$ **d)** -2
e) If the number of packs of pansies goes up by 1 then the number of packs of snapdragons goes down by 2.

82. a) $x =$ the number of pencils, $y =$ the number of pens
b)

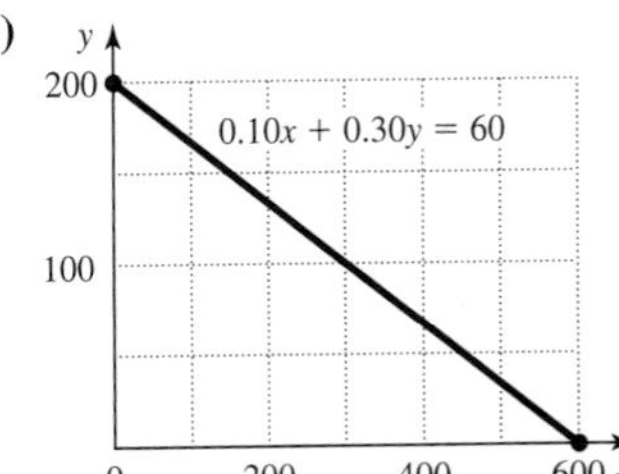

c) $y = -\frac{1}{3}x + 200$ **d)** $-\frac{1}{3}$
e) If the number of pencils increases by 3 then the number of pens goes down by 1.

83.

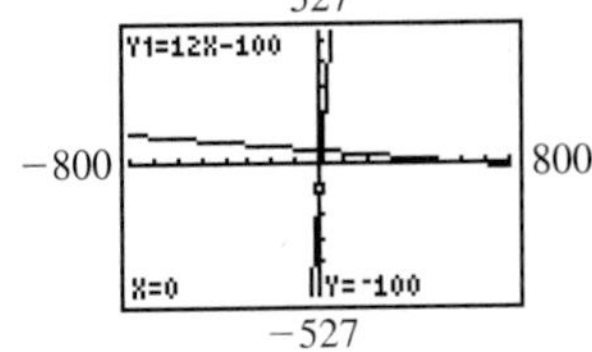

84.

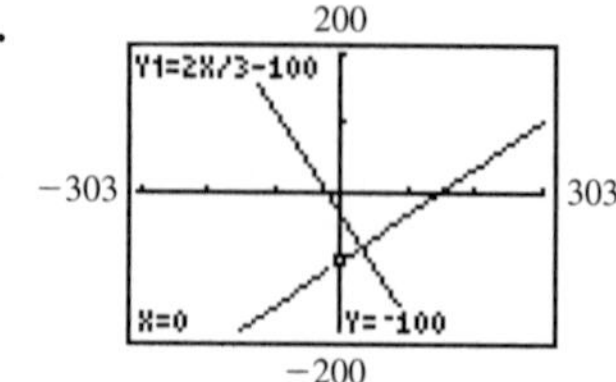

Section 3.4 Warm-ups F F T T F T T T T T

1. Point-slope form is $y - y_1 = m(x - x_1)$.
2. If we know any point and the slope of a line we can use point-slope form to write the equation.
3. If you know two points on a line, find the slope. Then use it along with a point in point-slope form to write the equation of the line.
4. Rewrite any equation in slope-intercept form to find the slope of the line.
5. Nonvertical parallel lines have equal slopes.
6. If lines with slopes m_1 and m_2 are perpendicular, then $m_1 = \frac{-1}{m_2}$.

7. $y = 5x + 11$ **8.** $y = -3x + 15$ **9.** $y = \frac{3}{4}x - 20$
10. $y = -\frac{2}{3}x + 30$ **11.** $y = \frac{2}{3}x + \frac{1}{3}$
12. $y = -\frac{1}{2}x - \frac{7}{15}$ **13.** $y = 3x - 1$ **14.** $y = 4x - 3$
15. $y = \frac{1}{2}x + 3$ **16.** $y = \frac{1}{2}x + 4$ **17.** $y = \frac{1}{3}x + \frac{7}{3}$
18. $y = \frac{1}{4}x + \frac{15}{4}$ **19.** $y = -\frac{1}{2}x + 4$ **20.** $y = -\frac{1}{3}x$
21. $y = -6x - 13$ **22.** $y = -8x - 13$
23. $2x - y = 7$ **24.** $3x + y = 1$ **25.** $x - 2y = 6$
26. $x - 3y = -15$ **27.** $2x - 3y = 2$
28. $3x - 2y = -10$ **29.** $2x - y = -1$ **30.** $4x - y = 3$
31. $x - y = 0$ **32.** $x + y = 0$
33. $3x - 2y = -1$ **34.** $2x - y = 1$
35. $3x + 5y = -11$ **36.** $2x - 3y = 7$ **37.** $x - y = -2$
38. $3x + 5y = 15$ **39.** $y = -x + 4$ **40.** $y = -x - 4$
41. $y = \frac{5}{3}x - 1$ **42.** $y = \frac{4}{3}x - 4$ **43.** $y = -\frac{1}{3}x + 5$
44. $y = -\frac{1}{2}x + 2$ **45.** $y = x + 3$ **46.** $y = -x + 3$
47. $y = -\frac{2}{3}x + \frac{5}{3}$ **48.** $y = -5x - 4$
49. $y = -2x - 5$ **50.** $y = \frac{3}{2}x + 4$ **51.** $y = \frac{1}{3}x + \frac{7}{3}$
52. $y = -2x + 4$ **53.** $y = 2x - 1$ **54.** $y = \frac{1}{2}x + \frac{7}{2}$
55. $y = 2$ **56.** $x = 3$ **57.** $y = \frac{2}{3}x$ **58.** $y = -\frac{3}{2}x$
59. $y = -x$ **60.** $y = x$ **61.** $y = 50$ **62.** $y = -40$
63. $y = -\frac{3}{5}x - 4$ **64.** $y = \frac{2}{3}x - 1$

65. a) Slope 0.9 means that the number of ATM transactions is increasing by 0.9 billion per year.
b) $y = 0.9x + 10.6$ **c)** 23.2 billion
66. a) $y = 3x + 20$ **b)** About 2017 **c)** 2017
67. a) $y = 1.6x + 53.1$
b) x = years since 1990, y = GDP in thousands of dollars
c) \$77,100
d)

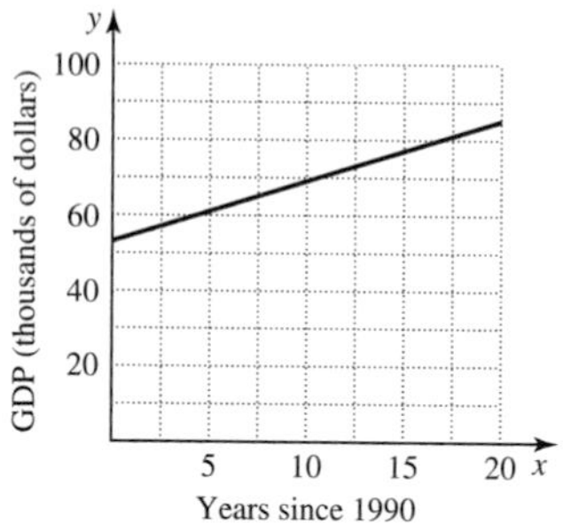

68. a) $y = 0.12x + 23.9$
b) x = the number of years since 1990, y = median age at first marriage
c) Median age increases 0.12 year each year or approximately 1 year in 8 years.
d) 2041
e)

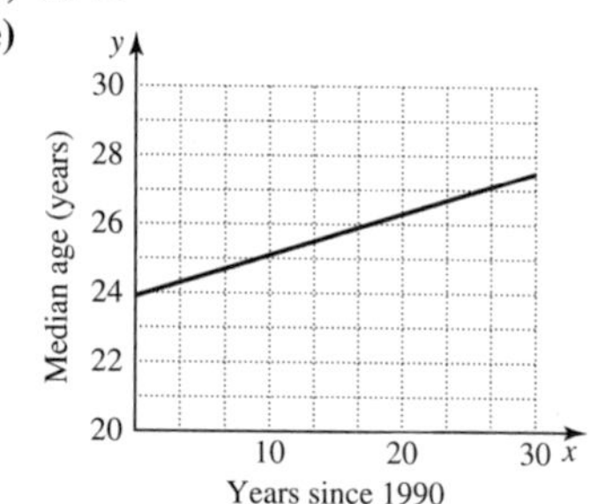

69. $C = 20n + 30$, \$170 **70.** $S = 180n - 360$, 1080°
71. $S = 3L - \frac{41}{4}$, 8.5 **72.** $F = \frac{9}{5}C + 32$, 113°F
73. $v = 32t + 10$, 122 ft/sec **74.** $C = 0.03n + 9$, \$81
75. a) $w = -\frac{1}{120}t + \frac{3}{2}$ **b)** $\frac{5}{6}$ inch **c)** 60°F
76. a) $P = 2L + 15$ **b)** 95 in. **c)** 100 in. **d)** 7.5 in.
77. $A = 0.6w$, 3.6 in.
78. $v = -32t + 100$, 4 ft/sec
79. a) $a = 0.08c$ **b)** 0.24 **c)** 6.25 mg/ml
80. $B = 13w + 715$, 1406.6 calories
81. 2, 3, $-\frac{2}{3}$; 4, -5, $\frac{4}{5}$; $\frac{1}{2}$, 3, $-\frac{1}{6}$; 2, $-\frac{1}{3}$, 6
82. $m = -\frac{A}{B}$
83. a)

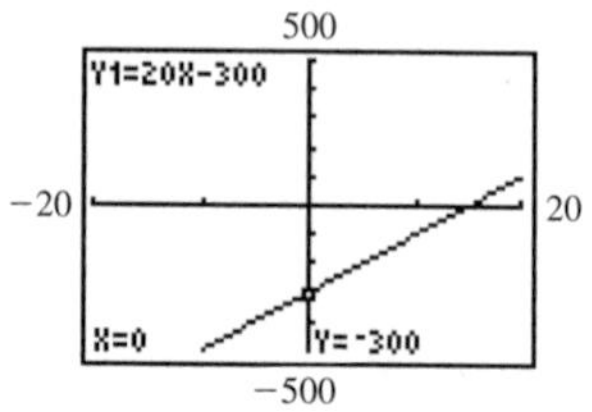

b)

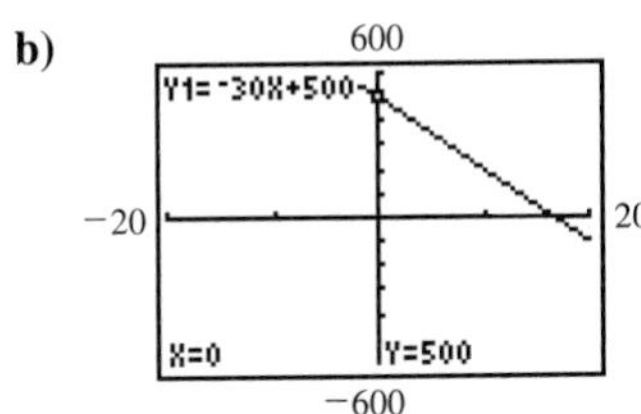

c)

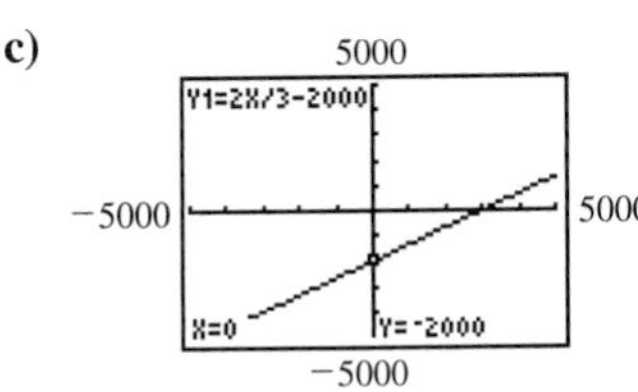

84. They look parallel, but they are not.
85. $-1 \le x \le 1,\ -1 \le y \le 1$
86. They will look perpendicular in the right window.

Section 3.5 Warm-ups T F T F F T T T F F

1. If y varies directly as x, then there is a constant k such that $y = kx$.
2. A variation constant is the constant k in the formulas $y = kx$ or $y = \frac{k}{x}$.
3. If y is inversely proportional to x, then there is a constant k such that $y = \frac{k}{x}$.
4. If y varies jointly as x and z, then there is a constant k such that $y = kxz$.
5. $T = kh$ **6.** $m = kp$ **7.** $y = \frac{k}{r}$ **8.** $u = \frac{k}{n}$
9. $R = kts$ **10.** $W = kuv$ **11.** $i = kb$
12. $p = kx$ **13.** $A = kym$ **14.** $t = \frac{k}{e}$ **15.** $y = \frac{5}{3}x$
16. $m = 2w$ **17.** $A = \frac{6}{B}$ **18.** $c = \frac{10}{d}$ **19.** $m = \frac{198}{p}$
20. $s = \frac{12}{v}$ **21.** $A = 2tu$ **22.** $N = 120pq$
23. $T = \frac{9}{2}u$ **24.** $R = 5p$ **25.** 25 **26.** 104
27. 1 **28.** 3 **29.** 105 **30.** 20
31. $\left(\frac{1}{2}, 600\right)$, (1, 300), (30, 10), $\left(900, \frac{1}{3}\right)$, inversely
32. $\left(\frac{1}{5}, 2500\right)$, (1, 500), (50, 10), $\left(1500, \frac{1}{3}\right)$, inversely
33. $\left(\frac{1}{3}, \frac{1}{4}\right)$, (8, 6), (12, 9), (20, 15), directly
34. $\left(\frac{1}{2}, \frac{1}{3}\right)$, (3, 2), (9, 6), (21, 14), directly
35. Directly, $y = 3.5x$ **36.** Directly, $y = 0.5x$
37. Inversely, $y = \frac{20}{x}$ **38.** Inversely, $y = \frac{500}{x}$
39. (1, 65), (2, 130), (3, 195), (4, 260)
40. (5, 8), (10, 16), (15, 24), (20, 32)
41. (20, 20), (40, 10), (50, 8), (200, 2)
42. (1, 20), (2, 10), (4, 5), (10, 2)
43. 1600, 12, 12 **44.** 300,000, 90, 100
45. 100.3 pounds **46.** \$24 **47.** 50 minutes **48.** 4 cm^3
49. \$17.40 **50.** 27 cents **51.** 80 mph **52.** 28 inches
53. k, (0, 0), no, $y = kx$

Enriching Your Mathematical Word Power

1. d **2.** a **3.** b **4.** c **5.** b **6.** a
7. c **8.** c **9.** d **10.** b **11.** c **12.** d

Review Exercises

1. Quadrant II **2.** Quadrant III **3.** x-axis
4. Quadrant I **5.** y-axis **6.** y-axis **7.** Quadrant IV
8. Quadrant II **9.** (0, −5), (−3, −14), (4, 7)
10. (9, −17), (3, −5), (−1, 3)
11. $\left(0, -\frac{8}{3}\right), \left(3, -\frac{2}{3}\right), \left(-6, -\frac{20}{3}\right)$
12. $\left(0, \frac{1}{2}\right), \left(-2, \frac{3}{2}\right), \left(2, -\frac{1}{2}\right)$

13.

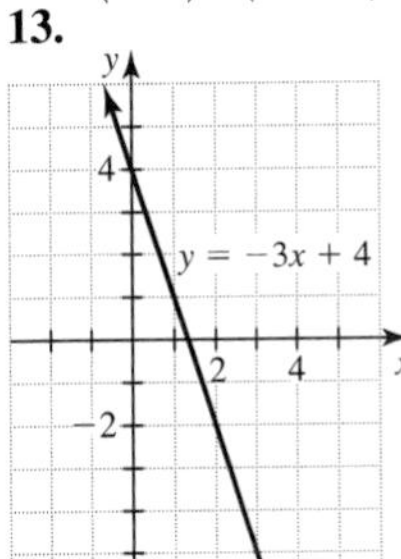

14.

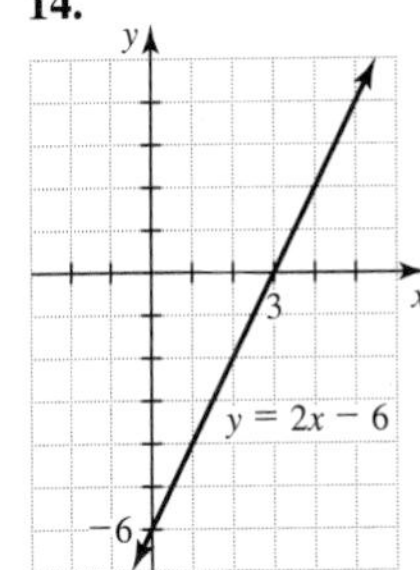

15.

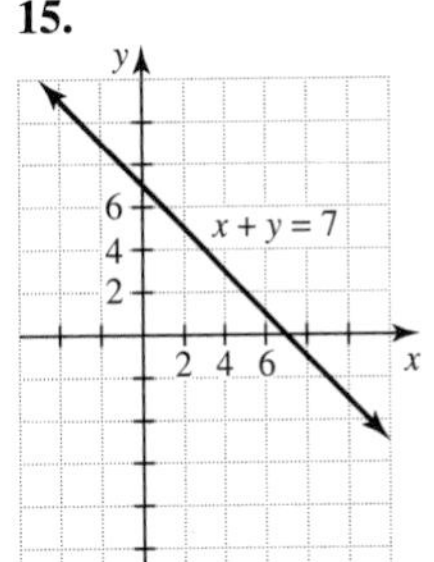

16.

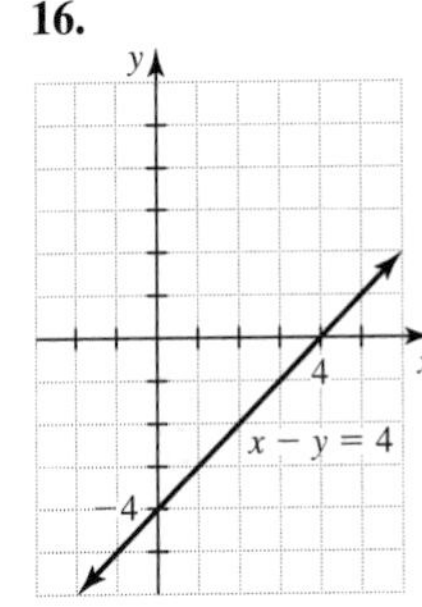

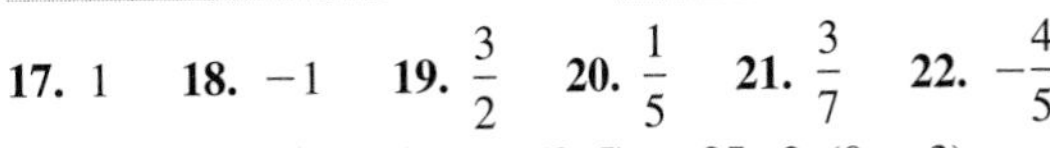

17. 1 **18.** −1 **19.** $\frac{3}{2}$ **20.** $\frac{1}{5}$ **21.** $\frac{3}{7}$ **22.** $-\frac{4}{5}$
23. 3, (0, −18) **24.** −1, (0, 5) **25.** 2, (0, −3)

26. $\frac{1}{2}, \left(0, -\frac{1}{2}\right)$ **27.** 2, (0, −4) **28.** $-\frac{3}{5}$, (0, −2)

29.

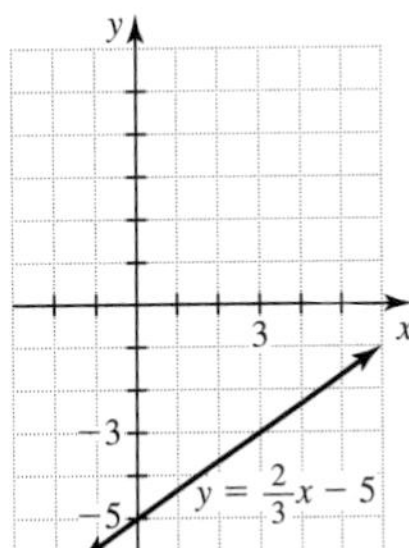

30.

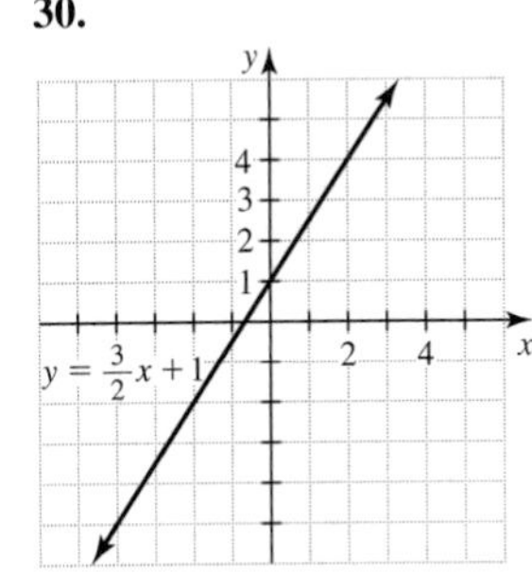

31.

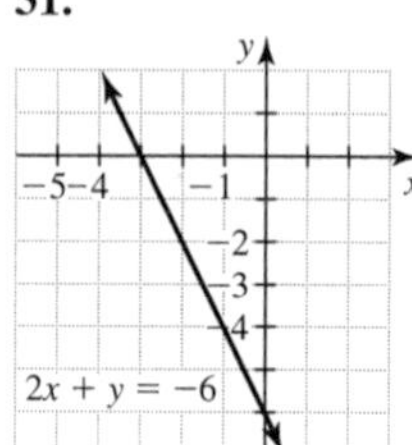

32.

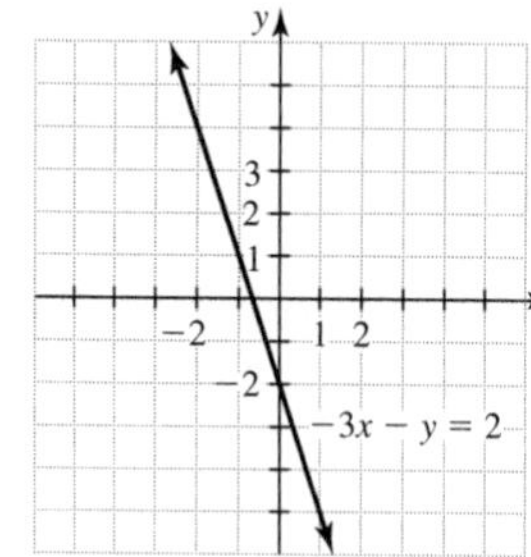

33.

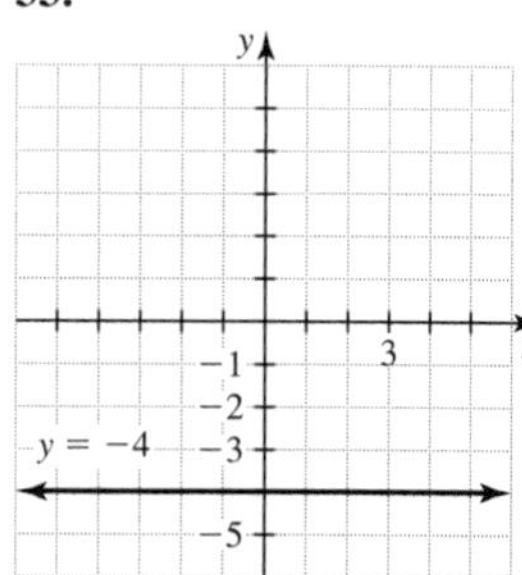

34.

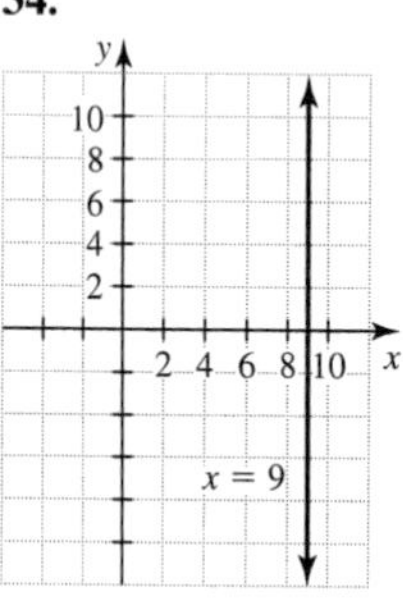

35. $x - 3y = -12$ **36.** $3x + 4y = -6$
37. $x + 2y = 0$ **38.** $3x + 5y = 45$ **39.** $y = 5$
40. $x = -2$ **41.** $y = \frac{2}{3}x + 7$ **42.** $y = -6x + 4$
43. $y = \frac{3}{7}x - 2$ **44.** $y = -x + 1$ **45.** $y = -\frac{3}{4}x + \frac{17}{4}$
46. $y = -\frac{2}{5}x - \frac{36}{5}$ **47.** $y = -2x - 1$ **48.** $y = \frac{1}{2}x - \frac{9}{2}$
49. $y = \frac{6}{5}x + \frac{17}{5}$ **50.** $y = \frac{5}{7}x - \frac{20}{7}$ **51.** $y = 3x - 14$
52. $y = x - 4$ **53.** $C = 32n + 49$, \$177
54. $h = 4t + 74$, 114 **55. a)** $q = 1 - p$ **b)** 1
56. $b = 500a - 24{,}000$ **57.** $y = 0.1x + 0.6$
58. $y = -4x + 28$ **59.** 132 **60.** 39 **61.** 2
62. 3 **63.** 60 **64.** 125
65. a) $C = 0.75T$ **b)** \$15 **c)** Increasing
66. a) $h = \frac{120}{n}$ **b)** 24 hours **c)** Decreasing

Chapter 3 Test

1. Quadrant II **2.** x-axis **3.** Quadrant IV
4. y-axis **5.** 1 **6.** $-\frac{5}{6}$ **7.** 3 **8.** 0 **9.** Undefined
10. $\frac{2}{3}$ **11.** $y = -\frac{1}{2}x + 3$ **12.** $y = \frac{3}{7}x - \frac{11}{7}$
13. $x - 3y = 11$ **14.** $5x + 3y = 27$
15. $y = \frac{1}{2}x - 3$ **16.** $2x - 3y = 6$

17.

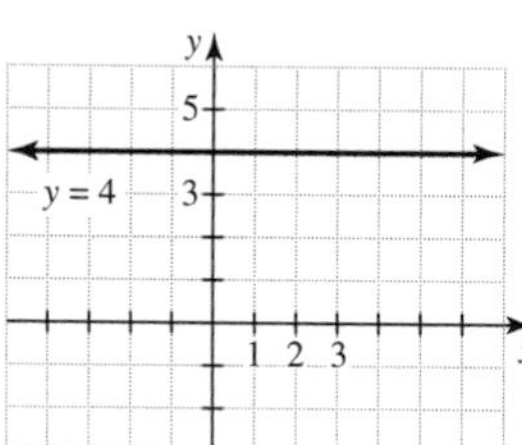

18.

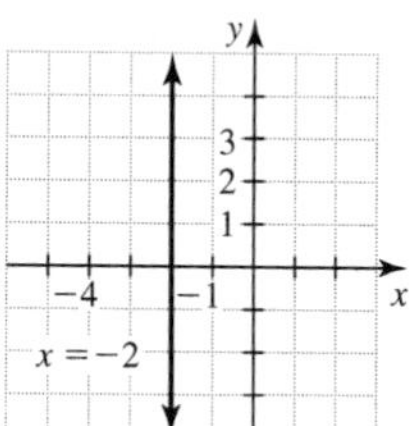

19. $S = 0.75n + 2.50$ **20.** $P = 3v + 20$, 80 cents
21. \$2.80 **22.** 18.75 days, decreases **23.** \$770

Making Connections Chapters 1–3

1. −1 **2.** −34 **3.** 1 **4.** 72 **5.** −4 **6.** −28
7. $-\frac{7}{2}$ **8.** 0.4 **9.** $\frac{1}{10}$ **10.** 15 **11.** $13x$
12. $3x - 36$ **13.** $\left\{\frac{5}{2}\right\}$ **14.** $\left\{\frac{7}{3}\right\}$ **15.** $\frac{1}{6}$ **16.** $\frac{5}{12}$
17. {2} **18.** {−4} **19.** $2x - 4$ **20.** $x + 2$ **21.** {5}
22. {3} **23.** ∅ **24.** All real numbers **25.** $y = \frac{t - 2}{3\pi}$
26. $y = mx + b$ **27.** $y = x - 4$ **28.** $y = 6$
29. $y = \frac{4}{5}$ **30.** $y = 200$
31. a) $\frac{2}{15}$ **b)** $\frac{1}{5}$ **c)** About 13% per year
d) \$276,000 saved, \$12,000 per year

Chapter 4

Section 4.1 Warm-ups T F F F T T T T T F
1. A system of equations is a pair of equations.
2. A solution to a system is a point or ordered pair that satisfies both equations.
3. In this section systems of equations were solved by graphing.
4. An independent system is a system with exactly one solution.
5. A dependent system is one for which the two equations are equivalent.
6. An inconsistent system is one that has no solution.
7. (3, −2) **8.** (2, 3) **9.** All three
10. All three **11.** None **12.** None **13.** (−2, 3)
14. (3, −1) **15.** (2, 4) **16.** (1, 3) **17.** (1, 2)
18. (2, −1) **19.** (0, −5) **20.** (2, 2) **21.** (−2, 3)
22. (1, −3) **23.** (0, 0) **24.** (0, 0) **25.** (2, 3)
26. (−1, −2) **27.** No solution, inconsistent
28. $\{(x, y) \mid x - y = 3\}$, dependent
29. (1, −2), independent **30.** (2, 0), independent
31. $\{(x, y) \mid 2x + y = 3\}$, dependent
32. $\{(x, y) \mid x - 2y = 8\}$, dependent
33. $\{(x, y) \mid y = x\}$, dependent
34. $\{(x, y) \mid y = -3x + 1\}$, dependent
35. (−4, −3), independent **36.** (−3, 1), independent
37. (−1, 0), independent **38.** No solution, inconsistent
39. Inconsistent **40.** Dependent **41.** Independent
42. Independent **43.** Inconsistent **44.** Independent

45. a) (5, 20)
b) For 5 toppings the cost is \$20 at both restaurants.
46. a) 100, 910; 400, 760; 700, 610; 1900, 10
b)

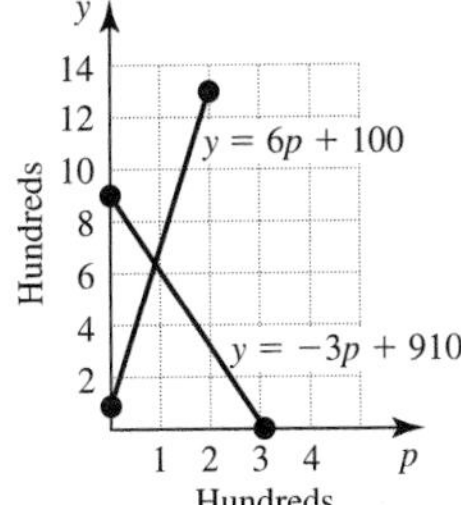

c) \$90

47. a) 800, 500; 1050, 850; 1300, 1200; 1800, 1900
b)

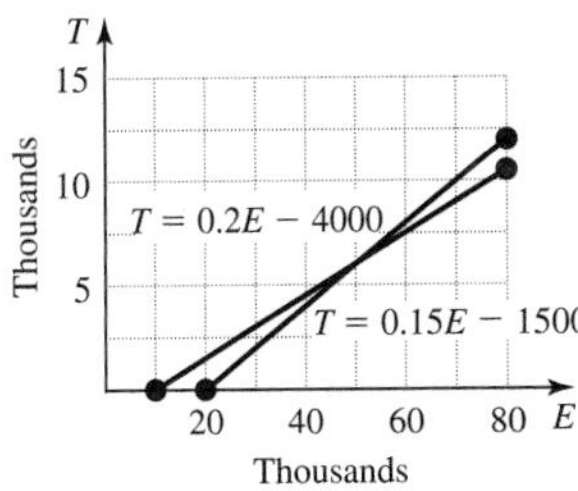

c) 15,000
d) Panasonic

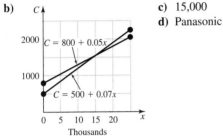

48. \$50,000, Schneider's plan

49. It is a dependent system.
50. $x + y = 8, x - y = -2$ **51.** $x + y = 1, x + y = 5$
52. (3.8, 3.2) **53.** (−2.4, −522.7) **54.** (1.4, −0.1)
55. (27.3, 3.3)

Section 4.2 Warm-ups T T T T T T F T T T
1. In this section we learned the substitution method.
2. Graphing is not accurate enough.
3. A dependent system is one in which the equations are equivalent.
4. An inconsistent system is one with no solution.
5. Using substitution on a dependent system results in an equation that is always true.
6. Using substitution on an inconsistent system results in a false equation.
7. (2, 5) **8.** (6, 1) **9.** (2, 3) **10.** (3, 5) **11.** (−2, 9)
12. $\left(\frac{10}{29}, \frac{2}{29}\right)$ **13.** (−5, 5) **14.** (3, −3) **15.** $\left(\frac{1}{3}, \frac{2}{3}\right)$
16. $\left(\frac{4}{3}, -\frac{2}{3}\right)$ **17.** $\left(\frac{1}{2}, \frac{1}{3}\right)$ **18.** $\left(\frac{1}{2}, -\frac{1}{4}\right)$
19. $\left(3, \frac{5}{2}\right)$, independent **20.** $\left(-\frac{3}{5}, \frac{14}{5}\right)$, independent
21. No solution, inconsistent
22. $\{(x, y) \mid 3x - y = 5\}$, dependent
23. No solution, inconsistent **24.** No solution, inconsistent
25. $\left(\frac{7}{3}, 0\right)$, independent **26.** $\left(\frac{11}{5}, \frac{3}{25}\right)$, independent
27. (−1, 1), independent **28.** $\{(x, y) \mid y = x\}$, dependent
29. (3, 2) **30.** (3, 3) **31.** (3, 1) **32.** (2, 1)
33. No solution **34.** No solution
35. \$12,000 at 10%, \$8,000 at 5%
36. \$18,000 at 10%, \$12,000 at 9%
37. Titanic \$601 million, Star Wars \$461 million
38. Length 13 yd, width 9 yd **39.** Lawn \$12, sidewalk \$7
40. Burger \$0.75, fries \$0.45
41. Left rear 288 pounds, left front 287 pounds, no
42. Left front 306, left rear 294, right front 282, right rear 318 pounds
43. \$2.40 per pound
44. Dum 121 lbs, Dee 120 lb
45. a) 69.2 years, 76.8 years
b)

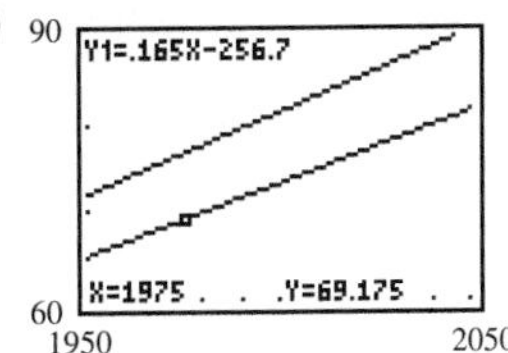

c) No **d)** 1614

Section 4.3 Warm-ups F T T T T F F F F T
1. In this section we learned to solve systems by the addition method.
2. The three methods presented are graphing, substitution, and addition.
3. In addition and substitution we eliminate a variable and solve for the remaining variable.
4. It is sometimes necessary to use the multiplication property of equality before adding the equations.
5. Eliminate the variable that is easiest to eliminate.
6. In the addition method inconsistent systems result in a false equation and dependent systems result in an equation that is always true.
7. (3, −1) **8.** (2, 3) **9.** (−3, 5) **10.** (3, 2)
11. (−4, 3) **12.** (−2, 1) **13.** $\left(\frac{1}{7}, \frac{9}{7}\right)$ **14.** (0, 0)
15. (8, 31) **16.** (30, −5) **17.** (−2, −3) **18.** (−4, 1)
19. (−1, 4) **20.** (1, −1) **21.** (−1, 2) **22.** (−2, −3)
23. (3, −1) **24.** (2, −3) **25.** (1, 2) **26.** (−1, 4)
27. No solution, inconsistent **28.** (4, 1), independent
29. $\{(x, y) \mid x + y = 5\}$, dependent **30.** (2, 0), independent
31. $\{(x, y) \mid 2x = y + 3\}$, dependent
32. No solution, inconsistent **33.** $\{(x, y) \mid x + 3y = 3\}$, dependent
34. (0, 2), independent **35.** No solution, inconsistent
36. (6, 2), independent **37.** (12, 18), independent
38. (10, 2), independent **39.** (40, 60), independent
40. (40, 60), independent **41.** (0.1, 0.1), independent
42. (2, 0.3), independent **43.** (1.5, −2.8)
44. (123, 55) **45.** (4, 3) **46.** (4, 4) **47.** (4, 1)

48. (3, 1) **49.** No solution **50.** No solution
51. 150 cars, 100 trucks **52.** 16 dimes, 14 nickels
53. 6 adults, 24 children **54.** Coffee \$0.45, doughnut \$0.35
55. 180 men, 120 women
56. 180 hours regular time, 30 hours overtime
58. a) $y = \dfrac{a_2c_1 - a_1c_2}{a_2b_1 - a_1b_2}$ **b)** $x = \dfrac{b_2c_1 - b_1c_2}{b_2a_1 - b_1a_2}$ **c)** (2, 1)

Section 4.4 Warm-ups T T T F F F T F T F

1. A linear inequality has the same form as a linear equation except that an inequality symbol is used.
2. An ordered pair satisfies a linear inequality if the inequality is correct when the variables are replaced by the coordinates of the ordered pair.
3. If the inequality symbol includes equality, then the boundary line is solid; otherwise it is dashed.
4. We shade the side that satisfies the inequality.
5. In the test point method we test a point to see which side of the boundary line satisfies the inequality.
6. With the test point method you can use the inequality in any form.

7. $(-3, -9)$ **8.** $(-2, 6)$ **9.** (3, 0), (1, 3)
10. (2, 0), $(-3, 9)$ **11.** (2, 3), (0, 5) **12.** $(-3, -4)$

13.

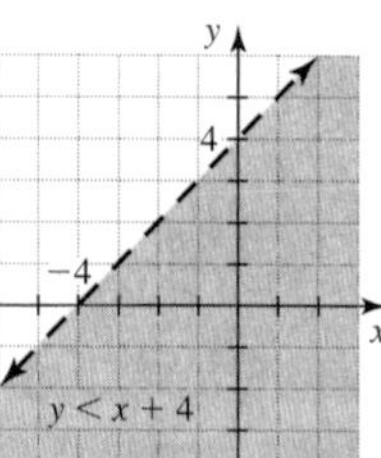

14.

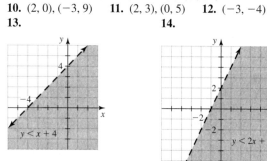

15.

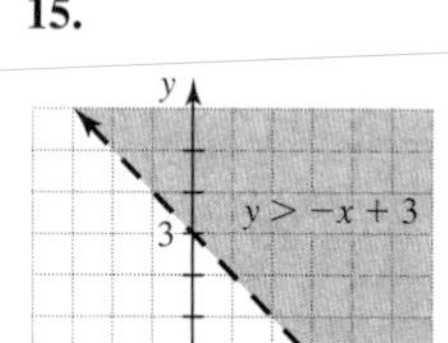
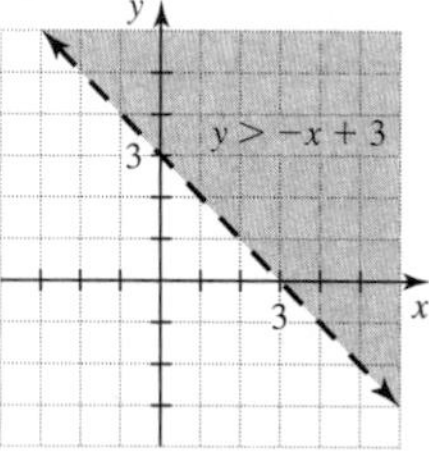

16.

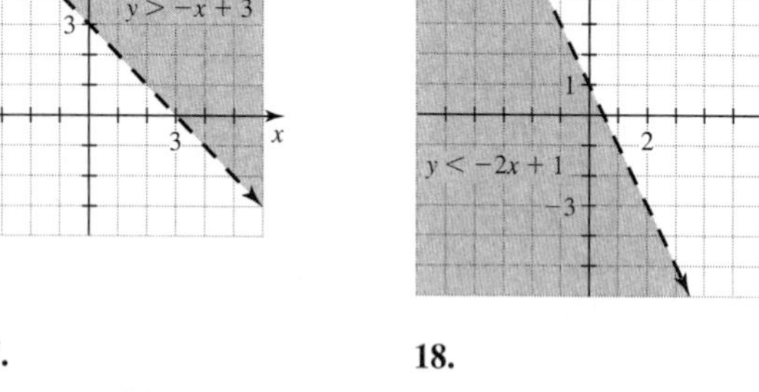

17.

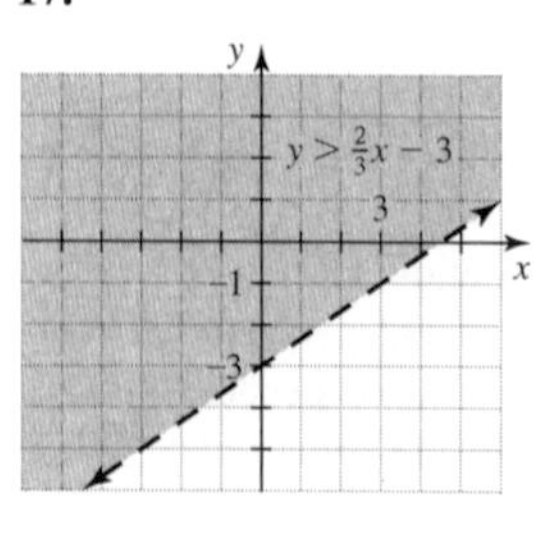

18.

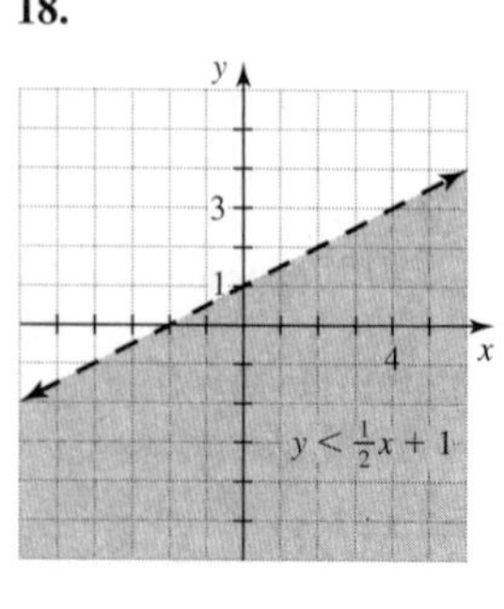

19.

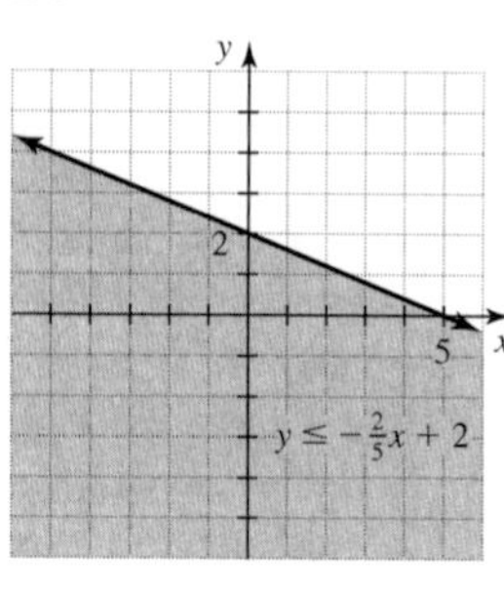

20.

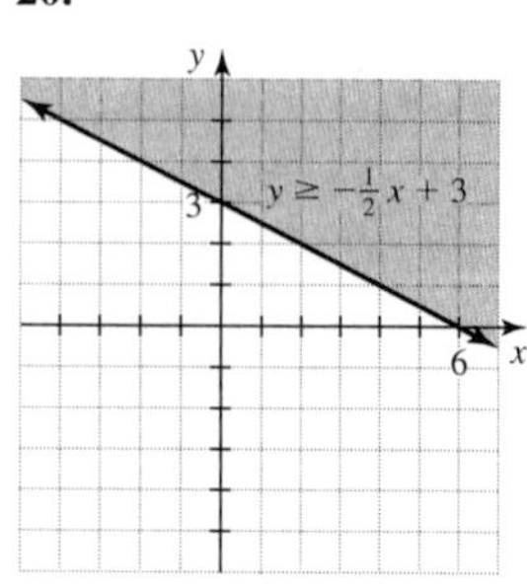

21.

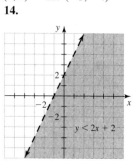

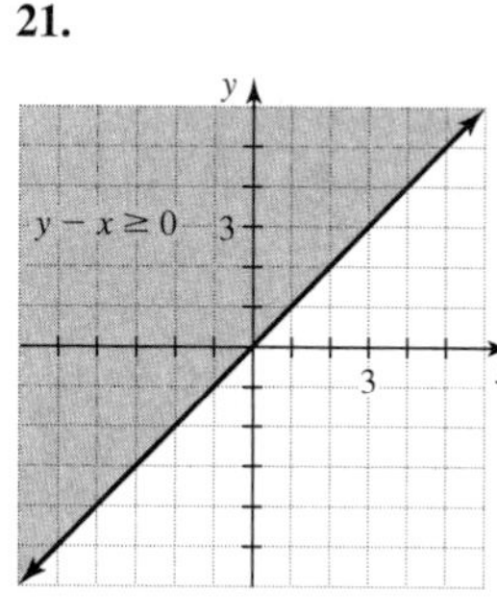

22.

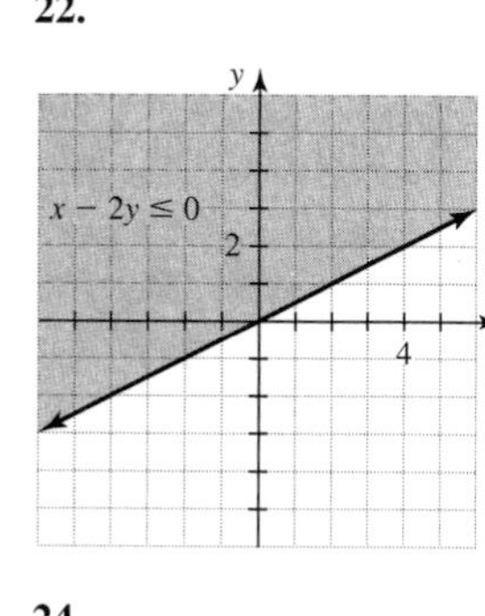

23.

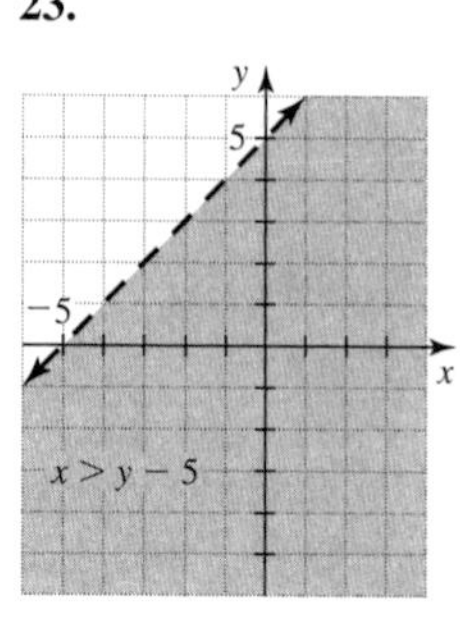

24.

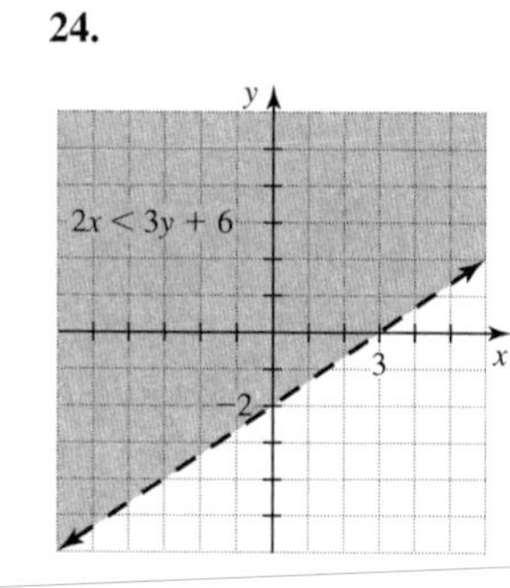

25.

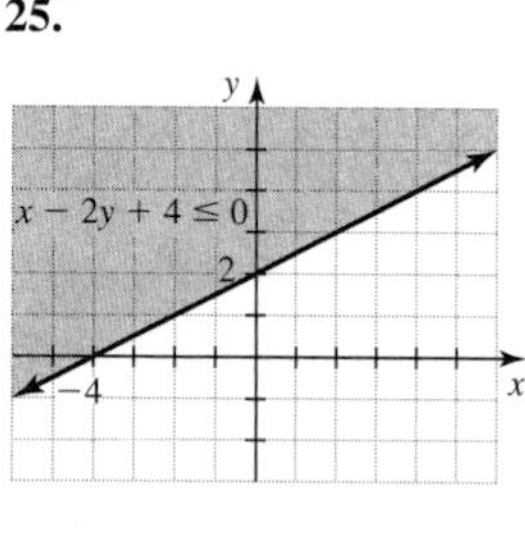

26.

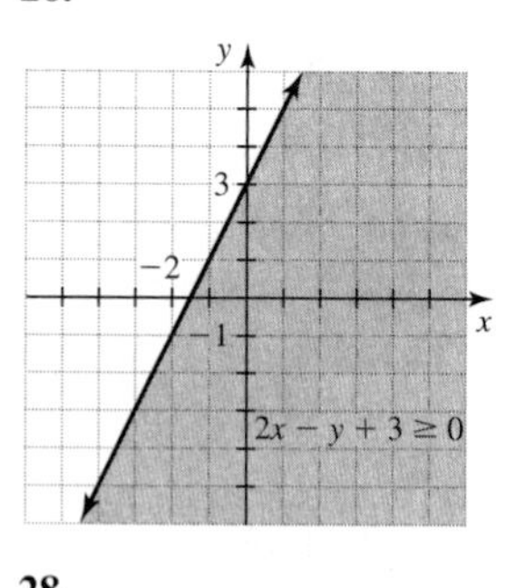

27.

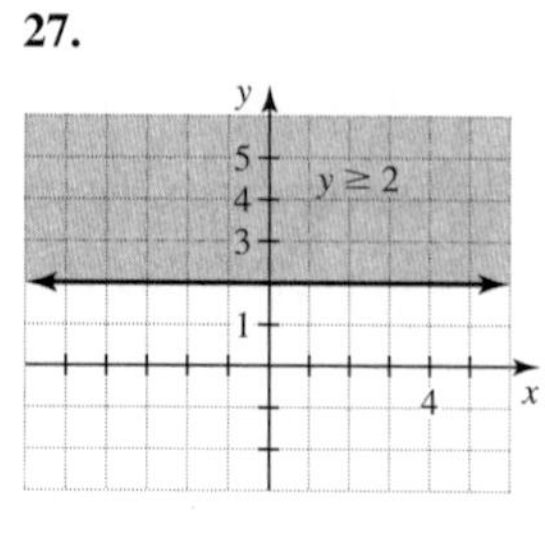

28.

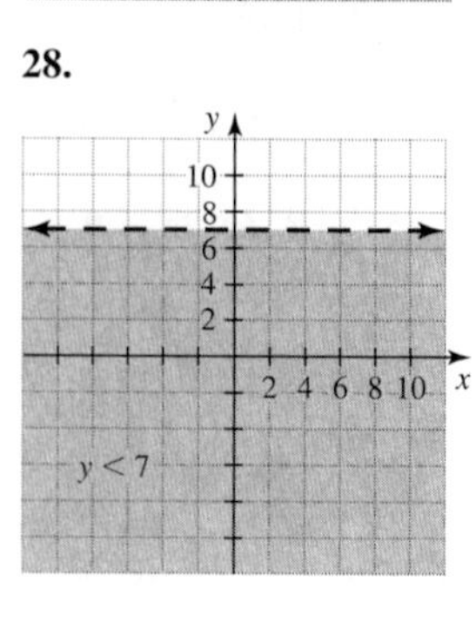

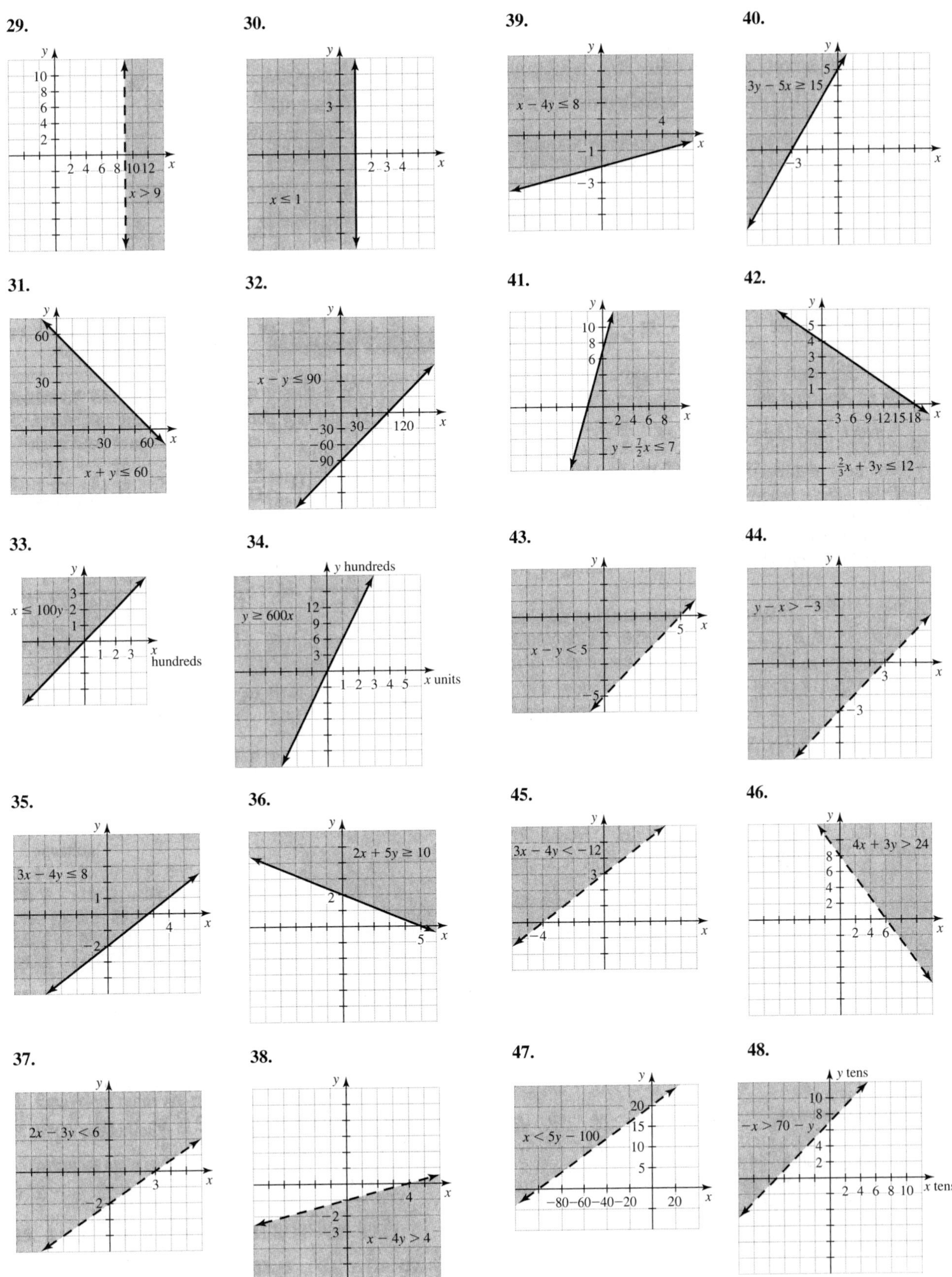
29.
x > 9
30.
x ≤ 1
39.
x − 4y ≤ 8
40.
3y − 5x ≥ 15
31.
x + y ≤ 60
32.
x − y ≤ 90
41.
y − 7/2 x ≤ 7
42.
2/3 x + 3y ≤ 12
33.
x ≤ 100y
x hundreds
34.
y hundreds
y ≥ 600x
x units
43.
x − y < 5
44.
y − x > −3
35.
3x − 4y ≤ 8
36.
2x + 5y ≥ 10
45.
3x − 4y < −12
46.
4x + 3y > 24
37.
2x − 3y < 6
38.
x − 4y > 4
47.
x < 5y − 100
48.
y tens
−x > 70 − y
x tens

49. $5x + 7y \leq 770$

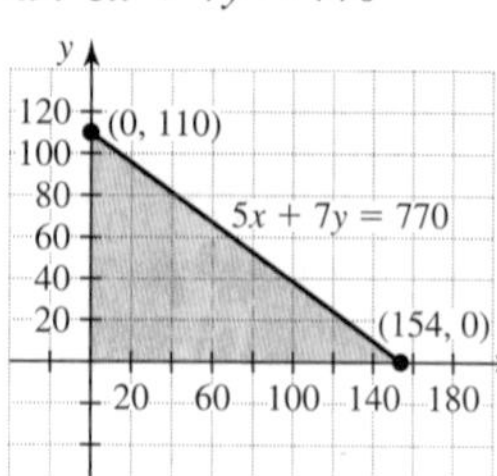

50. $5x + 4y \leq 1000$

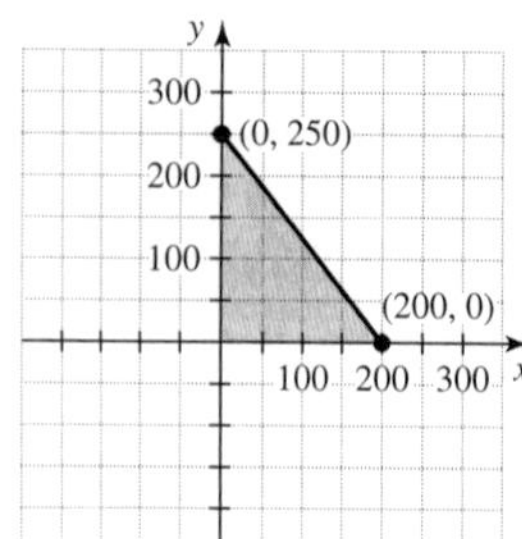

51.

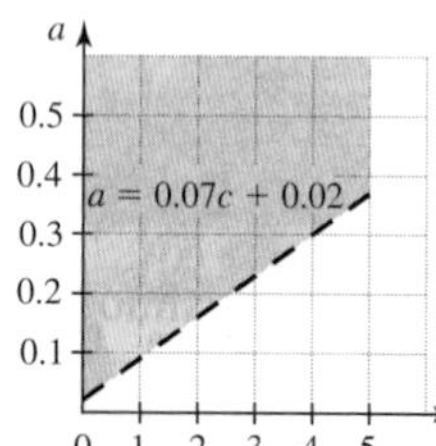

Section 4.5 Warm-ups F T T T F F T F T T

1. A system of linear inequalities in two variables is a pair of linear inequalities in two variables.

2. To see if an ordered pair satisfies a system of inequalities we can check to see if it satisfies both inequalities.

3. The solution set to a system of inequalities is usually described with a graph.

4. The boundary lines are solid if the inequality symbol includes equality; otherwise they are dashed.

5. To use the test point method, select a point in each region determined by the graphs of the boundary lines.

6. Any point will work as a test point, but it is usually simplest to select points on the axes.

7. (4, 3) **8.** (1, 1), (0, −1) **9.** (3, 6) **10.** (−3, 8)

11. (9, −5) **12.** (0, −7)

13.

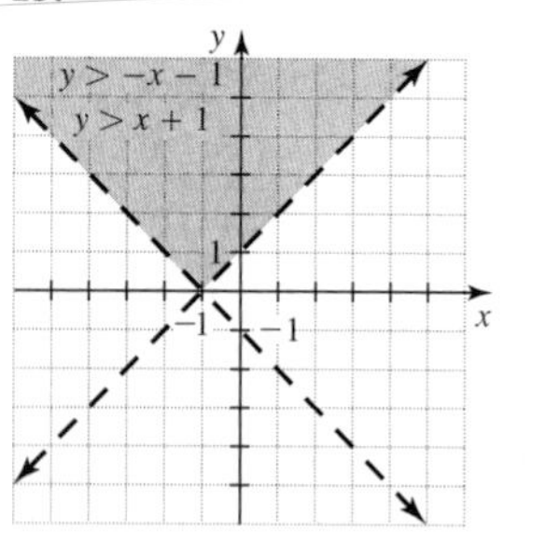

14.

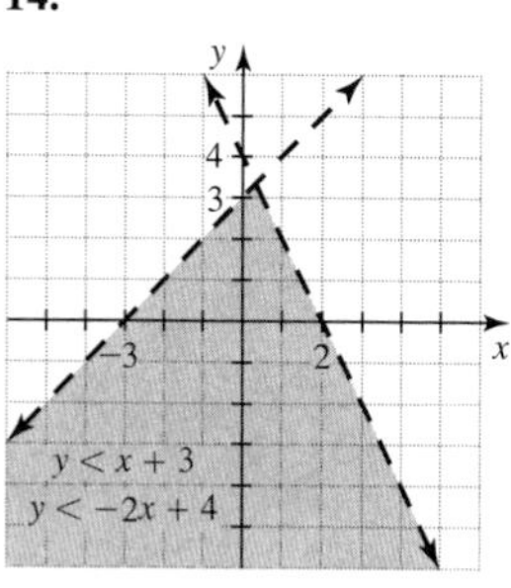
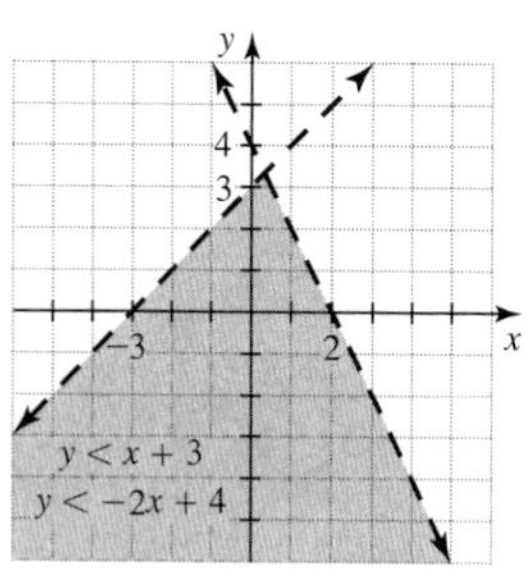

15.

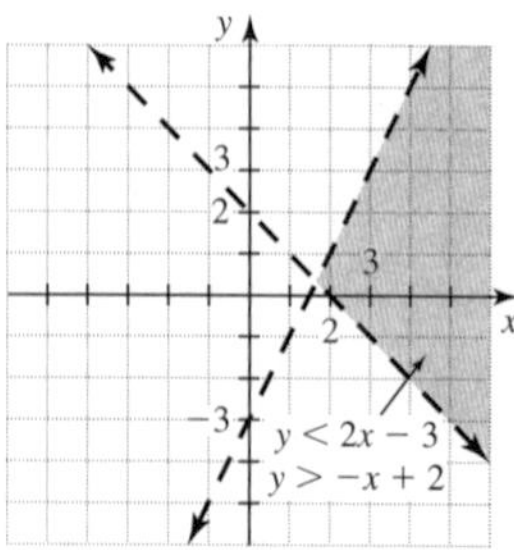

16.

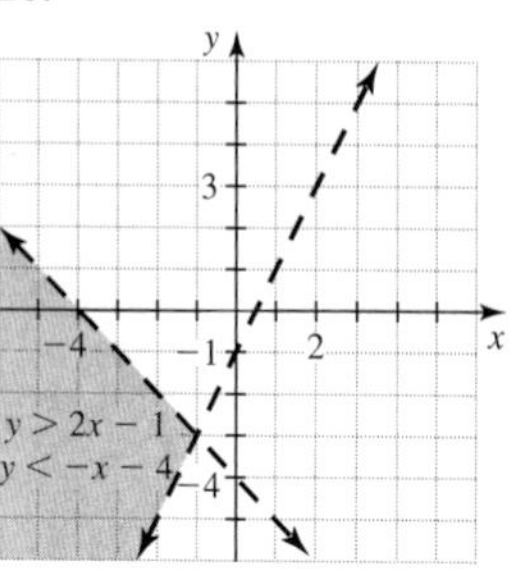

17.

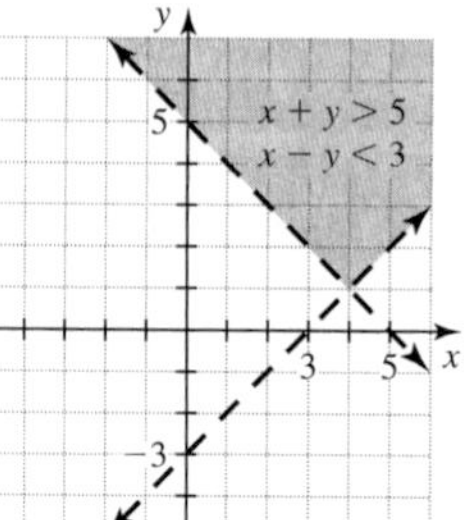

18.

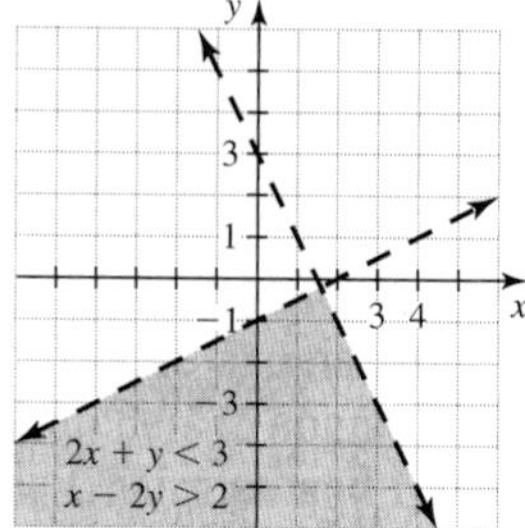

19.

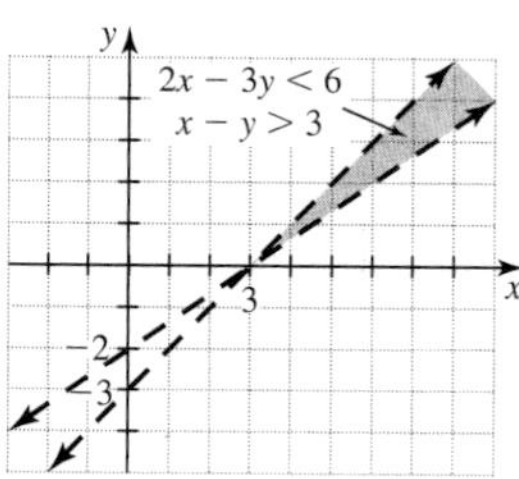

20.

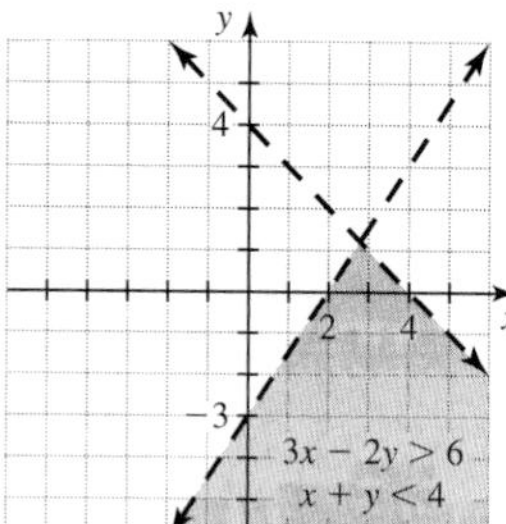

21.

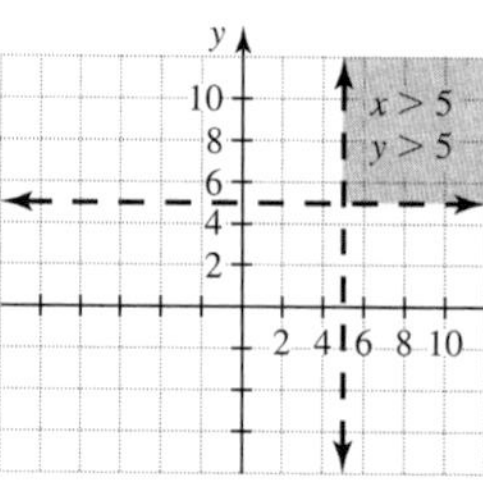

22.

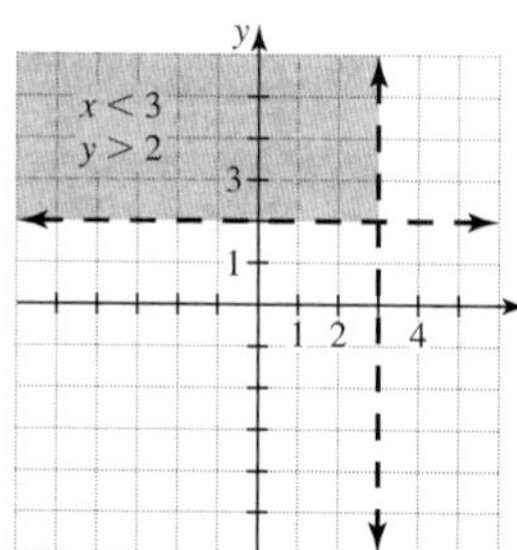

23.

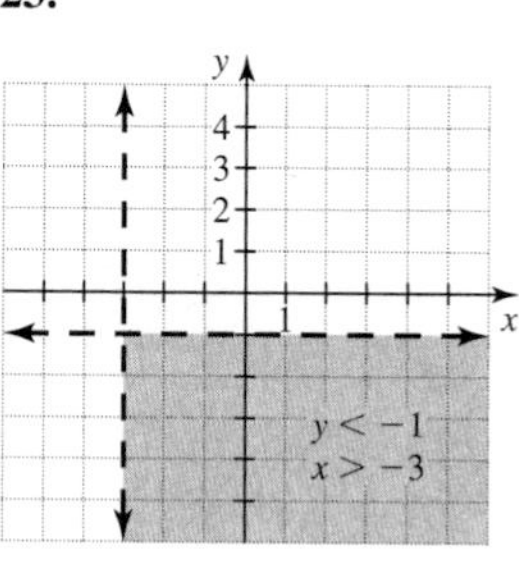

24.

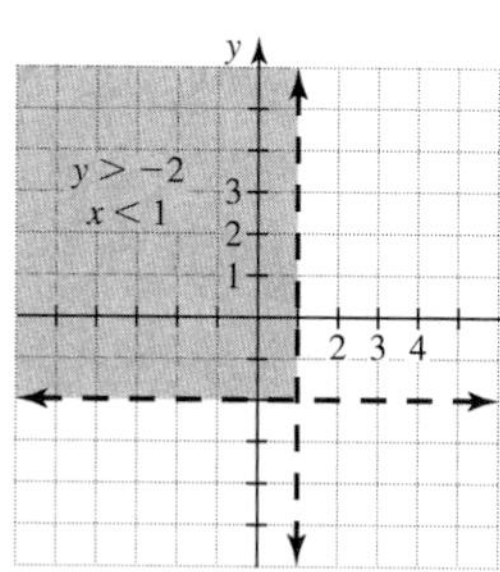

25.

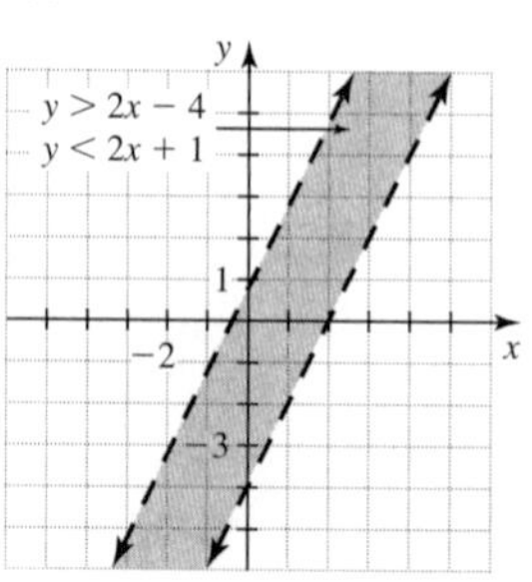

26.

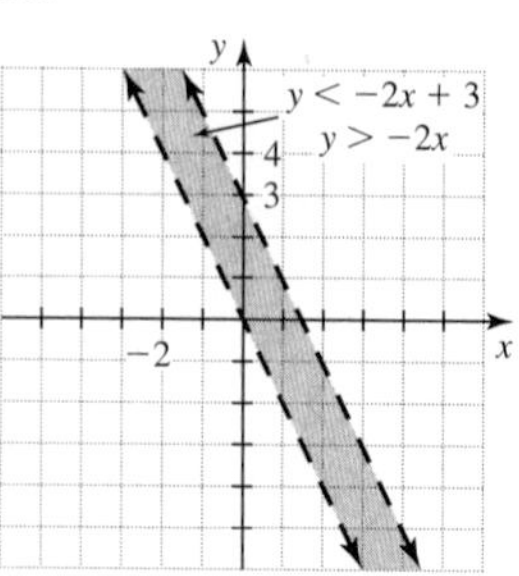

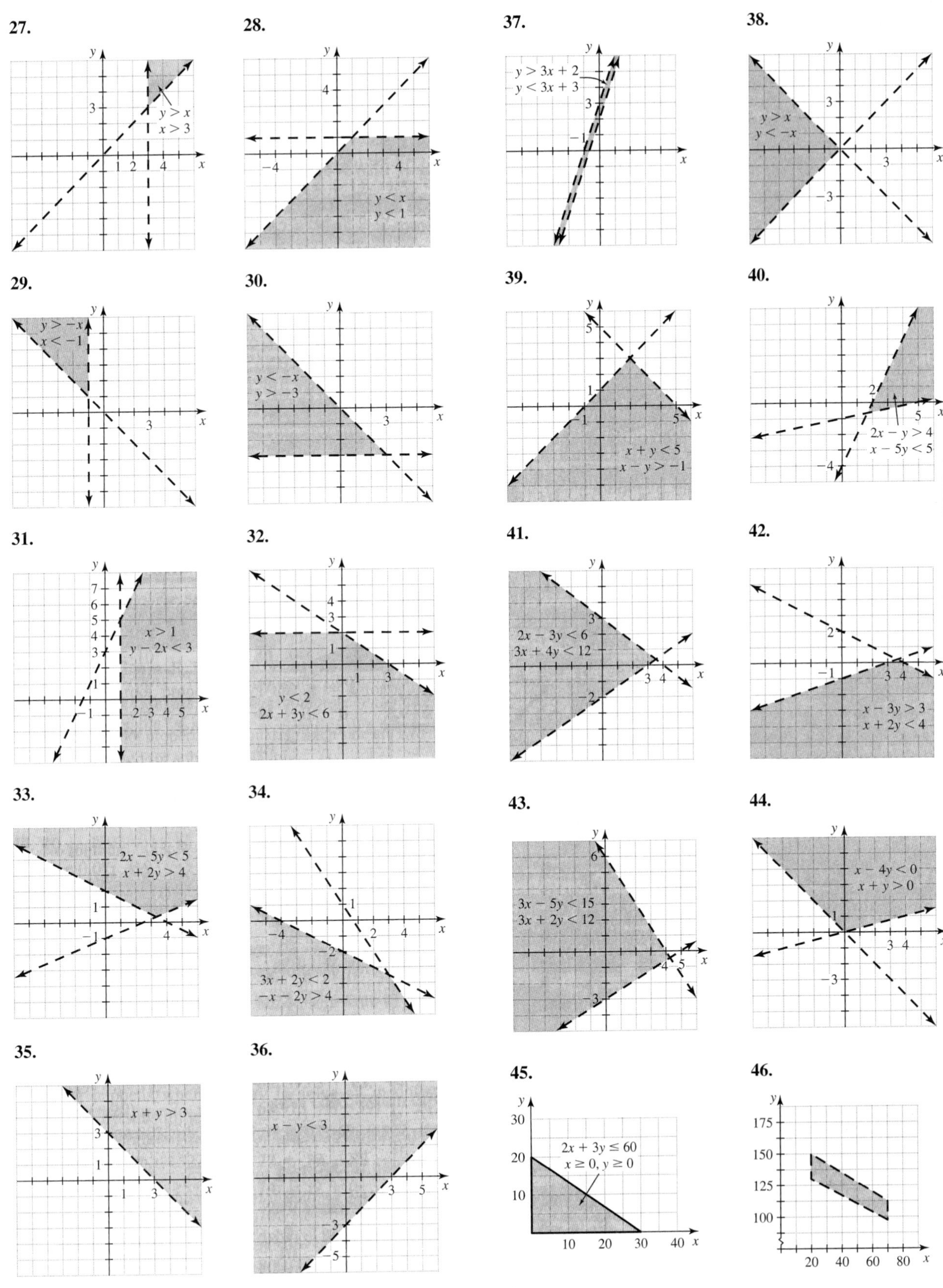
27.
y > x
x > 3
28.
y < x
y < 1
37.
y > 3x + 2
y < 3x + 3
38.
y > x
y < −x
29.
y > −x
x < −1
30.
y < −x
y > −3
39.
x + y < 5
x − y > −1
40.
2x − y > 4
x − 5y < 5
31.
x > 1
y − 2x < 3
32.
y < 2
2x + 3y < 6
41.
2x − 3y < 6
3x + 4y < 12
42.
x − 3y > 3
x + 2y < 4
33.
2x − 5y < 5
x + 2y > 4
34.
3x + 2y < 2
−x − 2y > 4
43.
3x − 5y < 15
3x + 2y < 12
44.
x − 4y < 0
x + y > 0
35.
x + y > 3
36.
x − y < 3
45.
2x + 3y ≤ 60
x ≥ 0, y ≥ 0
46.

47. $5x + 8y \le 800$
$5x + 7y \le 770$

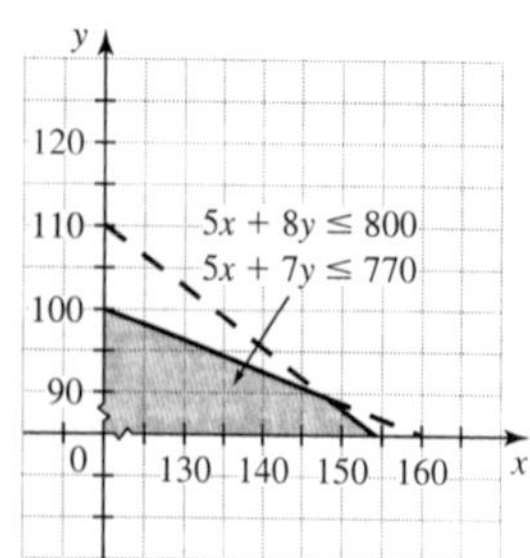

48. $5x + 8y \le 80$
$2x + 3y \le 30$

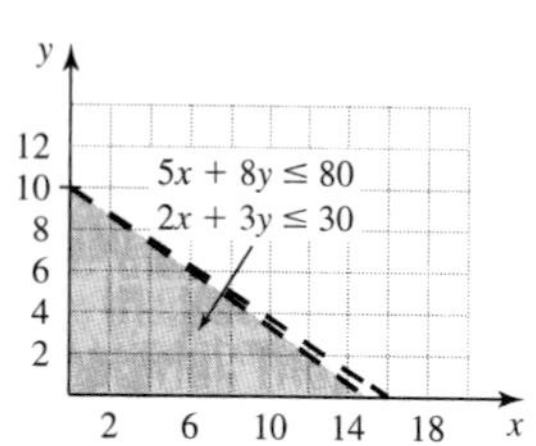

Enriching Your Mathematical Word Power

1. c **2.** a **3.** a **4.** d **5.** b **6.** c **7.** d **8.** b **9.** a **10.** b

Review Exercises

1. (1, 3) **2.** No solution **3.** (−1, 1) **4.** (−2, 8) **5.** (2, 6) **6.** (−1, 4) **7.** (−8, −3) **8.** (2, 1) **9.** (3, −1), independent **10.** (1, −3), independent **11.** $\{(x, y) \mid x - 2y = 4\}$, dependent **12.** No solution, inconsistent **13.** (1, −2), independent **14.** (−2, 3), independent **15.** (0, 0), independent **16.** No solution, inconsistent **17.** $\{(x, y) \mid x - y = 6\}$, dependent **18.** (0, 0), independent **19.** No solution, inconsistent **20.** (2, −3), independent

21.

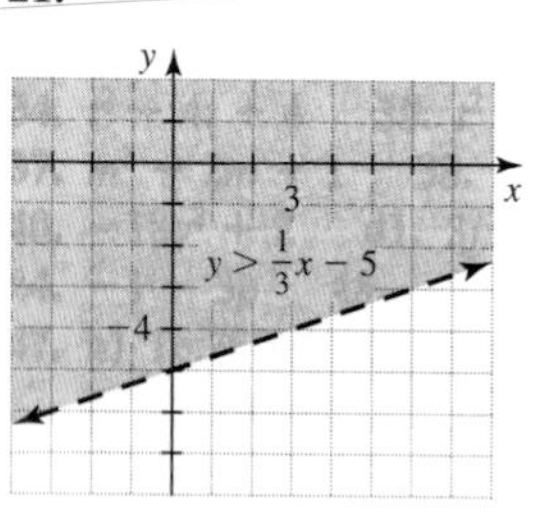

22.

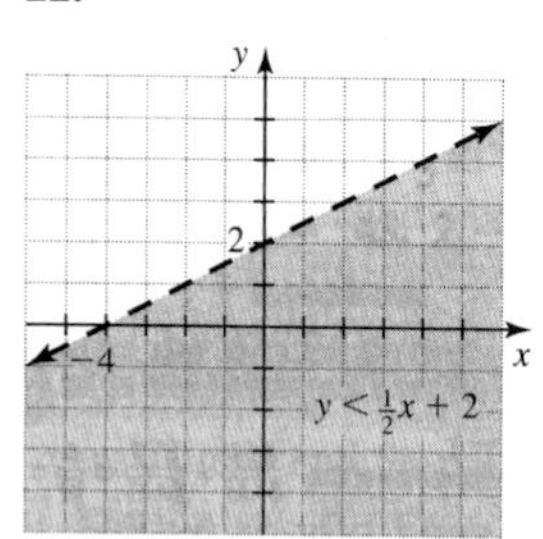

23.

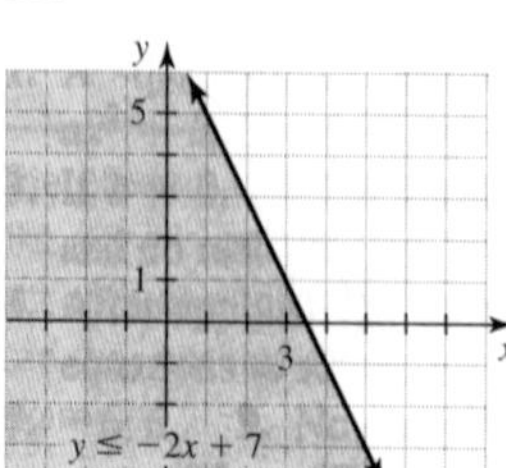

24.

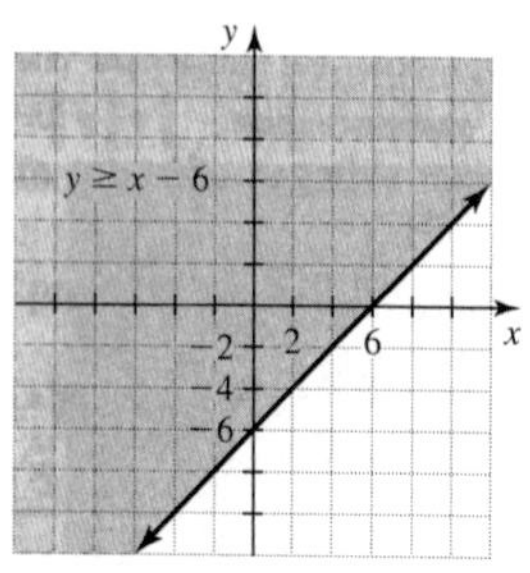

25.

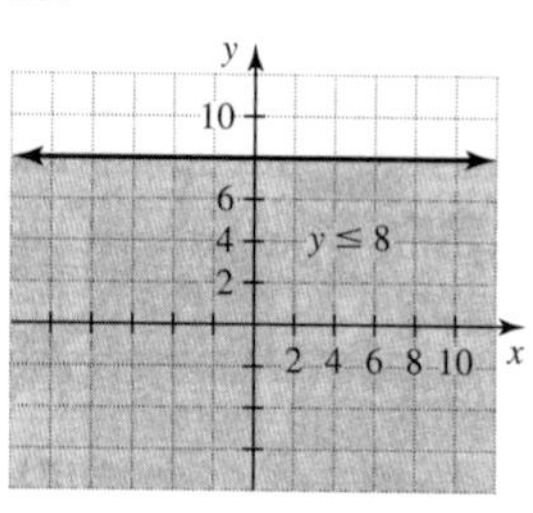

26.

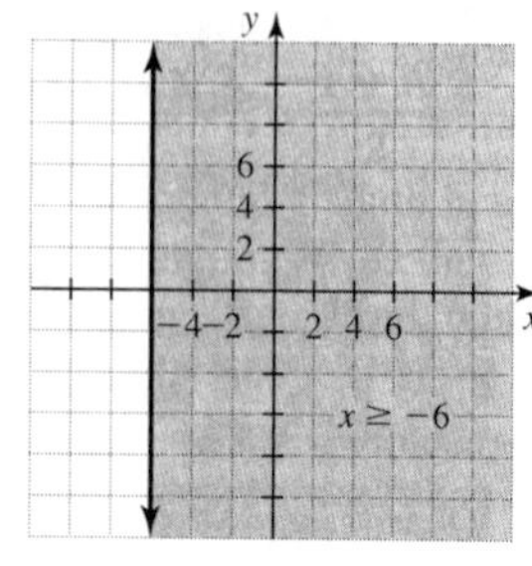

27.

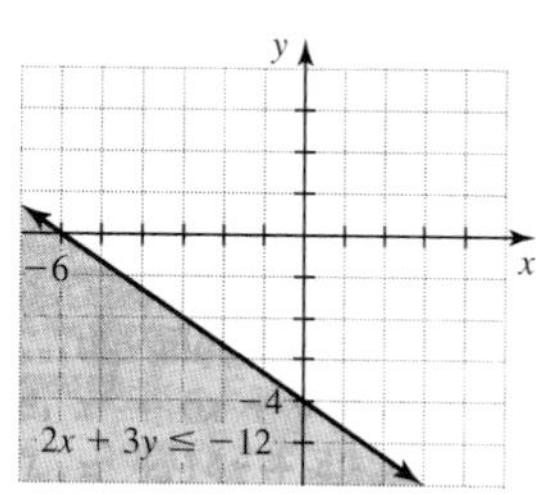

28.

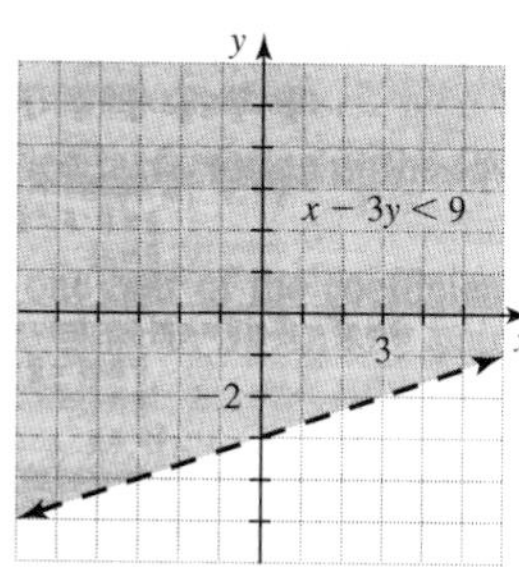

29.

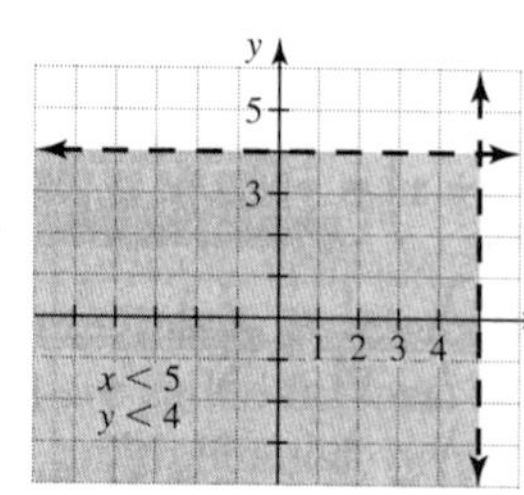

30.

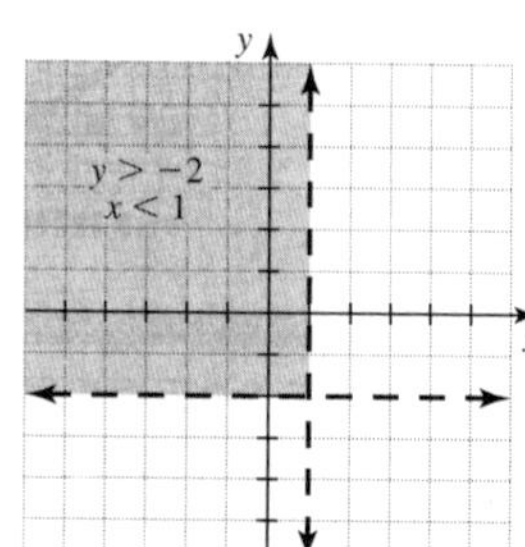

31.

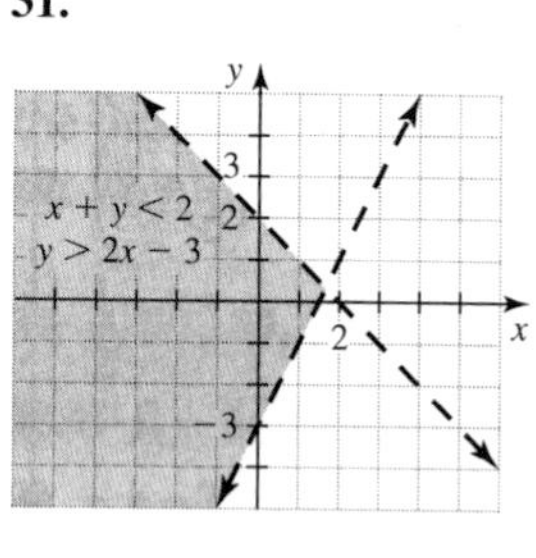

32.

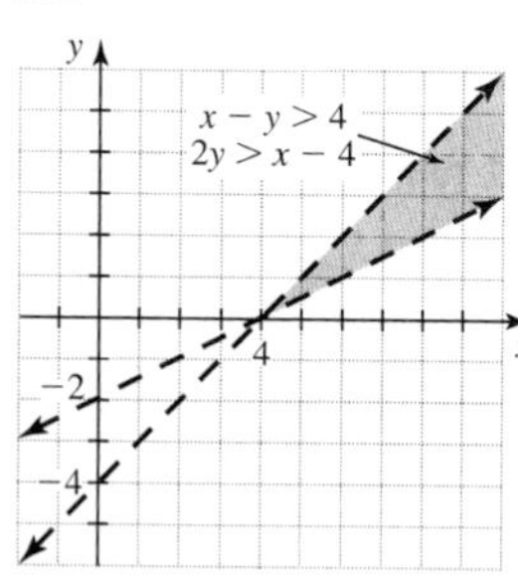

33.

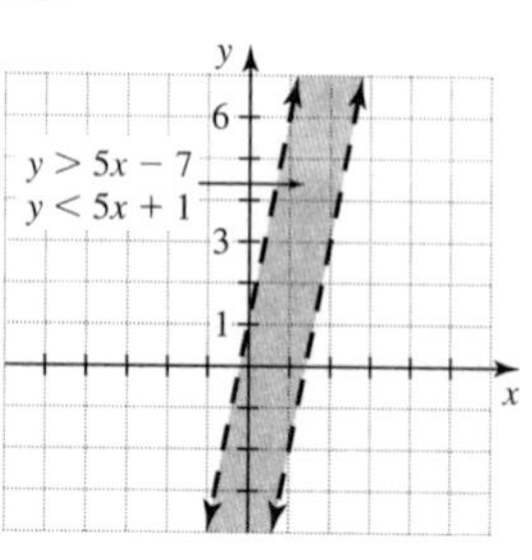

34.

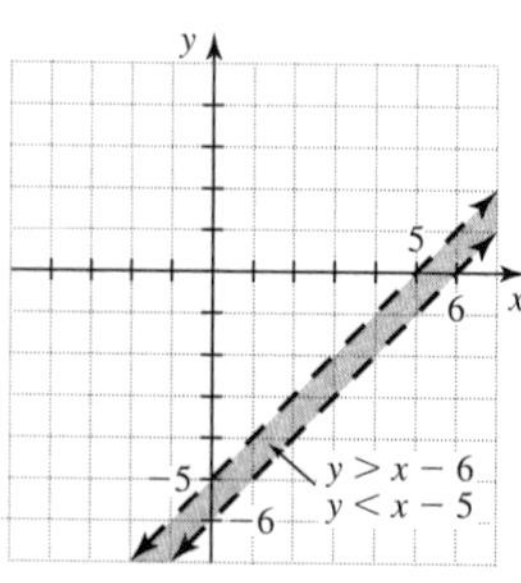

35.

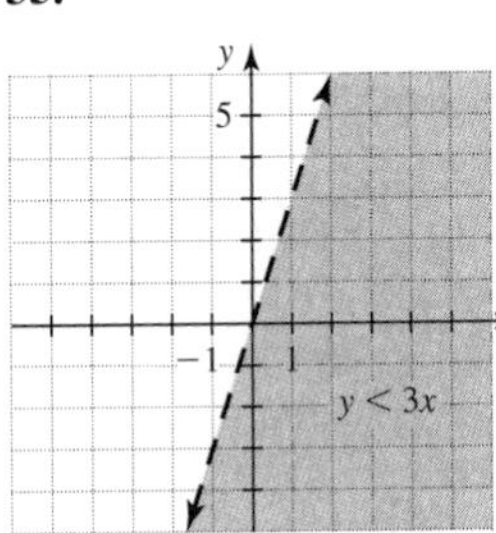

36.

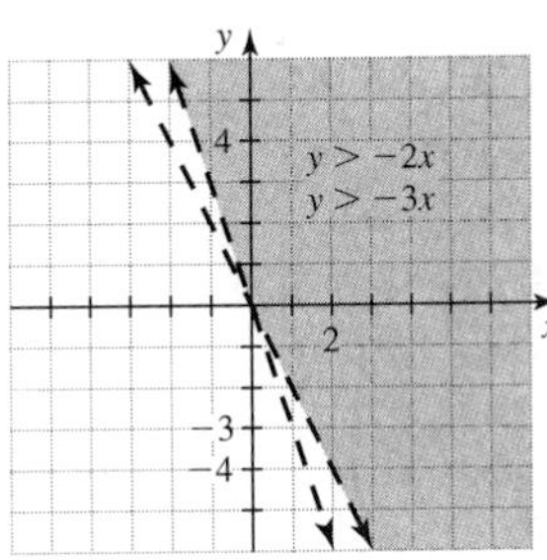

37. Apple \$0.45, orange \$0.35
38. Small \$0.55, medium \$0.65 **39.** 32 fives, 22 tens
40. \$6000 at 8%, \$4000 at 10%
41. 4 servings green beans, 3 servings chicken soup
42. 30-second \$800, 60-second \$1500

Chapter 4 Test

1. $(-1, 3)$ **2.** $(2, 1)$ **3.** $(3, -1)$ **4.** $(2, 3)$ **5.** $(4, 1)$
6. Inconsistent **7.** Dependent **8.** Independent
9.

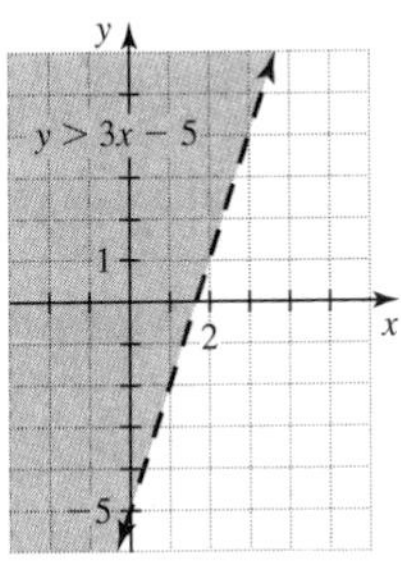

10.

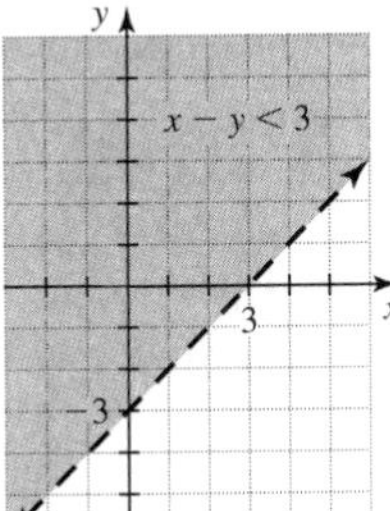

11.

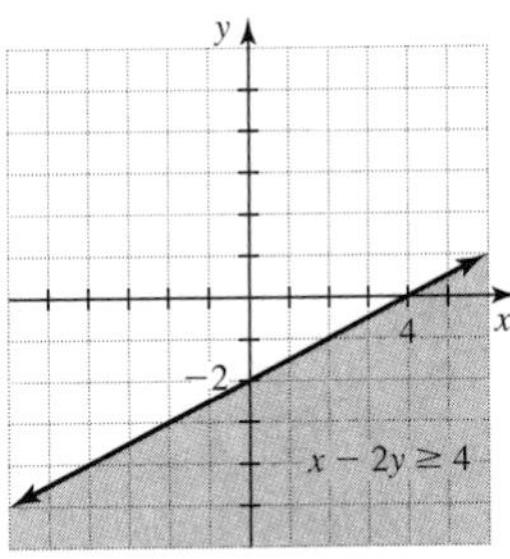

12.

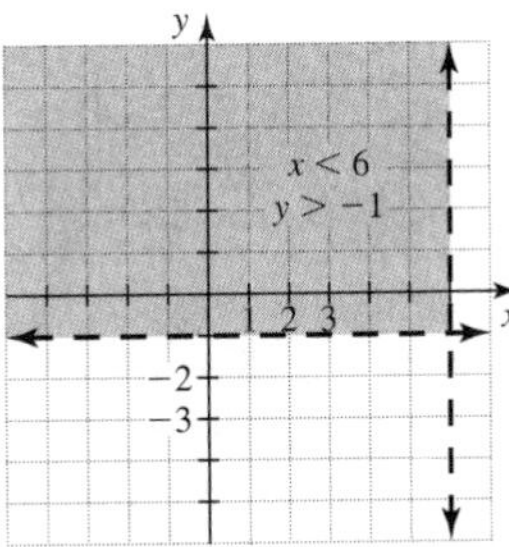

13.

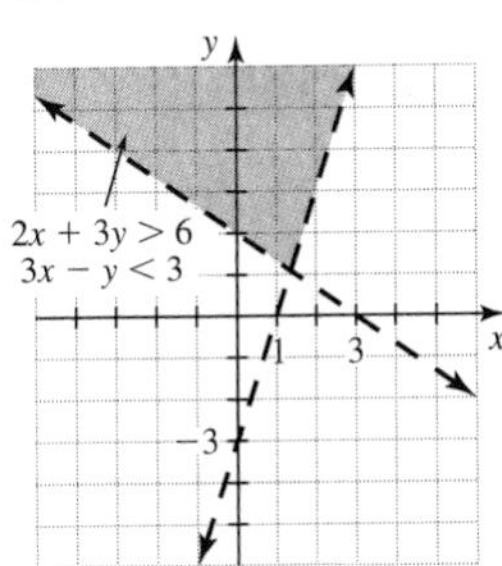

14.

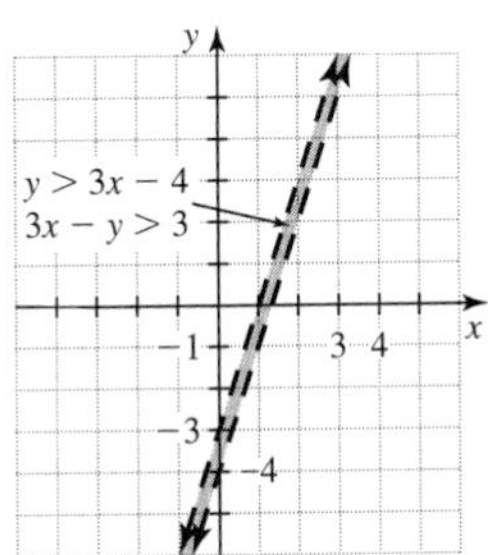

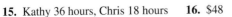

15. Kathy 36 hours, Chris 18 hours **16.** \$48

Making Connections Chapters 1–4

1. $\{7\}$ **2.** $\left\{\frac{5}{3}\right\}$ **3.** $\{12\}$ **4.** $\{1000\}$ **5.** $\varnothing$
6. All real numbers
7. $x > 4$

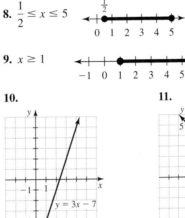

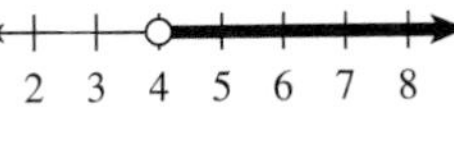

8. $\frac{1}{2} \leq x \leq 5$

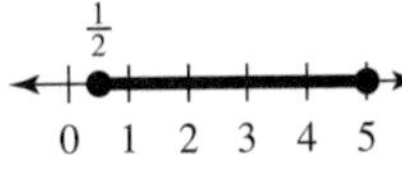

9. $x \geq 1$

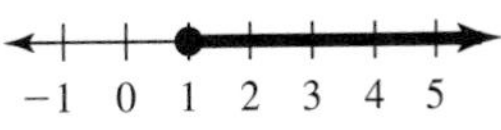

10.

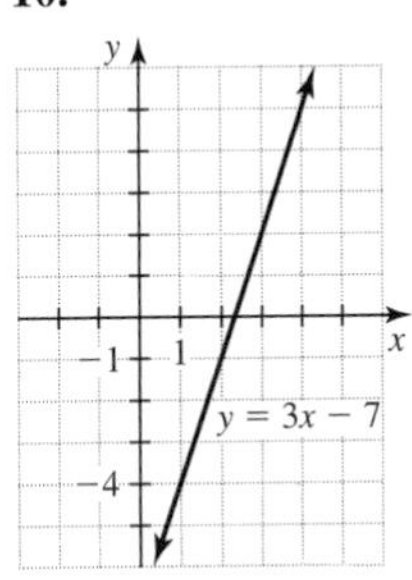

11.

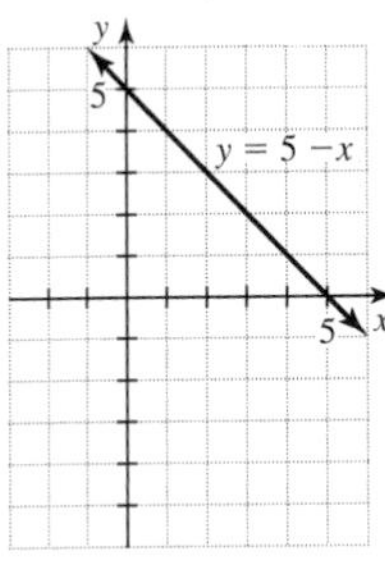

12.

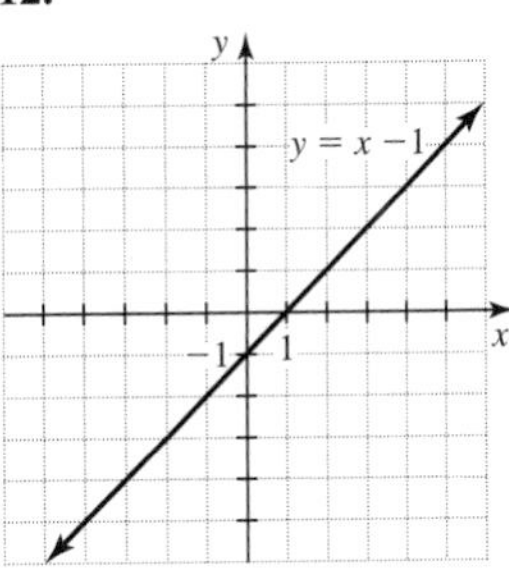

13.

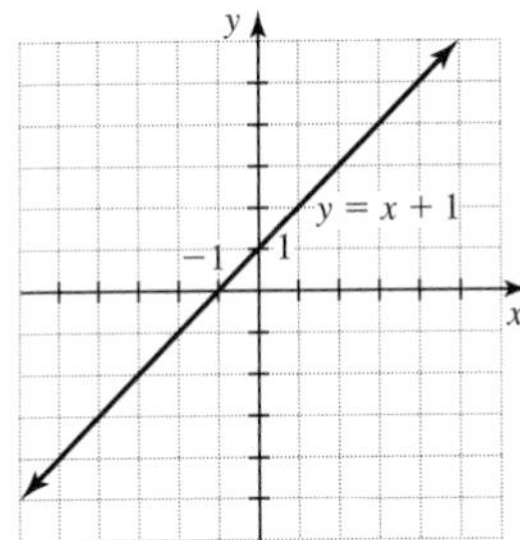

14.

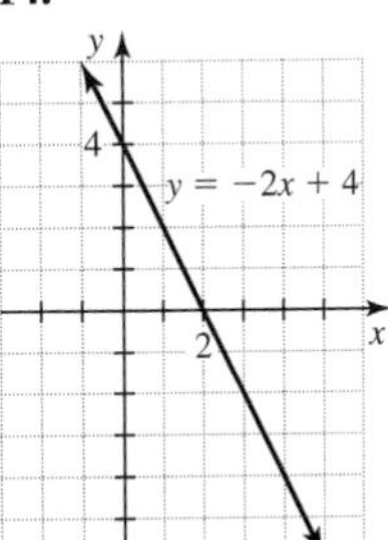

15.

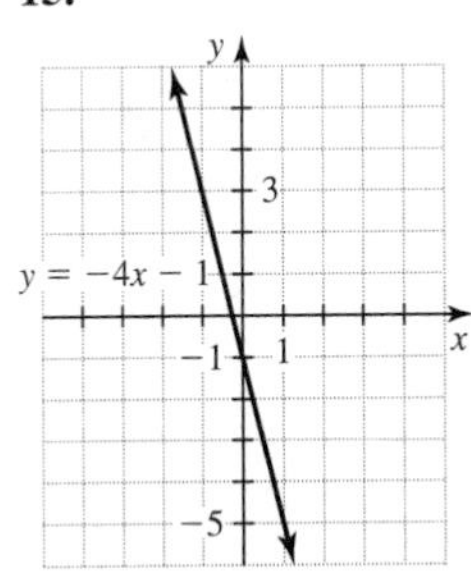

16.

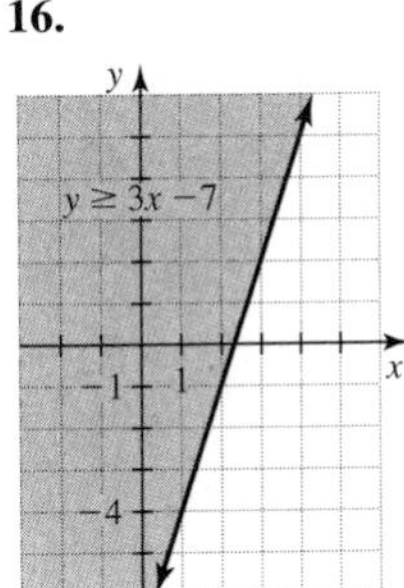

17.

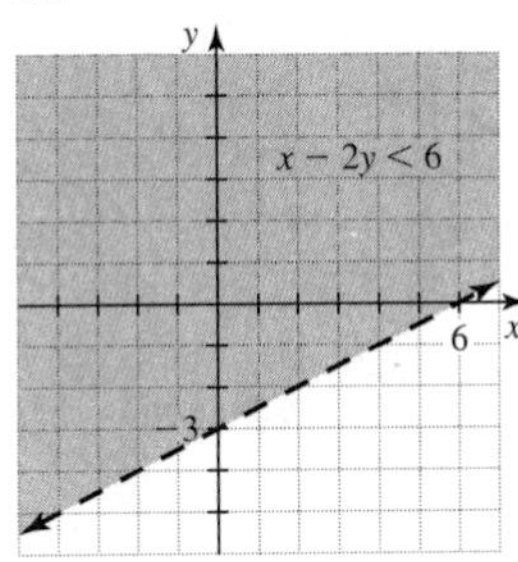

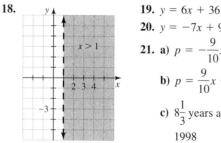

18.

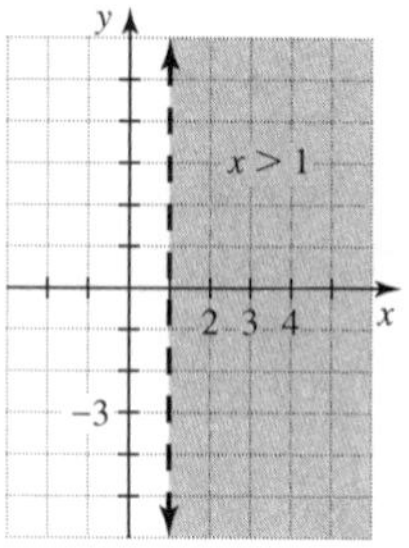

19. $y = 6x + 36$
20. $y = -7x + 95$
21. a) $p = -\frac{9}{10}x + 25$
b) $p = \frac{9}{10}x + 10$
c) $8\frac{1}{3}$ years after 1990 or 1998

Chapter 5

Section 5.1 Warm-ups F F T T F T F T T F

1. A term is a single number or the product of a number and one or more variables raised to powers.
2. A polynomial is a single term or a finite sum of terms.
3. The degree of a polynomial in one variable is the highest power of the variable in the polynomial.
4. The value of a polynomial is the number obtained when the variable is replaced by a number.
5. Polynomials are added by adding the like terms.
6. Polynomials are subtracted by subtracting like terms.

7. $-3, 7$ **8.** $10, -1$ **9.** $0, 6$ **10.** $-1, 0$
11. $\frac{1}{3}, \frac{7}{2}$ **12.** $\frac{1}{2}, -\frac{1}{4}$ **13.** Monomial, 0
14. Monomial, 0 **15.** Monomial, 3 **16.** Monomial, 8
17. Binomial, 1 **18.** Binomial, 1 **19.** Trinomial, 10
20. Trinomial, 6 **21.** Binomial, 6 **22.** Binomial, 2
23. Trinomial, 3 **24.** Trinomial, 2 **25.** 6 **26.** 16
27. $\frac{5}{8}$ **28.** $-\frac{1}{2}$ **29.** -85 **30.** -43 **31.** 71
32. -137 **33.** -4.97665 **34.** -42.608 **35.** $4x - 8$
36. $2x + 1$ **37.** $2q$ **38.** $2q + 10$ **39.** $x^2 + 3x - 2$
40. $2x^2 - 3$ **41.** $x^3 + 9x - 7$ **42.** $x^2 - x - 1$
43. $3a^2 - 7a - 4$ **44.** $2w^2 - 4$ **45.** $-3w^2 - 8w + 5$
46. $a^3 - 4a^2 - 6a + 6$ **47.** $9.66x^2 - 1.93x - 1.49$
48. $9.5x^2 - 1.12x - 11.65$ **49.** $-4x + 6$ **50.** $-2x - 6$
51. -5 **52.** -4 **53.** $-z^2 + 2z$ **54.** $-4z^2 - z$
55. $w^5 + w^4 - w^3 - w^2$ **56.** $w^6 - w^3 + w^2 - w$
57. $2t + 13$ **58.** $-4t^2 - 3t + 9$ **59.** $-8y + 7$
60. $y^3 - y^2 - 2y + 2$ **61.** $-22.85x - 423.2$
62. $288.96x - 780.4$ **63.** $4a + 2$ **64.** $3w - 5$
65. $-2x + 4$ **66.** $2x - 6$ **67.** $2a$ **68.** $2s - 7$
69. $-5m + 7$ **70.** $-8n + 6$ **71.** $4x^2 + 1$
72. $2x^2 + 3x - 11$ **73.** $a^3 - 9a^2 + 2a + 7$
74. $-3b^3 + 7b^2 + 4b - 7$ **75.** $-3x + 9$
76. $-2x^4 - x^2 + 2$ **77.** $2y^3 + 7y^2 - 4y - 14$
78. $-2q^2 - 11q + 14$ **79.** $-3m + 3$ **80.** $8m - 12$
81. $-11y - 3$ **82.** $-14y$ **83.** $2x^2 - 6x + 12$
84. $-8x^2 - 13x + 10$ **85.** $-5z^4 - 8z^3 + 3z^2 + 7$
86. $-v^4 - 3v^2 + v + 6$
87. $P(x) = 100x + 500$ dollars, \$5500
88. a) $T(x) = 0.4x^2 + 450x + 1350$ b) \$326,350
c) \$275,550, \$50,800
89. $P(x) = 6x + 3$, $P(4) = 27$ meters
90. $P(x) = 10x + 8$, $P(4) = 48$ feet
91. $5x + 40$ miles, 140 miles **92.** $5x + 30$ miles, 255 miles
93. 800 feet, 800 feet
94. a) $16t$ feet
b) 32 feet
c) When green ball hits ground
95. $0.17x + 74.47$ dollars, \$244.47
96. $-0.1x + 16.5$ milliliters, 11.5 milliliters
97. 1321.39 calories **98.** 1957.8 calories
99. Yes, yes, yes **100.** No, yes, yes
101. The highest power of x is 3. **102.** b and d

Section 5.2 Warm-ups F F T F T T T T T F

1. The product rule for exponents says that $a^m \cdot a^n = a^{m+n}$.
2. The sum of two monomials can be a binomial if the terms are not like terms.
3. To multiply a monomial and a polynomial we use the distributive property.
4. To multiply two binomials we use the distributive property twice.
5. To multiply any two polynomials we multiply each term of the first polynomial by every term of the second polynomial.
6. To find the opposite of a polynomial change the sign of each term in the polynomial.

7. $27x^5$ **8.** $15x^{12}$ **9.** $14a^{11}$ **10.** $15y^{27}$ **11.** $-30x^4$
12. $-16x^7$ **13.** $27x^{17}$ **14.** $16x^{11}$ **15.** $-54s^2t^2$
16. $-36qs^2$ **17.** $24t^7w^8$ **18.** $5h^9k^3$ **19.** $25y^2$
20. $36x^2$ **21.** $4x^6$ **22.** $9y^{10}$ **23.** $4y^7 - 8y^3$
24. $6t^8 + 18t^5$ **25.** $-18y^2 + 12y$ **26.** $-9y^3 + 9y$
27. $-3y^3 + 15y^2 - 18y$ **28.** $7x^5 - 35x^4 - 7x^2$
29. $-xy^2 + x^3$ **30.** $ab^3 - a^3b$
31. $15a^4b^3 - 5a^5b^2 - 10a^6b$ **32.** $24c^3d^3 - 8cd^5 + 8cd^2$
33. $-2t^5v^3 + 3t^3v^2 + 2t^2v^2$ **34.** $2m^3n^5 - m^3n^4 + 4m^2n^3$
35. $x^2 + 3x + 2$ **36.** $x^2 + 9x + 18$ **37.** $x^2 + 2x - 15$
38. $y^2 + 2y - 8$ **39.** $t^2 - 13t + 36$ **40.** $w^2 - 8w + 15$
41. $x^3 + 3x^2 + 4x + 2$ **42.** $x^3 - 1$
43. $6y^3 + y^2 + 7y + 6$ **44.** $4y^3 + 15y^2 + 13y + 3$
45. $2y^8 - 3y^6z - 5y^4z^2 + 3y^2z^3$
46. $18m^7n^2 - 23m^5n^4 - 3m^9 - 4m^3n^6$
47. $2a^2 + 7a - 15$ **48.** $2w^2 + 4w - 30$
49. $14x^2 + 95x + 150$ **50.** $15x^2 + 51x + 42$
51. $20x^2 - 7x - 6$ **52.** $8x^2 - 18x - 18$
53. $2am - 6an + mb - 3nb$ **54.** $3ax + 7a - 6xb - 14b$
55. $x^3 + 9x^2 + 16x - 12$ **56.** $-x^3 + 10x^2 - 26x + 35$
57. $-4a^4 + 9a^2 - 8a - 12$ **58.** $15x^3 - 7x^2 - 20x + 12$
59. $x^2 - y^2$ **60.** $a^4 - b^4$ **61.** $x^3 + y^3$ **62.** $8w^3 - v^3$
63. $u - 3t$ **64.** $3t + u$ **65.** $-3x - y$ **66.** $3y - x$
67. $3a^2 + a - 6$ **68.** $-3b^2 + b + 6$
69. $-3v^2 - v + 6$ **70.** $3t^2 - t + 6$
71. $-6x^2 + 27x$ **72.** $3x - 2$ **73.** $-6x^2 + 27x + 2$

74. $30 - 12x$ **75.** $-x - 7$ **76.** $-5x + 11$ **77.** $36x^{12}$
78. $9a^6b^2$ **79.** $-6a^3b^{10}$ **80.** $-32s^2tx^2$
81. $25x^2 + 60x + 36$ **82.** $25x^2 - 60x + 36$
83. $25x^2 - 36$ **84.** $4x^2 - 81$ **85.** $6x^7 - 8x^4$
86. $4a^4b^3$ **87.** $m^3 - 1$ **88.** $a^3 + b^3$
89. $3x^3 - 5x^2 - 25x + 18$ **90.** $-18y^3 + 21y^2 + 37y - 35$
91. $x^2 + 4x$ square feet, 140 square feet
92. $2x^2 + 3x - 2$ square meters, 63 square meters
93. $A(x) = x^2 + \frac{1}{2}x$, $A(5) = 27.5$ square feet
94. $V(x) = 6x^3 - 10x^2$, $V(3) = 72$ cubic inches
95. $x^2 + 5x$ **96.** $9x - x^2$
97. $8.05x^2 + 15.93x + 6.12$ square meters
98. $4.33x^2$ square inches
99. 30,000, \$300,000, $40{,}000p - 1000p^2$
100. $2000p - 100p^2$, 0, \$0, \$10
101. $10x^5 + 10x^4 + 10x^3 + 10x^2 + 10x$, \$67.16
102. $10x^5 + 20x^4 + 30x^3 + 40x^2 + 50x$, \$180.35

Section 5.3 Warm-ups F T T T T F T F F F

1. We use the distributive property to find the product of two binomials.
2. FOIL stands for first, outer, inner, and last.
3. The purpose of FOIL is to provide a faster method for finding the product of two binomials.
4. The maximum number of terms obtained in multiplying binomials is four.

5. $x^2 + 6x + 8$ **6.** $x^2 + 8x + 15$ **7.** $a^2 - a - 6$
8. $b^2 + b - 2$ **9.** $2x^2 - 5x + 2$ **10.** $2y^2 - 9y + 10$
11. $2a^2 - a - 3$ **12.** $3x^2 + 7x - 20$
13. $w^2 - 60w + 500$ **14.** $w^2 - 50w + 600$
15. $y^2 - ay + 5y - 5a$ **16.** $3a + 3t - ay - ty$
17. $5w - w^2 + 5m - mw$ **18.** $ab - hb + at - ht$
19. $10m^2 - 9mt - 9t^2$ **20.** $2x^2 - 3xy - 5y^2$
21. $45a^2 + 53ab + 14b^2$ **22.** $11x^2 + 47xy + 12y^2$
23. $x^4 - 3x^2 - 10$ **24.** $y^4 - y^2 - 2$
25. $h^6 + 10h^3 + 25$ **26.** $y^{12} - 3y^6 - 4$
27. $3b^6 + 14b^3 + 8$ **28.** $5n^8 + 14n^4 - 3$
29. $y^3 - 2y^2 - 3y + 6$ **30.** $x^3 - x^2 - x + 1$
31. $6m^6 + 7m^3n^2 - 3n^4$ **32.** $36y^8 - 30y^4z^2 + 6z^4$
33. $12u^4v^2 + 10u^2v - 12$ **34.** $10y^6w^4 + 17y^3w^2z + 3z^2$
35. $b^2 + 9b + 20$ **36.** $y^2 + 12y + 32$
37. $x^2 + 6x - 27$ **38.** $m^2 - m - 56$
39. $a^2 + 10a + 25$ **40.** $t^2 - 8t + 16$
41. $4x^2 - 4x + 1$ **42.** $9y^2 + 24y + 16$
43. $z^2 - 100$ **44.** $9h^2 - 25$ **45.** $a^2 + 2ab + b^2$
46. $x^2 - 2xy + y^2$ **47.** $a^2 - 3a + 2$
48. $b^2 - 9b + 8$ **49.** $2x^2 + 5x - 3$
50. $3y^2 - 4y - 15$ **51.** $5t^2 - 7t + 2$
52. $4t^2 - 8t + 3$ **53.** $h^2 - 16h + 63$
54. $h^2 - 14hw + 49w^2$ **55.** $h^2 + 14hw + 49w^2$
56. $h^2 - 49q^2$ **57.** $4h^4 - 4h^2 + 1$
58. $9h^4 + 6h^2 + 1$ **59.** $8a^2 + a - \frac{1}{4}$
60. $18b^2 + 3b - \frac{2}{9}$ **61.** $\frac{1}{8}x^2 + \frac{1}{6}x - \frac{1}{6}$
62. $\frac{1}{3}t^2 - \frac{11}{24}t + \frac{1}{8}$ **63.** $-12x^6 - 26x^5 + 10x^4$
64. $24x^3y^3 - 4x^2y^4 - 4xy^5$ **65.** $x^3 + 3x^2 - x - 3$
66. $a^3 - 4a^2 - 17a + 60$ **67.** $9x^3 + 45x^2 - 4x - 20$
68. $81x^3 - 486x^2 - 16x + 96$ **69.** $2x + 10$ **70.** $k - 15$
71. $2x^2 + 5x - 3$ square feet
72. $6x^2 + 13x + 6$ square meters
73. $5.2555x^2 + 0.41095x - 1.995$ square meters
74. $2.9998x^2 + 4.8981x + 1.9994$ square meters
75. 12 ft², $3h$ ft², $4h$ ft², h^2 ft², $h^2 + 7h + 12$ ft², $(h + 3)(h + 4) = h^2 + 7h + 12$
76. a^2, ab, ab, b^2, $a^2 + 2ab + b^2$, $(a + b)(a + b) = a^2 + 2ab + b^2$

Section 5.4 Warm-ups F T T T F T T T F F

1. The special products are $(a + b)^2$, $(a - b)^2$, and $(a + b)(a - b)$.
2. $(a + b)^2 = a^2 + 2ab + b^2$
3. It is faster to do by the new rule than with FOIL.
4. In $(a + b)(a - b)$ the inner and outer products have a sum of zero.
5. $(a + b)(a - b) = a^2 - b^2$
6. Higher powers of binomials are found by using the distributive property.

7. $x^2 + 2x + 1$ **8.** $y^2 + 4y + 4$ **9.** $y^2 + 8y + 16$
10. $z^2 + 6z + 9$ **11.** $9x^2 + 48x + 64$
12. $4m^2 + 28m + 49$ **13.** $s^2 + 2st + t^2$
14. $x^2 + 2xz + z^2$ **15.** $4x^2 + 4xy + y^2$
16. $9t^2 + 6tv + v^2$ **17.** $4t^2 + 12ht + 9h^2$
18. $9z^2 + 30kz + 25k^2$ **19.** $a^2 - 6a + 9$
20. $w^2 - 8w + 16$ **21.** $t^2 - 2t + 1$ **22.** $t^2 - 12t + 36$
23. $9t^2 - 12t + 4$ **24.** $25a^2 - 60a + 36$
25. $s^2 - 2st + t^2$ **26.** $r^2 - 2rw + w^2$
27. $9a^2 - 6ab + b^2$ **28.** $16w^2 - 56w + 49$
29. $9z^2 - 30yz + 25y^2$ **30.** $4z^2 - 12wz + 9w^2$
31. $a^2 - 25$ **32.** $x^2 - 36$ **33.** $y^2 - 1$ **34.** $p^2 - 4$
35. $9x^2 - 64$ **36.** $36x^2 - 1$ **37.** $r^2 - s^2$
38. $b^2 - y^2$ **39.** $64y^2 - 9a^2$ **40.** $16u^2 - 81v^2$
41. $25x^4 - 4$ **42.** $9y^4 - 1$ **43.** $x^3 + 3x^2 + 3x + 1$
44. $y^3 - 3y^2 + 3y - 1$ **45.** $8a^3 - 36a^2 + 54a - 27$
46. $27w^3 - 27w^2 + 9w - 1$
47. $a^4 - 12a^3 + 54a^2 - 108a + 81$
48. $16b^4 + 32b^3 + 24b^2 + 8b + 1$
49. $a^4 + 4a^3b + 6a^2b^2 + 4ab^3 + b^4$
50. $16a^4 - 96a^3b + 216a^2b^2 - 216ab^3 + 81b^4$
51. $a^2 - 400$ **52.** $1 - x^2$ **53.** $x^2 + 15x + 56$
54. $x^2 - 4x - 45$ **55.** $16x^2 - 1$ **56.** $81y^2 - 1$
57. $81y^2 - 18y + 1$ **58.** $16x^2 - 8x + 1$

59. $6t^2 - 7t - 20$ **60.** $6t^2 + 7t - 20$
61. $4t^2 - 20t + 25$ **62.** $4t^2 + 20t + 25$
63. $4t^2 - 25$ **64.** $9t^2 - 16$ **65.** $x^4 - 1$ **66.** $y^6 - 1$
67. $4y^6 - 36y^3 + 81$ **68.** $9z^8 - 48z^4 + 64$
69. $4x^6 + 12x^3y^2 + 9y^4$ **70.** $16y^{10} + 16y^5w^3 + 4w^6$
71. $\frac{1}{4}x^2 + \frac{1}{3}x + \frac{1}{9}$ **72.** $\frac{4}{9}y^2 - \frac{2}{3}y + \frac{1}{4}$
73. $0.04x^2 - 0.04x + 0.01$ **74.** $0.01y^2 + 0.1y + 0.25$
75. $a^3 + 3a^2b + 3ab^2 + b^3$
76. $8a^3 - 36a^2b + 54ab^2 - 27b^3$
77. $2.25x^2 + 11.4x + 14.44$
78. $11.9025a^2 - 15.87a + 5.29$
79. $12.25t^2 - 6.25$ **80.** $20.25h^2 - 32.49$
81. $x^2 - 25$ square feet, 25 square feet smaller
82. 4 square feet **83.** $3.14b^2 + 6.28b + 3.14$ square meters
84. $6\pi t + 9\pi$ square centimeters **85.** $v = k(R^2 - r^2)$
86. $C = 1.2\pi[2rw + w^2]$, \$307,624.75
87. $P + 2Pr + Pr^2$, \$242 **88.** $P + Pr + \frac{Pr^2}{4}$, \$220.50
89. \$20,230.06 **90.** \$26,619.83
91. The first is an identity and the second is a conditional equation.
92. A sum or difference can be squared with the distributive property, FOIL, or the special product rules. It is easier with the special product rules.

Section 5.5 Warm-ups F F T F T F T T T T
1. The quotient rule is used for dividing monomials.
2. The zero power of an nonzero real number is 1.
3. When dividing a polynomial by a monomial the quotient should have the same number of terms as the polynomial.
4. The terms of a polynomial should be written in descending order of the exponents.
5. The long division process stops when the degree of the remainder is less than the degree of the divisor.
6. Insert a term with zero coefficient for each missing term when doing long division.
7. 1 **8.** 1 **9.** 1 **10.** 1 **11.** 1 **12.** -2 **13.** 1
14. 1 **15.** x^6 **16.** y^6 **17.** $\frac{3}{a^5}$ **18.** $\frac{10}{b^4}$ **19.** $\frac{-4}{x^4}$
20. $\frac{2}{y^5}$ **21.** $-y$ **22.** $-a$ **23.** $-3x$ **24.** $2hk$
25. $\frac{-3}{x^3}$ **26.** $\frac{6}{z^6}$ **27.** $x - 2$ **28.** $-y + 2$
29. $x^3 + 3x^2 - x$ **30.** $2y^4 - 3y^2 + 4$ **31.** $4xy - 2x + y$
32. $3 + 2a^2b$ **33.** $y^2 - 3xy$ **34.** $-2h^3 + 3k$
35. $2, -1$ **36.** $-3, 10$ **37.** $x + 5, 16$ **38.** $x - 7, 30$
39. $x + 2, 7$ **40.** $x, 6$ **41.** $2, -10$ **42.** $5, 5$
43. $a^2 + 2a + 8, 13$ **44.** $w^2 + 4w + 8, 13$ **45.** $x - 4, 4$
46. $3x - 3, 3$ **47.** $h^2 + 3h + 9, 0$ **48.** $w^2 - w + 1, 0$
49. $2x - 3, 1$ **50.** $b + 6, -9$ **51.** $x^2 + 1, -1$
52. $a^2 - a + 2, 0$ **53.** $3 + \frac{15}{x - 5}$ **54.** $2 + \frac{2}{x - 1}$
55. $-1 + \frac{3}{x + 3}$ **56.** $-3 + \frac{3}{x + 1}$ **57.** $1 - \frac{1}{x}$
58. $1 - \frac{5}{a}$ **59.** $3 + \frac{1}{x}$ **60.** $2 + \frac{1}{y}$ **61.** $x - 1 + \frac{1}{x + 1}$
62. $x + 1 + \frac{1}{x - 1}$ **63.** $x - 2 + \frac{8}{x + 2}$
64. $x + 1 + \frac{2}{x - 1}$ **65.** $x^2 + 2x + 4 + \frac{8}{x - 2}$
66. $x^2 - x + 1 + \frac{-2}{x + 1}$ **67.** $x^2 + \frac{3}{x}$ **68.** $x + \frac{2}{x}$
69. $-3a$ **70.** $2x^5$ **71.** $\frac{4t^4}{w^5}$ **72.** $\frac{-3y^4}{z^7}$ **73.** $-a + 4$
74. $2 - z$ **75.** $x - 3$ **76.** $x^2 + 3x - 5$
77. $-6x^2 + 2x - 3$ **78.** $3x^2 - x + 5$ **79.** $t + 4$
80. $b + 7$ **81.** $2w + 1$ **82.** $z + 6$ **83.** $4x^2 - 6x + 9$
84. $4y^2 + 2y + 1$ **85.** $t^2 - t + 3$ **86.** $2u^2 + u - 1$
87. $v^2 - 2v + 1$ **88.** $y^2 + 2y + 2$ **89.** $x - 5$ meters
90. $2x + 3$ yards
91. $x^8 + x^7 + x^6 + x^5 + x^4 + x^3 + x^2 + x + 1$
92. $a^7 + a^6b + a^5b^2 + a^4b^3 + a^3b^4 + a^2b^5 + ab^6 + b^7$
93. $10x \div 5x$ is not equivalent to the other two.

Section 5.6 Warm-ups F F F F F F T F F T
1. The product rule says that $a^m a^n = a^{m+n}$.
2. The quotient rule says that $a^m/a^n = a^{m-n}$ if $m > n$ and $a^m/a^n = \frac{1}{a^{n-m}}$ if $n > m$.
3. These rules do not make sense without identical bases.
4. The power rule for exponents says that $(a^m)^n = a^{mn}$.
5. The power of a product rule says that $(ab)^n = a^n b^n$.
6. The power of a quotient rule says that $\left(\frac{a}{b}\right)^n = \frac{a^n}{b^n}$.
7. 128 **8.** x^{13} **9.** $6u^{10}$ **10.** $-18r^6$ **11.** a^4b^{10}
12. x^5y^7 **13.** $\frac{-1}{2a^4}$ **14.** $\frac{-1}{2t^9}$ **15.** $\frac{2a^6}{5b^4}$ **16.** $\frac{3y^3}{4x}$
17. 200 **18.** 4000 **19.** x^6 **20.** y^8 **21.** $2x^{12}$
22. $3y^{17}$ **23.** $\frac{1}{t^2}$ **24.** $\frac{1}{r^7}$ **25.** $\frac{1}{2}$ **26.** $\frac{1}{2y^4}$
27. x^3y^6 **28.** w^6y^{12} **29.** $-8t^{15}$ **30.** $-27r^9$
31. $-8x^6y^{15}$ **32.** $-27y^6z^9$ **33.** $a^9b^2c^{14}$ **34.** $\frac{2b^6c^{11}}{a^7}$
35. $\frac{x^{12}}{64}$ **36.** $\frac{y^6}{8}$ **37.** $\frac{16a^8}{b^{12}}$ **38.** $\frac{81r^6}{t^{10}}$ **39.** $-\frac{x^6}{8y^3}$
40. $\frac{81y^{24}}{16z^4}$ **41.** $\frac{4z^{12}}{x^8}$ **42.** $\frac{-125s^{21}}{t^9}$
43. 45 **44.** 4 **45.** 81 **46.** 16 **47.** -19 **48.** 91
49. -1 **50.** 343 **51.** $\frac{8}{125}$ **52.** $\frac{27}{64}$ **53.** 200
54. 27,000 **55.** 128 **56.** 1,000,000 **57.** $\frac{1}{16}$ **58.** $\frac{1}{81}$
59. $15x^{11}$ **60.** $-6y^4$ **61.** $-125x^{12}$ **62.** $64z^9$
63. $-27y^6z^{19}$ **64.** $4a^{13}b^7$ **65.** $\frac{3v}{u}$ **66.** $-4a$
67. $-16x^9t^6$ **68.** $-81a^{11}b^7$ **69.** $\frac{8}{x^6}$ **70.** $9y^6$

71. $\frac{-32a^{15}b^{20}}{c^{25}}$ 72. $\frac{-32c^5}{b^{20}}$ 73. $\frac{y^5}{32x^5}$ 74. z^{10}
75. $P(1 + r)^{15}$ 76. $P(1 + r)^6$

Section 5.7 Warm-ups T F F T F T T T T F

1. A negative exponent means "reciprocal," as in $a^{-n} = \frac{1}{a^n}$.
2. The operations can be evaluated in any order.
3. The new quotient rule is $a^m/a^n = a^{m-n}$ for any integers m and n.
4. Convert from scientific notation by multiplying by the appropriate power of 10.
5. Convert from standard notation by counting the number of places the decimal must move so that there is one nonzero digit to the left of the decimal point.
6. Numbers between 1 and 10 are not written in scientific notation.

7. $\frac{1}{3}$ 8. $\frac{1}{27}$ 9. $\frac{1}{16}$ 10. $\frac{1}{81}$ 11. $-\frac{1}{16}$ 12. $-\frac{1}{16}$
13. 4 14. $\frac{4}{9}$ 15. $\frac{8}{125}$ 16. $\frac{9}{16}$ 17. $\frac{1}{3}$ 18. $\frac{3}{4}$
19. 1250 20. 400 21. 82 22. 702 23. x 24. y^2
25. $-\frac{16}{x^4}$ 26. $-\frac{30}{y^2}$ 27. $\frac{6}{a^5}$ 28. $\frac{1}{b^8}$ 29. $\frac{1}{u^8}$
30. $\frac{1}{w^{10}}$ 31. $-4t^2$ 32. $\frac{2}{w}$ 33. $2x^{11}$ 34. $-3y^{15}$
35. $\frac{1}{x^{10}}$ 36. $\frac{1}{y^8}$ 37. a^9 38. b^{10} 39. $\frac{x^{12}}{16}$ 40. $\frac{y^2}{9}$
41. $\frac{y^6}{16x^4}$ 42. $\frac{s^2}{6t^4}$ 43. $\frac{x^2}{4y^6}$ 44. $27a^6b^9$ 45. $\frac{a^{16}}{16c^8}$
46. $\frac{w^4x^6}{9}$ 47. $\frac{1}{6}$ 48. $\frac{5}{6}$ 49. $\frac{3}{2}$ 50. $\frac{2}{7}$ 51. $-14x^6$
52. $\frac{2a^2}{b^2}$ 53. $\frac{2a^4}{b^2}$ 54. $6x^{12}y^3$ 55. 9,860,000,000
56. 40,070 57. 0.00137 58. 0.000093 59. 0.000001
60. 0.3 61. 600,000 62. 8,000,000 63. 9×10^3
64. 5.298×10^6 65. 7.8×10^{-4} 66. 2.14×10^{-4}
67. 8.5×10^{-6} 68. 5.67×10^9 69. 5.25×10^{11}
70. 3.4×10^{-11} 71. 6×10^{-10} 72. 8×10^{14}
73. 2×10^{-38} 74. 3×10^2 75. 5×10^{27}
76. 2.5×10^{-16} 77. 9×10^{24} 78. 8×10^{-15}
79. 1.25×10^{14} 80. 2×10^{-15} 81. 2.5×10^{-33}
82. 3.6×10^{23} 83. 8.6×10^9 84. 1.6×10^{14}
85. 2.1×10^2 86. 3×10^3 87. 2.7×10^{-23}
88. 2×10^{-1} 89. 3×10^{15} 90. 2×10^{17}
91. 9.135×10^2 92. 3.758×10^{13} 93. 5.715×10^{-4}
94. 7.359×10^9 95. 4.426×10^7 96. 6.337×10^{-57}
97. 1.577×10^{182} 98. 1.1×10^{100}
99. 4.910×10^{11} feet 100. 8.3 minutes
101. 4.65×10^{-28} hours 102. 1.735×10^{-15} cm^2
103. 9.040×10^8 feet 104. 6.5×10^{-13} meters
105. a) \$1 per pound and 1%
b) \$1,000,000 dollars per pound. c) No
106. a) \$1 per pound and 1% b) \$100 per pound and 0.1%
107. \$10,727.41 108. \$963.83
109. a) $w < 0$ b) m is odd c) $w < 0$ and m odd
110. e

Enriching Your Mathematical Word Power

1. a 2. d 3. b 4. c 5. d 6. b 7. a 8. b
9. c 10. a 11. a 12. c

Review Exercises

1. $5w - 2$ 2. $y - 5$ 3. $-6x + 4$ 4. $-2x^2 + 2x - 5$
5. $2w^2 - 7w - 4$ 6. $-t^2 - 4t - 2$ 7. $-2m^2 + 3m - 1$
8. $-n^4 + 2n^3 - n^2 + 4$ 9. $-50x^{11}$ 10. $6h^5t^7$
11. $121a^{14}$ 12. $144b^6$ 13. $-4x + 15$ 14. $-3x + 36$
15. $3x^2 - 10x + 12$ 16. $4x^3 - 20x^2 + 5$
17. $15m^5 - 3m^3 + 6m^2$ 18. $-4a^6 - 8a^5 - 16a^4$
19. $x^3 - 7x^2 + 20x - 50$ 20. $x^3 + 8$
21. $3x^3 - 8x^2 + 16x - 8$ 22. $5x^3 - 22x^2 + 5x + 12$
23. $q^2 + 2q - 48$ 24. $w^2 + 17w + 60$
25. $2t^2 - 21t + 27$ 26. $25r^2 + 15r + 2$
27. $20y^2 - 7y - 6$ 28. $11y^2 + 23y + 2$
29. $6x^4 + 13x^2 + 5$ 30. $2x^6 - 7x^3 - 49$ 31. $z^2 - 49$
32. $a^2 - 16$ 33. $y^2 + 14y + 49$ 34. $a^2 + 10a + 25$
35. $w^2 - 6w + 9$ 36. $a^2 - 12a + 36$ 37. $x^4 - 9$
38. $4b^4 - 1$ 39. $9a^2 + 6a + 1$ 40. $1 - 6c + 9c^2$
41. $16 - 8y + y^2$ 42. $81 - 18t + t^2$
43. $-5x^2$ 44. $3x^2$ 45. $\frac{-2a^2}{b^2}$ 46. $\frac{-3t^3}{h^2}$
47. $-x + 3$ 48. $y - 7$ 49. $-3x^2 + 2x - 1$
50. $-4x^2y^3 + 2xy^2 - y$ 51. -1 52. -1
53. $m^3 + 2m^2 + 4m + 8$ 54. $x^3 + x^2 + x + 1$
55. $m^2 - 3m + 6, 0$ 56. $4x^2 - 2x - 9, 0$
57. $b - 5, 15$ 58. $r - 2, 3$ 59. $2x - 1, -8$
60. $3y^2 - 2y + 2, -4$ 61. $x^2 + 2x - 9, 1$
62. $y^2 - 10y + 13, -19$ 63. $2 + \frac{6}{x - 3}$ 64. $3 + \frac{12}{x - 4}$
65. $-2 + \frac{2}{1 - x}$ 66. $-3 + \frac{15}{5 - x}$ 67. $x - 1 - \frac{2}{x + 1}$
68. $x + 6 + \frac{19}{x - 3}$ 69. $x - 1 + \frac{1}{x + 1}$
70. $-2x - 6 + \frac{-18}{x - 3}$ 71. $6y^{30}$ 72. $-15a^8$ 73. $\frac{-5}{c^6}$
74. $-2y^7$ 75. b^{30} 76. y^{40} 77. $-8x^9y^6$
78. $81a^{16}b^{24}$ 79. $\frac{8a^3}{b^3}$ 80. $\frac{27y^3}{8}$ 81. $\frac{8x^6y^{15}}{z^{18}}$
82. $\frac{a^4}{16b^{16}}$ 83. $\frac{1}{32}$ 84. $-\frac{1}{16}$ 85. $\frac{1}{1000}$ 86. $\frac{1}{5}$
87. $\frac{1}{x^3}$ 88. $\frac{1}{a^{12}}$ 89. a^4 90. a^{14} 91. a^{10} 92. b^4
93. $\frac{1}{x^{12}}$ 94. $\frac{1}{x^{50}}$ 95. $\frac{x^9}{8}$ 96. $\frac{9}{y^{10}}$ 97. $\frac{9}{a^2b^6}$
98. $125a^6b^3$ 99. 5×10^3 100. 9×10^{-5}
101. 340,000 102. 0.000000057 103. 4.61×10^{-5}

104. 4.4×10^4 **105.** 0.00000569 **106.** 5,500,000,000
107. 7×10^{-4} **108.** 1.8×10^{30} **109.** 1.6×10^{-15}
110. -2.7×10^{16} **111.** 8×10^1 **112.** 1.5×10^{15}
113. 3.2×10^{-34} **114.** 1.25×10^{32}
115. $x^2 + 10x + 21$ **116.** $k^2 + 9k + 20$
117. $t^2 - 7ty + 12y^2$ **118.** $t^2 + 13tz + 42z^2$ **119.** 2
120. 1 **121.** $-27h^3t^{18}$ **122.** $81y^6c^8$
123. $2w^2 - 9w - 18$ **124.** $6x^2 - 8x - 30$
125. $9u^2 - 25v^2$ **126.** $81x^4 - 4$ **127.** $9h^2 + 30h + 25$
128. $16v^2 - 24v + 9$ **129.** $x^3 + 9x^2 + 27x + 27$
130. $k^3 - 30k^2 + 300k - 1000$ **131.** $14s^5t^6$
132. $-10w^7r^{10}$ **133.** $\frac{k^8}{16}$ **134.** $\frac{81y^{12}}{h^{16}}$
135. $x^2 - 9x - 5$ **136.** $-5x^2 - x - 15$
137. $5x^2 - x - 12$ **138.** $2x^2 - 5x$
139. $x^3 - x^2 - 19x + 4$ **140.** $2x^3 - x^2 - 17x + 12$
141. $x + 6$ **142.** $a + 2$
143. $P(w) = 4w + 88, A(w) = w^2 + 44w, P(50) = 288$ ft, $A(50) = 4700$ ft^2
144. $P(x) = 4x - 48, A(x) = x^2 - 24x, P(44) = 128$ ft, $A(44) = 880$ ft^2
145. $R = -15p^2 + 600p$, \$5040, \$20
146. $R = -3q^2 + 900q$, 0

Chapter 5 Test

1. $7x^3 + 4x^2 + 2x - 11$ **2.** $-x^2 - 9x + 2$
3. $-2y^2 + 3y$ **4.** -1 **5.** $x^2 + x - 1$
6. $15x^5 - 21x^4 + 12x^3 - 3x^2$ **7.** $x^2 + 3x - 10$
8. $6a^2 + a - 35$ **9.** $a^2 - 14a + 49$
10. $16x^2 + 24xy + 9y^2$ **11.** $b^2 - 9$ **12.** $9t^4 - 49$
13. $4x^4 + 5x^2 - 6$ **14.** $x^3 - 3x^2 - 10x + 24$
15. $2 + \frac{6}{x - 3}$ **16.** $x - 5 + \frac{15}{x + 2}$ **17.** $-35x^8$
18. $12x^5y^9$ **19.** $-2ab^4$ **20.** $15x^5$ **21.** $\frac{-32a^5}{b^{10}}$
22. $\frac{3a^4}{b^2}$ **23.** $\frac{3}{t^{16}}$ **24.** $\frac{1}{w^2}$ **25.** $\frac{s^6}{9t^4}$ **26.** $\frac{-8y^3}{x^{18}}$
27. 5.433×10^6 **28.** 6.5×10^{-6} **29.** 4.8×10^{-1}
30. 8.1×10^{-27} **31.** $x - 2, 3$ **32.** $-2x^2 + x + 15$
33. $A(x) = x^2 + 4x, P(x) = 4x + 8, A(4) = 32$ ft^2, $P(4) = 24$ ft
34. $R = -150q^2 + 3000q$, \$14,400

Making Connections Chapters 1–5

1. 8 **2.** 32 **3.** 41 **4.** -2 **5.** 32 **6.** 32
7. -144 **8.** 144 **9.** $\frac{5}{8}$ **10.** $\frac{1}{9}$ **11.** 64 **12.** 34
13. $\frac{5}{6}$ **14.** $\frac{5}{36}$ **15.** 899 **16.** -1 **17.** $x^2 + 8x + 15$
18. $4x + 15$ **19.** $-15t^5v^7$ **20.** $5tv$ **21.** $x^2 + 9x + 20$
22. $x^2 + 7x + 10$ **23.** $x + 3$ **24.** $x^3 + 13x^2 + 55x + 75$
25. $3y - 4$ **26.** $6y^2 - 4y + 1$
27. $\left\{-\frac{1}{2}\right\}$ **28.** $\{7\}$ **29.** $\left\{\frac{14}{3}\right\}$ **30.** $\left\{\frac{7}{4}\right\}$ **31.** $\{0\}$
32. All real numbers **33.** $\left(-\frac{1}{2}, 0\right)$ **34.** $(0, -7)$ **35.** 2
36. $\frac{2}{3}$ **37.** $\frac{14}{3}$ **38.** $-\frac{1}{2}$
39. $\frac{2.25n + 100{,}000}{n}$, \$102.25, \$3.25, \$2.35, It averages out to 10 cents per disk.

Chapter 6

Section 6.1 Warm-ups F F F T T T T F F T
1. To factor means to write as a product.
2. A prime number is an integer greater than 1 that has no factors besides itself and 1.
3. You can find the prime factorization by dividing by prime factors until the result is prime.
4. The GCF for two numbers is the largest number that is a factor of both.
5. The GCF for two monomials consists of the GCF of their coefficients and every variable that they have in common raised to the lowest power that appears on the variable.
6. You can check all factoring by multiplying the factors.
7. $2 \cdot 3^2$ **8.** $2^2 \cdot 5$ **9.** $2^2 \cdot 13$ **10.** $2^2 \cdot 19$
11. $2 \cdot 7^2$ **12.** $2^2 \cdot 5^2$ **13.** $2^2 \cdot 5 \cdot 23$ **14.** $3 \cdot 5 \cdot 23$
15. $2^2 \cdot 3 \cdot 7 \cdot 11$ **16.** $3^2 \cdot 5 \cdot 13$ **17.** 4 **18.** 6
19. 12 **20.** 14 **21.** 8 **22.** 5 **23.** 4 **24.** 6
25. 1 **26.** 1 **27.** $2x$ **28.** $4x^2$ **29.** $2x$ **30.** $3y^3$
31. xy **32.** a^2x **33.** $12ab$ **34.** $15x^2yz^3$ **35.** 1 **36.** 1
37. $6ab$ **38.** $8z$ **39.** $9(3x)$ **40.** $3y(17)$ **41.** $8t(3t)$
42. $3u(6u)$ **43.** $4y^2(9y^3)$ **44.** $3z^2(14z^2)$ **45.** $uv(u^3v^2)$
46. $x^2y(x^3y^2)$ **47.** $2m^4(-7n^3)$ **48.** $4z^3(-2y^3z)$
49. $-3x^3yz(11xy^2z)$ **50.** $-12ab^3c^3(8a^2bc^2)$
51. $x(x^2 - 6)$ **52.** $10y^2(y^2 - 3)$ **53.** $5a(x + y)$
54. $3w(2z + 5a)$ **55.** $h^3(h^2 + 1)$ **56.** $y^5(y + 1)$
57. $2k^3m^4(-k^4 + 2m^2)$ **58.** $3h^3t^2(-2h^2 + t^4)$
59. $2x(x^2 - 3x + 4)$ **60.** $6x(x^2 + 3x + 4)$
61. $6x^2t(2x^2 + 5x - 4t)$ **62.** $3xy(5xy - 3y + 2x)$
63. $(x - 3)(a + b)$ **64.** $(y + 4)(3 + z)$
65. $(x - 5)(x - 1)$ **66.** $(a - 3)(a + 1)$
67. $(m + 1)(m + 9)$ **68.** $(x - 2)(x - 1)$
69. $(a + b)(y + 1)^2$ **70.** $(w + 8)(w + 2)^2$
71. $8(x - y), -8(-x + y)$ **72.** $2(a - 3b), -2(-a + 3b)$
73. $4x(-1 + 2x), -4x(1 - 2x)$
74. $5x(-x + 2), -5x(x - 2)$ **75.** $1(x - 5), -1(-x + 5)$
76. $1(a - 6), -1(-a + 6)$ **77.** $1(4 - 7a), -1(-4 + 7a)$
78. $1(7 - 5b), -1(-7 + 5b)$
79. $8a^2(-3a + 2), -8a^2(3a - 2)$
80. $15b^3(-2b + 5), -15b^3(2b - 5)$
81. $6x(-2x - 3), -6x(2x + 3)$
82. $4b(-5b - 2), -4b(5b + 2)$
83. $2x(-x^2 - 3x + 7), -2x(x^2 + 3x - 7)$
84. $2x^2(-4x^2 + 3x - 1), -2x^2(4x^2 - 3x + 1)$
85. $2ab(2a^2 - 3ab - 2b^2), -2ab(-2a^2 + 3ab + 2b^2)$
86. $3u^2v^3(4u^3v^3 + 6 - 5u^2v^2), -3u^2v^3(-4u^3v^3 - 6 + 5u^2v^2)$
87. $x + 2$ hours **88.** $x + 50$ cm
89. **a)** $S = 2\pi r(r + h)$ **b)** $S = 2\pi r^2 + 10\pi r$ **c)** 3 in.

90. **a)** $A = P(1 + rt)$ **b)** \$12,035
91. The GCF is an algebraic expression.

Section 6.2 Warm-ups F T F F T T T F F T T

1. A perfect square is a square of an integer or an algebraic expression.
2. $a^2 - b^2 = (a + b)(a - b)$
3. A perfect square trinomial is of the form $a^2 + 2ab + b^2$ or $a^2 - 2ab + b^2$.
4. A prime polynomial is a polynomial that cannot be factored.
5. A polynomial is factored completely when it is a product of prime polynomials.
6. Always factor out the GCF first.
7. $(a - 2)(a + 2)$ **8.** $(h - 3)(h + 3)$ **9.** $(x - 7)(x + 7)$
10. $(y - 6)(y + 6)$ **11.** $(y + 3x)(y - 3x)$
12. $(4x - y)(4x + y)$ **13.** $(5a + 7b)(5a - 7b)$
14. $(3a - 8b)(3a + 8b)$ **15.** $(11m + 1)(11m - 1)$
16. $(12n - 1)(12n + 1)$ **17.** $(3w - 5c)(3w + 5c)$
18. $(12w - 11a)(12w + 11a)$ **19.** Perfect square trinomial
20. Neither **21.** Neither **22.** Difference of two squares
23. Perfect square trinomial **24.** Neither **25.** Neither
26. Perfect square trinomial **27.** Difference of two squares
28. Neither **29.** Perfect square trinomial
30. Perfect square trinomial **31.** $(x + 6)^2$ **32.** $(y + 7)^2$
33. $(a - 2)^2$ **34.** $(b - 3)^2$ **35.** $(2w + 1)^2$
36. $(3m + 1)^2$ **37.** $(4x - 1)^2$ **38.** $(5y - 1)^2$
39. $(2t + 5)^2$ **40.** $(3y - 2)^2$ **41.** $(3w + 7)^2$
42. $(12x + 1)^2$ **43.** $(n + t)^2$ **44.** $(x - y)^2$
45. $5(x - 5)(x + 5)$ **46.** $3(y - 3)(y + 3)$
47. $-2(x - 3)(x + 3)$ **48.** $-5(y - 2)(y + 2)$
49. $a(a - b)(a + b)$ **50.** $y(x - 1)(x + 1)$
51. $3(x + 1)^2$ **52.** $3(2a + 3)^2$ **53.** $-5(y - 5)^2$
54. $-2(a + 4)^2$ **55.** $x(x - y)^2$ **56.** $xy(x + y)^2$
57. $-3(x - y)(x + y)$ **58.** $-8(a - b)(a + b)$
59. $2a(x - 7)(x + 7)$ **60.** $2y(4x - y)(4x + y)$
61. $3a(b - 3)^2$ **62.** $-2b(a - 2)^2$ **63.** $-4m(m - 3n)^2$
64. $10a(a - b)^2$ **65.** $(b + c)(x + y)$ **66.** $(3 + a)(x + z)$
67. $(x - 2)(x + 2)(x + 1)$ **68.** $(x - 1)(x + 1)^2$
69. $(3 - x)(a - b)$ **70.** $(x - 4)(a - b)$
71. $(a^2 + 1)(a + 3)$ **72.** $(y^2 + 8)(y - 5)$
73. $(a + 3)(x + y)$ **74.** $(x + 3)(x^2 + a)$
75. $(c - 3)(ab + 1)$ **76.** $(a + t)(x + b)$
77. $(a + b)(x - 1)(x + 1)$ **78.** $(m + n)(a - b)(a + b)$
79. $(y + b)(y + 1)$ **80.** $(c + w^2)(a + m)$
81. $6ay(a + 2y)^2$ **82.** $2b^3c(2b - c)^2$
83. $6ay(2a - y)(2a + y)$ **84.** $3bc(3b - 2c)(3b + 2c)$
85. $2a^2y(ay - 3)$ **86.** $9x^2y(x - 2y)$
87. $(b - 4w)(a + 2w)$ **88.** $(3m - n)(a + 6)$
89. $h = -16(t - 20)(t + 20)$, 6336 feet
90. **a)** $R = p(-0.08p + 300)$ **b)** $-0.08p + 300$
c) 60 pools **d)** \$2000, 140 pools **e)** \$280,000
f) \$0 or \$3800
91. $y - 3$ inches **92.** $k = 12$

Section 6.3 Warm-ups T T F F T F T F F F

1. We factored $ax^2 + bx + c$ with $a = 1$.
2. You can check all factoring by multiplying the factors.
3. If there are no two integers that have product of c and a sum of b, then $x^2 + bx + c$ is prime.
4. A sum of two squares with no common factor is prime.
5. A polynomial is factored completely when all of the factors are prime polynomials.
6. Always look for the GCF first.
7. $(x + 3)(x + 1)$ **8.** $(y + 5)(y + 1)$
9. $(x + 3)(x + 6)$ **10.** $(w + 2)(w + 4)$
11. $(a - 3)(a - 4)$ **12.** $(m - 2)(m - 7)$
13. $(b - 6)(b + 1)$ **14.** $(a + 6)(a - 1)$
15. $(y + 2)(y + 5)$ **16.** $(x + 3)(x + 5)$
17. $(a - 2)(a - 4)$ **18.** $(b - 3)(b - 5)$
19. $(m - 8)(m - 2)$ **20.** $(m - 16)(m - 1)$
21. $(w + 10)(w - 1)$ **22.** $(m + 8)(m - 2)$
23. $(w - 4)(w + 2)$ **24.** $(m - 8)(m + 2)$
25. Prime **26.** Prime **27.** $(m + 16)(m - 1)$
28. $(y + 5)(y - 2)$ **29.** Prime **30.** Prime
31. $(z - 5)(z + 5)$ **32.** $(p - 1)(p + 1)$
33. Prime **34.** Prime **35.** $(m + 2)(m + 10)$
36. $(m + 1)(m + 20)$ **37.** Prime **38.** Prime
39. $(m - 18)(m + 1)$ **40.** $(h + 9)(h - 4)$
41. Prime **42.** Prime **43.** $(t + 8)(t - 3)$
44. $(t - 12)(t + 2)$ **45.** $(t - 6)(t + 4)$
46. $(t + 12)(t + 2)$ **47.** $(t - 20)(t + 10)$
48. $(t + 20)(t + 10)$ **49.** $(x - 15)(x + 10)$
50. $(x - 15)(x - 10)$ **51.** $(y + 3)(y + 10)$
52. $(z + 3)(z + 15)$ **53.** $(x + 3a)(x + 2a)$
54. $(a + 2b)(a + 5b)$ **55.** $(x - 6y)(x + 2y)$
56. $(y + 4t)(y - 3t)$ **57.** $(x - 12y)(x - y)$
58. Prime **59.** Prime **60.** $(x - 8s)(x + 3s)$
61. $w(w - 8)$ **62.** $x^3(x - 1)$ **63.** $2(w - 9)(w + 9)$
64. $6w^2(w - 3)(w + 3)$ **65.** $x^2(w^2 + 9)$
66. $a^2b(a^2 + b^2)$ **67.** $(w - 9)^2$ **68.** $(w + 3)(w + 27)$
69. $6(w - 3)(w + 1)$ **70.** $w(3 - w)(3 + w)$
71. $2x^2(4 - x)(4 + x)$ **72.** $20(w^2 + 5w + 2)$
73. $3(w + 3)(w + 6)$ **74.** $w(w - 6)(w + 3)$
75. $w(w^2 + 18w + 36)$ **76.** $3a(a^2 + 6a + 12)$
77. $8v(w + 2)^2$ **78.** $3t(h + 1)^2$
79. $6xy(x + 3y)(x + 2y)$ **80.** $3xy^2(x^2 - x + 1)$
81. $(a + 6)(a - 2)$ **82.** $(a - 4)(a + 3)$ **83.** $(a - 3)(a)$
84. $(a - 3)(a + 3)$ **85.** $(a + 6)^2$ **86.** $(a - 6)^2$
87. $a^2(a + 12)(a - 3)$ **88.** $a^3(a + 4)(a - 3)$
89. $x + 4$ feet **90.** $2x + 4$ meters **91.** 3 feet and 5 feet
92. Height 1 foot smaller, width 3 feet larger
93. d **94.** c

Section 6.4 Warm-ups T F T F T F F F F T

1. We factored $ax^2 + bx + c$ with $a \neq 1$.
2. In the *ac* method we find two integers whose product is equal to *ac* and whose sum is *b*, and then we use factoring by grouping.
3. If there are no two integers whose product is *ac* and whose sum is *b*, then $ax^2 + bx + c$ is prime.
4. In trial and error, we make an educated guess at the factors and then check by FOIL.
5. 2 and 10 **6.** -2 and -18 **7.** -6 and 2
8. 8 and -1 **9.** 3 and 4 **10.** 2 and 15

11. −2 and −9 12. −4 and −15
13. −3 and 4 14. −20 and 3 15. $(2x + 1)(x + 1)$
16. $(2x + 1)(x + 5)$ 17. $(2x + 1)(x + 4)$
18. $(2h + 1)(h + 3)$ 19. $(3t + 1)(t + 2)$
20. $(3t + 5)(t + 1)$ 21. $(2x - 1)(x + 3)$
22. $(3x + 2)(x - 1)$ 23. $(3x - 1)(2x + 3)$
24. $(3x - 1)(7x + 3)$ 25. Prime 26. Prime
27. $(2x - 3)(x - 2)$ 28. $(3a - 5)(a - 3)$
29. $(5b - 3)(b - 2)$ 30. $(7y - 5)(y + 3)$
31. $(4y + 1)(y - 3)$ 32. $(7x + 1)(5x - 1)$
33. Prime 34. Prime 35. $(4x + 1)(2x - 1)$
36. $(4x + 1)(2x - 3)$ 37. $(3t - 1)(3t - 2)$
38. $(9t - 4)(t + 1)$ 39. $(5x + 1)(3x + 2)$
40. $(5x + 1)(3x - 2)$ 41. $(2a + 3b)(2a + 5b)$
42. $(5x + y)(2x + 3y)$ 43. $(3m - 5n)(2m + n)$
44. $(3a - 7b)(a + 3b)$ 45. $(x - y)(3x - 5y)$
46. $(m - 3n)(3m - 4n)$ 47. $(5a + 1)(a + 2)$
48. $(3y + 7)(y + 1)$ 49. $(2w + 3)(2w + 1)$
50. $(2z + 1)(3z + 5)$ 51. $(5x - 2)(3x + 1)$
52. $(15x - 2)(x + 1)$ 53. $(4x - 1)(2x - 1)$
54. $(4x - 1)(2x - 5)$ 55. $(15x - 1)(x - 2)$
56. $(15x + 1)(x + 2)$ 57. Prime 58. Prime
59. $2(x^2 + 9x - 45)$ 60. $(x + 2)(3x + 5)$
61. $(3x - 5)(x + 2)$ 62. $(3x - 2)(x - 5)$
63. $(5x + y)(2x - y)$ 64. $(4x + y)(2x - y)$
65. $(6a - b)(7a - b)$ 66. $(5a - b)(2a - 5b)$
67. $(x + 2)(3x + 1)$ 68. $(x - 3)(2x + 5)$
69. $(5x + 1)(x + 2)$ 70. $(4x + 1)(x - 5)$
71. $(3a - 1)(2a - 5)$ 72. $(2b - 5)(2b - 3)$
73. $w(9w - 1)(9w + 1)$ 74. $w^2(81w - 1)$
75. $2(2w - 5)(w + 3)$ 76. $2(x - 7)^2$
77. $3(2x + 3)^2$ 78. $12(y + 1)^2$
79. $(3w + 5)(2w - 7)$ 80. $(2y - 5)(4y + 3)$
81. $3z(x - 3)(x + 2)$ 82. $b(a + 5)(a - 3)$
83. $3x(3x^2 - 7x + 6)$ 84. $-2x(4x^2 - 2x + 1)$
85. $(a + 5b)(a - 3b)$ 86. $a^2(b - 5)(b + 3)$
87. $y^2(2x^2 + x + 3)$ 88. $6(3x^2 - x + 1)$
89. $-t(3t + 2)(2t - 1)$ 90. $-6(3t + 2)(2t - 1)$
91. $2t^2(3t - 2)(2t + 1)$ 92. $2t(3t + 2)(2t + 1)$
93. $y(2x - y)(2x - 3y)$ 94. $3(3x - y)(x + 3y)$
95. $-1(w - 1)(4w - 3)$ 96. $-1(5w - 1)(6w + 1)$
97. $-2a(2a - 3b)(3a - b)$ 98. $-3b(3a - b)(4a - b)$
99. $h = -8(2t + 1)(t - 3)$, 0 feet
100. a) $N(t) = -t(3t + 1)(t - 8)$ b) 150 c) 5 hr d) 250 components
101. a) −4, 4 b) ±8, ±16 c) ±1, ±7, ±13, ±29
102. a) −2, −6 b) 1, −8 c) 1, −9

Section 6.5 Warm-ups F F T T F T F T T F
1. If there is no remainder, then the dividend factors as the divisor times the quotient.
2. If you divide $a^3 - b^3$ by $a - b$, there will be no remainder.
3. If you divide $a^3 + b^3$ by $a + b$, there will be no remainder.
4. A sum of two cubes is of the form $a^3 + b^3$.
5. $a^3 + b^3 = (a + b)(a^2 - ab + b^2)$
6. $a^3 - b^3 = (a - b)(a^2 + ab + b^2)$
7. $(x + 4)(x - 3)(x + 2)$ 8. $(x - 1)(x + 3)(x - 2)$
9. $(x - 1)(x + 3)(x + 2)$ 10. $(x + 2)(x - 3)(x - 4)$
11. $(x - 2)(x^2 + 2x + 4)$ 12. $(x + 3)(x^2 - 3x + 9)$
13. $(x + 5)(x^2 - x + 2)$ 14. $(x - 3)(2x^2 + x + 2)$
15. $(x + 1)(x^2 + x + 1)$ 16. $(x + 3)(x - 2)(x + 1)$
17. $(m - 1)(m^2 + m + 1)$ 18. $(z - 3)(z^2 + 3z + 9)$
19. $(x + 2)(x^2 - 2x + 4)$ 20. $(y + 3)(y^2 - 3y + 9)$
21. $(a + 5)(a^2 - 5a + 25)$ 22. $(b - 6)(b^2 + 6b + 36)$
23. $(c - 7)(c^2 + 7c + 49)$ 24. $(d + 10)(d^2 - 10d + 100)$
25. $(2w + 1)(4w^2 - 2w + 1)$
26. $(5m + 1)(25m^2 - 5m + 1)$
27. $(2t - 3)(4t^2 + 6t + 9)$
28. $(5n - 2)(25n^2 + 10n + 4)$
29. $(x - y)(x^2 + xy + y^2)$
30. $(m + n)(m^2 - mn + n^2)$
31. $(2t + y)(4t^2 - 2ty + y^2)$
32. $(u - 5v)(u^2 + 5uv + 25v^2)$
33. $2(x - 3)(x + 3)$ 34. $3x(x - 2)(x + 2)$
35. $4(x + 5)(x - 3)$ 36. $3(x + 3)^2$
37. $x(x + 2)^2$ 38. $a(a - 2)(a - 3)$
39. $5am(x^2 + 4)$ 40. $3bm(w - 2)(w + 2)$
41. Prime 42. Prime 43. $(3x + 1)^2$
44. $3(3x^2 + 2x + 1)$ 45. $y(3x + 2)(2x - 1)$
46. $y^2(5x - 6)(x + 1)$ 47. Prime
48. Prime 49. $3(4a - 1)^2$ 50. $2(2b + 3)^2$
51. $2(4m + 1)(2m - 1)$ 52. $2(2a + 1)(8a - 3)$
53. $(3a + 4)^2$ 54. $3(x - 8)(x + 2)$
55. $2(3x - 1)(4x - 3)$ 56. $2(2x^2 - 3x - 6)$
57. $3a(a - 9)$ 58. $a(a - 25)$ 59. $2(2 - x)(2 + x)$
60. $x(x + 3)^2$ 61. Prime 62. Prime
63. $x(6x^2 - 5x + 12)$ 64. $(x - 1)(x + 1)(x + 2)$
65. $ab(a - 2)(a + 2)$ 66. $2(m - 30)(m + 30)$
67. $(x - 2)(x + 2)^2$ 68. $a(m + a)^2$
69. $2w(w - 2)(w^2 + 2w + 4)$
70. $mn(m + n)(m^2 - mn + n^2)$ 71. $3w(a - 3)^2$
72. $4a(2a^2 + 1)$ 73. $5(x - 10)(x + 10)$
74. $(5x - 4y)(5x + 4y)$ 75. $(2 - w)(m + n)$
76. $(w + 5)(a - b)$ 77. $3x(x + 1)(x^2 - x + 1)$
78. $3a^2(a - 3)(a^2 + 3a + 9)$ 79. $4(w^2 + w - 1)$
80. $(2w + 5)(2w - 1)$ 81. $a^2(a + 10)(a - 3)$
82. $y^3(2y - 5)(y + 4)$ 83. $aw(2w - 3)^2$
84. $bn(3n - 2)(3n + 7)$ 85. $(t + 3)^2$ 86. $t(t + 6)^2$
87. Length $x + 5$ cm, width $x + 3$ cm
88. Length $y - 3$ cm, height $y - 4$ cm
89. $(-1 + 1)^3 = (-1)^3 + 1^3$, $(1 + 2)^3 \neq 1^3 + 2^3$

Section 6.6 Warm-ups F F T T T F T T T F
1. A quadratic equation has the form $ax^2 + bx + c = 0$ with $a \neq 0$.

2. A compound equation is two equations connected with the word or.
3. The zero factor property says that if $ab = 0$ then $a = 0$ or $b = 0$.
4. Quadratic equations are solved by factoring in this section.
5. Dividing each side by a variable is not usually done because the variable might have a value of zero.
6. A triangle is a right triangle if and only if the sum of the squares of the legs is equal to the square of the hypotenuse.

7. $-4, -5$ **8.** $-6, -5$ **9.** $-\frac{5}{2}, \frac{4}{3}$ **10.** $\frac{8}{3}, -\frac{3}{4}$
11. 2, 7 **12.** $-9, 3$ **13.** $-4, 6$ **14.** $-6, 3$
15. $-1, \frac{1}{2}$ **16.** $-1, \frac{3}{2}$ **17.** $0, -7$ **18.** $0, -5$
19. $-5, 4$ **20.** $-7, 6$ **21.** $\frac{1}{2}, -3$ **22.** $1, \frac{7}{3}$
23. $0, -8$ **24.** $-4, 8$ **25.** $-\frac{9}{2}, 2$ **26.** $\frac{11}{3}, -2$
27. $\frac{2}{3}, -4$ **28.** $-\frac{1}{2}, \frac{1}{4}$ **29.** 5 **30.** 4 **31.** $\frac{3}{2}$
32. $-\frac{2}{5}$ **33.** $0, -3, 3$ **34.** $-5, 0, 5$ **35.** $-4, -2, 2$
36. $-2, -1, 1$ **37.** $-1, 1, 3$ **38.** $-5, -1, 5$
39. 0, 4, 5 **40.** $-3, 0, 1$ **41.** $-4, 4$ **42.** $-6, 6$
43. $-3, 3$ **44.** $-5, 5$ **45.** $0, -1, 1$ **46.** $-2, 0, 2$
47. $-3, -2$ **48.** $-4, 3$ **49.** $-\frac{3}{2}, -4$ **50.** $-3, \frac{1}{3}$
51. $-6, 4$ **52.** $-2, \frac{3}{2}$ **53.** $-1, 3$ **54.** $-1, 2$
55. $-4, 2$ **56.** 3, 6 **57.** $-5, -3, 5$
58. $-10, -1, 0, 10$ **59.** Length 12 ft, width 5 ft
60. Width 3 in., length 4 in. **61.** Width 5 ft, length 12 ft
62. 6 m and 8 m **63.** 2 and 3, or -3 and -2
64. -6 and -4, or 4 and 6 **65.** 5 and 6
66. Anita 4, Molly 8
67. **a)** 25 sec **b)** last 5 sec **c)** increasing
68. 4 sec **69.** 6 sec **70.** 2 sec
71. Base 6 in., height 13 in. **72.** 2 m and 3 m
73. 20 ft by 20 ft **74.** 9 in. **75.** 80 ft **76.** 40 m
77. 3 yd by 3 yd, 6 yd by 6 yd **78.** 4 km^2, 9 km^2, 25 km^2
79. 12 mi **80.** 10 paces **81.** 25% **82.** 50%

Enriching Your Mathematical Word Power

1. a **2.** d **3.** c **4.** a **5.** c **6.** b **7.** c **8.** a
9. d **10.** c

Review Exercises

1. $2^4 \cdot 3^2$ **2.** 11^2 **3.** $2 \cdot 29$ **4.** $2^2 \cdot 19$
5. $2 \cdot 3 \cdot 5^2$ **6.** $2^3 \cdot 5^2$ **7.** 18 **8.** 6 **9.** $4x$
10. $3ab$ **11.** $3(x + 2)$ **12.** $x(7x + 1)$
13. $-2(-a + 10)$ **14.** $-a(-a + 1)$ **15.** $a(2 - a)$
16. $3(3 - b)$ **17.** $3x^2y(2y - 3x^3)$ **18.** $a^3b^2(b^3 + 1)$
19. $3y(x^2 - 4x - 3y)$ **20.** $a(2a - 4b^2 - b)$
21. $(y - 20)(y + 20)$ **22.** $(2m - 3)(2m + 3)$
23. $(w - 4)^2$ **24.** $(t + 10)^2$ **25.** $(2y + 5)^2$
26. $2(a^2 - 2a - 1)$ **27.** $(r - 2)^2$ **28.** $3(m - 5)(m + 5)$
29. $2t(2t - 3)^2$ **30.** $(t - 3w)(t + 3w)$ **31.** $(x + 6y)^2$
32. $(3y - 2x)^2$ **33.** $(x - y)(x + 5)$ **34.** $(x + a)(x + y)$
35. $(b + 8)(b - 3)$ **36.** $(a - 7)(a + 5)$
37. $(r - 10)(r + 6)$ **38.** $(x + 8)(x + 5)$
39. $(y - 11)(y + 5)$ **40.** $(a + 10)(a - 4)$
41. $(u + 20)(u + 6)$ **42.** $(v - 25)(v + 3)$
43. $3t^2(t + 4)$ **44.** $-4m^2(m^2 + 9)$
45. $5w(w^2 + 5w + 5)$ **46.** $-3t(t^2 - t + 2)$
47. $ab(2a + b)(a + b)$ **48.** $y^2(2x - y)(3x + y)$
49. $x(3x - y)(3x + y)$ **50.** $h^2(h - 10)(h + 10)$
51. $(7t - 3)(2t + 1)$ **52.** $(5x + 1)(3x - 5)$
53. $(3x + 1)(2x - 7)$ **54.** $(x + 2)(2x - 5)$
55. $(3p + 4)(2p - 1)$ **56.** $(p - 1)(3p + 5)$
57. $-2p(5p + 2)(3p - 2)$ **58.** $-2(3q + 5)(q + 5)$
59. $(6x + y)(x - 5y)$ **60.** $(5a - 2b)(2a + b)$
61. $2(4x + y)^2$ **62.** $2(2a + 5b)^2$ **63.** $5x(x^2 + 8)$
64. $(w + 3)^2$ **65.** $(3x - 1)(3x + 2)$ **66.** $ax(x^2 + 1)$
67. $(x + 2)(x - 1)(x + 1)$ **68.** $(8x + 3)(2x - 1)$
69. $xy(x - 16y)$ **70.** $-3(x - 3)(x + 3)$ **71.** $(a + 1)^2$
72. $-2(w + 3)^2$ **73.** $(x^2 + 1)(x - 1)$
74. $9y^2(x - 1)(x + 1)$ **75.** $(a + 2)(a + b)$
76. $(2m + 5)^2$ **77.** $-2(x - 6)(x - 2)$
78. $3(2x - 3)(x + 5)$ **79.** $(m - 10)(m^2 + 10m + 100)$
80. $(2p + 1)(4p^2 - 2p + 1)$ **81.** $(x + 2)(x^2 - 2x + 5)$
82. $(x - 3)(x^2 + 3x + 4)$ **83.** $(x + 4)(x + 5)(x - 3)$
84. $(x - 5)(x^2 + x + 2)$ **85.** 0, 5 **86.** $-3, -2$
87. 0, 5 **88.** $-8, 7$ **89.** $-\frac{1}{2}, 5$ **90.** $-\frac{3}{4}, \frac{1}{3}$
91. $-4, -3, 3$ **92.** $-5, -1, 1$ **93.** $-2, -1$
94. $-1, 2$ **95.** $-\frac{1}{2}, \frac{1}{4}$ **96.** $\frac{2}{3}, \frac{3}{2}$ **97.** 5, 11
98. $-6, -5, -4$ or 4, 5, 6 **99.** 6 in. by 8 in.
100. 4 and 4.5 **101.** $v = k(R - r)(R + r)$
102. **a)** $V = \frac{4}{3}\pi(R - r)(R^2 + Rr + r^2)$ **b)** 1.5 cm
103. 6 ft **104.** 40 ft

Chapter 6 Test

1. $2 \cdot 3 \cdot 11$ **2.** $2^4 \cdot 3 \cdot 7$ **3.** 16 **4.** 6 **5.** $3y^2$
6. $6ab$ **7.** $5x(x - 2)$ **8.** $6y^2(x^2 + 2x + 2)$
9. $3ab(a - b)(a + b)$ **10.** $(a + 6)(a - 4)$
11. $(2b - 7)^2$ **12.** $3m(m^2 + 9)$ **13.** $(a + b)(x - y)$
14. $(a - 5)(x - 2)$ **15.** $(3b - 5)(2b + 1)$
16. $(m + 2n)^2$ **17.** $(2a - 3)(a - 5)$
18. $z(z + 3)(z + 6)$ **19.** $(x - 1)(x - 2)(x - 3)$
20. $\frac{3}{2}, -4$ **21.** $0, -2, 2$ **22.** $-2, \frac{5}{6}$
23. Length 12 ft, width 9 ft
24. -4 and 8

Making Connections Chapters 1–6

1. -1 **2.** 2 **3.** -3 **4.** 57 **5.** 16
6. 7 **7.** $2x^2$ **8.** $3x$ **9.** $3 + x$ **10.** $6x$ **11.** $24yz$

12. $6y + 8z$ **13.** $4z - 1$ **14.** t^6 **15.** t^{10} **16.** $4t^6$

17. $x < -9$

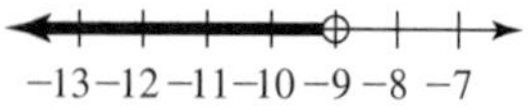

18. $x \geq 3$

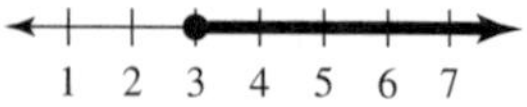

19. $x > 12$

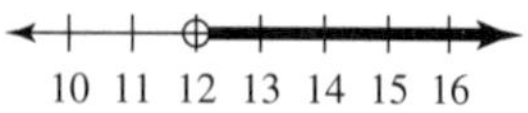

20. $x < 600$

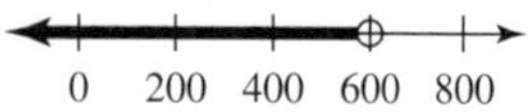

21. $\frac{3}{2}$ **22.** $-\frac{1}{2}$ **23.** $3, -5$ **24.** $\frac{3}{2}, -\frac{1}{2}$ **25.** $0, 3$

26. $0, 1$ **27.** All real numbers **28.** No solution

29. 10 **30.** 40 **31.** $-3, 3$ **32.** $-5, \frac{3}{2}$

33. Length 21 ft, width 13.5 ft

Chapter 7

Section 7.1 Warm-ups F T T F F T T F F T

1. A rational number is a ratio of two integers with the denominator not 0.
2. A rational expression is a ratio of two polynomials with the denominator not 0.
3. A rational number is reduced to lowest terms by dividing the numerator and denominator by the GCF.
4. A rational expression is reduced to lowest terms by dividing the numerator and denominator by the GCF.
5. The quotient rule is used in reducing ratios of monomials.
6. The expressions $a - b$ and $b - a$ are opposites.

7. -3 **8.** 1 **9.** 5 **10.** -2 **11.** $-0.6, 9, 401, -199$

12. $0, 4, -57.95, 3001.999$ **13.** -1 **14.** 7 **15.** $\frac{5}{3}$

16. $\frac{3}{2}$ **17.** $4, -4$ **18.** $-2, 3$

19. Any number can be used. **20.** Any number can be used.

21. $\frac{2}{9}$ **22.** $\frac{2}{3}$ **23.** $\frac{7}{15}$ **24.** $\frac{7}{9}$ **25.** $\frac{2a}{5}$ **26.** $\frac{7y}{5}$

27. $\frac{13}{5w}$ **28.** $\frac{17}{11y}$ **29.** $\frac{3x + 1}{3}$ **30.** $\frac{w + 1}{w}$ **31.** $\frac{2}{3}$

32. $\frac{m + 3w}{m - 2w}$ **33.** $w - 7$ **34.** $a + b$ **35.** $\frac{a - 1}{a + 1}$

36. $\frac{x - y}{x + y}$ **37.** $\frac{x + 1}{2x - 2}$ **38.** $\frac{2x + 4}{3x - 9}$ **39.** $\frac{x + 3}{7}$

40. $x + 1$ **41.** x^3 **42.** y^3 **43.** $\frac{1}{z^5}$ **44.** $\frac{1}{w^3}$

45. $-2x^2$ **46.** $\frac{-2}{y^6}$ **47.** $\frac{-3m^3n^2}{2}$ **48.** $\frac{-3v^5}{2}$

49. $\frac{-3}{4c^3}$ **50.** $\frac{-3}{2x^5y^2}$ **51.** $\frac{5c}{3a^4b^{16}}$ **52.** $\frac{5n^2}{8m^2p}$

53. $\frac{35}{44}$ **54.** $\frac{14}{15}$ **55.** $\frac{11}{8}$ **56.** $\frac{3}{2}$ **57.** $\frac{21}{10x^4}$

58. $\frac{8}{9y^3}$ **59.** $\frac{33a^4}{16}$ **60.** $\frac{18b^{63}}{11}$ **61.** -1

62. -1 **63.** $-h - t$ **64.** $-r - s$ **65.** $\frac{-2}{3h + g}$

66. $\frac{-5}{2b + a}$ **67.** $\frac{-x - 2}{x + 3}$ **68.** $-\frac{a + 1}{a + 2}$ **69.** -1

70. $-\frac{5}{3}$ **71.** $\frac{-2y}{3}$ **72.** $\frac{4 - y}{2}$ **73.** $\frac{x + 2}{2 - x}$

74. $\frac{x - 2}{2x + 4}$ **75.** $\frac{-6}{a + 3}$ **76.** $\frac{-2b - 4}{b - 1}$

77. $\frac{x^4}{2}$ **78.** $\frac{2}{x^7}$ **79.** $\frac{x + 2}{2x}$ **80.** $\frac{1 + 2x}{2}$ **81.** -1

82. $\frac{b - 2}{b + 2}$ **83.** $\frac{-2}{c + 2}$ **84.** $\frac{2}{t - 2}$ **85.** $\frac{x + 2}{x - 2}$

86. $\frac{3}{x - 2}$ **87.** $\frac{-2}{x + 3}$ **88.** $\frac{-2}{x - 2}$ **89.** q^2

90. $\frac{2s^7}{3s - 4}$ **91.** $\frac{u + 2}{u - 8}$ **92.** $\frac{v - 3}{v + 6}$ **93.** $\frac{a^2 + 2a + 4}{2}$

94. $\frac{2}{w + 3}$ **95.** $y + 2$ **96.** $\frac{x + y}{m - 6}$ **97.** $\frac{300}{x + 10}$ hr

98. $\frac{40}{x}$ mph **99.** $\frac{4.50}{x + 4}$ dollars/lb **100.** $\frac{x}{9}$ dollars/lb

101. $\frac{1}{x}$ pool/hr **102.** $\frac{1}{x - 3}$ lawn/hr

103. **a)** \$0.75 **b)** \$0.75, \$0.63, \$0.615 **c)** Approaches \$0.60

104. **a)** \$50,000, \$100,000 **b)** \$1,000,000 **c)** Gets larger and larger without bound

Section 7.2 Warm-ups T T T F T F F T T T

1. Rational numbers are multiplied by multiplying their numerators and their denominators.
2. Rational expressions are multiplied by multiplying their numerators and their denominators.
3. Reducing can be done before multiplying rational numbers or expressions.
4. To divide rational expressions, invert the divisor and multiply.

5. $\frac{7}{9}$ **6.** $\frac{2}{7}$ **7.** $\frac{18}{5}$ **8.** $\frac{5}{6}$ **9.** $\frac{42}{5}$ **10.** $\frac{21}{2}$ **11.** $\frac{a}{44}$

12. $\frac{5}{2p}$ **13.** $\frac{-x^5}{a^3}$ **14.** $\frac{wy^2}{6z^6}$ **15.** $\frac{18t^8y^7}{w^4}$ **16.** $\frac{4x^7}{z^3}$

17. $\frac{2a}{a - b}$ **18.** $\frac{2b^2 + 2}{b + 1}$ **19.** $3x - 9$ **20.** $12x + 30$

21. $\frac{8a + 8}{5a^2 + 5}$ **22.** 2 **23.** 30 **24.** 128 **25.** $\frac{2}{3}$

26. $\frac{1}{10}$ **27.** $\frac{10}{9}$ **28.** $\frac{22}{81}$ **29.** $\frac{7x}{2}$ **30.** $\frac{10uv^5}{7}$

31. $\frac{2m^2}{3n^6}$ **32.** $\frac{p^3}{6q^8}$ **33.** -3 **34.** $\frac{-3}{5a + 20}$

35. $\frac{2}{x + 2}$ **36.** $\frac{a^2 + a}{3a - 3}$ **37.** $\frac{1}{4t - 20}$ **38.** $\frac{1}{w^2 + 3w}$

39. $x^2 - 1$ **40.** $9y^2 - 12y + 4$ **41.** $2x - 4y$
42. $\frac{m + 2n}{2}$ **43.** $\frac{x + 2}{2}$ **44.** $\frac{a^2 + 1}{a + 1}$ **45.** $\frac{x^2 + 9}{15}$

46. $\frac{1}{4a - 12}$ **47.** $9x + 9y$ **48.** $x^2 + 5x + 4$ **49.** -3

50. $-\frac{2}{3}$ **51.** $\frac{a + b}{a}$ **52.** $-\frac{1}{5}$ **53.** $\frac{2b}{a}$ **54.** $\frac{2g}{3}$

55. $\frac{y}{x}$ **56.** $\frac{4}{81}$ **57.** $\frac{-a^6b^8}{2}$ **58.** $\frac{-8}{9a^2}$ **59.** $\frac{1}{9m^3n}$

60. r^2 **61.** $\frac{x^2 + 5x}{3x - 1}$ **62.** $\frac{x^4 + 5x^3}{3}$ **63.** $\frac{a^3 + 8}{2a - 4}$

64. $\frac{1}{w + 1}$ **65.** 1 **66.** $\frac{x^3 - 2x^2 + x}{9x^2 + 9}$

67. $\frac{(m + 3)^2}{(m - 3)(m + k)}$ **68.** $\frac{2}{x - w}$ **69.** $\frac{13.1}{x}$ mi

70. $\frac{840}{x + 2}$ magazines **71.** 5 square meters **72.** 4 yd^2

73. **a)** $\frac{1}{8}$ **b)** $\frac{4}{3}$ **c)** $\frac{2x}{3}$ **d)** $\frac{3x}{4}$

74. **a)** $R = \frac{3}{5}, R = \frac{11}{23}, H = \frac{3}{5}, H = \frac{11}{23}$

b) R reduces to H.

Section 7.3 Warm-ups F F T T F F F F T T

1. We can build up a denominator by multiplying the numerator and denominator of a fraction by the same nonzero number.
2. To build up the denominator of a rational expression, we can multiply the numerator and denominator by the same polynomial.
3. For fractions the LCD is the smallest number that is a multiple of all of the denominators.
4. For polynomial denominators the LCD consists of every factor that appears, raised to the highest power that appears on the factor.

5. $\frac{9}{27}$ **6.** $\frac{14}{35}$ **7.** $\frac{14x}{2x}$ **8.** $\frac{24y}{4y}$ **9.** $\frac{15t}{3bt}$ **10.** $\frac{7z}{2ayz}$

11. $\frac{-36z^2}{8awz}$ **12.** $\frac{-42y^2t^2}{18xyt}$ **13.** $\frac{10a^2}{15a^3}$ **14.** $\frac{21bc^3}{36c^8}$

15. $\frac{8xy^3}{10x^2y^5}$ **16.** $\frac{15x^2y^2z^2}{24x^5z^3}$ **17.** $\frac{-20}{-8x - 8}$

18. $\frac{-6}{2n - 2m}$ **19.** $\frac{-32ab}{20b^2 - 20b^3}$ **20.** $\frac{-15x^2}{18x^2 + 27x}$

21. $\frac{3x - 6}{x^2 - 4}$ **22.** $\frac{a^2 - 3a}{a^2 - 9}$ **23.** $\frac{3x^2 + 3x}{x^2 + 2x + 1}$

24. $\frac{-14x^2 + 21x}{4x^2 - 12x + 9}$ **25.** $\frac{y^2 - y - 30}{y^2 + y - 20}$

26. $\frac{z^2 - 11z + 30}{z^2 - 2z - 15}$ **27.** 48 **28.** 84 **29.** 180
30. 240 **31.** $30a^2$ **32.** $180x^2y$ **33.** $12a^4b^6$
34. $36m^6n^5w^8$ **35.** $(x - 4)(x + 4)^2$ **36.** $(x - 3)(x + 3)^2$
37. $x(x + 2)(x - 2)$ **38.** $y(y - 5)(y + 2)$

39. $2x(x - 4)(x + 4)$ **40.** $3y(y - 3)$ **41.** $\frac{4}{24}, \frac{9}{24}$

42. $\frac{25}{60}, \frac{9}{60}$ **43.** $\frac{9b}{252ab}, \frac{20a}{252ab}$ **44.** $\frac{28b^2}{525ab}, \frac{30}{525ab}$

45. $\frac{2x^3}{6x^5}, \frac{9}{6x^5}$ **46.** $\frac{9c}{24a^3b^9c}, \frac{20ab^9}{24a^3b^9c}$

47. $\frac{4x^4}{36x^3y^5z}, \frac{3y^6z}{36x^3y^5z}, \frac{6xy^4z}{36x^3y^5z}$ **48.** $\frac{35b^2}{84a^6b^3}, \frac{18a^3b^4}{84a^6b^3}, \frac{42a^5}{84a^6b^3}$

49. $\frac{2x^2 + 4x}{(x - 3)(x + 2)}, \frac{5x^2 - 15x}{(x - 3)(x + 2)}$

50. $\frac{2a^2 + 4a}{(a - 5)(a + 2)}, \frac{3a^2 - 15a}{(a - 5)(a + 2)}$

51. $\frac{4}{a - 6}, \frac{-5}{a - 6}$ **52.** $\frac{8}{2x - 2y}, \frac{-5x}{2x - 2y}$

53. $\frac{x^2 - 3x}{(x - 3)^2(x + 3)}, \frac{5x^2 + 15x}{(x - 3)^2(x + 3)}$

54. $\frac{5x^2 - 5x}{(x + 1)(x - 1)^2}, \frac{-4x - 4}{(x + 1)(x - 1)^2}$

55. $\frac{w^2 + 3w + 2}{(w - 5)(w + 3)(w + 1)}, \frac{-2w^2 - 6w}{(w - 5)(w + 3)(w + 1)}$

56. $\frac{z^2 + 2z - 3}{(z + 2)(z + 4)(z + 3)}, \frac{z^2 + 5z + 4}{(z + 2)(z + 4)(z + 3)}$

57. $\frac{-5x - 10}{6(x - 2)(x + 2)}, \frac{6x}{6(x - 2)(x + 2)}, \frac{9x - 18}{6(x - 2)(x + 2)}$

58. $\frac{3b}{b(2b - 3)(2b + 3)}, \frac{4b^3 - 6b^2}{b(2b - 3)(2b + 3)}, \frac{-10b - 15}{b(2b - 3)(2b + 3)}$

59. $\frac{2q + 8}{(2q + 1)(q - 3)(q + 4)}, \frac{3q - 9}{(2q + 1)(q - 3)(q + 4)}, \frac{8q + 4}{(2q + 1)(q - 3)(q + 4)}$

60. $\frac{-3p + 12}{(2p - 3)(p + 5)(p - 4)}, \frac{p^2 + 5p}{(2p - 3)(p + 5)(p - 4)}, \frac{4p - 6}{(2p - 3)(p + 5)(p - 4)}$

61. Identical denominators are needed for addition and subtraction.
62. c

Section 7.4 Warm-ups F T T T T F T F T F

1. We can add rational numbers with identical denominators as follows: $\frac{a}{c} + \frac{b}{c} = \frac{a + b}{c}$.
2. Rational expressions with identical denominators are added in the same manner as rational numbers.
3. The LCD is the smallest number that is a multiple of all denominators.
4. We use the least common denominator to keep the addition process as simple as possible.

5. $\frac{1}{5}$ **6.** $\frac{1}{2}$ **7.** $\frac{3}{4}$ **8.** $\frac{1}{3}$ **9.** $-\frac{2}{3}$ **10.** $-\frac{5}{4}$
11. $-\frac{3}{4}$ **12.** $-\frac{3}{5}$ **13.** $\frac{5}{9}$ **14.** $\frac{13}{12}$ **15.** $\frac{103}{144}$
16. $\frac{43}{30}$ **17.** $-\frac{31}{40}$ **18.** $-\frac{17}{60}$ **19.** $\frac{5}{24}$
20. $-\frac{2}{35}$ **21.** $\frac{1}{x}$ **22.** $\frac{1}{y}$ **23.** $\frac{5}{w}$ **24.** $\frac{4x}{y}$
25. 3 **26.** -4 **27.** -2 **28.** $\frac{8-2a}{3}$
29. $\frac{3}{h}$ **30.** $\frac{1}{t-3}$ **31.** $\frac{x-4}{x+2}$ **32.** $\frac{x+2}{x+6}$
33. $\frac{3}{2a}$ **34.** $\frac{7}{3w}$ **35.** $\frac{5x}{6}$ **36.** $\frac{3y}{4}$ **37.** $\frac{6m}{5}$
38. $\frac{9y}{4}$ **39.** $\frac{17}{10a}$ **40.** $\frac{11}{24y}$ **41.** $\frac{w}{36}$
42. $\frac{y+19}{35}$ **43.** $\frac{b^2-4ac}{4a}$ **44.** $\frac{7by+3}{7b}$
45. $\frac{2w+3z}{w^2z^2}$ **46.** $\frac{b^2-5a^4}{a^5b^3}$ **47.** $\frac{-x-3}{x(x+1)}$
48. $\frac{2-a}{a(a-1)}$ **49.** $\frac{3a+b}{(a-b)(a+b)}$
50. $\frac{5x-1}{(x+1)(x-1)}$ **51.** $\frac{15-4x}{5x(x+1)}$
52. $\frac{15-2a}{5a(a+3)}$ **53.** $\frac{a^2+5a}{(a-3)(a+3)}$
54. $\frac{4x+3}{(x-1)(x+1)}$ **55.** 0 **56.** $\frac{1}{3-x}$
57. $\frac{7}{2a-2}$ **58.** $\frac{11}{2x-4}$
59. $\frac{-2x+1}{(x-5)(x+2)(x-2)}$
60. $\frac{5x^2-7x}{(x-3)(x+3)(x+1)}$
61. $\frac{7x+17}{(x+2)(x-1)(x+3)}$
62. $\frac{2x^2+x-18}{(x+2)(x+3)(x-4)}$
63. $\frac{2x^2-x-4}{x(x-1)(x+2)}$ **64.** $\frac{2a^2+5a-1}{a(a-1)(a+1)}$
65. $\frac{a+51}{6a(a-3)}$ **66.** $\frac{-7c+19}{6c(2c+1)}$ **67.** g **68.** b
69. f **70.** c **71.** e **72.** d **73.** h **74.** a
75. $\frac{11}{x}$ feet **76.** $\frac{13}{6x}$ meters
77. $\frac{315x+600}{x(x+5)}$ hours, 5 hours
78. $\frac{440x-4000}{x(x-20)}$ hours, 5 hours
79. $\frac{4x+6}{x(x+3)}$ job, $\frac{5}{9}$ job
80. $\frac{3}{2x}$ barn, $\frac{3}{10}$ barn

Section 7.5 Warm-ups F T F F F F F T T T

1. A complex fraction is a fraction that has fractions in its numerator, denominator, or both.
2. You can multiply the numerator and denominator by the LCD or you can simplify the numerator and denominator, and then divide.

3. $-\frac{10}{3}$ **4.** $\frac{1}{6}$ **5.** $\frac{22}{7}$ **6.** $-\frac{7}{30}$ **7.** $\frac{14}{17}$ **8.** $\frac{13}{11}$
9. $\frac{45}{23}$ **10.** $-\frac{47}{37}$ **11.** $\frac{3a+b}{a-3b}$ **12.** $\frac{4-6x}{3x+4}$
13. $\frac{5a-3}{3a+1}$ **14.** $\frac{4y+3}{y-2}$ **15.** $\frac{x^2-4x}{6x^2-2}$
16. $\frac{6a+5a^2}{9a-9}$ **17.** $\frac{10b}{3b^2-4}$ **18.** $-\frac{102}{11}$ **19.** $\frac{y-2}{3y+4}$
20. $\frac{2a-7}{3a-10}$ **21.** $\frac{x^2-2x+4}{x^2-3x-1}$ **22.** $\frac{x^2-2x+6}{x^2-2x-15}$
23. $\frac{5x-14}{2x-7}$ **24.** $\frac{x^2-5x-2}{4x-5}$ **25.** $\frac{a-6}{3a-1}$
26. $\frac{2x-6}{x-15}$ **27.** $\frac{-3m+12}{4m-3}$ **28.** $\frac{3y+12}{y-3}$
29. $\frac{-w+5}{9w+1}$ **30.** $\frac{-2x-3}{5x+13}$ **31.** -1 **32.** $-\frac{x}{2}$
33. $\frac{6x-27}{4x-6}$ **34.** $\frac{5a-25}{4a+8}$ **35.** $\frac{2x^2}{3y}$ **36.** $\frac{3a}{2}$
37. $\frac{a^2+7a+6}{a+3}$ **38.** $\frac{y-6}{(y+2)^2}$ **39.** $1-x$ **40.** $\frac{1}{2}$
41. $\frac{32}{95}, \frac{11}{35}$ **42.** $\frac{11}{25}$

Section 7.6 Warm-ups F F F F F T T T T T

1. The first step is usually to multiply each side by the LCD.
2. In adding rational expressions we build up each expression to get the LCD as the common denominator.
3. An extraneous solution is a number that appears when we solve an equation, but it does not check in the original equation.
4. Extraneous solutions occur because we multiply by an expression involving a variable and that variable expression might have a value of zero.

5. 12 **6.** 30 **7.** 30 **8.** 6 **9.** 5 **10.** 8 **11.** 4
12. 2 **13.** 4 **14.** 8 **15.** 4 **16.** 3 **17.** 3 **18.** 4
19. 2 **20.** 5 **21.** $-5, 2$ **22.** $-4, 3$ **23.** $-3, 2$
24. $-1, 6$ **25.** 2, 3 **26.** $\frac{5}{3}, 4$ **27.** $-3, 3$ **28.** $-4, 5$
29. 2 **30.** 6 **31.** No solution **32.** No solution

33. No solution **34.** No solution **35.** 3 **36.** 1
37. 10 **38.** $\frac{18}{5}$ **39.** 0 **40.** 0 **41.** $-5, 5$
42. $-3, 3$ **43.** 3, 5 **44.** $-4, 2$ **45.** 1 **46.** -3
47. 3 **48.** 4 **49.** 0 **50.** 6 **51.** 4 **52.** 3
53. -20 **54.** 9 **55.** 3 **56.** 3 **57.** 3 **58.** 2
59. $54\frac{6}{11}$ mm **60.** 250.03 mm

Section 7.7 Warm-ups T F F T T T T F F F T

1. A ratio is a comparison of two numbers.
2. The ratio of a to b is also written as $\frac{a}{b}$ or $a : b$.
3. Equivalent ratios are ratios that are equivalent as fractions.
4. A proportion is an equation that expresses equality of two ratios.
5. In the proportion $\frac{a}{b} = \frac{c}{d}$ the means are b and c and the extremes are a and d.
6. The extremes-means property says that if $\frac{a}{b} = \frac{c}{d}$ then $ad = bc$.

7. $\frac{5}{7}$ **8.** $\frac{4}{1}$ **9.** $\frac{8}{15}$ **10.** $\frac{1}{16}$ **11.** $\frac{7}{2}$ **12.** $\frac{8}{3}$
13. $\frac{9}{14}$ **14.** $\frac{6}{5}$ **15.** $\frac{5}{2}$ **16.** $\frac{8}{9}$ **17.** $\frac{15}{1}$
18. $\frac{16}{1}$ **19.** 3 to 2 **20.** 9 to 16 **21.** 9 to 16
22. 1 to 4 **23.** 31 to 1 **24.** 8 to 5 **25.** 2 to 3
26. 3 to 10 **27.** 6 **28.** 6 **29.** $-\frac{2}{5}$ **30.** $-\frac{9}{4}$
31. $-\frac{27}{5}$ **32.** $-\frac{20}{3}$ **33.** 5 **34.** -3 **35.** $-\frac{3}{4}$
36. -7 **37.** $\frac{5}{4}$ **38.** 0 **39.** 108 **40.** 356
41. 176,000 **42.** 4.2 million **43.** Lions 85, Tigers 51
44. Length $\frac{16}{3}$ ft, width $\frac{10}{3}$ ft
45. 40 luxury cars, 60 sports cars
46. 45 rabbits, 10 foxes
47. 84 in. **48.** $\frac{28}{3}$ yd **49.** 15 min **50.** 2330 m
51. 536.7 mi **52.** $\frac{126}{19}$ days **53.** 3920 lb, 2000 lb
54. 390 lb **55.** 6000 **56.** 2500
57. a) 3 to 17 **b)** 14.4 lbs **58.** 17 to 33, 157.2 lbs
59. 4074 **60.** d

Section 7.8 Warm-ups T T F T T T F F T F T

1. $y = 2x - 5$ **2.** $y = -2x + 10$ **3.** $y = -\frac{1}{2}x - 2$
4. $y = -\frac{1}{2}x - 4$ **5.** $y = mx - mb - a$
6. $y = ax + ak + h$ **7.** $y = -\frac{1}{3}x - \frac{1}{3}$
8. $y = -\frac{3}{4}x - \frac{5}{4}$ **9.** $C = \frac{B}{A}$ **10.** $A = PC + PD$
11. $p = \frac{a}{1 + am}$ **12.** $m = \frac{3f}{2 + ft}$ **13.** $m_1 = \frac{r^2F}{km_2}$
14. $r = \frac{mv^2}{F}$ **15.** $a = \frac{bf}{b - f}$ **16.** $R = \frac{R_1R_2}{R_1 + R_2}$
17. $r = \frac{S - a}{S}$ **18.** $R = \frac{E - Ir}{I}$ **19.** $P_2 = \frac{P_1V_1T_2}{T_1V_2}$
20. $T_1 = \frac{P_1V_1T_2}{P_2V_2}$ **21.** $h = \frac{3V}{4\pi r^2}$ **22.** $S = 2\pi rh + 2\pi r^2$
23. $\frac{5}{12}$ **24.** 3 **25.** $-\frac{6}{23}$ **26.** $-\frac{8}{9}$ **27.** $\frac{128}{3}$
28. $\frac{15}{16}$ **29.** -6 **30.** $\frac{15}{2}$ **31.** $\frac{6}{5}$ **32.** $-\frac{9}{5}$
33. Marcie 4 mph, Frank 3 mph **34.** $\frac{15}{7}$ mph
35. Bob 25 mph, Pat 20 mph
36. Smith 15 mph and Jones 20 mph, or Smith 30 mph and Jones 35 mph
37. 5 mph **38.** 5 mph **39.** 6 hours **40.** 66 hours
41. 40 minutes **42.** 1 hour 20 minutes
43. 1 hour 36 minutes **44.** $\frac{40}{13}$ days
45. Bananas 8 pounds, apples 10 pounds **46.** 12
47. 80 gallons **48.** Master 2 hours, apprentice 6 hours

Enriching Your Mathematical Word Power

1. b **2.** a **3.** a **4.** d **5.** a **6.** b **7.** d **8.** a
9. c **10.** d

Review Exercises

1. $\frac{6}{7}$ **2.** $\frac{7}{3}$ **3.** $\frac{c^2}{4a^2}$ **4.** $\frac{13x^5}{5}$ **5.** $\frac{2w - 3}{3w - 4}$
6. $-\frac{3}{4}$ **7.** $-\frac{x + 1}{3}$ **8.** $\frac{3 - 3x}{5}$ **9.** $\frac{1}{2}k$
10. $\frac{1}{3}a^2b^4c$ **11.** $\frac{2x}{3y}$ **12.** $8a^5b$ **13.** $a^2 - a - 6$
14. $x + 1$ **15.** $\frac{1}{2}$ **16.** $\frac{-2y}{x^2 + xy}$ **17.** 108
18. 210 **19.** $24a^7b^3$ **20.** $180u^4v^5$ **21.** $12x(x - 1)$
22. $24a(a + 1)$ **23.** $(x + 1)(x - 2)(x + 2)$
24. $(x - 3)(x + 3)^2$ **25.** $\frac{15}{36}$ **26.** $\frac{6a}{45}$ **27.** $\frac{10x}{15x^2y}$
28. $\frac{18xy^7z}{42x^3y^8}$ **29.** $\frac{-10}{12 - 2y}$ **30.** $\frac{6}{2t - 4}$ **31.** $\frac{x^2 + x}{x^2 - 1}$
32. $\frac{t^2 + 5t}{t^2 + 2t - 15}$ **33.** $\frac{29}{63}$ **34.** $-\frac{1}{35}$ **35.** $\frac{3x - 4}{x}$
36. $\frac{2b + 3a}{2b}$ **37.** $\frac{2a - b}{a^2b^2}$ **38.** $\frac{10x + 9}{12x^3}$
39. $\frac{27a^2 - 8a - 15}{(2a - 3)(3a - 2)}$ **40.** $\frac{-2x + 19}{(x - 2)(x + 3)}$

41. $\frac{3}{a-8}$ **42.** $\frac{1}{x-14}$ **43.** $\frac{3x+8}{2(x+2)(x-2)}$

44. $\frac{-2x^2}{(x-3)(x+3)(x+1)}$ **45.** $-\frac{3}{14}$ **46.** $\frac{31}{3}$

47. $\frac{6b+4a}{3a-18b}$ **48.** $\frac{90-10x}{5y-18x}$ **49.** $\frac{-2x+9}{3x-1}$

50. $\frac{4a+1}{-3a-2}$ **51.** $\frac{x^2+x-2}{-4x+13}$ **52.** $\frac{4a+9}{a+4}$

53. $-\frac{15}{2}$ **54.** $\frac{14}{3}$ **55.** 9 **56.** No solution

57. -3 **58.** 6 **59.** $\frac{21}{2}$ **60.** $-4, 4$

61. 5 **62.** $\frac{27}{2}$ **63.** 8 **64.** 1155

65. 56 cups water, 28 cups rice **66.** $\frac{24}{25}$ pints

67. $y = mx + b$ **68.** $a = \frac{2A - hb}{h}$

69. $m = \frac{1}{F - v}$ **70.** $r = \frac{m}{1 - mt}$

71. $y = 4x - 13$ **72.** $y = -\frac{1}{3}x + \frac{7}{3}$

73. 200 hours **74.** Leon 45 mph, Pat 55 mph
75. Bert 60 cars, Ernie 50 cars **76.** 1000
77. 27.83 million tons
78. 230 million tons, 25.07 million tons

79. $\frac{10}{2x}$ **80.** $\frac{2}{a}$ **81.** $\frac{-2}{5-a}$ **82.** $\frac{1}{7-a}$

83. $\frac{3x}{x}$ **84.** $\frac{2ab}{b}$ **85.** $2m$ **86.** $5x^2$

87. $\frac{1}{6}$ **88.** $\frac{1}{2x}$ **89.** $\frac{1}{a+1}$ **90.** $\frac{x+3}{x^2-9}$

91. $\frac{5-a}{5a}$ **92.** $\frac{3b-14}{7b}$ **93.** $\frac{a-2}{2}$

94. $\frac{1-a}{a}$ **95.** $b - a$ **96.** -1 **97.** $\frac{1}{10a}$

98. $6a$ **99.** $\frac{3}{2x}$ **100.** $\frac{2}{3}$ **101.** $\frac{4+y}{6xy}$

102. $\frac{3}{x(x-1)}$ **103.** $\frac{8}{a-5}$ **104.** No solution

105. $-1, 2$ **106.** $\frac{6}{7}$ **107.** $-\frac{5}{3}$ **108.** $\frac{50}{3}$

109. 6 **110.** $\frac{x+2}{4}$ **111.** $\frac{1}{2}$ **112.** $-\frac{4}{5}$

113. $\frac{3x+7}{(x-5)(x+5)(x+1)}$

114. $\frac{a+7}{2(a+1)(a-1)}$

115. $\frac{-5a}{(a-3)(a+3)(a+2)}$

116. $\frac{-7a+1}{(a+2)(a-2)(a-1)}$ **117.** $\frac{2}{5}$

118. No solution

Chapter 7 Test

1. $-1, 1$ **2.** $\frac{2}{3}$ **3.** 0 **4.** $-\frac{14}{45}$ **5.** $\frac{1+3y}{y}$

6. $\frac{4}{a-2}$ **7.** $\frac{-x+4}{(x+2)(x-2)(x-1)}$ **8.** $\frac{2}{3}$

9. $\frac{-2}{a+b}$ **10.** $\frac{a^3}{18b^4}$ **11.** $-\frac{4}{3}$ **12.** $\frac{3x-4}{-2x+6}$

13. $\frac{15}{7}$ **14.** 2, 3 **15.** 12 **16.** $y = -\frac{1}{5}x + \frac{13}{5}$

17. $c = \frac{3M - bd}{b}$ **18.** 29 **19.** 7.2 minutes

20. Brenda 15 mph and Randy 20 mph, or Brenda 10 mph and Randy 15 mph
21. $72 billion

Making Connections Chapters 1–7

1. $\frac{7}{3}$ **2.** $-\frac{10}{3}$ **3.** -2 **4.** No solution **5.** 0

6. $-4, -2$ **7.** $-1, 0, 1$ **8.** $-\frac{15}{2}$ **9.** $-6, 6$

10. $-2, 4$ **11.** 5 **12.** 3 **13.** $y = \frac{c - 2x}{3}$

14. $y = \frac{1}{2}x + \frac{1}{2}$ **15.** $y = \frac{c}{2-a}$ **16.** $y = \frac{AB}{C}$

17. $y = 3B - 3A$ **18.** $y = \frac{6A}{5}$ **19.** $y = \frac{8}{3-5a}$

20. $y = 0$ or $y = B$ **21.** $y = \frac{2A - hb}{h}$ **22.** $y = -\frac{b}{2}$

23. 64 **24.** 16 **25.** 49 **26.** 121 **27.** $-2x - 2$

28. $2a^2 - 11a + 15$ **29.** x^4 **30.** $\frac{2x+1}{5}$ **31.** $\frac{1}{2x}$

32. $\frac{x+2}{2x}$ **33.** $\frac{x}{2}$ **34.** $\frac{x-2}{2x}$ **35.** $-\frac{7}{5}$ **36.** $\frac{3a}{4}$

37. $x^2 - 64$ **38.** $3x^3 - 21x$ **39.** $10a^{14}$ **40.** x^{10}
41. $k^2 - 12k + 36$ **42.** $j^2 + 10j + 25$

43. -1 **44.** $3x^2 - 4x$ **45.** $P = \frac{r+2}{(1+r)^2}$, $1.81, $7.72

Chapter 8

Section 8.1 Warm-ups T F T F T T F F T T
1. If $b^n = a$, then b is an nth root of a.
2. The principal root is the positive even root of a positive number.

3. If $b^n = a$, then b is an even root provided n is even or an odd root provided n is odd.
4. The nth root of a is written as $\sqrt[n]{a}$.
5. The product rule for radicals says that $\sqrt[n]{a} \cdot \sqrt[n]{b} = \sqrt[n]{ab}$ provided all of these roots are real.
6. The quotient rule for radicals says that $\sqrt[n]{a}/\sqrt[n]{b} = \sqrt[n]{a/b}$ provided all of these roots are real.

7. 6 **8.** 7 **9.** 2 **10.** 3 **11.** 10 **12.** 2
13. Not a real number **14.** Not a real number
15. 0 **16.** 1 **17.** -1 **18.** 0 **19.** 1 **20.** 1
21. Not a real number **22.** Not a real number
23. 2 **24.** 2 **25.** -2 **26.** -5 **27.** -10
28. -6 **29.** Not a real number **30.** Not a real number
31. m **32.** m^3 **33.** y^3 **34.** m^2 **35.** y^5 **36.** m^4
37. m **38.** x **39.** 27 **40.** 4 **41.** 32 **42.** 2^{33}
43. 125 **44.** 1,000,000 **45.** 10^{10} **46.** 10^9
47. $3\sqrt{y}$ **48.** $4\sqrt{n}$ **49.** $2a$ **50.** $6n$ **51.** x^2y
52. w^3t **53.** $m^6\sqrt{5}$ **54.** $z^8\sqrt{7}$ **55.** $2\sqrt[3]{y}$
56. $3\sqrt[3]{z^2}$ **57.** $-3w$ **58.** $-5m^2$ **59.** $2\sqrt[4]{s}$
60. $3\sqrt[4]{w}$ **61.** $-5a^3y^2$ **62.** $-3zw^5$ **63.** $\frac{\sqrt{t}}{2}$
64. $\frac{\sqrt{w}}{6}$ **65.** $\frac{25}{4}$ **66.** $\frac{1}{4}$ **67.** $\frac{\sqrt[3]{t}}{2}$ **68.** $\frac{\sqrt[3]{a}}{3}$
69. $\frac{-2x^2}{y}$ **70.** $\frac{-3y^{12}}{10}$ **71.** $\frac{2a^3}{3}$ **72.** $\frac{3a}{7b^2}$
73. $\frac{\sqrt[4]{y}}{2}$ **74.** $\frac{\sqrt[4]{5w}}{3}$ **75.** 3.968
76. 0.914 **77.** -1.610 **78.** 0.257 **79.** 6.001
80. 3.256 **81.** 0.769 **82.** 6.850 **83.** 46
84. 7.8 in. **85.** **a)** 228 mi **b)** 7000 ft
86. **a)** 8.0 knots **b)** 20 ft
87. **a)** No **b)** Yes **c)** No **d)** Yes
88. n is odd.

Section 8.2 Warm-ups T F T F T F T F F F

1. We use the product rule to factor out a perfect square from inside a square root.
2. The perfect squares are 1, 4, 9, 16, 25, and so on.
3. To rationalize a denominator means to rewrite the expression so that the denominator is a rational number.
4. A square root in simplified form has no perfect squares or fractions inside the radical and no radicals in a denominator.
5. To simplify square roots containing variables use the same techniques as we use on square roots of numbers.
6. Any even power of a variable is a perfect square.

7. $2\sqrt{2}$ **8.** $2\sqrt{5}$ **9.** $2\sqrt{6}$ **10.** $5\sqrt{3}$ **11.** $2\sqrt{7}$
12. $2\sqrt{10}$ **13.** $3\sqrt{10}$ **14.** $10\sqrt{2}$ **15.** $10\sqrt{5}$
16. $7\sqrt{2}$ **17.** $5\sqrt{6}$ **18.** $2\sqrt{30}$ **19.** $\frac{\sqrt{5}}{5}$ **20.** $\frac{\sqrt{6}}{6}$
21. $\frac{3\sqrt{2}}{2}$ **22.** $\frac{4\sqrt{3}}{3}$ **23.** $\frac{\sqrt{6}}{2}$ **24.** $\frac{\sqrt{42}}{6}$
25. $\frac{-3\sqrt{10}}{10}$ **26.** $-\frac{4\sqrt{5}}{5}$ **27.** $\frac{-10\sqrt{17}}{17}$
28. $-\frac{3\sqrt{19}}{19}$ **29.** $\frac{\sqrt{77}}{7}$ **30.** $\frac{\sqrt{30}}{3}$ **31.** $3\sqrt{7}$
32. $4\sqrt{3}$ **33.** $\frac{\sqrt{6}}{2}$ **34.** $\frac{\sqrt{15}}{5}$ **35.** $\frac{\sqrt{10}}{4}$
36. $\frac{\sqrt{10}}{6}$ **37.** $\frac{\sqrt{15}}{5}$ **38.** $\frac{\sqrt{15}}{5}$ **39.** 5
40. 3 **41.** $\frac{\sqrt{6}}{2}$ **42.** $\frac{\sqrt{70}}{7}$ **43.** a^4 **44.** y^5
45. $a^4\sqrt{a}$ **46.** $t^5\sqrt{t}$ **47.** $2a^3\sqrt{2}$ **48.** $3w^4\sqrt{2w}$
49. $2a^2b^4\sqrt{5b}$ **50.** $2xy\sqrt{3y}$ **51.** $3xy\sqrt{3xy}$
52. $3x^2y\sqrt{5xy}$ **53.** $3ab^4c\sqrt{3a}$ **54.** $5xy^4z^2\sqrt{5xy}$
55. $\frac{\sqrt{x}}{x}$ **56.** $\frac{\sqrt{2x}}{2x}$ **57.** $\frac{\sqrt{6a}}{3a}$ **58.** $\frac{\sqrt{10b}}{2b}$
59. $\frac{\sqrt{5y}}{5y}$ **60.** $\frac{\sqrt{2x}}{2x}$ **61.** $\frac{\sqrt{6xy}}{2y}$ **62.** $\frac{\sqrt{30w}}{5w}$
63. $\frac{\sqrt{6xy}}{3x}$ **64.** $\frac{\sqrt{6xy}}{2y}$ **65.** $\frac{2x\sqrt{2xy}}{y}$
66. $\frac{2s^2\sqrt{2st}}{t}$ **67.** $4x\sqrt{5x}$ **68.** $3y^{40}\sqrt{10}$
69. $3y^4x^7\sqrt{yx}$ **70.** $4xy^3\sqrt{3y}$ **71.** $4x^3\sqrt{5x}$
72. $x^2y\sqrt{7x}$ **73.** $\frac{-11\sqrt{6pq}}{3q}$ **74.** $-10t^2\sqrt{3t}$
75. $a^3b^8c\sqrt{ac}$ **76.** $3n^4b^5c^3\sqrt{nb}$ **77.** $\frac{\sqrt{6y}}{3x^9y^4}$
78. $\frac{2\sqrt{3n}}{3m^2n^3}$ **79.** 0 **80.** 0 **81.** 0 **82.** 0
83. **a)** $E = \frac{\sqrt{2AIS}}{I}$ **b)** 51.2
84. **a)** $V = \frac{29\sqrt{LCS}}{CS}$ **b)** 113.4

Section 8.3 Warm-ups F T F F T F F F T T

1. Like radicals have the same index and same radicand.
2. We combine like radicals just like we combine like terms.
3. Radicals can be added, subtracted, multiplied, and divided.
4. We use the FOIL method for multiplying sums of radicals.
5. Radical expressions such as $\sqrt{a} + \sqrt{b}$ and $\sqrt{a} - \sqrt{b}$ are conjugates.

6. Rationalizing a denominator that contains a sum can be done by multiplying the numerator and denominator by the conjugate of the denominator.

7. $7\sqrt{5}$ **8.** $2\sqrt{2}$ **9.** $2\sqrt[3]{2}$ **10.** $-3\sqrt[3]{6}$

11. $8u\sqrt{11}$ **12.** $-3m\sqrt{5}$ **13.** $4\sqrt{3}-4\sqrt{2}$

14. $5\sqrt{6}+4\sqrt{2}$ **15.** $-4\sqrt{x}-\sqrt{y}$

16. $8\sqrt{7}-6\sqrt{a}$ **17.** $5x\sqrt{y}+2\sqrt{a}$

18. $6a\sqrt{b}+\sqrt{a}$ **19.** $5\sqrt{6}$ **20.** $5\sqrt{3}$

21. $-14\sqrt{3}$ **22.** $-2\sqrt{2}$ **23.** $-\sqrt{3a}$

24. $-2\sqrt{5w}$ **25.** $3x\sqrt{x}$ **26.** $13x\sqrt{3x}$

27. $\frac{2\sqrt{3}}{3}$ **28.** $\frac{4\sqrt{5}}{5}$ **29.** $\frac{7\sqrt{3}}{3}$ **30.** $\frac{13\sqrt{2}}{2}$

31. $\sqrt{77}$ **32.** $\sqrt{39}$ **33.** 36 **34.** 24

35. $-12\sqrt{10}$ **36.** $-24\sqrt{6}$ **37.** $2a^4\sqrt{3}$

38. $w^8\sqrt{3}$ **39.** 3 **40.** -5 **41.** $-2m$

42. $10m^2$ **43.** $2+\sqrt{6}$ **44.** $3-\sqrt{6}$

45. $12\sqrt{3}+6\sqrt{5}$ **46.** $6\sqrt{2}+12\sqrt{5}$

47. $10-30\sqrt{2}$ **48.** $12-6\sqrt{6}$ **49.** $-7-\sqrt{5}$

50. $12-5\sqrt{6}$ **51.** 2 **52.** 2 **53.** 3 **54.** -3

55. $28-\sqrt{5}$ **56.** $28+20\sqrt{2}$ **57.** $-42-\sqrt{15}$

58. $108+22\sqrt{21}$ **59.** $37+20\sqrt{3}$

60. $24+12\sqrt{3}$ **61.** $\sqrt{2}$ **62.** $\sqrt{7}$

63. $\frac{\sqrt{15}}{3}$ **64.** $\frac{\sqrt{6}}{2}$ **65.** $\frac{2\sqrt{30}}{9}$

66. $\frac{\sqrt{21}}{4}$ **67.** $\frac{5\sqrt{7}}{3}$ **68.** $\frac{2\sqrt{30}}{5}$

69. $1+\sqrt{2}$ **70.** $1+\sqrt{2}$ **71.** $-2+\sqrt{5}$

72. $-2+\sqrt{5}$ **73.** $\frac{2-\sqrt{5}}{3}$ **74.** $\frac{-2-\sqrt{3}}{2}$

75. $\frac{2+\sqrt{6}}{3}$ **76.** $1+\sqrt{3}$ **77.** $\frac{3+2\sqrt{3}}{6}$

78. $\frac{3+2\sqrt{2}}{3}$ **79.** $5\sqrt{3}+5\sqrt{2}$ **80.** $\frac{3\sqrt{6}-3\sqrt{2}}{4}$

81. $\frac{\sqrt{15}+3}{2}$ **82.** $\frac{-2+\sqrt{10}}{3}$ **83.** $\frac{13+7\sqrt{3}}{22}$

84. $\frac{\sqrt{6}-3+\sqrt{2}-\sqrt{3}}{2}$ **85.** $\frac{19-11\sqrt{7}}{27}$

86. $\frac{3\sqrt{10}+12\sqrt{2}+5+4\sqrt{5}}{13}$ **87.** $3\sqrt{5a}$

88. $6a^2\sqrt{2}$ **89.** $\frac{5\sqrt{2}}{2}$ **90.** $-3\sqrt{6}$ **91.** $70+30\sqrt{5}$

92. 1 **93.** $\frac{5\sqrt{5}}{3}$ **94.** $2\sqrt{2}+\sqrt{3}$ **95.** 1.866

96. 0.045 **97.** -1.974 **98.** -1.465

99. 12 ft², $10\sqrt{2}$ ft **100.** 9, $5\sqrt{3}+\sqrt{39}$ **101.** 6 m³

102. $2\sqrt{6}+4\sqrt{3}+6\sqrt{2}\,\text{m}^2$

Section 8.4 Warm-ups F T F F T F F T T T

1. The square root property says that $x^2 = k$ for $k > 0$ is equivalent to $x = \pm\sqrt{k}$.

2. We do not take the square root of each side of an equation.

3. The square root property and squaring each side are two new techniques used for solving equations.

4. The square root property and the zero factor property produce equivalent equations without doing the same thing to each side.

5. Squaring each side can produce extraneous roots.

6. Squaring each side does not always give equivalent equations.

7. $-4, 4$ **8.** $-7, 7$ **9.** $-2\sqrt{10}, 2\sqrt{10}$

10. $-2\sqrt{6}, 2\sqrt{6}$ **11.** $-\frac{\sqrt{6}}{3}, \frac{\sqrt{6}}{3}$ **12.** $-\frac{\sqrt{6}}{2}, \frac{\sqrt{6}}{2}$

13. No solution **14.** No solution **15.** $-1, 3$

16. $-6, 0$ **17.** $5-\sqrt{3}, 5+\sqrt{3}$

18. $6-\sqrt{5}, 6+\sqrt{5}$ **19.** -19 **20.** 0

21. 90 **22.** 13 **23.** No solution **24.** No solution

25. $-5, 5$ **26.** $-\sqrt{2}, \sqrt{2}$ **27.** 3 **28.** 9

29. 0, 1 **30.** 0, 2 **31.** 6 **32.** -1

33. No solution **34.** No solution **35.** $\frac{13}{18}$ **36.** 4

37. 3, 5 **38.** 2, 3 **39.** 3 **40.** 3 **41.** 6, 8

42. 2, 4 **43.** $r = \pm\sqrt{\frac{V}{\pi h}}$ **44.** $r = \pm\sqrt{\frac{3V}{4\pi h}}$

45. $b = \pm\sqrt{c^2 - a^2}$ **46.** $x = \pm\sqrt{\frac{y-c}{a}}$

47. $b = \pm 2\sqrt{ac}$ **48.** $t = \pm\sqrt{\frac{2s-2v}{g}}$

49. $t = \frac{v^2}{2p}$ **50.** $x = \frac{y^2}{2}$ **51.** $-\sqrt{2}, \sqrt{2}$

52. No solution **53.** No solution

54. No solution

55. $\frac{1+2\sqrt{2}}{2}, \frac{1-2\sqrt{2}}{2}$

56. $\frac{2-3\sqrt{2}}{3}, \frac{2+3\sqrt{2}}{3}$

57. $\frac{9}{2}$ **58.** $\frac{5}{3}$ **59.** 3 **60.** 9 **61.** $-4, 2$

62. $-2, 8$ **63.** $\frac{5}{2}$ **64.** $\frac{1}{3}$ **65.** $-1.803, 1.803$

66. $-5.506, 3.506$ **67.** 0.993 **68.** 5.343

69. 0.362, 4.438 **70.** $-1.995, 1.995$

71. $3\sqrt{2}$ or 4.243 ft **72.** $5\sqrt{3}$ or 8.660 mi

73. $3\sqrt{2}$ or 4.243 ft **74.** $\frac{\sqrt{2}}{2}$ or 0.707 yd

75. $\sqrt{2}$ or 1.414 ft **76.** $2\sqrt{10}$ or 6.325 meters
77. 5 ft **78.** 10 ft
79. 2.5 sec **80.** $90\sqrt{2}$ or 127.279 ft
81. $\frac{200\sqrt{13}}{3}$ or 240.370 ft **82.** $2\sqrt{41}$ or 13 inches

Section 8.5 Warm-ups T F F T T T T F T T
1. The nth root of a is $a^{1/n}$.
2. The mth power of the nth root of a is $a^{m/n}$.
3. The expression $a^{-m/n}$ means $\frac{1}{a^{m/n}}$.
4. All of the rules of exponents hold for rational exponents.
5. The operations can be performed in any order, but the easiest is usually root, power, and then reciprocal.
6. The expression $a^{-m/n}$ is a real number except when n is even and a is negative, or when $a = 0$.
7. $7^{1/4}$ **8.** $(cbs)^{1/3}$ **9.** $\sqrt[5]{9}$ **10.** $\sqrt{3}$ **11.** $(5x)^{1/2}$
12. $(3y)^{1/2}$ **13.** $\sqrt{a}$ **14.** $\sqrt[5]{-b}$ **15.** 5 **16.** 4
17. -5 **18.** -2 **19.** 2 **20.** 2
21. Not a real number **22.** Not a real number
23. $w^{7/3}$ **24.** $a^{5/2}$ **25.** $2^{-10/3}$ **26.** $a^{-2/3}$
27. $\sqrt[4]{\frac{1}{w^3}}$ **28.** $\sqrt[3]{\frac{1}{6^5}}$ **29.** $\sqrt{(ab)^3}$ **30.** $\sqrt[5]{\frac{1}{3m}}$
31. 25 **32.** 100 **33.** 125 **34.** 64 **35.** $\frac{1}{81}$ **36.** $\frac{1}{8}$
37. $\frac{1}{8}$ **38.** $\frac{1}{125}$ **39.** $-\frac{1}{3}$ **40.** $\frac{1}{16}$
41. Not a real number **42.** Not a real number
43. $x^{1/2}$ **44.** y **45.** $n^{1/6}$ **46.** $w^{7/20}$ **47.** $x^{3/2}$
48. $a^{1/6}$ **49.** $2t^{1/4}$ **50.** $\frac{2}{w^{1/12}}$ **51.** x^2 **52.** $\frac{1}{y^2}$
53. $5^{1/8}$ **54.** $\frac{1}{7^{9/2}}$ **55.** xy^3 **56.** tw^2 **57.** $\frac{x}{3y^4}$
58. $\frac{wt^2}{2}$ **59.** $\frac{3}{4}$ **60.** $\frac{1}{4}$ **61.** $\frac{1}{9}$ **62.** $\frac{1}{2}$ **63.** 27
64. $\frac{1}{125}$ **65.** $\frac{1}{12}$ **66.** $-\frac{8}{9}$ **67.** Yes, 9.2 m^2
68. 54 cm^2, 150 cm^2 **69.** 13.8% **70.** 1.2%
71. a) 15.9 **b)** Decreasing **c)** 28,000 pounds
72. 278, 294, 312, 330, 350, 371, 393, 416, 441, 467, 495 Hz
73. $a < 0$ and m and n are odd
74. $a < 0$ and n even, or $a = 0$
75. $3^{-8/9} < 3^{-2/3} < 3^{-1/2} < 3^0 < 3^{1/2} < 3^{2/3} < 3^{8/9}$, $7^{-9/10} < 7^{-3/4} < 7^{-1/3} < 7^0 < 7^{1/3} < 7^{3/4} < 7^{9/10}$, $m < n$

Enriching Your Mathematical Word Power
1. d **2.** b **3.** b **4.** b **5.** d **6.** b **7.** c **8.** d

Review Exercises
1. 2 **2.** -3 **3.** 10 **4.** 10 **5.** x^6 **6.** a^5 **7.** x^2
8. a^3 **9.** $2x$ **10.** $3y^2$ **11.** $5x^2$ **12.** $2y^4$ **13.** $\frac{2x^8}{y^7}$
14. $\frac{3y^4}{t^5}$ **15.** $\frac{w}{4}$ **16.** $\frac{a^2}{5}$ **17.** $6\sqrt{2}$ **18.** $4\sqrt{3}$
19. $\frac{\sqrt{3}}{3}$ **20.** $\frac{2\sqrt{5}}{5}$ **21.** $\frac{\sqrt{15}}{5}$ **22.** $\frac{\sqrt{30}}{6}$
23. $\sqrt{11}$ **24.** $\sqrt{10}$ **25.** $\frac{\sqrt{6}}{4}$ **26.** $\frac{1}{3}$
27. y^3 **28.** z^5 **29.** $2t^4\sqrt{6t}$ **30.** $2p^3\sqrt{2p}$
31. $2m^2t\sqrt{3mt}$ **32.** $3pq^3\sqrt{2pq}$ **33.** $\frac{\sqrt{2x}}{x}$
34. $\frac{\sqrt{5y}}{y}$ **35.** $\frac{a^2\sqrt{6as}}{2s}$ **36.** $\frac{x^3\sqrt{15xw}}{3w}$ **37.** $10\sqrt{7}$
38. $-2\sqrt{6}$ **39.** $-\sqrt{3}$ **40.** $8\sqrt{2}$ **41.** 30
42. -36 **43.** $-45\sqrt{2}$ **44.** $96\sqrt{6}$
45. $-15\sqrt{3} - 9$ **46.** $24\sqrt{2} + 16$
47. $-3\sqrt{2} + 3\sqrt{5}$ **48.** $2 - 2\sqrt{3}$ **49.** -22
50. -5 **51.** $26 - 4\sqrt{30}$ **52.** $24 + 12\sqrt{3}$
53. $38 - 19\sqrt{6}$ **54.** $-37 + 2\sqrt{15}$ **55.** $\frac{\sqrt{10}}{4}$
56. $\frac{\sqrt{15}}{2}$ **57.** $\frac{1 - \sqrt{5}}{5}$ **58.** $-3 + \sqrt{3}$
59. $-\frac{3 + 3\sqrt{5}}{4}$ **60.** $\frac{2\sqrt{3} - \sqrt{6}}{3}$ **61.** $-20, 20$
62. $-11, 11$ **63.** $-\frac{\sqrt{21}}{7}, \frac{\sqrt{21}}{7}$ **64.** $-\frac{\sqrt{21}}{3}, \frac{\sqrt{21}}{3}$
65. $4 - 3\sqrt{2}, 4 + 3\sqrt{2}$ **66.** $-1 - 2\sqrt{5}, -1 + 2\sqrt{5}$
67. 81 **68.** 400 **69.** 4 **70.** 1 **71.** 6 **72.** 4, 5
73. $t = \pm 2\sqrt{2sw}$ **74.** $t = -b \pm \sqrt{b^2 - 4ac}$
75. $t = \frac{9a^2}{b}$ **76.** $t = (a - w)^2$ **77.** $\frac{1}{125}$
78. $\frac{1}{243}$ **79.** 5 **80.** 27 **81.** $\frac{1}{8}$ **82.** $\frac{1}{25}$
83. $\frac{1}{x}$ **84.** $t^{1/6}$ **85.** $-\frac{1}{2}x^2$ **86.** $\frac{1}{9}x^6$
87. w^2 **88.** $m^{7/12}$ **89.** $\frac{t^3}{3s^2}$ **90.** $\frac{x^4y^2}{4}$ **91.** $\frac{4}{x^8y^{20}}$
92. $\frac{y^{3/2}t}{256}$ **93. a)** 18.6% **b)** First two years
94. 13.2% **95.** 0.4 cm **96.** $r = \sqrt{\frac{A}{\pi}}$
97. $4\pi\sqrt{10}$ or 40 in.^{2} **98.** $24\pi\sqrt{61}$ or 589 yd^2
99. 22 in. **100.** 34.9 in. by 19.6 in.

Chapter 8 Test
1. 6 **2.** 12 **3.** -3 **4.** 2 **5.** 2 **6.** $2\sqrt{6}$
7. $\frac{\sqrt{6}}{4}$ **8.** Not a real number **9.** $3\sqrt{2}$
10. $7 + 4\sqrt{3}$ **11.** 11 **12.** $\sqrt{7}$ **13.** $\frac{2\sqrt{15}}{3}$

14. $1 + \sqrt{2}$ **15.** 81 **16.** $3\sqrt{2} - 3$ **17.** $y^{3/4}$

18. $3x^{2/3}$ **19.** xy^3 **20.** $\frac{1}{5u^4w}$ **21.** $\frac{\sqrt{3t}}{t}$

22. $2y^3$ **23.** $2y^4$ **24.** $3t^3\sqrt{2t}$ **25.** $-9, 3$

26. 18 **27.** $-\frac{\sqrt{10}}{5}, \frac{\sqrt{10}}{5}$ **28.** $\frac{4}{3}$ **29.** 11

30. $r = \pm\sqrt{\frac{S}{\pi h}}$ **31.** $b = \sqrt{c^2 - a^2}$

32. $\frac{5\sqrt{2}}{2}$ meters **33.** 178.4 meters

Making Connections Chapters 1–8

1. $-\frac{3}{2}$ **2.** $\frac{3}{2}$

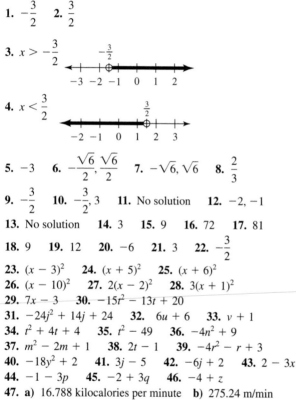

3. $x > -\frac{3}{2}$

4. $x < \frac{3}{2}$

5. -3 **6.** $-\frac{\sqrt{6}}{2}, \frac{\sqrt{6}}{2}$ **7.** $-\sqrt{6}, \sqrt{6}$ **8.** $\frac{2}{3}$

9. $-\frac{3}{2}$ **10.** $-\frac{3}{2}, 3$ **11.** No solution **12.** $-2, -1$

13. No solution **14.** 3 **15.** 9 **16.** 72 **17.** 81

18. 9 **19.** 12 **20.** -6 **21.** 3 **22.** $-\frac{3}{2}$

23. $(x - 3)^2$ **24.** $(x + 5)^2$ **25.** $(x + 6)^2$

26. $(x - 10)^2$ **27.** $2(x - 2)^2$ **28.** $3(x + 1)^2$

29. $7x - 3$ **30.** $-15t^2 - 13t + 20$

31. $-24j^2 + 14j + 24$ **32.** $6u + 6$ **33.** $v + 1$

34. $t^2 + 4t + 4$ **35.** $t^2 - 49$ **36.** $-4n^2 + 9$

37. $m^2 - 2m + 1$ **38.** $2t - 1$ **39.** $-4r^2 - r + 3$

40. $-18y^2 + 2$ **41.** $3j - 5$ **42.** $-6j + 2$ **43.** $2 - 3x$

44. $-1 - 3p$ **45.** $-2 + 3q$ **46.** $-4 + z$

47. a) 16.788 kilocalories per minute **b)** 275.24 m/min **c)** Increasing

Chapter 9

Section 9.1 Warm-ups T F F T T F F T F F

1. A quadratic equation is an equation of the form $ax^2 + bx + c = 0$, where $a \neq 0$.
2. If $b = 0$, a quadratic equation can be solved by the square root property.
3. If $b = 0$, we can get the square root of a negative number and no real solution.
4. If $b \neq 0$, some quadratics can be solved by factoring.
5. After applying the zero factor property we will have linear equations to solve.
6. This section reviews material about quadratic equations that was given earlier in this text.

7. $-6, 6$ **8.** $-9, 9$ **9.** No real solution

10. No real solution **11.** $-\sqrt{10}, \sqrt{10}$ **12.** $-\sqrt{2}, \sqrt{2}$

13. $-\frac{\sqrt{15}}{3}, \frac{\sqrt{15}}{3}$ **14.** $-\frac{\sqrt{35}}{5}, \frac{\sqrt{35}}{5}$

15. $-\frac{2\sqrt{6}}{3}, \frac{2\sqrt{6}}{3}$ **16.** $-\frac{2\sqrt{15}}{5}, \frac{2\sqrt{15}}{5}$ **17.** 1, 5

18. $-8, -2$ **19.** $2 - 3\sqrt{2}, 2 + 3\sqrt{2}$

20. $5 - 2\sqrt{5}, 5 + 2\sqrt{5}$ **21.** $-\frac{3}{2}, -\frac{1}{2}$ **22.** $\frac{3}{2}, \frac{1}{2}$

23. $\frac{2 - \sqrt{2}}{2}, \frac{2 + \sqrt{2}}{2}$ **24.** $\frac{-4 - \sqrt{2}}{2}, \frac{-4 + \sqrt{2}}{2}$

25. $\frac{-1 - \sqrt{2}}{2}, \frac{-1 + \sqrt{2}}{2}$ **26.** $\frac{1 - \sqrt{6}}{2}, \frac{1 + \sqrt{6}}{2}$

27. 11 **28.** -45 **29.** $-3, 5$ **30.** $-3, 4$

31. -3 **32.** -5 **33.** $-1, 2$ **34.** $-6, 5$ **35.** 0, 2

36. 0, 2 **37.** $0, \frac{3}{2}$ **38.** $0, \frac{5}{2}$ **39.** $-7, \frac{3}{2}$ **40.** $2, \frac{1}{2}$

41. 5 **42.** 2 **43.** $-\frac{5}{2}, 6$ **44.** $-\frac{1}{5}, \frac{1}{3}$ **45.** 4, 6

46. $-9, \frac{3}{2}$ **47.** $-\sqrt{6}, \sqrt{6}$ **48.** $-\frac{\sqrt{2}}{2}, \frac{\sqrt{2}}{2}$ **49.** 3, 9

50. $-3, 2$ **51.** 3, 4 **52.** 5, 6 **53.** $\frac{5\sqrt{2}}{2}$ meters

54. $5\sqrt{2}$ meters **55.** $4\sqrt{5}$ blocks **56.** $s = \frac{d\sqrt{2}}{2}$

57. $2\sqrt{61}$ feet **58.** 9 feet **59.** 6.3% **60.** 2.5%

61. a) 2 sec and 3 sec **b)** 5 sec **62.** 2 sec **63.** 9

64. January 16 **65.** c

66. a) 6 **b)** 10 **c)** $\frac{1}{2}n^2 - \frac{1}{2}n$ **d)** 16

Section 9.2 Warm-ups F F F F F T T T T T T

1. Not every quadratic can be solved by factoring or the square root property.
2. Any quadratic can be solved by completing the square.
3. The last term is the square of one-half of the coefficient of the middle term.
4. If $a \neq 1$, divide each side by a.

5. $x^2 + 6x + 9 = (x + 3)^2$ **6.** $x^2 - 4x + 4 = (x - 2)^2$

7. $x^2 + 14x + 49 = (x + 7)^2$

8. $x^2 + 16x + 64 = (x + 8)^2$

9. $x^2 - 16x + 64 = (x - 8)^2$

10. $x^2 - 14x + 49 = (x - 7)^2$

11. $t^2 - 18t + 81 = (t - 9)^2$

12. $w^2 + 18w + 81 = (w + 9)^2$

13. $m^2 + 3m + \frac{9}{4} = \left(m + \frac{3}{2}\right)^2$

14. $n^2 - 5n + \frac{25}{4} = \left(n - \frac{5}{2}\right)^2$

15. $z^2 + z + \frac{1}{4} = \left(z + \frac{1}{2}\right)^2$

16. $v^2 - v + \frac{1}{4} = \left(v - \frac{1}{2}\right)^2$

17. $x^2 - \frac{1}{2}x + \frac{1}{16} = \left(x - \frac{1}{4}\right)^2$

18. $y^2 + \frac{1}{3}y + \frac{1}{36} = \left(y + \frac{1}{6}\right)^2$

19. $y^2 + \frac{1}{4}y + \frac{1}{64} = \left(y + \frac{1}{8}\right)^2$

20. $z^2 - \frac{4}{3}z + \frac{4}{9} = \left(z - \frac{2}{3}\right)^2$

21. $(x + 5)^2$ 22. $(x - 3)^2$ 23. $(m - 1)^2$

24. $(n + 2)^2$ 25. $\left(x + \frac{1}{2}\right)^2$ 26. $\left(y - \frac{1}{2}\right)^2$

27. $\left(t + \frac{1}{6}\right)^2$ 28. $\left(v - \frac{1}{3}\right)^2$ 29. $\left(x + \frac{1}{5}\right)^2$

30. $\left(y - \frac{1}{8}\right)^2$ 31. $-5, 3$ 32. $-6, 4$

33. $-3, 7$ 34. $-2, 6$ 35. -3 36. 5

37. $\frac{1}{2}, 1$ 38. $-\frac{1}{2}, 2$ 39. $\frac{3}{2}, 2$

40. $\frac{3}{4}, -2$ 41. $-1, \frac{1}{3}$ 42. $-\frac{1}{3}, 3$

43. $-1 - \sqrt{7}, -1 + \sqrt{7}$

44. $-2 - 2\sqrt{2}, -2 + 2\sqrt{2}$

45. $-3 - 2\sqrt{2}, -3 + 2\sqrt{2}$

46. $3 - 2\sqrt{3}, 3 + 2\sqrt{3}$

47. $\frac{1 - \sqrt{13}}{2}, \frac{1 + \sqrt{13}}{2}$

48. $\frac{-1 - \sqrt{5}}{2}, \frac{-1 + \sqrt{5}}{2}$

49. $\frac{-3 - \sqrt{21}}{2}, \frac{-3 + \sqrt{21}}{2}$

50. $\frac{3 - \sqrt{5}}{2}, \frac{3 + \sqrt{5}}{2}$

51. $\frac{1 - \sqrt{33}}{4}, \frac{1 + \sqrt{33}}{4}$

52. $\frac{-1 - \sqrt{5}}{4}, \frac{-1 + \sqrt{5}}{4}$

53. $5 - \sqrt{7}, 5 + \sqrt{7}$ 54. $-4, 3$

55. $-\frac{\sqrt{15}}{3}, \frac{\sqrt{15}}{3}$

56. No real solution 57. No real solution

58. $-3 - \sqrt{2}, -3 + \sqrt{2}$ 59. $4 - 2\sqrt{2}, 4 + 2\sqrt{2}$

60. $-6, \frac{7}{2}$ 61. $\frac{7}{2}$ 62. $-\sqrt{7}, \sqrt{7}$

63. $-3 - 2\sqrt{5}, -3 + 2\sqrt{5}$ 64. $-6, 0$ 65. $2 \pm \sqrt{2}$

66. $2 \pm \sqrt{3}$ 67. $-0.6, 0.4$ 68. $0.3, 0.6$

69. $5 - \sqrt{3}$ in. and $5 + \sqrt{3}$ in.

70. Length $3 + \sqrt{3}$ in., width $3 - \sqrt{3}$ in.

71. $6 - \sqrt{2}$ and $6 + \sqrt{2}$ 72. $4 - \sqrt{5}$ and $4 + \sqrt{5}$

73. 12 years old

74. a) \$27,200 b) \$10 and \$15 c) \$12.50

75. a) $x^2 - 10x + 22$ b) $5 \pm \sqrt{3}$ c) $x^2 - 3x + \frac{7}{4} = 0$

Section 9.3 Warm-ups T F F F T F F T T T

1. The quadratic formula solves any quadratic equation.
2. The quadratic formula is $x = \frac{-b \pm \sqrt{b^2 - 4ac}}{2a}$.
3. The quadratic formula is used to solve $ax^2 + bx + c = 0$.
4. The discriminant is $b^2 - 4ac$.
5. If $b^2 - 4ac < 0$ then there are no real solutions.
6. We have solved quadratic equations by factoring, the square root property, completing the square, and the quadratic formula.

7. $-5, 3$ 8. $6, -3$ 9. -5 10. 6 11. $-2, \frac{3}{2}$

12. $-3, \frac{5}{2}$ 13. $-\frac{3}{2}, \frac{1}{2}$ 14. $-\frac{3}{2}, -\frac{1}{2}$

15. $\frac{3 - \sqrt{3}}{2}, \frac{3 + \sqrt{3}}{2}$ 16. $\frac{-3 - \sqrt{3}}{3}, \frac{-3 + \sqrt{3}}{3}$

17. $\frac{-2 - \sqrt{2}}{2}, \frac{-2 + \sqrt{2}}{2}$ 18. $2 - \sqrt{2}, 2 + \sqrt{2}$

19. No real solution 20. No real solution 21. $0, \frac{1}{2}$

22. $-1, 0$ 23. $-\frac{\sqrt{15}}{5}, \frac{\sqrt{15}}{5}$ 24. $\frac{-2\sqrt{7}}{7}, \frac{2\sqrt{7}}{7}$

25. $-7 + 4\sqrt{3}, -7 - 4\sqrt{3}$ 26. $12 - 2\sqrt{33}, 12 + 2\sqrt{33}$

27. 0, one 28. 0, one 29. -47, none 30. -59, none

31. 181, two 32. 76, two 33. 0, one 34. 0, one

35. -15, none 36. -7, none 37. -59, none

38. 25, two 39. $-2, \frac{1}{2}$ 40. $-\frac{1}{2}, 4$ 41. $0, 3$ 42. $-2, 5$

43. $-\frac{6}{5}, 4$ 44. $-\frac{3}{5}, 2$ 45. $-4, -2$ 46. $-\frac{1}{2}, 3$

47. $0, 9$ 48. $-3, 3$ 49. $-8, -2$ 50. $\frac{1}{3}$

51. No real solution 52. $\frac{4}{3}, 5$

53. $\frac{3 - \sqrt{33}}{2}, \frac{3 + \sqrt{33}}{2}$ 54. $\frac{5 - \sqrt{41}}{2}, \frac{5 + \sqrt{41}}{2}$

55. $0, \frac{3}{2}$ 56. $-\sqrt{5}, \sqrt{5}$ 57. $-0.79, 3.79$

58. $-0.73, 2.73$ 59. $-1.36, 2.36$

60. $0.88, 3.42$ 61. 0.30 62. $-2.11, 0.91$

63. **a)** \$150,000 **b)** \$0 and \$40 **c)** \$17.50 or \$22.50 **d)** \$20
64. \$6.91 and \$18.09

Section 9.4 Warm-ups F T F F T T T F T T

1. Width $-1 + \sqrt{11}$ meters, length $1 + \sqrt{11}$ meters
2. $-2 + 2\sqrt{5}$ cm and $2 + 2\sqrt{5}$ cm **3.** $4\sqrt{2}$ feet
4. 6 meters by 8 meters
5. Height $-3 + \sqrt{19}$ inches, base $3 + \sqrt{19}$ inches
6. $4 + 2\sqrt{3}$ and $4 - 2\sqrt{3}$
7. $3 + \sqrt{5}$ or 5.24 hours
8. $\dfrac{5 + \sqrt{17}}{2}$ or 4.56 days
9. $16 + 4\sqrt{26}$ or 36.40 hours
10. $\dfrac{-3 + \sqrt{409}}{2}$ or 8.61 days
11. $\dfrac{15 + \sqrt{241}}{8}$ or 3.82 seconds
12. $\dfrac{25 + \sqrt{655}}{4}$ or 12.65 seconds.
13. 1.5 seconds
14. $\dfrac{-15 + 5\sqrt{137}}{16}$ or 2.7 seconds
15. $2\sqrt{19}$ or 8.72 mph
16. Gladys 36 mph, Bonita 45 mph
17. $\dfrac{-13 + \sqrt{409}}{2}$ or 3.61 feet
18. $\dfrac{-5 + 5\sqrt{5}}{2}$ or 3.09 feet

Section 9.5 Warm-ups T T T T F F F F F T

1. A complex number is a number of the form $a + bi$, where a and b are real and $i = \sqrt{-1}$.
2. The real numbers together with the imaginary numbers make up the set of complex numbers.
3. Complex numbers are added or subtracted like binomials in which i is the variable.
4. The complex conjugate of $a + bi$ is $a - bi$.
5. If $b > 0$, then $\sqrt{-b} = i\sqrt{b}$
6. In the complex number system all quadratic equations have at least one solution.
7. $5 + 9i$ **8.** $9 + 5i$ **9.** 1 **10.** $-5 + 4i$
11. $2 - 8i$ **12.** $-4 - 8i$ **13.** $-1 - 4i$
14. $-2 - 5i$ **15.** -1 **16.** 8 **17.** $\dfrac{3}{4} + \dfrac{1}{2}i$
18. $\dfrac{5}{12} - \dfrac{1}{2}i$ **19.** $6 - 9i$ **20.** $-12 + 8i$
21. -36 **22.** -9 **23.** -36 **24.** -9
25. $21 - i$ **26.** $6 - 27i$ **27.** $21 - 20i$
28. $-7 + 24i$ **29.** 25 **30.** 34 **31.** 2
32. 10 **33.** 29 **34.** 25 **35.** 52 **36.** 53
37. 13 **38.** 17 **39.** 1 **40.** 4 **41.** $1 - 3i$
42. $1 - 2i$ **43.** $-1 + 4i$ **44.** $-2 + 3i$ **45.** $-3 + 2i$
46. $-8 - 3i$ **47.** $\dfrac{8}{13} + \dfrac{12}{13}i$ **48.** $\dfrac{20}{41} + \dfrac{25}{41}i$ **49.** $\dfrac{3}{5} + \dfrac{4}{5}i$
50. -1 **51.** $-4i$ **52.** $-\dfrac{15}{13} + \dfrac{23}{13}i$ **53.** $5 + 3i$
54. $6 + 4i$ **55.** $-3 - i\sqrt{7}$ **56.** $2 - i\sqrt{3}$
57. $-1 + i\sqrt{3}$ **58.** $-2 + i\sqrt{2}$ **59.** $2 + \dfrac{1}{2}i\sqrt{5}$
60. $-3 - i\sqrt{6}$ **61.** $-\dfrac{2}{3} + \dfrac{1}{3}i\sqrt{7}$
62. $1 - \dfrac{1}{2}i\sqrt{5}$ **63.** $\dfrac{1}{5} - i$ **64.** $\dfrac{1}{3} - i$
65. $-9i, 9i$ **66.** $-10i, 10i$ **67.** $-i\sqrt{5}, i\sqrt{5}$
68. $-i\sqrt{6}, i\sqrt{6}$ **69.** $-i\dfrac{\sqrt{6}}{3}, i\dfrac{\sqrt{6}}{3}$ **70.** $-i\dfrac{\sqrt{15}}{5}, i\dfrac{\sqrt{15}}{5}$
71. $2 - i, 2 + i$ **72.** $3 - i, 3 + i$ **73.** $3 - 2i, 3 + 2i$
74. $2 - 5i, 2 + 5i$ **75.** $2 - i\sqrt{3}, 2 + i\sqrt{3}$
76. $5 - i\sqrt{2}, 5 + i\sqrt{2}$ **77.** $\dfrac{2 - i}{3}, \dfrac{2 + i}{3}$
78. $\dfrac{1}{2} - \dfrac{1}{2}i, \dfrac{1}{2} + \dfrac{1}{2}i$ **79.** $\dfrac{1 - i\sqrt{3}}{2}, \dfrac{1 + i\sqrt{3}}{2}$
80. $\dfrac{5}{2} - \dfrac{1}{2}i\sqrt{2}, \dfrac{5}{2} + \dfrac{1}{2}i\sqrt{2}$ **81.** $\dfrac{2 - i\sqrt{5}}{2}, \dfrac{2 + i\sqrt{5}}{2}$
82. $\dfrac{2}{3} - \dfrac{1}{3}i\sqrt{6}, \dfrac{2}{3} + \dfrac{1}{3}i\sqrt{6}$ **83.** $-6 - 24i$
84. $-17 + 20i$ **85.** 0 **86.** 0 **87.** $x^2 - 12x + 37$
88. $x^2 - 6x + 58$ **89.** $x^2 + 9 = 0$
90. $x^2 - 4x + 29 = 0$
91. Natural: 54, integers: 54, rational: 54, $-3/8$, irrational: $3\sqrt{5}$, real: 54, $-3/8$, $3\sqrt{5}$, imaginary: $6i$, $\pi + i\sqrt{5}$, complex: all
92. Real: b, imaginary: a, c, d, complex: all

Section 9.6 Warm-ups T F T T F T T T T F

1. The graph of $y = mx + b$ is a straight line.
2. The graph of $y = ax^2 + bx + c$ with $a \neq 0$ is a parabola.
3. The graph of $y = ax^2 + bx + c$ opens upward if $a > 0$.
4. The vertex is the highest point on a parabola that opens downward or the lowest point on a parabola that opens upward.
5. The x-coordinate of the vertex is $-b/(2a)$.
6. Use $x = -b/(2a)$ to find x and then use the original equation to find y.
7. $(3, -6), (4, 0), (-3, 0)$ **8.** $(0, 1), (-4, -3), (2, -3)$
9. $(4, -128), (0, 0), (2, 0)$ **10.** $(-2, 1), (-1, 2), (-3, 2)$

11.

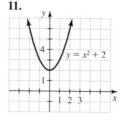

$y = x^2 + 2$

12.

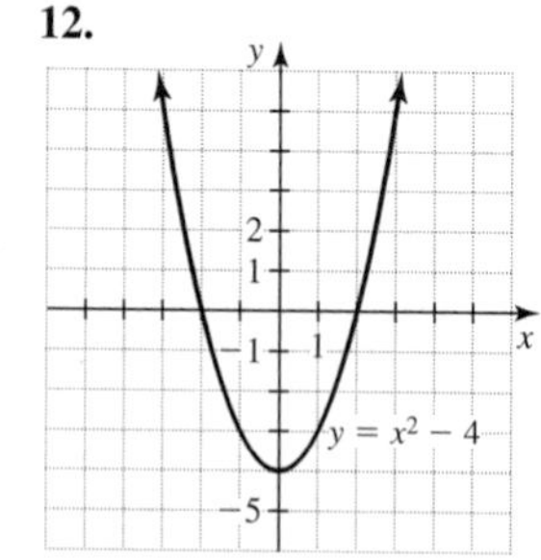

13.

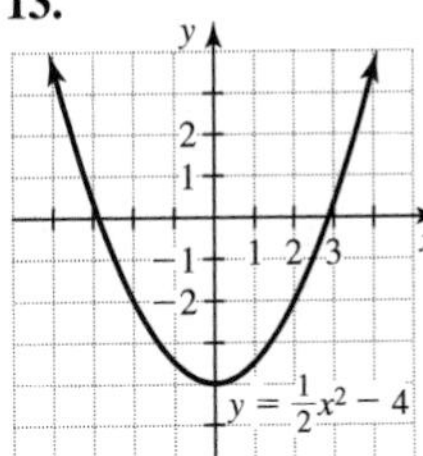

14.

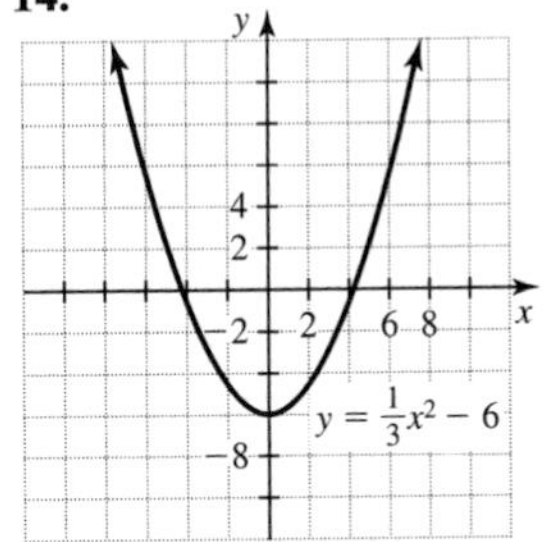

15.

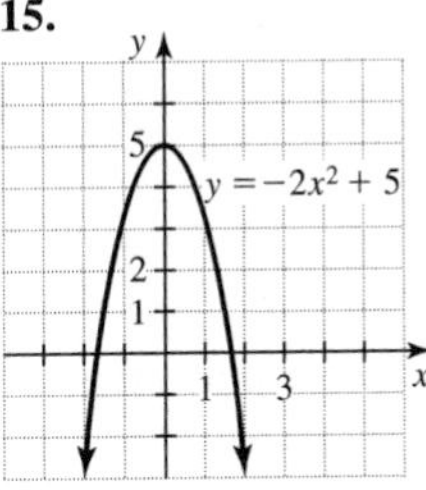

16.

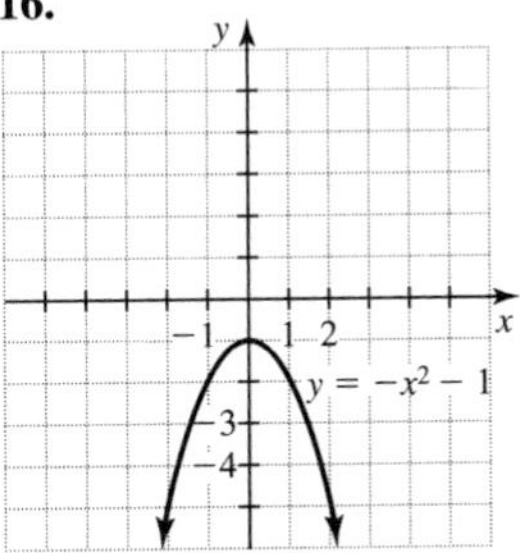

17.

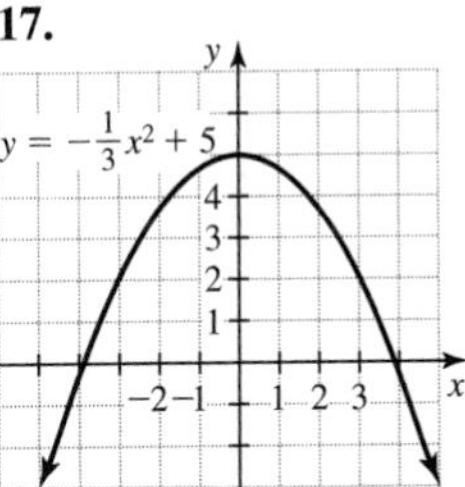

18.

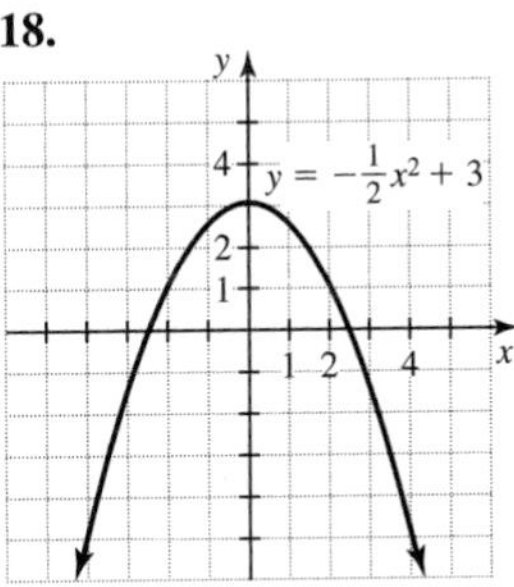

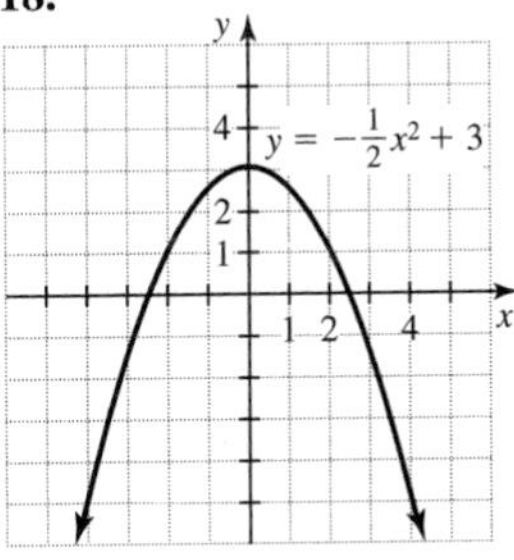

19.

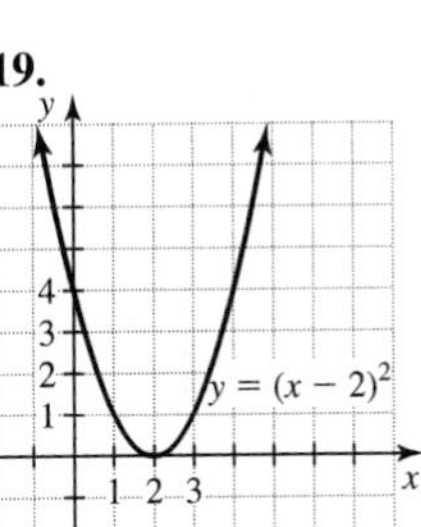

20.

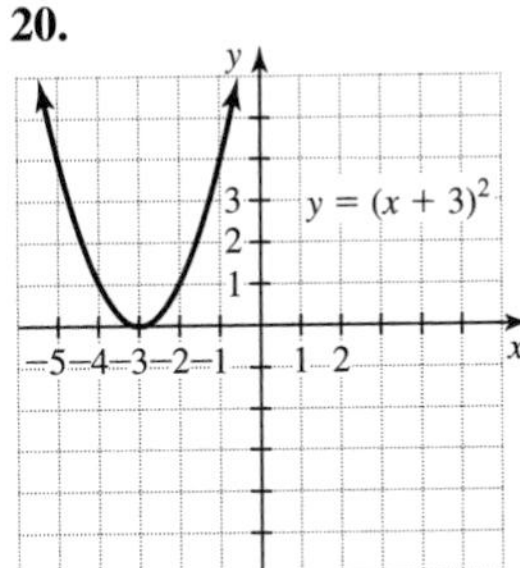

21. Vertex $\left(\frac{1}{2}, -\frac{9}{4}\right)$, intercepts $(0, -2)$, $(-1, 0)$, $(2, 0)$

22. Vertex $(-1, -4)$, intercepts $(0, -3)$, $(-3, 0)$, $(1, 0)$

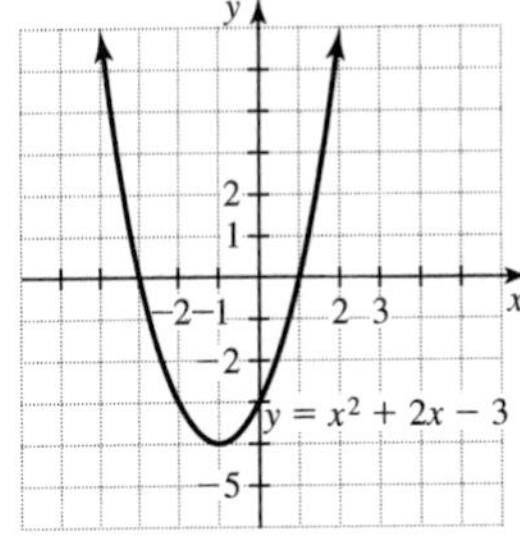

23. Vertex $(-1, -9)$, intercepts $(0, -8)$, $(-4, 0)$, $(2, 0)$

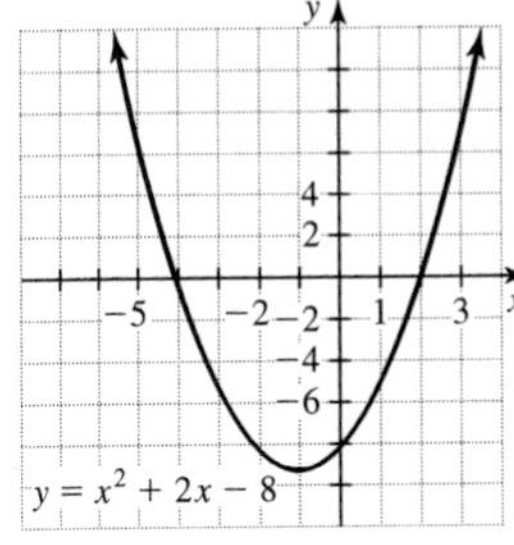

24. Vertex $\left(-\frac{1}{2}, -\frac{25}{4}\right)$, intercepts $(0, -6)$, $(-3, 0)$, $(2, 0)$

25. Vertex $(-2, 1)$, intercepts $(0, -3)$, $(-1, 0)$, $(-3, 0)$

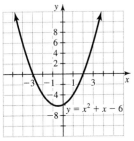

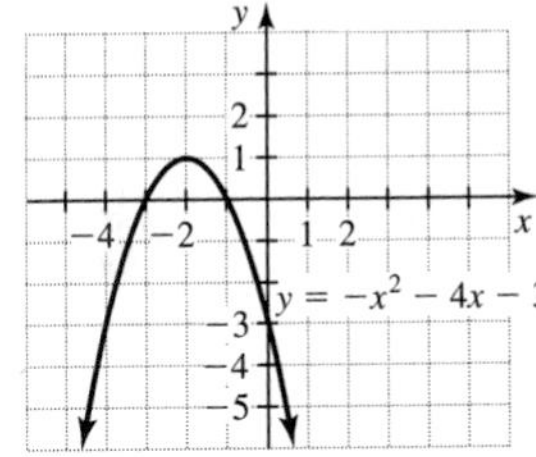

26. Vertex $\left(-\frac{5}{2}, \frac{9}{4}\right)$, intercepts $(0, -4)$, $(-1, 0)$, $(-4, 0)$

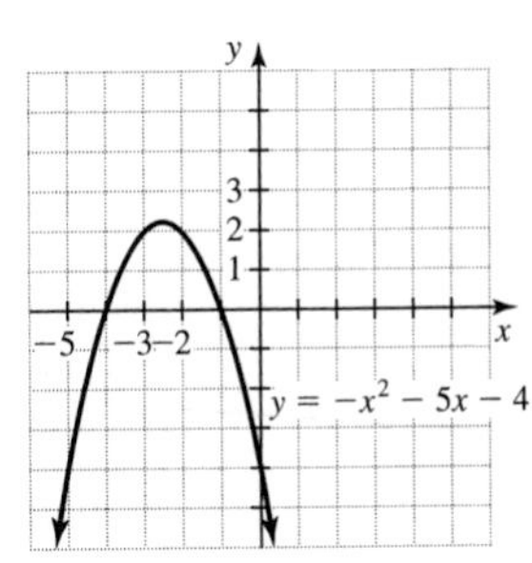

27. Vertex $\left(\frac{3}{2}, \frac{25}{4}\right)$, intercepts (0, 4), (4, 0), (−1, 0)

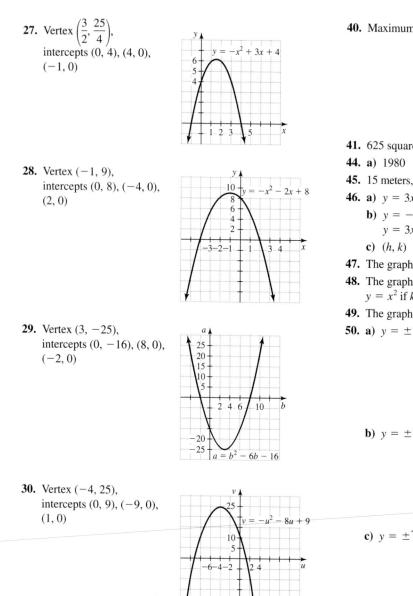

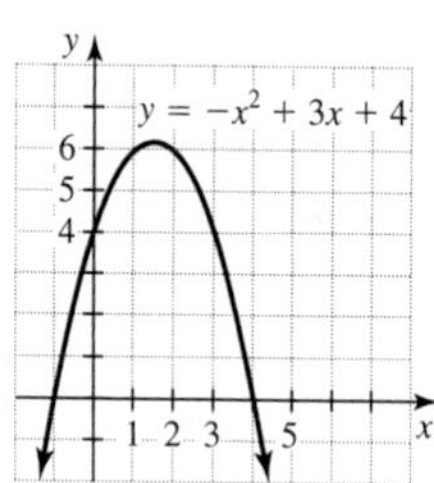

28. Vertex (−1, 9), intercepts (0, 8), (−4, 0), (2, 0)

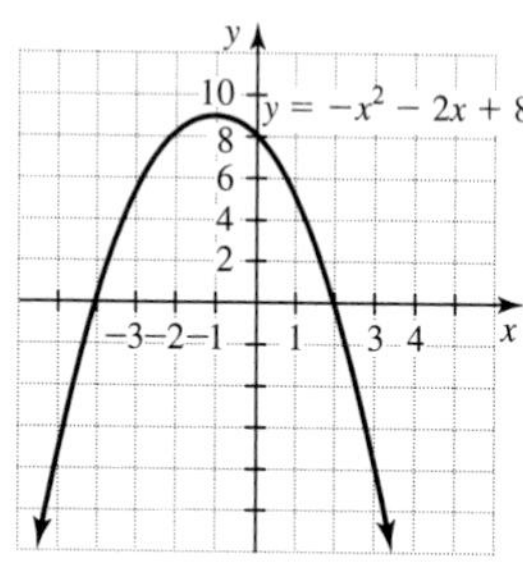

29. Vertex (3, −25), intercepts (0, −16), (8, 0), (−2, 0)

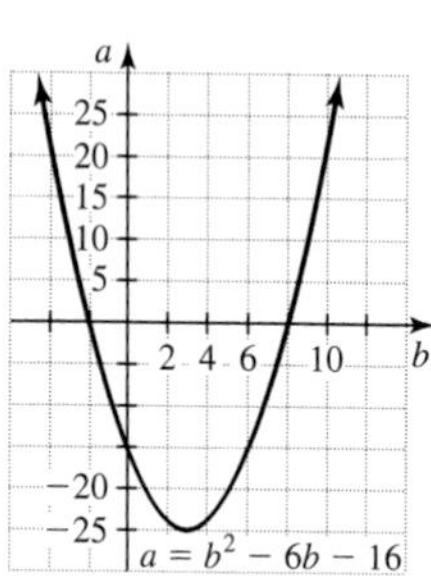

30. Vertex (−4, 25), intercepts (0, 9), (−9, 0), (1, 0)

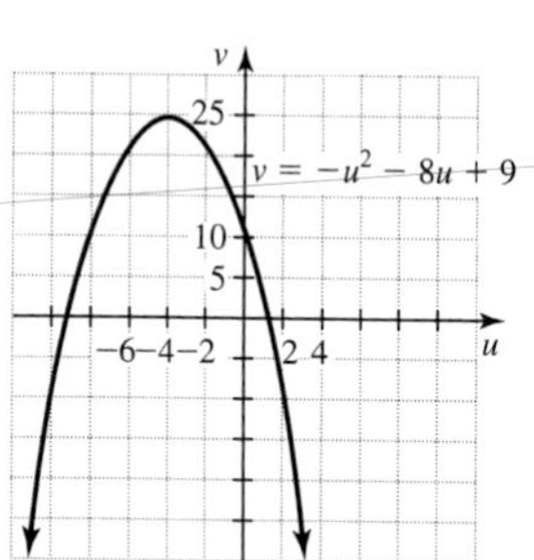

31. Minimum −8 **32.** Maximum 33 **33.** Maximum 14
34. Minimum 6 **35.** Minimum 2 **36.** Minimum 4
37. Maximum 2 **38.** Maximum 48
39. Maximum 64 feet

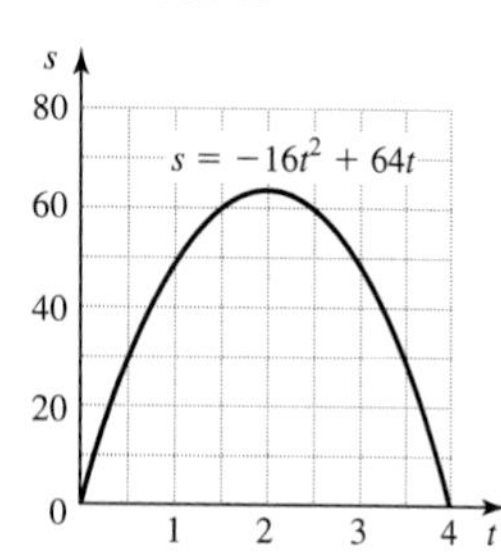

40. Maximum 16 feet

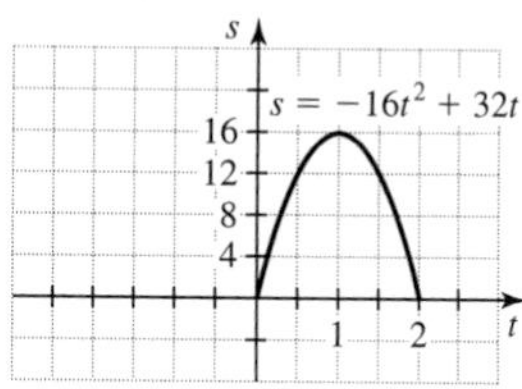

41. 625 square meters **42.** 85, \$5.50 **43.** 2 P.M.
44. **a)** 1980 **b)** 1981 **c)** 3.83 **d)** stable population
45. 15 meters, 25 meters
46. **a)** $y = 3x^2 - 12x + 18$, (2, 6)
b) $y = -4x^2 + 40x - 109$, (5, −9), $y = 3x^2 + 12x + 6$, (−2, −6)
c) (h, k)
47. The graph of $y = kx^2$ gets narrower as k gets larger.
48. The graph of $y = (x - k)^2$ lies k units to the right of $y = x^2$ if $k > 0$ and $-k$ units to the left of $y = x^2$ if $k < 0$.
49. The graph of $y = x^2$ has the same shape as $x = y^2$.
50. **a)** $y = \pm\sqrt{x + 1}$

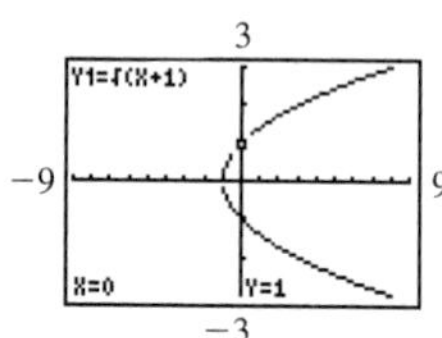

b) $y = \pm\sqrt{-x}$

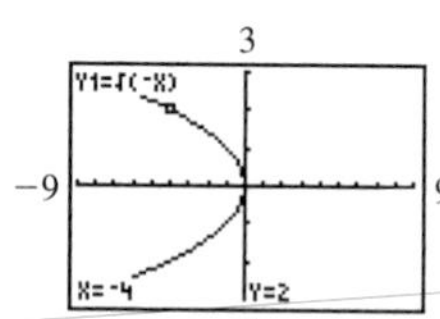

c) $y = \pm\sqrt{4 - x^2}$

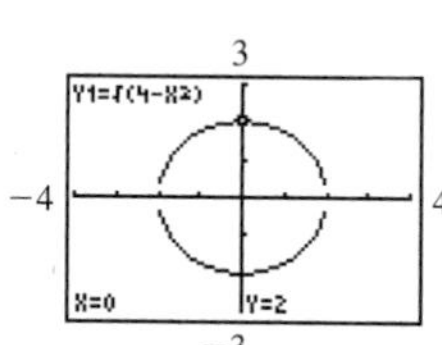

51. **a)** Vertex (0.84, −0.68), x-intercepts (1.30, 0), (0.38, 0)
b) Vertex (5.96, 46.26), x-intercepts (12.48, 0), (−0.55, 0)

Section 9.7 Warm-ups F T F T T T T T F T

1. A function is a set of ordered pairs in which no two have the same first coordinate and different second coordinates.

2. Functions can be expressed by means of a verbal rule, a formula, a table, a list, or a graph.

3. All descriptions of functions involve ordered pairs that satisfy the definition.

4. A graph is the graph of a function if no vertical line crosses the graph more than once.

5. The domain is the set of all first coordinates of the ordered pairs.

6. In function notation x represents the first coordinate and $f(x)$ represents the second coordinate.

7. $C = 0.50t + 5$ 8. $C = 40s + 20{,}000$

9. $T = 1.09S$ 10. $R = 0.05A$

11. $C = 2\pi r$ 12. $C = \pi d$

13. $P = 4s$ 14. $P = 2L + 20$

15. $A = 5h$ 16. $A = 11h$

17. Yes 18. Yes 19. Yes 20. Yes

21. No 22. No 23. Yes 24. Yes

25. Yes 26. No 27. Yes 28. Yes

29. No 30. Yes 31. Yes 32. Yes

33. Yes 34. Yes 35. No 36. No

37. Yes 38. Yes 39. No 40. Yes

41. Yes 42. Yes 43. Yes 44. Yes

45. Yes 46. Yes 47. No 48. No

49. Yes 50. Yes 51. No 52. No

53. No 54. Yes

55. $\{1, 2, 3\}, \{3, 5, 7\}$ 56. $\{-1, 3\}, \{4, 5\}$

57. $\{0, 2, 4\}, \{1\}$ 58. $\{4, 5, 7, 8\}, \{-2\}$

59. R (real numbers), R

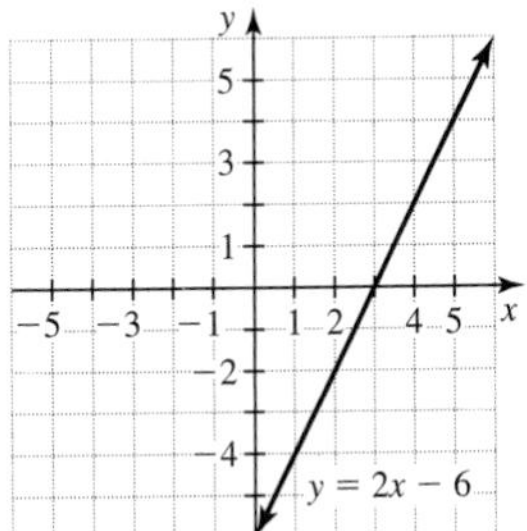

60. R, R

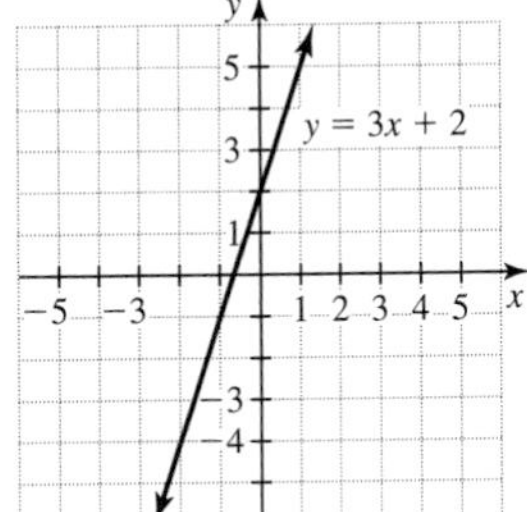

61. R, R

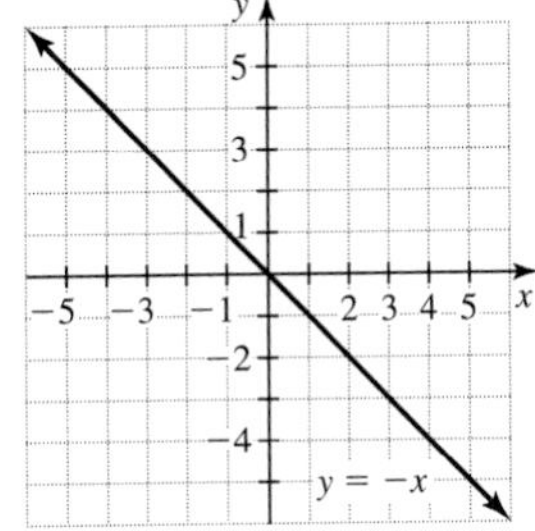

62. R, R

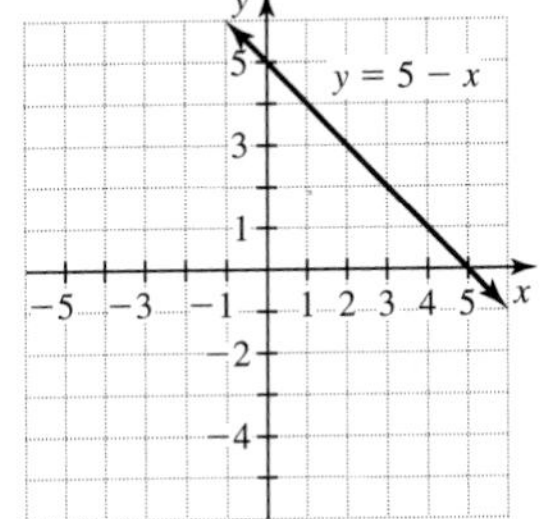

63. $R, \{y \mid y \geq -2\}$

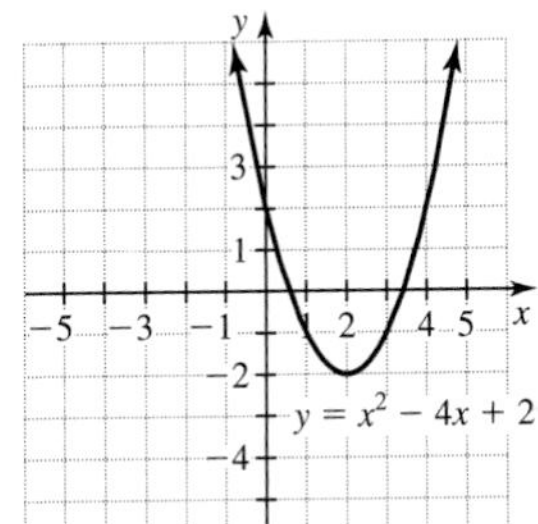

64. $R, \{y \mid y \geq -4\}$

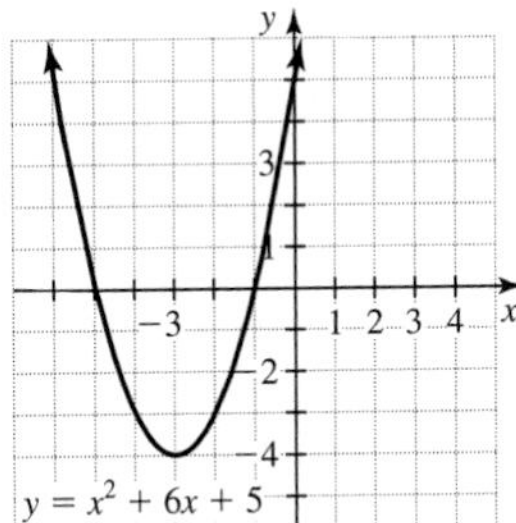

65. $R, \{y \mid y \leq 7\}$

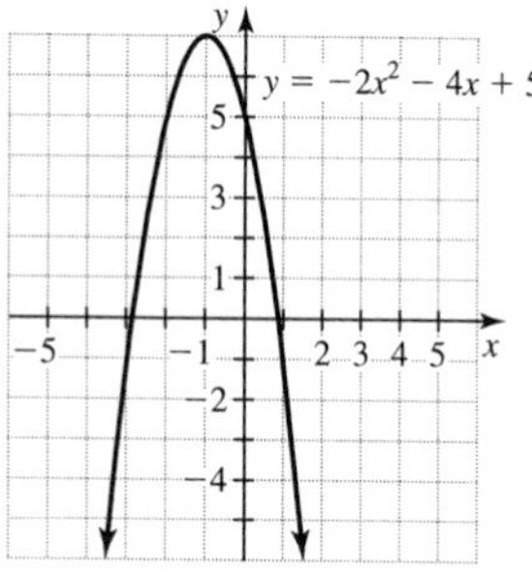

66. $R, \{y \mid y \leq 4\}$

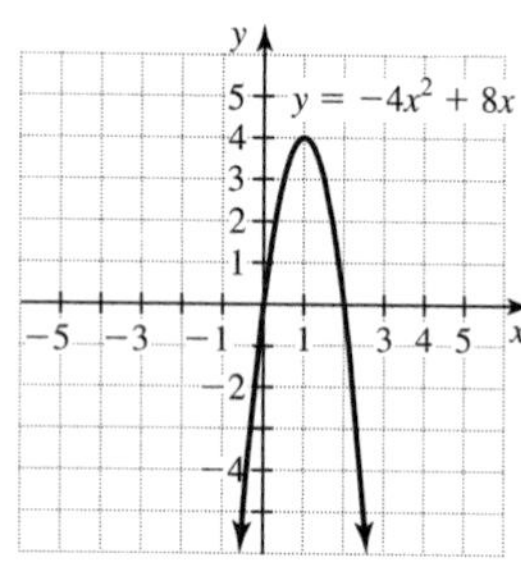

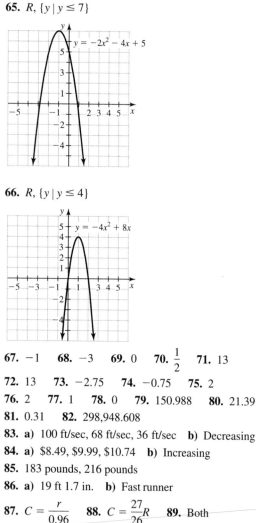

67. -1 68. -3 69. 0 70. $\frac{1}{2}$ 71. 13
72. 13 73. -2.75 74. -0.75 75. 2
76. 2 77. 1 78. 0 79. 150.988 80. 21.39
81. 0.31 82. 298,948.608
83. a) 100 ft/sec, 68 ft/sec, 36 ft/sec b) Decreasing
84. a) \$8.49, \$9.99, \$10.74 b) Increasing
85. 183 pounds, 216 pounds
86. a) 19 ft 1.7 in. b) Fast runner
87. $C = \frac{r}{0.96}$ 88. $C = \frac{27}{26}R$ 89. Both
90. b is a function of a. 91. Neither 92. Neither
93. Neither 94. Both 95. b is a function of a.
96. b is a function of a.

Section 9.8 Warm-ups T T F F F F T T T T

1. The basic operations of functions are addition, subtraction, multiplication, and division.
2. We perform the operation with functions by adding, subtracting, multiplying, or dividing the expressions that define the functions.
3. In the composition function the second function is evaluated on the result of the first function.
4. Since each operation is a function, the order of operations determines the order in which the functions are composed.

5. $x^2 + 2x - 3$ 6. $-x^2 + 6x - 3$
7. $4x^3 - 11x^2 + 6x$ 8. $\frac{4x - 3}{x^2 - 2x}$ 9. 12
10. 5 11. -30 12. -19 13. -21 14. -88
15. $\frac{13}{8}$ 16. $-\frac{11}{8}$ 17. $y = 6x - 20$
18. $y = -6x + 11$ 19. $y = x + 2$
20. $y = x$ 21. $y = x^2 + 2x$ 22. $y = x^2 + x - 1$
23. $y = x$ 24. $y = x$ 25. -2 26. -7 27. 5
28. 28 29. 7 30. 3 31. 5 32. 0 33. 5 34. 0
35. 4 36. $\frac{1}{2}$ 37. 22.2992 38. 23.761
39. $4x^2 - 6x$ 40. $\frac{x^2 + 12x + 27}{4}$ 41. $2x^2 + 6x - 3$
42. $\frac{x^2 + 3x + 3}{2}$ 43. x 44. x 45. $4x - 9$
46. $x^4 + 6x^3 + 12x^2 + 9x$ 47. $\frac{x + 9}{4}$ 48. $8x - 21$
49. $F = f \circ h$ 50. $N = h \circ f$ 51. $G = g \circ h$
52. $P = f \circ g$ 53. $H = h \circ g$ 54. $M = f \circ f$
55. $J = h \circ h$ 56. $R = f \circ h \circ g$
57. $K = g \circ g$ 58. $Q = f \circ g \circ h$
59. 112.5 in.^{2}, $A = \frac{d^2}{2}$ 60. $P = 4\sqrt{A}$
61. $P(x) = -x^2 + 20x - 170$ 62. $A = \frac{(32 + 3\pi)x^2}{8}$
63. $J = 0.025I$ 64. $A = \pi\frac{M}{4}$
65. a) 397.8 b) $D = \frac{1.116 \times 10^7}{L^3}$ c) decreases
66. a) 15.97 b) $S = 14{,}400d^{-2/3}$ c) decreases

Enriching Your Mathematical Word Power

1. b 2. a 3. d 4. c 5. b 6. a 7. c 8. a
9. b 10. c 11. a 12. d 13. c 14. a 15. b

Review Exercises

1. $-3, 3$ 2. $-1, 1$ 3. 0, 9 4. 0, 1 5. $-1, 2$
6. $-1, 10$ 7. $9 - \sqrt{10}, 9 + \sqrt{10}$
8. $-5 - \sqrt{14}, -5 + \sqrt{14}$ 9. $\frac{3}{2}$ 10. $-\frac{1}{3}$ 11. 4, 5
12. 1, 3 13. $-7, 2$ 14. 5 15. $-4, \frac{1}{2}$ 16. $-\frac{3}{2}, 1$
17. $-2 - \sqrt{11}, -2 + \sqrt{11}$ 18. $-3 - 2\sqrt{3}, -3 + 2\sqrt{3}$
19. $-7, 4$ 20. $-2, 3$ 21. $\frac{-3 - \sqrt{29}}{2}, \frac{-3 + \sqrt{29}}{2}$
22. $\frac{-2 - \sqrt{7}}{3}, \frac{-2 + \sqrt{7}}{3}$ 23. $-5, \frac{1}{2}$
24. $\frac{-3 - \sqrt{19}}{2}, \frac{-3 + \sqrt{19}}{2}$ 25. 0, one
26. -24, none 27. -44, none 28. 49, two

29. 76, two **30.** 0, one **31.** $-\frac{2}{3}, \frac{1}{2}$ **32.** $-\frac{2}{3}, \frac{5}{2}$

33. $\frac{1-\sqrt{17}}{2}, \frac{1+\sqrt{17}}{2}$ **34.** $1-\sqrt{5}, 1+\sqrt{5}$

35. $\frac{3-\sqrt{14}}{5}, \frac{3+\sqrt{14}}{5}$ **36.** $3-\sqrt{5}, 3+\sqrt{5}$

37. $0, \frac{5}{3}$ **38.** $-\frac{5}{2}, 3$ **39.** 13 meters

40. $20\sqrt{2}$ or 28.284 yards

41. $\frac{-2+\sqrt{46}}{2}$ or 2.391 meters and $\frac{2+\sqrt{46}}{2}$ or 4.391 meters

42. Width $\frac{-3+\sqrt{89}}{2}$ or 3.217 feet, height $\frac{3+\sqrt{89}}{2}$ or 6.217 feet

43. Height $-2+2\sqrt{11}$ or 4.633 inches, base $2+2\sqrt{11}$ or 8.633 inches

44. Base $\frac{1+\sqrt{33}}{2}$ or 3.372 meters, height $\frac{-1+\sqrt{33}}{2}$ or 2.372 meters

45. $3+\sqrt{2}$ or 4.414 and $3-\sqrt{2}$ or 1.586.

46. Width $4-\sqrt{3}$ or 2.268 feet, length $4+\sqrt{3}$ or 5.732 feet

47. New printer $7+\sqrt{65}$ or 15.062 hours, old printer $9+\sqrt{65}$ or 17.062 hours

48. Blake $\frac{15+3\sqrt{17}}{2}$ or 13.685 hours, Cassie $\frac{9+3\sqrt{17}}{2}$ or 10.685 hours.

49. $7-3i$ **50.** $-7-9i$ **51.** $-3+7i$ **52.** $-2i$

53. $15-25i$ **54.** $-18+8i$ **55.** $-55+48i$

56. 29 **57.** $-1-i\sqrt{2}$ **58.** $2-i\sqrt{6}$

59. $\frac{3}{37}+\frac{19}{37}i$ **60.** $\frac{9}{73}+\frac{24}{73}i$ **61.** $\frac{19}{17}+\frac{9}{17}i$

62. $2-3i$ **63.** $-11i, 11i$ **64.** $-2i\sqrt{30}, 2i\sqrt{30}$

65. $8-i, 8+i$ **66.** $5-i\sqrt{3}, 5+i\sqrt{3}$

67. $\frac{3-3i\sqrt{7}}{4}, \frac{3+3i\sqrt{7}}{4}$ **68.** $\frac{3-i\sqrt{3}}{3}, \frac{3+i\sqrt{3}}{3}$

69. Vertex $(3, -9)$, intercepts $(0, 0)$, $(6, 0)$

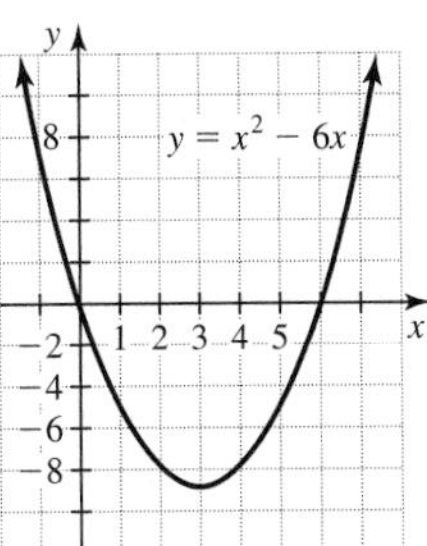

70. Vertex $(-2, -4)$, intercepts $(0, 0)$, $(-4, 0)$

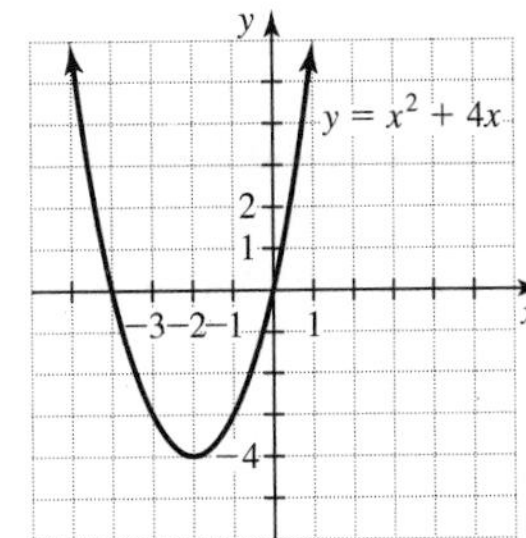

71. Vertex $(2, -16)$, intercepts $(0, -12)$, $(-2, 0)$, and $(6, 0)$

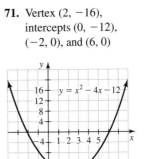

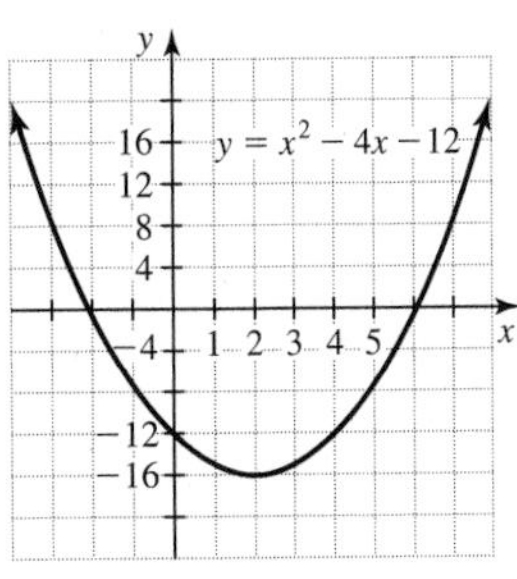

72. Vertex $(-1, -25)$, intercepts $(0, -24)$, $(-6, 0)$, $(4, 0)$

73. Vertex $(2, 8)$, intercepts $(0, 0)$, $(4, 0)$

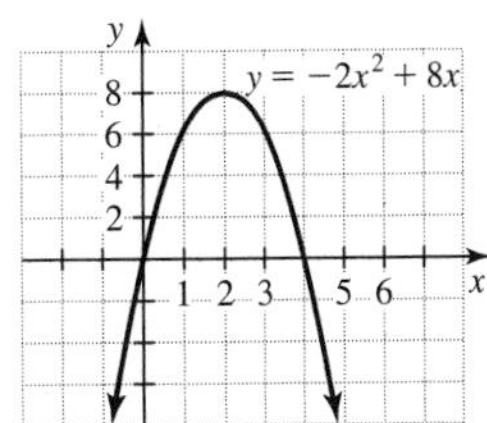

74. Vertex $(1, 3)$, intercepts $(0, 0)$, $(2, 0)$

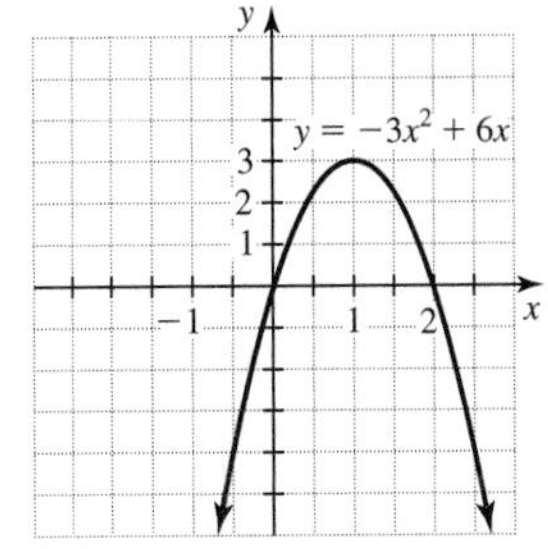

75. Vertex $(1, 4)$, intercepts $(0, 3)$, $(-1, 0)$, $(3, 0)$

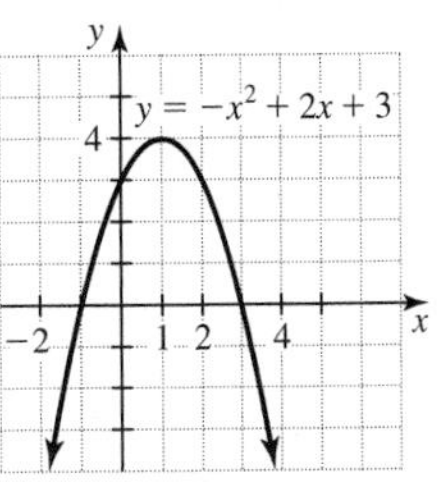

76. Vertex $\left(-\frac{3}{2}, \frac{1}{4}\right)$, intercepts $(0, -2)$, $(-2, 0)$, $(-1, 0)$

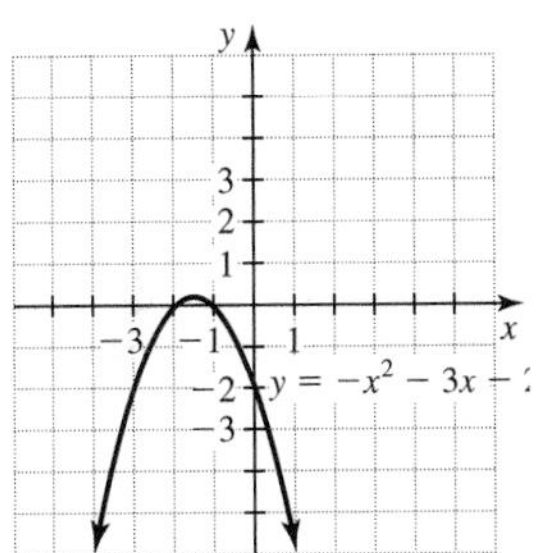

77. \$20.40, 400 **78.** \$99,800, 750 **79.** Yes **80.** No

81. No **82.** Yes **83.** Yes **84.** Yes **85.** No

86. No **87.** $\{1, 2, 3\}, \{0, 2\}$ **88.** $\{2, 4, 6\}, \{3\}$

89. Domain R, range $\{y \mid y \geq -3\}$

90. Domain R, range $\{y \mid y \geq -7\}$

91. Domain R, range $\{y \mid y \leq 4.125\}$

92. Domain R, range $\left\{y \mid y \leq \frac{22}{3}\right\}$

93. -4 **94.** -3 **95.** $\sqrt{2}$ **96.** 319 **97.** 99

98. $9x^2+24x+15$ **99.** 17 **100.** $-x^2+5x+5$

101. $3x^3-x^2-10x$ **102.** -8 **103.** 20

104. $9x+20$ **105.** $F = f \circ g$ **106.** $G = g \circ f$

107. $H = g \circ h$ **108.** $K = h \circ g$ **109.** $I = g \circ g$
110. $J = g \circ h \circ h$

Chapter 9 Test

1. 0, one **2.** -31, none **3.** 17, two **4.** $-1, \frac{3}{5}$
5. $\frac{2 - \sqrt{10}}{2}, \frac{2 + \sqrt{10}}{2}$ **6.** $-7, 3$
7. $\frac{-3 - \sqrt{29}}{2}, \frac{-3 + \sqrt{29}}{2}$ **8.** $-5, 4$ **9.** 3, 25
10. $-\frac{1}{2}, 3$ **11.** $10 + 3i$ **12.** $-6 + 7i$ **13.** -36
14. $42 - 2i$ **15.** 68 **16.** $2 - 3i$ **17.** $-1 + i\sqrt{3}$
18. $-2 + i\sqrt{2}$ **19.** $\frac{3}{5} + \frac{4}{5}i$ **20.** $-3 - i\sqrt{3}, -3 + i\sqrt{3}$
21. $\frac{3}{5} - \frac{4}{5}i, \frac{3}{5} + \frac{4}{5}i$
22. Domain R, range $\{y \mid y \le 16\}$
23. Domain R, range $\left\{y \mid y \ge -\frac{9}{4}\right\}$

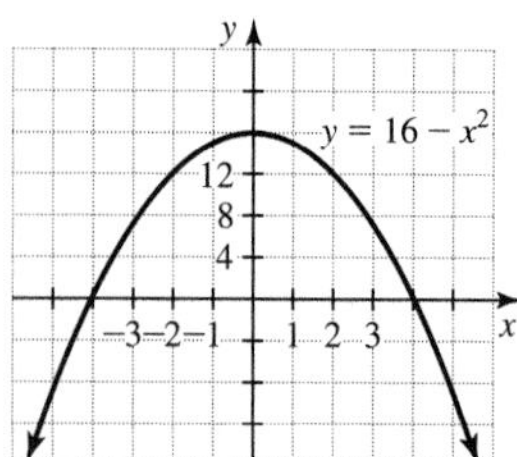

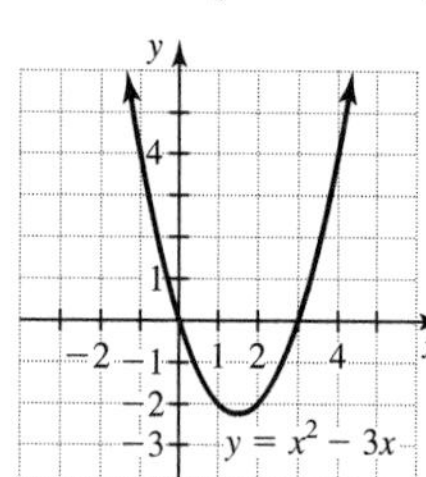

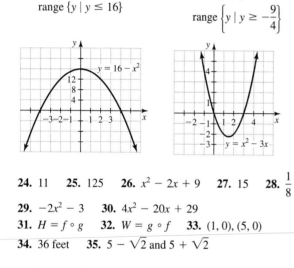

24. 11 **25.** 125 **26.** $x^2 - 2x + 9$ **27.** 15 **28.** $\frac{1}{8}$
29. $-2x^2 - 3$ **30.** $4x^2 - 20x + 29$
31. $H = f \circ g$ **32.** $W = g \circ f$ **33.** (1, 0), (5, 0)
34. 36 feet **35.** $5 - \sqrt{2}$ and $5 + \sqrt{2}$

Making Connections Chapters 1–9

1. $\frac{1}{2}$ **2.** 1 **3.** $-\frac{\sqrt{2}}{2}, \frac{\sqrt{2}}{2}$ **4.** $\frac{1 - 2\sqrt{2}}{2}, \frac{1 + 2\sqrt{2}}{2}$
5. $\frac{2 - \sqrt{6}}{2}, \frac{2 + \sqrt{6}}{2}$ **6.** 0, 2 **7.** $-1, \frac{1}{2}$ **8.** 5
9. $1 - i\sqrt{14}, 1 + i\sqrt{14}$ **10.** $-1, 2$ **11.** $y = \frac{5}{4}x - 2$
12. $y = 3x - 9$ **13.** $y = \frac{2}{3}x + \frac{16}{3}$ **14.** $y = -\frac{b}{a}$
15. $y = \frac{-b \pm \sqrt{b^2 - 4ac}}{2a}$ **16.** $y = -\frac{2}{3}x + 7$
17. $y = -\frac{4}{3}x + \frac{2}{9}$ **18.** $y = \pm\sqrt{a^2 - x^2}$ **19.** $P = 4s$
20. $A = s^2$ **21.** $P = 2d\sqrt{2}$ **22.** $d = \sqrt{2A}$
23. $(2, -3)$ **24.** (3, 10)
25.

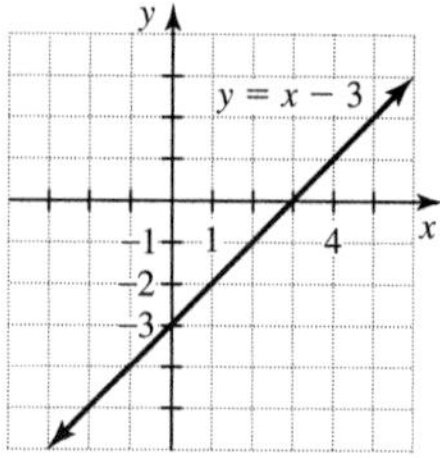

26.

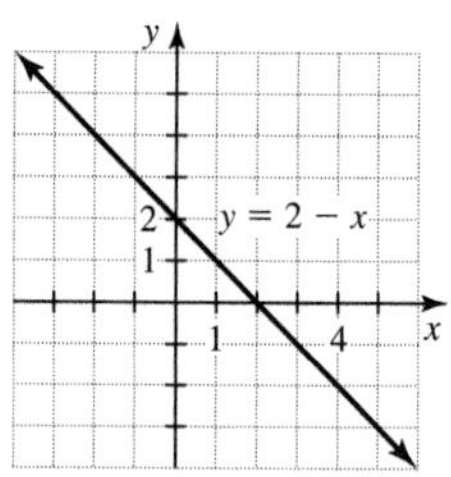

27.

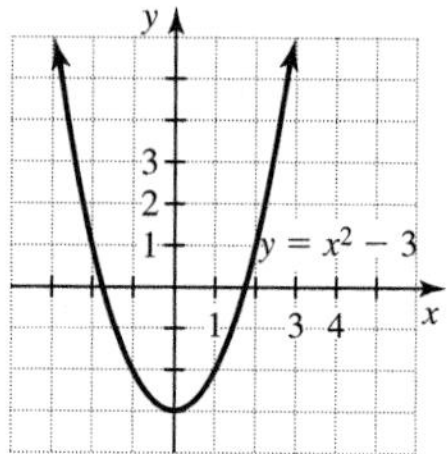

28.

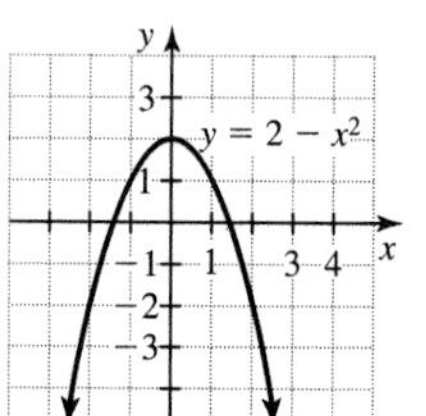

29. $y = \frac{2}{3}x - 4$

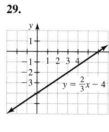

30.

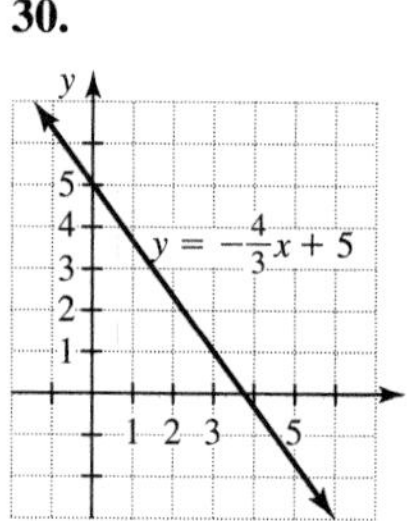

31. a) $s = -250p + 10{,}500$ **b)** Increased
c) $R = -250p^2 + 10{,}500p$ **d)** \$21

Appendix A

Geometry Review Exercises

1. 12 in. **2.** 24 ft^2 **3.** 60° **4.** 6 ft
5. 20 cm **6.** 144 cm^2 **7.** 24 ft^2 **8.** 13 ft
9. 30 cm **10.** No **11.** 84 yd^2 **12.** 30 in.
13. 32 ft^2 **14.** 20 km **15.** 7 ft, 3 ft^2
16. 22 yd **17.** 50.3 ft^2 **18.** 37.7 ft
19. 150.80 cm^3 **20.** 879.29 ft^2
21. 288 in.^{3}, 288 in.^{2} **22.** 4 cm
23. 100 mi^2, 40 mi **24.** 20 km
25. 42.25 cm^2 **26.** 33.510 ft^3, 50.265 ft^2
27. 75.4 in.^{3}, 100.5 in.^{2} **28.** 56°
29. 5 cm **30.** 15 in. and 20 in.
31. 149° **32.** 12 km **33.** 10 yd

INDEX

E

F

Q

R

S

T

U

V

W

X

Y

Z

DEFINITIONS, RULES, AND FORMULAS

Subsets of the Real Numbers

Natural Numbers $= \{1, 2, 3, \ldots\}$

Whole Numbers $= \{0, 1, 2, 3, \ldots\}$

Integers $= \{\ldots -3, -2, -1, 0, 1, 2, 3, \ldots\}$

Rational $= \left\{\frac{a}{b} \,\middle|\, a \text{ and } b \text{ are integers with } b \neq 0\right\}$

Irrational $= \{x \mid x \text{ is not rational}\}$

Properties of the Real Numbers

For all real numbers a, b, and c

$a + b = b + a;\ a \cdot b = b \cdot a$ Commutative

$(a + b) + c = a + (b + c);\ (ab)c = a(bc)$ Associative

$a(b + c) = ab + ac;\ a(b - c) = ab - ac$ Distributive

$a + 0 = a;\ 1 \cdot a = a$ Identity

$a + (-a) = 0;\ a \cdot \frac{1}{a} = 1\ (a \neq 0)$ Inverse

$a \cdot 0 = 0$ Multiplication property of 0

Absolute Value

$$|a| = \begin{cases} a & \text{for } a \geq 0 \\ -a & \text{for } a < 0 \end{cases}$$

Order of Operations

No parentheses or absolute value present:

1. Exponential expressions
2. Multiplication and division
3. Addition and subtraction

With parentheses or absolute value:

First evaluate within each set of parentheses or absolute value, using the order of operations.

Exponents

$a^0 = 1$ $\qquad$ $a^{-1} = \frac{1}{a}$

$a^{-r} = \frac{1}{a^r} = \left(\frac{1}{a}\right)^r$ $\qquad$ $\frac{1}{a^{-r}} = a^r$

$a^r a^s = a^{r+s}$ $\qquad$ $\frac{a^r}{a^s} = a^{r-s}$

$(a^r)^s = a^{rs}$ $\qquad$ $(ab)^r = a^r b^r$

$\left(\frac{a}{b}\right)^r = \frac{a^r}{b^r}$ $\qquad$ $\left(\frac{a}{b}\right)^{-r} = \left(\frac{b}{a}\right)^r$

Roots and Radicals

$a^{1/n} = \sqrt[n]{a}$ $\qquad$ $a^{m/n} = (\sqrt[n]{a})^m = \sqrt[n]{a^m}$

$\sqrt[n]{ab} = \sqrt[n]{a} \cdot \sqrt[n]{b}$ $\qquad$ $\sqrt[n]{\frac{a}{b}} = \frac{\sqrt[n]{a}}{\sqrt[n]{b}}$

Factoring

$a^2 + 2ab + b^2 = (a + b)^2$

$a^2 - 2ab + b^2 = (a - b)^2$

$a^2 - b^2 = (a + b)(a - b)$

$a^3 - b^3 = (a - b)(a^2 + ab + b^2)$

$a^3 + b^3 = (a + b)(a^2 - ab + b^2)$

Rational Expressions

$\frac{a}{b} + \frac{c}{b} = \frac{a + c}{b}$ $\qquad$ $\frac{a}{b} - \frac{c}{b} = \frac{a - c}{b}$

$\frac{ac}{bc} = \frac{a}{b}$ $\qquad$ $\frac{a}{b} + \frac{c}{d} = \frac{ad + bc}{bd}$

$\frac{a}{b} \cdot \frac{c}{d} = \frac{ac}{bd}$ $\qquad$ $\frac{a}{b} \div \frac{c}{d} = \frac{a}{b} \cdot \frac{d}{c}$

If $\frac{a}{b} = \frac{c}{d}$, then $ad = bc$.